W0018582

Supercritical Fluid Technology for Drug Product Development

DRUGS AND THE PHARMACEUTICAL SCIENCES

Executive Editor

James Swarbrick
PharmaceuTech, Inc.
Pinehurst, North Carolina

Advisory Board

Larry L. Augsburger
University of Maryland
Baltimore, Maryland

Harry G. Brittain
Center for Pharmaceutical Physics
Milford, New Jersey

Jennifer B. Dressman
Johann Wolfgang Goethe-University
Frankfurt, Germany

Anthony J. Hickey
University of North Carolina School of Pharm
Chapel Hill, North Carolina

Jeffrey A. Hughes
University of Florida College of Pharmacy
Gainesville, Florida

Ajaz Hussain
U.S. Food and Drug Administration
Frederick, Maryland

Trevor M. Jones
The Association of the
British Pharmaceutical Industry
London, United Kingdom

Hans E. Junginger
Leiden/Amsterdam Center
for Drug Research
Leiden, The Netherlands

Vincent H. L. Lee
University of Southern California
Los Angeles, California

Stephen G. Schulman
University of Florida
Gainesville, Florida

Jerome P. Skelly
Alexandria, Virginia

Elizabeth M. Topp
University of Kansas School of Pharmacy
Lawrence, Kansas

Geoffrey T. Tucker
University of Sheffield
Royal Hallamshire Hospital
Sheffield, United Kingdom

Peter York
University of Bradford School of Pharmacy
Bradford, United Kingdom

DRUGS AND THE PHARMACEUTICAL SCIENCES

A Series of Textbooks and Monographs

Supercritical Fluid Technology for Drug Product Development

edited by
Peter York
University of Bradford
Bradford, United Kingdom

Uday B. Kompella
University of Nebraska
Omaha, Nebraska, U.S.A.

Boris Y. Shekunov
Ferro Corporation
Independence, Ohio, U.S.A.

informa
healthcare

FIRST INDIAN REPRINT, 2014

Informa Healthcare USA, Inc.
52 Vanderbilt Avenue
New York, NY 10017

© 2008 by Informa Healthcare USA, Inc. (original copyright 2004 by Marcel Dekker, Inc.)
Informa Healthcare is an Informa business

Printed and bound in India by Bhavish Graphics.

International Standard Book Number-10: 0-8247-4805-0 (Hardcover)
International Standard Book Number-13: 978-0-8247-4805-0 (Hardcover)

This book contains information obtained from authentic and highly regarded sources. Reprinted material is quoted with permission, and sources are indicated. A wide variety of references are listed. Reasonable efforts have been made to publish reliable data and information, but the author and the publisher cannot assume responsibility for the validity of all materials or for the consequences of their use.

No part of this book may be reprinted, reproduced, transmitted, or utilized in any form by any electronic, mechanical, or other means, now known or hereafter invented, including photocopying, microfilming, and recording, or in any information storage or retrieval system, without written permission from the publishers.

For permission to photocopy or use material electronically from this work, please access www.copyright. com (http://www.copyright.com/) or contact the Copyright Clearance Center, Inc. (CCC) 222 Rosewood Drive, Danvers, MA 01923, 978-750-8400. CCC is a not-for-profit organization that provides licenses and registration for a variety of users. For organizations that have been granted a photocopy license by the CCC, a separate system of payment has been arranged.

Trademark Notice: Product or corporate names may be trademarks or registered trademarks, and are used only for identification and explanation without intent to infringe.

Visit the Informa Web site at
www.informa.com

and the Informa Healthcare Web site at
www.informahealthcare.com

FOR SALE IN SOUTH ASIA ONLY

Preface

Proprietary pharmaceutical product development is driven by continuous innovations in drug discovery, drug polymorph preparation, dosage form design, and process engineering, while meeting rigorous regulatory standards. Innovations in all these areas are feasible with the application of technologies utilizing supercritical fluids (SCFs). Following a brief explanation of the SCF technologies, this preface introduces the objectives and the content of this book.

Under supercritical conditions, that is, above a substance's critical pressure and temperature, materials exhibit liquid-like solvent properties and gas-like viscosity, thermal conductivity, and diffusivity. While the solvent properties are beneficial in drug solubilization, polymer plasticization, and extraction of organic solvents or impurities, the gas-like properties significantly enhance mass transfer and promote extraction or reaction selectivity. High compressibility of fluids in the near-critical region offers the advantage of tunable solvent power within microseconds, which is utilized in fractionation and separation processes. These unique and complementary physical characteristics allow the development of efficient and versatile processes based on supercritical fluids. However, SCFs should not be considered as universal "super-solvents". Such a statement would be correct if high temperatures were permissible for pharmaceuticals. The transition into supercritical state for most polar liquids occurs at high temperatures, which are generally prohibitive for pharmaceutical manufacturing. Requirements for the mild processing conditions, purity, safety, and good economics explain why most of the current SCF applications employ supercritical carbon dioxide. However, many drug substances, reagents, and excipients have very low solubility in supercritical CO_2, and in attempting to overcome this major limitation, much research has been carried out to increase the utility of CO_2 as a solvent by employing polar organic co-solvents, novel surfactants, and different heterogeneous systems for extractions or reactions. There is also an increasing tendency to utilize other SCFs such as hydrofluorocarbons, propane, ethane, and dimethyl ether in specific applications for extraction and particle formation.

The original concept of supercritical fluids as a "green" alternative to traditional organic solvents in the 1980s in many important food and nutraceutical applications has now evolved into wider processing applications including extraction, fractionation, separation, cleaning, chemical reactions, and polymer processing. High-pressure engineering has advanced to afford the design of safe, large-scale SCF plants with improved economics. Significant progress has been achieved in the area of analytical and preparative supercritical fluid chromatography. Within the last decade, supercritical fluids have also made rapid inroads into pharmaceutical development and provided several innovations as reflected by hundreds of patents granted in this area.

All the advances in SCF applications have been supported by increased fundamental and mechanistic understanding of SCF processes, theoretical modeling, and optimization. It is interesting to see how limitations in one area of SCF technology opened up new research directions in others. For example, low solubility of most solid pharmaceuticals in pure CO_2 made it an ideal "anti-solvent" for precipitation and crystallization processes, a direct way of producing composite drug-excipient particles as well as fine powders in relatively pure physical forms. SCF technology already has a special place in particle engineering for drug delivery. Many drug delivery systems, including injectable, respiratory, and oral formulations, require uniform fine particles with defined solid-state properties and predictable drug release profile. However, manufacturing of such particles still represents significant challenges to the pharmaceutical industry. These challenges include the quality and consistency of particulate products and scale-up. The requirement for more consistent particle engineering was probably the most important driving force in the development of SCF particle technology. Additional process parameters—pressure and fluid density—enable the achievement of crystallization conditions not possible with conventional batch crystallization techniques, thus facilitating production of new polymorphic forms, materials with different crystallinity, and surface energetics. The enclosed SCF systems make possible the processing of potent and heat-, light-, or shear-sensitive compounds with an additional benefit of a sterile environment suitable for cGMP manufacturing.

This field is still very young, and SCF technologies are not yet widespread in the pharmaceutical industry. However, many innovative drug formulations, which involve SCF processing, are passing through different evaluation and development stages. Of course, SCF technologies are not going to replace existing and well-established techniques such as solvent coating, granulation, spray-drying, lyophilization, or liquid extraction. However there are certainly many niche areas of pharmaceutical applications where SCF technology is beneficial for product quality, batch consistency, and the reduction of manufacturing complexity and cost. Perhaps one of the major strengths of SCF particle technology is not in its universality but in the versatility and flexibility in offering alternative processing approaches. There is little doubt that new uses of supercritical fluids or compressed gases will be found, particularly in the area of nanotechnology, composite and porous materials, disperse systems such as liposomes and emulsions, and biological substance formulation.

Progress in SCF technology requires a multidisciplinary approach, which is captured in the structure of this book. While the fundamentals of supercritical fluids are advanced by the scientists in the areas of basic physicochemical sciences and engineering, scientists at the interface of these areas and pharmaceutical sciences are enabling the application of supercritical fluid technologies for the development of pharmaceuticals. These applications are spurred by various pharmaceutical developmental needs, which were partly alluded above: (a) synthesis of solid forms of drugs with better biopharmaceutical properties such as solubility, stability, and absorption, (b) engineering of drug and drug-excipient particles with uniform size distribution and unique surface features, enhanced yield, and reduced residual solvent content, to enable precise and/or targeted drug delivery, to improve process efficiency, and for better safety and compliance with regulatory agencies, (c) extraction and/or purification of drugs from natural, synthetic, and biological matrices for drug discovery and/or analysis. Several non-pharmaceutical researchers, who primarily contributed to the fundamentals of supercritical fluid technologies, have progressed to apply their technologies to pharmaceutical applications. However, a unifying platform to better educate the pharmaceutical scientists in the fundamentals of supercritical fluid technologies and the non-pharmaceutical scientists working on the fundamentals of supercritical fluids in the applications of this technology to pharmaceutical research is thus far lacking. Therefore, one purpose of this book is to provide a unifying compendium that brings together the fundamentals, process engineering and pharmaceutical applications of supercritical fluid technologies, in order to better inform the pharmaceutical scientists. Another purpose of this book is to provide a snapshot of the current state and scientific achievements and to stimulate further research in this area. This book will also be a valuable teaching aid to graduate and undergraduate students in physical pharmaceutics, pharmaceutical technology, chemical engineering, and related areas. It will also provide a valuable resource for scientists involved in drug product development. In addition, scientists and engineers with innovative supercritical fluid technologies will find this book useful in identifying new applications for their technologies.

This book is divided into three complimentary sections. Part I provides the fundamentals of supercritical fluid technology. Fundamentals of particle formation are included in greater detail because preparation of particulate drug and drug delivery systems is a relatively new and active area of pharmaceutical application of SCFs. Chapter 1 focuses on the molecular-level design of CO_2-soluble surfactants, co-solvents, and covalent modification, which would make biomaterials and pharmaceuticals solubilized in supercritical CO_2. Thermodynamics of the processes (Chapter 2) provides the basis for understanding of phase behavior and drug solubility, important for all SCF processes. Chapter 3 gives a comprehensive review of, and highlights new findings related to, the non-equilibrium processes of fluid-dynamics, mass-transfer, and precipitation kinetics including general mechanistic and scale-up principles. The research challenge here is intimately linked with the design of reliable in-line measurements in SCFs, in order to provide means for scientifically driven process development, monitoring, and control. Chapter 4 makes a logical continuation of this topic, discussing chemical engineering approaches to different particle formation processes. It is shown that SCFs offer a wide range of complementary techniques to produce particles for pharmaceutical applications.

Part II is dedicated to drug delivery applications. It starts with a general discussion of the interfacial and colloid phenomena (Chapter 5), which creates a link between the fundamentals of interactions in the carbon dioxide–based systems, and specific applications for pharmaceutical processing and drug delivery system design. Production of powders for respiratory applications (Chapter 6) is one of the most promising applications of SCF particle technology and may serve as an example of SCF processing that can significantly improve the drug delivery characteristics of dry-powder and metered-dose inhalers. This chapter also gives insight into characterization and optimization of respiratory particles. Control of the drug physical form, in terms of crystallinity, solid solution, hydration, and polymorphism, is considered in Chapter 7. Here the emphasis is placed on vigorous solid-state characterization which makes possible the design of low-risk, controllable, and predictable formulation processes, as required for pharmaceutical development and manufacture. Polymeric materials and their interactions with supercritical fluids are detailed in Chapter 8 because SCFs appear attractive for the preparation and impregnation of monolithic drug delivery matrixes, bioactive scaffolds, contact lenses, and some other biomedical products. Chapter 9 considers potential drug products for controlled drug release, which include nano- and micro- particles, films, and foams as well as use of SCF for extraction of residual organic solvent from drug delivery systems. Biological materials (Chapter 10) entail special consideration in formulation and processing, with a detailed assessment of their chemical and physical stability in solution and solid-state form. This technological area is driven by a search for more efficient processing and stabilization techniques as well as development of alternatives to solution injection drug delivery techniques including inhaled and controlled or sustained release formulations.

Part III of this book presents various important developments that will either directly or indirectly impact the development of pharmaceutical products. Performance of chemical reactions in supercritical fluids offers unique advantages, and it holds tremendous potential in the synthesis of drugs and excipients intended for pharmaceutical products. The preparation of chiral drugs is discussed in Chapter 11. This review shows that asymmetric catalysis, whether by enzyme, heterogeneous or homogeneous catalysis, can be performed in SCF with significantly enhanced reaction selectivity, rate, and increased catalyst stability. Developments in supercritical fluid chromatography (SFC), an integral part of drug analysis and purification process, are discussed in Chapter 12. It is shown that there is a significant economical incentive for greater penetration of SFC in discovery and development of small drug molecules. Supercritical fluid extraction (Chapter 13) is one of the most developed technological areas which is now consolidated into a powerful tool with such different objectives as the extraction of active compounds or oils used in the pharmaceutical preparations or for monitoring of levels of drugs and their metabolites in biological tissues and fluids, and also for the extraction of active compounds from food. Food industry pioneered the industrial level applications of supercritical fluids. Major advances are being made in the design and scale-up of supercritical fluid processes for nutraceuticals (Chapter 14), which can be a rich resource for pharmaceutical product development. It is noted here, as well as mentioned in other chapters, that the same technological platforms that employ supercritical fluids can also be applied with liquid compressed gases (for example, liquid CO_2 in combination with organic solvents and also pressurized liquids such as water and ethanol above their boiling point). This allows the extension of the applications of "real" SCF because these solvents share some advantages with SCF in terms of mass-transfer and also enable better solubility for certain extracts. Several engineering marvels fail during scale-up and upon imposition of rigorous regulatory challenges associated with pharmaceutical product development. However, supercritical fluid technologies have surpassed several hurdles, and the processes are scaleable and comply with good manufacturing practices, as described in Chapter 15.

The preparation of this book would not be possible without the enthusiastic and timely contributions of various experts in supercritical fluid processes. The authors have summarized various areas of research while sharing their own experience in the field. We are also indebted to the expert staff at Marcel Dekker, Inc., for their assistance in the preparation of this book. Finally, we are grateful for the support of our colleagues and collaborators at Ferro Corporation, Nektar Therapeutics, Bradford University, and the University of Nebraska Medical Center. We hope that this compendium will serve some of your pharmaceutical product development needs, and we look forward to your feedback.

Peter York
Uday B. Kompella
Boris Y. Shekunov

Contents

PART III.   OTHER APPLICATIONS AND SCALE-UP ISSUES

Contributors

Elisabeth Badens Laboratoire de Procédés Propres et Environment, Université d'Aix–Marseille, Aix-en-Provence, France

Jerzy Baldyga Department of Chemical and Process Engineering, Warsaw University of Technology, Warsaw, Poland

Nagesh Bandi GlaxoSmithKline, Parsippany, New Jersey, U.S.A.

Eric J. Beckman Department of Chemical and Petroleum Engineering, University of Pittsburgh, Pittsburgh, Pennsylvania, U.S.A.

Terry A. Berger Berger Instruments, Newark, Delaware, U.S.A.

Olivier Boutin Laboratoire de Procédés Propres et Environment, Université d'Aix–Marseille, Aix-en-Provence, France

Gérard Charbit Laboratoire de Procédés Propres et Environment, Université d'Aix–Marseille, Aix-en-Provence, France

Albert H. L. Chow Division of Pharmaceutical Sciences, School of Pharmacy, The Chinese University of Hong Kong, Shatin, Hong Kong

Jean-Yves Clavier Lavipharm, East Windsor, New Jersey, U.S.A., and Separex 5, Champignuelles, France

Sandro R. P. DaRocha   Department of Chemical Engineering, The University of Texas at Austin, Austin, Texas, U.S.A.

Ram B. Gupta   Department of Chemical Engineering, Auburn University, Auburn, Alabama, U.S.A.

Marek Henczka   Department of Chemical and Process Engineering, Warsaw University of Technology, Warsaw, Poland

Philip G. Jessop   Department of Chemistry, University of California, Davis, California, U.S.A.

Keith P. Johnston   Department of Chemical Engineering, The University of Texas at Austin, Austin, Texas, U.S.A.

A. Jurado-López   Analytical Chemistry Department, University of Córdoba, Córdoba, Spain

Sergei G. Kazarian   Department of Chemical Engineering and Chemical Technology, Imperial College, London, United Kingdom

Jerry W. King   Supercritical Fluid Facility, Chemistry Division, Los Alamos National Laboratory, Los Alamos, New Mexico, U.S.A.

B. L. Knutson   Department of Chemical and Materials Engineering, University of Kentucky, Lexington, Kentucky, U.S.A.

Uday B. Kompella   Department of Pharmaceutical Sciences, University of Nebraska Medical Center, Omaha, Nebraska, U.S.A.

C. Ted Lee   Department of Chemical Engineering, The University of Texas at Austin, Austin, Texas, U.S.A.

Ge Li   Department of Chemical Engineering, The University of Texas at Austin, Austin, Texas, U.S.A.

M. D. Luque de Castro   Analytical Chemistry Department, University of Córdoba, Córdoba, Spain

J. L. Luque-Garcia   Analytical Chemistry Department, University of Córdoba, Córdoba, Spain

Mamata Mukhopadhyay   Chemical Engineering Department, Indian Institute of Technology, Bombay, India

Janice L. Panza   Department of Chemical and Petroleum Engineering, University of Pittsburgh, Pittsburgh, Pennsylvania, U.S.A.

Michel Perrut   Lavipharm, East Windsor, New Jersey, U.S.A., and Separex 5, Champignuelles, France

Christopher B. Roberts   Department of Chemical Engineering, Auburn University, Auburn, Alabama, U.S.A.

Marazban Sarkari   RX Kinetix, Louisville, Colorado, U.S.A.

Boris Y. Shekunov   Pharmaceutical Technologies, Ferro Corporation, Independence, Ohio, U.S.A.

Henry H. Y. Tong   Division of Pharmaceutical Sciences, School of Pharmacy, The Chinese University of Hong Kong, Shatin, Hong Kong

Matthew Z. Yates   Department of Chemical Engineering, The University of Texas at Austin, Austin, Texas, U.S.A.

1

Chemistry and Materials Design for CO_2 Processing

Janice L. Panza and Eric J. Beckman
Department of Chemical and Petroleum Engineering,
University of Pittsburgh, Pittsburgh, Pennsylvania, U.S.A.

Carbon dioxide has many properties that make it an attractive solvent, including low toxicity, nonflammability, and an environmentally benign nature. It has been proposed as a "green" alternative to traditional organic solvents because it is neither regulated as a volatile organic chemical (VOC) nor restricted in food or pharmaceutical applications. Both liquid and supercritical CO_2 have been exploited as solvents; however, it has been suggested that supercritical fluids have the additional benefit of "solvent tunability" (1). Small changes in pressure in the supercritical region lead to considerable changes in fluid density, which in turn, lead to changes in solvent properties. Under supercritical conditions, CO_2 like all supercritical fluids, offers mass transfer advantages over conventional organic solvents owing to its gaslike diffusivities, low viscosity, and vanishing interfacial tension.

Considerable research is being done to increase the utility of CO_2 as a solvent. In the area of pharmaceutical applications, supercritical fluids are useful as solvents in the production of particulate drugs, the extraction and separation of active ingredients, and the preparation of microemulsions and sustained drug delivery systems (2). A major disadvantage, insofar as solvent

1

behavior is concerned, is that CO_2 is a poor solvent for many polar and nonpolar compounds, although it is miscible with many small, nonpolar (and hence volatile) molecules at moderate pressures. Whereas organic compounds are usually classified as either hydrophilic or lipophilic, it is often the case that neither hydrophilic nor lipophilic molecules exhibit appreciable solubility in carbon dioxide at pressures less than 500 bar. The term "CO_2-philic" has thus been coined to describe molecules that exhibit high solubility in CO_2 (when solid–fluid phase equilibria govern the situation) at moderate pressures or, where liquid–liquid phase behavior is concerned, complete miscibility at moderate pressures (3). It is unlikely, however, that most compounds to be used in pharmaceutical and/or biomaterial applications will demonstrate appreciable solubility or complete miscibility in CO_2 at economically tractable pressures.

Although exactly what governs solubility in CO_2 is still not entirely clear, this chapter reviews the properties of CO_2 and the important characteristics that dictate what can and cannot readily be mixed with carbon dioxide. We present methods to solubilize CO_2-phobic or essentially insoluble compounds such as pharmaceuticals and biomaterials in CO_2. Finally, taking hints from detailed studies on solubility in CO_2, we predict what characteristics are necessary to design materials for solubility in CO_2.

1. REVIEW OF SOLUBILITY IN CO_2

1.1. Early Work and the Discovery of Fluorinated CO_2-philes

When examining the phase behavior of compounds in carbon dioxide, it is important to note that both solid–fluid phase behavior (solubility) and liquid–liquid phase behavior (miscibility) are reported in the literature. In general, the designation of one material as more or less "CO_2-philic" than another refers to whether solubility or miscibility pressures (for a given concentration) are relatively lower or higher than those of the comparison material.

CO_2 is in some ways an elusive solvent. Defining exactly the molecular characteristics that govern solubility/miscibility in both liquid and supercritical CO_2 continues to create controversy in the scientific literature. Indeed, what was originally hypothesized about CO_2's behavior in earlier work no longer holds true today. Although the solvent properties of CO_2 are not entirely clear, the identification of new CO_2-philic compounds continues to occur. The progression of our understanding of the governing solvent properties of CO_2 deemed important through the years is presented in this section.

Our review begins in 1954 with the seminal publication of Francis (4). Although the phase behavior of different organic or inorganic compounds in liquid CO_2 had been published prior to this work, Francis reported the phase behavior of 261 compounds (binary mixtures) and 464 ternary systems that included liquid CO_2 as one of the components. At that time, no other reports of ternary systems in liquid CO_2 had been published. Francis found that at moderate concentrations (up to 40% by weight), CO_2 acts as a dissolved gas and cosolvent, and hence promotes mixing in ternary systems. At higher concentrations (60–90%), CO_2 is a relatively poor solvent for many of the same compounds. That CO_2 is a rather weak solvent for most compounds still holds true today.

Initially, the Hildebrand solubility parameter [δ, defined as the square root of the cohesive energy density (CED) divided by molar volume] was used as an indication of what ought be soluble/miscible in CO_2, where compounds that have a similar δ to CO_2 would be soluble. In 1969 Giddings compared the elution power of a series of dense gases with that of a series of liquids based on δ (5). Giddings suggested that the solubility parameter was a good indicator of elution power and could serve as a guide in the selection of a fluid to use in chromatography. According to Giddings's approximation, δ was proportional to the square root of the critical pressure ($P_c^{1/2}$):

$$\delta = \frac{1.25 P_c^{1/2} \rho_{rg}}{\rho_{rl}}$$

where ρ_{rg} and ρ_{rl} are the reduced density of the supercritical gas and the reduced density of the liquid at its normal boiling point, respectively. Based on values of δ_s calculated by Giddings, the solvent power of CO_2 should be comparable to that of pyridine; clearly the use of a simple model for δ could not be directly applied to carbon dioxide.

Allada later approximated δ as follows for liquids:

$$\delta = \left(\frac{\Delta H_{vap} - RT}{V} \right)^{1/2}$$

where ΔH_{vap} is the heat of vaporization at temperature T and V is the molar volume (6). In this approximation, the numerator can be replaced by the change in internal energy (ΔU) as one moves from the dense fluid state to zero density (the ideal gas state). Based on this approximation, at temperatures near 25°C and pressures between 100 and 300 bar, the δ of CO_2 was calculated to be like that of an alkane. The available experimental data on compounds in CO_2, however, did not really support this hypothesis.

In 1984 Hyatt performed a comprehensive study on the solubility of many organic compounds in liquid and supercritical CO_2, (7), generalizing the solvent behavior of CO_2 based on his own work and the work of Francis (4), Gouw (8), and Alwani (9):

1. Liquid CO_2 behaves like a hydrocarbon solvent with the exception of methanol miscibility.
2. Liquid CO_2 *does not* interact with organic weak bases (e.g., anilines, pyrroles, pyridines) but forms salts with aliphatic primary and secondary amines.
3. Many pairs of immiscible or partially miscible liquids form a single phase in liquid CO_2.
4. Aliphatic hydrocarbons having chain lengths less than C_{20} and small aromatic hydrocarbons dissolve in liquid CO_2.
5. Halocarbons, aldehydes, esters, ketones, and low molecular weight alcohols are soluble in CO_2.
6. Poly(hydroxy aromatics) are insoluble.
7. Polar compounds such as amides, ureas, urethanes, and azo dyes show low solubility in CO_2.
8. Few compounds with molecular weights above 500, regardless of structure, are soluble in liquid CO_2.

Based on visible and IR spectroscopy studies of organic solutes in CO_2, Hyatt concluded that CO_2 behaves like a hydrocarbon solvent with very low polarizability. Hyatt's study was important in that it demonstrated that CO_2 could dissolve (or was miscible with) most low molecular weight organic compounds; however, some of his hypotheses do not appear to hold true today. In particular, the notion that CO_2 had solvent properties similar to hydrocarbons (as also indicated by Allada) was misleading, but lingered for years. CO_2 could not dissolve many of the compounds that would dissolve in low carbon alkanes, and vice versa. In addition, statement 2 was proved incorrect, as discussed later in this section, when CO_2 was shown to interact with bases (10).

Iezzi et al. also originally subscribed to the notion that CO_2 had solvent properties similar to those of alkanes (11). Pentane, isoctane, and perfluorohexane were used as low pressure screening fluids for high pressure CO_2 based on the similarity of dipole moments, solubility parameters, and polarizability/dipolarity parameters of the hydrocarbons and carbon dioxide. In this study, fluorination of an alkyl segment of a compound was found to increase its solubility in CO_2. This was one of the earlier studies showing an apparent favorable interaction (and hence ready miscibility) between fluorinated compounds and CO_2.

During this period, Consani and Smith performed an extensive, qualitative study on the solubility of many commercially available surfactants in CO_2 (12). Not surprisingly, most ionic surfactants were found to be relatively insoluble in CO$_2$, whereas fluorinated analogues were observed to dissolve in CO$_2$, in line with the work of Iezzi regarding the effect of fluorination.

Given the work by Iezzi and Consani, Hoefling et al. investigated how the incorporation of fluoroether functional groups into polymers and surfactants enhanced solubility in CO$_2$ (13). The polymer poly(hexafluoropropylene oxide) displayed high miscibility in CO$_2$ (MW 13,000 soluble up to 10 wt % polymer in CO$_2$ at 295 K and approximately 17 MPa). Furthermore, those surfactants that incorporated fluoroether functionalities, hydroxyaluminum bis[poly(hexafluoropropylene oxide) carboxylate], poly(hexafluoropropylene oxide) carboxylic acid, and sodium poly(hexafluoropropylene oxide) carboxylate, exhibited complete miscibility in CO$_2$ at pressures less than 20 MPa.

Further studies on fluoroether-functional amphiphiles were performed by Newman et al. (14), who noted that several competing structural factors appeared to determine miscibility pressures in CO$_2$. Normally, increasing the molecular weight of a compound tends to increase the cloud point pressure; Newman found that increasing the molecular weight via the addition of CO$_2$-philic fluoroether groups (and branching of CO$_2$-philic tails) decreased the cloud point pressure. Thus miscibility pressures of such compounds in CO$_2$ depend upon a balance between the enthalpy and entropy of mixing. Increasing the number of CO$_2$-philic groups in a compound renders it more soluble in CO$_2$ owing to an enhanced enthalpy of mixing, but eventually increasing the size of the molecule leads to less solubility because the entropy of mixing is decreased. Silicone-based amphiphiles [poly(dimethylsiloxane)] were also investigated for their solubility in CO$_2$, since like fluoroethers, dimethylsiloxane has a low solubility parameter (15). As demonstrated by Hoefling et al., the trends in the balance of the enthalpy and entropy of mixing also holds true for silicone-based surfactants (16); however, silicone-based surfactants are much less CO$_2$-philic in CO$_2$ than fluoroethers (15).

In the early 1990s, DeSimone's group showed that poly(1,1-dihydroperfluorooctyl acrylate) [or poly(FOA)] exhibits extremely low miscibility pressures in CO$_2$, regardless of molecular weight (17). These investigators synthesized Poly(FOA) (homogeneously) in supercritical CO$_2$ to a high molecular weight of ~2.7 × 10^5 g/mol. The investigators attributed the high solubility of poly(FOA) to specific solute–solvent interactions between electronegative fluorines and the electron-poor carbon in CO$_2$. Small-angle neutron scattering (SANS) was later used to monitor the conformation of poly(FOA) and other known CO$_2$-philes in CO$_2$ (18). The second virial coefficient, which describes the interaction between the polymer and solvent, was found to be positive for poly(FOA) in CO$_2$, indicating that CO$_2$ is a good solvent for the polymer. By contrast, CO$_2$ was found to be a relatively poor solvent for fluoroether polymers and silicones.

Dardin et al. used ^{1}H-NMR and ^{19}F-NMR techniques to demonstrate that proton and fluorine nuclei behave differently in supercritical CO_2 (19,20). Whereas protons are susceptible to only the bulk characteristics of CO_2, fluorine nuclei experienced an additional contribution due to magnetic shielding attributed to van der Waals interactions between the fluorine and CO_2. However, neither IR work by Yee et al. (21) nor first-principles calculations (22) suggested that fluoroalkyl groups and CO_2 interact specifically. Recent work by McHugh et al. (23) suggests that fluorinated materials interact specifically with CO_2 only if the presence of the fluorines leads to creation of a substantial dipole–moment (thus creating dipole–quadrupole interactions with carbon dioxide). How and with what carbon dioxide exhibits favorable specific interactions continues to be an interesting target for research.

For the most part, compounds containing fluorine and silicone functionalities were assumed to be compatible with CO_2 based on the low solubility parameters of both. Although solubility parameter (δ) calculations (using the equation of state) of CO_2 indicate that this gas has solvent power similar to that of low carbon ($< C_6$), alkanes, approximately 20% of δ may be due to the large quadrupole moment of CO_2 leading to an inflated value (24). Recognizing that δ of CO_2 was inflated, Johnston and colleagues suggested that polarizability per volume (α/v) was a better measure of the solvent power of CO_2 (25). Calculations of α/v show CO_2 to be a feeble solvent.

Fluorination of a compound is a useful stategy in certain situations, but it *does not* guarantee solubility in CO_2 at room temperature and moderate pressures. McHugh et al. demonstrated that a completely fluorinated trifluoroethylene–hexafluoropropylene copolymer (FEP_{19}) (Figure 1A) requires temperatures over 190°C and pressures greater than 1000 bar for solubility in CO_2 (26,27). In contrast, a partially fluorinated vinylidene fluoride–hexafluoropropylene copolymer (Fluorel) (Figure 1B) was soluble in CO_2 at more moderate (but still extreme) conditions: temperatures of 100°C and pressures below 1000 bar (26). The difference between these two polymers is in their respective polarities; Fluorel is polar, which allows it to interact favorably with carbon dioxide through its quadrupole moment. Since CO_2 has such a low polarizability, its quadrupole moment plays a major role in solubility. CO_2 is hence a good solvent for Fluorel, most likely a result of specific interactions between CO_2 and the vinylidene fluoride segments (28).

Like Fluorel, poly(vinylidene fluoride) (PVDF) (Figure 1C) was expected to exhibit favorable interactions with CO_2 owing to the dipole moment associated with the CH_2 and the CF_2 groups (29). However, higher temperatures and pressures were required to dissolve PVDF in CO_2 relative to Fluorel, even though PVDF is more polar. This indicated that polymer architecture also plays a role in solubility in CO_2. Fluorel has a larger free

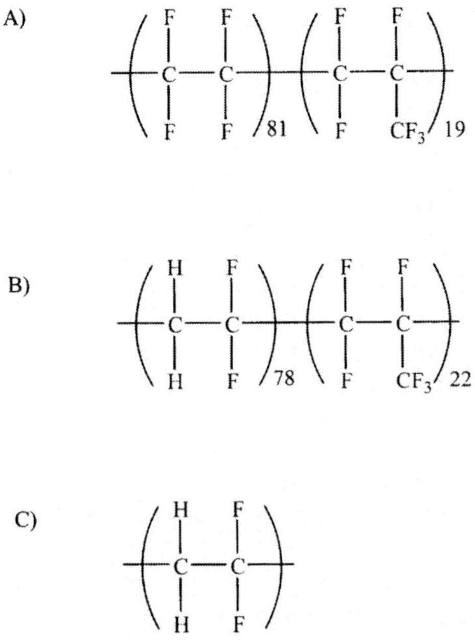

A)

B)

C)

FIGURE 1 Structure of (A) FEP$_{19}$, (B) Fluorel, and (C) PVDF.

volume (as shown by its much lower glass transition temperature), which makes it easier to dissolve in CO$_2$ owing to enhanced entropy of mixing. The stiffer the main chain, generally the more difficult (higher temperatures, pressures) it is to dissolve in CO$_2$ (26). In summary, CO$_2$-philicity represents the sum of all the effects of structural parameters on the mixture thermodynamics, specific interactions with CO$_2$, solute self-interactions, and entropic contributions.

Eastoe et al. studied the ability of fluorinated surfactants to form microemulsions in CO$_2$ (30). Replacing the terminal F for H in the surfactant $C_8F_{17}COO^-Na^+$ to form $HC_8F_{16}COO^-Na^+$ resulted in a permanent dipole moment at the hydrophobic tip of the chain, making the surfactant "bipolar" (31). The replacement of F with H in the terminal methyl group reduced packing in micelles, consistent with the dipolar interactions, and increased the critical micelle concentration (cmc) by a factor of 4. Most likely, the tail with the H in the terminal methyl group is more attractive to CO$_2$ as a result of the dipole moment, which increases the cmc and reduces the extent of molecular packing in micelles. Eastoe and colleagues suggested using the limiting air–water surface tensions at the critical micelle concentration, γ_{cmc}, as an indicator of surfactant solubility in CO$_2$—the lower the value, the more soluble the molecule ought to be (32). In their studies, the γ_{cmc} decreased with an increase in chain length from $-(CF_2)_4-$ to $-(CF_2)_6-$ and γ_{cmc} increased with the replacement of F with H in the terminal methyl group of a series of fluorinated analogues to the oft-used surfactant sodium bis(2-ethyl-1-hexyl)sulfo-succinate (AOT) (33,34).

From this significant body of work, one can surmise that although incorporating fluorine into a compound may enhance its solubility in CO_2, it is also necessary for the compound to be somewhat polar if high solubility in CO_2 is to be obtained. Partial fluorination of a compound can lead to creation of dipoles, which enhance the solubility in CO_2 owing to specific interactions with CO_2. In addition, increases to polymer-free volume, based on choice of materials with substantial backbone flexibility, increases solubility in CO_2.

1.2. Nonfluorous CO_2-Philes

Kazarian et al. showed that other functional groups do interact specifically with CO_2 (35). These investigators used Fourier transform IR spectroscopy to show that CO_2 exhibits specific interactions with polymers possessing electron-donating functional groups, such as carbonyls, most likely of a Lewis acid/base nature. Their results suggest that CO_2 acts as the electron acceptor rather than the electron donor, for example, with poly(methylmethacrylate) (PMMA), while noting that CO_2 may also act as the electron donor in other cases. Likewise, Quinn et al. have shown that CO_2 interacts as an electron acceptor with the oxygen atom of water in salt hydrates (36). Meredith et al. demonstrated specific interactions between CO_2 and Lewis bases such as triethylamine, pyridine, and tributyl phosphate (10). The authors speculate that Lewis base groups add specific interactions that although weak, could raise solubility in CO_2. As early as the 1980's, Harris et al. recognized that specific interactions with Lewis bases were important to solubility in CO_2; however, the work was published in a patent and did not receive great exposure in the scientific community (37). The patent describes a method for increasing the viscosity of CO_2 for oil recovery through the use of soluble polymers. The inventors claimed that a suitable polymer should contain a certain amount of electron donor capacity to interact (in a donor–acceptor fashion) with carbon dioxide yet still exhibit an overall low cohesive energy density. Whereas a basic molecular design for such polymers was proposed, no examples were given, and hence this work remained more a theoretical than a practical construct.

O'Neill and coworkers approached the problem of CO_2 solubility from a different perspective, taking low cohesive energy density (and hence weak self-interactions) to be the key parameter (38). Solubility studies on diverse polymers such as polyethers, polyacrylates, polysiloxanes, and many block copolymers demonstrated that a decrease in the cohesive energy density

(CED) of a polymer, closer to that of CO$_2$, increased the solubility (lowered the miscibility pressures) in CO$_2$. O'Neill theorized that solubility is thus *not controlled* by polymer–CO$_2$-specific interactions, but mainly by poor polymer–polymer interactions, and that a low cohesive energy density (reflected by a low surface tension) of a polymer primarily determines its solubility in CO$_2$. In a U.S. patent issued in 1998, Johnston et al. demonstrated that surfactants based on poly(propylene oxide) and poly(butylene oxide) were also quite soluble in CO$_2$ (39). The invention was significant for demonstrating that nonfluorinated surfactants can exhibit appreciable solubility in CO$_2$.

Eastoe et al. discovered that incorporation of trimethyl functional groups onto the hydrocarbon end of an AOT analogue surfactant rendered the molecule CO$_2$ soluble (Figure 2) (40). Increasing methylation decreases the value of γ_{cmc}; the maximum water solubilization capacity of these micelles was not reported. Similarly, Lagalante et al. studied the effect of hydrocarbon branching on the solubility of metal chelates in CO$_2$ (41). Substitution of *tert*-butyl groups for methyl groups on the end of metal chelates rendered the molecules more soluble in CO$_2$. This is consistent with the approach described by O'Neill (38), in that the inclusion of branched alkanes tends to lower the cohesive energy density, as proposed earlier by Dandge and colleagues (42).

Furthermore, studies by Liu et al. demonstrated that other nonfluorous and non-silicone-containing surfactants were soluble in CO$_2$. Dynol-604, an acetylenic glycol-based nonionic surfactant, showed high solublility in CO$_2$ (5 wt % at 313.15 K and ~20 MPa) and formed water-in-CO$_2$ microemulsions capable of dissolving the water-soluble dye methyl orange (43). The water loading within the micelles, defined as the molar ratio of water to surfactant, increased with an increase in pressure and decreased with an increase in temperature. The nonionic surfactant tetraethylene glycol *n*-laurel ether (C$_{12}$E$_4$) was less soluble than Dynol-604 (2.9 wt % at 313.15 K and ~20 MPa); however, not surprisingly the addition of the cosolvent *n*-pentanol enhanced the solubility of C$_{12}$E$_4$ in CO$_2$ (44). The nonionic surfactants Ls-36 and Ls-37, which are based on ethylene oxide and propylene oxide, were also soluble at about 4 wt % at similar temperatures and pressure to Dynol-604 and C$_{12}$E$_4$ (45). These surfactants were also able to form water-in-CO$_2$ microemulsions, although the maximum water solubilization was found to be low (W$_0$ < 10).

Based on the fundamental work of other investigators through the 1990s, Fink et al. proposed a template for the design of nonfluorinated CO$_2$-philic materials (46). These researchers investigated the use of side chain functionalization attached to silicone oligomers to improve CO$_2$ solubility. The authors speculated that highly flexible chains (using low glass transition temperatures as an indication of chain flexibility) would lead to a more favorable entropy of mixing with CO$_2$. As suggested by O'Neill (38), Meredith (10), Kazarian (35), and Rindfleisch (26), Lewis base groups were included to

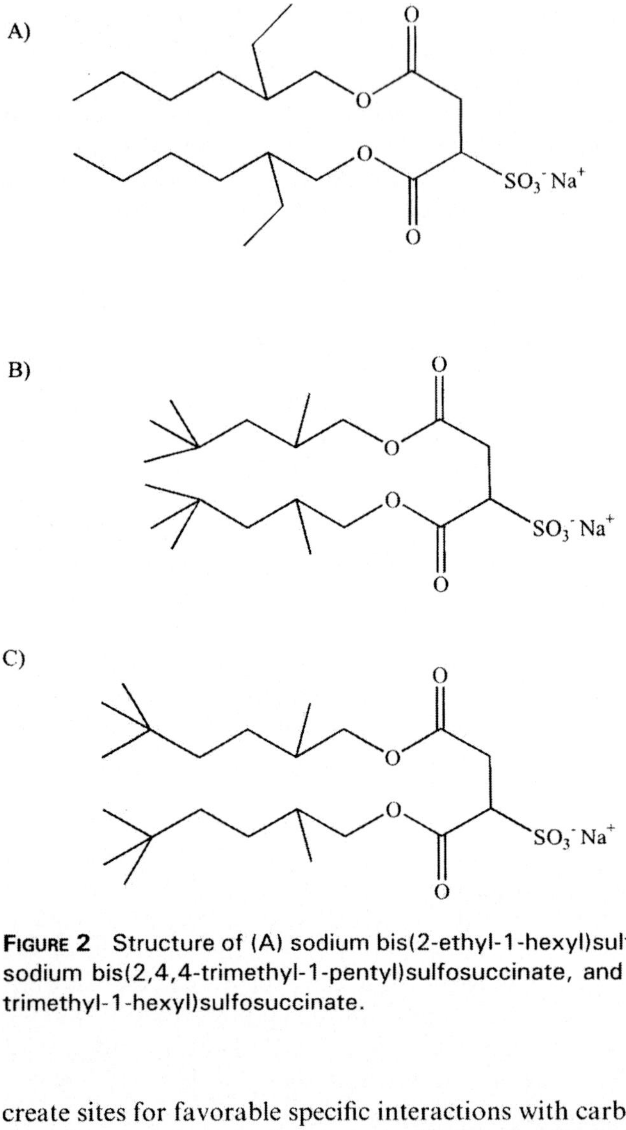

A)

B)

C)

FIGURE 2 Structure of (A) sodium bis(2-ethyl-1-hexyl)sulfosuccinate (AOT), (B) sodium bis(2,4,4-trimethyl-1-pentyl)sulfosuccinate, and (C) sodium bis(3,5,5-trimethyl-1-hexyl)sulfosuccinate.

create sites for favorable specific interactions with carbon dioxide. Although the subjects were treated separately, Fink and colleagues recognized that the entropy and enthalpy of mixing are coupled, and any plan to render the enthalpy of mixing more favorable through the addition of more specific interactions with CO_2 will eventually reach a point of diminishing return, with the entropy of mixing becoming less favorable. Based on their hypothesis, Fink et al. demonstrated that functionalization of a silicone oligomer (Figure 3) with only five ester-functional side chains lowered the cloud point

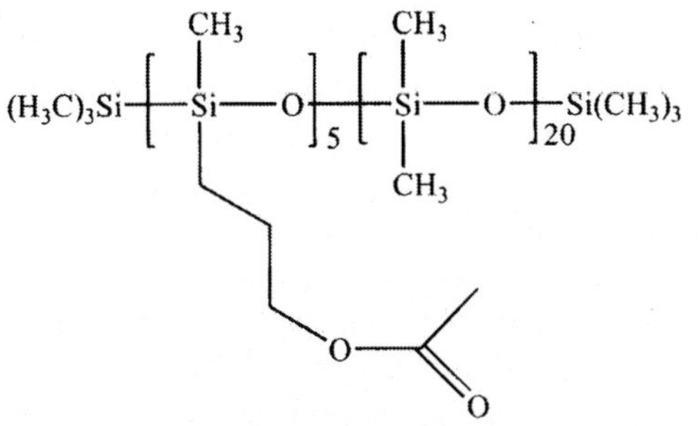

FIGURE 3 Functionalized silicone oligomer.

curve at 22°C by 2500 psi (close to the cloud point of a perfluoropolyether). The investigators concluded that the specific interactions of terminal carbonyls of the grafted side chains and CO$_2$ improved the solubility of the silicone oligomer. This was the first indication that hydrocarbon segments could render a molecule CO$_2$-philic.

Sarbu et al. hypothesized that a CO$_2$-philic copolymer would contain two monomers: one (M$_1$) that provides high flexibility and high free volume to enhance the entropy of mixing and a low cohesive energy density (weak solute-solute interactions), and a second (M$_2$) that contains Lewis base groups to interact with CO$_2$, but lacks both donors and acceptors of hydrogen bonds (47). Finally, these investigators suggested that the enthalpic interactions between M$_1$ and M$_2$ should not be favorable, which would promote the dissolution of the material in any solvent, including carbon dioxide. To create the model copolymer, carbonyls (M$_2$) were incorporated into a polymer via copolymerization of CO$_2$ with propylene oxide (PO) (M$_1$) (Figure 4). A copolymer of 250 repeat units with 15.4% carbonate units exhibited lower miscibility pressures in CO$_2$ than a fluoroether whose chain length is much lower (175 repeats), proving the copolymer to be CO$_2$-philic. Sarbu et al. showed that functionalization of polyethers with carbonyl groups (acetate) in the side chain or incorporation of carbonyls (carbonate) into the polyether backbone also results in a cloud point pressure in CO$_2$

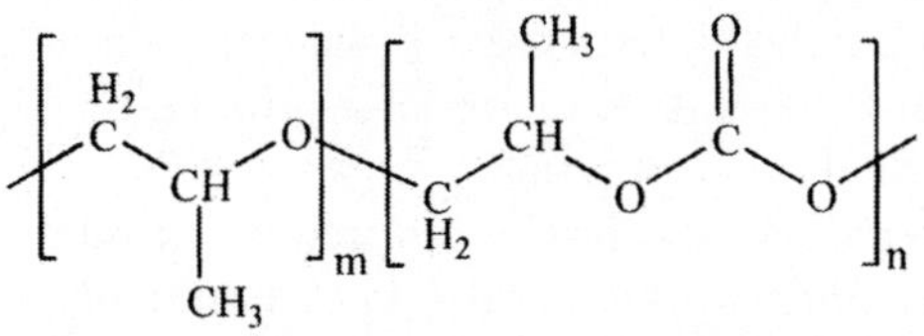

FIGURE 4 Structure of a CO$_2$-philic poly(ether–carbonate) copolymer.

lower than that found in the homopolymer (48). Whether it is better to have the carbonyls in the backbone or side chain is currently being investigated.

Lastly, based on ab initio calculations that revealed that methyl acetate exhibits a favorable *two-point* interaction with CO_2, Raveendran et al. suggested that acetylation of hydroxyl groups might be a possible method to solubilize hydroxylated compounds into CO_2 (49). In theory, a hydrogen atom attached to the carbonyl carbon (α carbon) forms a weak yet cooperative C—H $\cdots$ O hydrogen bond between the hydrogen and one of the oxygen atoms of CO_2. To test their hypothesis, the investigators explored the solubility of three acetylated carbohydrates in CO_2; the peracylated sugars exhibited low miscibility pressures at concentrations up to 30 wt % in supercritical CO_2 (~100–110 bar, 40°C).

To summarize, much of the research regarding CO_2 solubility has been directed at the solubility of fluorinated compounds in CO_2, where it appears that three structural variables might govern the low miscibility pressures of fluorinated compounds in CO_2. First, CO_2 might interact specifically with the electronegative fluorine. Second, fluorinated molecules have weak self-association as indicated by low solubility parameters and low interfacial surface tensions, lowering miscibility pressures in low cohesive energy density CO_2. Finally, fluorination of a compound enhances solubility in CO_2 only if the compound is also somewhat polar. Likewise, these reasons also are important to the solubility of nonfluorinated compounds in CO_2. Specific interactions, weak self-association, and polarity also play a dominant role in governing the phase behavior of nonfluorous compounds in CO_2.

2. MODELING SOLUBILITY IN CO_2

Investigators have attempted to devise mathematical models to predict the phase behavior of compounds in CO_2 by means of solute chemical structure alone. Equations of state often fall short of accurate prediction owing to lack of experimentally determined quantities (such as vapor pressure) and other physicochemical properties of the solute (50). Ashour et al., for example, surmised that no single cubic equation of state exists that is appropriate for the prediction of solubility in all supercritical fluid mixtures (51). To further complicate the issue, more than 40 different forms of equations of state and 15 different types of mixing rules have been evaluated vis-à-vis phase behavior in carbon dioxide (52); choosing the correct equation to model solubility in CO_2 for a specific system can be a challenging undertaking.

Investigators have tried in the past to develop a comprehensive model that both is predictive and does not require experimental data to assign binary interaction parameters. Politzer et al. related solubility in CO_2 to the total variance of the electrostatic potential on the molecular surface and the molecular volume (53,54). Solubility was found to vary inversely with the variance of the surface potential, which is consistent with solute–solute interactions dominating over solute–solvent interactions in determining solubility in low density supercritical solutions. This theoretical work hence agrees with the experimental work of O'Neill et al., who found that polymer–polymer interactions dominate polymer–CO$_2$ reactions in determining CO$_2$ solubility (38).

In 1997, Engelhardt and Jurs used a quantitative structure–property relationship (QSPR) to predict the solubilities of a diverse group of 58 organic compounds in supercritical CO$_2$ (55). Numerical descriptors encoding information about the topological, geometric, and electronic properties of each compound were calculated from their molecular structures. Topological descriptors included the number of atoms and bonds, the number of paths (topology), and the molecular weight. Electronic descriptors provided information about fine molecular structure, including charge on the most positive and negative atoms. Geometric descriptors provided information based on the three-dimensional size and shape of a compound. The best model for the prediction of solubility in supercritical CO$_2$ found by Engelhardt and Jurs includes descriptors that may provide insight into the factors that influence solubility in CO$_2$. The seven descriptors to which solubility was found to be most sensitive were included in the final model; Table 1 summarizes the seven descriptors and their definitions. Two of these descriptors encode size information (S3C8 and SHDW5). The next two descriptors listed likely provide information on hydrogen bonding (MCHG0 and CTAA0). The last three descriptors possibly give information on dispersion interactions, di-

TABLE 1 Descriptors Used for Prediction of Solubility in Supercritical CO$_2$

Descriptor	Definition
S3C8	Simple cluster molecular connectivity for paths of length 3
SHDW5	Area of the molecular projected onto the XZ plane divided by a box with dimensions corresponding to the maximum dimension of the molecule in the X and Z planes
MCHG0	Maximum charge difference between a hydrogen-bonding donor and a hydrogen-bonding acceptor atom
CTAA0	Number of hydrogen-bonding acceptor atoms in each molecule
PPSA3	Sum of the surface areas of the positively charged atoms in a structure times their respective charges
RNCG1	Charge on the most negative atom divided by the total negative charge of the molecule
PVOL	Polarization volume of the molecule (polarizability) divided by its molecular volume

Source: Ref. 53.

pole–induced-dipole interactions, and electrostatic interactions (PPSA3, RNCG1, and PVOL). Three of the descriptors identified by Engelhardt and Jurs are consistent with the characteristics that have been identified experimentally to be important in dictating solubility in CO_2. Both MCHG0 and CTAA0 describe specific interactions (10,35–37), while PVOL describes the polarizability per volume, previously deemed an important indicator for solubility in CO_2 (25).

Of course, one would be in favor of a simple method by which to predict CO_2 solubility if it did so with some accuracy. Vetere, for example, developed a shortcut method for predicting the solubility of solids in CO_2 (56). This empirical method requires only the input of two properties of the pure components, the molecular weight and the melting temperature (which reflects the strength of solute–solute interactions and is hence relevant). However, the absolute average percentage deviation of calculated from experimental values ranged from 4.0 to 78.4% for nonpolar solids and 5.8 to 118.0% for polar solids, indicating that the predictive ability of the method varies immensely depending on the system being studied. Jouyban et al. developed a method for using a minimum number of experiments to predict solubility in CO_2 (57). In this case, though, correlative empirical equations based on independent variables such as temperature, pressure, and density were used after the model had been "trained," by generating a minimum number of experimental data points that were fitted to the model to allow computation of the model constants. Jouyban used nicotinic acid (NA) and *p*-acetoxyacetanilide (PAA) to train the model equation and generate the constants. The average absolute deviation of this trained model equation was about 17% when compared with the literature data of 401 compounds, although the lack of molecular inputs makes it unclear how reliable this model would be to predict fine effects of structure on phase behavior.

3. IMPORTANT CHARACTERISTICS FOR SOLUBILITY IN CO_2

In summary, we will attempt to pinpoint the specific characteristics that are necessary for a compound to be soluble in or miscible with CO_2 at moderate pressures. This identification of the important characteristics that govern solubility in CO_2 should aid in the design of CO_2-philic compounds for biomaterial and pharmaceutical applications. The characteristics and their importance are summarized in Table 2.

The first important characteristic is the presence of functional groups within a compound that can interact specifically with CO_2 in interactions of the Lewis acid/base type. These functional groups may be electron donating, such as carbonyls, to interact with CO_2, which would be the electron acceptor.

TABLE 2   Important Characteristics for the Design of CO$_2$-philic Compounds

Characteristic	Importance	Ref.
Electron-donating functional groups	Specific solute–solvent interactions (Lewis acid/base interactions, CO$_2$ electron acceptor)	35–37
Electron-accepting functional groups	Specific solute–solvent interactions (Lewis acid/base interactions, CO$_2$ electron donor)	10
Polarity/partial dipoles	Interact with the quadrupole moment of CO$_2$	26–29, 31
Molecular architecture	Free volume and flexibility as indicated by low glass transition temperatures	26, 29, 46, 47, 48
Low CED (low surface tension)	Weak intramolecular interactions more significant than compound–CO$_2$ interactions	38
Balance between entropy and enthalpy of mixing	A point of diminishing return will be reach (viz-, too many favorable interactions will lead to unfavorable entropy)	47, 48

On the other hand, the compound can have electron-accepting groups (Lewis base) to interact with CO$_2$, which would then be a Lewis acid. An important point is that the functional groups should not have both hydrogen-bonding donors and acceptors, which would result in the presence of overly strong solute–solute interactions.

A second potential attribute is simply polarity. A compound designed for solubility in CO$_2$ should contain some polar character, (e.g., dipoles capable of interacting with CO$_2$ through its strong quadrupole moment). An important limitation here is that the compound should not be so polar such that solute–solute interactions dominate.

A third feature for solubility is molecular architecture, especially in the case of large compounds such as polymers. The polymer should exhibit large free volume with a high degree of flexibility. A good indication of polymer flexibility and high free volume is a low glass transition temperature.

A low cohesive energy density as signified by low surface tension is a fourth significant trait for solubility in CO$_2$. Low CED, close to that of CO$_2$, denotes weak solute–solute interactions, apparently a dominant factor for CO$_2$ solubility. The compound must not interact strongly with itself; robust interaction of CO$_2$ with the compound is less essential.

Solubility is a fine balance between the enthalpy of mixing and the entropy of mixing. The compound must contain the specific interactions for favorable enthalpy of mixing, but there is a point of diminishing return that will be reached with too many specific interactions. At that point, the unfavorable entropy of mixing will prevail.

4. SOLUBILIZATION OF INSOLUBLE COMPOUNDS INTO CO_2: POTENTIAL APPLICATIONS IN THE PHARMACEUTICAL INDUSTRY

Many biomaterials and pharmaceuticals will *not* dissolve in CO_2 at moderate pressures; however, that does not mean that CO_2 cannot be used as the process solvent. Methods for overcoming limited solubility in CO_2 include use of surfactants, use of cosolvents, and covalent modification of the insoluble compound with CO_2-philic "ponytails."

4.1. Surfactants

Traditionally, a surfactant, or surface active agent, is an amphiphilic molecule containing both hydrophilic and lipophilic segments. Surfactants are capable of forming micelles, aggregates arranged so that the hydrophilic segment interacts with the aqueous phase and the lipophilic segment is oriented to interact with the organic phase. In the case of CO_2-soluble surfactants, the surfactant would instead contain CO_2-philic and CO_2-phobic segments. The CO_2-philic segment interacts with the continuous CO_2 phase and the CO_2-phobic segment can be chosen from either hydrophilic or lipophilic molecules, based on the application of the surfactant (Figure 5). The CO_2-philic material may be created from any of the compounds already discussed, including fluorinated polymers, silicone-based compounds, and even CO_2-philic hydro-

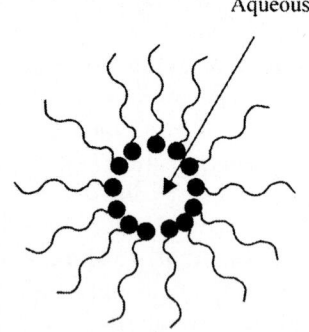

Aqueous or organic core

CO_2 phase

FIGURE 5 Representation of a micelle in CO_2: solid circles are hydrophilic or lipophilic head groups; wavy lines are CO_2-philic tails.

carbons. Surfactants reduce interfacial tension and aid in the solubilization of CO_2-phobic compounds, such as polymers and biological molecules in CO_2.

Surfactants have also been used to overcome the solubility limitation of synthetic polymers in CO_2 (most common synthetic polymers would be considered to be CO_2-phobic). For example, surfactants have been used to aid in the dispersion polymerization of poly(methylmethacrylate) (PMMA) in CO_2 (58–60). The surfactants used in the polymerizations of PMMA are more accurately referred to as stabilizers. The CO_2-phobic region acts as anchor to the growing polymer, either by physical adsorption or by chemical grafting. The CO_2-philic region sterically stabilizes the growing polymer particles, preventing flocculation and precipitation. When a biopolymer is not soluble in CO_2, specific surfactants may be designed to aid in the solubilization of the polymer into CO_2.

Biocompatible polymers have widespread applications to medicine. A common biopolymer used in sutures, poly(lactide-*co*-glycolide) (PLGA), dissolves in CO_2, where solubility is strongly dependent on the glycolide content in the backbone (61). PLGA with 50 repeat units is poorly soluble in CO_2 at pressures up to 3000 bar; therefore CO_2 is not usually considered to be a useful process solvent for this polymer. However, the liquid–liquid phase boundary of a polymer–CO_2 binary is highly asymmetric (Figure 6) owing to the large size disparity of the solute and solvent. This means that while dissolving small amounts of polymer in CO_2 might require impractically high pressures, the dissolution of CO_2 in the polymer (i.e., CO_2 as diluent and plasticizer) is far more tractable. Indeed, it is known that CO_2 swells many

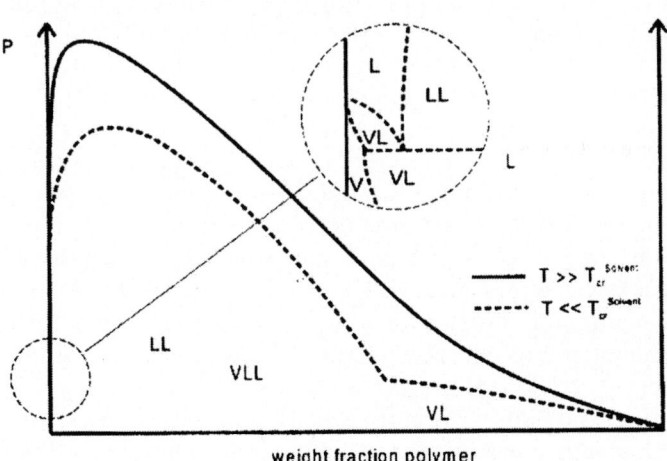

weight fraction polymer

FIGURE 6 Generic phase diagram of a polymer–CO_2 mixture.

amorphous polymers to a great extent ($\leq 30\%$ CO_2) at mild pressures (~ 100 bar), significantly depressing the glass transition temperature. In the case of biocompatible polymers such as poly(lactide-glycolide), this allows one to create foams for tissue engineering (without the use of organic solvents) and to use solvents or high temperature melt processes in mixing biologically active molecules with the polymer (62,63).

Proteins and enzymes play a significant role in pharmaceutical applications from the pharmaceutical products themselves to the specialized catalysts for the production of pharmaceuticals. CO_2 is an interesting solvent for bioextractions and as a reaction medium for biological molecules owing to its nontoxicity and the ease of product recovery; however, surfactants are necessary to solubilize the large biological molecules. Johnston et al. were the first to show that a protein, bovine serum albumin (BSA), can be solubilized in a water-in-CO_2 microemulsion, where an ammonium carboxylate PFPE surfactant (MW 740) was used to form the microemulsions (64). Fluorescence was used to monitor the solubilization of BSA (MW 67,000) labeled with acrylodan (BSA-Ac) in a stable aqueous environment in CO_2. The fluorescence of BSA-Ac in this water-in-CO_2 microemulsion when the PFPE surfactant (1.4 wt %) was used was similar to that of native BSA in buffer, pH 7.0. Ghenciu et al., using fluoroether-functional anionic and nonionic surfactants, showed that a protein can be extracted from either buffer or cell-free culture media with retention of most of the protein's activity following recovery by depressurization (65).

Finally, DeSimone et al. designed and synthesized amphiphilic copolymers comprising a perfluoroacrylate polymer, poly(1,1,2,2-tetrahydroperfluorodecyl acrylate) [poly(TAN)] and poly(ethylene glycol) (PEG) to form a poly(TAN-*co*-PEG) copolymer to be used in bioextractions (66). Poly (TAN-*co*-PEG) was able to extract BSA from an aqueous solution into CO_2.

Holmes et al. reported the first enzyme catalyzed reactions in water-in-CO_2 microemulsions (67). Two reactions, a lipase-catalyzed hydrolysis and a lipoxygenase-catalyzed peroxidation, were demonstrated in water-in-CO_2 microemulsions using the surfactant di(1*H*,1*H*,5*H*-octafluoro-*n*-pentyl) sodium sulfosuccinate (di-HCF4). A major concern of enzymatic reactions in CO_2 is the pH of the aqueous phase, which is approximately 3 when there is contact with CO_2 at elevated pressures. Holmes et al. examined the ability of various buffers to maintain the pH of the aqueous solution in contact with CO_2. The biological buffer 2-(*N*-morpholino)ethanesulfonic acid sodium salt (MES) was the most effective, able to maintain a pH of 5, depending on the pressure, temperature, and buffer concentration. The activity of the enzymes in the water-in-CO_2 microemulsions was comparable to that in a water-in-heptane microemulsion stabilized by the surfactant AOT, which contains the same head group as di-HCF4.

Another potential pharmaceutical application of CO_2 is as the reaction medium for organic synthesis. To this end, organic syntheses have been conducted in water-in-CO_2 microemulsions formed with an ammonium carboxylate PFPE surfactant (68). Nucleophilic substitution reactions occurred between hydrophilic nucleophiles and CO_2-soluble reactants. The reaction yields and rate constants were an order of magnitude greater than traditional water-in-oil microemulsions under similar conditions (except pressure), likely owing to lower microviscosity of the water-in-CO_2 microemulsions. Water-in-CO_2 emulsions also exhibited higher yields than water-in-oil emulsions, a property attributed to lower interfacial tension and viscosity of the water–CO_2 interface than the water–oil interface and higher diffusivity in CO_2 (69). Greater yields were obtained from organic synthesis in water-in-CO_2 emulsions than in water-in-CO_2 microemulsions owing to the larger amount of water in an emulsion, which allowed for greater excess of the hydrophilic nucleophile in these reactions (69).

4.2. Cosolvents

Cosolvents can enhance solubility of compounds in CO_2, a topic beyond the scope of this chapter. It is useful, however, to point out some details on cosolvents in CO_2. McHugh et al. demonstrate that a cosolvent can provide the specific interactions that are necessary to solubilize a compound in CO_2 (70,71). For instance, butyl acrylate (BA) and ethyl hexyl acrylate (EHA) decrease the cloud point pressure of acrylate polymers owing to the specific polar interactions between the cosolvent with the acrylate backbone of the polymer (70). Addition of ethyl methacrylate (EMA) and butyl methacrylate (BMA) reduces the pressure needed to solubilize poly(ethyl methacrylate) (PEMA) and poly(butyl methacrylate) (PBMA) in CO_2 (71).

While addition of cosolvents facilitates the dissolution of most materials in CO_2, these substances also tend to detract from the sustainability (and perhaps biocompatibility) of the process. Use of certain cosolvents (owing to their toxicity) could entirely counterbalance the benefits accrued through use of a benign solvent such as CO_2. Hence one must weigh the benefits versus the problems that accompany the use of cosolvents in carbon dioxide.

4.3. Covalent Modification

Another method for solubilizing an essentially insoluble compound into CO_2 is to covalently attach a CO_2-philic compound to it, which is generally how CO_2-philic surfactants were prepared. The CO_2-philic segment, referred to as a CO_2-philic "ponytail," is attached to the CO_2-phobic segment, thus rendering the entire molecule soluble in CO_2. CO_2-philic ponytails have been covalently attached to biological molecules (CO_2-phobic segment), forming amphiphilic molecules that are soluble in CO_2 without the need for other surfactants. Such molecules have pharmaceutical applications in bioextractions and biocatalysis in CO_2.

Ghenciu and Beckman designed an affinity surfactant containing the ligand biotin for the extraction of avidin (Figure 7A) (72). The surfactants were prepared both with and without a hydrophilic PEG spacer, and the CO_2-philic tail was composed of PFPE. The phase behavior of the surfactant was a function of both the overall molecular weight and the ratio of the number of CO_2-philic to hydrophilic groups. The surfactant that contained a PEG spacer was able to extract more avidin than the surfactant without the spacer, probably because the material with the PEG spacer had better surface activity. An inverse emulsion (20:80 liquid CO_2 to avidin solution) and three-phase emulsion (40:60 liquid CO_2 to avidin solution), both using the PFPE 7500/PEG 600 biotin surfactant, were compared on their abilities to

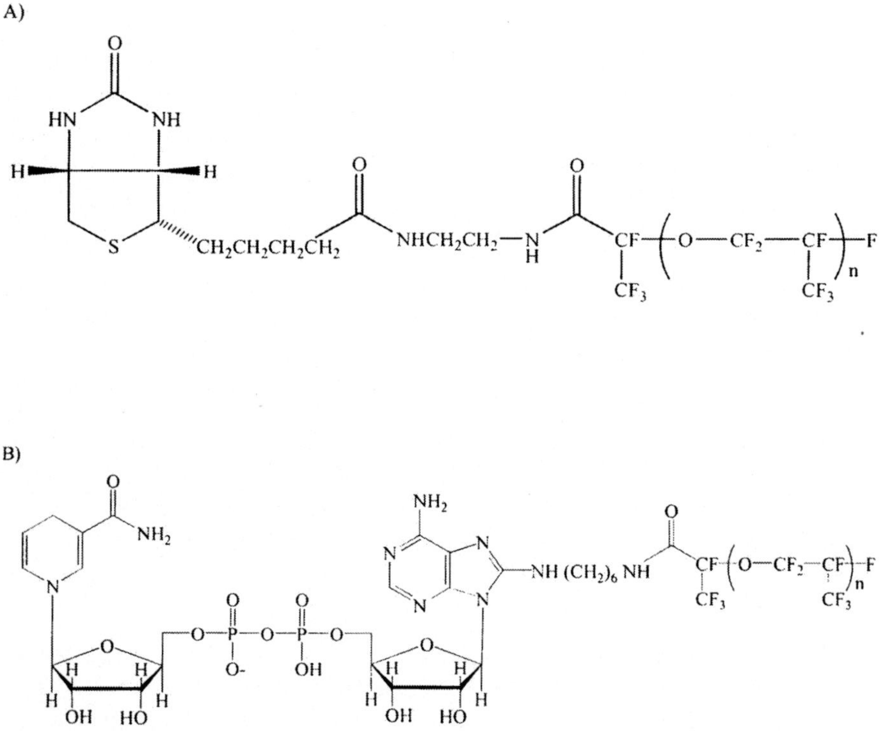

A)

B)

FIGURE 7   Biological molecules modified with CO_2-philic ponytails. structure of (A) fluoroether–biotin, (B) FNAD.

extract avidin. The three-phase emulsion extracted more than double the amount of protein obtained from the inverse emulsion, possibly because the former was better able to partition the surfactant–protein aggregates. To recover the proteins, the three-phase emulsion was stripped with liquid CO$_2$ until the emulsion broke (owing to the continuous removal of the surfactant–protein complex).

Panza et al. synthesized a CO$_2$-philic amphiphile from the coenzyme nicatinamide adenine dinucleotide (MW 664) and a covalently attached perfluoropolyether (MW 2500) (Figure 7B) (73). The fluorofunctional co-enzyme (FNAD) was soluble up to 5 mM in CO$_2$ at room temperature and 1400 psi. The CO$_2$-soluble FNAD was able to participate in a cyclic oxidation/reduction reaction catalyzed by the enzyme horse liver alcohol dehydrogenase (HLADH) in CO$_2$ at room temperature and 2600 psi.

5. CONCLUSION

The design of CO$_2$-philic molecules, especially in the case of biomaterials and pharmaceutical application, is a challenging feat. The main obstacle is that small changes in molecular structure can have significant effect on CO$_2$-solubility. CO$_2$ is an enigmatic solvent and much is still unknown; however, several characteristics seem to be important for solubility in CO$_2$. Compounds that contain one or more of the following features may have enhanced solubility in CO$_2$: functional groups that interact with CO$_2$, slight polarity or partial dipoles, molecular architecture with high flexibility and high free volume, and low cohesive energy density. Incorporating one or more of these traits into a compound to form a CO$_2$-soluble molecule is a trial-and-error process, since a fine balance between the enthalpy and entropy of mixing must be present for the compound to be soluble in CO$_2$. Biomaterials and pharmaceuticals that are insoluble in CO$_2$ may be solubilized in CO$_2$ using CO$_2$-soluble surfactants, cosolvents, and covalent modification with CO$_2$-philic ponytails.

REFERENCES

1. McHugh M, Krukonis V. Supercritical Fluid Extraction. 2d ed. Boston: Butter-worth-Heinemann, 1994.
2. Dondeti P, Desai Y. Supercritical fluid technology in pharmaceutical research. In: Swarbrick J, Boylan JC, eds. Encyclopedia of Pharmaceutical Technology. Vol. 18. New York: Marcel Dekker, 219–248.
3. DeSimone JM, Maury EE, Menceloglu YZ, McClain JB, Romack TJ, Combes JR. Dispersion polymerizations in supercritical carbon dioxide. Science 1994; 265:356–359.

4. Francis AW. Ternary systems of liquid carbon dioxide. J Phys Chem 1954; 58: 1099–1114.
5. Giddings JC, Myers MN, McLaren L, Keller RA. High pressure gas chromatography of nonvolatile species. Science 1968; 162:67–73.
6. Allada SR. Solubility parameters of supercritical fluids. Ind Eng Chem Process Design Dev 1984; 23:344–348.
7. Hyatt JA. Liquid and supercritical carbon dioxide as organic solvents. J Org Chem 1984; 49:5097–5101.
8. Gouw T. Solubility of C6–C8 aromatic hydrocarbons in liquid carbon dioxide. J Chem Eng Data 1969; 14:473–474.
9. Alwani Z. Löslichkeitsverhalten von schwerflüchtigen biochemischen Stoffen in Komprimiertem Kohlendioxid. Angew Chem Int Ed Engl 1980; 19:633–634.
10. Meredith JC, Johnston KP, Seminario JM, Kazarian SG, Eckert CA. Quantitative equilibrium constants between CO_2 and Lewis bases from FTIR spectroscopy. J Phys Chem 1996; 100:10837–10848.
11. Iezzi A, Bendale P, Enick RM, Turber M, Brady J. Gel formation in carbon dioxide–semifluorinated alkane mixtures and phase equilibria of carbon dioxide–perfluorinated alkane mixture. Fluid Phase Equilibria 1989; 52:307–317.
12. Consani KA, Smith RD. Observations on the solubility of surfactants and related molecules in carbon dioxide at 50°C. J Supercrit Fluids 1990; 3:51–65.
13. Hoefling T, Stofesky D, Reid M, Beckman E, Enick RM. The incorporation of a fluorinated ether functionality into a polymer or surfactant to enhance CO_2-solubility. J Supercrit Fluids 1992; 5:237–241.
14. Newman DA, Hoefling TA, Beitle RR, Beckman EJ, Enick RM. Phase behavior of fluoroether-functional amphiphiles in supercritical carbon dioxide. J Supercrit Fluids 1993; 6:205–210.
15. Hoefling TA, Beitle RR, Enick RM, Beckman EJ. Design and synthesis of highly CO_2-soluble surfactants and chelating agents. Fluid Phase Equilibria 1993; 83: 203–212.
16. Hoefling TA, Newman DA, Enick RM, Beckman EJ. Effect of structure on the cloud-point curves of silicone-based amphiphiles in supercritical carbon dioxide. J Supercrit Fluids 1993; 6:165–171.
17. DeSimone JM, Guan Z, Elsbernd CS. Synthesis of fluoropolymers in supercritical carbon dioxide. Science 1992; 257:945–947.
18. McClain JB, Londono D, Combes JR, Romack TJ, Canelas DA, Betts DE, Wignall GD, Samulski ET, DeSimone JM. Solution properties of a CO_2-soluble fluoropolymer via small angle neutron scattering. J Am Chem Soc 1996; 118:917–918.
19. Dardin A, Cain JB, DeSimone JM, Johnson CS Jr, Samulski ET. High-pressure NMR of polymers dissolved in supercritical carbon dioxide. Macromolecules 1997; 30:3593–3599.
20. Dardin A, DeSimone JM, Samulski ET. Fluorocarbons dissolved in supercritical carbon dioxide. NMR evidence for specific solute–solvent interactions. J Phys Chem B 1998; 102:1775–1780.

21. Yee GG, Fulton JL, Smith RD. Fourier transform infrared spectroscopy of molecular interactions of heptafluoro-1-butanol or 1-butanol in supercritical carbon dioxide and supercritical ethane. J Phys Chem 1992; 96:6172–6181.

22. Diep P, Jordan KD, Johnson JK, Beckman EJ. CO_2–fluorocarbon and CO_2–hydrocarbon interactions from first principle calculations. J Phys Chem A 1998; 102:2231–2236.

23. McHugh MA, Park IH, Reisinger JJ, Ren Y, Lodge TP, Hillmyer MA. Solubility of CF_2-modified polybutadiene and polyisoprene in supercritical carbon dioxide. Macromolecules 2002; 35:4653–4657.

24. McFann GJ, Johnston KP, Howdle SM. Solubilization in nonionic reverse micelles in carbon dioxide. AIChE J 1994; 40:543–555.

25. Harrison K, Goveas J, Johnston KP, O'Rear EA. Water-in-carbon dioxide microemulsions with a fluorocarbon–hydrocarbon hybrid surfactant. Langmuir 1994; 10:3536–3541.

26. Rindfleisch F, DiNoia TP, McHugh MA. Solubility of polymers and copolymers in supercritical CO_2. J Phys Chem 1996; 100:15581–15587.

27. Mertdogan CA, Byun HS, McHugh MA, Tuminello WH. Solubility of poly (tetrafluoroethylen-*co*-19 mol% hexafluoropropylene) in supercritical CO_2 and halogenated supercritical solvents. Macromolecules 1996; 29:6548–6555.

28. Mertdogan CA, DiNoia TP, McHugh MA. Impact of backbone architecture on the solubility of fluorocopolymers in supercritical CO_2 and halogenated supercritical solvents: comparison of poly(vinylidene fluoride-*co*-22 mol% hexafluoropropylene) and poly(tetrafluoroethylene-*co*-19 mol% hexafluoropropylene). Macromolecules 1997; 30:7511–7515.

29. DiNoia TP, Conway SE, Lim JS, McHugh MA. Solubility of vinylidene fluoride polymers in supercritical CO_2 and halogenated solvents. J Polym Sci B: Polym Phys 2000; 38:2832–2840.

30. Eastoe J, Cazelles BMH, Steytler DC, Holmes JD, Pitt AR, Wear TJ, Heenan RK. Water-in-CO_2 microemulsions studied by small-angle neutron scattering. Langmuir 1997; 13:6980–6984.

31. Downer A, Eastoe J, Pitt AR, Penfold J, Heenan RK. Adsorption and micellisation of partially- and fully-fluorinated surfactants. Colloids Surf A Physicochem Eng Aspects 1999; 156:33–48.

32. Eastoe J, Paul A, Downer A, Steytler DC, Rumsey E. Effects of fluorocarbon surfactant chain structure on stability of water-in-carbon dioxide microemulsions. Links between aqueous surface tension and microemulsion stability. Langmuir 2002; 18:3014–3017.

33. Eastoe J, Downer A, Paul A, Steytler DC, Rumsey E, Penfeld J, Heenan RK. Fluoro-surfactants at air/water and water/CO_2 interfaces. Phys Chem Chem Phys 2000; 2:5235–5242.

34. Steytler DC, Rumsey E, Thorpe M, Eastoe J, Paul A, Heenan RK. Phosphate surfactants for water-in-CO_2 microemulsions. Langmuir 2001; 17:7948–7950.

35. Kazarian SG, Vincent MF, Bright FV, Liotta CL, Eckert CA. Specific intermolecular interaction of carbon dioxide with polymers. J Am Chem Soc 1996; 118:1729–1736.

36. Quinn R, Appleby JB, Pez GP. Salt hydrates: new reversible absorbents for carbon dioxide. J Am Chem Soc 1995; 117:329–335.

37. Harris TV, Irani CA, Pretzer WR. Enhanced oil recovery using CO_2 flooding. US patent 4,913,235, 1990.

38. O'Neill ML, Cao Q, Fang M, Johnston KP, Wilkinson SP, Smith CD, Kerschner JL, Jureller SH. Solubility of homopolymers and copolymers in carbon dioxide. Ind Eng Chem Res 1998; 37:3067–3079.

39. Johnston KP, Wilkinson SP, O'Neill ML, Robeson LM, Mawson S, Bott RH, Smith CD, Surfactants for hetergeneous processes in liquid or supercritical CO_2. US patent 5,733,964, 1998.

40. Eastoe J, Paul A, Nave S, Steytler DC, Robinson BH, Rumsey E, Thorpe M, Heenan RK. Micellization of hydrocarbon surfactants in supercritical carbon dioxide. J Am Chem Soc 2001; 123:988–989.

41. Lagalante AF, Hansen BN, Bruno TJ, Sievers RE. Solubilities of copper(II) and chromium(III) β-diketonates in supercritical carbon dioxide. Inorg Chem 1995; 34:5781–5785.

42. Dandge DK, Heller JP, Wilson KV. Structure–solubility correlations: organic compounds and dense carbon dioxide binary systems. Ind Eng Chem Prod Res Dev 1985; 24:162–166.

43. Liu J, Han B, Li G, Zhang X, He J, Liu Z. Investigation of nonionic surfactant Dynol-604 based reverse microemulsions formed in supercritical carbon dioxide. Langmuir 2001; 17:8040–8043.

44. Liu J, Han B, Li G, Liu Z, He J, Yang G. Solubility of the non-ionic surfactant tetraethylene glycol n-laurel ether in supercritical CO_2 with n-pentanol. Fluid Phase Equilibria 2001; 187–188:247–254.

45. Liu J, Han B, Wang Z, Zhang J, Li G, Yang G. Solubility of Ls-36 and Ls-45 surfactants in supercritical CO_2 and loading water in the CO_2/water/surfactant systems. Langmuir 2002; 18:3086–3089.

46. Fink R, Hancu D, Valentine R, Beckman EJ. Toward the development of "CO_2-philic" hydrocarbons. 1. Use of side-chain functionalization to lower the miscibility pressure of polydimethylsiloxanes in CO_2. J Phys Chem B 1999; 103:6441–6444.

47. Sarbu T, Styranec TJ, Beckman EJ. Non-fluorous polymers with very high solubility in supercritical CO_2 down to low pressure. Nature 2000; 405:165–168.

48. Sarbu T, Styranec TJ, Beckman EJ. Design and synthesis of low cost, sustainable CO_2-philes. Ind Eng Chem Res 2000; 39:4678–4683.

49. Raveendran P, Wallen SL. Sugar acetates as novel, renewable CO_2-philes. J Am Chem Soc 2002; 124:7273–7275.

50. Johnston KP, Peck DG, Kim S. Modeling supercritical mixtures: how predictive is it? Ind Eng Chem Res 1989; 28:1115–1125.

51. Ashour I, Almehaideb R, Fateen SE, Aly G. Representation of solid fluid phase equilibria using cubic equations of state. Fluid Phase Equilibria 2000; 167:41–61.

52. Pfohl O, Petkov S, Brunner G. Manual for a program to calculate phase equilibria. www.tu-harburg.de/vt2/pe2000.

53. Politzer R, Lane P, Murray JS, Brinck T. Investigation of relationships between

solute molecule surface electrostatic potentials and solubilities in supercritical fluids. J Phys Chem 1992; 96:7938–7943.

54. Politzer R, Murray JS, Lane P, Brinck T. Relationships between solute molecular properties and solubility in supercritical CO_2. J Phys Chem 1993; 97:729–732.

55. Englehart HL, Jurs PC. Prediction of supercritical carbon dioxide solubility of organic compounds from molecular structure. J Chem Inf Comput Sci 1997; 37: 478–484.

56. Vetere A. A short-cut method to predict the solubilities of solids in supercritical carbon dioxide. Fluid Phase Equilibria 1998; 148:83–93.

57. Jouyban A, Rehman M, Shekunov BY, Chan HK, Clark BJ, York P. Solubility prediction in supercritical CO_2 using minimum number of experiments. J Pharm Sci 2002; 91:1287–1295.

58. Hsiao YL, Maury EE, DeSimone JM, Mawson S, Johnston KP. Dispersion polymerization of methyl methacrylate stabilized with poly(1,1-dihydroperfluorooctyl acrylate) in supercritical carbon dioxide. Macromolecules 1995; 28: 8159–8166.

59. Shaffer KA, Jones TA, Canelas DA, DeSimone JM. Dispersion polymerizations in carbon dioxide using siloxane-based stabilizers. Macromolecules 1996; 29:2704–2706.

60. Lepilleur C, Beckman EJ. Dispersion polymerization of methyl methacrylate in supercritical CO_2. Macromolecules 1997; 30:745–756.

61. Conway SE, Byun HS, McHugh MA, Wang JD, Mandel FS. Poly(lactide-*co*-glycolide) solution behavior in supercritical CO_2, CHF_3, and $CHClF_2$. J Appl Polym Sci 2001; 80:1155–1161.

62. Sparacio D, Beckman EJ. Generation of microcellular biodegradable polymers using supercritical carbon dioxide. ACS Symposium Series 713. In: Solvent-Free Polymerization and Processes. Washington, DC: American Chemical Society, 1998:181–193.

63. Mandel FS, Wang JD, Howdle SM, Shakesheff KM. PCT Int Appl 2002; WO 0220624.

64. Johnston KP, Harrison KL, Clarke MJ, Howdle SM, Heitz MP, Bright FV, Carlier C, Randolph TW. Water-in-carbon dioxide microemulsions: An environment for hydrophiles including proteins. Science 1996; 271:624–626.

65. Ghenciu EG, Russell AJ, Beckman EJ, Steele L, Becker NT. Solubilization of subtilisin in CO_2 using fluoroether-functional amphiphiles. Biotechnol Bioeng 1998; 58:572–580.

66. DeSimone JM, Crette SA, LeClerc JM, Kendall JL, Carbonell RG. Bioextractions with carbon dioxide. Proccedings of Int 5th Meeting on Supercritical Fluids, 1998:813–819.

67. Holmes JD, Steytler DC, Rees GD, Robinson BH. Bioconversions in a water-in-CO_2 microemulsion. Langmuir 1998; 14:6371–6376.

68. Jacobsen GB, Lee CT Jr, Johnston KP. Organic synthesis in water/carbon dioxide microemulsions. J Org Chem 1999; 64:1201–1206.

69. Jacobsen GB, Lee CT Jr, daRocha SRP, Johnston KP. Organic synthesis in water/carbon dioxide emulsions. J Org Chem 1999; 64:1207–1210.

70. McHugh MA, Rindfleisch F, Kuntz PT, Schmaltz C, Buback M. Cosolvent effect of alkyl acrylates on the phase behavior of poly(alkyl acrylates)–supercritical CO_2 mixtures. Polymer 1998; 39:6049–6052.
71. Byun HS, McHugh MA. Impact of "free" monomer concentration on the phase behavior of supercritical carbon dioxide–polymer mixtures. Ind Eng Chem Res 2000; 39:4658–4662.
72. Ghenciu EG, Beckman EJ. Affinity extraction into carbon dioxide. 1. Extraction of avidin using a biotin-functional fluoroether surfactant. Ind Eng Chem Res 1997; 36:5366–5370.
73. Panza JL, Russell AJ, Beckman EJ. Synthesis of fluorinated NAD as a soluble coenzyme for enzymatic chemistry in fluorous solvents and carbon dioxide. Tetrahedron 2002; 58:4091–4104.

2

Phase Equilibrium in Solid–Liquid–Supercritical Fluid Systems

Mamata Mukhopadhyay
Chemical Engineering Department, Indian Institute of Technology, Bombay, India

The bioavailability of solid drugs and the efficacy of their delivery systems are often constrained by size, morphology, and size distribution, since these factors decide the solubilities of the drugs in the aqueous media of our body fluid systems. The lower the particle size of these solid drugs, the higher their dissolution rates and their bioavailability in the body fluids. Micronized drugs with uniform size and morphology are functionally most effective for controlled delivery to the target organs such as heart, lung, tissues, and bones. The pharmaceutical industry today needs a promising method of producing micronized drugs with a narrow size distribution for utilizing the new drug delivery routes, such as dry powder inhalers, needle-free injections, and controlled-release devices. Like micronization, two other processes, micro-impregnation and microencapsulation of the drugs, are also adopted for improvement of functional and aesthetic values. There is great potential for utilizing supercritical carbon dioxide ($scCO_2$) as the medium in all these processes, because the pharmaceutical compounds are thermally labile and the micronized drugs should be free from any organic solvent residues and artifacts. Investigators consider $scCO_2$ to be an attractive solvent or antisolvent in these processes owing to the tunability of its solvent power with small variations in pressure and temperature and to its excellent transport properties (e.g., viscosity and diffusivity); in addition, $scCO_2$ in nontoxic, nonflammable, and environmentally benign.

The major advantages of these $scCO_2$ methods include single-step operation, a mild operating temperature, and a very narrow size distribution of particles with controlled morphology, unlike the conventional techniques. Conventional methods include spray drying, recrystallization using solvent evaporation or liquid antisolvent (1) and ball milling and sieving (2,3). These methods produce particles in the range of 5 to 50 μm, whereas pneumatic jet grinding can produce particles in the range of 1 to 10 μm. These conventional techniques have the common disadvantages of multiple steps, and poor control of the size, morphology, and size distribution of the drug powders, in addition to certain other disadvantages specific to each process.

The general principle of the $scCO_2$ technique involves dissolution of the pharmaceutical substance in the $scCO_2$ itself as a solvent, or in a suitable organic solvent, followed by a fast decrease of its solvent power in either case. This results in very high supersaturation/nucleation of the solute, extremely rapid phase change, and subsequently formation of ultrafine solid particles. When the $scCO_2$ is used as the solvent, the solvent power is reduced by rapid depressurization, as in the rapid expansion of supercritical solution (RESS) or in the process for particles from gas-saturated solutions (PGSS). This is termed depressurization crystallization. When an organic solvent is used as the solvent and CO_2 as antisolvent, the dissolution of gaseous antisolvent (GAS) or supercritical (CO_2) antisolvent (SAS) in the solution reduces the solvent power. This is known as the antisolvent crystallization process and is similar to precipitation with compressed antisolvent (PCA). Supercritical CO_2 is highly soluble in an organic solvent and vice versa. The mechanism of the two-way diffusion of CO_2 to solvent and solvent to CO_2, combined with the fast rate of mixing and reduction of solution droplets by atomization, as in the case of solution-enhanced dispersion by supercritical fluid (SEDS), enables formation of monodispersed nanoparticles (4,5). These processes using $scCO_2$ as the processing medium have been found promising for large-scale commercial applications (6). This chapter reviews the basic thermodynamic principles involved and the modeling of the multiphase equilibrium behavior in the binary ($scCO_2$–solid) and ternary ($scCO_2$–solvent–solid) systems needed in the design of the crystallization processes.

1. SOLID SOLUBILITY AND SUPERSATURATION

Solubility is defined as the maximum amount of solute that can be dissolved in a given amount (mass or moles or volume) of the solution under the equilibrium condition at a given temperature and density/pressure. A solution is normally referred to as a homogeneous mixture of a solvent [which is a liquid or a supercritical fluid (SCF)] and a solute (which is a solid to be crystallized). The solubility of a solid solute in a solvent generally increases with temperature or density, though the extent of the increase depends on the interaction between the solute and the solvent. The solubility may also decrease with increasing temperature for a sparingly soluble solute or at pressures less than crossover pressures in the case of the SCF solvent. A phase diagram representing the phase behavior of solid–liquid, gas–liquid, solid–SCF, solid–liquid–SCF, or SCF/liquid–liquid–solid equilibrium normally describes the solubility of a substance in a particular system. In other words, the solubility is defined in terms of the concentration of the solute in the saturated solution at which there is equilibrium between the solid and the solution phases.

A solution in which the concentration of the solid solute exceeds the equilibrium solubility or the saturated concentration at a given temperature is known as a supersaturated solution. The supersaturation is defined as the concentration difference Δx_i (in mol fraction/wt fraction), where

$$\Delta x_i = x_i - x_i^* \tag{1}$$

The degree of supersaturation S is given by the ratio of the concentrations as

$$S = \frac{x_i}{x_i^*} \tag{2}$$

where x and x_i^* are the actual (supersaturated) and the equilibrium (saturated) mole fractions, respectively, of solute i at the given temperature and pressure condition. Thermodynamically, this ratio S is related to the difference in the chemical potential μ_i and the activity coefficient γ_i, since we know that

$$\bar{\mu}_i - \bar{\mu}_i^* = RT\ln\frac{\gamma_i x_i}{\gamma_i^* x_i^*} \tag{3}$$

$$\ln S = \frac{\bar{\mu}_i - \bar{\mu}_i^*}{RT} - \ln\frac{\gamma_i}{\gamma_i^*} \tag{4}$$

It is understood that crystallization is a transient process associated with a phase change leading to crystal formation. Its rate depends on the level of supersaturation, which is the driving force for crystallization (7). Creating a high degree of supersaturation induces crystallization. "Precipitation," also often used to mean rapid crystallization, generally refers to the simultaneous and relatively rapid occurrence of nucleation, crystal growth, and agglomeration. As a result, a large number of crystals with relatively small sizes, typically between 0.1 and 10 μm, can be produced. The high degree of supersaturation needed for precipitation normally results from a chemical reaction, and such precipitation is often referred to as reactive crystallization. Often antisolvent crystallization is also referred to as precipitation, since it involves very rapid phase change and formation of a solid phase from a homogeneous liquid solution under a relatively high degree of supersaturation. Supersaturation is thus the most crucial parameter for controlling the precipitation process and physical characteristics of the particles formed.

2. NUCLEATION AND SPINODAL CURVE

As mentioned earlier, a supersaturated solution is not in the equilibrium condition. Crystallization moves the solution toward equilibrium by relieving its supersaturation. A supersaturated solution is thus not stable. There is a maximum degree of supersaturation for a solution before it becomes unstable. The region between this unstable boundary and the equilibrium (binodal) curve is termed the metastable zone, and it is here that the crystallization process occurs. The absolute limit of the metastable zone, known as the spinodal curve (8), is given by the locus of the maximum limit of supersaturation at which nucleation occurs spontaneously. Thermodynamically, the spinodal curve within the two-phase region is defined by the criterion

$$\frac{\partial^2 g}{\partial x_i^2} = 0 \tag{5}$$

and the binodal curve, the boundary between the single-phase region and the two-phase region, is characterized by the criterion

$$\frac{\partial^2 g}{\partial x_i^2} > 0 \tag{6a}$$

whereas the unstable region of the solution is given by the condition

$$\frac{\partial^2 g}{\partial x_i^2} < 0 \tag{6b}$$

where g is the free energy of the solution and is given by $\sum x_i \bar{\mu}_i$. In other words, the spinodal curve in the two-phase region demarcates the unstable region and the metastable regions, which are described later (Figure 7b).

Supersaturated homogeneous solutions are thus metastable, and the metastability decreases with increasing supersaturation. In other words, nucleation starts after the solute concentration has exceeded a definite value, and then solute molecules form aggregates or clusters. If, however, the crystals are placed in a supersaturated solution, the solution eventually attains equilibrium after the growth of the crystals upon crystallization.

Crystallization is thus a two-step process: nucleation and growth. The first step is nucleation, which occurs spontaneously when the supersaturation attains the metastable limit at the spinodal curve. Knowledge of the spinodal curve is useful in understanding the mechanism of crystallization. However, the actual metastable limit may often exceed the spinodal curve in certain systems, and in such cases the phase transition mechanism is explained in terms of spinodal decomposition (9) or the solution history, impurities present, or rate of increase of supersaturation.

Thermodynamically, the Gibbs free energy change for nucleation from a homogeneous solution is given by the sum of the free energy gain (which is positive) for the formation of the nucleus (crystal) solid surface and the loss of free energy (which is negative) for phase transition. The following Arrhenius type of expression gives the rate of nucleation:

$$B_0 = A \exp\left(-\frac{\Delta g_{cr}}{kT} \right) \tag{7}$$

where A is a preexponential parameter and Δg_{cr} is the free energy change for formation of a nucleus of a critical size. The rate of nucleation increases with increasing supersaturation and temperature, and decreases with the increase in the surface energy. Mullins (10) reports that instantaneous nucleation of water vapor: an induction time of, say, 0.1 s is possible only at a degree of supersaturation of around 4. Thus knowledge of the solid solubility is needed to find the degree of supersaturation for the design of a crystallization process. Accordingly, to find the rate of nucleation and growth in all crystallization processes, the phase equilibrium behavior or the binodal curve must be known for the sparingly soluble solid solute in $scCO_2$ or the solvent plus CO_2 as a function of pressure, temperature, and composition.

3. PHASE BEHAVIOR

For the rational design of commercial process plants, it is essential to understand the fluid phase equilibrium behavior of binary systems with CO_2 as the solvent for RESS and PGSS, and that of ternary systems with CO_2 as the antisolvent for GAS, SAS, PCA, and SEDS. This, in addition to knowledge of the kinetics of nucleation, growth, aggregation, and aging, enables manipulation of the phase transition pathways desired for the production of monodispersed submicrometer particulate solids with controlled structure, morphology, size, and shape. While RESS involves phase transition from supercritical solution phase to particulate solids, PGSS, GAS, SAS, PCA, and SEDS involve phase transition from $scCO_2$-dissolved liquid solutions to solid precipitates.

The phase diagrams for some systems are quite simple, like the solubility curve for binary mixtures, whereas for some complex mixtures, it is difficult to define the degree of supersaturation owing to the large number of crystallizing components present. So it is necessary to make an approximation and represent a complex system on the basis of a ternary phase diagram of only three prominent components. The phase boundaries obviously depend on the number of components present, and these are explained in the sections that follow.

3.1. Single-Component Phase Boundary

The equilibrium phase boundaries between solid and liquid (i.e., melting curve), solid and gas (i.e., sublimation curve), and liquid and gas (i.e., vaporization curve) for a single component are represented by a line on a pressure–temperature diagram as shown in Figure 1. The state of coexistence of all three phases in equilibrium is represented by the triple point. The highest point on the vaporization curve, which is unique to a given substance, is termed the critical point, and the supercritical fluid is defined as the state at which both pressure and temperature exceed the critical point values for the corresponding substance. The fluid now acquires unique properties of a good solvent, strong pressure and temperature dependence of solubility of certain

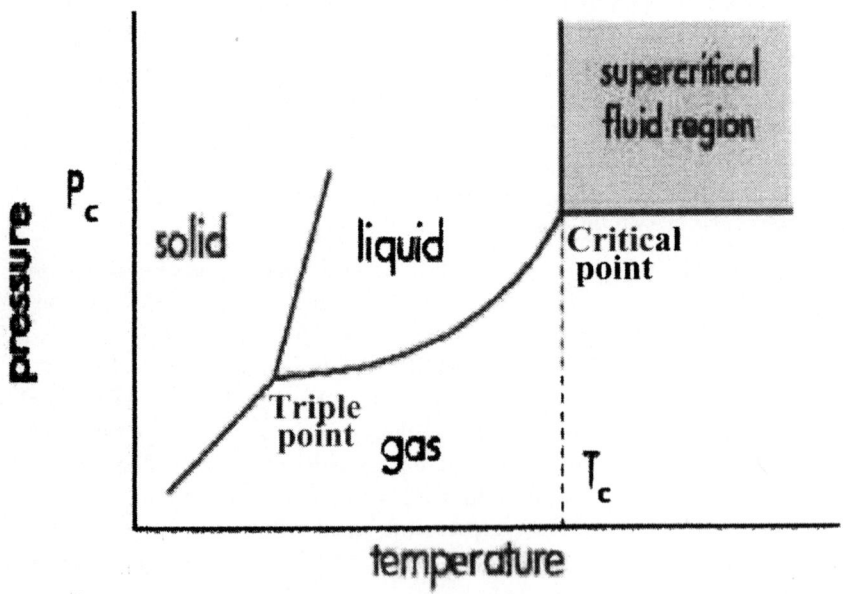

FIGURE 1   Pressure–temperature phase diagram for a single component.

solid solutes, and good extraction characteristics. The degree of freedom of the variables is 1 on the phase boundary line and 0 on the triple point and critical point, whereas outside these lines it is 2 (e.g., pressure and temperature) according to the Gibbs phase rule. It is to be noted that some energy is required to cross over the phase boundary, and the energy required for phase transformation vanishes as the critical point is approached. For attaining the SCF state, it is thus common practice to first lower the temperature of gaseous CO_2 at a near-critical subcritical pressure to condense it and cool it further, to ensure that it is pumped to the supercritical pressure and heated to the supercritical temperature.

3.2. Two-Component Phase Boundary

The degree of freedom for a binary mixture on the vapor–liquid (V-L) boundary is 2. The phase boundary is represented on a T-x or a P-x diagram and the locus of the critical points, (V = L) of all binary mixtures on a P-T space. The mixture's critical point (CP) is defined as the point where the two phases are identical, (i.e., L = V). The critical point of a binary mixture is designated as either mixture critical pressure or mixture critical temperature. As shown in Figure 2a, the mixture critical pressure is the highest pressure on an isothermal P-x diagram of the binary mixture beyond which there is no two-phase region at a particular temperature (11). A binary mixture is supercritical at all pressures above its mixture critical pressure and compositon such that the mole fraction of the SCF component is above its critical value on the isothermal P-x diagram. However, the two components are completely miscible for all compositions at any pressure above the mixture critical pressure.The solubility of a liquid solvent in the fluid phase passes through a minimum value and then increases with pressure, at high pressures near the mixture critical pressure, provided the temperature is higher than the critical temperature of the lighter component, CO_2. This is similar to the behavior of the solubility of a solid solute in $scCO_2$. On the other hand, as can be seen from Figure 2b the two-phase vapor–liquid boundary may extend at a temperature above the mixture critical temperature on a T-x diagram for a particular pressure. Thus a binary mixture is not necessarily supercritical above its mixture critical temperature and critical composition, and hence a binary P-T phase diagram needs to be considered involving all three parameters, namely, pressure, temperature, and composition.

3.3. Classification of Binary P-T Phase Diagrams

3.3.1. Liquid–Fluid Systems

To ascertain the nature of phase behavior of the binary system of interest with a minimal quantum of experimental effort, it is desirable to have prior knowl-

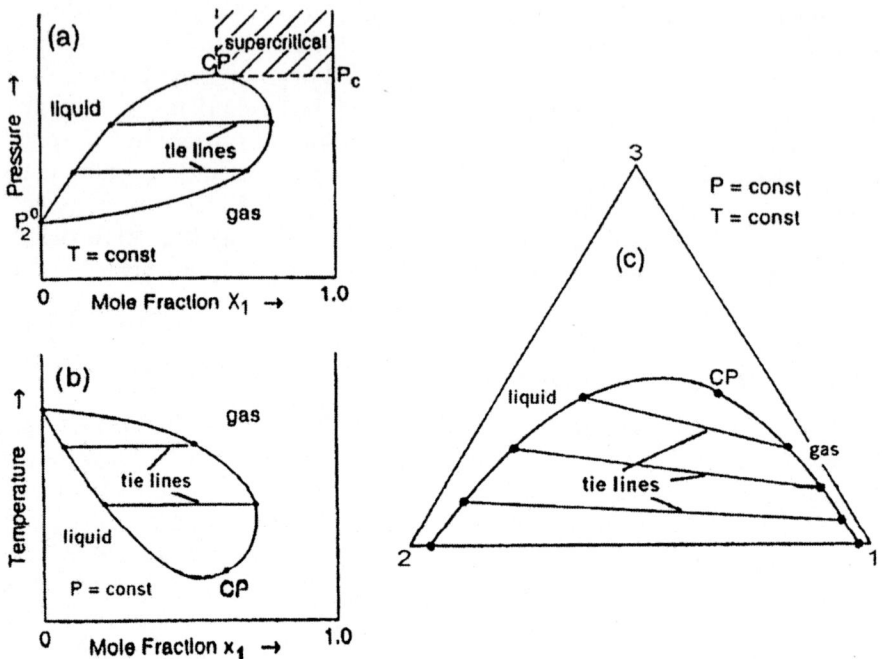

FIGURE 2 Critical points in binary [(a) isothermal, (b) isobaric] and (c) ternary mixtures.

edge of its characteristic features according to the classification of the various possible phase diagrams on a *P-T* plane. Ekart et al. (12) cite that van Konynenberg and Scott had grouped all possible phase diagrams of liquid–fluid phase behavior into six major schematic classes (I–VI) on the basis of *P-T* projections of the critical (L = V) locus and three-phase (L-L-V) lines. The distinct features of these six classes are illustrated in Figure 3. According to this representation, a *P-T* surface for a binary system represents 2 degrees of freedom, whereas the solid lines depicting equilibrium between two phases for a pure component represent 1 degree of freedom. The dashed lines depicting the mixture critical curve and the dashed-dotted lines depicting the L-L-V equilibrium for a binary mixture also represent 1 degree of freedom. A point for binaries corresponds to zero degrees of freedom according to the Gibbs phase rule. The two important types of point are as follows: (a) the critical points of the pure components (circles in Figure 3) and (b) the critical end points for the binaries (triangles), where two phases critically merge to form a single phase in the presence of another phase. For liquid–fluid systems, this

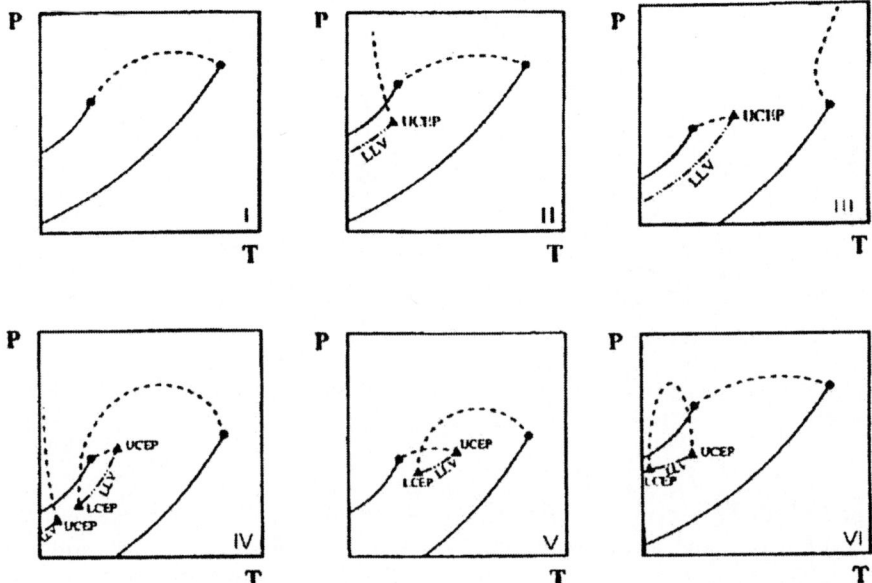

FIGURE 3 Van Konynenberg and Scott's classification of binary L-V phase behavior into types I–VI (12).

is formed by the intersection of a critical (L = V) line with the three-phase (L-L-V) line.

3.3.2. Solid–Fluid Systems

Two possible types of solid–fluid P-T phase diagram are commonly observed, as reported in the literature.

Type I (Solid–Fluid) System. The distinguishing feature of systems of this type (where components of the mixture are similar) is that the critical (L = V) mixture curve runs continuously between the critical points of the lighter and the heavier components, in addition to the occurrence of a continuous S-L-V line (13). The S-L-V line begins at the triple point of the heavy component, bends toward the lower temperature as the pressure is increased, and ends at a temperature less than the critical temperature of the lighter component as described in Figure 4a. The melting point of the heavier component (solid) decreases with increasing pressure in the presence of an SCF component, and this depression is analogous to the freezing point depression of ice in the presence of a salt. In other words, at low pressures and at a temperature less than the critical temperature of the heavier component, the S-V equilibrium is observed until the three-phase S-L-V line is ap-

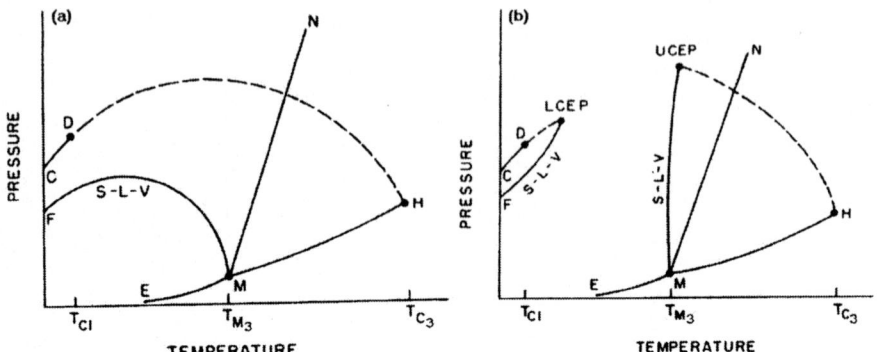

FIGURE 4   Binary solid–fluid phase behavior of (a) type I and (b) type II systems (13).

proached. Thus, for a type I system, at a low pressure there is a region of S-V equilibrium, and as pressure increases, the S-L-V line may be encountered, resulting a transition from the S-V to the S-L region if the composition of the mixture is rich in the solid component.

Type II (Solid–Fluid) System.   In type II systems (when the solid and the SCF component are very dissimilar in molecular size, structure, and polarity), the S-L-V line is no longer continuous, and the critical (L = V) mixture curve also is not continuous. The branch of the three-phase S-L-V line starting with the triple point of the solid solute does not bend as much toward lower temperature with increasing pressure as it does in the case of type I system. This is because the SCF component is not very soluble in the heavy molten solute. The S-L-V line rises sharply with pressure and intersects the upper branch of the critical mixture (L = V) curve at the upper critical end point (UCEP), and the lower temperature branch of the S-L-V line intersects the critical mixture curve at the lower critical end point (LCEP). Between the two branches of the S-L-V line there exists S-V equilibrium only (13).

As described in Figure 4b the phase behavior of a type II binary system is depicted by the vapor pressure (L-V boundary) curves for the pure components, sublimation (S-V boundary) and melting (S-L boundary) curves for the solid component, and especially the S-L-V line on the *P-T* space. For an organic solid drug solute, the triple-point temperature is sufficiently higher than the critical temperature of the SCF solvent. The (L = V) critical locus has two branches and is intersected by two S-L-V lines at UCEP and LCEP, respectively, in the presence of the solid phase. The S-L-V line indicates that the melting of the solid is lowered in the presence of the SCF solvent component as it is dissolved in the molten (liquid) phase. The S-L-V line may exhibit a minimum temperature with an increase in pressure and at a temperature less than this minimum, an S-V equilibrium exists at all pressures. At a temperature between the UCEP and the minimum temperature, an isotherm intersects the S-L-V line at two points at which either S-L or V-L equilibrium is observed. The S-V equilibrium is again observed on increasing the pressure. However, the LCEP usually occurs very close to the critical point of the pure SCF component. If temperature is slightly increased above the LCEP temperature, only S-V equilibrium is observed at all pressures, since the L-L-V line ends at LCEP. There is a relatively large solubility enhancement with pressure near LCEP, although the solubility is quite low (14).

Although P-T projections are useful in classifying the phase diagrams, it is necessary to study the three-dimensional P-T-x diagrams for a complete overview of the phase behavior of the binary system. The dew point curve (P-x) finds an inflection at LCEP, and at any mixture critical (V = L) point, the P-x diagram must exhibit zero slope: that is, $(\partial P/\partial x)_T = 0$. As a result, a large solid solubility enhancement is observed with a slight increase in pressure if the temperature is increased slightly above the LCEP temperature.

At temperatures slightly below the UCEP temperatures, a large solid solubility enhancement is observed in the vicinity of the UCEP pressure. The inflection in the solubility isotherm near the UCEP temperature is much greater, and as a result the solubility of the solid solute in the SCF phase is very high near the UCEP temperature (13). However the shape of the S-L-V line controls the solubility behavior of a solid solute, and less dramatic solubility enhancements are observed at lower temperatures, which are further removed from the UCEP temperature. At pressures much higher than the UCEP pressures, the SCF is less compressible and so the solubility attains a limiting value.

3.4. Three-Component Phase Boundary

3.4.1. Liquid–Liquid–Fluid Systems

The phase boundary between vapor and liquid or supercritical fluid and liquid and the occurrence of the critical point (CP) for a ternary mixture are represented on a triangular diagram (Figure 2c) at a fixed temperature and pressure, since the degree of freedom is 3. Mc Hugh and Krukonis report that Elgin and coworkers classified the phase behavior of the ternary systems consisting of an SCF component with two liquid components to study the effect of the SCF component on miscibility or phase splitting (13). The transformation of isothermal ternary diagrams with pressure is schematically represented by the horizontal sections of a prism such that the three sides of the prism would depict the isothermal P-x diagrams for the constituent binaries respectively. According to this ternary (liquid–liquid–fluid) classification, ternary phase diagrams are grouped into three different classes based on the appearance of the L-L-V and L-L regions with increasing pressures.

Type I Ternary (Liquid–Liquid–Fluid).   The ternary phase behavior is characterized by the absence of an L-L-V immiscibility region when the two liquid components (e.g., ethanol and water) are infinitely miscible. One of the liquid components (e.g., ethanol) becomes miscible with the SCF component (i.e., CO_2) at higher pressures (greater than the mixture critical pressure of the binary). An example of type I ternary phase behavior is depicted by the phase boundary for ethanol–water–CO_2 system, as illustrated in the triangular diagram (Figure 5a) in terms of the three concentrations (mole fractions) at a pressure of 120 bar, at which ethanol–CO_2 is supercritical (15,16). The phase boundary between the single-phase and two-phase regions is the locus of the end points of the tie lines and is known as the binodal curve. All three components are distributed in the equilibrated phases, namely the fluid phase and the liquid phase. The relative dissolution of the component in the two phases is characterized by the distribution coefficient. This is given by the ratio of the concentration (mole fraction) of the component in the fluid phase to that in the liquid phase, as represented by the two end points of a tie line. The mixture critical point of a ternary mixture, also known as the plait point, is identified as the concurrence of the two end points of a tie line at which all three components are completely miscible. This is not observed in Figure 5b for a lower pressure of 69 bar and 308 K, where the (ethanol and CO_2) binary mixture is not supercritical. To observe the plait point, therefore, it is necessary to have two (out of three) binaries in a ternary system completely miscible at the given temperature and pressure.

Type II Ternary (Liquid–Liquid–Fluid).   Type II ternary phase behavior is characterized by the appearance within the P-x prism of the L-L-V region (13), which does not extend to the sides of the prism, each side depicting the P-x behavior of a constituent binary, as schematically shown in Figure 5c. At pressures below the critical pressure of the SCF component, two liquid phases appear along with an L-L-V region and expand considerably with increasing pressure. At pressures above the mixture critical pressure of the SCF component and one liquid component, the L-L-V region disappears and the phase behavior becomes identical with that for a type I ternary system.

Type III Ternary (Liquid–Liquid–Fluid).   Type III ternary phase behavior is characterized by liquid–liquid immiscibility, which is observed even at low pressures. A large L-L-V region exists along with two single–liquid phase regions and three possible two-phase regions as schematically described in Figure 5d. Upon increasing pressure to a value greater than the mixture

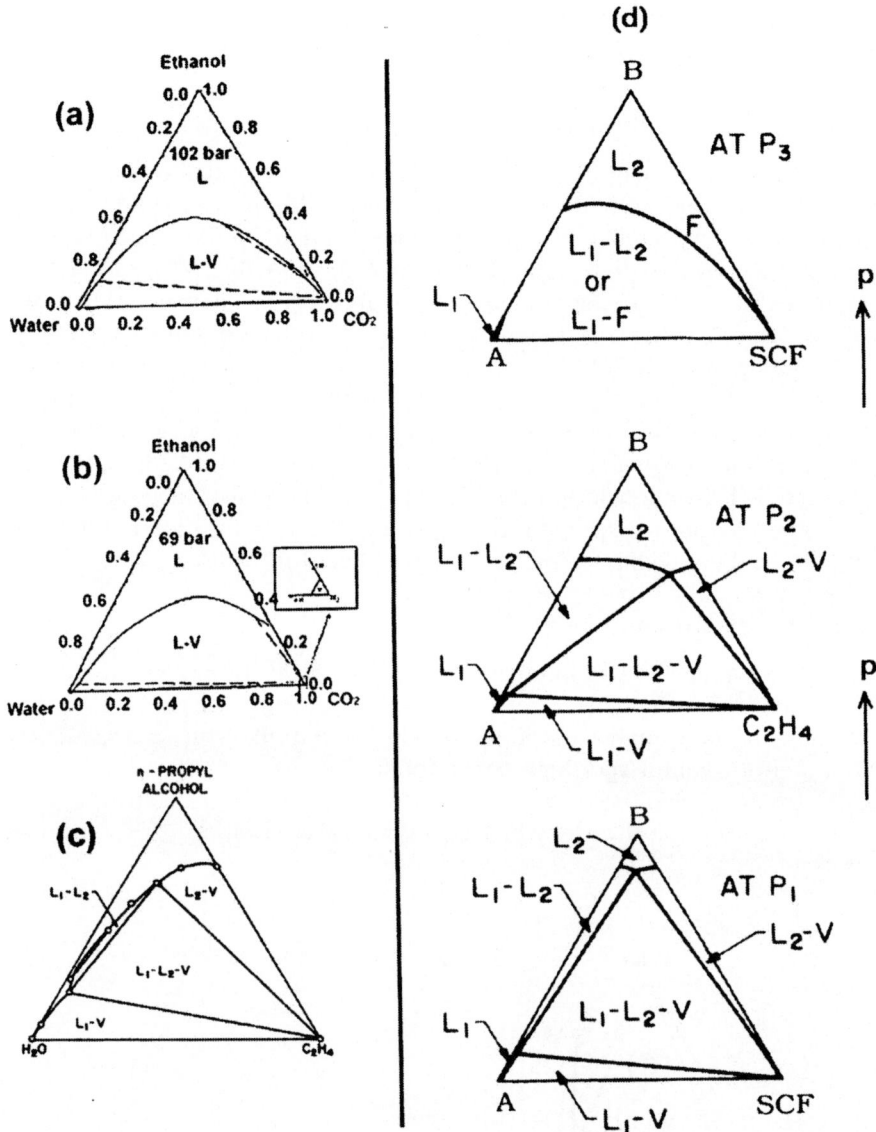

FIGURE 5 (a) Type I ternary (L-L-V) phase behavior at 120 bar (16). (b) Type I ternary (L-L-V) phase behavior at 69 bar (16). (c) Ternary phase behavior of a type II (L-L-V) system (13). (d) Effect of pressure on a type III (L-L-V) system (13).

critical pressure, the LLV region disappears and only a single liquid phase, a single fluid phase, and a liquid–fluid region appear (13).

3.4.2. Solid–Liquid–Fluid Systems

The phase behavior of a (solid–liquid–fluid) ternary system containing one solid component and one liquid component in addition to the SCF component is also very similar to type III ternary (liquid–liquid–fluid) phase behavior inasmuch as all three binaries are partially miscible. There could be as many as four phase boundaries for six possible regions, namely (a) homogeneous liquid (L) phase, (b) fluid–liquid (V-L), (c) solid–liquid (S-L), (d) fluid–liquid–solid (S-L-V), (e) fluid–solid (S-V), and (f) homogeneous vapor (V) phases, as shown in Figure 6a. These are reduced to four regions with the disappearance of S-L-V and L-V regions, as shown in Figure 6b at a pressure above the mixture critical pressure of the binary mixture of the liquid solvent and the SCF component (15). A thorough understanding of these multiple phase boundaries is required for monitoring the crystallization pathways to obtain the desired characteristics of the particles formed (17).

3.5. Crystallization Pathways

The RESS and PGSS techniques involve two steps: continuous dissolution of the solid solute in the SCF solvent or vice versa, and rapid depressurization of the solution through a nozzle and particle formation by nucleation induced by a uniform and a very high degree of supersaturation. However GAS is operated on a semibatch mode in which the solution is taken in a high-pressure crystallizer through which CO_2 is passed continuously for some time,

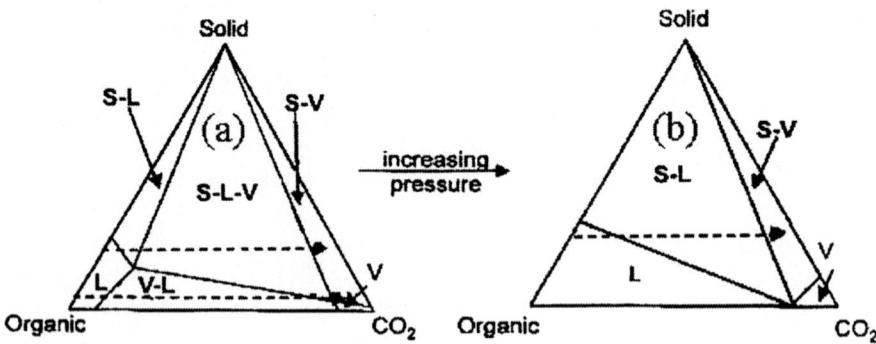

FIGURE 6 Effect of pressure on (a) ternary (S-L-V) and (b) binary (S-L) phase behavior (15).

following which the particles are separated from the filtrate and dried in flowing CO_2. An advantage of the GAS process is the sequential relative solubility reduction of the dissolved solutes with increasing CO_2 dissolution on increasing pressure. This often results in fractional crystallization and plays a significant role on the purification and recovery of the desired bioactive component in the product precipitated from the solution.

On the other hand, SAS and SEDS are operated on a continuous mode in which continuous addition of SC CO_2 along with a cosolvent (in SEDS) with simultaneous spraying of the liquid solution through a nozzle enables complete removal of the residual organic solvent after the particles are formed. The two-way diffusion of CO_2 and solvent, respectively, to and from the ultrafine droplet causes rapid crystallization followed by complete evaporation of the solvent in the flowing CO_2 stream.

Consequently, the phase behavior of a binary system encountered in RESS/PGSS, and a ternary system in SAS/GAS, is quite complex. As illustrated in Figure 7a, in the RESS/PGSS process the solid solute and the fluid separates out from the homogeneous phase very rapidly, thus resulting in the formation of solid microspheres or fibers. In the case of gaseous antisolvent (GAS) or supercritical antisolvent (SAS) precipitation, or precipitation with compressed antisolvent (PCA), an organic solvent is present and the pathway enters from the liquid (L) solution region into the solid–fluid (S-V) region, via the liquid–solid (L-S) and liquid–fluid (L-V) two-phase regions and the solid–liquid–fluid (S-L-V) three-phase region, crossing a series of tie lines. For micronization from solid–liquid homogeneous (L) mixtures by GAS, SAS, or SEDS, a large number of phase changes may be encountered at pressures below the mixture critical condition, as described in Figure 7a. The organic solvent is still present at the interface between the solution and the CO_2 continuum, and after crystallization this needs to be removed by flushing with CO_2. However this phase diagram changes with pressure, temperature, and the thermodynamic state. For example, at pressures above the mixture critical condition, the transformation in the SAS or SEDS process may occur entirely within the S-L region, and because there is no interface between the droplet and the SC CO_2 continuum, the crystallization pathway traverses from the S-L to S-V regions, as shown earlier in Figure 6b. As a consequence, the mechanisms of two-way mass transfer for the supersaturation and crystallization pathways are considered to be different below and above the mixture critical pressures.

The schematic diagram depicting the phase behavior of a polymer–solvent–CO_2 system as presented in Figure 7b affords insight into the crystallization pathways indicated by arrows. The diagram shows the binodal and spinodal curves for phase separation. As explained in Figure 7b, the solvent evaporates into CO_2 along path AA' with too little CO_2 influx, thus avoiding

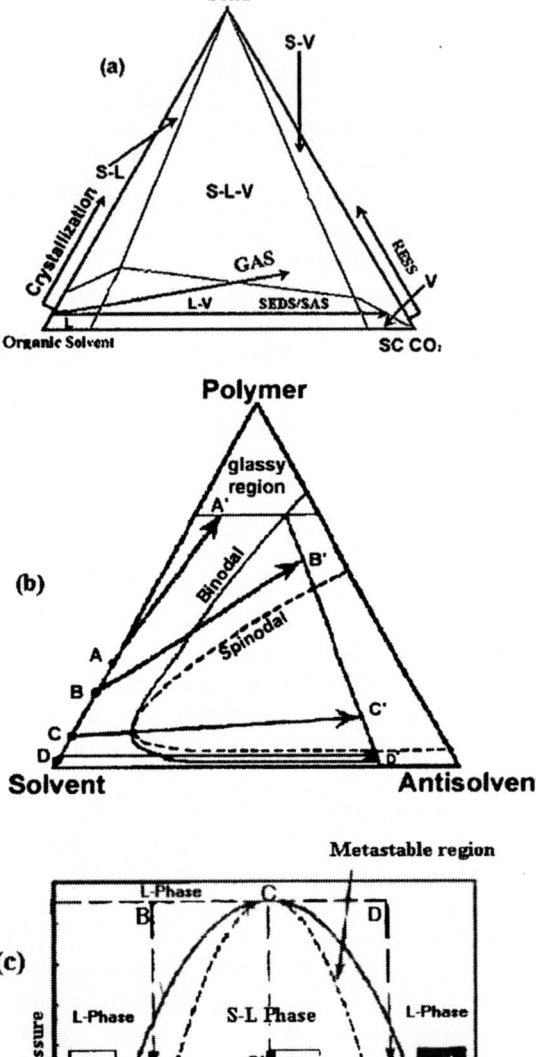

FIGURE 7   (a) Different pathways for crystallization in binary and ternary systems
(17). (b) Pathways for antisolvent crystallization in a polymer–solvent–CO_2
system (17). (c) Effect of concentration on polymer morphology (8).

phase separation, and the path does not enter the two-phase region. As a result, the polymer vitrifies as a denser particle (17). In path *BB′* at the polymer-rich side of the plait point the pathway enters the two-phase region outside the spinodal curve as enough CO_2 enters the droplet. As a result, solvent voids nucleate and grow in the polymer-rich phase until the polymer vitrifies into a porous solid. Path *CC′* enters the spinodal region, which is characterized by bicontinuous phases and often results in rapid spinodal decomposition that produces fine structures, rather than discrete droplets. Path *DD′* is on the polymer-lean side of the binodal curve when the influx of CO_2 is much greater than the evaporation of the solvent. This trend may reverse after the solution droplet has been saturated with CO_2 when diffusion of CO_2 reverses and the evaporation of the solvent is more significant. Simultaneously, solid nucleates and grows within a solvent continuous phase. As a result, the porosity of the polymer increases as the pathway approaches the polymer lean phase. Figure 7c shows that the crystallization pathway either enters the glassy region or the high porosity region from the homogeneous single-phase region, depending on the initial polymer concentration. The morphology is thus decided by the crystallization pathway, or the mode by which the solution loses its solvent power. These pathways highlight the importance of phase behavior along with mass transfer for complete understanding and control of the particle size and morphology.

4. EQUATIONS OF STATE FOR MODELING PHASE EQUILIBRIUM

Different cubic equations of states have so far been used to model the solid–liquid–fluid phase equilibrium and the solubility of an organic solid solute in the fluid or liquid phase. These equations are listed elsewhere (11), along with the procedures for evaluating the binary parameters and adjustable interaction constants in them. These cubic equations of state (EOS) can be summarized as having the following general form:

$$P = \frac{RT}{v - b} - \frac{a(T)}{v^2 + v(b + c) - (bc + d^2)} \tag{8}$$

For the Peng-Robinson (18) equation of state, $b = c$ and $d = 0$. For the Redlich–Kwong–Soave equation of state (see Ref. 19), $c = d = 0$. For the three-parameter Trebble–Bishnoi–Salim (20) equation of state, $d = v_c/3$. For the Patel–Teja (21) equation, $d = 0$. In Eq. (8) the temperature-dependent parameter $a(T)$ is expressed as follows:

$$a(T) = a_c \cdot \alpha(T) \tag{9}$$

The temperature dependence of the attraction parameter $\alpha(T)$ may be obtained from the original references. Equation (8) may be applied to mixtures with the classical van der Waals mixing and combining rules:

$$a = \sum_{i}^{n} \sum_{j}^{n} x_i x_j a_{ij} \quad \text{with} \quad a_{ij} = a_{ji} = (a_{ii} \, a_{jj})^{1/2} (1 - k_{ij}) \tag{10}$$

$$b = \sum_{i}^{n} \sum_{j}^{n} x_i x_j b_{ij} \quad \text{with} \quad b_{ij} = b_{ji} = \frac{b_i + b_j}{2} (1 - l_{ij}) \tag{11}$$

$$c = \sum_{i}^{n} \sum_{j}^{n} x_i x_j c_{ij} \quad \text{with} \quad c_{ij} = c_{ji} = \frac{cc_i + c_j}{2} (1 - m_{ij}) \tag{12}$$

and

$$d = \sum_{i=1}^{n} x_i d_{ij} \tag{13}$$

For the Redlich–Kwong–Soave (R-K-S) and Peng–Robinson (P-R) equations of state, the composition-dependent combining rule for a_{ij} was given by Panagiotopoulos and Reid (22) as

$$a_{ij} = (a_{ii} \, a_{jj})^{1/2} \left[1 - n_{ij} + (n_{ij} - n_{ji}) x_i \right] \tag{14}$$

with

$$b = \sum_{i=1}^{n} b_i x_i$$

The accuracy of an EOS essentially depends on the mixing rule and its interaction constants. The most commonly used equation of state for supercritical fluid phase equilibrium calculations is the P-R EOS with one or two interaction constants evaluated from the binary equilibrium data. In the absence of experimental data, the interaction constants in the P-R or the R-K-S EOS were predicted by the group contribution approach based on the modified Huron and Vidal mixing rule and the modified UNIFAC model for the excess Gibbs energy (8,11). However they yielded limited success in representing high pressure L-L-V phase equilibrium. From now on, for the sake of consistency, the supercritical fluid (F), vapor (V), and gaseous (G) phases are all designated by V.

5. SOLUBILITY OF DRUGS IN SUPERCRITICAL CARBON DIOXIDE

To select the best method for particle formation, it is first necessary to have knowledge of the solubility of the solid solute in scCO$_2$ as well as in the organic solvents. The solubility of a high molecular weight solid drug substance in scCO$_2$ depends on temperature, pressure, and cosolvent concentration in addition to physicochemical characteristics. A few selected references for the solubility data reported in the literature for several drug substances are listed in Table 1.

It is seen that solubility increases with density and that the effect of the density on solubility sharply increases at higher densities (Figure 8). This implies that at higher densities, molecular interactions are enhanced and so is solubility. Solubility also increases with a decrease in temperature at a pressure less than the crossover pressure of the substance, beyond which it increases with temperature (Figure 9). This retrograde phenomenon is attributed to the two opposite effects of an increase in temperature on the solubility of a solid solute in an SCF solvent: (a) a decrease in solubility due to a rise in temperature that decreases the density of the SCF solvent, and also (b) an increase in solubility owing to the increase in the volatility of the substance with temperature (11). The effect of addition of a cosolvent in scCO$_2$ is utilized in achieving an enhancement in solubility at lower pressure due to the specific molecular interactions. The degree to which the polarity of the supercritical fluid phase is modified depends on the amount and the nature of the cosolvent added (66) (Figures 10 and 11). The effects of all three parameters along with the physicochemical characteristics of the substance can be combined by an EOS approach or by an empirical correlation for modeling the drug solubility in scCO$_2$.

5.1. Modeling Binary Solid–Fluid (S-V) Equilibrium by the EOS Method

The solubility y_3 of solid solute 3 at equilibrium conditions in SCF solvent 1 at pressure P and temperature T can be expressed as

$$y_3 = \frac{P_3^s}{P\overline{\phi}_3^V}\exp\left[\frac{v_3^s(P - P_3^s)}{RT}\right] \tag{15}$$

where P_3^s is the sublimation pressure, v_3^s is the molar volume of the solid, and the fugacity coefficient, $\overline{\phi}_3^V$ is given as

$$RT\ln\overline{\phi}_3^V = \int_\infty^v\left[\left(\frac{\partial P}{\partial n_3}\right)_{T,v,n_j} - \frac{RT}{v}\right]dV - RT\ln Z \tag{16}$$

where Z is the compressibility factor of the mixture. The most commonly used EOS for solubility calculations of pharmaceutical solids (23) in scCO$_2$ has been the P-R EOS with two adjustable interaction parameters, k_{ij} and l_{ij} in a_{ij}

TABLE 1 Selected References for Solubility of Solid Drugs in $scCO_2$

Solute	Temperature K	Pressure (bar)	Ref.
Epicatechin	313	80–120	69
α-Tetralol	303–333	125–258	70
Sulfathiazole	298–353	100–200	71
9,10-Anthraquinone	308–318	84–306	72
p-Quinone	308–318	86–292	
Ascorbyl palmitate	308–313	130–200	73
Butylhydroxyanisole			
Dodecyl gallate			
1,8-Bis(prop-2'-enyloxy)-9,	318–348	121–405	74
10-anthraquinone			
Caffeine	313–353	199–349	75
β-Carotene	313–353	97–260	
Theobromine	313–353	193–345	
α-Theophylline	313–353	199–349	
β-Carotene	310–340	97–260	76
o,m,p-Coumeric acid	308–323	85–250	77
Coumarin	308–323	85–250	78
Hydroxycoumarin	308–323	85–250	
Methoxycoumarin	308–323	85–250	
Methyl coumarin	308–323	85–250	
Dihydroxyxanthone	305–348	74–355	79
Eicosanic acid	308–328	34–211	80
1-Eicosanal	308–328	36–412	
Endrin	313–333	100–220	81
Lindane	313–333	100–220	
Methoxychlor	313–333	100–220	
Flavone	308–318	91–253	82
Hydroxyflavone	308–318	91–253	
o-hydroxybenzoic acid	318–328	101–203	83
p-Hydroxybenzoic Acid			
Ketoprofen	312–332	100–220	84
Nimesulfide	312–332	100–220	
Menthol	308–328	64–115	85
Naproxen	313–333	90–193	66
Nifedipine	313–333	90–193	68
Nitrendipine	313–333	90–193	
Paclitaxel	310–329	141–345	86
Squalene	303–328	79–275	87
Sulfamethoxine	313–333	131–488	88
Sulfamerazine	313–333	151–474	
Sulfamethazine	313–333	136–476	
Taxol	308–318	207–483	89
Vanillic acid	313–328	86–250	90

TABLE 1 Continued

Solute	Temperature K	Pressure (bar)	Ref.
Vitamin D_2, D_3, K_1	313–353	200–350	91
Aspirin			92
Caffeine			93
Cholesterol			94
Dipalmitoylphosphatidyl-choline (DPPC)			95
β-Estradiol			96
Ibuprufen			97
Griseofulvin			98
Salicylic acid			
Stigmasterol			99

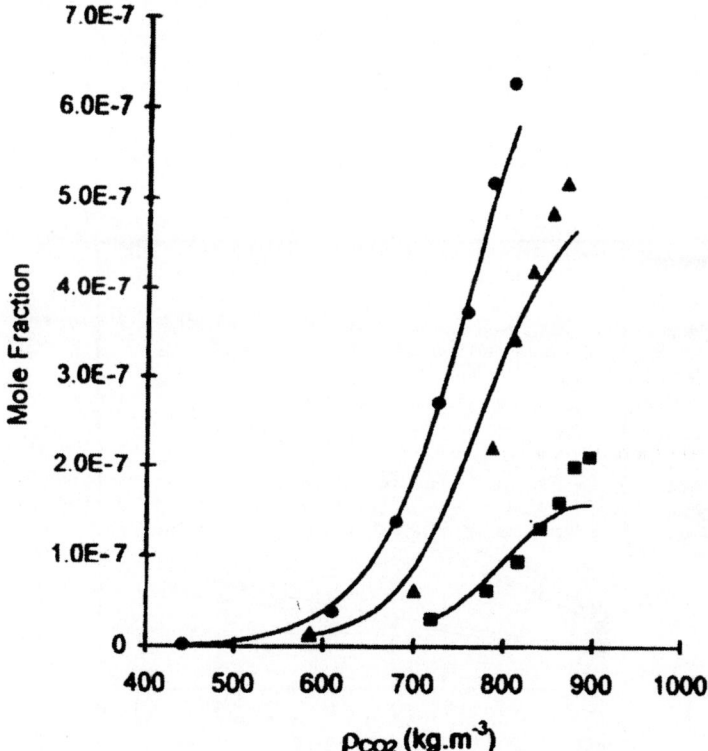

FIGURE 8 Effect of density on solid (β-carotene) solubility in $scCO_2$: experimental data at 40°C (■), 50°C (▲), 60°C (●); calculated values (solid curves) by the PR EOS (65).

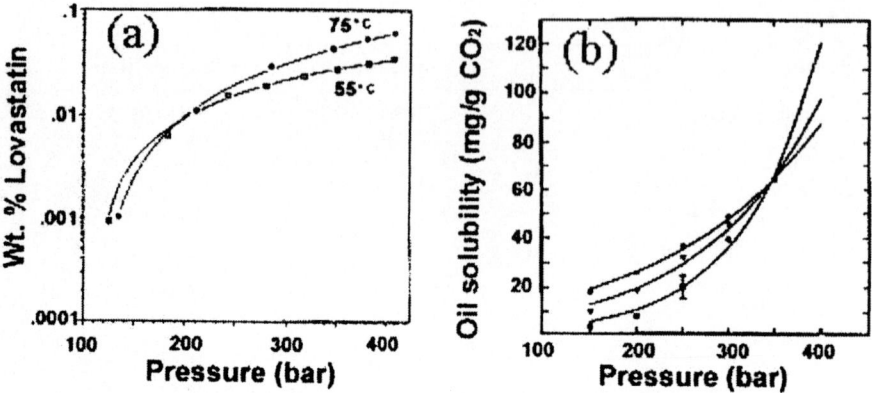

FIGURE 9   Effect of Temperature on (a) solid (lovastatin) solubility in $scCO_2$ (67) at 75°C (●) and 55°C (□).(b) Oil solubility in $scCO_2$ (11) at 40°C(●), 50°C (▲), and 60°C(■).

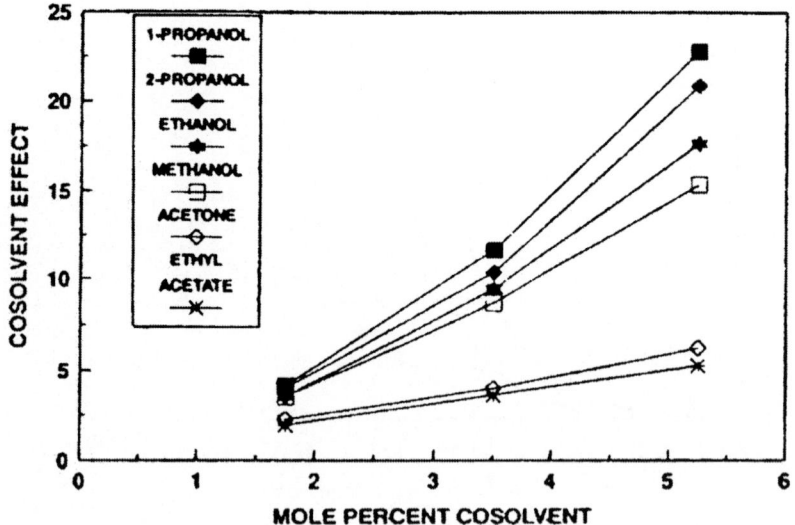

FIGURE 10   Effect of different cosolvents on solid (naproxen) solubility enhancement in $scCO_2$ at 333 K and 179.3 bar (66).

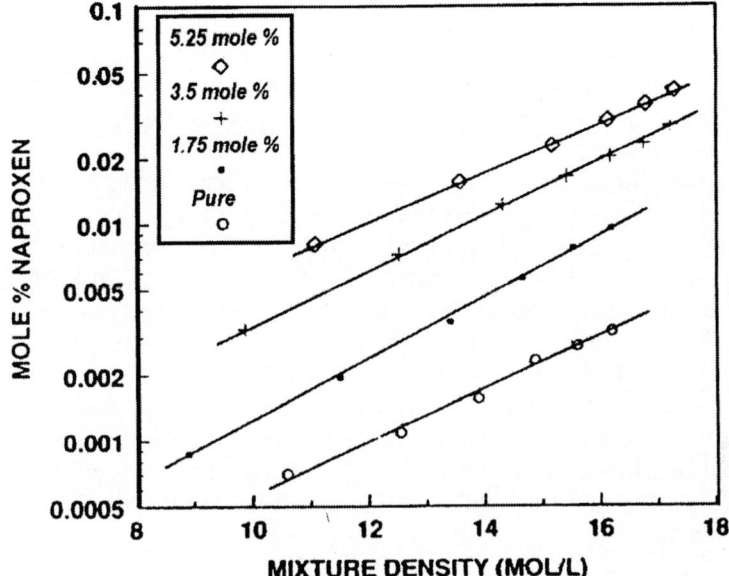

FIGURE 11 Effect of cosolvent (methanol) concentration on naproxen solubility enhancement in scCO$_2$ at 333 K (66).

and b_{ij}, respectively. The P-R EOS is expressed as

$$P = \frac{RT}{v - b} - \frac{a}{v(v + b) + b(v - b)} \tag{17}$$

The fugacity coefficient, $\overline{\phi}_3^v$ is calculated from:

$$\ln \overline{\phi}_3^V = \frac{b_3}{b}\left(\frac{Pv}{RT} - 1\right) - \ln\left[\frac{P(v^V - b)}{RT}\right] - \frac{a}{2\sqrt{2}bRT}$$
$$\times \left[\frac{2\sum_k x_k a_{3k}}{a} - \frac{b_3}{b}\right] \ln\left[\frac{v^V + 2.414b}{v^V - 0.414b}\right] \tag{18}$$

The solute properties, needed for determination of solid solubility, can be estimated by the group contribution method of Lyderson (see Ref. 24).

The interaction constants k_{ij} and l_{ij} may be regressed from the experimental solubility data. There is much uncertainty in both the estimated and reported experimental data on the sublimation pressure of the solid solute. It is therefore necessary to assess the accuracy of the saturation pressure. Alternatively, P_3^s may be regressed from experimental data and l_{ij} may be neglected. There are a few new mixing rules, which have been employed for prediction of solute solubility in SC CO_2 and are listed elsewhere (11). These rules enable reduction of the sensitivity of the interaction parameter k_{ij}, as a result of which it is possible to correlate them in terms of the pure component properties, making the method a predictive one. The uncertainties in modeling S-V equilibrium have been analyzed by Xu et al. (25). The P-R EOS generally results in good correlation with the experimental solubility data as illustrated in Figure 8 for β-carotene solubility (65) in scCO₂, although both P-R EOS and R-K-S EOS resulted good agreement for naproxen solubility (66) in SC CO_2 with and without a cosolvent. However Gordillo et al. (26) modeled the solubility of penicillin G, using the R-K-S EOS with good agreement (Figure 12). The experimental data on solubility of a few pure solid pharmaceuticals listed in Table 2 allow a glimpse of its variation with temperature and density.

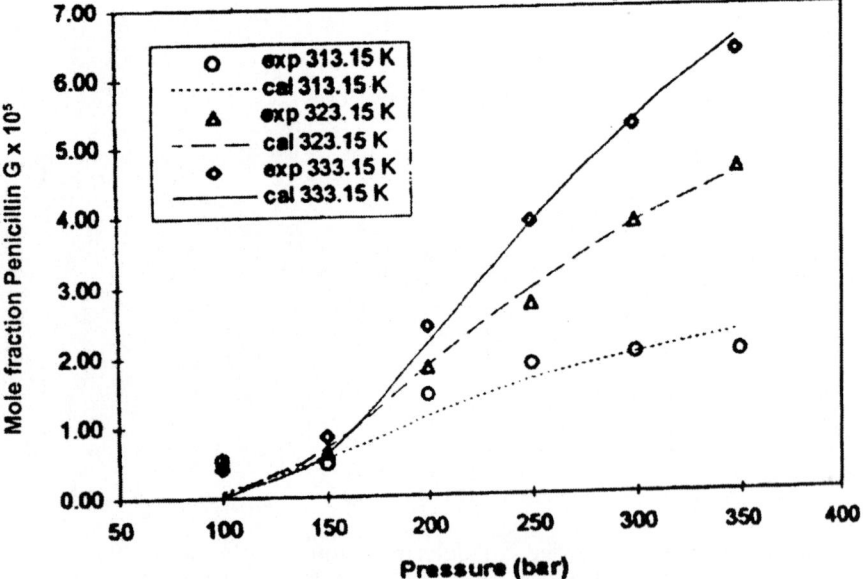

FIGURE 12   Modeling of solubility of penicillin G in scCO₂ (26) by means of the R-K-S EOS.

TABLE 2 Solubility of Drugs in scCO$_2$

Drug	Temperature (K)	Pressure (bar)	Density (g/cm^3)	Solubility (mole fraction × 10^4)	Ref.
Progesterone	313	115	0.705	1.64	100
		205	0.846	5.99	
		240	0.874	7.20	
	333	115	0.398	0.09	
		205	0.733	4.86	
		240	0.777	7.37	
Methoxyprogesterone acetate	313	110	0.686	0.29	100
		180	0.821	1.01	
		220	0.859	1.63	
		150	0.606	0.331	
	333	200	0.725	1.10	
		220	0.754	1.32	
Nifedipene	339	135	0.560	0.0508	68
		210	0.734	0.1506	
		296	0.827	0.3059	
	353	212	0.6207	0.1877	
		250	0.686	0.2532	
		280	0.777	0.3803	
Lovastatin	328	125		9 (wt % × 10^{-4})	67
		280		187	
		382		310	
		409		341	
	348	134		10 (wt % × 10^{-4})	
		285		284	
		381		526	
		409		600	

5.2. Modeling Solid Solubility in Supercritical Fluid by Empirical Correlation

There are several empirical correlations for estimating the solubility of solid pharmaceuticals in scCO$_2$. The logarithm of the solubility is found to be a linear function of the logarithm of the density over a pressure range, as given by Chrastil's equation (27)

$$\ln s = A_1 \ln \rho + \frac{A_2}{T} + A_3 \tag{19}$$

where s is the solubility (g/L), ρ is the solvent density (g/mL), and T is the temperature (K); A_1 has the physical significance of an association number representing the average number of CO_2 molecules in the solvated complex, A_2 depicts the vaporization and solvation enthalpies of the solute, and A_3 depends on the molecular weight of the solid. The numerical values of the constants A_1, A_2, and A_3 are reported in the literature (28). This equation was further modified by Mendez-Santiago and Teja (29) to correlate the solid solubility with SCF density ρ in terms of the enhancement factor E with respect to the ideal gas solubility, as

$$T \ln E = A + B\rho \qquad (20)$$

where A and B are two adjustable parameters. To achieve more accurate predictions, however, del Valle and Aguilera (30) added one more term:

$$\ln s = B_0 + \frac{B_1}{T} + B_2 \ln \rho + \frac{B_3}{T^2} \qquad (21)$$

Equations (20) and (21) do not include any term involving pressure; hence the pressure effect is combined with density, rather than the isothermal pressure effect. And hence these equations are not expected to produce good accuracy for isothermal solubility data at different pressures.

Yu et al. (31) proposed an empirical correlation, as given by Eq. (22), relating the solute solubility, y_3 (mole fraction) to the pressure P (bar) and the temperature T(K):

$$y_3 = C_0 + C_1 P + C_2 P^2 + C_3 PT(1 - y_3) + C_4 T + C_5 T^2 \qquad (22)$$

However Gordillo et al. (26) modified this correlation to improve the accuracy for calculation of the solute solubility in $scCO_2$, in the form of Eq. (23) and evaluated the empirical constants C_0–C_5.

$$y_3 = C_0 + C_1 P + C_2 P^2 + C_3 PT + C_4 T + C_5 T^2 \qquad (23)$$

Jouyban et al. (32) recently provided a new correlation for the solid solubility in $scCO_2$, which is given as

$$\ln y_2 = M_0 + M_1 P + M_2 PT + \frac{M_4 T}{P} + M_5 \ln \rho \qquad (24)$$

and evaluated the empirical constants M_0 to M_5 based on a large database for several drug substances. The calculation of the density of pure $scCO_2$ requires a cubic equation of state such as the P-R EOS. However, the following empirical equation was also proposed by Jouyban et al. (32) for $scCO_2$:

$$\ln \rho = -27.091 + 0.609\sqrt{T} + \frac{3966.170}{T} - \frac{3.445P}{T} + 0.401\sqrt{P} \qquad (25)$$

To calculate the saturated liquid density of CO_2 by Eq. (25), the pseudo–vapor pressure of liquid CO_2, up to $T_r = 1.2$, can be obtained from the following relation (33):

$$\ln\left(\frac{f^{sat}CO_2}{73.77}\right) = 8.83 - \frac{9.18}{T_r} - 4.01 \ln T_r \tag{26}$$

Garcia-Gonzalez et al. (34) used the correlation given by Bartle et al. (28) to calculate the solubility of hydroquinone and p-quinone in $scCO_2$ as

$$\ln\left(\frac{Py}{P_{ref}}\right) = \frac{a+b}{T} + C(\rho - \rho_{ref}) \tag{27}$$

where ρ_{ref} and P_{ref} correspond to the reference density and the reference pressure, respectively. Sovova et al. (35) gave a correlation for enhancement of β-carotene solubility in $scCO_2$ modified with a cosolvent as

$$y - y_0 = 0.012(y_0 w_e)^{0.5} \quad \text{for ethanol}$$

$$y - y_0 = 0.003(y_0 w_o)^{0.5} \quad \text{for vegetable oil}$$

where y_0 is the solubility of β-carotene without cosolvent; w_o and w_e are weight fractions of vegetable oil and ethanol, respectively, in the mixture (g/g). The solubility correlations for various pharmaceuticals in $scCO_2$ have been compiled and are available in the literature (36).

6. CHARACTERIZATION OF DEPRESSURIZATION CRYSTALLIZATION IN BINARY SYSTEMS

In both processes of depressurization crystallization, namely, particles from gas-saturated solutions (PGSS) and rapid expansion from supercritical solutions (RESS), the solid solute and the SCF solvent constitute the binary system. Supersaturation is attained as a result of a drastic reduction of solubility by rapid depressurization, and subsequently the two-phase (S-V) system is formed. The process chosen depends on the solubility of the solid solute and its melting point.

The PGSS technique utilizes the principle of melting point depression of the heavy solute with dissolution of $scCO_2$ at high pressures and saturation of the resultant molten phase of the solute with the SCF solvent to form the so-called gas-saturated solution. By rapid depressurization of such a solution, the $scCO_2$ is removed from the melt, which leads to either a dramatic increase of the melting point of the solute or an increase in the glass transition temperature of the polymer, resulting in plasticization or swelling of the polymer. Since the temperature is also lowered along with pressure by the Joule–Thomson effect or by evaporation cooling, a very rapid and high supersaturation is attained, causing nucleation and formation of ultrafine solid particles.

PGSS has the advantage of processing a concentrated solution of solute, enabling high productivity. At a temperature higher than the melting point of the solid solute, the binary mixture forms two phases: one rich in the molten solute and the other rich in the SCF component with a low content of molten solute. In PGSS the solute-rich liquid phase is depressurized through a nozzle. On the other hand, RESS has the advantage of having an operating temperature much lower than the melting point of the solid solute, and the lower temperature of operation minimizes thermal degradation of the solid solute. In this process the dilute supercritical solution is depressurized through a nozzle for precipitation of the particulate solids without entering into the liquid phase, as described earlier (Figure 4b).

One of the key parameters for process selection is solid solubility in the SCF solvent. As a rule of thumb, RESS is normally employed for micronization when the solute solubility in the SCF solvent is higher than 1000 ppm; if it is less, the solute is dissolved in an organic solvent for processing by SAS or GAS. On the other hand, if the melting point of the solid is relatively low or if it has little thermal lability, PGSS is preferred for micronization (37). For example, micronization of the practically insoluble calcium channel blocker nifedipine, a cardiovascular drug, was investigated by PGSS over a pressures of 100 to 200 bar and temperatures of 170 to 200°C, and it was possible to obtain a reduction in particle size from 48.6 µm to 15.4 µm with 50% enhancement in drug solubility (38). It was, however, reported (39) that the final size of the micronized drug particles was dependent on the preexpansion conditions and agglomeration, though with increasing pressure the particle size decreased. Simultaneous addition of a hydrophobic polymer PEG 4000 facilitated a substantial decrease in the melting point (50–70°C) with a view to avoiding agglomeration and thermal degradation of nifedipine at high temperatures (175–185°C); in addition, the dissolution rate of the coprecipitated drug was doubled in comparison to pure nifedipine (39).

In RESS, it is advantageous to operate near the UCEP owing to an increased pressure dependence on solubility. The solid solute solubility in the SCF solvent increases very sharply with pressure near the LCEP and UCEP. Since the temperature is greater and the slope of isothermal solubility (y-P) curve is higher at UCEP, the solute solubility is very high near UCEP. However, in RESS the S-L-V line needs to be avoided during depressurization. On the other hand, PGSS needs to be operated in the L-V region at a temperature higher than the S-L-V line, and depressurization enables transition from the L-V to the S-V region. The advantage of PGSS over RESS is that micronization can occur at a much lower pressure and a lower ratio of SCF solvent to the molten solute. Accordingly, for PGSS it is crucial to know the solubility of the SCF solvent in the liquid phase, which increases with pressure and decreases with temperature.

6.1. *P-T* Trace of S-L-V Equilibrium in Binary Systems for RESS/PGSS

For ascertaining the process conditions of RESS and PGSS, it is essential to have knowledge of the equilibrium solubility of the solute in dense gas (SCF phase) and vice versa, and also the *P-T* trace for the solid–liquid–vapor (S-L-V) phase transition of the drug substance. If all three phases coexist, there is only a single degree of freedom for a binary system, and a *P-T* trace of the S-L-V equilibrium is sufficient to determine the phase equilibrium compositions.

There are at least four variations in the *P-T* behavior of S-L-V phase transition (40) as described in Figure 13 and are listed here:

1. *Type I P-T trace*: This S-L-V line has a negative slope of *P* vs *T*. This type occurs when the SCF component has high solubility in the molten solute.
2. *Type II P-T trace*: The SCF component is slightly soluble in the molten solute. As a result, the melting point is increased owing to an increase in the hydrostatic pressure.

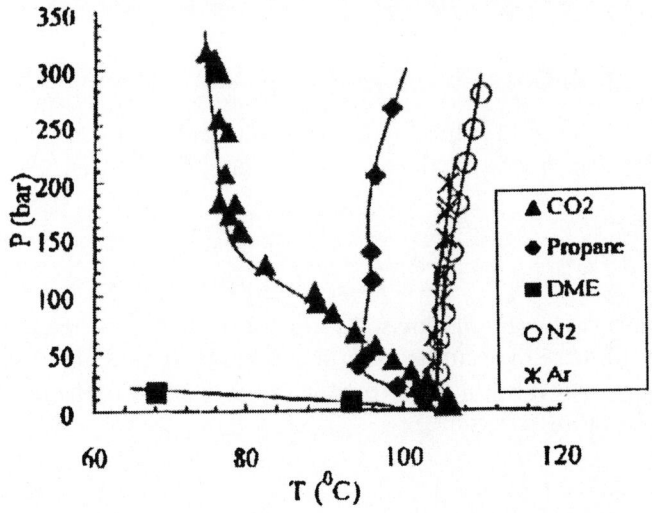

FIGURE 13 Variations in *P-T* behavior of binary (S-L-V) phase transitions for a vitamin K_3–gas [CO_2, propane, dimethyl ether (DME), argon, or nitrogen] system (40).

3. *Type III P-T trace*: There is a minimum temperature due to two competing opposite effects of increasing solubility of the SCF component in the molten solute and also increasing melting point with increasing hydrostatic pressure.
4. *Type IV P-T trace*: There exist both maximum and minimum temperatures.

Before the solubility isotherms of low-melting solid solutes can be predicted, it is necessary to ascertain whether there is a solid–liquid transition along the solubility isotherm due to the melting point depression. For RESS, one needs to study the S-V equilibrium and for PGSS it is required to study the L-V equilibrium at a temperature above the melting point.

6.2. Using the EOS Approach to Model Binary S-L-V Equilibrium

The thermodynamic modeling and calculation procedures for S-L-V equilibrium in binary systems for RESS/PGSS involve simultaneous solution of the phase equilibrium relations for the two components, namely, the SCF solvent, 1, and the solid solute, 3, for all three phases, S, L, and V, as given by the following equations:

$$f_3^S(T, P) = \bar{f}_3^V(T, P, y_3) \tag{28}$$

$$\bar{f}_3^V(T, P, y_3) = \bar{f}_3^L(T, P, x_3) \tag{29}$$

$$\bar{f}_1^V(T, P, y_1) = \bar{f}_1^L(T, P, x_1) \tag{30}$$

As shown earlier, the solid solubility in a supercritical fluid solvent is calculated by Eq. (15), which was obtained by rearranging Eq. (28) as

$$y_3 = \frac{\phi_3^S P_3^S \exp\left[v_3^S(P - P_3^S)/RT\right]}{P\bar{\phi}_3^V} \tag{15}$$

where ϕ_3^S is the fugacity coefficient of pure solid vapor at P_3^S and v_3^S is the solid molar volume. In the absence of accurate experimental data, the sublimation pressure of the solid, P_3^S may be estimated from the vapor pressure of the subcooled liquid P_3^L and heat of fusion as

$$\ln P_3^S = \ln P_3^L - \frac{\Delta H_3^f}{RT_{tp}}\left(\frac{T_{tp}}{T} - 1\right) \tag{31}$$

This is based on the fact that the solid state fugacity of pure solute 3 at the triple-point pressure P_{tp} is related to the liquid state fugacity of the pure solute with the temperature correction, neglecting the difference in the heat capacities in the solid and liquid states (8), as

$$\ln f_3^S(T, P_{tp}) = \ln f_3^L(T, P_{tp}) - \frac{\Delta H_3^f}{RT_{tp}}\left(\frac{T_{tp}}{T} - 1\right) \tag{32}$$

To take into account the effect of pressure (41), these fugacities can be written, respectively, as

$$\ln f_3^S(T, P) = \ln f_3^S(T, P_{tp}) + \frac{v_3^S(P - P_{tp})}{RT} \tag{33}$$

$$\ln f_3^L(T, P) = \ln f_3^L(T, P_{tp}) + \frac{v_3^L(P - P_{tp})}{RT} \tag{34}$$

Combining these three equations one can obtain Eq. (35)

$$\ln f_3^S(T, P) = \ln f_3^L(T, P) + \frac{\left(v_3^S - v_3^L\right)\left(P - P_{tp}\right)}{RT} - \frac{\Delta H_3^f}{RT_{tp}}\left(\frac{T_{tp}}{T} - 1\right) \tag{35}$$

where v_3^S and v_3^L are respectively the molar volumes of the solid and liquid phases, T_{tp} is the triple point temperature, and P_{tp} is the reference (triple point) pressure of the solute.

For the P-T trace of the S-L-V equilibrium, two more equations are obtained by expressing Eq. (29) and (30), respectively, as

$$y_3\,\overline{\phi}_3^V = x_3\,\overline{\phi}_3^L \tag{36}$$

$$y_1\,\overline{\phi}_1^V = x_1\,\overline{\phi}_1^L \tag{37}$$

The S-L-V equilibria for a number of binary systems were modeled by using this approach with the help of the P-R EOS with one (k_{ij}) or two (k_{ij} and l_{ij}) interaction constants regressed from the experimental solute solubility data in the SCF phase. These interaction constants were subsequently used for calculation of the P-T trace of S-L-V equilibrium and compositions (13). Though reasonably good agreement was obtained in correlating the P-T trace of the S-L-V equilibrium with one interaction constant, the calculated UCEP pressure was much higher than the experimentally observed UCEP pressures (41). Better agreement was reported when two interaction constants were employed.

Knez and Skerget (40) reported good agreement with two interaction constants in the P-R EOS for modeling the S-L-V line for vitamin K$_3$, but agreement for vitamins D$_2$ and D$_3$ was not as good. The significant deviation in the latter systems was attributed to the temperature dependency of the interaction constants. It is however important to employ the most appropriate approach for evaluation of the interaction constants.

Kikic et al. (41) employed only the values of P and T at the UCEP to evaluate these two constants (k_{ij} and l_{ij}). The location of the UCEP was estimated from the experimental data by locating the intersection of the S-L-V line and ($L = V$) critical locus curve. For a type I P-T trace (with no temperature minimum with increasing pressure, e.g., a naphthalene–ethylene system), the solubility isotherm at T_{UCEP} provides an inflection point at $P = P_{UCEP}$ (13). By setting the first and second derivatives to zero at this point, one can obtain the two equations needed for the two binary interaction parameters, k_{ij} and l_{ij}, respectively. When this approach was used for the inter-

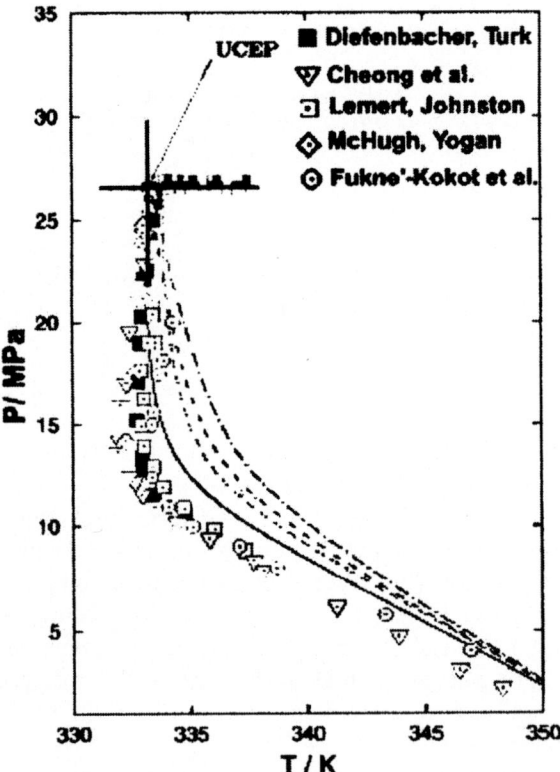

FIGURE 14 Comparison of predicted and experimental binary (S-L-V) phase trace for a naphthalene–CO_2 system: (——) R-K-S EOS with van der Waals mixing rule, (——)R-K-S EOS with Panagiotopoulos–Reid mixing rule, (- - -) P-R EOS with van der Waals mixing rule, and (———) P-R EOS with Panagiotopoulos–Reid mixing rule. (From Ref. 44.)

action constants, the P-T trace as well as the location of UCEP and the solubility data were predicted well for a naphthalene–ethylene system (41). On the other hand, for a type III P-T trace (with a temperature minimum with increasing pressure, e.g., a biphenyl–CO_2 system), the solubility isotherm at T_{UCEP} presents a horizontal peak at P_{UCEP} and thus the approach is slightly different. The two interaction constants were regressed, respectively, by setting the first derivative to zero from the left side as well as from the right side. The P-T trace of S-L-V equilibrium, the UCEP position, and the S-L-V equilibrium compositions of biphenyl–CO_2 thus obtained showed remarkable improvement in agreement, in comparison with the results of Paulaitis et al. obtained by using only one constant (42,43). However, Diefenbacher and Türk (44) observed that the R-K-S EOS yields a better description of the S-L-V line than the P-R EOS and that no improvement in the calculated S-L-V line was obtained by employing the composition-dependent mixing rules of Eq. (14). It was rather observed that the accuracy of prediction of the S-L-V line with two interaction constants for a_{ij} was worse than that obtained without an additional interaction parameter for b_{ij}. Figure 14 compares the experimentally determined S-L-V line for a CO_2–naphthalene (for a type III P-T trace) system with the predicted S-L-V line calculated by the R-K-S and P-R EOS (44) with the two mixing rules. It may be noted here that the two binary interaction parameters were regressed from the UCEP pressure and temperature and not regressed from the experimental solubility (P-y) data. Accordingly, the nature of P-T trace of the S-L-V equilibrium is crucial for deciding the modeling approach, which is strongly dependent on the solubility of the SCF solvent in the molten phase.

7. CHARACTERIZATION OF ANTISOLVENT CRYSTALLIZATION IN TERNARY SYSTEMS

It is necessary to find the appropriate solvent for the given solute and the optimum temperature and pressure conditions for instantaneous crystallization of the solute from its solution, at which the antisolvent, CO_2 either is totally miscible in the solvent or has a very high solubility in it. The solvent and CO_2 are completely miscible as the binary mixture reaches the mixture critical point. There is an exponential increase in the total volume of the solution with increasing dissolution of CO_2. Originally, the relative total volume expansion (RTVE) of the solvent was the criterion for the GAS crystallization, which was defined (45) as

$$\text{RTVE} = \frac{\Delta V}{V} = \frac{V_L(T, P, X_1) - V_2(T, P_0)}{V_2(T, P_0)} \tag{38}$$

where V_1 is the total volume of the liquid solution, V_2 is the total volume of the pure solvent at the same temperature and reference pressure, P_0 (normally, atmospheric pressure). Upon increasing pressure, $\Delta V/V$ increases slowly at low pressures and rapidly at high pressures, as shown in Figure 15a. The isothermal plots of $\Delta V/V$ vs P are different for different temperatures. They are also different for different solvents although the behavior is the same (45). However, with the appearance of liquid–liquid immiscibility, RTVE remains invariant with increasing pressure. In addition, RTVE vs X_1 is not able to recognize the difference between the solvents for a given antisolvent, CO_2. Similarly, RTVE vs X_1 does not indicate any appreciable change with temperature, as shown in Figure 15b. This behavior may be explained by expressing RTVE in terms of molar volumes (46)

$$\frac{\Delta V}{V} = \frac{1}{1 - X_1} \cdot \frac{v_L(T, P, X_1)}{v_2(T, P_0)} - 1 \tag{39}$$

where v_L is the molar volume of the liquid solution and v_2 is the molar volume of the solvent, 2; X_1 denotes the mole fraction of CO_2 in the binary solution of the solvent and CO_2. Since the ratio, v_L/v_2 is close to unity, $\Delta V/V$ may be considered to be simply dependent on X_1, the key controlling parameter. RTVE was thus inappropriate as a criterion for the antisolvent crystallization process, since it is unable to characterize the difference between the solvents as well as temperatures.

Subsequently, the relative molar volume expansion (RMVE) was considered (46) to be a more appropriate parameter, and a minimum in the variation of RMVE vs pressure or X_1, was suggested as the criterion for

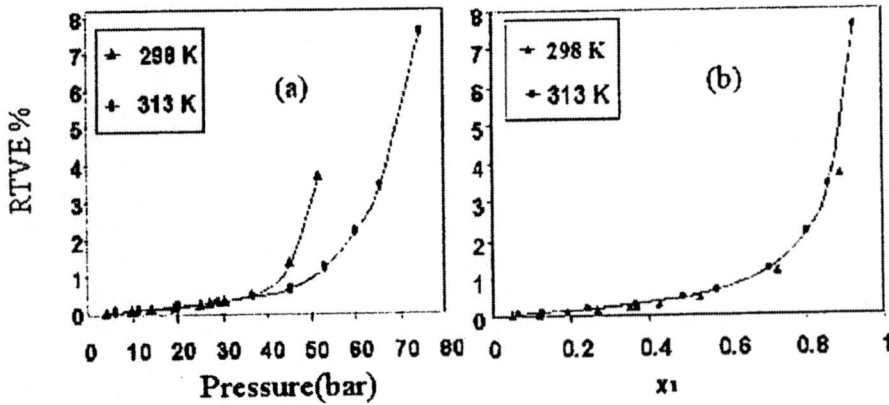

FIGURE 15   RTVE of Hexane: (a) effect of pressure and (b) effect of CO_2 dissolution (45).

ascertaining the optimum condition for the antisolvent crystallization process. The relative molar volume expansion was defined as

$$RMVE = \frac{\Delta v}{v} = \frac{v_L(T, P, X_1)}{v_2(T, P_0)} - 1 \tag{40}$$

However, its value is negative because the molar volume of solution v_L decreases with CO_2 dissolution. So there is no expansion as such, inasmuch as the molar volume of the solution v_L is mostly less than that of the pure solvent v_2, except for a very high value of X_1 in any organic solvent irrespective of whether $v_2 > v_L$ or $v_2 < v_L$. The behavior of RMVE for a hexane–CO_2 system at two temperatures is shown in Figure 16.

It may be noted that the occurrence of a minimum value of the molar volume of the binary (solvent–CO_2) solution is due to the reduction of the partial molar volume of the solvent $\bar{v}_2$, which is the actual indicator for lowering its solvent power for the solid solute, rather than RTVE or RMVE per se (47). With increasing CO_2 dissolution, the liquid molar volume of solution v_L or simply v decreases, since $v_L < \bar{v}_2$. It passes through a minimum, when $v = \bar{v}_1 = \bar{v}_2$, and then increases if $\bar{v}_1 > \bar{v}_2$, as follows:

$$\left(\frac{\partial v}{\partial X_1} \right)_{P.T} = (\bar{v}_1 - \bar{v}_2) \tag{41}$$

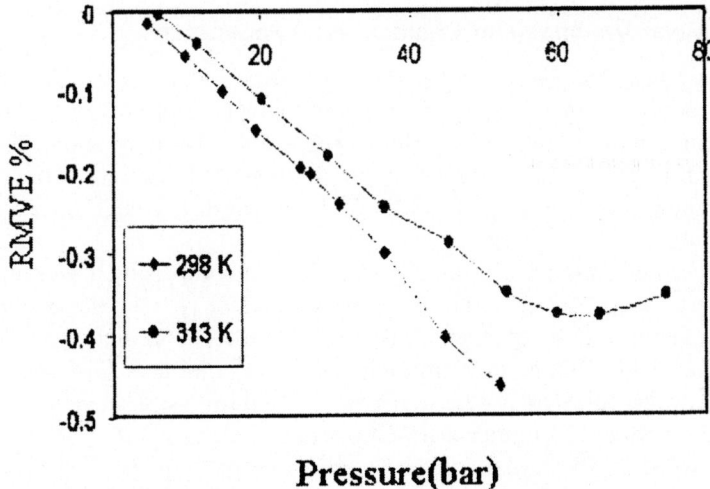

FIGURE 16 Effect of pressure on RMVE of hexane (46).

Also

$$\frac{dv}{dX_1} = \left(\frac{\partial v}{\partial P}\right)_{X_1,T} \frac{dP}{dX_1} + \left(\frac{\partial v}{\partial X_1}\right)_{P,T} \tag{42}$$

where

$$\bar{v}_i = \left(\frac{\partial V}{\partial n_i}\right)_{P,T,n_j} \tag{43}$$

and

$$v = X_1\bar{v}_1 + X_2\bar{v}_2 \tag{44}$$

According to the Gibbs–Duhem equation

$$\bar{v}_2 = \left[v - X_1\left(\frac{\partial v}{\partial X_1}\right)_{P,T}\right] \tag{45}$$

The first term on the right-hand side of Eq. (42) is negative and small, and the second term is initially negative and later positive. Accordingly, the liquid molar volume initially decreases and later increases with CO_2 dissolution owing to the increase of the second term at high pressures, close to the vapor pressure of CO_2 at a subcritical temperature or near the mixture critical pressure of the solvent–CO_2 system at supercritical temperatures. It may be recalled that the RTVE behavior also shows an exponential increase with pressure at such pressures.

7.1. Liquid Molar Volumes and Characteristic Parameters

It is customary to use the P-R EOS for the calculation of the partial molar volumes and molar volumes of binary systems of CO_2, and solvent (11). Values of the interaction constant k_{12} for several CO_2–solvent systems (48) are listed in Table 3. The behavior of molar volumes for different binary systems as a function of CO_2 mole fraction X_1 is shown in Figure 17 at two temperatures (49).

It may be noted that even in the case of a system comprising a solvent and a less dense antisolvent (e.g., ethanol–CO_2), that is, for $v_1 > v_2$, the liquid molar volume of solution v first marginally decreases with X_1 and then exceeds v_2 at a high value of X_1. That is, a minimum in the molar volume occurs where $v = \bar{v}_1 = \bar{v}_2$. However this may not be sharp at all. Even for a system in which the organic liquid solvent is lighter than CO_2 (e.g., hexane, where $v_1 < v_2$), the minimum value in the molar volume is also not prominent; rather, it flattens out as X_1 approaches unity. As a consequence, RMVE is insensitive to CO_2 mole fraction X_1 for several systems. RMVE was earlier suggested by

TABLE 3 Interaction Constants in the P-R Equation

System	k_{ij}	l_{ij}	Ref.
CO_2 + hexane at 298 K	0.112	0	48
CO_2 + hexane at 313 K	0.125	0	48
CO_2 + acetonitrile at 298 K	0.07	0	45
CO_2 + 1,4-dioxane at 298 K	−0.05	−0.05	45
CO_2 + Ethyl acetate at 298 K	−0.02	0.01	
CO_2 + Dimethyl sulfoxide	0.015	−0.025	45
CO_2 + N,N-Dimethylformamide	0.017	−0.043	45
CO_2 + Ethanol at 298 K	0.089		45
CO_2 + methanol at 323 K	0.079		101
Methanol + water at 323 K	−0.0966		101
CO_2 + water at 323 K	−0.019		101
CO_2 + toluene at 298 K	−0.09		56
CO_2 + 1-propanol at 298 K	−0.0488	−0.0044	58
CO_2 + butanol at 298 K	−0.10	−0.035	50
CO_2 + acetone at 308.318 K	−0.072	−0.078	53

Badilla et al. as the criterion for characterizing the antisolvent crystallization process (46).

Subsequently another new criterion has been introduced (47) in terms of the relative partial molar volume reduction (RPMVR) of solvent. This is based on the fact that the partial molar volume of solvent, $\bar{v}_2$ slowly decreases with increasing CO_2 mole fraction X_1 and drastically reduces at high values of

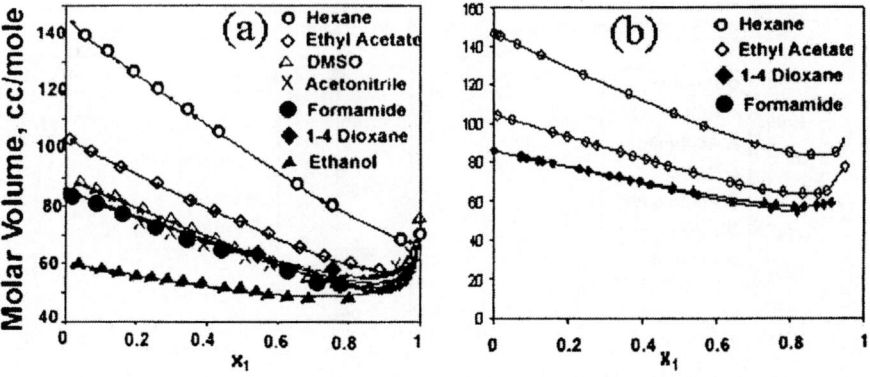

FIGURE 17 Molar volumes of different solvent–CO_2 systems at (a) 298 K and (b) 313 K (49).

CO_2 mole fraction. This can be observed from Figure 18 for a large number of solvents at 298 and 313 K. It can be noticed that $\bar{v}_2$ decreases with temperature for a given value of X_1.

At a very high CO_2 dissolution in organic solvents at pressures lower or higher than the mixture critical pressure, the significant reduction in the partial molar volume of the solvent is attributed to the clustering of CO_2 molecules around the solvent molecule. This phenomenon of clustering causes the solvent molecule to lose its affinity for the solid solute molecule, with consequent lowering of its solvent power. This causes aggregation of the solid solute molecules, leading to nucleation and crystallization. (49). It is the dissolution of CO_2, that is the crucial parameter for attaining high supersaturation and nucleation. The value of X_1 at which $\bar{v}_2$ sharply decreases is the appropriate condition for antisolvent crystallization because it causes drastic reduction of solute solubility and thereby supersaturation. Accordingly, the solid solute solubility X_3 (solute mole fraction on a CO_2-free basis) in the (solvent–CO_2) binary mixed solvent (49) was considered to be proportional to the partial molar volume, $\bar{v}_2$ and was expressed as:

$$X_3(T, P) = \frac{\bar{v}_2(T, P, X_1)}{\bar{v}_2(T, P_0, X_{10})} X_{30}(T, P_0) \tag{46}$$

where X_1, X_{10} are the mole fractions of CO_2 (on a solute-free basis) in the liquid phase at system pressure P and reference pressure P_0, respectively, and $\bar{v}_2$ is the partial molar volume of the solvent in the binary (solvent–CO_2)

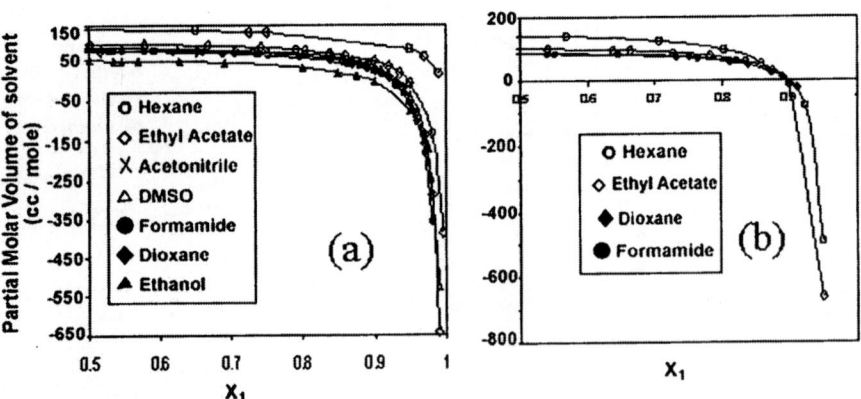

FIGURE 18 Partial molar volumes of different solvents with CO_2 mole fraction at (a) 298 K and (b) 313 K (49).

mixture. This is similar to the equation by Chang and Randolph (50) for solid solubility S (mg/mL) in a solvent in which the effect of pressure on the volume of the solution was not considered. A similar equation was also reported by Berends et al. (51) for solid solubility w (kg/kg), where CO_2 solubility in the solution was not considered.

Equation (46) implies that the fraction of solute crystallized is proportional to the reduction in the partial molar volume of the solvent. Accordingly, the characteristic parameter for the antisolvent crystallization is the relative partial molar volume reduction (RPMVR), which is defined as

$$RPMVR \equiv -\frac{\Delta \bar{v}_2}{v_2} = 1 - \frac{\bar{v}_2}{v_2} \tag{47}$$

From the behavior of RPMVR of different solvents at 298 K, as shown in Figure 19, it can be noted that the value of RPMVR is always positive and is different for different solvents. It can be observed that RPMVR for hexane is lower than that for other solvents and slowly increases with X_1. So fractional crystallization can be performed better in hexane because control of X_1 is easier. Figure 20 shows RPMVR of ethyl acetate and hexane at two temperatures. It can be observed that RPMVR is higher at 313 K than at 298 K. So a lower temperature of 298 K is more appropriate for fractional crystallization, which is needed during the purification process, whereas a

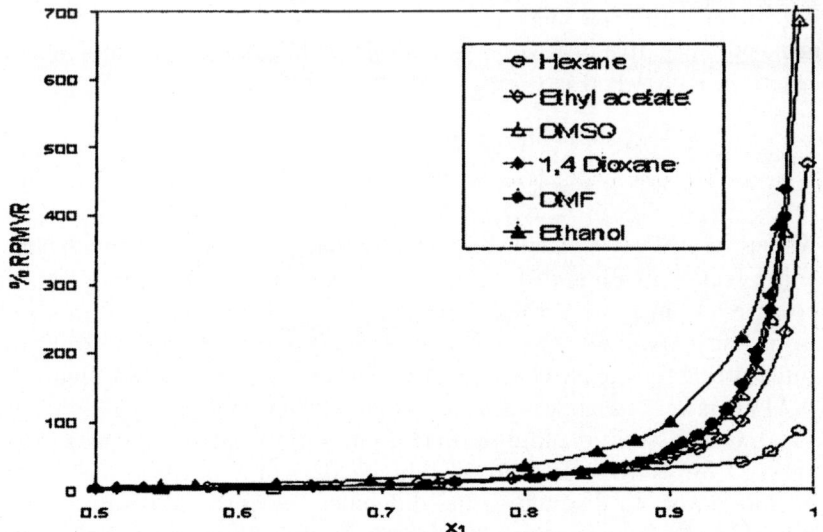

FIGURE 19 RPMVR of different solvents at 298 K (49).

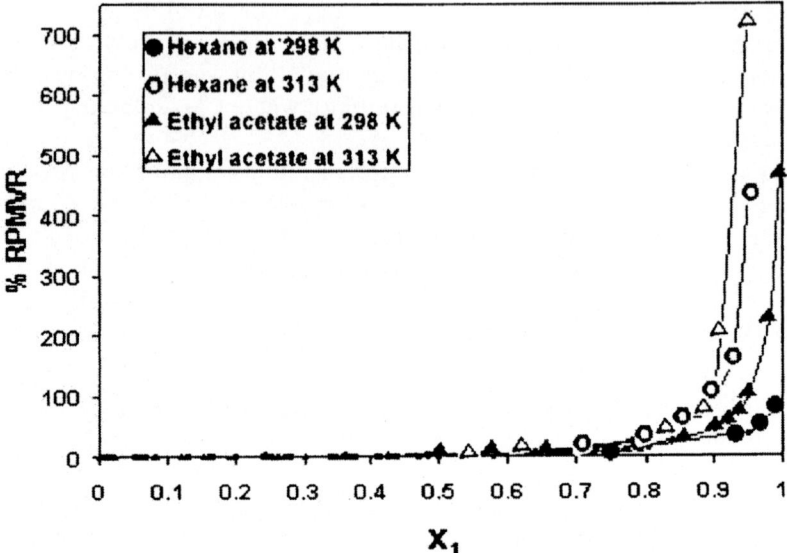

FIGURE 20 RPMVR of two solvents at 298 and 313 K.

higher temperature is appropriate for micronization, which requires very rapid supersaturation.

Another parameter, the relative molar volume reduction (RMVR), was defined as the negative of RMVE [Eq. (40)] for the sake of comparison with the criterion, RPMVR [Eq. (47)]:

$$\text{RMVR} = -\frac{\Delta v}{v} = 1 - \frac{v_L(T, P, X_1)}{v_2(T, P_0)} \tag{48}$$

The values of RMVR and RPMVR are compared for two solvents hexane and ethyl acetate in Figure 21. As can be seen, both RMVR and RPMVR are positive and increase with X_1. However, RMVR increases with X_1 with no prominent maximum for the hexane–CO_2 system, whereas it exhibits a maximum value for the ethyl acetate–CO_2 system. It is clear from Figure 21 that RMVR is very much less sensitive (than RPMVR) in its variation with X_1 at high CO_2 dissolution, and the maximum is not at all prominent. On the contrary, RPMVR increases sharply at high values of X_1 for both systems at 298 K and is able to distinguish the differences in both temperatures and solvents. Thus the most appropriate parameter for characterizing the anti-solvent crystallization is RPMVR, since it gives a direct measure of the degree of supersaturation.

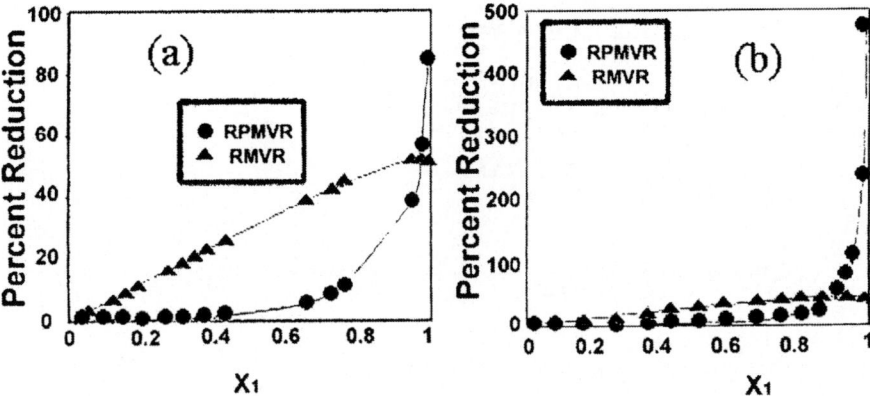

FIGURE 21 Comparison of RPMVR and RMVR at 298 K for (a) hexane and (b) ethyl acetate.

7.2. Antisolvent Effects on Molar and Partial Molar Volumes

The variation in the molar volume of the solution at a high CO_2 dissolution is relatively less sensitive than the partial molar volume of the solvent. The plots in Figure 22 compare the molar volume v and the partial molar volumes of solvent and CO_2 for hexane–CO_2 and ethyl acetate–CO_2 systems. It can be observed that the plots of $\bar{v}_1$, v, and $\bar{v}_2$ intersect at a high value of X_1, at which v tends to pass through a minimum, though not prominently, since all three values are equal at this value of X_1. In other words, the value of X_1 corresponding to the minimum molar volume can be obtained graphically only from the point of intersection of these three plots.

At pressures close to the vapor pressure of CO_2 or the mixture critical pressure, the liquid phase comprises mostly CO_2 and as a consequence, CO_2 molecules cluster around the solvent molecules, causing a reduction in $\bar{v}_2$, which in turn is due to the effect of pressure on CO_2 volume. The partial molar volume of the solvent, $\bar{v}_2$, decreases with an increase in temperature as shown in Figure 23 because a higher pressure is required to retain the same amount of CO_2 dissolved in the solution at a higher temperature. As a result, there is a greater pressure effect on CO_2 molecules surrounding the solvent molecule, and $\bar{v}_2$ decreases with temperature.

7.3. Modeling Ternary S-L Equilibrium and Predicting the Solute Mole Fraction

It is well known that CO_2 solubility in a solvent increases with pressure and decreases with temperature. Also the solute solubility X_3 in a solvent increases

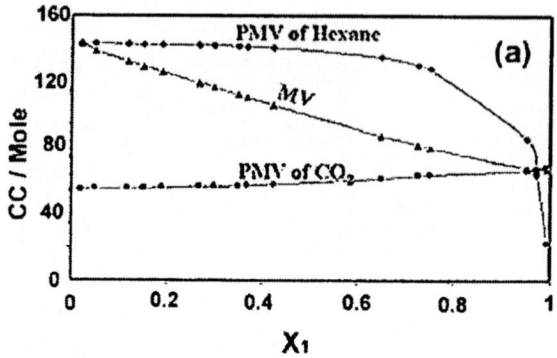

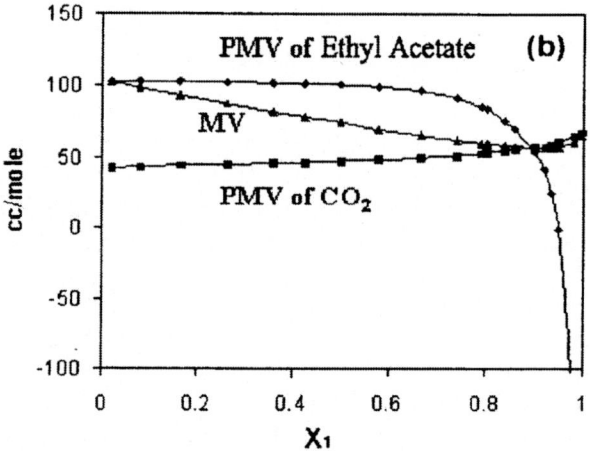

FIGURE 22 Antisolvent effect on molar volume and partial molar volumes at 298 K for (a) a hexane–CO_2 system and (b) an ethyl acetate–CO_2 system (49).

with temperature and decreases with CO_2 dissolution at any temperature or pressure. So X_1 and temperature are the controlling parameters for the solute mole fraction x_3 in the (CO_2–solvent–solid) ternary liquid phase, where $x_3 = X_3(1 - x_1)$ and $x_1 = X_1(1 - x_3)$. Both solute–solvent and solvent–antisolvent interactions are considered in the terms X_3 and $\bar{v}_2$, respectively, in Eq. (46) for the ternary S-L equilibrium. The value of $\bar{v}_2$ is calculated from Eq. (45) for the binary mixture of solvent and CO_2 (i.e., with no solute present in the solution). As shown earlier, $\bar{v}_2$ decreases with temperature, causing X_3 to decrease with

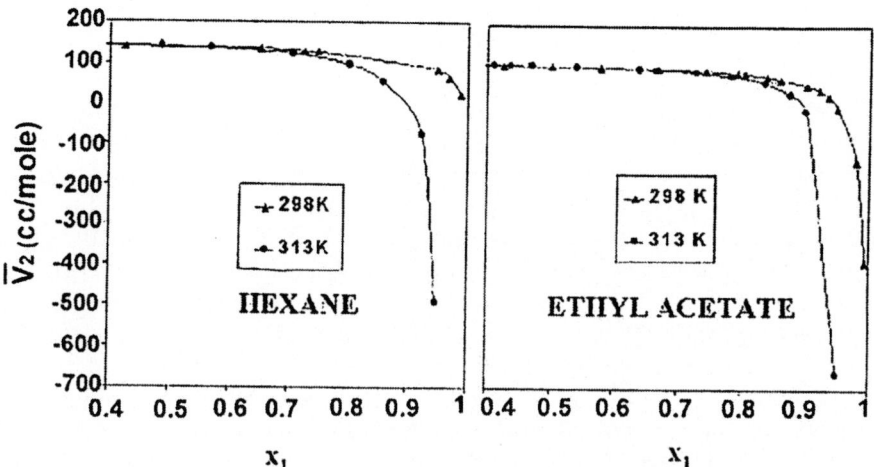

FIGURE 23 Effect of temperature on PMV of hexane and ethyl acetate (49).

temperature. On the other hand, X_3 increases with temperature with or without CO_2, as X_1 decreases with temperature. So temperature has two opposite effects, which are taken into consideration in terms of $\bar{v}_2$ and X_1.

The solubilities of pure β-carotene in ethyl acetate and cholesterol in acetone, predicted by using Eq. (46), compare well with the corresponding experimental data reported in the literature (52,53). The behavior of the curve for X_3 vs X_1 is similar to that for $\bar{v}_2$ vs X_1 in that both are drastically reduced at high values of X_1, although both remain almost invariant at lower values of X_1. This was further validated by experimental data (54) from comparisons of the antisolvent effects on the reduction of solubility of pure β-carotene in hexane and in ethyl acetate. This trend was also observed for a lecithin–hexane system (55). The solute solubility is negligible at zero or negative values of $\bar{v}_2$, which occur at a very high CO_2 dissolution.

However the agreement of the predicted solubility, X_3 by Eq. (46), with the corresponding experimental data for naphthalene and phenanthrene in toluene (56) was found to be inadequate. The deviation is significant at high values of both X_3 and X_1. To overcome this limitation, a new model has been developed. For this purpose, first a molar volume correction is incorporated (57), as shown in Eq. (49), to predict the binary solid solubility X_3, and then x_1 is replaced with X_1, leading to Eq. (50), to predict x_3 from binary data, since $X_1 = x_1/(1 - x_3)$ and $X_3 = x_3/(1 - x_1)$.

$$X_3(T, P) = \frac{\bar{v}_2(T, P, X_1)/v(P, X_1)}{\bar{v}_2(T, P_0, X_{10})/v(P_0, X_{10})} X_{30}(T, P_0) \tag{49}$$

$$x_3(T, P) = \frac{(1 - X_1)\bar{v}_2(T, P, X_1)/v(P, X_1)}{(1 - X_{10})\bar{v}_2(T, P_0, X_{10})/v(P_0, X_{10})} x_{30}(T, P_0) \qquad (50)$$

where P_0 and X_{10} are, respectively, the reference pressure and reference CO_2 solubility on a solute-free basis; X_1 is the CO_2 mole fraction on a solute-free basis.

It was thus postulated that the liquid phase mole fraction x_3 of the solid solute in the ternary system is proportional to the "partial molar volume fraction" (PMVF) of the solvent in the binary solution. In other words, the solvent capacity of the CO_2-diluted solution is due to the solvent's contribution to the molar volume of the solution. This new term, PMVF, defined as $[(1 - X_1)\bar{v}_2/v]$, is thus a characteristic parameter depicting the solvent capacity of the binary (CO_2-diluted) solvent mixture for the solid solute, and it varies between 0 and 1. Figure 24 shows that the PMVF of the solvent decreases with increasing CO_2 mole fraction, as does the solvent capacity for the solid solute (57). The retention of solid molecules in the liquid phase is primarily attributed to the capacity of the partially surrounded solvent molecule to retain its affinity for the solute molecules. In other words, the number of solid molecules that can be accommodated in the vicinity of the solvent molecules in the CO_2-diluted solution phase is attributed to the relative strength of the clustered solvent molecules, which are only partially available for the solid solute molecules.

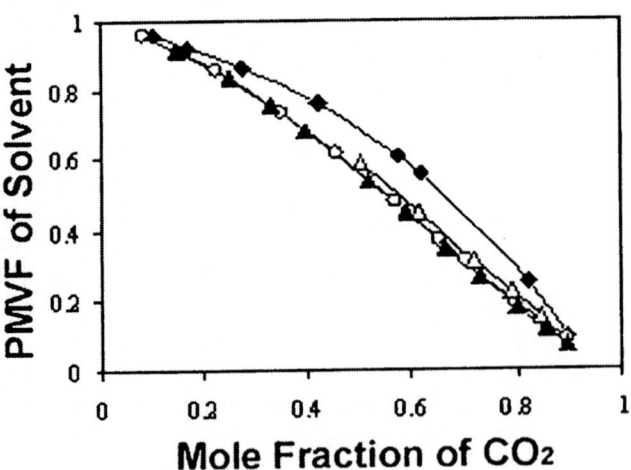

FIGURE 24   PMVF of different solvents in binary mixtures with CO_2 dissolution in toluene at 298 K (♦), acetone at 318 K (□), acetone at 308 K (), n-butanol at 298 K (○) (57).

The solubilities of several pure solids substances in a number of organic solvents have been calculated by means of Eq. 50, using $\bar{v}_2$ and v at various values of X_1. For comparison of the predicted values with the experimental data, the reported experimental solubility data in milligrams per milliliter must be converted to the mole fraction of solid solute in the ternary liquid phase with due pressure correction of the liquid volume. Figure 25 shows that the calculated values of the liquid phase composition, x_3 in the CO_2–solvent–solid solute systems, are in very good agreement with the corresponding experimental data for a large number of solids. These include naphthalene and phenanthrene in toluene (56), acetaminophen in butanol (50), β-carotene in toluene (50), cholesterol in acetone (53), salicylic acid in 1-propanol (58), and catechin in ethanol (59). This validates the concept of the PMVF of solvent as a measure of the capacity of the CO_2-diluted solvent. It is interesting to note that the ternary process phase diagram (x_1-x_2-x_3) for isothermal S-L equilibrium at different pressures can be generated (60) from knowledge of the PMVF of solvent in the binary (solvent–CO_2) solution and the solid solubility at some reference pressure P_0, as illustrated in Figure 26. It may be noted that X_1 or pressure, solution temperature, and x_3 at the reference pressure P_0 are the three input parameters needed for this calculation corresponding to the three degrees of freedom for the ternary S-L equilibrium.

8. MODELING TERNARY S-L-V EQUILIBRIUM FOR SAS/GAS

Solvent selection for the micronization of drugs is crucial because the molecules may be multifunctional and polar with a tendency toward hydrogen bonding, resulting in specific solute–solvent interactions (61). However, it is generally believed that the solute–antisolvent interaction is negligible compared with the solvent–antisolvent interaction for the S-L-V equilibrium in antisolvent crystallization systems. Thus the solvent–CO_2 interaction is solely responsible for the reduction of solvent power of the mixed solvent. Accordingly, the PMVF of solvent in a binary (solvent–antisolvent) mixture depicts the solute mole fraction in the ternary (solute–solvent–antisolvent) liquid phase.

It is imperative to know the S-L-V equilibrium compositions for the ternary (CO_2–solvent–solid) system, for these give the concentrations at the interface, which are needed for calculating the two-way mass transfer rates of CO_2 and solvent in the antisolvent crystallization processes and for the selection of operating conditions for the desired crystallization pathways. Three kinds of data are usually generated for ternary (solute–solvent–antisolvent) systems: (a) the liquid phase compositions for S-L equilibrium at a fixed

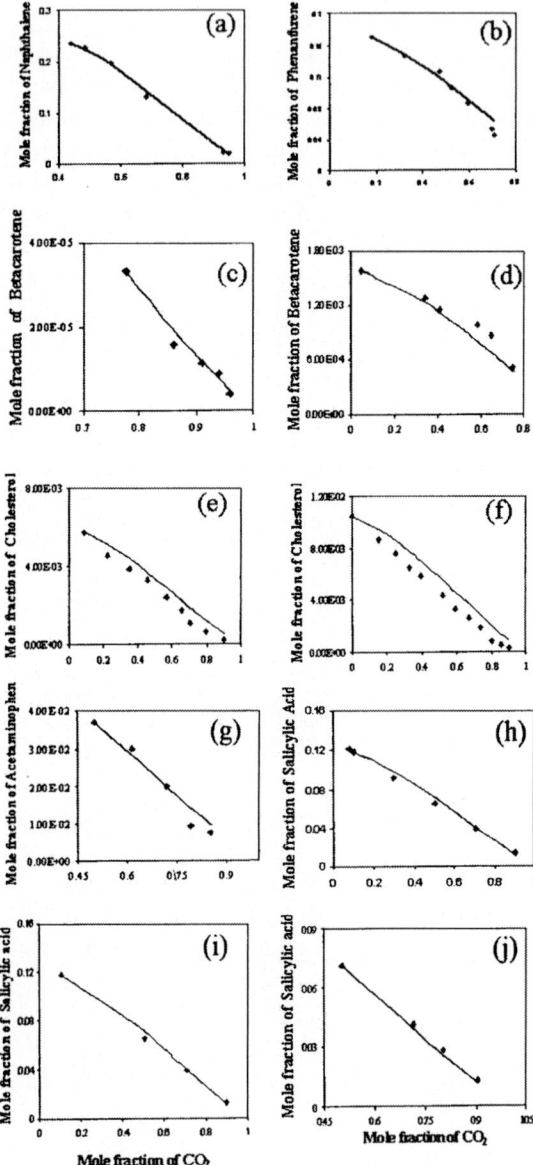

Naphthalene

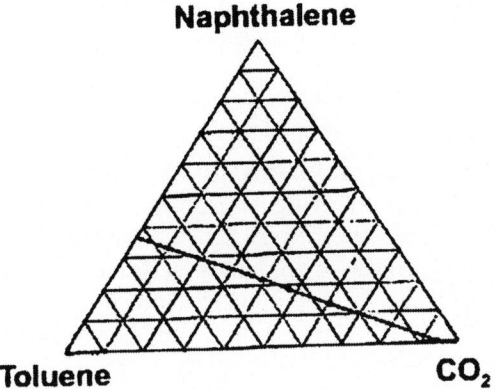

Toluene   **CO$_2$**

FIGURE 26   Ternary process diagram of solution for S-L equilibrium at 298K from binary data.

temperature and different pressures (i.e., P-x_1-x_2-x_3 on a prism), (b) the P-T trace for S-L or L-V equilibrium for a fixed solute concentration in liquid phase on antisolvent-free basis, and (c) the P-T trace at S-L-V equilibrium and also the liquid and vapor phase compositions.

Different approaches to modeling the ternary S-L-V equilibrium reported so far essentially differ in the calculation procedure adopted for the S-L equilibrium. For example, the liquid phase composition in the ternary S-L equilibrium for different pressures at any temperature may be calculated by means of (a) the expanded liquid EOS and activity coefficient model, (b) the EOS model, or (c) the PMVF of solvent. Subsequently, the isofugacity criterion for the L-V equilibrium is considered to predict the bubble point pressure and the vapor phase composition to ensure that all three (S-L-V) phases will coexist. Clearly, the ternary liquid and vapor phase mole fractions can be

FIGURE 25   Comparison of predicted liquid mole fraction x_3 of different solid solutes for ternary S-L equilibrium with CO$_2$ dissolution from PMVF of solvent in binary mixtures by Eq. (50) with experimental mole fractions of (a) naphthalene in toluene solution at 298 K (56), (b) phenanthrene in toluene solution at 298 K (56), (c) β-carotene in ethyl acetate solution at 298 K (52), (d) β-carotene in toluene solution at 298 K (50), (e) cholesterol in acetone at 308 K (53), (f) cholesterol in acetone solution at 318K (53), (g) acetaminophen in n-butanol solution at 298 K (50), (h) salicylic acid in 1-propanol solution at 303 K (X_{SA} = 0.132), experimental (58), (i) salicylic acid in 1-propanol solution at 288 K (X_{SA} = 0.132) (58), (j) salicylic acid in 1-propanol solution with at 288 K (X_{SA} = 0.144) (58).

predicted from the binary (solvent–CO_2) data in the third approach. However, in the absence of experimental solid solubility data at any reference condition, it is imperative to predict this S-L equilibrium composition by adopting a molecular thermodynamic model. The approach to calculating the L-V equilibrium, however, remains more or less the same for all approaches to the modeling of a multiphase equilibrium for a ternary (GAS/SAS) system. The computational procedure for SEDS involving a cosolvent as the fourth component is very similar to that used for the ternary S-L-V equilibrium, except that one additional equation is employed for the mole fraction of the second liquid component.

8.1. Three-Phase Equilibrium

In pursuing an accurate thermodynamic description of the three-phase, three-component system, the phase equilibrium compositions can be calculated after pressure and temperature have been fixed, since it is known from the Gibbs phase rule that there are only 2 degrees of freedom. There are five unknown compositions, assuming that the solid is crystalline and pure and that its solubility in the vapor/fluid phase is negligible. Two of these unknown mole fractions are eliminated by the constraints that the mole fractions in each phase sum up to unity. To find these three unknown mole fractions, namely, x_1, x_3, and y_2, only three equilibrium relations are required.

The relevant phase equilibrium equations for the three components are

$$\bar{f}_1^V(T, P, y_1) = \bar{f}_1^L(T, P, x_1) \tag{51}$$

$$\bar{f}_2^V(T, P, y_2) = \bar{f}_2^L(T, P, x_2) \tag{52}$$

$$f_3^S(T, P) = \bar{f}_3^L(T, P, x_3) \tag{53}$$

8.1.1. Expanded Liquid Model

Dixon and Johnston (56) employed the isofugacity criteria along with an expanded liquid EOS model, combining the regular solution theory for the liquid phase activity coefficients and the P-R EOS for the fluid phase fugacity coefficients. According to this model, it is necessary to know all three binary interaction constants along with the solubility parameters of the solvent and solute. For the vapor–liquid equilibrium of antisolvent 1 and solvent 2, the equilibrium equations [Eqs. (51) and (52)] are expressed in terms of the respective fugacity coefficients, and for the solid–liquid equilibrium of solid 3, Eq. (53) is expressed in terms of the activity coefficient as:

$$y_1 \bar{\phi}_1^V = x_1 \bar{\phi}_1^L \tag{54}$$

$$y_2 \overline{\phi}_2^V = x_2 \overline{\phi}_2^L \tag{55}$$

$$\overline{f}_3^L = \gamma_3 x_3 f_3^L \tag{56}$$

$$f_3^S(T, P) = P_3^S \phi_3^S \exp\left[\frac{v_3^S(P - P_3^S)}{RT}\right] \tag{57}$$

It was reported by Dixon and Johnston (56) that up to a moderate pressure, the liquid phase activity coefficient of the solid component γ_3 could be obtained by any activity coefficient model or, more conveniently, from the regular solution theory, as

$$RT \ln \gamma_3 = v_3^L (\delta_3 - \overline{\delta})^2 \tag{58}$$

where $\overline{\delta}$ is the volume fraction averaged solubility parameter as given by Eqs. (59) and (60).

$$\psi_i = \frac{(x_i v_i^L)}{\sum_j x_j v_j^L} \tag{59}$$

$$\overline{\delta} = \sum \psi_i \delta_i \tag{60}$$

Combining Eqs. (56) and (57) according to Eq. (53) and expressing f_3^L by a pressure correction on f_3^{0L}, one can get

$$P_3^S \exp\left[\frac{v_3^S(P - P_3^S)}{RT}\right] = x_3 \gamma_3(P_0, x_1^0) f_3^{0L} \left[\frac{\overline{\phi}_3^L(P, x_1) P}{\overline{\phi}_3^L(P^0, x_1^0) P^0}\right] \tag{61}$$

where, P^0 is the reference pressure or atmospheric pressure at which the regular solution theory is valid, f_3^{0L} is the fugacity of the subcooled pure liquid solute at P^0, and the term in the square bracket on the right-hand side corrects for the effect of pressure on the activity coefficient of 3, the solid solute in the liquid phase.

This expanded liquid EOS model in conjunction with the regular solution theory model predicted the solute solubilities of naphthalene and phenanthrene in toluene with CO_2 reasonably well at both high and low pressures (56).

8.1.2. The EOS Model

In the EOS model, as suggested by Kikic et al. (41), the solid state fugacity of the pure solute [left-hand side of Eq. (53)] may be calculated from Eq. (35). This is computed from the heat of fusion at the triple point, the triple-point temperature, the triple-point pressure (reference pressure), and the fugacity of the pure solute at the fictitious subcooled liquid state at the given temperature and pressure (19). This liquid phase fugacity of the pure solute may be calculated by using the P-R EOS. The fugacity of the solute component in the liquid phase [right-hand side of Eq. (53)] is calculated in terms of the fugacity coefficients by using the P-R EOS and the solute mole fraction x_3 in the liquid phase, from Eq. (18). This model has been tested by Kikic et al. (41) for naphthalene, phenanthrene, and β-carotene in toluene with CO_2, and good agreement was obtained when the liquid phase compositions were correlated.

For the case in which the solute mole fraction in the vapor phase y_3 cannot be neglected, the S-L-V equilibrium calculations also include the S-V equilibrium for the solid solute in addition to Eqs. (51) to as follows:

$$f_3^S(T, P) = \bar{f}_3^V(T, P, y_3) \tag{28}$$

Accordingly,

$$y_3 = \frac{P_3^S \phi_3^S \exp[v_3^S P/RT]}{P \bar{\phi}_3^V} \tag{15}$$

as

$$\bar{f}_3^V = P \bar{\phi}_3^V y_3 \tag{15a}$$

$$\ln P_3^S = \ln P_{tp} - \frac{\Delta H^S}{R} \left(\frac{1}{T} - \frac{1}{T_{tp}} \right) \tag{62}$$

where P_3^S, the sublimation (S-V) pressure of the solid solute, is calculated from the heat of sublimation, ΔH^S at the triple point; P_{tp} is the triple-point pressure, and T_{tp} is the triple-point temperature.

The experimentally determined S-L-V equilibrium data for salicylic acid (2-hydroxy-benzoic acid)-1-propanol–CO_2 were correlated by using the Stryjek–Vera modification of the Peng–Robinson EOS in conjunction with Eq. (35) for the solid state fugacity of the solute (58,62), as described earlier. This procedure also yielded good agreement of the liquid phase compositions of salicylic acid in the temperature and pressure ranges of 273 to 367 K and 1.0 to 12.5 MPa. The P-T traces of S-L and L-V equilibria were calculated for a fixed solute concentration on CO_2-free basis, and subsequently the P-T trace for the S-L-V equilibrium was found from the point of intersection of these two lines. The liquid phase compositions of the solute as a function of pressure at a constant temperature at the condition of S-L-V equilibrium were calculated to assess the effect of pressure or addition of antisolvent on solute crystallization. It was reported that two isobaric points of the CO_2 mole fraction could be observed on the curve of the S-L equilibrium temperature vs the CO_2 mole fraction at constant temperature as it passes through a minimum at a particular CO_2 mole fraction (Figure 27a). The two opposite effects of temperature on the ternary S-L equilibrium also were mentioned in earlier (Section 7.3). Most of the dissolved solids can be crystallized out in a narrow pressure range in the vicinity to this minimum. The agreement of the predicted values of the P-T traces for the S-L and L-V boundaries with the corresponding experimental data is demonstrated in Figure 27b, for a fixed solute mole fraction on a CO_2-free basis.

The method suggested by Badilla et al. (46) for modeling the S-L-V equilibrium is similar to that by Kikic et al. (41) except that the solid state fugacity of the solute f_3^s (T, P) was calculated by Eq. (57), for which the sublimation pressure P_3^S was predicted from the triple-point pressure and the heat of sublimation by means of Eq. (62).

8.1.3. The PMVF Model

A recently developed model for the S-L-V equilibrium (63) utilizes the PMVF of the solvent in a binary mixture and the solute solubility at a reference pressure. This approach uses Eq. (50) to predict the ternary liquid mole fractions for the S-L equilibrium at different CO_2 mole fractions corresponding to different pressures and at a fixed temperature. Next the pressure is adjusted to satisfy the isofugacity criterion for the L-V equilibrium, to permit the prediction of the vapor phase composition at which all three (S-L-V) phases coexist. This is repeated for other temperatures to obtain the P-T trace of the S-L-V equilibrium. The P-T trace for the constant liquid phase composition of the

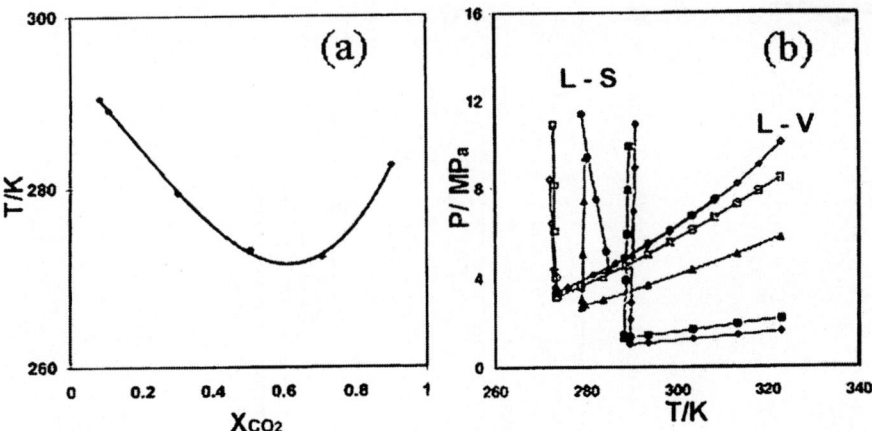

FIGURE 27   (a) Isobaric S-L equilibrium temperature vs CO_2 mole fraction (for $X_3 = 0.16$) at 7 MPa. (b) Prediction of P-T trace of ternary S-L and L-V equilibria in CO_2–propanol–salicylic acid for $x_3 = 0.132$ (58).

TABLE 4 Comparison of Predicted Data with Experimental S-L-V Data (56) of CO_2–Toluene–Naphthalene System at 298 K

Pressure (bar)		Liquid Mole Fractions			Vapor Mole Fractions			
Experimental	Calculated[a]	x_1	x_2	x_3	y_1	y_2	y_3	ARD (%)
32.9	30.83	0.334	0.429	0.237	0.99826	0.0173	0.000018	6.3
36.7	35.372	0.373	0.398	0.229	0.99830	0.00163	0.000019	3.62
43.4	42.679	0.456	0.346	0.198	0.99827	0.00171	0.000024	1.66
49.9	51.600	0.591	0.277	0.132	0.99800	0.00196	0.000033	3.40
57.1	58.151	0.907	0.0688	0.0242	0.99793	0.00200	0.000068	1.84
57.7	58.761	0.928	0.0511	0.0209	0.99809	0.00183	0.000082	1.84

% AARD = percent absolute average relative deviation.

[a] k_{12} = 0.100, k_{13} = 0.11, k_{23} = 0.0, l_{12} = 0.0, l_{13} = 0.0, l_{23} = 0.0.

TABLE 5 CO_2–Toluene–Phenanthrene System at 298 K

Pressure (bar)		Liquid mole fractions			Vapor mole fractions			ARD (%)
Experimental	Calculated[a]	x_1	x_2	x_3	y_1	y_2	y_3	
14.9	14.899	0.147	0.682	0.171	0.99685	0.00314	0.00001	0.006
26.6	26.783	0.275	0.579	0.146	0.99774	0.00226	0.00001	0.688
35.7	35.137	0.378	0.426	0.126	0.99792	0.00208	0.00000	1.597
42.3	43.200	0.470	0.424	0.106	0.99787	0.00213	0.00000	2.129
52.7	53.920	0.676	0.2787	0.0453	0.99755	0.00245	0.00000	2.310
59.6	60.045	0.964	0.03454	0.00146	0.99816	0.00184	0.00000	0.754
62.4	62.755	0.991	0.00849	0.00051	0.99918	0.00080	0.0000	0.569

% AARD = 1.1503.

[a] $k_{12} = 0.100$, $k_{13} = 0.12$, $k_{23} = 0.0$, $l_{12} = 0.0$, $l_{13} = 0.0$, $l_{23} = 0.0$.

solute at the S-L-V equilibrium is obtained from the isothermal plots of P vs $(x_3/1 - x_3)$. This method has yielded good agreement for both liquid phase compositions and isothermal bubble point pressures (60) for naphthalene and phenanthrene in (toluene–CO_2) mixed solvent at the S-L-V equilibrium, as illustrated in Tables 4 and 5. The computed isothermal ternary equilibrium compositions compare well with the corresponding experimental data for the S-L-V equilibrium, with the % AARD in the bubble point pressures of 3.11 and 1.15, respectively, for the two systems at 298 K over the pressure range of 14.9 to 62.4 bar. This method has the advantage of not requiring estimation of x_3 by means of the isofugacity criterion (41), in which an EOS is used for calculating the hypothetical liquid phase fugacity of pure subcooled solid solute, or an activity coefficient model is used with a pressure correction.

It is also observed that the presence of the nonvolatile solute like phenanthrene has negligible effect on the vapor phase mole fractions of toluene and CO_2, inasmuch as the latter for the ternary system are very much close to those for the binary (toluene and CO_2) system at the same pressure and 298 K (60). The P-T trace for the constant liquid phase compositions at the S-L-V equilibrium is obtained from the isothermal plots of P vs $(x_3/1 - x_3)$ at three temperatures (Figure 28). It can be seen that the pressure required for attaining the S-L-V equilibrium increases with temperature at the same composition and increases with x_3 at constant temperature. This trend is similar to that reported by Kikic et al. (41). Figure 29 depicts the effects of variations in pressure and temperature on isothermal and isobaric triangular

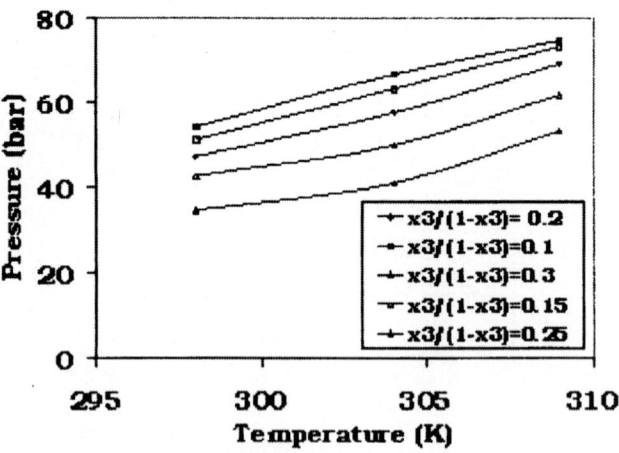

FIGURE 28 P-T trace predicted by the PMVF model for S-L-V equilibrium for a CO_2–toluene–naphthalene system at constant solute compositions.

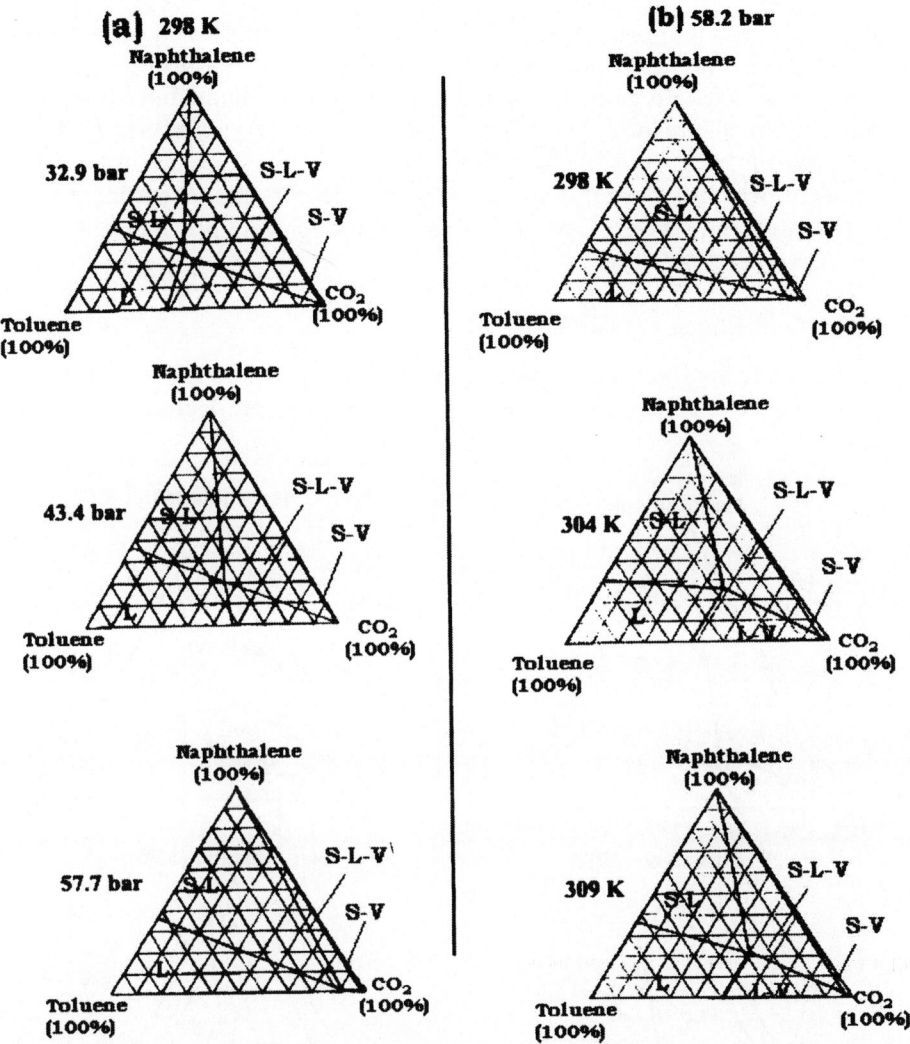

FIGURE 29 Ternary phase diagrams of a toluene–CO_2–Naphthalene system predicted by the PMVF model: (a) effect of pressure at 298 K and (b) effect of temperature at 58.2 bar (60).

phase diagrams for a naphthalene–toluene–CO_2 system, which were generated by using this new model. It can be seen that with increasing pressure, the isothermal S-L-V region shrinks and is completely eliminated above the mixture critical pressure. This indicates that the solid crystallization pathway directly traverses from the S-L to S-V regions above the mixture critical pressure, which is desirable for the solvent-free particles. With increasing temperature, however, S-L-V region expands at a constant pressure below the mixture critical pressure.

8.2. Four-Phase Equilibrium

In case there are four phases, namely, S, L_1, L_2, and V, and the solid phase is pure (in a three-component system), there is only a single degree of freedom.

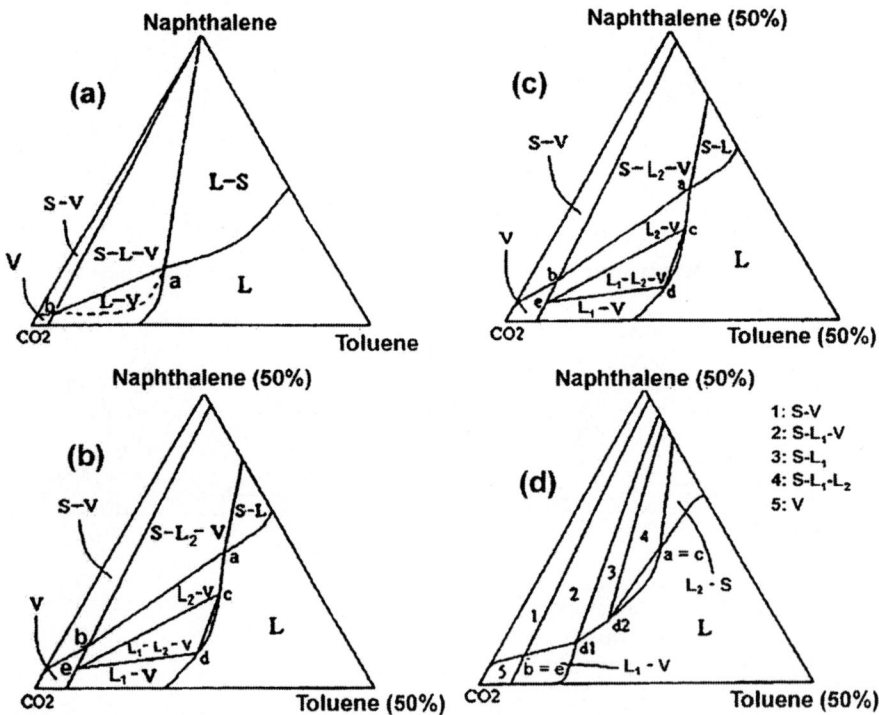

FIGURE 30 Effect of ternary compositions on phase transitions for a toluene–CO_2–naphthalene system: (a) S-L-V equilibrium at a given P and T, (b) S-L_1-V and L_1-L_2-V equilibrium at $315 < T < 325$ K, (c) S-L_1-L_2-V at $315 < T < 320$ K and $82 < P < 91$ bar, and (d) S-L_1-V and S-L_1-L_2 equilibrium at $315 < T < 320$ K and P above that S-L_1-L_2-V equilibrium at each temperature (41).

An equilibrium equation can be written for each component between L_1 and V or L_1 and L_2 or S and L_1. In all, seven equations can be written for S-L_1-L_2-V equilibrium, and one can compute P-T data: $L_1 = L_2 - V$ critical locus and $L_1 - L_2 = V$ critical locus (64). Figure 30 demonstrates the effect of compositions on the phase transition and occurrence of a number of phases for CO_2–toluene–naphthalene system (41). It can be noted that a second liquid phase appears at a temperature between 315 and 325 K, and as P increases a second liquid phase appears, depending on the overall composition at the $L_1 - L_2$ critical point.

9. CONCLUDING REMARKS

Micronization by means of $scCO_2$ is of great interest in the pharmaceutical industries owing to the ability of the process to produce uniform size and controlled morphology of the micronized particles. In conventional thermal crystallization, nonuniform temperature propagation causes nonuniform supersaturation and hence nonuniform crystallization. As a result, it gives rise to broad product size distribution. In contrast, both depressurization crystallization and antisolvent crystallization propagate rapidly and give rise to an essentially uniform and very high degree of supersaturation, leading to ultrafine particles with narrow size distribution.

The P-T trace of the solid–liquid–fluid boundaries and the mixture critical locus are important to locate the UCEP because there is a remarkable solubility enhancement in its vicinity. Both RESS and PGSS are operated to get the advantage of this phenomenon, with the distinction that RESS is operated such that the S-L-V line is avoided, while in PGSS the molten solute is saturated with the $scCO_2$ before depressurization. In GAS and SAS it is the solvent–CO_2 interaction that plays the key role, and temperature and CO_2 dissolution are the controlling process parameters. It is also established that the partial molar volume reduction of the solvent induced by the dissolution of CO_2 is attributed to the clustering of the solvent molecule by the CO_2 molecules in the liquid phase. Hence RPMVR is the true indicator of lowering of the solvent power for the solute to attain rapid supersaturation. The partial molar volume fraction, PMVF of solvent in the binary CO_2–solvent mixture, decreases with CO_2 mole fraction and is proportional to the solid mole fraction in the solution diluted with CO_2. This chapter illustrates different approaches to thermodynamic modeling in binary (RESS and PGSS) and ternary (GAS and SAS) systems to enable one to predict the P-T trace and phase compositions of V-L, S-V, S-L, and S-L-V equilibria and to ascertain the phase diagrams needed for the process design. It is clearly understood that no single equation of state or model will work for all situations. Equations with more accurate representation of molecular interactions are expected to improve the ability to model complex phase behavior. Accordingly the mixing rules and the method of evaluation of the interaction constants need careful selection for proper representation of the highly non ideal asymmetric binary and ternary mixtures.

NOMENCLATURE

a, b	constants in the EOS
g	free energy of the solution
k_{12}, l_{12}	binary interaction constant between CO_2 (1) and the solvent (2)
P	pressure
T	temperature
V	total volume of the liquid solution
v	molar volume of the solution
v_1	molar volume of CO_2 (component 1)
v_2	molar volume of the solvent (component 2)
$\bar{v}_1$	partial molar volume of component 1 in a binary mixture of 1 and 2
$\bar{v}_2$	partial molar volume of component 2 in a binary mixture of 1 and 2
X_1	mole fraction of CO_2 in the liquid phase on a solute-free basis
X_2	mole fraction of the solvent in the liquid phase on a solute-free basis
X_3	mole fraction of the solute (component,3) in the liquid phase on a CO_2-free basis
x_3	mole fraction of the solute in the ternary liquid phase
y_3	mole fraction of the solute in the vapor–fluid phase

REFERENCES

1. Nass R. Pharmaceutical suspensions. In: Lieberman HA, Rigger M, Banker G, eds. Pharmaceutical Dosage Forms. New York: Marcel Dekker, 1988:151.
2. Hixon L, Prior M, Prem H, Van Cleef J. Sizing materials by crushing and grinding. Chem Eng 1990; 97:99.
3. Van Cleef J. Powder technology. Am Sci 1991; 79:304.
4. Chattopadhyay P, Gupta RB. Protein nanoparticles formation by SAS with enhanced mass transfer. AIChE J 2002; 48(2):235–244.
5. Knutson BL, Debenedetti PG, Tom JW. Cohen S, Bernstein H, eds. Microparticulate Systems for the Delivery of Proteins and Vaccines. Drugs and Pharmaceutical Sciences Series. New York: Marcel Dekker, 1996:89–125.

6. Reverchon E. Supercritical antisolvent precipitation: its applications to micro-particle generation and products fractionation. Proceedings of the 5th Meeting on Supercritical Fluids, Materials, and Natural Products Processing. Vol. 1. Nice: France, 1998:221–236.

7. Bristow S, Shekunov T, Shekunov BY, York P. Analysis of the super-saturation and precipitation process with supercritical CO_2. J Supercrit Fluids 2001; 21:257–271.

8. Prausnitz JM, Lichtenhaler RN, Gomez de Azevedo E. Molecular Thermodynamics of Fluid Phase Equilibria. 3rd ed. Englewood Cliffs, NJ: Prentice Hall, 1999.

9. Myerson AS, Ginde R. Crystals, crystal growth, and nucleation, Chapter 2. In: Myerson Allen S, ed. Handbook of Industrial Crystallization. Woburn, MA: Butterworth Heinmann, 2002.

10. Mullins JW. Ind Chem 1960; 36:272.

11. Mukhopadhyay M. Natural Extracts Using Supercritical CO_2. Boca Raton, FL: CRC Press, 2000:11–82, Chapter 2.

12. Ekart MP, Brennecke JF, Eckert CA. Molecular analysis of phase equilibria in supercritical fluids. Chapter 3. In: Bruno TJ, Ely JF, eds. Supercritical Fluid Technology. Florida, Boca Raton, FL: CRC Press, 1991:163–192.

13. McHugh MA, Krukonis VJ. Supercritical Fluid Extraction: Principles and Practice. Stoneham, MA: Butterworth-Heinemann, 1994.

14. Diepen GAM, Scheffer FEC. (a) On critical phenomena of saturated solutions in binary systems. J Am Chem Soc 1948; 70:4081 (b) The solubility of naphthalene in supercritical ethylene. J Am Chem Soc 1948; 70:4085.

15. Palakodaty S, York P. Phase behavioral effects on particle formation processes using supercritical fluids. Pharma Res 1999; 16:7.

16. Yao S, Guan Y, Zhu Z. Investigation of phase equilibrium for ternary systems containing ethanol, water and carbon dioxide at elevated pressures. Fluid Phase Equilibria 1994; 99:249–259.

17. Dixon DJ, Johnston KP, Bodmeier RA. Polymeric materials formed by pre-cipitation with a compressed fluid antisolvent. AIChE J 1993; 39(1):127–139.

18. Peng DY, Robinson DB. A new two-constant equation of state. Ind Eng Chem Fundam 1976; 15:59.

19. Reid RC, Prausnitz JM, Poling BE. The Properties of Gases and Liquids. 4th ed. New York: McGraw-Hill, 1987.

20. Salim PH, Trebble MA. A modified Trebble–Bishnoi equation of state: Thermodynamic consistency revisited. Fluid Phase Equilibria 1991; 65:59–71.

21. Patel NC, Teja AS. A new cubic equation of state for fluids and fluid mixtures. Chem Eng Sci 1982; 37:463–473.

22. Panagiotopoulos AZ, Reid RC. New mixing rule for cubic EOS for highly polar, asymmetric systems. ACS Symp Ser 1986; 300:571–582.

23. Werling JO, Debenedetti PG. Numerical modeling of mass transfer in the supercritical antisolvent process. J Supercrit Fluids 1999; 16:167–181.

24. Baum EJ. Chemical Property Estimation. Theory and Application, ASPEN MAX Release 1.1, Aspen Tech Europe. Boca Raton, FL: Lewis Publishers, 1998.

25. Xu G, Scurto AM, Castier M, Brennecke JF, Stadtherr MA. Reliable computation of high pressure solid–fluid equilibrium. Ind Eng Chem Res 2000; 39:1624–1636.

26. Gordillo MD, Blanco MA, Molero A, de la Martinez, Ossa E. Solubility of the antibiotic penicillin G in supercritical carbon dioxide. J Supercrit Fluids 1999; 15:183–190.

27. Chrastil J. Solubility of solids in supercritical fluids using fusion properties. J Phys Chem 1982; 86:3016.

28. Bartle KD, Clifford AA, Jafar SA, Shilstone GF. Solubilities of solids and liquids of low volatility in supercritical carbon dioxide. J Phys Chem Ref Data 1991; 20(4):713–756.

29. Mendez-Santiago, Teja AS. Proceedings of the 8th International Conference on Properties and Phase Equilibria, Noordwijkerhout, Netherlands, 1998:123.

30. Del Valle JM, Aguilera JM. An improved equation for predicting the solubility of vegetable oils in supercritical CO_2. Ind Eng Chem Res 1988; 27: 1551.

31. Yu Z, Singh B, Rizvi SSH, Zollewg JA. Solubilities of fatty acids, fatty acid esters, triglycerides, and fats and oils in supercritical carbon dioxide. J Supercrit Fluids 1994; 7:51.

32. Jouyban A, Chan H-K, Foster NR. Mathematical representation of solute solubility in supercritical carbon dioxide using empirical expressions. J Supercrit Fluids 2002; 24(1):19–35.

33. Sandro RPdR, Olivera JVd, Avila SGD. A three-phase ternary model for CO_2–solid–liquid equilibrium at moderate pressures. J Supercrit Fluids 1996; 9: 1–5.

34. Garcia-Gonzalez J, Molina MJ, Rodriguez F, Mirada F. Solubilities of hydroquinone and p-quinone in supercritical carbon dioxide. Fluid Phase Equilibria 2002; 4995:1–9.

35. Sovova H, Stateva RP, Galushko AA. Solubility of β-carotene in supercritical-CO_2 and the effect of entrainers. J Supercrit Fluids 2001; 21:195–203.

36. Kikic I, Sist P. Applications of SCFs to pharmaceuticals: controlled drug release systems. 2nd NATO ASI on Supercritical Fluids, Kemer, Turkey, 1998.

37. Alessi P, Cortesi A, Kikic I, Foster NR, Macnaughton SJ, Colombo I. Particle production of steroid drugs using supercritical fluid processing. Ind Eng Chem Res 1996; 35:4718–4726.

38. Weidner E, Knez Z Novak Z. PGSS (particles from gas-saturated solutions): a new process for powder generation. Proceedings of the 3rd International Symposium on Supercritical fluids, ISASF 1994:229.

39. Kerc J, Srcic S, Knez Z, Sencar-Bozic P. Micronisation of drugs using supercritical carbon dioxide. Int J Pharm 1999; 182:33–39.

40. Knez Z, Skerget M. Phase equilibria of the vitamins D_2, D_3 and K_3 in binary systems with CO_2 and propane. J Supercrit Fluids 2001; 20:131–144.

41. Kikic I, Lora M, Bertucco A. Thermodynamic analysis of three phase equilibria in binary and ternary systems for applications in rapid expansion of a supercritical solution (RESS) particles from gas-saturated solutions (PGSS) and supercritical antisolvent (SAS). Ind Eng Chem Res 1997; 36:5507–5515.

42. Paulaitis ME, McHugh MA, Chai CP. Solid solubilities in supercritical fluids at elevated pressures. Chemical Engineering at Supercritical Conditions. Ann Arbor, MI: Ann Arbor Science, 1983:139–158.

43. Paulaitis ME, Krukonis VJ, Kurnik RT, Reid RC. Supercritical fluid extraction. Rev Chem Eng 1983; 1:179.

44. Diefenbacher A, Türk M. Phase equilibria of organic solid solutes and supercritical fluids with respect to the RESS Process. J Supercrit Fluids 2002; 22(3): 175–184.

45. Kordikowski A, Schenk AP, Nielen RMV, Peters CJ. Volume expansions and V-L-E of binary mixtures of a variety of polar solvents and certain near critical solvents. J Supercrit Fluids 1995; 8(3):205–216.

46. Badilla JCDF, Peters CJ, Arons JDS. Volume expansion in relation to the gas antisolvent process. J Supercrit Fluids 2000; 17(1):13–23.

47. Mukhopadhyay M. Purification of phytochemicals by selective crystallization using carbon dioxide as antisolvent. Proceedings of the 10th International Symposium on Supercritical Fluid Chromatography, Extraction and Processing, Myrtle Beach, SC, August 2001.

48. Singh S. Dense CO_2 as antisolvent for purification of lecithin. MTech. dissertation, Indian Institute of Technology, Bombay, India, 2001.

49. Mukhopadhyay M. Partial molar volume reduction of solvent for solute crystallization using carbon dioxide as antisolvent. J Supercrit Fluids 2003; 25(3): 213–223.

50. Chang CJ, Randolph AD. Solvent expansion and solute solubility predictions in GAS–expanded liquids. R & D notes. AIChE J 1990; 36(6):939–942.

51. Berends EM, Bruinsma OSL, de Graauw J, Van Rosmalein GM. Crystallisation of phenanthrene from toluene with carbon dioxide by the GAS process. AIChE J 1996; 14(2):1431–1439.

52. Cocero MJ, Ferrero S, Vicente S. Gas crystallization of β-carotene from ethyl acetate solutions using CO_2 as antisolvent. Proceedings of the 5th International Symposium on Supercritical Fluids, Atlanta, April 13-18, 2000.

53. Liu JW, Song L, Yang G, Han B. Study of the phase behavior of the cholesterol–Acetone–CO_2 system and recrystallisation of cholesterol by antisolvent CO_2. J Supercrit Fluids January 2002. In Press.

54. Mukhopadhyay M, Patel CR. Gas Antisolvent crystallisation at ambient temperature for enrichment of phytochemicals. Proceedings of the National Seminar and Workshop on Advanced Separation Processes, IIT-Kharagpur, India, August 2002.

55. Mukhopadhyay M, Singh S. Refining of crude lecithin using dense carbon dioxide as antisolvent. Accepted for publication in J Supercrit Fluids 2003.

56. Dixon DJ, Johnston KP. Molecular thermodynamics of solubilities in gas antisolvent crystallization. AIChE J 1991; 37(10):1441–1449.

57. Mukhopadhyay M, Dalvi SV. Solid solubility prediction from partial molar volume fraction of solvent in antisolvent–solvent mixture. Proceedings of Super Green-2002, 1st International Symposium on Supercritical Fluid Technology for Energy and Environmental Applications, Suwon, Korea, Nov. 3–6, 2002.

58. Shariati A, Peters CJ. Measurement and modeling of the phase behavior of ternary systems of interest for the GAS Process I. The system carbon dioxide + 1-propanal + salicylic acid. J Supercrit Fluids 2002; 25:195.

59. Berna A, Chafer A, Monton JB, Subirats S. High pressure solubility data of system ethanol (1) + catechin (2) + CO_2 (3). J Supercrit Fluids 2001; 20:157–162.

60. Dalvi SV. Mathematical modeling of antisolvent crystallization using carbon dioxide. MTech. dissertation, Indian Institute of Technology, Bombay, India. January 2003.

61. Kolar P, Shen J, Tsuboi A, Ishikawa T. Solvent selection for pharmaceuticals. Fluid Phase Equilibria 2002; 194–197:771–782.

62. Yeo Sang-Do, Kim Min-Su, Lee Jong-Chan. Recrystallization of sulfathiazole and chlorpropamide using the supercritical fluid antisolvent process. J Supercrit Fluids. In Press.

63. Mukhopadhyay M, Dalvi SV. A new thermodynamic method for solid–liquid–vapor equilibrium in Ternary systems from binary data for antisolvent crystallization. Proceedings of the 6th International Symposium, France, April 2003.

64. Hong SP, Luks KD. Multiphase equilibria of the mixture carbon dioxide–naphthalene–toluene. Fluid Phase Equilibria 1992; 74:133.

65. Mendes RL, Nobre BP, Coelho JP, Palavra AF. Solubility of β-carotene in supercritical carbon dioxide and ethane. J Supercrit Fluids 1999; 16:99–106.

66. Ting SST, Macnaughton SJ, Tomasko DL, Foster NR. Solubility of naproxen in supercritical carbon dioxide with and without cosolvents. Ind Eng Chem Res 1993; 32:1471–1481.

67. Mohamed RS, Halverson DS, Debenedetti PG, Prud'homme RR. Solids formation after the expansion of supercritical mixtures. In: Johnston KP, ed. Supercritical Fluid Science and Technology, ACS Symposium Series, Chapter 23. Washington, DC: American Chemical Society, 1994:355–378.

68. Knez M, Skerget P, Sencar-Bozic, Rizner A. Solubility of nifedepine and nitrendipine in supercritical CO_2. J Chem Eng Data 1995; 40:216–220.

69. Chafer A, Berna A, Monton JB, Munoz R. High pressure solubility data of system ethanol (1) + catechin (2) + CO_2 (3). J Supercrit Fluids 2001; 20:157–162.

70. Borg P, Jaubert JN, Denet F. Solubility of α-tetralol in pure carbon dioxide and in mixed solvent formed by ethanol and CO_2. Fluid Phase Equilibria 2001; 191:59–69.

71. Kordikowski A, Siddiqi M, Palakodaty S. Phase equilibria for the CO_2 + methanol + sulfathiazole system at high pressure. Fluid Phase Equilibria 2001; 4829:1–13.

72. Coutsikos P, Magoulas K, Tassios D. Solubilities of p-quinone and 9,10-anthraquinone in supercritical carbon dioxide. J Chem Eng Data 1997; 42:463.

73. Cortesi A, Kikic I, Alessi P, Turtoi G, Garnier S. Effect of chemical structure on the solubility of antioxidants in supercritical carbon dioxide: Experimental data and correlation. J Supercrit Fluids 1999; 140:139.

74. Fat'hi HR, Yamini Y, Sharghi H, Shamsipur M. Solubilities of some recently synthesized 1,8-dihydroxy-9,10-anthraquinone derivatives in supercritical carbon dioxide. Talanta 1999; 48:951.

75. Johannsen M, Brunner G. Solubilities of xanthines, caffeine, theophylline and theobromine in supercritical carbon dioxide. Fluid Phase Equilibria 1994; 95: 215.
76. Subra P, Castellani S, Ksibi H, Garrabos Y. Contribution to the determination of the solubility of β-carotene in supercritical carbon dioxide and nitrous oxide: experimental data and modeling. Fluid Phase Equilibria 1997; 131:269.
77. Choi ES, Noh MJ, Yoo KP. Solubilities of *o*-, *m*-, and *p*-coumaric acid in supercritical carbon dioxide at 308.15–323.15 K and 8.5–25 MPa. J Chem Eng Data 1998; 43:6.
78. Choi ES, Noh MJ, Yoo KP. Effect of functional groups on the solubilities of coumarin derivatives in supercritical carbon dioxide. In: Abraham MA, Sunol AK, eds. Supercritical Fluids: Extraction and Pollution. Washington, DC: American Chemical Society, 1997:110–118.
79. Ghiasvand AR, Hossenl M, Sharghi H, Yamini Y, Shamsipur M. Solubilities of some hydroxyxanthone derivatives in supercritical carbon dioxide. J Chem Eng Data 1999; 44:1135.
80. Yau JS, Tsai FN. Solubilities of 1-eicosanol and eicosanoic acid in supercritical carbon dioxide from 308.2 to 328.2 K at pressures to 21–26 MPa. J Chem Eng Data 1994; 39:827.
81. MacNaughton SJ, Kikic I, Rovedo D, Foster NR, Alessi P. Solubility of chlorinated pesticides in supercritical carbon dioxide. J Chem Eng Data 1995; 40: 593.
82. Uchiyama H, Mishima K, Oka S, Ezawa M, Ide M, Takai T, Park PW. Solubilities of flavone and 3-hydroxyflavone in supercritical carbon dioxide. J Chem Eng Data 1997; 42:570.
83. Lucien FP, Foster NR. Influence of matrix composition on the solubility of hydroxybenzoic acid isomers in supercritical carbon dioxide. Ind Eng Chem Res 1996; 35:4686.
84. MacNaughton SJ, Kikic I, Foster NR, Alessi P, Cortesi A, Colombo I. Solubility of anti-inflammatory drugs in supercritical carbon dioxide. J Chem Eng Data 1996; 41:1083.
85. Sovova, Jez J. Solubility of menthol in supercritical carbon dioxide. J Chem Eng Data 1994; 39:840.
86. Vandana K, Teja AS. The solubility of paclitaxel in supercritical carbon dioxide and N₂O. Fluid Phase Equilibria 1995; 135:83.
87. Sovova H, Jez J, Khachaturyan M. Solubility of squalane, dionyl phthalate and glycerol in supercritical carbon dioxide. Fluid Phase Equilibria 1997; 137: 185.
88. Hampson JW, Maxwell RJ, Shadwell SHRJ. Solubility of three veterinary sulfonamides in supercritical carbon dioxide by a recirculating equilibrium method. J Chem Eng Data 1999; 44:1222.
89. Nalesink KA, Hansen BN, Hsu JT. Solubility of pure taxol in supercritical carbon dioxide. Fluid Phase Equilibria 1998; 146:315.
90. Stassi A, Bettini R, Gazzaniga A, Giordano F, Schiraldi A. Assessment of solubility of ketoprofen and vanillic acid in supercritical carbon dioxide under the dynamic conditions. J Chem Eng Data 2000; 45:61.

91. Johansson M, Brunner G. Solubilities of the fat-soluble vitamins A, D, E, and K in supercritical carbon dioxide. J Chem Eng Data 1997; 42:106.

92. Domingo C, Berends E, Van Rosmalen GM. J Supercrit Fluids 1997; 10:39–55.

93. Subra P, Boissinot P, Benzaghau S. Proceedings of the 5th Meeting of Supercritical Fluids. Vol 1. France, 1998:307–312.

94. Krober H, Teipel U, Krause H. Manufacture of submicron particles via expansion of supercritical fluids. Chem Eng Technol 2000; 23:763–765.

95. Sievers RE, Hybertson B, Hausen B. European patent EP 0627910B1, 1993.

96. Krukonis V. Supercritical fluid nucleation of difficult-to-comminute solids. Proceedings of the AIChE Meeting, San Francisco, 1984.

97. Charoenchaitrakol F, Dehghani F, Foster NR. Micronization with supercritical fluids by RESS. Proceedings of the 5th Meeting on Supercritical Fluids. Atlanta, April 8–12, 2000.

98. Reverchon E, Della Porta G, Taddeo R. Solubility and micronization of griseofulvin in supercritical carbon dioxide and CHF_3. Ind Eng Chem Res 1995; 34:4087–4091.

99. Ohgaki K, Kobayashi H, Katayama T, Hirokawa N. Formation of jet supercritical solution. J Supercrit Fluids 1990; 3:103.

100. Alessi P, Cortessi A, Kikic I, Foster NR, MacNaughton SJ, Colombo I. Particle production of steroid drugs using supercritical fluid processing. Ind Eng Chem Res 1996; 35(12):4718–4726.

101. Palakodaty S, York P, Pitchard JP. Supercritical fluid processing of materials from aqueous solutions: the application of SEDS to lactose as a model substance. Pharma Res 1998; 15(12):1835–1843.

3

Fluid Dynamics, Mass Transfer, and Particle Formation in Supercritical Fluids

Jerzy Baldyga and Marek Henczka

Warsaw University of Technology, Department of Chemical and Process Engineering, Warsaw, Poland

Boris Y. Shekunov

Ferro Corporation, Independence, Ohio, U.S.A.

Supercritical fluids (SCFs) are involved in numerous industrial processes and have a potentially wide field of new applications. Current applications include extraction processes, reaction chemistry and polymerization, food fractionation, waste recycling, soil remediation, cleaning of electronic and optical equipment parts, impregnation, dry powder coating, aerogels, nanotechnology, and crystallization and particle formation of pharmaceuticals and many other powdered products (1). Industrial processes related to energy production as well as liquid rocket, diesel, and the gas turbine engines operate at supercritical conditions for the injected fuel (2). In this chapter we are interested in the mechanisms and phenomena affecting particle formation processes entailing the use of supercritical fluids, with particular reference to the production of fine particles for pharmaceutical applications. A description of different methods for particle formation by means of supercritical fluids can be found elsewhere in this volume, (Chapter 4 by Charbit et al.) and in Refs. 3

to 5. Most of these methods are based on crystallization or precipitation of a solid phase from supersaturated product solutions, whereas supersaturation is created in the system by means of supercritical fluids. Particle size distribution, particle morphology, and many physical properties of the solid state depend on the spatial distribution of the supersaturation and the timing of its evolution. To understand how solute supersaturation is created and how to predict its structure and evolution in time, one needs to consider balances of momentum (flow), species (mixing), energy (heating and cooling), and population (particle size distribution, PSD) and to apply the flux and source terms defined for the supercritical fluids. The physics of fluid behavior in the supercritical regime is different from that observed in gases or liquids, because SCFs exhibit some unique properties that must be taken into account during theoretical modeling, analytical studies, and the design of processing equipment.

The material in this chapter is organized as follows. In Section 1 we give a brief overview of experimental techniques used in our research as the basis for modeling and data interpretation; in Section 2 we review briefly specific properties of supercritical fluids with emphasis on molecular level transitions observed by means of X-ray scattering and their effect on mass transfer parameters and transition mechanisms; in Sections 3 and 4 we present aspects of fluid mechanics, mass transfer, and mixing in supercritical fluids. In Section 5 we review the kinetics of the precipitation process affecting particle size and morphology, including aggregation effects. Section 6 summarizes the most salient conclusions to be derived from the precipitation models. Finally Section 7 considers the most important theoretical aspects of process scale-up.

1. IN SITU ANALYSIS OF MIXING AND PRECIPITATION IN SCF

Particle formation, solvent dispersion, and mixing are essentially nonequilibrium dynamic processes that therefore require in situ measurements of such time- and scale-dependent parameters as fluid velocity, concentration of species, and droplet and particle size. Optical techniques are noninvasive, lend themselves well to quantitative analysis, and are particularly advantageous for dynamic measurements. However, for high pressure compressible solvents such as CO_2, there are significant experimental problems associated with the intricate cell design, reproducibility, and complexity of such measurements. In particular, as shown in the following sections, even the most uniform precipitation conditions with minimal nozzle back pressure are accompanied by significant gradients of both temperature and species concentration, which can obscure optical measurements in the supercritical flow. A high number concentration of very small particles is also typical for dispersion and precipitation in SCFs, imposing severe restrictions on particle size analysis. Therefore several complementary techniques have been developed and applied for such measurements, as summarized in Table 1 (6–11). Table 1 illustrates the physical principles and range of experimental conditions used. The most important parameter is the working pressure P in relationship to the mixture critical pressure P_m, which is the transition point between single- and two-phase mass transfer mechanisms (see the Section 2.3). Generally, large concentration and temperature gradients created in two-phase fluids make several experimental techniques unsuitable for quantitative analysis in such disperse systems.

Laser interferometry is an important example of a technique developed in our group for SCF measurements (9–12). This method is based on a simple principle whereby the changes of fluid composition or concentration gradients are detected as variations of the fluid refractive index, shown by the curvature or distance of the interferometric fringes observed. A spatial phase shift interferometric method (Figure 1a) was designed on the basis of a Mach–Zehnder interferometer. An optical flow cell is inserted in an object laser beam, whereas a second laser beam is used as a reference. Images (interferograms) are detected by using a multichannel optical decoder and digital video cameras. Parallel computer processing of these pictures allows one to calculate the amplitude and phase shift of the fringes and to transform this information into concentration or temperature gradients within the optical cell (11). This technique provides very fast measurement of mixing or heat transfer in SCF flow or visualization of the mass transfer mechanisms between liquid droplets and an SCF, explained in Sections 3 and 4. A confocal configuration of this interferometer also enables imaging of individual solid particles and particle size measurements (13).

The most crucial part of any of the optical systems listed in Table 1 is the high pressure optical cell, which is specifically designed or modified for each experimental method. For example, the cell shown in Figure 1b has a volume of about 3 cm^3 and could be used either with high quality sapphire windows (~7 mm thick) or with two 0.5 mm thin windows made of natural diamonds, giving an unsupported view area of about 3 mm. Single-crystal diamond is the only material that is transparent at the wavelength used in the small-angle X-ray scattering (SAXS) measurements. Such cells also have been used for phase Doppler and laser diffraction measurements, which require a plug flow configuration and high flow velocity. A cell for particle imaging velocimetry (PIV) was equipped with three high quality fused silica windows, which provided an undistorted view of a relatively wide (20 mm) measuring zone at 90° relative to the flow. All these cells were equipped with thermostatic aluminum jackets and designed to withstand working pressures up to 300 bar. The cells were also built on a modular principle whereby the view area could be moved relative to the nozzle by using different lengths of connecting tube sections,

TABLE 1 Experimental In Situ Optical Techniques Applied to the Analysis of Fluid Dynamics and Particle Formation in Supercritical Fluids

Method	Principles	Applications
1. UV on-line spectrometry	Concentration of solute is analyzed before or after particle precipitation.	Measurement of the equilibrium solubility and/or effluent concentration to calculate the supersaturation in single-phase solutions at $P > P_m$ (6,7).
2. Small-angle X-ray scattering (SAXS)	Forward scattering of synchrotron radiation is defined by fluid local inhomogeneity on a scale below 0.1 µm.	Analysis of correlation length and density fluctuations in pure and solvent-modified SCFs, in particular around the mixture critical curve ($P \approx P_m$). Also measurement of the mean nuclei size and microscale mixing segregation (8).
3. Interferometric microscopy	An image of a pendant solvent droplet is superimposed on an interferogram of the diffusion boundary layer around the droplet.	Combined measurements of the surface tension, concentration gradient, and mass transfer coefficients on the liquid–vapor interface ($P < P_m$) and also measurements of the solution refractive index in the vapor or liquid phase (9,10).
4. Phase shift laser interferometry	Interferometric images are processed online to obtain phase information on the refractive index distribution in the flow	Measurements of the concentration and temperature gradients during jet mixing in the single-phase solutions ($P > P_m$) (11).

TABLE 1   Continued

Method	Principles	Applications
5. Particle imaging velocimetry (PIV)	A sheet of double-pulsed laser light is scattered by small particles (0.1–10 μm) within a chosen flow cross section. The images obtained are analyzed by means of correlation algorithms.	Studies of the flow patterns, flow velocity, nucleation time constants and, more qualitatively, particle size and number particle concentration. Applicable to solid particles in single-phase solutions ($P > P_m$) or for liquid droplets ($P < P_m$).
6. Laser diffraction	Intensity measurements of forward-scattered laser light correspond to particle size in the range of 0.1 to 100 μm.	Size distribution of particles in single-phase flow ($P > P_m$).
7. Phase Doppler anemometry	Time-dependent fluctuations of optical phase and intensity are caused by droplets (1–100 μm size range) passing through a small measuring volume created by two laser beams.	Size distribution, number concentration, and velocity of liquid droplets in two-phase flow ($P < P_m$).

thus permitting investigators to observe different flow areas. Different inserts were used to modify the flow pattern (e.g., by eliminating the backflow), or to reduce the optical pass length.

2.   PROPERTIES OF SUPERCRITICAL FLUIDS

2.1.   Inhomogeneity on the Molecular Scale

According to classical thermodynamics (see Chapter 2 by Mukhopadhyay, this volume), the supercritical state of a substance is achieved when the temperature and the pressure are raised over the critical values T_c and P_c at the critical point (cp). It is often said that SCFs exhibit some properties of both liquids and gases: having liquidlike density, they can have gaslike viscosity,

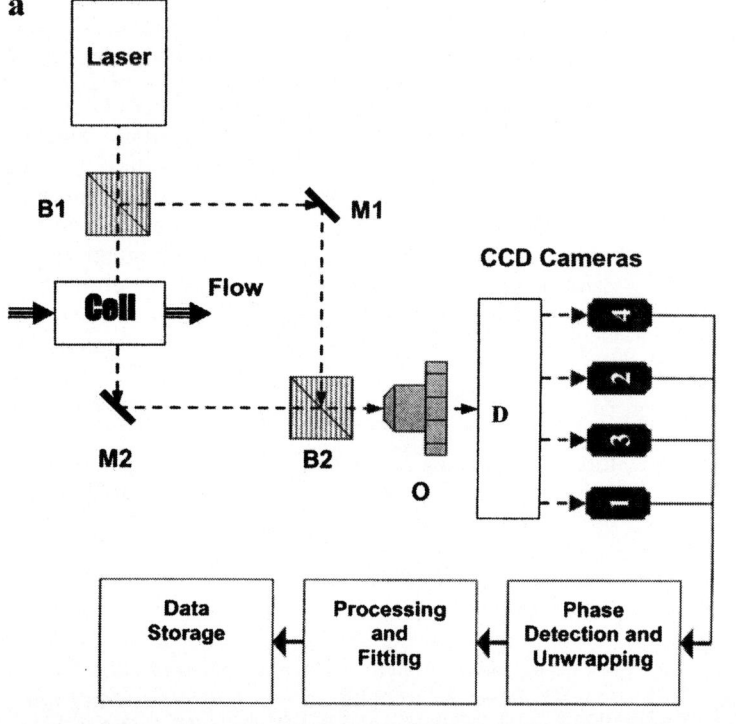

FIGURE 1 (a) Schematic of a laser interferometer. Mirrors (M1 and M2) and cubic beam splitters (B1 and B2) constitute the Mach–Zehnder configuration. The interferometric fringes overlap the image of the flow or droplet within the optical cell created by the objective (O). The optical decoder (D) provides four parallel phase shift images, which are processed and stored by a computer to extract information on the dynamics of mass or heat transfer. (b) Optical cell, used for interferometry, laser diffraction, phase Doppler anemometry, or small-angle X-ray scattering (SAXS) measurements, manufactured by Thar Technologies Inc. (Pittsburgh, PA). This cell is equipped with single-crystal diamond windows for SAXS.

conductivity, and diffusivity. It is important to acknowledge however that supercritical properties result not from the density itself, but from the inhomogeneity of molecular distribution in the SCF. On a characteristic microscale between about 10 and 100 Å, the supercritical state is defined by statistical clusters of augmented density with a structure resembling that of liquids, surrounded by less dense and more chaotic regions of compressed gas. The number and dimensions of these clusters vary dramatically with pressure and temperature, resulting in high compressibility near the critical point. The dissolving power of a fluid depends on its density, which means that solubility in an SCF can be varied continuously within a narrow pressure range. When an organic cosolvent is mixed with an SCF, the behavior of the mixture becomes more complex. Even within the region of a pressure–temperature–mole fraction (P-T-x) diagram that can be thermodynamically considered to be a supercritical state, some regions will be "more supercritical" than others in terms of both the molecular structure and the resulting mass transfer properties. Therefore it is important to identify quantitatively the boundaries of the processing conditions under which the critical phenomena take place.

For example, Figure 2 shows the P-T mixture critical curve formed by the mixture critical points (mcp) of ethanol–CO_2 system. A mixture becomes supercritical at its mcp with increasing pressure and temperature for a given solvent mole fraction. In the supercritical state, the more dense, "liquidlike" state SCF(L) can be identified above this curve, whereas the vaporlike state SCF(V) belongs to the range of conditions characterized by lower pressure and higher temperature. Structural changes in such mixtures have been studied by using small-angle X-ray scattering (8). The mean cluster size is defined by the Ornstein–Zernike correlation length ξ, and the number of clusters is described by the mean-square number density fluctuation $\langle (\Delta N)^2 \rangle / \langle N \rangle$, which is directly related to the thermodynamic isothermal compressibility (14). Figure 3 plots the behavior of these parameters as a function of pressure for pure carbon dioxide and an ethanol–CO_2 mixture. Like ξ, the $\langle (\Delta N)^2 \rangle / \langle N \rangle$ function has a maximum. For pure CO_2 this maximum coincides with a ridge along the extension of the gas–liquid coexistence curve as shown by Nishikawa and coworkers (14). This ridge is also very close to the critical isochore (i.e., where SCF density equals critical density). However this maximum shifts toward lower pressures and becomes more pronounced with addition of cosolvent. The ethanol mole fraction required to produce maximum fluctuation varies with SCF density. It was shown by Shekunov and Sun (8) that at $P > P_m$, the ethanol mole fraction corresponding to this maximum is smaller than that at the mcp, changing the entire character of the fluctuation. In other words, during mixing between an SCF and a liquid cosolvent, the mixture will exhibit supercritical properties only at a certain intermittent composition, which depends on the pressure and temperature. As discussed by Lockemann

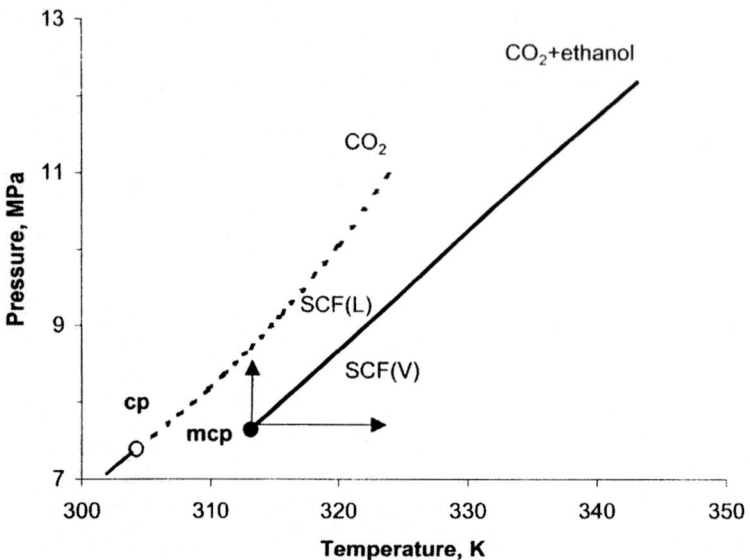

FIGURE 2 *P-T* mixture critical curve (solid line) for ethanol–CO_2 modeled from the Peng–Robinson equation of state, also shown in relation to the extension of the gas–liquid coexistence curve for pure CO_2 (dashed line) (8).

and Schlünder (15), the correlation length ξ is a measure of the interfacial mixing zone in the supercritical state, where the fluctuations in SCF lead to a significant resistance to mass transfer. Addition of cosolvent at concentrations above that at the mixture critical composition will transform the fluid into a liquid subcritical state. A pressure increase of about 3 MPa (Figure 3) will transform the physical properties of an SCF from those characteristic of compressed gases at lower pressures to those of a liquid state at higher pressures.

2.2. Phase Transition

A very important aspect of phase behavior in a system consisting of a volatile organic solvent, such as ethanol, and a supercritical fluid, such as CO_2, is that the mixture critical pressure P_m coincides with the liquid–vapor phase transition. This means that above P_m a single phase exists for all solvent compositions, whereas the (ethanol-rich and CO_2-rich) two-phase region lies below this curve. This fact has important implications for the mass transfer and precipitation mechanisms. Complete miscibility of fluids above P_m means that there is no defined or stable vapor–liquid or liquid–liquid interface, and the surface tension is reduced to zero and then thermodynamically becomes

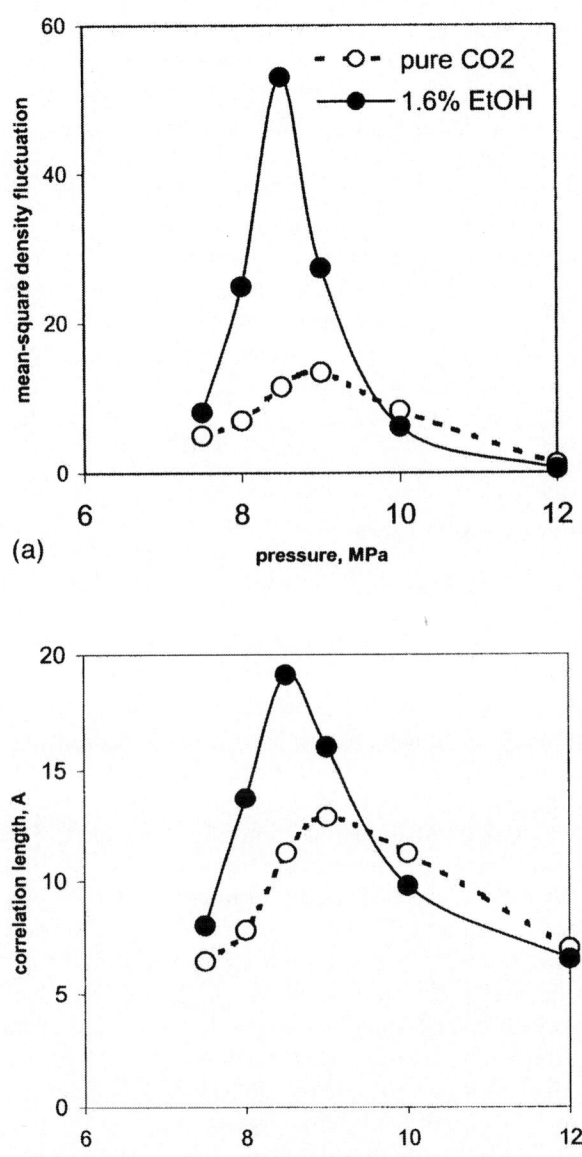

FIGURE 3 (a) Mean-square density fluctuation and (b) correlation length as functions of pressure at constant temperature 313.1 K for pure carbon dioxide and a mixture containing 1.6 mol % of ethanol.

negative (10). This effect can be observed by means of interferometry, as shown in Figure 4. Relatively large droplets below P_m (Figure 4a) decrease significantly in a transient region in the vicinity of P_m (Figure 4b) and eventually become smaller than the outside capillary diameter above P_m (Figure 4c), at which point the droplet shape or curvature can no longer be identified. The mass transfer below P_m proceeds through the diffusion boundary layer, which looks similar to droplet evaporation during spray drying. The transient region between these mechanisms corresponds to the mcp curve in Figure 3, taking into account, however, that dynamic surface tension in the boundary layer tends to increase P_m (10). Above P_m a convective mixing occurs, as in the case of subcritical miscible liquids, (e.g., ethanol and water). Quantitatively, however, there is a significant difference when mass transfer occurs between a liquid and supercritical fluid in both cases as shown in the section that follows.

2.3. Mass and Heat Transfer Parameters

The viscosities, diffusivities, and thermal conductivities of SCFs are intermediate to the properties of liquids and gases. This can be seen from the corresponding dimensionless numbers. The Schmidt number ($Sc = v/D_m$), represents the relative importance of momentum diffusion to the mass diffusion, where D_m and v denote the molecular diffusivity and kinematic viscosity, respectively. The Schmidt number for an SCF takes values similar to the values characteristic for gas mixtures (of order 1 or larger), up to $Sc \approx 10$. For $Sc > 1$, fluid deformation is faster than molecular diffusion, which means that spots, threads, and slabs of unmixed component can persist in single-phase systems. This phenomenon is observed in liquids ($Sc \gg 1$), and also in supercritical fluids when $Sc > 1$. The Prandtl number ($Pr = v/D_t$) is a measure of the importance of momentum diffusion to the heat diffusion (conduction). For supercritical fluids $Pr > 1$ and is larger than in a gas phase ($Pr \approx 1$). This means that the heat diffusion in an SCF is slightly slower than momentum diffusion and some hot spots may be created during mixing. The relative importance of heat diffusion to the mass diffusion is expressed by the Lewis number $Le = D_t/D_m$. In gases $Le \approx 1$, whereas in liquids Le is between 10 and 100, indicating faster diffusion of heat than mass. Harstad and Bellan (16) have studied the Lewis problem under supercritical conditions. They

---

FIGURE 4 Behavior of ethanol droplets in pure CO_2 at 363.15 K and pressures of (a) 8, (b) 12, and (c) 13.5 MPa. The interferometric fringes show the diffusion boundary layer (a, b) or mixing flow (c). The droplet diameter in (a) is about 1.6 mm.

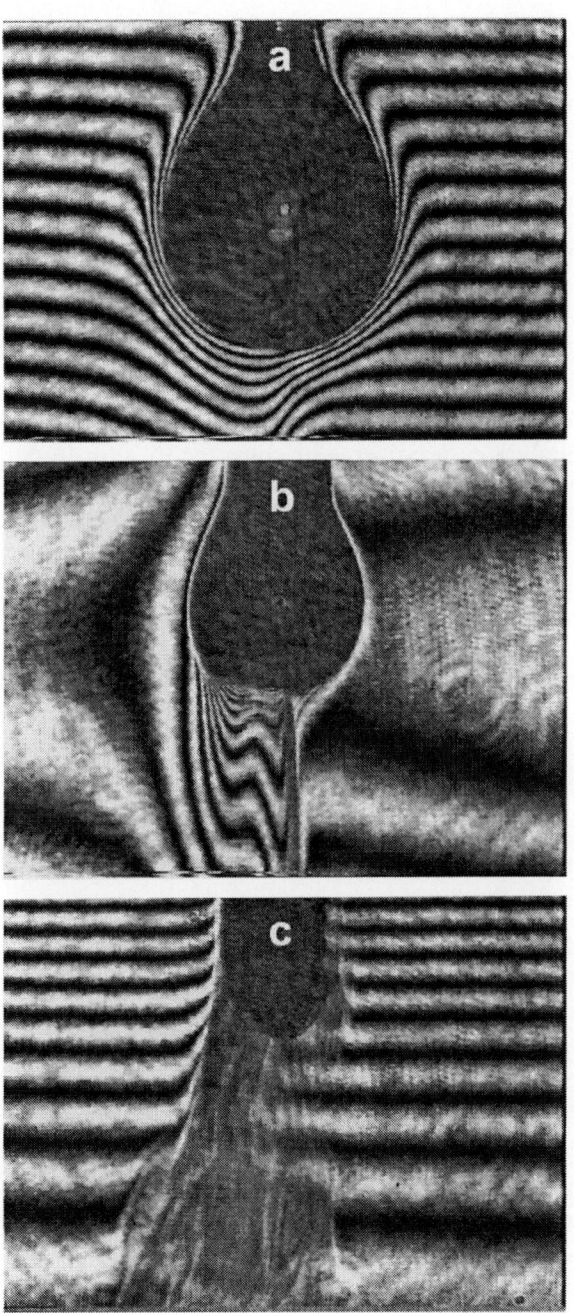

have shown that the effective value of the Lewis number can be 2 to 40 times larger than the traditionally calculated Lewis number, owing to the combined effects of the small mass diffusion factor and the difference between the specific enthalpies of the two mixed species. The mass diffusion factor α_d is a thermodynamic parameter that acts to alter the effective diffusivity ($D_{eff} = D_{m\alpha_d}$) in the Fickian mass diffusion flux. We will come back to application of the mass diffusion factor when we consider "nonequilibrium" thermodynamic effects. As shown by Miller (17) and Lou and Miller (18), "nonequilibrium" effects, which can be significant in supercritical fluids, can result in the generation of concentration gradients; similarly, diffusion of energy can result from concentration gradients (Soret and Dufour effects).

There is anomalous variation of fluid thermal physical properties near the critical state. When the critical point is approached, all thermometric and transport properties of the fluid either diverge or converge to zero. In particular, the latent heat, the surface tension, and the heat and mass diffusion coefficients vanish at the critical point, whereas the thermal conductivities and viscosities as well as the heat capacities are enhanced. Nevertheless, heat transfer can be very fast near the critical point. Such enhanced heating results from the piston effect- at constant volume, there is critical divergence of the thermal expansion coefficient. Thus, even a very small inhomogeneity of temperature can generate strong convection and large fluxes of energy (19).

2.4. Differences Between Subcritical and Supercritical Fluids

It follows from the foregoing considerations that the differences between sub- and supercritical fluids are defined by

Reduction of surface tension followed by phase transition around the mcp

Inhomogeneity of molecular structure (solution clustering) in mixtures, most pronounced in the P-T area between the mcp and liquid–vapor continuation curve for a pure SCF

The magnitude and behavior of the mass and heat transfer parameters

Supercritical conditions above ($P > P_m$) or below ($P < P_m$) the mixture critical curve lead to different crystallization and aggregation mechanisms (7,20). In the first case there is no interfacial tension, and mass transfer is determined by the flow–molecular diffusion interactions during mixing, defined here as jet mixing. In the second case, there is interfacial tension between the phases, and the mixing mechanism is based on atomization to small droplets while energy and species are being transferred between the drops and their environment. We shall term these processes "droplet dispersion" and "mass transfer". One should consider separately behavior of the compressible homogeneous SCF phase and the two-phase system, where such phenomena as destabilization of the interface and drop breakage are of importance. The transition of one mechanism into another occurs within a very narrow interval of pressures and temperatures and has a pronounced effect on particle formation. On molecular level, the density fluctuations in an SCF produce inhomogeneities of supersaturation and diffusion flux and may result in an increase of the nuclei density. Transport of solute molecules to nuclei can be reduced in SCFs, leading to production of nonagglomerated and small particles as shown in Section 5.

3. JET MIXING

3.1. Isothermic Mixing

Consider the flow of a single-phase SCF mixture, which can be observed above the mixture critical pressure P_m, as seen in Figure 2. Because processes of particle formation using SCFs are usually carried out in turbulent flow systems, we consider turbulent flows. In this section we assume isothermic conditions to distinguish the pure effects of fluid flow. Under typical supercritical conditions, the SCF mixture is a dense fluid, and the local fluid density is strongly dependent on local pressure and composition. To predict the effects of pressure on fluid density, the Peng–Robinson equation of state (EOS) (21) is often used (16,17,22) because of its accurate density predictions at high pressure and its relatively simple form

$$P = \frac{RT}{V - B} - \frac{A}{V^2 + 2VB - B^2} \tag{1}$$

with

$$A = 0.457236 \frac{R^2 T_C^2}{P_C}\left[1 + C\left(1 - \sqrt{\frac{T}{T_C}}\right)^2\right], \quad B = 0.077796 \frac{RT_C}{P_C} \tag{2}$$

$$C = 0.37464 + 1.54226\omega - 0.26992\omega^2 \tag{3}$$

where R is the universal gas constant and ω is the acentric factor of the substance. For mixtures, the two parameters specifying the EOS are

$$A_m = \sum_\alpha \sum_\beta X_\alpha X_\beta A_{\alpha\beta}, \quad \text{and} \quad B_m = \sum_\alpha \sum_\beta X_\alpha X_\beta B_{\alpha\beta} \tag{4}$$

In Eq. (4), the summation represents sums over both species. The diagonal elements of the matrices $A_{\alpha\alpha}$ and $B_{\alpha\alpha}$ are equal to their pure substance counterparts. The Peng–Robinson parameters $A_{\alpha\beta}$ and $B_{\alpha\beta}$ are provided by an appropriate set of mixing rules that vary for different equations of state and may have many variations even for the same EOS. The recommended mixing rules are:

$$A_{\alpha\beta} = \sqrt{A_\alpha A_\beta}(1 - k_{\alpha\beta}), \qquad \text{and} \qquad B_{\alpha\beta} = \frac{1}{2}(B_\alpha + B_\beta) \tag{5}$$

The binary interaction parameter $k_{\alpha\beta}$ is a function of the species being considered and is taken to be $k_{\alpha\alpha} = 0$.

The flow of an SCF under isothermic conditions is governed by the general conservation equation

$$\frac{\partial \rho}{\partial t} + \frac{\partial}{\partial x_j}(\rho u_j) = 0 \tag{6}$$

where u_j is the jth Cartesian component of fluid velocity, t is the time, and ρ is the fluid density, $\rho = M_i x_i + M_j x_j / V$. The repeated indices imply summation over all coordinate directions, and V is the molar volume of the mixture. The momentum conservation equation reads:

$$\frac{\partial}{\partial t}(\rho u_i) + \frac{\partial}{\partial x_j}(\rho u_i u_j) = -\frac{\partial p}{\partial x_i} + \frac{\partial}{\partial x_j}\left[\mu\left(\frac{\partial u_i}{\partial x_j} + \frac{\partial u_j}{\partial x_i} - \frac{2}{3}\frac{\partial u_k}{\partial x_k}\delta_{ij}\right)\right] \tag{7}$$

where P is the static pressure, μ denotes the molecular viscosity, and δ_{ij} is the Kronecker delta. To apply relatively simple computational fluid dynamics (CFD) models (k-ε model, Reynolds stress model) and include effects of variable density, Favre averaging can be employed. Favre averaging is based on weighting the averaged quantities by the local, instantaneous value of density. The local Favre average $\tilde{y}$ of the field variable y reads

$$\tilde{y} = \frac{\overline{\rho y}}{\overline{\rho}} \tag{8}$$

where an overbars denotes a conventional Reynolds average and a tilde denotes a Favre average. The field variables can be thus decomposed as follows:

$$y = \tilde{y} + y'' = \overline{y} + y' \tag{9}$$

where primes denote fluctuations with respect to the Reynolds average and double primes denote fluctuations with respect to the Favre average. The Favre average conservation of mass equation reads

$$\frac{\partial \overline{\rho}}{\partial t} + \frac{\partial \rho}{\partial x_j}(\overline{\rho}\tilde{u}_j) = 0 \tag{10}$$

and the Favre average momentum conservation equation take

$$\frac{\partial}{\partial t}(\overline{\rho}\widetilde{u}_i) + \frac{\partial}{\partial x_j}(\rho\widetilde{u}_i\widetilde{u}_j) - \frac{\partial}{\partial x_j}(\tau_{ij}) = -\frac{\partial \widetilde{p}}{\partial x_i} \tag{11}$$

where the Favre-averaged Reynolds stress is defined as

$$\tau_{ij} = -\overline{\rho}\,\widetilde{u_i''u_j''} = -\overline{\rho u_i'u_j'} \tag{12}$$

Equations (11) and (12) are unclosed; application of the k-ε model to close this set of partial differential equations enables effective computation of the flow field. Figure 5b shows sample computations illustrating the velocity and density distribution of a CO_2–ethanol mixture in a mixing nozzle (SEDS process, see Chapter 4 by Charbit et al.) at $P > P_m$. Figure 5a shows that for both the nozzle chamber and the particle formation vessel in the region of the jet, there is intensive circulation. Figure 5b illustrates the variation of average density resulting from the mixing of supercritical carbon dioxide with ethanol (nozzle chamber) and the pressure variation in the nozzle and the particle formation vessel. Figure 6 shows the experimental velocity field obtained for such a nozzle by means of the PIV technique (Table 1). The circulation around the jet is clearly seen. This flow plays an important role for establishment of in supersaturation profile as shown in Section 5.

We have considered the situation of only one phase for any mixture composition; this means that there is no surface tension and the fluid behavior is completely characterized by the turbulent flow described by the mass and momentum balance equations. To solve these equations, one needs to model the diffusional mixing of the species present in the system and to identify local values of the thermodynamic and transport properties, as considered in Section 3.2. Here we just point out that once the methods for predicting local values of fluid density and viscosity have been worked out, one should be able to integrate Eqs. (10) and (11).

3.2. Mixing and Heat Transfer

Let us again consider single-phase mixing (i.e., for $P > P_m$). We would like to follow the progress of turbulent mixing of a chemically passive tracer with an SCF. The tracer is then redistributed over the mixing chamber, the concentration variance decreases owing to mixing on the molecular scale (micromixing), and the temperature varies owing to the effects of the enthalpy of mixing and energy dissipation, as well as to the Joule–Thomson effect. To model such a process one needs to start from the momentum and mass balances. Modeling is based on Favre averaging, Eq. (8), so we apply Eqs. (10) and (11) together with a balance for the kinetic energy of turbulence k and the

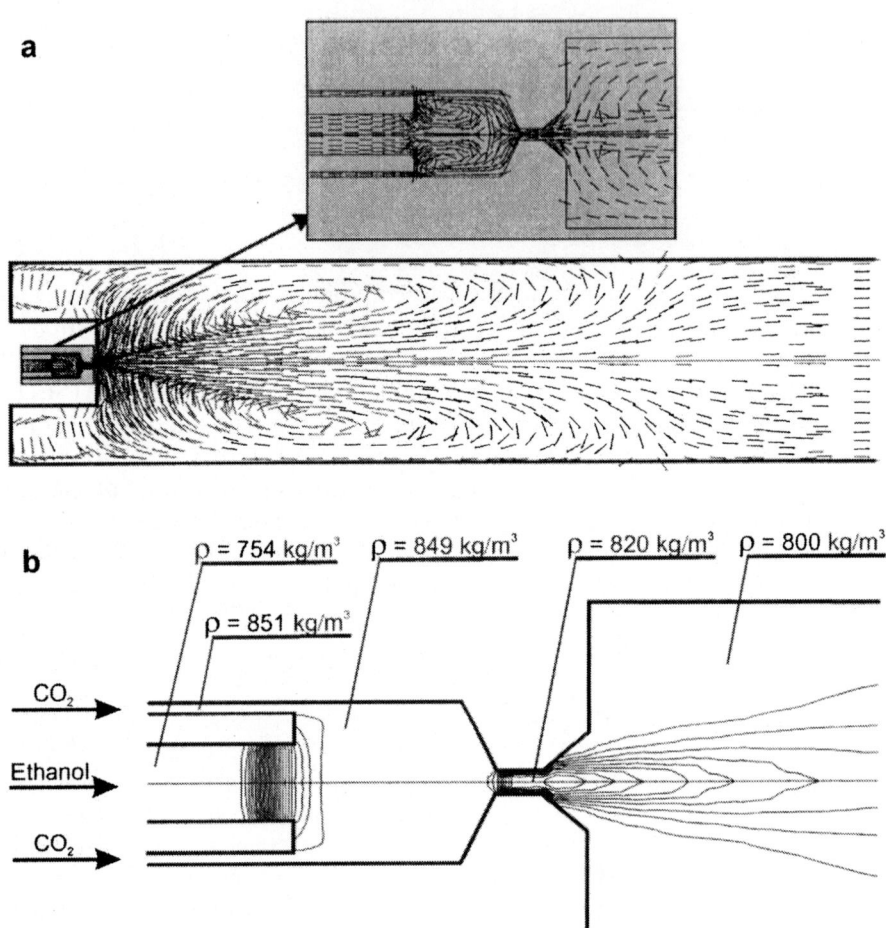

FIGURE 5 (a) Flow pattern in the precipitation vessel from a two-component mixing nozzle. (b) Distribution of fluid density in this system at 20 MPa and 323.1 K in the vessel.

a

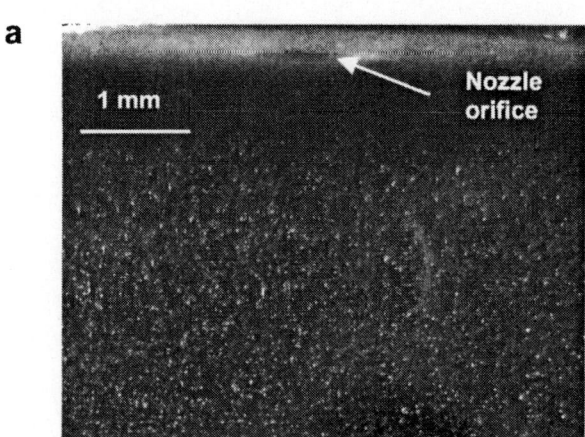

b

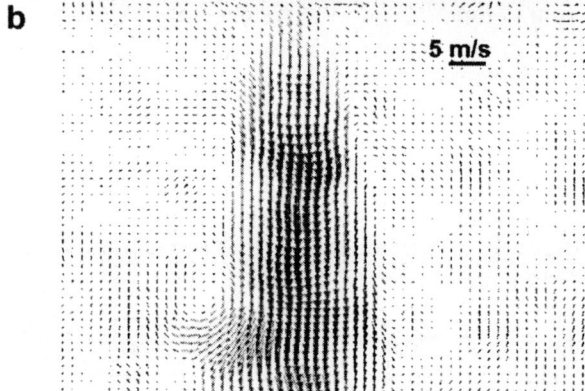

FIGURE 6   (a) Image of acetaminophen crystals (mean size ~3 μm) obtained by precipitation from ethanol solution with CO_2 and used as seeds. (b) Jet flow pattern obtained by application of autocorrelation PIV algorithm to this image. The conditions correspond to those in Figure 5.

rate of energy dissipation $\tilde{\varepsilon}$. Distributions of the mean mixture fraction $\tilde{f}$ and its variance σ^2 are calculated by means of the following equations:

$$\frac{\partial\left(\overline{\rho}\tilde{f}\right)}{\partial t} + \frac{\partial\left(\overline{\rho}\tilde{u}_j\tilde{f}\right)}{\partial x_j} = \frac{\partial}{\partial x_j}\left[\overline{\rho}(D_T + D_m)\frac{\partial\tilde{f}}{\partial x_j}\right] \tag{13}$$

$$\frac{\partial\left(\overline{\rho}\tilde{\sigma}^2\right)}{\partial t} + \frac{\partial\left(\overline{\rho}\tilde{u}_j\tilde{\sigma}^2\right)}{\partial x_j} = \frac{\partial}{\partial x_j}\left[\overline{\rho}(D_T + D_m)\frac{\partial\tilde{\sigma}^2}{\partial x_j}\right]$$

$$+ 2D_T\left(\frac{\partial\tilde{f}}{\partial x_j}\right)^2 - \frac{R\tilde{\varepsilon}}{\tilde{k}}\tilde{\sigma}^2 \tag{14}$$

where $R \approx 2$. Equation (14) is valid for $Sc \leq 1$ D_m and $D_T = C_\mu \tilde{k}^2/Sc_T\tilde{\varepsilon}$ denote the molecular and turbulent diffusivity, respectively. The mixture fraction f represents the concentration of the conserved scalar and can be interpreted as a mass fraction of fluid fed into the system from a chosen feed point

$$f = \frac{Y_\alpha - Y_{\alpha,1}}{Y_{\alpha,2} - Y_{\alpha,1}} \tag{15}$$

where Y_j is a mass fraction of the tracer α fed into the system through feed point 2. In the case of $Sc \gg 1$, the concentration variance can be divided into three parts according to the scale of inhomogeneity and the related mechanism of mixing. One can distinguish the inertial–convective, viscous–convective, and viscous–diffusive mechanisms of mixing and describe them by using several time constants for mixing (23–25). Such complex models enable one to include the effects of molecular viscosity and diffusivity on the rate of mixing on the molecular scale. The local values of such fluid properties as density, molecular viscosity, and molecular diffusivity depend on temperature, which means that the mixture fraction and concentration variance balances need to be solved together with the energy balance equation, which in the Favre-averaged forms reads

$$\frac{\partial\left(\overline{\rho}\tilde{h}\right)}{\partial t} + \frac{\partial\left(\overline{\rho}\tilde{u}_j\tilde{h}\right)}{\partial x_j} = \frac{\partial}{\partial x_j}\left[\frac{k_{\text{eff}}}{C_{\text{pm}}}\frac{\partial\tilde{h}}{\partial x_j}\right] + \tau_{ij}\frac{\partial\tilde{u}_i}{\partial x_j} \tag{16}$$

$$+ \frac{\partial}{\partial x_j}\left[\sum_{\alpha=1}^{N}\tilde{h}_\alpha\left\{\overline{\rho}D_{\text{eff}} - \frac{k_{\text{eff}}}{C_{\text{pm}}}\right\}\frac{\partial\tilde{Y}_\alpha}{\partial x_j}\right] + S_h$$

where k_{eff} is the effective conductivity ($k_{\text{eff}} = k + k_T$, where k_T is the turbulent thermal conductivity), h is the total enthalpy, h_α is the enthalpy for species α, and S_h represents additional sources. To introduce into Eq. (16) effects of the heat of mixing, one can use the so-called integral heat of solution $\overline{Q_m}$ (per unit mass of mixture) and $Q_{m\alpha} = \overline{Q_m}/Y_\alpha$ (per unit mass of component of interest). Thus variation of $\overline{Q_m}$ with concentration reads

$$\frac{\partial \overline{Q_m}}{\partial Y_\alpha} = \frac{\partial Q_m}{\partial Y_\alpha} Y_\alpha + Q_m \tag{17}$$

We calculate the dependence of Q_m on concentration from the Peng–Robinson equation of state. The resulting source of energy S_h then depends on the local gradient of concentration, local fluid velocity, and fluid density:

$$S_h = \overline{\rho \tilde{u}_i} \left(\frac{\partial Q_m}{\partial Y_\alpha} \tilde{Y}_\alpha + Q_m \right) \frac{\partial \tilde{Y}_\alpha}{\partial x_i} \tag{18}$$

Typical model predictions for the jet mixing process between ethanol and supercritical carbon dioxide are shown in Figure 7. The distribution of $\tilde{f}$ shows that there is very fast premixing of ethanol with supercritical carbon dioxide already in the mixing chamber (Figure 7a). However, mixing on the molecular scale is not perfect, and fluctuations of ethanol concentration are observed (Figure 7b). The intensity of segregation I_S is equal to unity when the fluid elements are just premixed on the macroscale without mixing on the molecular scale; I_S is zero when mixing on the molecular scale is perfect for given $\tilde{f}$ value. The root-mean-square fluctuation of the mixture fraction is represented by $\sqrt{I_S}$. Clearly, mixing on the molecular scale follows macro-mixing (i.e., turbulent convection and turbulent diffusion); macromixing decreases large-scale inhomogeneity, producing at the same time inhomogeneity on the molecular scale that is finally dissipated as a result of the effects of molecular diffusion (micromixing). Finally, Figure 7c shows the temperature distribution in the system; the effects of the enthalpy of mixing are clearly present in the nozzle chamber, and afterward the most spectacular decrease of temperature results from the Joule–Thomson effect.

These predictions have been confirmed by experimental results obtained by using phase shift interferometry (11). The experiments consisted of using different nozzles to mix ethanol and CO_2. At high Reynolds numbers (small nozzle diameters), ethanol distribution inside the jet was shown to be homogeneous; thus the refractive index gradients were entirely created by the temperature difference between the jet and the reservoir liquid. However at relatively small Reynolds numbers (Figure 8a), poor mixing within the nozzle is manifested in large concentration fluctuations in the jet and, when time-averaged, translates into inhomogeneity of ethanol concentration across the flow. This large-scale inhomogeneity is a result of insufficient turbulent dif-

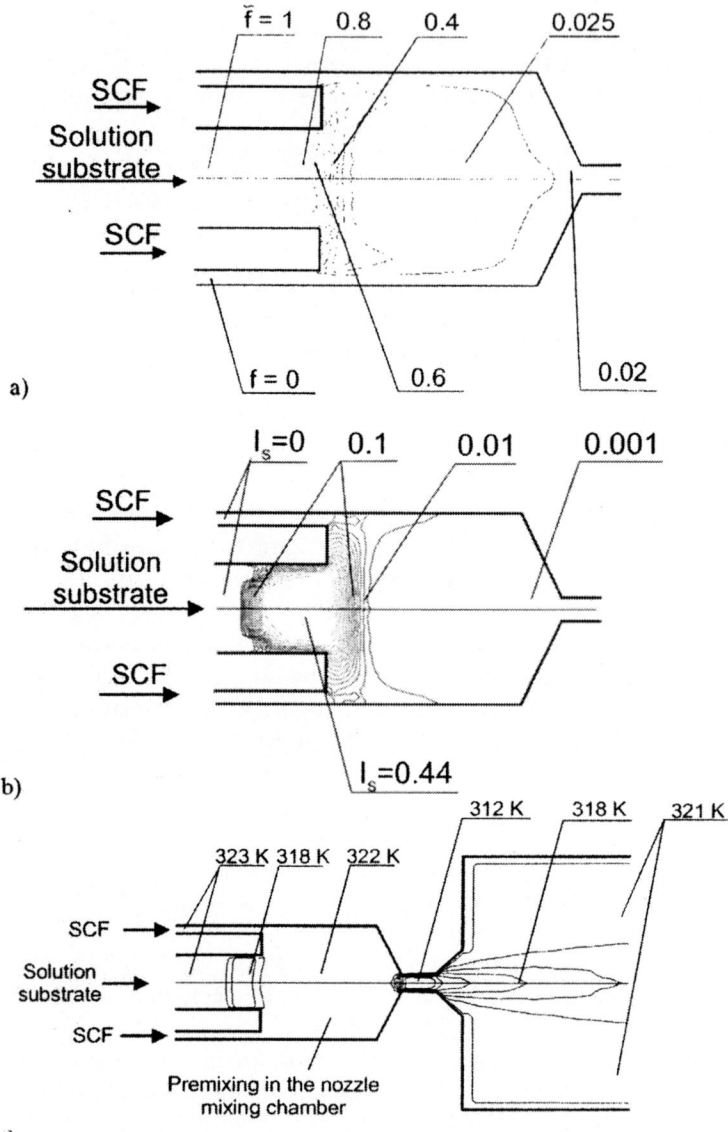

FIGURE 7 Effects of mixing ethanol and CO_2 in the nozzle for inlet temperature $T_f = 323.1$ K and pressure in the precipitation vessel $P = 20$ MPa; (a) distribution of the average mixture fraction $\tilde{f}$, (b) distribution of the mixture fraction concentration variance, and (c) distribution of temperatures in the system.

a

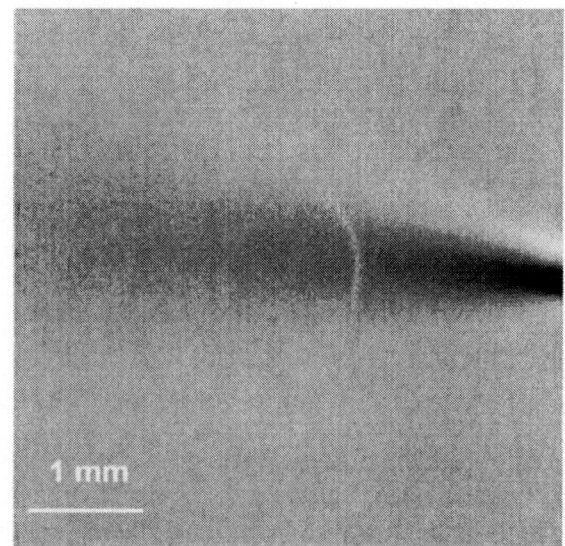

b

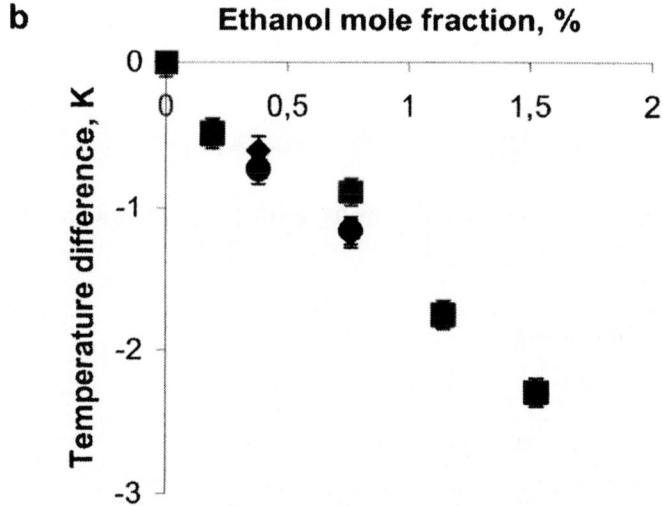

FIGURE 8   Effects of mixing ethanol and CO_2 observed by means of phase shift interferometry at T = 323.1 K and P = 20 MPa. (a) Time-averaged radial distribution of ethanol concentration in a nozzle with a large orifice (0.75 mm), equal to the diameter of the mixing chamber in Figure 7. Gray background corresponds to a well-mixed fluid, whereas dark or light deviations corresponds to CO_2-rich and ethanol-rich parts of the stream. (b) Temperature shift in the jet caused entirely by the enthalpy of the mixing effect and shown as a function of the ethanol mole fraction.

fusion in the nozzle chamber. The temperature drop caused by the enthalpy of mixing is significant (Figure 8b) and is comparable to the Joule–Thomson effect under these conditions (see Figure 7c). It should be noted that the enthalpy of mixing increases as the density of CO_2 decreases and pressure approaches P_m. Under these conditions, both the enthalpy and Joule–Thomson effects may increase by an order of magnitude. In conclusion, the nozzle geometry and the flow Reynolds numbers are crucial for uniform mixing, and changes in the temperature cannot be neglected.

3.3. Cross-Diffusion Effects During Mixing

Possible influences of nonequilibrium "cross-diffusion" effects on the mixing process were investigated by means of direct numerical simulations (DNS) of mass fraction fluctuations in stationary isotropic turbulence for binary mixtures under supercritical conditions (26,27). The authors have shown that after some time, the initially perfectly mixed species become segregated owing to the presence of temperature and pressure fluctuations and the resulting Soret mass cross-diffusion fluxes J_i^T and J_i^P, induced by temperature and pressure gradients. Based on DNS results (26,27), we propose a phenomenological model that predicts the rate of production of the concentration variance as

$$\left\langle J_i^P \frac{\partial Y_\alpha}{\partial x_i} \right\rangle \approx C_K C^P \langle \rho \rangle \frac{\left(\overline{Y_\alpha' Y_\alpha'}\right)^{1/2}}{(D\alpha_d)^{1/2}} \frac{\tilde{\varepsilon}^{5/6}}{l^{2/3}} \tag{19}$$

where C_K is the correlation coefficient for the mass fraction gradient with the pressure gradient [$C_K \approx 0.26$–0.51 (26)] and α_d is a mass diffusion factor related to the fugacity coefficient that alters the effective diffusivity ($D_{\text{eff}} = D_m \alpha_d$). When statistical balance is achieved between the Fickian micromixing rate and the variance production rate, Eq. (19), we get

$$\left(\overline{Y_\alpha' Y_\alpha'}\right)^{1/2} = C_X C_K C^P \frac{\tilde{\varepsilon}^{1/2}}{(D_m \alpha_d)^{1/2}} = C_X C_K \langle \rho \rangle \frac{D_m^{1/2} \tilde{\varepsilon}^{1/2}}{\alpha_d^{1/2}} \frac{\tilde{Y}_A \tilde{Y}_B}{R\tilde{T}}$$
$$\times \left(\frac{M_A M_B}{M_m}\right) \left(\frac{\partial V/\partial X_B}{M_B} - \frac{\partial V/\partial X_A}{M_A}\right) \tag{20}$$

Results of calculations (Figure 9) show that the concentration variance increases with increases in both the power input for mixing $\tilde{\varepsilon}$ and the molecular weight ratio of the species. The cross-diffusion effects can affect the process of particle production only in the case of fast mixers (e.g., nozzle mixers), characterized by extremely large values of the rate of energy dissipation. It should be noted particle nucleation is very sensitive to supersaturation (Section 5) and

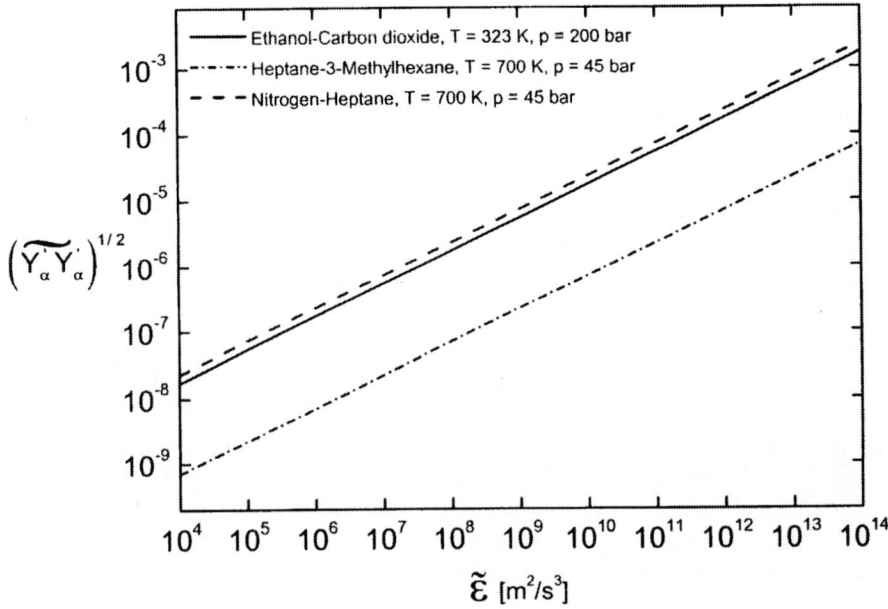

Figure 9 Effect of power input on fluctuations in concentration in supercritical fluids at steady state.

therefore these variances in concentration can be important factors in the mechanism of nuclei generation in supercritical solutions.

4. DROPLET DISPERSION AND MASS TRANSFER

4.1. Dispersion by a Single Liquid Nozzle

We consider now the two-phase system at $P < P_m$ where one phase is in a supercritical state and the second in a liquid state. There is a surface tension between both phases, and the fluid behavior is determined by turbulence and a complex phenomenon of momentum exchange between the phases. In practical applications fluid behavior is strongly affected by mass and heat exchange between the phases (16). Let us start from the phenomenon of jet breakup observed for steady injection of a liquid through a single nozzle into a quiescent gas or SCF. One then can recognize several breakup regimes and relate these regimes to the stability analysis of the liquid jets. Such an analysis of the growth of some small initial perturbations of the liquid surface that includes effects of liquid inertia, interfacial tension, and viscous and aerodynamic forces is presented in detail by Reitz and Bracco (28). With the jet Reynolds and Ohnesorge numbers defined by $Re_j = \rho_L u_0 d_0 / \mu_L$ and $Oh_j = \mu_L / \sqrt{\rho_L d_0 \sigma}$, respectively, the most important breakup regimes are presented in Figure 10. It is clearly seen that wind-induced breakup and atomization are achieved at lower injection velocities when the gas is compressed. Rayleigh (29) analyzed a specific case of $Oh_j = 0$ and $We_2 = \rho_g u_0^2 d_0 / \sigma = 0$ (i.e., an inviscid liquid jet at low velocity and low gas density). He calculated drop size from the wavelength of the axisymmetric surface disturbances with the maximum growth rate. Drops in the Rayleigh regime (also called the varicose regime) have diameters larger than the jet diameter and are created several nozzle diameter from the injection point; the jet breakup length increases proportionally to the jet velocity. When the jet velocity and/or the gas density is increased, the jet becomes affected by the surrounding gas. When these interactions have purely inertial character, as suggested by Weber (30), the strength of destabilization can be expressed by the Weber number We_2. In fact, at least at finite values of the Reynolds number, effects of gas viscosity are observed (31) because the fluctuating pressure of this gas related to the wave elevation is reduced (32). The maximum wave growth rate in this regime, called the first wind-induced regime, occurs at small wavenumbers, so droplet sizes are still of the order of the jet diameter and droplets still are created many nozzle diameters from the

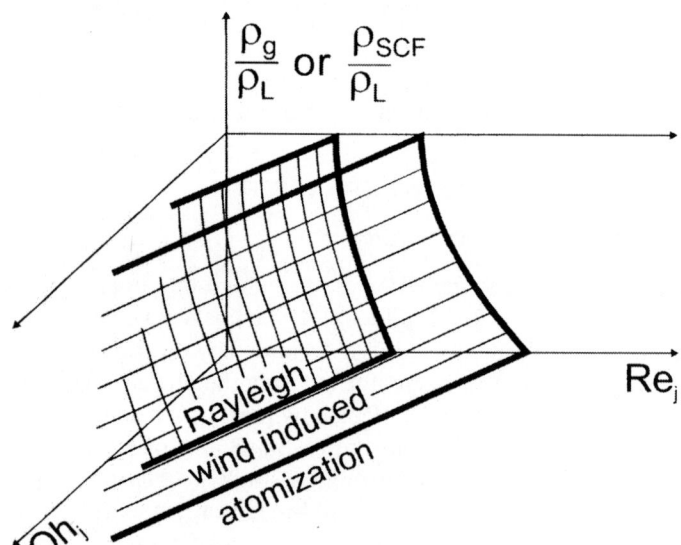

FIGURE 10 Schematic presentation of the liquid jet regime boundaries: effect of the density ratio.

nozzle. When, however, We_2 is further increased, interface tension forces start to stabilize the jet. The maximum growth of waves occurs at shorter wavelengths, which are induced by relative motion between the jet and surrounding gas. The onset of the related, second wind-induced breakup regime is observed at $We_2 \sim 12$ to 13 (33,34). Droplets much smaller than the jet diameter are observed at short wavelengths. The breakup starts a small distance downstream from the nozzle; Taylor (35) estimated the intact core length L_2 as

$$L_2 = C_1 a \left(\frac{\rho_L}{\rho_g} \right)^{1/2} \frac{1}{f(\Gamma)} \tag{21}$$

where C_1 is a constant of order unity and, $\Gamma = (\rho_L/\rho_g)(\sigma^2/\mu_L^2 u^2) = (\rho_L/\rho_g)$ $(Re_j/We_1)^2$, $We_1 = \rho_L u_0^2 d_0/\sigma$ and $f(\Gamma)$ is an increasing function of Γ that at high Γ reaches a constant value $f(\Gamma) = \sqrt{3}/6$ at $\Gamma \approx 10^3$. The intact surface length L_1 can be estimated (28) from

$$L_1 = \frac{C_2 L_2}{We_2} \tag{22}$$

which means that L_1 decreases with increasing We_2. In Eq. (22), C_2 is a constant that weakly depends on the drop size. Finally, in the atomization regime the breakup starts at the nozzle exit and the spray takes a cone shape. Droplet sizes are much smaller than the jet diameter. The spray angle was found to increase with gas density (36) and liquid viscosity. In this regime, the geometry of the nozzle strongly affects the process (28). The phenomenon is very complex and involves the superposition of many elementary mechanisms.

Assuming that atomization is caused by aerodynamic surface wave growth, Ranz (37) predicted the spreading angle, Θ, of the atomizing jet by relating the radial velocity of the unstable surface waves to the axial injection velocity

$$\tan \left(\frac{\Theta}{2} \right) = \frac{4\pi}{A} \left(\frac{\rho_g}{\rho_L} \right)^{1/2} f(\Gamma) \tag{23}$$

where A is the nozzle-dependent proportionality constant. The onset of atomization is observed when (28)

$$\left.
\begin{array}{ll}
\left(\dfrac{\rho_L}{\rho_p} \right)^{1/2} < K & \text{for } \Gamma > 1 \\[4mm]
\left(\dfrac{\rho_L We_1}{\rho_p Re_j} \right)^{1/2} < K & \text{for } \Gamma < 1
\end{array}
\right\} \tag{24}$$

where K is a geometry-dependent constant. The wind-induced breakup regions can be identified with the so-called sinous region (38). Another criterion for transition from a sinous regime to an atomization regime (39) reads

$$We_1 = 10^6 Re_j^{-0.45} \tag{25}$$

Most of the results just presented point out the strong effects of the gas phase density on jet destabilization, which should also have a significant effect on drop formation in SCFs.

Consider now the behavior of drops. Drop size is calculated directly from the original Rayleigh theory (29), so for low viscosity jets, we write

$$d = 1.89 d_j \tag{26}$$

where d_j represents jet diameter. For higher liquid viscosity (say, for $Oh_j > 0.1$), the Weber (30) theory predicts

$$d = 1.89 d_j \left(1 + \frac{3\mu_L}{\sqrt{\sigma \rho_L d_j}} \right)^{1/2} \tag{27}$$

In the Rayleigh or varicose breakup regime, the Duffie and Marshall equation (40), which is based on experimental data, predicts

$$d_G = 36 d_0^{0.56} Re_j^{-0.10} \tag{28}$$

where d_G is a geometric mean drop diameter. The drop size distributions are often expressed as the normal, log-normal, Rosin–Rammler and Nakiyana–Tanasawa distributions (41).

4.2. Drop Breakup and Dispersion by a Twin-Fluid Nozzle

The force balance or momentum balance equation describes the spatial motions of small particles or droplets in an unsteady, nonuniform flow. The acceleration of the particle mass increased by the added mass results from action of several forces on the particle. Assuming that there are not any nonlinear interactions between the various forces acting on the particle, one can write (42)

$$\rho_P V_p \left(1 + \frac{C_M}{\Psi} \right) \frac{d V_d}{dt} = F_d + F_g + F_L + F_S + F_H + F_W \tag{29}$$

where C_M is the added mass coefficient, Ψ denotes the ratio of particle (or drop) density to the continuous phase density, and V_p is the particle volume.

In Eq. (29), $\mathbf{V_d}$ represents the dispersed phase velocity, $\mathbf{F_D}$ is the drag force, $\mathbf{F_g}$ denotes the force of gravity, $\mathbf{F_L}$ is the lift force, $\mathbf{F_S}$ represents effects of the fluid stress gradients, $\mathbf{F_H}$ is the Basset history term, and $\mathbf{F_W}$ represents interactions with the wall. The review paper by Loth (42) presents and discusses all the forces present in Eq. (29). Here we limit ourselves to the most important effect of drag forces. In the case of spherical solid particles of diameter d, $\mathbf{F_D}$ can be expressed as

$$\mathbf{F_D} = -\frac{1}{8}\pi d^2 \rho_C C_D |\mathbf{V_d} - \mathbf{V_c}|(\mathbf{V_d} - \mathbf{V_c}) \tag{30}$$

where $\mathbf{V_d} - \mathbf{V_c}$ is the local difference of velocity between the particle phase and the continuous phase. For creeping flow (Stokes regime), the drag coefficient C_D is

$$C_D = \frac{24}{Re_p}, \quad Re_p \ll 1 \tag{31}$$

where

$$Re_p = \frac{|\mathbf{V_d} - \mathbf{V_c}|d\rho_c}{\mu_c} \tag{32}$$

and d is the particle diameter. At higher Reynolds number, inertial effects become influential and can be included by employing the correction factor f_{Re} with $C_D = (24/Re_p)f_{Re}$, where f_{Re} can be expressed by (43)

$$f_{Re} = A + BRe_p^m + CRe_p^n \tag{33}$$

with

$A = 1, B = 3/16, m = 1, C = 0$ for $Re_p < 1$

$A = 1, B = 0.1935, m = 0.6305, C = 0$ for $1 < Re_p < 285$

$A = 1, B = 0.015, m = 1, C = 0.2283, n = 0.427$ for $285 < Re_p < 2000$

$A = 0, B = 0.44/24, m = 1, C = 0$ for $2000 < Re_p < 3.5 \times 10^5$

A similar set of equations proposed in the context of atomization was given by Liu et al. (44). The correlations derived for spherical particles are a useful basis on which to develop drag expressions for nonspherical particles. Available methods are presented and compared by Chhabra et al. (45).

Consider now the situation when the particle is a liquid. Loth (42) discusses two regimes: $\Psi = \rho_L/\rho_c \gg 1$ (liquid drops in gas) and Ψ of order unity (immiscible liquid drops in a liquid or SCF) or less (gas bubbles in a liquid, $\Psi = \rho_g/\rho_c \ll 1$). In the case of $\Psi \gg 1$, internal circulation does not affect drag significantly and the drag coefficient expressions for a solid are often used, provided the droplet does not undergo significant deformation. Droplets of $\Psi \gg 1$ can be regarded as not deformed if $B < 0.14$, where B is the Bond or Eötvös number, $B = (|\rho_p - \rho_c|gd^2)/\sigma$. In quiescent flow with the velocity close to its terminal value, the shape of the drops is determined by the Bond and Morton number, $Mo = (g\mu_c^4|\rho_p-\rho_c|)/\rho_c^2\sigma^3$ (42,46), and so is the drag coefficient. Deformation is significant when $Mo < 1.2 \cdot \times 10^{-7}B^{8.15}$ for $B < 5$ or $Mo < 0.2 \times 10^{-7}B^{2.83}$ for $B \geq 5$ (42), with ellipsoidal shape for $B < 40$ and spherical cap for $B > 40$. Drop–environment interactions result in drop breakup and related creation of new surface area. In what follows we consider two kinds of drop breakup: the first resulting from the velocity difference between the drop and the continuous phase velocity and the second resulting from fluctuations in velocity and pressure. Consider drop breakup resulting from the velocity difference and resulting acceleration of droplets (47,48). Figure 11 illustrates basic phenomena involved in drop breakup. Disintegration of droplets becomes possible in some critical deformation stage (Figure 11a,b), while the deformation is due to the pressure difference generated by the flow. Atomization from the surface of large drops can result from presence of the boundary layer. As explained by Schlichting (49), some of the retarded fluid in the boundary layer (Figure 11c,d) is transported into the main stream when the phenomenon of "separation" occurs. This happens when a region with an adverse pressure gradient exists along the wall and the retarded fluid elements do not have enough kinetic energy to penetrate the region of increased pressure, whereupon the boundary layer is deflected from the surface. Then the boundary layer fluid can take some small drops from the liquid surface and convey them into the main stream. When the flow of a gas is parallel to the liquid surface, short waves are formed on the drop surface and liquid is stripped from the wave crests (Figure 11e). Probably the most important mechanism of drop breakage is related to the Rayleigh–Taylor instability (Figure 11f). Taylor (50) considered instabilities of the surface between two fluids in the case of two fluids that are accelerated in the direction perpendicular to this interface. He showed that this surface is unstable when the fluid of lower density accelerates the fluid of higher density. Figure 12, which shows how the above-mentioned mechanisms contribute to drop breakup, is a compilation of results and ideas presented by many authors (47,48,51). The catastrophic or explosive breakup presented in Figure 12 is related to the Rayleigh–Taylor instability just discussed. The forces required to breakup a drop are present until a velocity difference occurs between the drop and the surrounding fluid. Taylor (52) has given a criterion for drop breakup

$$\frac{\rho_L(\Delta V)^2 d}{4\sigma} > C_1 \tag{34}$$

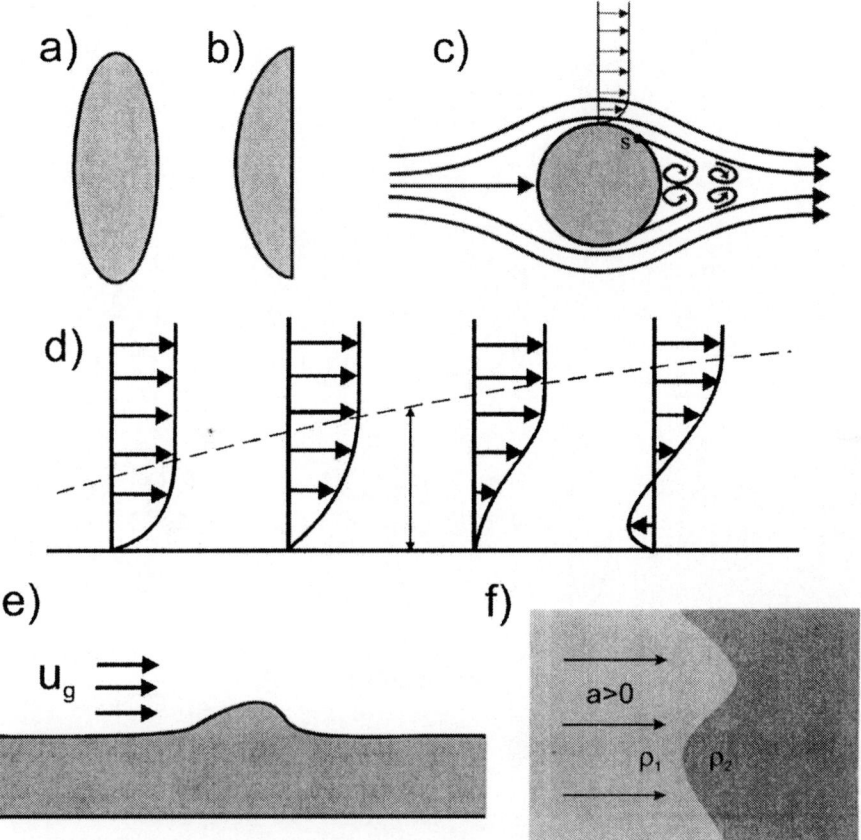

FIGURE 11 Phenomena connected with drop breakup: (a) deformation (splitting) of droplets, (b) deformation into lenticular shape, (c) development and separation of the boundary layer, (d) velocity distribution near the separation point of the boundary layer, (e) Kelvin–Helmholtz instability, and (f) Rayleigh–Taylor instability, $\rho_1 < \rho_2$.

with $C_1 \approx 2.7$ which gives the critical Weber number, above which drop breakup is roughly 12. Joseph et al. (48) proposed a criterion for drop diameter that is stable against the Rayleigh–Taylor instability

$$d < 2\pi \sqrt{\frac{\sigma}{\rho_L a}} \tag{35}$$

where a is the acceleration. When the average drop velocity is almost equal to the average velocity of the surroundings and the flow is turbulent, there still

a)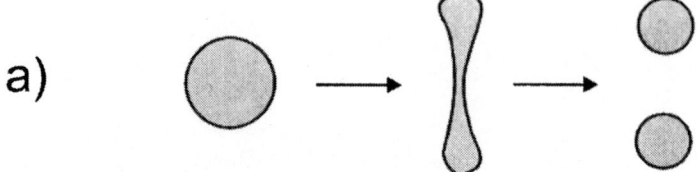

b)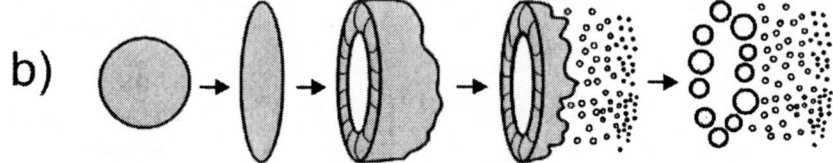

c)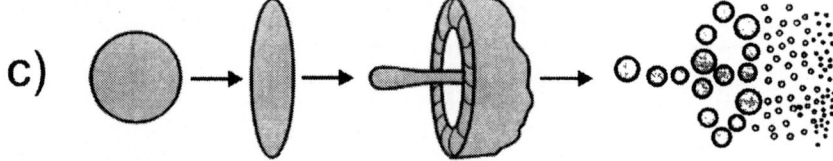

d)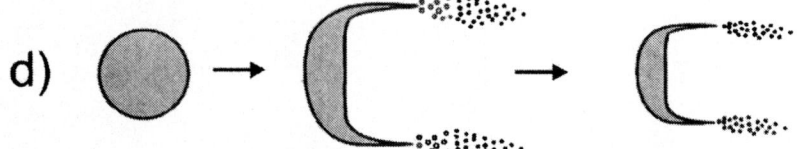

e)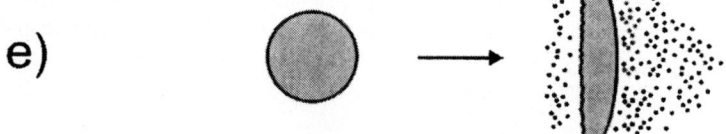

f)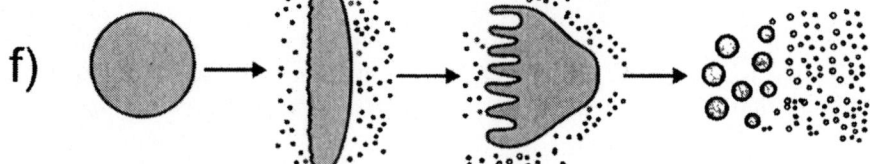

can be some drop breakage resulting from the action of turbulent stresses. The phenomenon of drop breakage by turbulent stresses has been a subject of many studies starting from fundamental papers by Kolmogorov (53) and Hinze (54). Fine-scale turbulence was interpreted in these studies by using the theory of Kolmogorov (55). According to this theory the equation for the maximum stable drop size for drops of low viscosity reads

$$d_{max} = \frac{C_X \sigma^{0.6}}{\varepsilon^{0.4} \rho_c^{0.6}} \tag{36}$$

Where ε is the rate of kinetic energy dissipation per unit mass and C_x is of order 1. Equation (36) is valid for drops whose diameter falls within the inertial subrange of turbulence, $\eta < d < L$, where L is the integral scale of turbulence and $\eta = v_c^{3/4}/\varepsilon^{1/4}$ is the Kolmogorov microscale of turbulence. The method proposed by Kolmogorov (53) and Hinze (54) was extended by Bałdyga and Podgórska (56) and Bałdyga and Bourne (57) to the case of more realistic intermittent turbulence, which was described by means of multi-fractal formalism. In this model drop size in the inertial subrange also depends on the integral scale of turbulence, which is related to the scale of the system. This new formalism predicts

$$d_{max} = C_X^{1.54} L \left(\frac{\sigma}{\rho_c \varepsilon^{2/3} L^{5/3}} \right)^{0.93} \tag{37}$$

for the maximum stable drop size and

$$g(d) = C_g \sqrt{\ln\left(\frac{L}{d}\right)} \varepsilon^{1/3} d^{-2/3} \int_{0.12}^{\alpha_Y} \left(\frac{d}{L}\right)^{[\alpha+2-3f(\alpha)]/3} \tag{38}$$

for breakup frequency. In Equation (38) C_g in the order of 10^{-2} to 10^{-3} is the proportionality constant, α is the scaling exponent, $f(\alpha)$ is the multifractal spectrum, and

$$\alpha_X = \frac{\frac{5}{2} \log \dfrac{L\varepsilon^{0.4} \rho_c^{0.6}}{C_x \sigma^{0.6}}}{\log \dfrac{L}{d}} - 1.5 \tag{39}$$

FIGURE 12 Mechanisms of droplet breakup and atomization: (a) vibrational breakup, (b) "bag", "hat", or "parachute" breakup, (c) "bag and stamen," "parachute and stamen," or "umbrella" breakup, (d) stripping of thin surface liquid layer, (e) wave crest stripping, and (f) "catastrophic" or "explosive" stripping.

Extensions of this theory to droplets of large viscosity and droplets smaller than the Kolmogorov microscale can be found in the literature (56,57).

Equation (35) for drop breakup caused by acceleration and Eqs. (37) to (39) to model drop breakup by turbulent stresses can be used to interpret drop behavior for the twin-fluid nozzle shown in Figure 7. The predicted size of the ethanol drops dispersed in supercritical carbon dioxide is compared with measured values in Figure 13. One can see that the model predicts well the

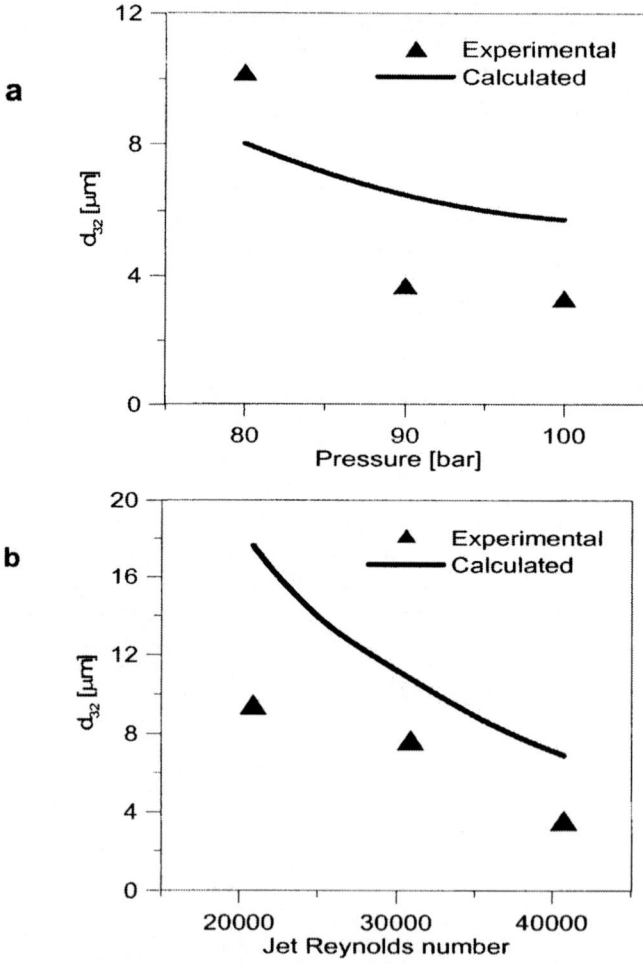

FIGURE 13 Drop breakup in the SEDS process: (a) effect of pressure in the particle formation vessel and (b) effect of the jet Reynolds number.

experimentally observed trends, although agreement is not perfect. An increase in pressure decreases drop size owing to decreased interfacial tension at higher pressure values (Figure 13a), and an increase of the Reynolds number of the jet, decreases the drop size by leading to increases in acceleration and in the rate of energy dissipation (Figure 13b).

4.3. Mass and Heat Transfer Between SCF Phase and Droplets

We are interested in the dispersion of a small amount of liquid solution in an SCF at $P < P_m$; that is, the system is far from thermodynamic equilibrium, with resulting heat and mass exchange between the drops and their supercritical environment. The simplest model that can be used in this case is the classical model of drop evaporation (58) that predicts a linear decrease of the squared droplet diameter in time and is often referred to as the "d^2 law":

$$d(t)^2 = d_{d0}^2 - K_V t \tag{40}$$

and

$$K_V = -\frac{d\left[d(t)^2\right]}{dt} = \frac{d_{d0}^2}{t_{vap}} = \frac{8\rho_{gs}D_{gs}}{\rho_L}\ln\left(1 + \frac{\langle C_P\rangle(T_\infty - T_{in})}{L}\right) \tag{41}$$

where d_{d0} is the initial droplet diameter, K_V is the evaporation rate constant (59,60), t_{vap} is droplet vaporization time, D_{gs} and ρ_{gs} represent the mass diffusivity in the gas phase at the droplet surface and the gas density at the droplet surface, respectively, $\langle C_P\rangle$ is an average value of the specific heat at constant pressure, and L is the specific latent heat of vaporization. Equations (40) and (41) are derived by assuming an isobaric, diffusion-controlled process, spherical symmetry (so, e.g., buoyancy effects are not included) and the process in quasi–steady state, which means that accumulation terms in the equations for the conservation of mass, momentum, and energy are neglected as small in comparison to the flux terms. Also the effects of the relative velocity between the drop and the fluid and effects of fluid deformation are neglected, although both are of importance in turbulent flows when droplets are large enough.

Equation (41) implies that mass transfer from a droplet is limited only by the heat of vaporization and is defined by the temperature difference $(T_\infty - T_{in})$ between the surrounding fluid and the droplet. For isothermal conditions, this simply means that mass transfer from the droplet is infinitely slow or heat transfer is infinitely fast. The model was used somewhat successfully to interpret experimental data for evaporating droplets as well as combusting with fuel supplied to the droplet surface (with the evaporation constant K_v replaced by combustion constant K_c). The success of such predictions can be explained by noting that after a short initial transient period, the drop reaches an almost steady temperature and in many cases droplet temperature is uniform (61,62). In the case of liquid–SCF interface, however, the d^2 law may not be applicable. The major differences between supercritical and subcritical behavior are related to mass transfer. Under supercritical conditions the effective length scales for heat and mass transport increase and decrease, respectively, in comparison to those at subcritical conditions. This means that the relative rate of mass transfer in relation to heat diffusion decreases under supercritical conditions, thus increasing the effective Lewis number.

An approach considered by Werling and Debenedetti (22) involved mass transfer between a liquid droplet and a supercritical phase on the basis of purely diffusive mechanisms under isothermal conditions. The authors solved the time-dependent conservation equations for species by assuming, spherical symmetry, equilibrium at the vapor–liquid interface, and stagnant droplets, with the only convective motion being that induced by diffusive mass transfer. The Peng–Robinson EOS was used to evaluate fugacity coefficients and the composition dependence of the gas and liquid densities. Application of the fugacity coefficients allowed simulation of the decrease to zero of the diffusion coefficient at the critical point. Computations were carried out for the carbon dioxide–toluene system. The results showed that the initial interfacial flux is always into the droplet, resulting in an initial increase in the droplet radius. Predicted droplet swelling was appreciable, with up to a fourfold increase in radius. After initial rapid diffusion of CO_2 into the droplets, the CO_2 concentration gradient decreases, and as the net diffusion becomes outward, the droplets start to shrink. Finally, mass transfer out of the 100% saturated droplets has been completed, and the drops shrink fast. This stage looks similar to droplet "evaporation"; however we use the term qualitatively here because there are considerable differences between evaporation into a gaseous and an SCF phase. It was pointed out (22) that swelling/evaporation behavior changes dramatically from one system to another owing to several competing factors, and so the phenomenon is case dependent. In addition, some fundamental limitations of this model related to the fluid dynamics and surface mass transfer may apply.

Let us consider experimental results obtained for an ethanol–CO_2 exchange. Figure 14a shows CO_2 bubbles injected into pure ethanol solvent. Compared with ethanol droplets injected into pure CO_2 under the same conditions (Figure 4a), the mass transfer is much more intense, and the concentration gradients are greater than those for a relatively slow evaporation of ethanol drops into the supercritical phase. In fact, this process is so intense that deformation and surface instability of the bubbles are clearly visible in

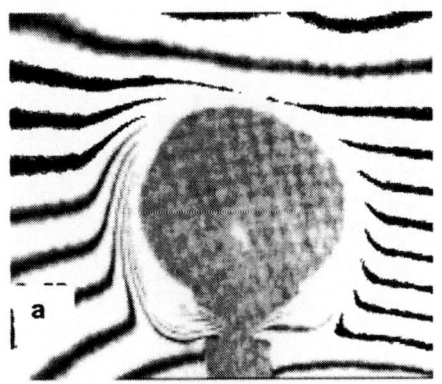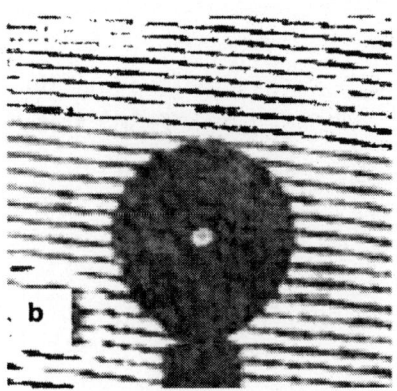

FIGURE 14 Images obtained by means of laser interferometry showing the behavior of CO_2 bubbles injected into ethanol for (a) pure solvent and (b) solvent saturated with CO_2; $T = 363.15$ K, $P = 8$ MPa, the bubble diameter is about 1.6 mm. The interferometric fringes show intensive mass transfer in the solvent phase.

Figure 14a, compared with the stable equilibrium bubble shape (Figure 14b) or the stable droplet shape during the evaporation stage in (Figure 4a). Thus the initial flux of CO_2 into ethanol at $P < P_m$ is much faster than the reverse process of ethanol evaporation, which can be explained by the higher equilibrium concentration of CO_2 in the ethanol-rich phase and also by the larger coefficient of internal mass transfer in comparison to the vapor phase. Clearly, fluid dynamics plays a very important role for both internal and external mass transfer, as illustrated by the very strong gravity convection (concentration plumes) clearly visible in Figures 4 and 14a.

Results of experimental investigations of the steady state mass exchange between an isolated droplet of ethanol vaporizing under normal gravity conditions and a quiescent supercritical carbon dioxide are shown in Figure 15. The droplets were hanging on the feeding capillary (as shown in Figure 4), and steady state conditions were achieved in such a way that the rate of ethanol feeding was equal to the rate of ethanol mass transfer to the supercritical carbon dioxide. This dynamic equilibrium was controlled by means of interferometry. The results show that mass exchange is dominated by the transfer of ethanol to the diffusion boundary layer of thickness δ_c. The first stage of ethanol "expansion" was completed even before the droplet had emerged from the capillary. It should be noted that the initial transient period in which droplets swell can be interpreted as a heating period when the droplet temperature quickly rises (62,63). No temperature gradient, however,

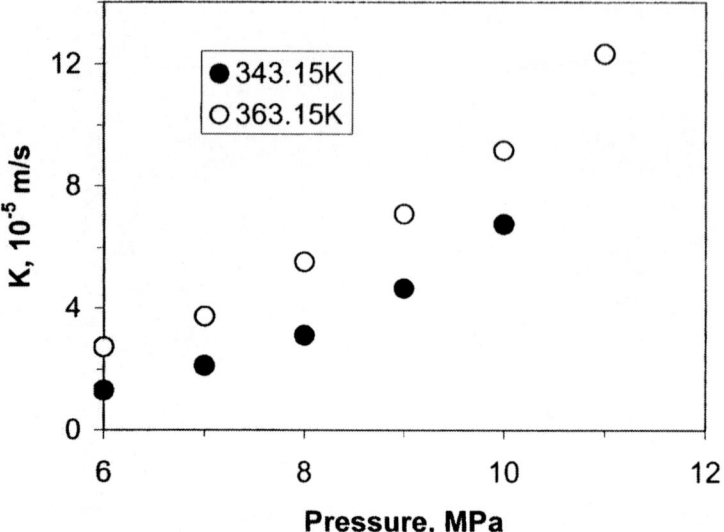

FIGURE 15 The overall mass transfer coefficients experimentally obtained for ethanol droplets suspended in pure CO_2, as a function of pressure at two different temperatures. The droplet diameter is about 1.6 mm.

was detected between the droplet and surrounding CO_2, indicating almost isothermal conditions for the second evaporative stage of mass exchange, consistent with predicted large Lewis numbers in supercritical CO_2. The overall mass transfer coefficient K, shown in Figure 15, increases with temperature, and also with pressure. This coefficient can be expressed by a series of mass transfer resistances comprising the interfacial mass transfer coefficient k_i, the mass transfer coefficient, k_L, effective within the liquid droplet, and the diffusion mass transfer resistance k_D within the vapor phase diffusion boundary layer

$$\frac{1}{K} = \left(\frac{1}{k_L} + \frac{1}{k_i} + \frac{1}{k_D M} \right) \tag{42}$$

where M is the surface distribution constant defined as $M = (M_L/M_g)(x_{gi}/x_{Li})(\rho_g/\rho_L)$, where x_{gi}, and x_{Li}, are the equilibrium ethanol mole fractions at the corresponding vapor and liquid sides of the interface, and ρ_g and ρ_L are the densities of the vapor and liquid phases. This equation is obtained by considering mass fluxes for each mass transfer step. For the conditions in Figure 15, the characteristic value of M is between 0.01 and 0.05.

The external mass transfer coefficient k_D can be defined by the Sherwood number Sh using the following equation:

$$Sh = \frac{k_D d}{D_V} = \frac{d}{\delta_c} \tag{43}$$

which gives the ratio between the droplet diameter and the characteristic thickness of diffusion layer. When very small droplets, of diameter smaller than the Kolmogorov microscale, $\lambda_K = (v^3/\varepsilon)^{1/4}$, are suspended in turbulent fluid, then assuming that they behave like small rigid spheres, one can calculate the external mass transfer coefficient from the theoretical relation developed by Batchelor (64):

$$Sh = 2 + 0.69 \left(\frac{d^2 \varepsilon^{1/2}}{D_V v^{1/2}} \right)^{1/3} \tag{44}$$

The term $(d^2\varepsilon^{1/2})/(D_V v^{1/2}) = d\Delta u/D_V$, where $\Delta u = (\varepsilon/v)^{1/2}d$, represents the Peclet number for fine-scale drop–turbulence interactions. In the case of larger droplets there is a significant velocity difference between the droplet and the continuous phase; then the approach presented in Section 4.2 and employing the relative velocity Reynolds number, Eq. (32), can be used, resulting in the following relation:

$$Sh = 2 + \frac{0.55 Re_p Sc}{\left(1.232 + Re_p Sc^{4/3}\right)^{1/2}} \tag{45}$$

The internal, liquid phase mass transfer coefficient k_L can also be calculated from classical relations. In the case of a completely stagnant spherical droplet, the analytical short-time solution (i.e., for Fourier number $Fo = D_L t/R^2 < 0.0293$) yields

$$Sh = \frac{k_L d}{D_L} = \sqrt{\frac{4}{\pi}} Fo^{-0.5} \tag{46}$$

The long-time solution is time independent and reads

$$Sh = \frac{k_L d}{D_L} = \frac{2}{3}\pi^2 \tag{47}$$

The dispersed phase mass transfer coefficient of a liquid drop with internal circulation (65) takes higher values than are observed in the case of the stagnant droplet. For example, the long-time solution takes the form

$$Sh = \frac{k_L d}{D_L} = 17.7 \tag{48}$$

Finally, there is the value of the interfacial mass transfer coefficient k_i. Sun and Shekunov (10), based on interferometric measurements of drop non-equilibrium surface concentrations proposed that at least for some solvents, there is a significant surface resistance to mass transfer. Experimental data on mass transfer surface resistance are rather scarce. In regular applications of chemical engineering it is usually assumed that the fluid phases are in equilibrium at the interface, so the resistance at the phase boundary is neglected. Also the model by Werling and Debenedetti (22) and most other theoretical work with SCF diffusion does not consider interfacial resistance. There are, however, indications that interfacial resistance can be at least as important as diffusive resistance in dense liquids or SCF systems. A classical interpretation of the interfacial resistance between liquid and vapor is presented by Sherwood et al. (66). Upon calculating the rate of collision of gas molecules with the liquid surface from the kinetic theory, it is possible to express the rate of evaporation and find the interfacial resistance. The coefficient k_i is usually very large for droplets in vacuum or low density gases, somewhat in the order between 10^{-3} and 10^{-2} m/s. It can, however, significantly decrease when surfactants are adsorbed at the surface or in a case of SCF–liquid mass transfer. Lockemann and Schlünder (15) measured interfacial resistance in the liquid phase between the supercritical CO_2 and either methyl myristate or methyl palmitate. They showed that this resistance can be expressed by means of the Ornstein–Zernike correlation length ζ (Section 2) which significantly increases near the mixture critical point where CO_2 solubility in the liquid phase is very large $[k_i \sim \exp(-\zeta)]$. If the Ornstein–Zernike correlation length ζ is increased by a factor of 4, the interfacial mass transfer coefficient k_i will be decreased by two orders of magnitude, whereupon the overall mass transfer coefficient is completely dominated by the interfacial resistance. Lockemann and Schlünder explain the interfacial resistance by citing the effect of the solvation of CO_2 molecules by liquid molecules, these then workers show that solvation at high pressures and the adsorption of surfactant at liquid interfaces are equivalent in their effects. It should be noted that their experiments showed that internal mass transfer coefficient (in the liquid phase) decreased with increased P_m. In the present experiments with ethanol droplets, external mass transfer into an SCF is more important than internal mass transfer. The behavior of the mass transfer coefficient in Figure 15 can be explained by using a model of molecular adsorption–desorption processes similar to those used for crystallization or dissolution (67). In this case there are solvation and energy activation barriers during the phase transition at the phase boundary:

$$k_i = B_T \exp\left[-\frac{(E_A + \Delta H)}{RT}\right] \tag{49}$$

Where E_A and ΔH are the corresponding activation energy and enthalpy of phase transition and the coefficient B_T defines the maximum probability that molecules will cross the interface between the liquid and SCF (vapor) phases. This simple relationship can explain the behavior of the mass transfer coefficient in Figure 15 when it is dominated by the interfacial resistance. Indeed, k_i increases with temperature T according to Eq. (49); also, both parameters E_A and ΔH should decrease with increase of pressure, since the structure and composition of the liquid and vapor phases become very similar to each other around the mixture critical point. The decrease of ΔH with pressure for the ethanol–CO_2 system has been confirmed by interferometric studies of jet mixing described in Section 3.2 and also by calorimetric measurements described by Cordray et al. (68). According to Eq. (43) the diffusion mass transfer coefficient k_D may also increase in parallel with k_i as a result of more intensive convection within the diffusion boundary layer.

The results of this section indicate that the models based on a purely diffusive mechanism of mass and heat transfer are incomplete because nonequilibrium surface concentration, interfacial mass transfer resistance, and the fluid dynamic regime around and within the droplets must be taken into account as described by Eqs. (42) to (49).

5. PARTICLE FORMATION

5.1. Precipitation In Jet Mixing

The process of crystallization or precipitation involves the nucleation and growth of crystals (particles) from a supersaturated solution and is often followed by the agglomeration of crystals. Supersaturation is the driving force for precipitation, which is defined by the difference between the real solute concentration c and equilibrium solute concentration $c_{eq}(P,T)$ at given pressure and temperature. The strict definition of supersaturation is linked to the difference of chemical potential $\Delta\mu$:

$$s_{aff} = \frac{-\Delta\mu}{kT} = \ln\frac{\gamma c}{\gamma_0 c_{eq}} \tag{50}$$

where c is the solute concentration, $c_{eq}(P,T,c_i)$ the solute equilibrium concentration, γ and γ_{eq} are the corresponding activity coefficients, and s_{aff} represents dimensionless affinity. In SCF solutions, the activity coefficient is a function of pressure, temperature, and mixture composition. For most antisolvent precipitation processes with CO_2, supersaturation is always generated by changing the concentration (c or c_{eq}) with only small variation of pressure and temperature. For very low concentrations of a nonvolatile solute, the $\gamma(c)$ dependence is often negligible, and $\gamma/\gamma_{eq} \approx 1$. Thus the simplified definition of supersaturation S

$$S = \frac{c}{c_{eq}} \approx S_a = \frac{a}{a_{eq}}$$

could be applied and directly calculated from experimental data. Supersaturation during the initial stages of precipitation can be characterized by the maximum attainable supersaturation S_m. This corresponds to the ideal case of CO_2–solution fresh feed being completely mixed on a molecular level and is calculated as (69)

$$S_m = \frac{c_0 Q_A}{c_{eq}(Q_A + Q)} \tag{51}$$

where Q_A and Q are the flow rates (mol/s) of the solution and CO_2, respectively, and c_0 is the molar feed concentration of solute.

The particles are created in supersaturated solution by nucleation events. There are various mechanisms of nucleation. In primary nucleation, creation of crystals is not affected by the presence of the crystallizing material itself. Primary nucleation may be homogeneous (when it occurs in the absence of any solid phase) or heterogeneous (when nucleation is facilitated by the presence of foreign solid particles). In secondary nucleation, formation of a solid phase is affected by the presence of the solid phase of the crystallizing material. In precipitation from an SCF, the supersaturation is typically high, so secondary nucleation either does not take place or has a negligible effect (70), and the most important mechanism is the primary homogeneous nucleation, which can be approximated by (69)

$$R_{N,homo} = R_{N,max} \exp\left\{-A[\ln(S)]^{-2}\right\} \tag{52}$$

where A is a constant proportional to the specific surface free energy of the nuclei and inversely proportional to T^3, whereas $R_{N,max}$ is the maximum possible nucleation rate (as $S \to \infty$). Heterogeneous nucleation that takes place on a foreign surface usually has much lower surface energy than that of solute crystals, which means that the A value is much smaller in the case of heterogeneous nucleation. Another difference between the two nucleation kinetics is that the heteronuclei are used up, thus limiting the maximum possible rate of heterogeneous nucleation $R_{N,max}$. The nucleation constant is defined as the time required to form a given number of nuclei: $\tau_N \sim 1/R_N$.

Crystal growth in a supersaturated solution is a complex process that occurs in many stages, including transport of solute from the bulk solution through the diffusion and adsorption layers, integration into the crystal lattice, and release and transport of crystallization heat. The simplest description of this complex process is based on diffusion reaction theories (67,70,71). According to these theories, the solute is transported to the crystal surface with driving force $c-c_i$ and then integrated into the surface with the driving force c_i-c_{eq}. These two stages can be represented by

$$N = k_D(c - c_i) = k_r(c_i - c_{eq})^g \tag{53}$$

where N is the molar flux (i.e., the number of moles diffusing and integrating per unit time and unit crystal area) k_D is a diffusion mass transfer coefficient k_r represents a rate constant for the surface integration step, and g is an order of surface reaction (often $g = 2$).

In practice, the rate of crystal growth can be expressed as the rate of displacement of a crystal surface or the rate of increase of a particle mass. Most frequently, the growth rate is defined as

$$G = \frac{k_a R_G}{3 k_V \rho_C} \tag{54}$$

where R_G is a mass deposition rate, k_V and k_a are the volume and surface shape factors, respectively, and ρ_c is the crystal density. The growth constant is defined as the time required to reach a certain mass of crystals: $\tau_R \sim 1/GA$, where A is the specific surface area of particles per unit volume of suspension (69). Precipitation results in the creation of a dispersed phase represented by a population of particles that, with time, change their spatial position as well as such properties as particle size, shape, and mass. The particles are created by nucleation and increase their size in the growth process; the identity of the particles in the population is then continuously destroyed and recreated by breakup and agglomeration phenomena. Because of the large number of particles, it is easier to consider not their individual histories but instead the properties of the population of particles. A predictive multidimensional particle distribution theory is based on population balance equations (72). For a single internal coordinate L (crystal size), the population balance takes the form

$$\frac{\partial \Psi}{\partial t} + \frac{\partial}{\partial x_j}\left(u_{pi}\Psi\right) + \frac{\partial}{\partial L}(G\Psi) = B - D \tag{55}$$

with

$$\Psi(L_0, \mathbf{x}, t) = \frac{R_N[S(\mathbf{x}, t)]}{G[S(\mathbf{x}, t)]} \tag{56}$$

where L_0 is a size of nuclei, $\mathbf{u}_p = (u_{p1}, u_{p2}, u_{p3})$ is a vector of particle velocity along external coordinates $\mathbf{x} = (x_1, x_2, x_3)$, and B and D are the birth and death functions representing the effects of agglomeration and breakage. To describe particle formation in turbulent flow, we use in what follows the moment transformation of the population balance for size independent growth

$$\frac{\partial m_i}{\partial t} + \frac{\partial}{\partial x_j}\left(u_{pj}m_i\right) = 0^i R_N + iGm_{i-1} + B_m - D_m \tag{57}$$

for $i = 0, 1, 2, 3, \ldots$ and with moments defined as

$$m_i(\mathbf{x}, t) = \int_0^\infty \Psi(L, \mathbf{x}, t)L^i dL \tag{58}$$

Equation (57) employs the local instantaneous values of velocity and concentration. To describe the effects on the process of turbulent fluctuations of the particle velocity as well as species and particle concentrations, one can use Reynolds-averaged form of the population balance

$$\frac{\partial \overline{m_i}}{\partial t} + \frac{\partial}{\partial x_j}(\overline{u_{pj}m_i}) = \frac{\partial}{\partial x_j}\left(D_{pT}\frac{\partial \overline{m_i}}{\partial x_j}\right)0^i\overline{R_N} + i\overline{Gm_{i-1}} + \overline{B_m} - \overline{D_m} \tag{59}$$

where D_{pT} is the local value of the particle turbulent diffusivity coefficient. The terms $\overline{R_N}$ and $\overline{Gm_{i-1}}$ can be calculated by using the concept of mixture fraction described in Section 3.2 as shown by Bałdyga and Orciuch (73). Of course any other probability density function (PDF) method can be used as well. Equation (59) should be solved together with the partial differential equations describing momentum, mass, and species balances and should be supplemented with specific kinetics for the nucleation, growth, agglomeration, and breakage of particles. One can then use moments $\overline{m_i}$ to calculate average particle size, for example $L_{32} = \overline{m_3}/\overline{m_2}$ or $L_{43} = \overline{m_4}/\overline{m_3}$.

5.2. Comparison of Experimental and Theoretical Data for Jet Precipitation

We now consider an example of precipitation by means of the solution enhanced dispersion by supercritical fluids (SEDS) process to the precipitation from an ethanol solution of model drug, acetaminophen. Details of the experimental method are presented in the literature (69). The aim of this study was to investigate the effects of flow rate through a coaxial nozzle (74) on the particle size and shape. There is an intensive mixing of fresh substrate solution with the supercritical antisolvent in the coaxial passages for the solution and the antisolvent, respectively. The mixing process was analyzed by means of CFD simulation (75,76). The k-ε model available in Fluent 6.0 and the mixing models discussed in Section 3 were applied in these computations. The Favre-averaged form of Eq. (59) was used in the modeling assuming, $\tilde{u}_{pj} = \tilde{u}_j$ and $D_{pT} = D_T$, which results from the fact that the velocity of very small particles differs hardly at all from the fluid velocity. Agglomeration and breakage phenomena were neglected, and supersaturation was calculated from the local concentrations of acetaminophen and ethanol, whereas the mean concentrations of the local species were calculated from species balances. For example, the acetaminophen concentration was calculated from

$$\frac{\partial(\bar{\rho}\tilde{c}_c)}{\partial t} + \frac{\partial}{\partial x_j}(\bar{\rho}\tilde{u}_j\tilde{c}_c) = \frac{\partial}{\partial x_j}\left[\bar{\rho}(D_m + D_T)\frac{\partial\tilde{c}_c}{\partial x_j}\right] - \bar{\rho}k_a\rho_p\frac{\widetilde{Gm_2}}{2M_c} \tag{60}$$

The results here are given for a fixed mixture composition (i.e., for the same proportion of mass fluxes of the carbon dioxide Q and acetaminophen solution in ethanol Q_A) but at different flow velocities, expressed through the Reynolds number $Re = u_0d_0/v$, where u_0 is the nozzle fluid velocity, d_0 is the nozzle diameter, and v represents kinematic viscosity. Examples of computations were shown in Figures 5 and 7. Figure 7a,b showed that the most intensive premixing occurs the nozzle chamber, and the highest supersaturation is produced there. Figure 5a showed that in the particle formation vessel there is intensive circulation in the region of the jet, which means that backmixing of the fresh fluid from the nozzle with the reservoir fluid is occurring in the system . This dilution of the supersaturated solution reduces supersaturation. A decrease in temperature results from the heat of mixing and from the Joule–Thomson effects (Figure 7c). It is interesting to note that supercritical carbon dioxide penetrates the core of the solution flow, so precipitation may start in the feeding tubes themselves. Computed distributions of the acetaminophen supersaturation S for $Re = 1000$ and $Re = 30,000$ are shown in Figure 16. An increase of the flow rate by a factor of 30 increases the maximum supersaturation from 8.2 to 11.5 and also increases the extent of the zone of high supersaturation but decreases the residence time in this region. Afterward the fluid is cooled in the nozzle owing to the Joule–Thomson effect, and this cooling further increases supersaturation.

Results of experimental investigations are compared with model predictions using CFD and also an analytical model based on different scales of inhomogeneity and time constants (69), as shown in Figure 17. The rapid decrease of particle size at $Re < 10,000$ is caused by the establishment of a supersaturation profile (in fact concentration and temperature profiles). Up to $Re \approx 10,000$ maximum, supersaturation increases in the system with increasing Re, which results in a decrease of particle size because the rate of nucleation is more sensitive to supersaturation than the rate of crystal growth. At $Re > 10,000$, the dominant effect is the decreasing the residence time of fluid elements in the region of high supersaturation, Hence decreasing the time available for nucleation in regions of very high supersaturation with increasing Re, and particle size increases again. Experimental data and model

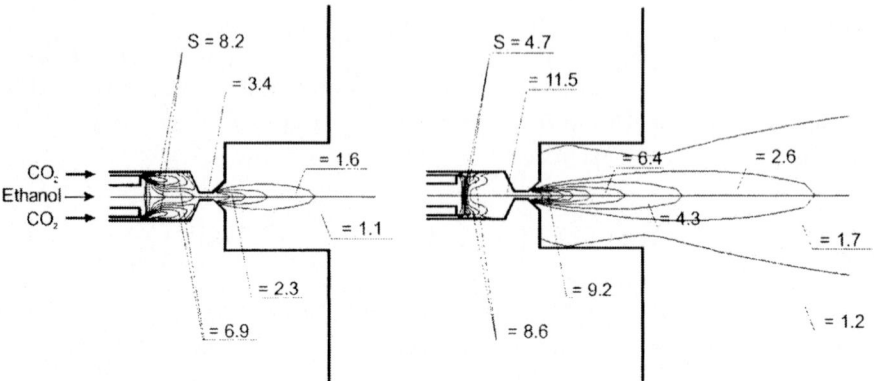

Figure 16 Supersaturation profiles during precipitation of acetaminophen by means of a two-component mixing nozzle at (a) Re = 1000 and (b) Re = 30,000. Pressure is 20 MPa, temperature 323.1 K. The volumetric flow ratio between CO_2 and the ethanol solution is constant at $Q/Q_A \approx 100$.

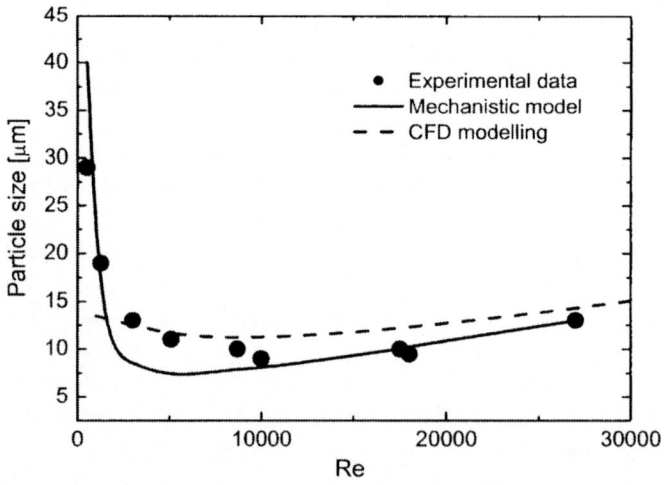

Figure 17 Comparison of experimental data with model predictions for mean particle size of acetaminophen crystals obtained by jet mixing, from direct CFD calculations of macromixing and supersaturation profile in Figure 16 and from mechanistic modeling (44–46), which allows for inclusion of the effects of molecular viscosity and diffusivity on rate of mixing on molecular scale (important for low Re).

predictions agree well with the foregoing discussion, although it is clear that the CFD model better predicts behavior at large Re, where macromixing effects dominate. The analytical model allows one to incorporate specific features of micromixing that are more important for small Re and therefore better predict the formation of large particles. Ideally, more complex CFD models can incorporate the effects of molecular viscosity and diffusivity on the rate of mixing on the molecular scale as discussed in Section 3.2, providing a complete solution for particle size distribution. Variation of Re also affects particles shape; at $Re \ll 10{,}000$, needlelike crystals are produced, whereas at $Re > 10{,}000$, more compact, isometric crystals are observed (69). This results from the fact that the initial level of supersaturation also affects particle morphology, as discussed in Section 6.

5.3. Precipitation in Droplets

We consider now precipitation in a two-phase system under supercritical conditions for one of the phases and $P < P_m$. Let us consider particle formation in which the substrate solution is sprayed into the supercritical fluid. In this case the nucleation and immediate postnucleation growth are confined within the ethanol-rich phase and supersaturation is generated within solution droplets as a result of droplet "expansion," but more importantly because of droplet "evaporation," as explained in Section 4.3. The evaporation stage is more important because the equilibrium concentration of CO_2 in the droplets is usually not high enough to initiate nucleation. For example, acetaminophen can precipitate at CO_2 concentrations in the ethanol-rich phase above 90% mole fraction (7). This is possible only very close to the mcp, where the droplets also become unstable (see Section 2). Near the nozzle, the Reynolds number of the droplets is large and the concentration of the solvent in the supercritical CO_2 is small; therefore the main mass transfer resistance for absorption of CO_2 is in the liquid phase. The ethanol evaporation is controlled by the combined mass transfer coefficient K, which includes the interfacial and diffusion transfer resistance. It is interesting to compare characteristic time scales for droplet saturation and evaporation. The characteristic time constant for droplet saturation with CO_2, τ_D, can be calculated by using the liquid phase mass transfer coefficient k_L.

For the completely stagnant droplets of diameter equal to 10 μm, the value $\tau_D = 0.4$ ms is obtained for $D_L = 10^{-8} m^2/s$ and $\tau_D = 4$ ms for $D_L = 10^{-9} m^2/s$. This agrees with an estimate of Wubbolts et al. (77), where for stagnant droplets of diameter 35 μm and $D_L = 2 \times 10^{-9} m^2/s$, the time necessary to obtain 90% saturation was in the order of 40 ms. When there is circulation within the droplet, the time constants τ_D would be decreased by a factor of 2.7.

Interfacial resistance is present in the thin layer, the coefficient K is almost constant, and one can neglect the effect of drop curvature on the total resistance. For a CO_2-saturated ethanol droplet of volume V and surface A the ethanol balance reads

$$\frac{d(Vc_L)}{dt} = -AKc_L \tag{61}$$

where c_L represents ethanol concentration within the droplet, and ethanol concentration in the bulk of CO_2 is assumed to be negligible. From Eq. (61) we obtain a relation for droplet shrinkage rate

$$\frac{d[d(t)]}{dt} = -2K \tag{62}$$

that results in a linear drop in shrinkage when the interfacial transfer resistance controls drop evaporation:

$$d(t) = d_{d0} - 2Kt \tag{63}$$

This linear shrinkage law differs this case from the "d_{d0}^2 law" represented by Eq. (40). Equation (63) shows that the droplet lifetime is equal to

$$t_{vap} = \frac{d_{d0}}{2K} \tag{64}$$

Moreover, for droplets of diameter $d_{d0} = 10\ \mu m$ and for $T = 353.15$ K, $P = 8$ MPa (so for mass transfer coefficient $K = 5 \times 10^{-5}$ m/s) results in $t_{vap} = 0.125$ s, which is much larger than the saturation time τ_D. Because the equilibrium concentration of CO_2 in the droplet is too small to create a supersaturated solution of the solute, the supersaturation of the nonvolatile solute is created by its concentration due to solvent evaporation from the droplet

$$\frac{dc_c}{dt} = -\frac{d \ln V}{dt} c_c = -\frac{3}{d}\frac{d[d(t)]}{dt} = \frac{6Kc_c}{d_{d0} - 2Kt} = \frac{3c_c}{t_{vap} - t} \tag{65}$$

which after an integration gives

$$c_c = c_{c0}\left(\frac{t_{vap}}{t_{vap} - t}\right)^3 \tag{66}$$

with t_{vap} given by Eq. (64).

Equations (65) and (66) show that the process of creating supersaturation within the droplets is much slower than in homogeneous mixtures (i.e., at pressures above the critical mixture pressure). Increase of the solute concentration from c_{c0} to $2c_{c0}$ is observed after $0.206 t_{vap}$, which for $t_{vap} = 0.125$ s gives 26 ms compared with micromixing time values of order of 0.01 ms (69). This observation is confirmed by optical measurements that point to relatively long evaporation times for droplets dispersed in a jet, shown by relatively long (in the order of 10 cm or more) spray patterns observed in supercritical CO_2 at $P < P_m$ (7). Assuming that the vapor phase diffusion boundary layer resistance is controlling and that the droplets are smaller than the Kolmogorov microscale during evaporation one gets for the same conditions considered earlier droplet evaporation times in the range of 0.025 to 0.050 s; again this is longer than the saturation time τ_D, but shorter than the time resulting from interfacial resistance and so is not considered in the order-of-magnitude analysis presented in this section. When, however, there is no interfacial resistance, the external diffusion boundary layer resistance controls the rate of evaporation.

Once supersaturation has been created, crystallization occurs by nucleation and growth of crystals within the droplets. At high supersaturation, nucleation can be homogeneous, in most cases, however, the solution contains many very fine foreign solid particles that become nucleation-active centers for heterogeneous nucleation that dominate the process at lower supersaturation. When the liquid is divided into so many small droplets that their number exceeds the number of foreign particles, some of the droplets will not contain impurities and the solute will crystallize within such droplets by an apparently homogeneous mechanism. This shows that in some fine droplets, nucleation may start later than in other droplets, with obvious consequences for the particle size distribution and morphology.

In some cases another kind of heterogeneous nucleation may be observed at the droplet surface. High solute concentration is created at the drop surface, and when the droplet is small enough, so that the characteristic time of the solute diffusion $\tau_{Dc} = [d(t)]^2/D_c$ is smaller than both the evaporation time and the nucleation induction time, then depending on the droplet size and supersaturation, the process can proceed either by the mononuclear mechanism with a single crystalline particle per droplet or by the polynuclear mechanism, resulting in polycrystalline particles formed from many nuclei created within the droplet. In the case of large droplets, the diffusion time τ_D for the solute can be much larger than the induction time, and then the mass transfer within the drop can control the process, resulting, for example, in hollow particles.

In the case of droplets of volume so small that their content can be regarded as uniform, the process can be modeled by applying to each droplet a macroscopic population balance (72) in the form

$$\frac{\partial \Psi}{\partial t} + \frac{\partial}{\partial L}(G\Psi) + \Psi \frac{d \ln V}{dt} = B - D \tag{67}$$

with

$$\frac{d \ln V}{dt} = -\frac{1}{t_{vap} - t}$$

and the boundary condition

$$\Psi(L_0) = \frac{R_{N,homo}(S) + R_{N,hetero}(S)}{G(S)}$$

where L_0 is the size of the nuclei. This equation should be solved together with the dynamic balance of the solute being crystallized

$$\frac{dc_c}{dt} = -c_c \frac{d \ln V}{dt} - \frac{k_a \rho_p G \int_0^\infty \Psi L^2 dL}{2M_c} = \frac{3c_c}{t_{vap} - t} - \frac{k_a \rho_p G \int_0^\infty \Psi L^2 dL}{2M_c}$$

(68)

and also with balances for the solvent and antisolvent.

At the initial evaporation period there is no crystallization within the droplet, and Eq. (68) is equivalent to Eq. (65), yielding the solute concentration given by Eq. (66). When crystallization starts, there is competition between the first term on the right-hand side of Eq. (68) (creation of supersaturation) and the second term (consumption of supersaturation). For smaller droplets of smaller evaporation time t_{vap}, supersaturation occurs more rapidly and at a higher level, which obviously promotes homogeneous nucleation and results in the creation of smaller primary particles created in initially smaller droplets. Because the creation of supersaturation in droplets is relatively slow, there is a good chance for heterogeneous nucleation to start in the droplets when there are many foreign particles. The concentration of impurities per unit volume is usually independent of droplet size, nor should the size of primary particles depend on droplet size in this case. For example, the experimental results given by Shekunov et al. (20) indicate that for acetaminophen, the primary particle size is a very weak function of both pressure and temperature at $P < P_m$. Since strong agglomerates are created mostly within droplets (high supersaturation is necessary to cement the primary particles), their mass decreases with decreasing the drop size, which depends on the amount of solute available in the droplet. The rate of creation of supersaturation by drop shrinkage can then affect the structure of agglomerates; for faster shrinkage, one can expect the agglomerates to have a higher fractal dimension and thus smaller size at the same mass. These observations have been confirmed by experimental results (7): the measured size of the aggregates increased with solute concentration and rapidly decreased with the nozzle Re number. The size of agglomerates in this work varied between 1 and 100 primary particles, which corresponded well to the predicted droplet size at different Reynolds number.

5.4. Particle Aggregation

Precipitation results in formation of a solid–liquid dispersion. Small particles forming this dispersion tend to aggregate (i.e., to form larger entities called agglomerates). The rate of particle agglomeration depends on particle size, shape, and number concentration, as well as composition of the solution and structure of the flow. Agglomeration is affected by attractive and repulsive colloidal forces, hydrodynamic forces, and the inertia of the particles. Smoluchowski (78) reported two limiting types of agglomeration for colloidal particles:

1. *Perikinetic, where agglomeration takes place as a result of Brownian motion.* The following equation was proposed for determining the collision rate of particles (78)

$$N_{ij,P} = 4\pi D_{ij}^{\infty}(a_i + a_j)c_i c_j \tag{69}$$

where $N_{ij,P}$ is the number of collisions of particles of radii a_i and a_j, respectively, per unit time and unit volume, c_i represents the number concentration of particles of radius a_i, and D_{ij}^{∞} is the mutual Brownian diffusion coefficient

$$D_{ij} = D_i + D_j = \frac{kT}{6\pi\mu}\left(\frac{1}{a_i} + \frac{1}{a_j}\right) \tag{70}$$

2. *Orthokinetic, where agglomeration is dominated by convection.* That is, the particles are transported by movement of the surrounding fluid (78). The analysis of Smoluchowski, which was limited to laminar shear flow, was extended to turbulent flow by Saffman and Turner (79)

$$N_{ij,O} = 1.3\left(\frac{\varepsilon}{v}\right)^{1/2}(a_i + a_j)^3 c_i c_j \tag{71}$$

where ε is the rate of energy dissipation per unit mass and v represents the kinematic viscosity.

Both these equations neglect the effects of interparticle forces. In the case of a static fluid, the effects of interparticle forces can be included by dividing the Smoluchowski relation [Eq. (69)] by the Fuchs' stability ratio W as given by Spielman (80). Thus W depends on properties of the solution including its composition as well as properties of particle surface. When the fluid is in motion, the stability ratio W becomes dependent on the flow as well

(73). In the limit of a very high rate of energy dissipation and for large enough particles (i.e., for large enough values of the Peclet number $Pe \gg 1$)

$$Pe = \frac{(\varepsilon/\nu)^{1/2}(a_i + a_j)^2}{D_{ij}} \tag{72}$$

one obtains $N_{ij,O} \approx N_{ij,P}/W$. When collisions of particles in fluids of high density are considered, effects of particle inertia are usually negligible. However, when particle density is large in comparison to fluid density, the effects of inertia must be included. This can be done by using theories explained in Refs. (79 to 82).

However, not every collision leads to agglomeration. When there is enough time during the collision and enough material (i.e., when the solution is supersaturated), a crystalline bridge can be built between the particles, forming agglomerates that are strongly bonded. Otherwise either the collision is unsuccessful or a weak aggregate bound by weak bonds (van der Waals forces) is formed. To express the rate of agglomeration β_{ij} in relation to the rate of collision N_{ij}, the concept of probability or efficiency of agglomeration, Pa, must be used as well

$$\beta_{ij} = N_{ij} \cdot Pa \tag{73}$$

The efficiency of agglomeration, Pa, depends on the supersaturation, collision geometry, and details of the flow (83–85), namely, Pa increases with increases in the supersaturation ratio S and decreases with increasing rate of energy dissipation ε.

Analysis of the kinetics of agglomeration just presented enables us to interpret the effects of supersaturation on particle size as shown in Figure 18. At low supersaturation, heterogeneous nucleation dominates. The rate of nucleation is then less sensitive to supersaturation than the rate of crystal growth, so increasing of initial supersaturation affects the rate of crystal growth more than the rate of nucleation, and larger crystals are produced. At higher initial supersaturation, homogeneous nucleation dominates and the rate of nucleation is very sensitive to the supersaturation. Thus more and smaller particles are produced if the initial supersaturation is continually increased. At still higher initial supersaturation, a large number of small particles [high c_i and c_j in Eq. (71)] results in large number of collisions. Owing to high supersaturation, agglomeration efficiency [Pa in Eq. (73)] is also high, and an increase of agglomerate size with increasing initial supersaturation is thus observed. Agglomeration of large particles in the orthokinetic regime is affected by hydrodynamics. Increasing the rate of energy dissipation increases the rate of collision but decreases agglomeration efficiency Pa, limiting the maximum attainable size of agglomerates. The isometric and amorphous par-

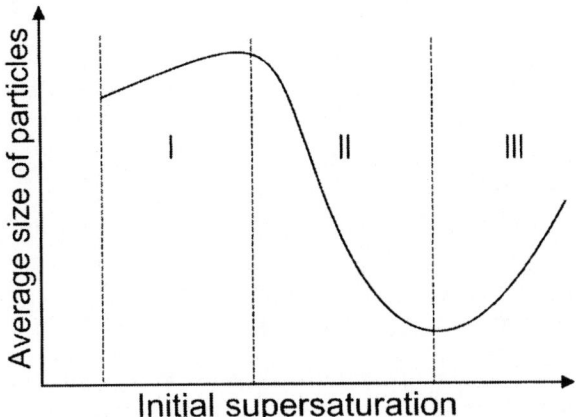

FIGURE 18 Effect of initial supersaturation on the average particle size: I, sub-range of heterogeneous nucleation; II, homogeneous nucleation dominates; III, agglomeration dominates.

ticles often agglomerate (region III in Figure 18), forming particles of a very complex structure. The aggregate shapes are often characterized by agglomerate fractal dimensions. A mass fractal dimension of an aggregate D_F is related to its mass m by

$$m \sim L^{D_{\mathrm{F}}} \tag{74}$$

where L is the aggregate size. The lower the fractal dimension, the more open the agglomerate structure. In practice, fractal dimensions, in the range of 1.8 to 3 are observed, with D_F strongly related to agglomeration mechanism (86).

Numerous examples of particle aggregation in SCF precipitation described in the literature. Chapter 6 by Shekunov shows how to optimize the size of respiratory particles by operating at the minimum particle size shown in Figure 18. Supersaturation during SCF precipitation is usually high, therefore particles can be produced on the either side of the minimum in Figure 18. As a result different dependencies of particle size on supersaturation (or more often on solute concentration) are obtained. Bristow et al. (7) showed that for small solute molecules like acetaminophen, particle size decreases at higher concentration for precipitation above P_m; Owing to the aggregation mechanism in the droplets, however, the size increases when precipitation occurs below P_m. There are several distinct situations in which the aggregation mechanism dominates. For example, precipitation of polymer or proteins above P_m, usually results in the formation of nanoparticles.

According to Eqs. (70) and (72), aggregation approximately follows the d^{-6} law in relation to particle diameter. Such strong dependence leads to the aggregation of nanoparticles, in particular materials with strong interparticle interactions, reflected in the high values of Pa number. As a result, most proteins and polymers exhibit an increase of particle size with solute concentration. This aggregation can usually be reduced by optimizing the fluid dynamics or by precipitation with additives or surfactants.

6. IMPORTANCE OF CHARACTERISTIC MATERIAL AND TIME CONSTANTS FOR PRECIPITATION

As shown earlier, particle size and shape produced by SCF precipitation depend on fluid thermodynamics and the mixing and dispersion processes as well as on material properties. In this section we summarize the most important aspects of this behavior.

One of the most significant material constants is the equilibrium solubility c_{eq}. A high solubility in SCF (the indicative threshold value for most applications can be taken to be $c_0 \approx 10^{-4}$ mole fraction resulting in supersaturation $S_m < 2$) means that the supersaturation is too small to initiate homogeneous nucleation, and very large particles, in small yield, will be produced. For most SCF precipitations, S_m is typically in the range between 4 and 50. The nucleation rate is a very sensitive function of supersaturation, and it is emphasized that values of c_{eq} and S_m must be measured in modified (e.g., CO_2 + ethanol) fluid because even under very fast mixing, precipitation occurs in the solvent-enriched SCF phase. The c_{eq} threshold defines what precipitation method is to be used: RESS is preferable for compounds with high solubility; however, most classes of pharmaceutical materials are only sparingly soluble or practically insoluble in CO_2, and therefore antisolvent techniques are indicated.

Other important material constants are the time constants for nucleation (τ_N) and growth (τ_G). A small τ_N means a high nucleation rate at relatively low supersaturation and, consequently, smaller particles. On the other hand, a large τ_G implies small particle growth rate (e.g., "arrested" growth when certain surfactants or additives are used), which also results in small particle size. For precipitation at $P > P_m$, these time constants should be compared with a characteristic time of mixing τ_M, which comprises macro-, meso-, and micromixing stages as defined by Shekunov et al. (69). If $\tau_M \ll \tau_N$, the mixing is very fast and more uniform small particles are produced, whereas the condition $\tau_M \gg \tau_N$ leads to larger particles and greater sensitivity of particle size and shape to fluid dynamic conditions. For precipitation in droplets, at $P < P_m$, the precipitation time constants should be compared with the characteristic time of droplet evaporation t_{vap}, which increases linearly with droplet diameter. Very often, a decrease in mixing or dispersion also results in a significant decrease of the product yield because the effluent fluid is extracted from the vessel without particle precipitation (69). For inefficient mixing at small Re, this effect is also predicted from the CFD computation in Figure 16. For the majority of SCF applications, supersaturation is high and τ_N is comparable to the characteristic mixing time. Therefore, even for very fast nozzle mixing, fluid velocity and nozzle design must be optimized, as shown, for example by Shekunov in Chapter 6.

The molecular structure of compounds plays an important role in precipitation. Most pharmaceutical applications, process are relatively large synthetic organic drug molecules or macromolecules (e.g., polymers or proteins) that form molecular bonds, with the notable exception of some salts in which ionic interactions are dominant. As an empirical rule, the simpler the molecule or structural symmetry, the higher the particle crystallinity, the smaller the nucleation constant τ_N, and also the smaller the growth constant τ_G. For example, the relatively small acetaminophen molecule forms a highly crystalline structure, and competition between its small τ_N and τ_G constants leads to production of micrometer-sized particles in the range of 5 to 30 μm by volume, depending on the mixing conditions (Figure 17). For such molecular crystals, the particle shape is often defined by periodic chains of the strongest intermolecular bonds (e.g., hydrogen bonds), leading to different relative growth rates of the most important crystal faces (67). Low supersaturation often results in blocking of more slowly growing faces, producing anisometric or needle-shaped particles. At high supersaturation, however, all crystal faces tend to grow at similar rates because the same kinetic mechanism and diffusion conditions are present, and thus small isometric crystals (prisms) or even rounded particles can be produced. Therefore an increase in supersaturation is desirable to produce more isometric particles, which are preferable for drug delivery systems. For certain supersaturated solution compositions colloids (positively or negatively charged sols) may be formed as well. At still higher initial supersaturation, an amorphous precipitate (the metastable solid phase) is usually formed and afterward transforms slowly into the crystalline solid phase (87).

Some materials can be produced only in amorphous or polycrystalline phase. For example, SCF precipitation of polymers such as poly(L-lactic acid)(PLLA) and proteins such as lysozyme and insulin have been widely investigated (see, e.g., Chapters 6, 9, and 10 in this book). These molecules form amorphous (proteins) or semicrystalline (PLLA) structures and typically have small values of both c_{eq} and τ_N, combined with a relatively large τ_G constant (small growth rate). Similarly, organic or inorganic salts usually have very small c_{eq} and τ_N constants. This explains why it is easy to produce very small (often submicrometer), spherical particles of these materials, al-

TABLE 2 Qualitative Description of the Effect of Different Process Parameters on the Size and Morphology of Particles Obtained by Antisolvent Precipitation[a]

Process parameter	Effect on particles
Pressure, P	Transition between single-phase and two-phase mass transfer occurs at $P = P_m$. Crystallization tends to be more consistent, of higher yield and purity, and providing less agglomerated particles at $P > P_m$. The smallest particle size can be obtained at pressures slightly above P_m. However further increase of pressure may result in particle size increases and a more acicular particle shape because of increasing solubility in modified CO_2. Most applications require pressures below 20 MPa.
Temperature, T	Similar to pressure, temperature causes transition of the mass transfer mechanism at $T = T_m$. $T < T_m$ corresponds to crystallization in a single phase, affording well-defined particles. $T > T_m$ may lead to agglomerates. At $P < P_m$, an increase of temperature may result in a particle size decrease because of the increased drop in mass transfer. At $P > P_m$ an increase of temperature causes an increase in solubility and formation of larger particles. Typical and most feasible applications are below 363 K.
Solute concentration	Particle size typically exhibits a minimum because of the competing effects of increasing nucleation rate and increasing aggregation rate at high supersaturation.
Solution flow rate	Particle size typically shows a minimum because of the competing effects of increasing concentration and increasing solubility with increasing flow rate. High solution flow rates result in reduced yield and long acicular or irregular particles or aggregates.
CO_2 flow rate	Increase of CO_2 flow rate typically results in smaller particle size because of increased supersaturation and mass transfer. However, high relative ratios between CO_2 and solution flow rates decrease the particle production rate and may prove economically unfavorable.

TABLE 2 Continued

Process parameter	Effect on particles
Nozzle flow velocity	Particle size typically has a minimum for mixing nozzles because of the competing effects on supersaturation of increasing mass transfer and increasing flow dilution at high flow velocities. For capillary nozzles, there is usually a progressive decrease of particle size with nozzle velocity.
Nature of solvent	In addition to phase behavioral changes (P_m and T_m), particle size decreases for solvents with a lower saturation limit because of increased supersaturation. More volatile solvents tend to produce smaller particles because of the enhanced mass transfer rates.
Nature of solute	Compounds with lower equilibrium solubility in solvent-modified CO_2 produce smaller particles because intermittent supersaturation during mixing is higher. Materials with a larger nucleation rate and a smaller growth rate produce smaller particles. Amorphous or semicrystalline materials (e.g., polymers or proteins) tend to produce smaller and more aggregated particles than crystalline materials because of their reduced growth rates.

[a]The process is compound specific. A more detailed analysis is given in the text.

though aggregation represents a significant problem for these compounds, as discussed in Section 5.

The effects of such precipitation conditions as pressure, temperature, and solute and solvent concentrations are listed in Table 2. These effects are related to the behavior of the maximum attainable supersaturation S_m, as described by Equation (51) and also to the phase transition at the mcp between the jet mixing and droplet dispersion mechanisms. As follows from our earlier discussions in Sections 2 and 5, the most uniform particles can be produced under conditions very close to the mcp (practically, just above P_m), where the supercritical properties are most pronounced, local values of supersaturation are high, and the constants c_{eq}, τ_N, and τ_G are the smallest. Figure 19a shows that particles produced below the mixture critical pressure are more agglomerated than to the individual and separate crystal particles obtained at high pressure. This is because the nuclei confined within the droplets at $P < P_m$ coalesce and fuse during further growth. In this respect,

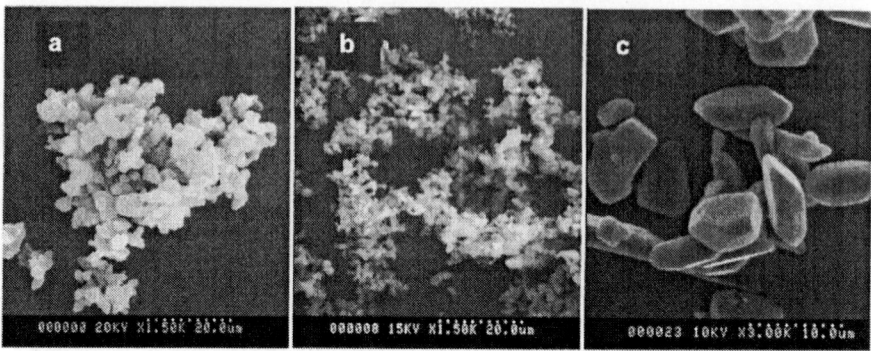

FIGURE 19 Acetaminophen particles produced (a) below, (b) in the vicinity of, and (c) above the mixture critical pressure.

the average size of the agglomerates formed decreases with the mean droplet size. Indeed, experiments with different nozzle sizes have shown that the average size of fused agglomerates decreases with increasing nozzle Reynolds number (7). In contrast, at $P > P_m$, there is rapid mixing between the drug solution and CO_2 before precipitation occurs. The volumetric flow of CO_2 is typically between one and two orders of magnitude greater than that of the drug solution; correspondingly, the agglomeration is less likely to proceed in the CO_2-rich media, although such mixing can produce larger primary particles, as shown in Figure 19c. The "best of both worlds" is achieved just above P_m: the local "hot spots" of drug concentration are still high, as in the droplets, whereas the mass transfer is as fast as in the completely miscible solvents, producing small, uniform, nonaggregated particles (Figure 19b). More detailed experimental descriptions of these phenomena can be found in the literature (7,69). Chapter 6 by Shekunov gives an example of optimization of particle size for a respiratory drug based on behavior of S_m, as a function of solution flow rate, solute concentration, and nozzle length.

7. GENERAL SCALE-UP PRINCIPLES FOR PRECIPITATION PROCESSES

Scaling up of chemical and pharmaceutical industry apparatus is a major task for chemical engineers, representing as it does the fundamental step in the realization of industrial plants. The objective of process development is the definition of chemical and physical processes in sufficient detail to build a plant that will operate with predictable output and product quality at a predictable cost that is as small as possible (88). This means that the apparatus should operate in a predictable fashion on the large, industrial scale, perhaps orders of magnitude larger than the laboratory scale. However, changing the scale affects the processes of mass, momentum, and heat transfer and the resulting concentration, stress, and temperature distributions, as well as features of the populations of entities present in the system (particles, drops, bubbles, microorganisms).

To scale up the process, one can use an empirical approach: that is, carry out the process at a number of scales, and collect enough information to be able to use the principles of similarity to make an empirical prediction of system performance on a larger scale (extrapolation). This method requires several empirical scale-up steps including laboratory, and pilot-scale experiments and extrapolation that is, despite these investigations, limited. The laboratory system does not need to be similar to the industrial plant, but it must be designed in a way that gives the best information. For example, the effects of fluid mechanics and mass transfer transport processes should be well defined or eliminated to identify the process kinetics. In the case of the antisolvent precipitation process, one should identify crystal nucleation and growth kinetics or at least determine the effects of supersaturation level and history on product particle size and morphology.

A pilot plant is often used to test the technology that will be used on an industrial scale to check the effects of phenomena that are not present or not influential at the laboratory scale. Extrapolation from the pilot scale to the industrial scale calls for the use of principles of similarity that can be obtained by means of the method of dimensional analysis. If the differential equations governing a considered process are given, dimensional analysis allows the determination of dimensionless parameters characterizing the process and helps to give some physical interpretation to those parameters. For example, from the Navier–Stokes equation [Eq. (7)], we can define the inertia force $\rho_0 u_0^2 / L_0$, the pressure force $\Delta P_0 / L_0$, and the viscous force $\mu_0 u_0 / L_0^2$, where L_0, u_0, ΔP_0, ρ_0, and μ_0 are the reference or characteristic values. Dividing the inertia force by the viscous force, one obtains the Reynolds number, $Re = u_0 L_0 \rho_0 / \mu_0$, as the similarity criterion for the flow. This means that two flows in geometrically similar systems I and II of the scale L_{0I} and L_{0II}, respectively, have similar flow patterns, that is,

$$\frac{u_{jI}(x_1/L_{0I}, x_2/L_{0I}, x_3/L_{0I})}{u_{0I}} = \frac{u_{jII}(x_1/L_{0II}, x_2/L_{0II}, x_3/L_{0II})}{u_{0II}}$$

if $Re_I = u_{0I} L_{0I} \rho_{0I} / u_{0I}$ is equal to $Re_{II} = u_{0II} L_{0II} \rho_{0II} / u_{0II}$. Thus such systems are regarded as dynamically similar. Similarly we can use Eq. (65) to define the characteristic time for convection L_0 / u_0, the characteristic times for molecular and turbulent diffusion L_0^2 / D_{m0} and L_0^2 / D_{T0}, respectively, and the characteristic time for particle growth $c_{c0}M_c/(\rho_p k_a m_2 G)$. The ratio of diffusion and convection time defines the Peclet numbers $Pe = u_0 L_0/D_{m0}$ and $Pe = u_0 L_0/D_{T0}$ for molecular and turbulent diffusional mixing, respectively. The identity of such Peclet numbers in systems of different scales yields similar concentration distributions if there is neither chemical reaction nor precipitation. In the case of precipitation also, the ratio of the characteristic time for crystal growth and the characteristic convection time should be the same in systems differing in scale. One should define more such scales, using all the balance equations [time scale for nucleation from Eq. (52), time scales for micromixing from Eq. (14), as well as time scales for energy transfer and energy generation from Eq. (16)] and using their ratios to define more dimensionless similarity criteria.

When there are no governing differential equations available, the Buckingham pi theorem can be used instead. This theorem states that any complete physical relationship (describing a process) can be expressed in terms of a set of independent dimensionless groups as discussed earlier and that the number of dimensionless groups i used to describe a process involving n variables is given by

$$i = n - r \tag{75}$$

where r is the number of fundamental dimensions (e.g., mass, length, time) that appear in each of the variables involved. Of course the pi theorem does not say how to select the significant variables; depending on the variables selected, different similarity criteria can be obtained. The proper choice of variables depends very much on personal knowledge, experience, and intuition to choose what is reasonable and what is not.

In geometrically similar systems there is complete similarity if all necessary dimensionless criteria derived either from differential equations or by using the pi theorem are equal. In complex precipitation processes such complete similarity is impossible. Moreover, because we want to obtain identical not similar products from the systems differing in scale, we usually want to reproduce the product quality (particle size, particle morphology), mixture composition, and structure of the suspension on a larger scale. We thus use limited similarity, which means that we lose several degrees of freedom (we cannot manipulate particle size, solution composition, viscosity, and diffusivity), and we obtain this way a reduced number of similarity criteria.

A more fundamental approach to scale-up is based on modeling with the goal of predicting its effect and understanding the process in sufficient detail. "Modeling approach" here means representing our particle production system by a set of equations. To this end, one can use empirical models, mechanistic models, semiempirical models, and CFD. In the case of purely empirical models that are based on fitting to experimental data, one chooses the form of equations (polynomials, neural networks, etc.) to represent how the dependent variables chosen (e.g., particle size) depend on such independent variables as concentration, temperature, pressure, and nozzle velocity. Such an approach is not recommended, since such models could give erroneous predictions outside the experimental region (i.e., in the larger scale system).

The recommended approach to modeling is to create models based on fundamental balances (of mass, species, energy, population) and basic kinetics and use them to build a complete model of the precipitator, as shown in earlier sections. Such a set of equations is known as a physical or a mechanistic model. Complete physical models are difficult to create and solve because they require identification in advance of all physical and chemical subprocesses, properties, and parameters. That is why the semiempirical models of a form similar to the complete physical models (but usually simpler) and with fewer equations are often used for scaling up. Parameters of such models are often given in lumped form, some of them fitted to available experimental data obtained from the small-scale system. Such a model can be useful for scaling up, but one cannot be sure that the scale-up will be completely correct because there is no guarantee that the model contains the complete mechanism (88). However, scale-up errors should be smaller than in the case of purely empirical models. CFD codes that are based on reasonable simplifications (closures) regarding their accuracy can be placed between the physical and semiempirical models; their application was demonstrated earlier.

Finally, there are empirical, conservative scale-up criteria such as holding constant during scale-up the power per unit mass (or volume), heat and mass transfer coefficients, nozzle velocity, or residence time. These criteria are often contradictory and should be used with care.

The following scale-up strategy is recommended. Before considering process modeling or empirical investigations for scaling up, one should perform an analysis of time constants. here it is important to know how to define and use the time constants, for as their comparison can show the subprocesses that are less important on the large scale, which will indicate any requirements related to such processes that can be neglected. For this analysis one can use crude estimates of media properties and process kinetics. One can use this analysis to create one or more solutions for the scaled-up system and then use the semiempirical model to choose one of the proposed solutions and to check the trends of variation of the most important variables with scale increase. Semiempirical modeling can be supported at this stage by CFD to model such hydrodynamic details as flow pattern, local concentration, and stresses, as well as by scale-up experiments. If possible instead of the semiempirical model, a physical model employing CFD should be applied to increase the accuracy of predictions of scale-up effects and finally to optimize the process.

This chapter mainly concerns mixing and precipitation in a jet. Therefore we illustrate some general scaleup principles by considering a nozzle scale-up based on the combined time constants for mixing (τ_M). nucleation (τ_N), and growth (τ_G), as defined in Section 5. One can classify the precipitation subprocesses from the point of view of their competition with mixing. Nucleation is instantaneous $\tau_M \gg \tau_N$, fast if $\tau_M \approx \tau_N$, and slow if $\tau_M \ll \tau_N$. The effect of mixing should be observed in the first two cases. Since only the mixing subprocesses (69) that affect the precipitation should be considered when one is scaling up, the foregoing criteria are helpful for discrimination of mixing subprocesses. When nucleation is very fast (homogeneous nucleation usually is), it can be affected by almost all the mixing subprocesses. The proper velocity scale in the considered system is the jet velocity u_0, and the proper length scale is the jet nozzle diameter d_0. The other length scales (e.g., vessel diameter, length) are proportional to d_0 in geometrically similar systems. Hence, the age of the fluid elements at position x on the axis scales with the characteristic convection time $\tau_C = d_0/u_0$. Therefore during the time history $t/\tau_C = f(x/d_0)$, fluid elements have the same dimensionless positions x/d_0 after the same dimensionless time periods t/τ_C. So keeping the same values of the mean residence time in both systems is equivalent to keeping the same τ_C values. The rate of energy dissipation, ε, appears in many equations describing both jet mixing and droplet dispersion. Generally the rate of energy dissipation scales as $\varepsilon \propto u_0^3/d_0$; this means that at the same points in dimensionless space the ratio $\varepsilon_1/\varepsilon_2$ remains the same. Ideally, process scale-up requires the same ε time history for fluid elements. (Then the same process parameters are "seen" by fluid elements after the same time periods in both systems.) Unfortunately, this leads to contradictory requirements:

$$\left(\frac{d_0}{u_0}\right)_1 = \left(\frac{d_0}{u_0}\right)_2 \tag{76}$$

and

$$\left(\frac{u_0^3}{d_0}\right)_1 = \left(\frac{u_0^3}{d_0}\right)_2 \tag{77}$$

So scale-up in which the fluid elements experience the same rate of energy dissipation in both systems at the same physical time (or at the same age) is impossible. Therefore compromises have to be considered:

1. Requirement of the same rate of energy dissipation, Eq. (77). Let us define parameter X

$$\frac{Q_2}{Q_1} = \frac{u_{02}d_{02}^2}{u_{01}d_{01}^2} = X \tag{78}$$

which is the ratio between the flow rates Q_1 and Q_2 on the different scale. From Eqs. (76) and (77) we have:

$$\left(\frac{u_{02}}{u_{01}}\right) = X^{1/7} \tag{79}$$

and

$$\left(\frac{d_{02}}{d_{01}}\right) = X^{3/7} \tag{80}$$

This gives a significant increase in nozzle diameter and a relatively small increase in nozzle velocity. As a result, the convection time significantly increases:

$$\tau_{C2} = \tau_{C1} \cdot X^{2/7} \tag{81}$$

The fluid elements thus pass the same trajectories and experience the same ε values, but move $X^{2/7}$ times more slowly in dimensionless space. This criterion can be identified, for example, with the same maximum stable droplet size in both systems, the same exchange coefficients for objects of the same size at the same dimensionless positions, the same mixing parameters, and so on. However, this means that mixing, nucleation, and growth of crystals take place at different values of ε when they are fast. As an example for nucleation time τ_N (or, more exactly, nucleation distance, $\tau_N u_0$), Eqs. (79) and (80) indicate that in the larger scale system the nucleation distance is $X^{1/7}$ times longer, whereas the ε profile is "elongated" $X^{3/7}$ times. This scale-up should be used for slow or "simple" processes, such as droplet breakup according to the Kolmogorov mechanism in Eq. (36), when the time factor is not important, only the final result. When the time factor is of importance (chemical reaction, crystallization, or even some droplet formation), this is not an appropriate method. For example, application of the multifractal mechanism Eqs. [(37–39)] for asymptotically stable droplets for $\alpha = 0.12$ results in $d_{max} \propto \varepsilon^{-0.617} d_0^{-0.54}$; so for the same ε, droplets are smaller in the larger scale system. This example also illustrates the importance of the definition of a mechanistic model that may lead to different scale-up criteria.

2. The requirement of the same time history in Eq. (76) leads to the following relations:

$$\frac{u_{02}}{u_{01}} = X^{1/3} \tag{82}$$

and

$$\frac{d_{02}}{d_{01}} = X^{1/3} \tag{83}$$

This gives similar increases in nozzle diameter and nozzle velocity. Because the rate of energy dissipation scales as $\varepsilon \propto u_0^3/d_0$; we obtain $\varepsilon_2 = \varepsilon_1 \cdot X^{2.3}$. This means that the rate of energy dissipation significantly increases during scale-up. In both systems the fluid elements pass the same way with the same velocity in dimensionless space. In physical space they will attain the maximum ε for the same period of time, but the profile of ε is everywhere $X^{2.3}$ times higher. The energy is larger in the large-scale system, so when energy should not be smaller than a certain value (e.g., de aggregation, emulsification), we are on the safe side here, although shear-sensitive materials such as proteins may not be good candidates for this criterion.

3. A pragmatic solution between (1) and (2): in a large-scale system, fluid elements move faster than in option 1 but experience smaller ε than in option 2. This means

$$\left(\frac{u_0^2}{d_0}\right)_1 = \left(\frac{u_0^2}{d_0}\right)_2 \tag{84}$$

and

$$\frac{u_{02}}{u_{01}} = X^{1/5} \tag{85}$$

$$\frac{d_{02}}{d_{01}} = X^{2/5} \tag{86}$$

Now $\varepsilon_2 = \varepsilon_1 \cdot X^{1/5}$ and $(d_2/u_0^2)/(d_1/u_0^1) = X^{1.5}$, so it is really a compromise between similarity in time and similarity in energy.

8. CONCLUSION

The major thrust of this chapter is the description of the particle formation mechanism during SCF antisolvent precipitation in a jet. In this case, the fluid dynamics of the process can be described by vigorous mathematical models. Our purpose was to show that different aspects of this process, such as thermodynamics, fluid dynamics, and precipitation kinetics, are complementary and can provide a comprehensive process description only when taken together, although clearly much experimental and theoretical work remains to complete the picture. Moreover, antisolvent precipitation with SCF cannot be represented by a single model and requires separate approaches for conditions below and above mixture critical point (mcp) with singularities of mass and heat transfer behavior at the mcp. The greatest potential of SCF in the chemical engineering and pharmaceutical industries lies in the unique thermodynamic properties and adjustability of SCF solvent parameters around the mcp. There is however a lack of experimental knowledge of mass and heat transfer coefficients and of the kinetics of precipitation and corresponding material constants. This challenge is intimately linked to the design of reliable in-line measurements in SCF, to provide means for scientifically driven process development, monitoring, and control. As shown in this chapter, a combination of different optical techniques can be used for this purpose. Further work in this area is indicated. In the area of theoretical modeling and prediction, CFD can offer a comprehensive solution for process optimization and scale-up. It is clear, however, that these efforts should be supported by correct physical models for molecular level processes of micromixing, nucleation, and growth.

The difference between well-known SCF antisolvent techniques such as GAS, PCA, and SEDS usually can be attributed to the specific nozzle mixing (or dispersing) technique involved. Enhanced mass and heat transfer can also be achieved by using mechanical and ultrasonic mixers and ultrafast jet expansion techniques. There are new developments for particle formation by means of dispersed systems such as emulsions, micelles, colloids, and polymer matrixes. It should be emphasized that all these processes involve the same fundamental aspects of mass and heat transfer phenomena between an SCF and a subcritical phase. Clearly the ultimate goal of all SCF particle technologies is to achieve predictable, consistent, and economical production of fine pharmaceuticals or chemicals. This is possible only on the basis of comprehensive mechanistic understanding and well-developed scale-up principles.

REFERENCES

1. Perrut M. Supercritical fluid applications: industrial development and economic issues. Ind Eng Chem 2000; 39:4531–4535.
2. Bellan J. Supercritical (and subcritical) fluid behavior and modeling: drops, streams, shear and mixing layers, jets and sprays. Prog Energy Combust Sci 2000; 26:329–366.
3. Jung J, Perrut M. Particle design using supercritical fluids: literature and patent survey—review. J Supercrit Fluids 2001; 20:179–219.
4. Marr R, Gamse T. Use of supercritical fluids for different processes including new developments—a review. Chem Eng Proc 2000; 39:19–28.
5. Subra P, Jestin P. Powder elaboration in supercritical media: comparison with conventional routes. Powder Technol 1999; 103:2–9.
6. Bristow SC, Shekunov BY, York P. Solubility analysis of drug compounds in supercritical carbon dioxide using static and dynamic extraction systems. Ind Eng Chem Res 2001; 40:1732–1739.
7. Bristow S, Shekunov T, Shekunov BY, York P. Analysis of the supersaturation and precipitation process with supercritical CO_2. J Supercrit Fluids 2001; 21: 257–271.
8. Shekunov BY, Sun Y. Analysis of mixing and crystallisation in supercritical

fluids using small angle X-ray scattering. Proceedings of the 6th International Symposium on Supercritical Fluids 2003; 3:1813–1818.

9.   Shekunov BY, Baldyga J, Sun Y, Astrakcharchik E, York P. Optical characterization and mechanism of antisolvent precipitation in turbulent flow. Proceedings of the 7th Meeting on Supercritical Fluids 2000; 1:65–70.

10. Sun Y, Shekunov BY. Surface tension between ethanol droplets and supercritical CO_2. J Supercrit Fluids 2003; 27:73–83.

11. Astrakharchik-Farrimond E, Shekunov BY, York P, Sawyer NBE, Morgan SP, Somekh MG, See CW. Dynamic measurements in supercritical flow using instantaneous phase-shift interferometry. Exp Fluids 2002; 33:307–314.

12. Sun Y, Shekunov BY, York P. Refractive index of supercritical CO_2–ethanol solvents. Chem Eng Commun 2003; 190:1–14.

13. Astrakharchik-Farrimond E, Shekunov BY, York P, Sawyer NBE, Morgan SP, Somekh MG, See CW. Particle imaging using wide-field phase confocal microscope. Part Sys Part Charact 2003; 20:104–110.

14. Nishikawa K, Tanaka I, Amemiya Y. Small-angle X-ray scattering of supercritical carbon dioxide. J Phys Chem 1996; 100:418–421.

15. Lockemann CA, Schlünder EU. High-pressure mass transfer coefficients in the liquid phase of the binary systems carbon dioxide–methyl myristate and carbon dioxide–methyl palmitate. Chem Eng Proc 1996; 35:121–129.

16. Harstad K, Bellan J. The Lewis number under supercritical conditions. Int J Heat Mass Transfer 1999; 42:961–970.

17. Miller RS. Long time mass fraction statistics in stationary compressible isotropic turbulence at supercritical pressure. Phys Fluids 2000; 128:2020–2032.

18. Lou H, Miller RS. On the scalar probability density function transport equation for binary mixing in isotropic turbulence at supercritical pressure. Phys Fluids 2001; 13(1):3386–3399.

19. Onuki A, Hao H, Ferell RA. Phys Rev A 1990; 41:2256.

20. Shekunov BY, Hanna M, York P. Crystallization process in turbulent supercritical flows. J Cryst Growth 1999; 198/199, 1345–1351.

21. Peng DY, Robinson DB. A new two-constants equation of state. Ind Eng Chem Fundam 1976; 15:59–63.

22. Werling JO, Debenedetti PG. Numerical modeling of mass transfer in the supercritical antisolvent process: miscible conditions. J Supercrit Fluids 2000; 18: 11–24.

23. Bałdyga J. Turbulent mixer model with application to homogeneous, instantaneous chemical reactions. Chem Eng Sci 1989; 44:1175–1182.

24. Bałdyga J, Henczka M. Turbulent mixing and parallel chemical reactions in a pipe. Recent Prog Genie Procedes 1997; 11:341–348.

25. Fox RO. The spectral relaxation model of the scalar dissipation rate in homogeneous turbulence. Phys Fluids 1995; 7:1082–1094.

26. Miller RS. Long time mass fraction statistics in stationary compressible isotropic turbulence at supercritical pressure. Phys Fluids 2000; 12:2020–2032.

27. Lou H, Miller RS. On the scalar probability density function transport equation for binary mixing in isotropic turbulence at supercritical pressure. Phys Fluids 2001; 13:3386–3398.

28. Reitz RD, Bracco FV. Mechanisms of breakup of round liquid jets. In: Cheremisinoff NP, ed. Encyclopedia of Fluid Mechanics. Houston, TX: Gulf Publishing, 1986:231–249.
29. Rayleigh WS. On the instability of jets. Proc Lond Math Soc 1878; 4:10.
30. Weber C. On the breakdown of a fluid jet. Z Angew Math Mech 1931; 11:136–154.
31. Fenn RW, Middleman S. Newtonian jet stability—the role of air resistance. AIChE J 1969; 15:379.
32. Benjamin TB. J Fluid Mech 1959; 6:161.
33. Stirling AM, Sleicher CA. The instability of capillary jets. J Fluid Mech 1975; 68:477.
34. Ranz WE. On Sprays and Spraying. Department of Engineering Research. Pennsylvania State University. Bulletin, 1956:655.
35. Taylor GI. Generation of Ripples by Wind Blowing over a Viscous Fluid. Collected Works of G.I. Taylor, Vol. 3, 1940.
36. Reitz RD, Bracco FV. Mechanism of atomization of a liquid jet. Phys Fluids 1982; 25:1730.
37. Ranz WE. Some experiments on orifice sprays. Can J Chem Eng 1958; 36:175.
38. Merrington AC, Richardson EG. The breakup of liquid jets. Proc Phys Soc 1947; 59:1–13, Part I, No. 331.
39. Ohnesorge G. Z Angew Math Mech 1936; 16:355.
40. Duffie JA, Marshall WR, Jr. Factors influencing the properties of spray-dried materials. Part I. Vol. 49. Chem Eng Proc, 1953; 8:417–423.
41. Narasimhamurty GSR, Parushothamen A, Sivaji K. Hydrodynamics of liquid drops in air. In: Cheremisinoff NP, ed. Encyclopedia of Fluid Mechanics. Houston, TX: Gulf Publishing, 1986:250–279.
42. Loth E. Numerical approaches for motion of dispersed particles, droplets and bubbles. Prog Energy Combust Sci 2000; 26:161–223.
43. Clift R, Grace JR, Weber ME. Bubbles, Drops and Particles. New York: Academic Press, 1978.
44. Liu H, Rangel RH, Lavernia EJ. Modeling of droplet–gas interactions in spray atomization of Ta-2.5W alloy. Mater Sci Eng 1995; A191:171–184.
45. Chhabra RP, Agarwal L, Sinha NK. Drag on non-spherical particles: an evaluation of available methods. Powder Technol 1999; 101:288–295.
46. Garner FH, Lihou DA. DA DECHEMA-Monograph 1965; 55:155–178.
47. Gelfand BE. Droplet breakup phenomena in flows with velocity lag. Prog Energy Combust Sci 1996; 22:201–265.
48. Joseph DD, Belanger J, Beavers GS. Breakup of a liquid drop suddenly exposed to a high-speed airstream. Int J Multiphase Flow 1999; 25:1263–1303.
49. Schlichting H. Boundary Layer Theory. New York: Mc Graw-Hill, 1979.
50. Taylor GI. The intstability of liquid surfaces when accelerated in a direction perpendicular to their planes. Part I. Proc R Soc Lond 1950; A 201:192–196; also in The Scientific Papers of G.I. Taylor. Vol. 3. In: Batchelor GK, eds. Cambridge: University Press, 1993.
51. Pilch M, Erdman C. Use of breakup time data and velocity history data to predict the maximum size of stable fragments for acceleration-induced break-up of a liquid drop. Int J Multiphase Flow 1987; 13:741–757.

52. Taylor GI. The shape and acceleration of a drop in a high-speed air stream. Advisory Council on Scientific Research and Technical Development. Ministry of Supply, AC 10647/Phys. C69; also in The Scientific Papers of G.I. Taylor Vol. 3. In: Batchelor GK, ed. Cambridge: University Press, 1993.

53. Kolmogorov AN. Disintegration of drops in turbulent flows. Dokl Akad Nauk SSSR 1949; 66:825 828.

54. Hinze JO. Fundamentals of the hydrodynamic mechanism of splitting in dispersion process. AIChE J 1955; 1:289–295.

55. Kolmogorov AN. The local structure of turbulence in incompressible viscous fluid for very large Reynolds number. Dokl Akad Nauk SSSR 1941; 30:301–305.

56. Bałdyga J, Podgórska W. Drop breakup in intermittent turbulence. Maximum stable and transient sizes of drops. Can J Chem Eng 1998; 76:456 470.

57. Bałdyga J, Bourne JR. Turbulent Mixing and Chemical Reactions. Chichester: John Wiley & Sons, 1999.

58. Godsave GAE. Burning of fuel droplets. Fourth Symposium (International) on Combustion. New York: Williams & Wilkins, 1953:818–830.

59. Spalding DB. Some Fundamentals of Combustion. London: Butterworth, 1955.

60. Arias-Zugasti M, Garcia-Ybarra PL, Castilo JL. Droplet vaporization at critical conditions: long-time convective-diffusive profiles along the critical isobar. Phys Rev 1999; 60:2930–2941.

61. Wiliams FA. On the assumption underlying droplet vaporization and combustion theories. J Chem Phys 1960; 33:133–144.

62. Givler SD, Abraham J. Supercritical droplet vaporization and combustion studies. Prog Energy Combust Sci 1996; 22:1–28.

63. Harstad K, Bellan J. An all-pressure fluid drop model applied to a binary mixture: heptane in nitrogen. Int J Multiphase Flow 2000; 26:1675–1706.

64. Batchelor GK. Mass transfer from small particles suspended in turbulent fluid. J Fluid Mech 1980; 98:609 623.

65. Kronig R, Brink JC. On the theory of extraction from falling droplets. Appl Sci Res 1950; A2:142–154.

66. Sherwood TK. Pigford RL. Wilke CR. Mass Transfer. New York: McGraw-Hill, 1975.

67. Chernov AA. Modern Crystallography III, Crystal Growth; Springer Series in Solid State Physics. Berlin: Springer-Verlag, 1984.

68. Cordray DR, Izatt RM, Christensen JJ, Oscarson JL. The excess enthalpies of (carbon dioxide + ethanol) at 308.15, 325.15, 373.15, 413.15, and 473.15 K from 5.00 to 14.91 MPa. J Chem Thermodyn 1988; 20:655–663.

69. Shekunov BY, Bałdyga J, York P. Particle formation by mixing with supercritical antisolvent at high Reynolds numbers. Chem Eng Sci 2001; 56:2421–2433.

70. Söhnel O, Garside J. Precipitation. Oxford: Butterworth-Heinemann, 1992.

71. Mullin JW. Crystallization. Oxford: Butterworth-Heinemann, 1993.

72. Randolph AD, Larson MA. Theory of Particulate Processes. New York and London: Academic Press, 1971.

73. Bałdyga J, Orciuch W. Some hydrodynamic aspects of precipitation. Powder Technol 2001; 121:9–19.

74. Hanna M, York P. World Patent WO 95/01221, 1994.
75. Bałdyga J, Henczka M, Czarnocki R, Shekunov BY. Scale-up of the supercritical antisolvent precipitation process. Inż Ap Chem 2002; 41:20–21.
76. Bałdyga J, Henczka M, Kubicki D, Shekunov BY. Modeling of turbulent mixing with application of supercritical antisolvent. Inż Ap Chem 2002; 41:22–23.
77. Wubbolts FE, Bruinsma OSL, Rosmalen GM. Dry-spraying of ascorbic acid or acetaminophen solutions with supercritical carbon dioxide. J Cryst Growth 1999; 198/199:767–772.
78. Smoluchowski M. Versuch einer mathematischen Theorie der Koagulations-Kinetik kolloider Lösungen. Z Phys Chem 1917; 92:129–168.
79. Saffman PG, Turner JS. On the collision of drops in turbulent clouds. J Fluid Mech 1956; 1:16–30.
80. Spielman LA. Viscous interactions in Brownian coagulation. J Colloid Interface Sci 1970; 33:562–571.
81. Abrahamson J. Collision rates of small particles in a vigorously turbulent fluid. Chem Eng Sci 1975; 30:1371–1379.
82. Kruis FE, Kusters KA. The collision rate of particles in turbulent flow. Chem Eng Commun 1997; 158:201–230.
83. David R, Marchal P, Klein JP, Villermaux J. Crystallization and precipitation engineering. III. A discrete formulation of the agglomeration rate of crystals in a crystallization process. Chem Eng Sci 1991; 46(1):205–213.
84. Hounslow MJ, Mumtaz HS, Collier AP, Barrick JP, Bramley AS. A micromechanical model for the rate of aggregation during precipitation from solutions. Chem Eng Sci 2001; 56:2543–2552.
85. Bałdyga J, Jasińska M, Krasiński A, Kubicki D, Franke D, Goesele W, Welker S, Zauner R. Mixing effects in precipitation–aggregation problem. Proceedings of 4th International Symposium on Mixing in Industrial Processes, Toulouse, France, paper 2001; 11:75.
86. Elimelech M, Gregory J, Jia X, Williams R. Particle Deposition and Aggregation. Oxford: Butterworth-Heinemann, 1995.
87. Dirksen JA, Ring TA. Fundamentals of crystallization: Kinetic effects on particle size distributions and morphology. Chem Eng Sci 1991;46), 2389–2427.
88. Rose LM. Chemical reactor design in practice. Chemical Engineering Monograph 13. Amsterdam: Elsevier, 1981.

4

Methods of Particle Production

**Gérard Charbit, Elisabeth Badens,
and Olivier Boutin**
Laboratoire de Procédés Propres et Environnement, Université
d'Aix–Marseille, Aix-en-Provence, France

Production of ultrafine particles with desired properties is the objective of many industries. Precise control of particle size is beneficial in the development of catalysts, adsorbents with high specific surface area, porous solids for chromatography, pigments, polymers, ion-exchange resins, explosives, nutraceuticals, and pharmaceuticals. This task is often more challenging in the pharmaceutical industry, which requires the processing of labile therapeutic entities. The pharmaceutical industry aims at producing particles of micrometer or submicrometer size for use in formulations administered by various routes including the parenteral, inhalation, and oral modes. In most cases, the requirement of small particle size is accompanied by the desirability of particle size distribution (PSD) that is as narrow as possible. For instance, to achieve optimal drug delivery to the lung, it is important to ensure that the drug is formulated into microparticles of the appropriate aerodynamic size, shape, and apparent density; in this particular case, the microparticles must have a mass median aerodynamic diameter of approximately 2 μm.

One purpose of size reduction is to increase the bioavailability of a drug by improving the drug dissolution rate. Since 60 to 70% of the molecules synthesized by pharmaceutical firms exhibit poor solubility and bioavailability, a major effort is being directed at overcoming this limitation. Such physicochemical properties of a drug as crystallinity and polymorphism greatly influence its solubility, stability, and bioavailability. Thus, the development of manufacturing processes that allow control over these properties would be ideal. Conventional processes for particle formation suffer from limitations in producing a desirable end product. Milling results in 10 to 50 μm particles with large PSD, and the process generates high local temperatures that are likely to modify or even damage the product. Spray drying yields particles of suitable size, but it exposes the product to locally high temperature. Lyophilization produces particles with a very broad PSD. Finally, precipitation from solution by the addition of an organic antisolvent (salting out), does not allow precise control of the particle size and results in high content of residual organic solvent, which must be removed. Conventional methods of forming controlled-release microspheres or microcapsules include the extrusion, prilling, spray cooling, spray drying, coacervation, interfacial polymerization, and hot melt approaches. These techniques fail to yield particles of suitable size or size distribution. In addition, some of these methods subject labile molecules to high thermal stress.

Thus, there is a need to identify alternative approaches to produce small particles for pharmaceutical use. Use of supercritical fluids (SFs) or compressed gases in general is an attractive alternative. Though historically the first report dates back more than a hundred years (1), it was in the 1980s that intensive research started in this field, giving renewed interest to supercritical technologies.

The following unique advantages offered by supercritical fluids and compressed gases acting as solvents or antisolvents allow a wide control of particle morphology:

> Solvent properties that can be tuned by changing pressure and/or temperature
> Reduced viscosities and increased mass transfer
> Dramatic modification (decrease) of solvent strength by dissolution of a dense gas
> Control of supersaturation via pressure, temperature and antisolvent concentration
> Low level of operating temperature (since in most cases carbon dioxide is the SF)

A number of novel and promising processes have been devised that can be divided into three groups

1. *Precipitation from supercritical solutions composed of supercritical fluid and solute(s).* Rapid expansion of supercritical solutions, the RESS

process, exemplifies this first group. In this method one dissolves the solute or solutes to be comminuted in a supercritical fluid, this mixture is then expanded by means of a restrictor, which causes the solid to precipitate. This technique calls for molecules that are fairly soluble in SCFs, which constitutes a limitation, since most drugs have a poor solubility in Supercritical carbon dioxide ($scCO_2$), a commonly used SCF for pharmaceuticals.

2. *Precipitation from solutions using SCFs or compressed gases as antisolvents.* Solids that are insoluble in SCFs or compressed gas can be micronized by means of this approach. The basic principle is to allow a solution of a substrate in a liquid primary solvent of interest to contact a supercritical fluid or a dense gas. The simultaneous transfers of CO_2 and primary solvent from one phase to the other lead to supersaturation and the precipitation of the solid. Several applications were developed on this basis, differing from one another in the contact mode of the two phases, in the dispersion device selected, in phase flow direction, and in mode (batch or semicontinuous). Precipitation with a compressed fluid antisolvent (PCA), gas antisolvent (GAS), supercritical antisolvent (SAS), aerosol solvent extraction system (ASES), and solution-enhanced dispersion by supercritical fluids (SEDS) are the designations proposed for techniques relevant to this group. As highlighted earlier, since most drugs cannot be operated by the RESS, the antisolvent techniques are effective for a very wide range of compounds.

3. *Precipitation from gas saturated solutions* (PGSS), (and related methods differs from groups). The last group, which consists of PGSS 1 and 2 in that the SCF does not act as either solvent or antisolvent. This process involves dissolving an SCF or a compressed gas in the molten material, then expanding the solution through a nozzle.

These versatile techniques combined with the pressing pharmaceutical need for particle design have been an impetus for intense research in this field, which has resulted in an exponential growth in articles devoted to these applications. A number of reviews have been published (2–14), clearly demonstrating that SCF technologies are viable approaches. However, a thorough review of the literature indicates that different acronyms were used to refer to the same technique. These discrepancies essentially concern the second group, the antisolvent techniques. Some examples of these discrepancies are summarized in the nonexhaustive list that follows.

Thiering et al. (10) sorted all antisolvent methods into two different categories: ASES and PCA. For Bungert et al. (4), SAS is only a fractionation process applied, for instance, to the recovery of polymers. For Jung and Perrut (11), GAS and SAS are similar and thus both are referred by GAS. The same authors refer to ASES as the process known as SAS in other issues. In the same way, Krober and Teipel (12) used PCA to refer to the application called SAS in many others papers. Reverchon et al. (13) include in a single acronym, SAS, processes described elsewhere as PCA, SAS, and GAS; in the same paper, differentiations were made between liquid batch SAS, gas batch SAS, and semicontinuous SAS.

Finally, depending on the authors, the same acronym may have different meanings. Thus meaning SEDS, "solution-enhanced dispersion by supercritical fluids," was first defined by the research group of Bradford University in its patents (15,16) filed in 1995 and 1996, but more recently edited book, the same acronym was used for "solution-enhanced dispersion of solids" (17).

The foregoing examples show that no absolute rule relates the acronym to the antisolvent process mode, and consequently, there is lot of confusion. Nevertheless, without any doubt, antisolvent techniques may be divided into two categories: first, those introducing the compressed gas into the liquid phase (GAS) and second, wherein the liquid phase is injected into a sub- or supercritical continuum (PCA, SAS, ASES, SEDS). Regarding the first category there is little ambiguity, since it includes only one process. The situation is somewhat more complex for the second class of processes, since the distinctions defined by the original papers were progressively erased. In current publications, almost no distinction is made between SAS, PCA, and ASES; SEDS is the only process, that is differentiated owing to the presence of a specific device for introducing the liquid solution. This is why in this chapter, after having pointed out the original references, we will classify the antisolvent processes into the following three groups:

GAS
PCA, SAS, ASES
SEDS

The GAS process, which is typically a batch process, was first devised by Gallagher et al. (18). In this method, the solution formed of the substrate and the primary solvent is placed in an autoclave and a compressed gas is progressively added until saturation is reached.

In the PCA process originally described by Dixon et al. (19), a small volume of a liquid solution is dispersed through a capillary in a chamber filled with a compressed gas, which can be either subcritical or supercritical. In the original paper, PCA appears as a batch process; nevertheless in most articles thereafter, "PCA" was mainly used when the continuum was subcritical. Subsequently, the use of a coaxial nozzle was suggested to improve the control of particle morphology (20).

Lim et al. extended the PCA concept, suggesting spraying the liquid solution continuously in a continuum of SF by means of a capillary; this new method was termed the SAS process (21,22). In this semicontinuous operating mode, the SCF and the primary solvent are continuously removed, while the solid remains in the autoclave.

Historically, the first report on ASES would seem to be attributable to Bleich et al. (23). ASES does not fundamentally differ from the SAS; indeed ASES uses a nozzle instead of a capillary to form smaller droplets and consequently smaller particles. As soon as atomization is reached at high liquid solution flow rate with capillaries, however, this difference disappears. Figure 1 shows the length breakup variation of a methylene chloride jet in $scCO_2$ vs the liquid flow rate. For these experiments, the internal diameter of the capillary was of 150 μm. As can be seen, (Figure 1e), at 180 mL/h, atomization is realized.

The SEDS process was first described by Hanna and York (15,16). It uses the compressed gas as an antisolvent, and additionally as a dispersing agent, to improve the extraction of the primary solvent. A nozzle with two or three coaxial channels is used to introduce the liquid solution(s) and the dense gas.

This first overview shows that supercritical fluid processing offers a wide range of tools for addressing the challenges faced by the pharmaceutical industry. The principles of these techniques as well as examples of applications in the field of drug manufacturing are detailed in the sections that follow.

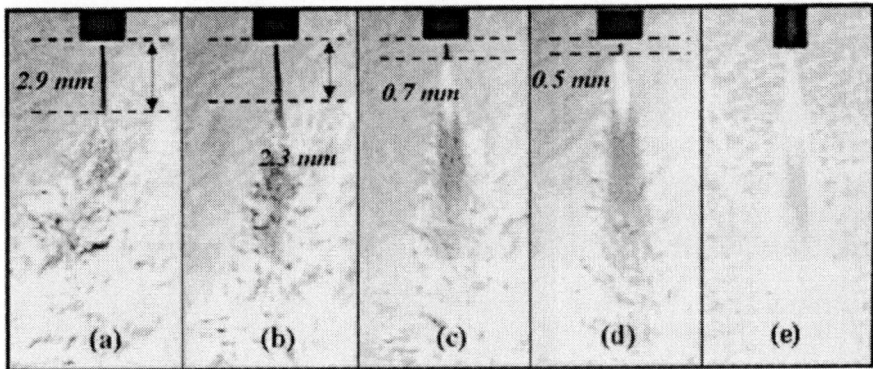

FIGURE 1 Variation of the jet breakup length with the liquid flow rate for five flow rates (mL/h) = (a) 15, (b) 30, (c) 45, (d) 60, and (e) 180. (From E. Carretier et al. Ind Eng Chem Res 2003; 42:331–338.)

1. RAPID EXPANSION OF SUPERCRITICAL SOLUTIONS (RESS)

1.1. Process Description

The RESS process relies on the solvent properties of carbon dioxide. Because CO_2 is a nonpolar molecule, this process will be mainly efficient and interesting for micronizing nonpolar molecules. For this reason, a preliminary study on the solubility of the compounds with pressure and temperature is necessary. As usual, the solvent polarity can be modified and enhanced by adding to the supercritical CO_2, small quantities of an organic cosolvent. This is primarily because the solvent power of an SCF is strongly dependent on its density, which can be adjusted by small variations of pressure and temperature (11).

The principle of the technique may be described as follows. The active substance to be micronized is partly solubilized in a continuous stream of pure $scCO_2$, in some cases with the addition of a cosolvent; the mixture so formed is then expanded. The pressure decrease causes the CO_2 to evaporate, leading to supersaturation and precipitation of the solid.

To obtain small enough particles with a uniform particle size distribution, the expansion must be fast ($<10^{-5}$ s) and uniform. This is possible because the pressure variation can be very rapid and travels at the speed of the sound, leading to uniform conditions within the expanding fluid (24).

Figure 2 shows a schematic of the RESS process, which is operated as follows. CO_2 is pumped (a) and raised to the desired pressure. When a cosolvent is used, it is pumped in the same way (b) and introduced into the CO_2 flow. This flow is then heated to the desired temperature (c) and allowed to enter a tank loaded with the active substance (d) for extraction. In this part of the process, the solvent power is strong because of the high pressure and because of the possible presence of a cosolvent. This mixture is then depressurized in an expansion vessel (e) by means of a capillary or a nozzle, with a typical inner diameter of 50 to 60 μm. The restrictor must be heated to avoid plugging by solid precipitation. The expansion chamber is generally at or near atmospheric pressure. A frit filter is placed at the exit of the expansion chamber to keep the particles formed in the expansion vessel. A cyclone (f) separates the solvent from the CO_2, which can be recycled (g).

The list that follows, though not exhaustive, gives the key factors to be controlled in this process if the desired particle size and size distribution are to be obtained (11):

Temperature and pressure in the saturator and in the expansion vessel
Solubility of the active substance in the CO_2 mixture
Nozzle diameter
Dimensions of the expansion vessel

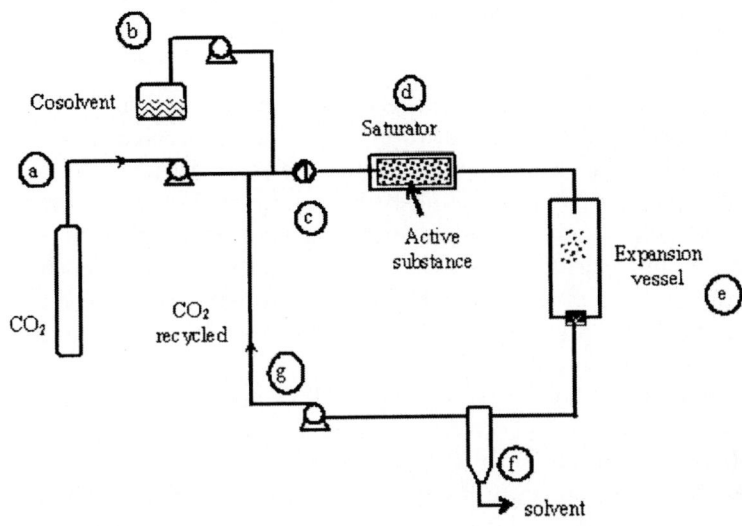

Cosolvent

Saturator

Active substance

Expansion vessel

CO$_2$

CO$_2$ recycled

solvent

△ pump

⊕ heat exchanger

▪ frit filter

FIGURE 2 Schematic drawing of the RESS process; for (a)–(g), see text.

Some investigators have reported different models of the process, especially for phenomena occurring in the expansion chamber in the capillary nozzle (24) and (25).

As for other supercritical processes, it is possible to micronize one substrate or more—for instance, to encapsulate an active substance in a biocompatible polymer. In such a case, the purpose is to obtain a controlled-release system of the active substance, and both materials (the active ingredient and the polymer) must be dissolved in the CO$_2$ mixture. Depending on the relative solubility of the two materials in the supercritical medium, one can obtain an encapsulated product or a composite matrix of the two products (26–28).

The RESS process, without the use of a cosolvent, is very attractive and probably the most easily managed supercritical process. Unfortunately many molecules are not soluble in pure scCO$_2$, and these require the use of a cosolvent. However, such an addition makes the RESS process less favorable, since it loses its main advantage of being system free of organic solvent. Any residual solvent in the product must be removed.

1.2. Application of the RESS for Processing Pharmaceuticals

Historically, Krukonis (29), in 1984, was the first to describe the application of the concept of supercritical micronization organic and inorganic materials. Later on, researchers of the Battelle Institute developed the first processes based on this concept (30,31). These investigators primarily focused on the formation of particles from various polymers. As mentioned, this process is undoubtedly the simplest supercritical technique, and that is why it has been tested on a number of pharmaceuticals including those that have a low solubility in pure carbon dioxide. In the 1990s many other processes emerged, as detailed in the sections that follow.

Although particles are obtained by the RESS process recently nanometric particle production has been reported (32–39), in most cases micrometer- and submicrometer-sized (Table 1) (40–71).

The more recent results are interesting because drug particles of less than 500 nm exhibit improved bioavailability. While most of the studies employed expansion into a CO_2 atmosphere, some of these reports employed expansion in a vessel filled with water (33,37). Obviously, this process is applicable to products insoluble in water. Usually, a surfactant is added to avoid agglomeration and flocculation, thereby stabilizing small particles (37).

2. GAS ANTISOLVENT PROCESS (GAS)

2.1. Process Description

Gallagher et al. (18) first propose the GAS process to overcome limitations encountered with the RESS process. Indeed, many materials, more particularly drugs, are polar compounds that cannot dissolve to an "appreciable" extent in supercritical fluids. Furthermore, the applicability of the RESS remains problematic at large scale owing to the high consumption of supercritical fluids and the high pressures involved. Inspired by classical antisolvent techniques applied to liquid phases (i.e., "salting out"), the authors suggested using supercritical fluids or gases near their vapor pressure as antisolvents.

This attractive technique was first tested to micronize explosives (18,72–74). Mechanical methods such as milling generate high local temperatures, and therefore are useless in this particular case; the use of $scCO_2$ as an antisolvent allowed comminution to proceed at mild temperatures. As already noted, the GAS process is a batch technique, which entails the gradual introduction of a compressed gas into a liquid solution of the solute of interest in a primary organic solvent. This method is based on the ability of liquids to solubilize large amounts of gases. This solubilization generally induces large volumetric expansions of the liquid phase (severalfold) and a decrease of its density up to a factor 2. Figure 3 shows experimental expansion behavior of a binary mixture consisting of CO_2 and Ethanol (75). The initial part of the curve (low pressure levels) shows a linear increase of the liquid volume; a sharp variation appears in the middle part of the curve (at 4 MPa and 318 K). The last part of the curve is almost vertical, indicating the transition from a two-phase to a single-phase system. One can note that at $T = 318$ K and $P = 7.8$ MPa, the initial volume of the liquid phase was multiplied by a factor 7.5. Other common binary systems involving carbon dioxide were studied by Kordikowski et al. (76).

When a solid has been solubilized in the liquid prior to the introduction of the compressed gas, the volumetric expansion is accompanied by a decrease of the liquid solvent strength, which causes the solid to precipitate as ultra fine particles. The physicochemical properties of the solute of interest strongly influence the choice of a solvent/antisolvent pair. The antisolvent should have appreciable mutual solubility with the solvent and should have little or no affinity for the solute. As will be seen, the solute–solvent affinity is also an effective factor that can strongly influence the morphology of the end product.

The main advantages of the GAS process over the RESS are as follows: (a) lower pressure levels (most applications reported involve pressures that do not exceed 10 Mpa, a low level of pressure that is often accompanied by milder conditions of temperature than in the RESS, (b) wider range of compounds that can be processed, and (c) greater number of process parameters to control particle size and morphology.

On the other hand, the GAS process exhibits some drawbacks in comparison to the RESS.

1. The use of a primary organic solvent can be problematic if residual traces remain in the final product; this can constitute a serious limitation in drug processing.
2. The progressive addition of a compressed gas leads to variable supersaturation ratios in the liquid phase, which can result in broader particle size distribution.
3. Generally, the GAS process yields larger particles than the RESS because lower levels of supersaturation ratios are reached in the former.
4. Particles are mainly produced in a liquid phase, requiring an additional stage of drying, which is carried out by flushing pure fluid during an appreciable time.
5. Because conditions of pressure are transient, the process scale-up can be somewhat difficult.

TABLE 1   RESS Experiments for Pharmaceutical Products

Substrate	Solvent (+ cosolvent)	P (MPa)[a]	T (K)[a]	Morphology / particle size (μm)	Ref.
α-Asarone	CO_2	—	—	—	Chen (40)
Benzoïc acid	CO_2	20–28	308–328	—	Tavana (42)
Benzoïc acid	CO_2	16–28	308–338	1–2	Berends (43)
Benzoïc acid	CO_2	13–30	350–420	0.2–1.3	Cihlar (44)
Benzoïc acid	CHF_3	13–20	350–415	0.8–1.2	Türk (45)
Benzoïc acid	CO_2	16–20	373–403	Needles, 2–3	Domingo (46)
Benzoïc acid	CO_2	20–30	350–397	0.2– 1.4	Helfgen (47)
Benzoïc acid	CO_2	13–30	300–350	0.2–0.7	Helfgen (38)
Benzoïc acid	CO_2	20	360–420	0.2–0.5	Türk (39)
β-Carotene	CO_2	17.5–25	378–330	0.3–20	Peters (48)
β-Carotene	C_2H_4 + toluene	31	343		Chang (49)
β-Estradiol	CO_2	34.5	328	<1	Krukonis (29)
β-Sitosterol	CO_2	20	360–420	0.15–0.25	Türk (39)
Caffeine	CO_2			1–10	Reverchon (50,51)
Caffeine	CO_2	15	369–383	Needles, 3–5	Subra (52)
Carbapenem antibiotic + docusate acid	CO_2 + ethyl acetate	8	343	Needdles, 8	Merrifiel (36)
Cholesterol	CO_2	19.3	308	Microspheres, 2–3	Sievers (41)
Cholesterol	CO_2			5–50	Frederiksen (53)
Cholesterol	CO_2	20–30	353–422	<0.35	Türk (45)
Cholesterol	CO_2, CHF_3	20–30	350–387	0.2	Helfgen (47)
Cholesterol	CO_2	—	—	0.4	Kröber (54)

Compound	Fluid				Reference
Cholesterol	CO_2	13–30	300–350	0.2	Helfegen (47)
Cyclosporine	CO_2	14	333	Expansion in water	Pace (33)
Cyclosporine	CO_2	345	333	Expansion in water	Young (37)
Cyclosporine	CO_2	—	—	0.02–0.09	Godinas (55)
Denbufylline	CO_2	—	—	—	King (56)
Dipalmitoylphosphatidylcholine (DPPC)	CO_2 + EtOH	19.3	308	—	Sievers (41)
Eudragit	CO_2 + EtOH	—	—	—	Mishima (58)
Fenofibrate	CO_2	21	—	Expansion in water	Pace (33)
Fenofibrate	CO_2	—	—	0.2	Godinas (55)
Flavone	CO_2 + EtOH	25	308	+ PEG, 1–5	Mishima (28)
3-hydroxyflavone	CO_2 + EtOH	—	—	+ PEG, Microspheres, 10	Mishima (57)
Griseofulvin	CHF_3	18–22	333	1–10	Reverchon (58)
Griseofulvin	CHF_3	13–30	300–350	0.25	Helfegen (38)
Griseofulvin	CHF_3	20	360–420	0.15–0.25	Türk (39)
Hydrogenated palm oil	CO_2	11	353	5	Peirico (59)
Ibuprofen	CO_2	13–19	308	<2	Charoenchaitrakool (60)
Lazaroid	CO_2 + EtOH	19.3	308	—	Sievers (41)
Lidocaine	CO_2	—	—	0.01	Frank (35)
Lidocaine	C_2H_4 + toluene	37.9	—	0.13–0.25	Mohamed (64,65)
Mevinolin	CO_2	—	—	1	Krukonis (29)
Mevinolin	CO_2	—	—	10–50	Larson (63)

TABLE 1 Continued

Substrate	Solvent (+ cosolvent)	P (MPa)[a]	T (K)[a]	Morphology / particle size (μm)	Ref.
Nabumetone	CO_2	4	323	Microspheres, 0.02	Merrifiel (36)
Naproxen	CO_2	12.5–40	328	+ PLA, 4–12	Kim (27)
Naproxen	CO_2	60	343	1–3	Stahl (66)
Phenacetin	CO_2-CHF_3	—	—		Loth (67)
Phenatren	CO_2	16–20	373–403	2–8	Domingo (46)
Progesterone	CO_2	—	—	Microspheres, 2–5	Coffey (32)
Salicylic acid	CO_2	—	—	Needles, 1–170	Reverchon (50,51)
Salicylic acid	CO_2	22.3	318	<4	Subra (68)
Salicylic acid	CO_2	16–20	373–403	1–5	Domingo (69)
Steroids	CO_2	13–25	313–333	1–10	Alessi (70)
Stigmasterol	CO_2	10–15	373	Whiskerlike crystal, 0.05–2	Ohgaki (71)
Ropivacaine + lidocaine	CO_2 + H_2O	—	—	0.1–1	Brodin (62)
Testosterone	CO_2	—	—	<0.4	Coffey (32)
Theophylline	CO_2	22.5	338	Microspheres, 2–5	Subra (68)
Tropic acid ester	CO_2	—	—		Peirico (59)

[a] Pressure and temperature given for preexpansion conditions.

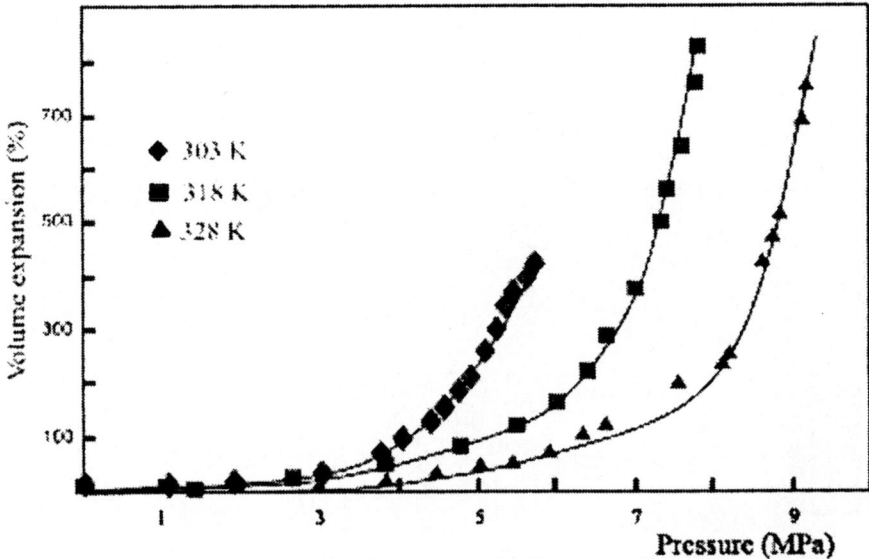

FIGURE 3 Volumetric expansion of CO_2–ethanol mixtures with pressure and temperature.

In the schematic of the GAS process shown in Figure 4, the compressed gas in tank 1 is pumped with a high pressure pump 2 and introduced into a buffer vessel 3. It can be fed to the crystallization vessel 4 either through the top 5 or most commonly through the bottom 6, so that it bubbles in the liquid. The autoclave is equipped with a circulating water jacket to maintain the working temperature during the entire process. The autoclave is also provided with a stainless steel frit filter 7 and a stirring device 8. At the end of the precipitation step, the fluid content of the precipitator is flushed to atmospheric pressure in a separation vessel 9, where the gas 10 and the liquid phase 11 are separated.

2.2. Influence of Process Parameters

Upon careful review of the literature, despite certain possible discrepancies, some general trends become apparent regarding the influence of operating parameters on particle size and morphology. Since the driving force of crystallization is supersaturation, any operating parameter influencing that state, will act on both morphology and particle size. Temperature, mixing conditions, and solute concentration are of course relevant parameters, but

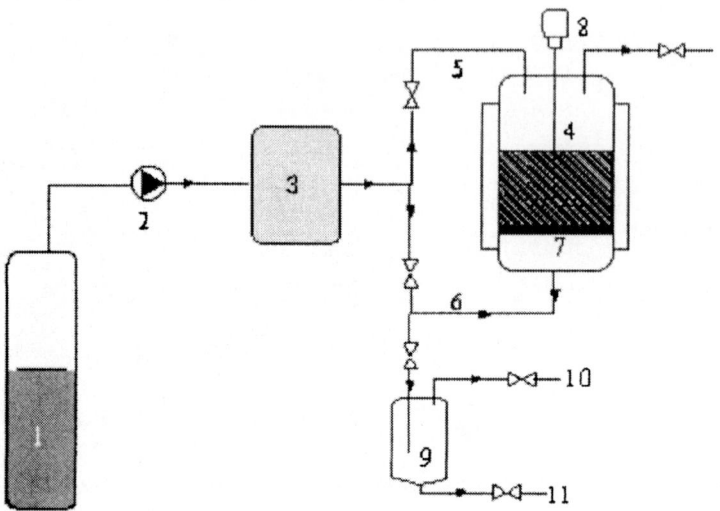

FIGURE 4 Schematic flow diagram of the GAS process; for 1–11, see text.

very marked effects are induced by pressure variations. The mode of adding the compressed gas governs the pressure modification inside the precipitator and therefore, the supersaturation profile. For this reason, several pressurization patterns were proposed to afford better control of the properties of the end product.

In most cases, pressure is linearly increased up to its final value, which is kept constant during a certain time to favor the crystal growth. Next, a filtration and a washing step are carried out and the vessel is depressurized. In this mode, the slopes of the initial part of the pressure profile as well as the duration of the crystal growth period are the control parameters. In modifications of this basic method suggested by Berends et al. (77), pressure was raised linearly until saturation and turbidity were reached. A "pulse period" was then performed to favor nuclei creation. During the "growth period," pressure was increased up to its highest value, which remained constant during a "holding period" involving the filtration and the washing steps.

Kitamura et al. proposed three different ways to manage the pressure profile (78). In the "stepwise method," carbon dioxide was added gradually in 5-bar increments until the maximum pressure was reached. In the "rapid pressurization method" the maximum pressure was reached as rapidly as possible. Finally the so-called two-step pressurization involved an initial increase in pressure as rapidly as possible, followed by reduction to a "decompression point" to control the crystal growth; this second step is similar to the "pulse period" suggested by Berends et al. (77), even though the goal was different.

It is noteworthy that the addition of the compressed gas is accompanied by a temperature increase, which can influence the particle size. Because rapid introduction of gas results in a fast increase of pressure, under such conditions, efficient heat transfer must be ensured via the jacket. Wubbolts et al. suggested the use of liquid CO_2 instead of $scCO_2$, with the micronization processed under helium pressure (79).

Finally, it should be pointed out that a constant pressurization rate does not result in a linear volumetric expansion of the liquid phase.

Most of the studies reported evidence that increasing the rate of the gas entry results in the production of smaller and more uniform particles (18, 22,78,80–84). However one can find some exceptions to this rule (79,85). Besides the rate of pressure increase, the level of final pressure is also of great importance (78,86).

The temperature trends are not marked as those with pressure. Some reports show that an increase in temperature leads to bigger particles (74,85,87); this increase was generally attributed to agglomeration of small particles. However, for some molecules, no effect was detected (76,88,89).

The solute concentration in the liquid solution is of course a major factor in the process. In most cases, high initial concentration resulted in larger particles owing to strong solute–solvent interaction (79,82,85,88–90), while at low concentration fine particles were formed but were generally agglomerated.

Finally, it is commonly admitted that stirring the liquid phase enhances the mass transfer rate and therefore favors production of smaller particles. However, it was pointed out that a high stirrer frequency may be responsible of significant attrition (77).

This short review shows that it is rather difficult at present to connect the process parameters to of micronization to the efficiency of the undertaking. This is due to the strong interrelation between the kinetic and thermodynamic phenomena involved in the particle production.

Results related to drug micronization by GAS are summarized in Table 2, which also includes some data dealing with biocompatible polymers often used as carriers in sustained delivery devices (91–95). Unless otherwise noted, the antisolvent used in these studies is carbon dioxide. The GAS technique was successfully carried out to achieve the fractionation of various mixtures (96,97).

A related method, termed depressurization of an expanded liquid organic solution (DELOS), was recently developed by Ventosa et al. (98). This technique was carried out in an apparatus very similar to the one used for the GAS. In the study reported, depressurization of the expanded liquid phase

TABLE 2 GAS Experiments for Pharmaceutical Products[a]

Substrates	Solvent	Solute concentration	P (Mpa)	T(K)	Particle features	Observations	References
Acetaminophen	Ethanol	7 wt %	Atmospheric to 7.1 atm	293	—	Almost no product after washing	Wubbolts (79)
Acetaminophen	Ethanol	7 wt %	Atmospheric to 7.1 atm, in 7.5 min	293	Platelets		Wubbolts (79)
Acetaminophen	Ethanol	14wt %	Atmospheric to 7.1 atm	293	Platelets		Wubbolts (79)
ACP	DMSO	0.1–5wt %	Final pressure: 8–12 atm P gradient: 0.3–2 MPa/min	298–333	Spheres 0.1–1 μm		Pallado (91)
ALAFF	DMSO	0.1–5wt %	Final pressure: 8–12 atm P gradient: 0.3–3 MPa/min	298–333	Spheres 0.1–1 μm		Pallado (91)
Carbamazepine	Acetone	—	Final pressure: 7 atm	313	Needles, 31 μm	Untreated CBZ: prismatic	Moneghini (90)
Carbamazepine	Acetone Ethyl acetate DCM	—	Final pressure: 5–6 atm	313	Needles	Best results with ethyl acetate	Kikic (92)
Composite CBZ-PEG 4000	Acetone	—	Final pressure: 7 atm	313	Needles and few agglomerates	CBZ/PEG : 5:1	Moneghini (90)
Composite CBZ-PEG 4000	Acetone		Final pressure: 7 atm	313	Needles and agglomerates	CBZ/PEG : 1/1.5	Moneghini (90)
Composite CBZ-PEG 4000	Acetone		Final pressure: 7 atm	313	Agglomerates	CBZ/PEG : 1/11	Moneghini (90)
Chlorpropamide	Acetone	20 wt %	0.02 atm and 1.75 MPa/min	303	Columnar habit		Yeo (95)
Chlorpropamide	Ethyl acetate	5 wt %	0.02 atm and 1.75 Pa/min	303	Platy crystals		Yeo (95)

Compound	Solvent	Concentration	Conditions	Temperature	Morphology	Notes	Reference
Cholesterol	Acetone	3.5 wt. %	1.0 mL CO_2/min	308	Needles		Liu (86)
Cholesterol	Acetone	3.5 wt. %	2.5 mL CO_2/min	308	Tablets		Liu (86)
Cholesterol	Acetone	3.5 wt. %	2.5 mL CO_2/min	308	Large tablets		Liu (86)
Cu2(indomethacin)	DMF		Final pressure: 5.8 atm	298–313	Rhombic crystals 15 μm		Warwick (93)
Cu2(indomethacin)	DMF		Slow expansion Final pressure: 5.8 atm	298	Rhombic, 100 μm, bipyramidal, 10 μm		Warwick (93)
Cu2(indomethacin)	DMF	5–200 mg/g	Rapid expansion Final pressure: 5.9 atm reached in 10 or 120 min	298–313	Rhombic and bipyramidal		Warwick (84)
Cu2 (indomethacin)	DMSO	5–200 mg/g	Final pressure: 5.8 atm	298–313	Clusters of needles, 50–100 μm		Warwick (84)
Cu2 (indomethacin)	NMP	5–200 mg/g	Final pressure: 5.8 atm	298–313	Clusters of needles	Warwick (84)	Warwick (84)
p-HBA	Methanol	5 wt %	—	298–318	Spheres, 1 μm, platelets, 10–20 μm	Increasing T produces smaller particles	Thiering (82)
p-HBA	Acetone	35 wt %	—				Thiering (82)
p-HBA	Methanol	5 wt %	9 MPa/h	308	Dendritic Needles, 1.0 mm	Amorphous particles at high pressurization rate	Thiering (82)
HYAFF-11	DMSO	0.5, 1.0, 1.5 wt %	2–75 bar/h Final pressure: 10 atm P gradient: 0.5–2 MPa/min	308–323	Spheres Spherical Mean size: 0.35 μm	Particles solubilized during washing for some drugs	Bertucco (89)
HYAFF-11–drugs	DMSO	1.0 wt %	Final pressure: 10 atm P gradient: 0.5–2 Mpa/min	308–313	Spheres, 0.32–0.4 μm		Bertucco (89)
HYAFF-7 HYAFF-11	DMSO	0.1–5 wt %	Final pressure: 8–12 atm P gradient: 0.3–2 MPa/min				

TABLE 2 Continued

Substrates	Solvent	Solute concentration	P (Mpa)	T(K)	Particle features	Observations	References
Insulin	DMSO	5 mg/cm³	Final pressure: 8.62 atm 0.057 MPa/min	308	Spheres 4 µm		Yeo (22)
Insulin	DMSO	5.1 mg/cm³	0.004–3.5 Mpa/min	308–323	1.4–8 µm	Spheres at low T Nonspherical at high T Aggregates at high pressure increase	Thiering (85)
Insulin	Methanol	(saturated)	0.03–3.5 Mpa /min	308	0.1–1 µm	Aggregates at high pressure increase	Thiering (85)
Insulin	Methanol	12.5–100% of saturation		308	0.05–0.6 µm and aggregates	Aggregates at low concentration	Thiering (85)
Composite/ Insulin–PEG/PLA	—	—		—	Spheres, 0.4–0.6 µm	Better loading with low MW PEG	Elvassore (94)
Lysozyme	DMSO	0.3–2.5 mg/cm³	0.004–3.5 MPa/min	291–318	Spheres, 0.05–0.3 µm	Non-effect of T Aggregates at high pressure increase	Thiering (85)
Myoglobin	DMSO		0.07 MPa/min	308	0.05–0.3 µm	Monodisperse powders	Thiering (85)
PLGA	Acetone	37.5 g/L		298	Spheres, 50 nm	Tightly agglomerated	Dillow (87)
	Acetone	37.5 g/L		273–288	Larger particles	Less aggregation	
	Acetone	104 g/L	Final pressure: 6 atm	299	Non particles	Formation of a film	
Sulfothiazole	Ethanol	0.0143 mole/L	Final pressure: 5 and 5.8 atm; increments of 0.5 MPa/min	298	Needles, 2–6 mm		Kitamura (78)

Sulfothiazole	Ethanol	0.0143 mole/L	Final pressure: 5–5.5–5.8 atm Rapid pressurization	298	Needles 1.5–4 mm		Kitamura (78)
Sulfothiazole	Ethanol	0.0143 mole/L	Two-step pressurization	298	45–280 µm		Kitamura (78)
Sulfothiazole	Acetone	1 wt %	First pressure: 6–6.2–6.5 atm, 0.02 Mpa/min	303–313	Tablets	Particles smaller than untreated for high injection rate and bigger	Yeo (95)
Sulfothiazole	Methanol	1 wt %	1.75 MPa/min 0.02 MPa/min	303–313	Tablets	No significant size modification at high injection	Yeo (95)
Drug	Ethanol		1.75 MPa/min 10–220 mL/min	278–323	Needles 0.24–10 µm	Amorphous spherical particles, more or less agglomerated	Muller (83)
Drug	Acetone Acetonitrile		10–220 mL/min	278–323	0.24–10 µm	Needlelike and prismatic crystals	Muller (83)

[a] Abbreviations: ACP, cross-linked polysaccharide of hyaluronic acid; ALAFF, polymer of an ester of alginic acid; CBZ, carbamazepine; PEG, polyethylene glycol; DCM, dichloromethane; DMF, dimethylformamide; DMSO, dimethyl sulfoxide; NMP, N-methylpyrrolidone; P-HBA, p-hydroxybenzoic acid; HYAFF-1, hyaluronic acid benzylic ester; PEG-PLA, poly(ethylene glycol)–poly(L-lactide); PLGA, poly(lactic-co-glycolide) acid.

was performed in two steps, and the apparatus was equipped with two different frit filters, to collect separately the particles formed in the two depressurization steps. Thus, the influence of the supersaturation ratio on particle size could be studied, and it was shown that low supersaturation results in much smaller particles. For a given work temperature, the initial composition of the liquid phase, the molar fraction of CO_2 in the expanded phase, and the pressure levels of depressurization determine domains in which the GAS or the DELOS techniques is appropriate.

3. PRECIPITATION WITH A COMPRESSED FLUID ANTISOLVENT (PCA), A SUPERCRITICAL ANTISOLVENT (SAS), AND THE AEROSOL SOLVENT EXTRACTION SYSTEM (ASES) PROCESSES

3.1. Principle

Dozens of drugs have been recrystallized by means of the precipitation with a compressed fluid antisolvent (PCA) and the supercritical antisolvent (SAS) and aerosol solvent extraction system (ASES) processes.

These three precipitation processes involve the dispersion of a solution of the substrate of interest, dissolved in an organic solvent, through a capillary or a nozzle and into a continuous supercritical or subcritical antisolvent phase, which generally sweeps the vessel. In most cases, the experimental procedure is as follows (see Figure 5). The antisolvent is first introduced into the precipitation vessel. When the experimental pressure has been reached, an outlet valve is opened to maintain a constant pressure and a constant antisolvent flow rate. Then, there are two alternatives. Pure solvent is introduced upon reaching and the steady state (constant ratio of antisolvent and solvent), the liquid phase turns from pure solvent to the liquid solution of the substrate. The other alternative is to introduce the solution of the substrate directly into the pure antisolvent phase. In that case, the supercritical phase composition evolves and a certain time is required before the steady state is reached. In both cases, the solution is introduced through a capillary or a nozzle. The dispersed liquid phase can form a liquid jet or can be atomized in fine droplets depending on the injection device (geometry, orifice diameter), the liquid flow rate, the operating pressure and temperature, and so on. During the process, the precipitation results from two phenomena: the fast diffusion of the antisolvent into the liquid phase and the evaporation of the organic solvent into the continuous phase, generally in a supercritical state. Both transfers rapidly yield supersaturation, which causes the substrate to precipitate in the form of nano- or microparticles. The fluid mixture

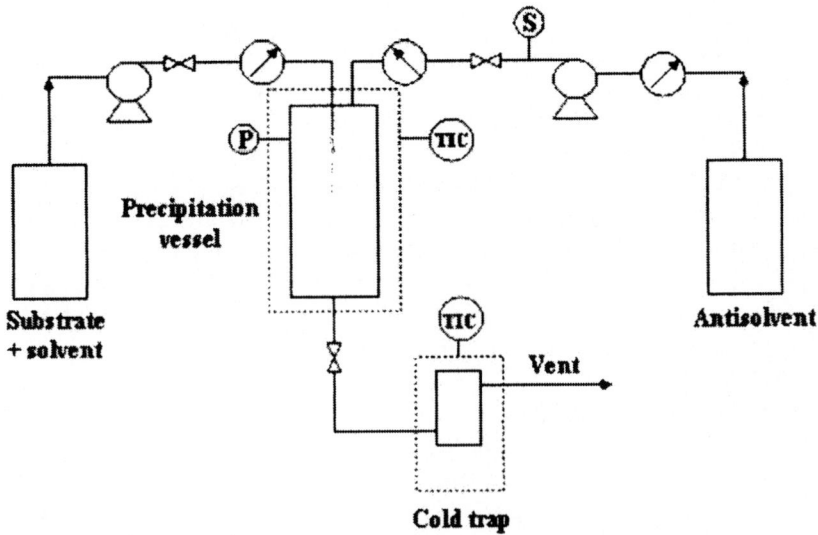

FIGURE 5 Apparatus for PCA, SAS, and ASES processes: TIC, temperature indicator/controller.

(antisolvent + solvent) flows to a cold trap at atmospheric pressure, where the separation of gas and liquid occurs.

When the injection of liquid is finished, a washing step is carried out to remove the organic solvent and to prevent it from condensing during the depressurization step. For this purpose, the feed of pure antisolvent is maintained, to renew the vessel content. The vessel pressure is then reduced to atmospheric pressure, and solid particles are collected on a filter at the bottom of the vessel and/or on the walls of the vessel; in some cases, metallic or polymeric baskets are used for the particle harvesting.

Authors differ with respect to the duration of each stage of the process, the process mode (batch or semibatch), the capillary/nozzle, and so on. The nozzle orifice diameter can vary from 20 to 760 μm. As detailed next, some authors have modified the introduction device to improve the dispersion and to enhance mass transfer.

3.2. Publications and Patents

Reverchon (6) and Jung (11) have reviewed patents and publications dealing with supercritical precipitation processes.

The first papers on these antisolvent processes applied to drugs are related to the precipitation of pure components. The required goals were to reduce drug particle size to permit use for aerosol delivery or to enhance a drug's dissolution rate and/or bioavailability (21,22,79,99–123). Papers reported later dealt with the coprecipitation of a drug and a biocompatible polymer (112,124–129). Some authors compared the characteristics of drugs or composites formed by antisolvent processes to those formed by conventional precipitation processes. The effect of process parameters on the morphology, particle size, and particle size distribution are often investigated. Recent papers presented the improvements brought to traditional SAS process to enhance mass transfer and avoid agglomeration (100,106,111, 113,117,125).

Table 3 lists different drugs precipitated by the SAS, ASES, and PCA processes. Almost all authors used carbon dioxide as the compressed antisolvent.

Yeo and Debenedetti (22,26) were the first to use a supercritical antisolvent process to micronize proteins. The purpose was to obtain small particles suitable for inhaler formulation. Indeed, a particle diameter lower than 5 μm is required for an effective drug delivery to the lungs with a dry powder inhaler, a metered dose inhaler or a nebulizer. Micrometer-sized particles of insulin and catalase were formed from organic solutions. The activity of processed insulin was found to be indistinguishable from that of the unprocessed material. Previous investigations (115) on lysozyme, trypsin, and insulin showed that the supercritical process results in changes in the secondary structure with respect to the commercial materials. However, characterization of the redissolved precipitates shows that the changes are reversible. This work tended to show that the particles formed were suitable for direct aerosol delivery and long-term storage at ambient conditions.

In 1933 Lim and Debenedetti (107) patented the formation of protein microparticles by antisolvent precipitation. Since that date, other patents dealing with the formation of protein by the same processes have been filed. Recently, Niu (100) patented the formation of insulin and albumin using an ultrasonic nozzle.

Steckel et al. (105) used the ASES process to micronize several steroids and particles with the required size were obtained. No chemical decomposition was observed. The wettability of the active products was improved upon addition of a physiological surfactant to the initial solution. This is advantageous for suspensions of the drug particles in water, hydrophilic solvents, or propellants. Some modifications were brought to the conventional process. First, once the organic solution spraying has been done, the antisolvent injection is stopped and a sedimentation phase is allowed to take place: this phase is claimed to increase the yield. Second, the removal of the residual solvent is improved by using an adsorber filled with active carbon. In 1998, Steckel and Muller (110) characterized Fluticasone-17-proponiate microparticles for their particle size, morphology, and crystallinity. The particle size distribution appeared to be very narrow. Slight differences in the Debye–Scherrer diagrams of the ASES processed product and of the raw material were observed. A pseudopolymorphic modification of fluticasone-17-propionate was suggested following the inclusion of methylene chloride. Twin impinger experiments with metered dose inhaler formulation yielded information about the aerodynamic behavior of the aerosol droplets. In studies of the influence of surfactant type and micronization technique, the ASES process were shown to efficiently micronize this anti-inflammatory drug to a particular shape, with a surfactant coating.

Reverchon et al. micronized several antibiotics by the SAS process by using different organic solvents (101–103,120). These authors investigated the influence of solvent type, operating pressure and temperature, and the organic solution concentration on particle size, particle size distribution, and morphology. It was demonstrated that choice of solvent is a key factor for successful precipitation using the SAS process (101). Two antibiotics were completely extracted by the supercritical phase with two given solvents, and they could be successfully micronized following a change of the organic solvents (see Table 3). The effects of different process parameters on particle size, morphology, and particle size distribution were studied for the tetracycline–N-methylpyrrolidone system. The coalescence of particles decreased with increasing operating pressure, while the diameter of the aggregates increased with increasing tetracycline concentration in the initial solution.

Amoxicillin was micronized from dimethyl sulfoxide (DMSO) and N-methylpyrrolidone solutions (102). From the DMSO solutions, two different morphologies were obtained in the same experiment, while solutions of N-methylpyrrolidone yielded spherical and nonagglomerated particles. Concentration appeared to have the major influence on the particle size, and the weight fraction of the organic solvent must be below a certain value to avoid sintering.

Salbutamol (121) was precipitated from DMSO solutions; the influence of the initial solution concentration on the dimension of the salbutamol particles was investigated. Interesting results were obtained, since salbutamol formed "balloons," hollow spheres, and rodlike microparticles with a tendency to connect together. With increasing the concentration, the organization of the rodlike particles changed. To interpret the formation of the balloons, Reverchon described the dispersion of the liquid phase. The liquid droplets formed at the injector are subject to a very fast expansion owing to antisolvent diffusion, when saturation is reached on the droplet surface, an outer skin of solute is formed.

TABLE 3 PCA, SAS, and ASES Experiments for Pharmaceutical Products

Substrate	Solvent	Antisolvent	P (MPa)	T (K)	Nozzle/capillary dimension (μm)	Morphology/particle size[a]	Reference
Acetaminophen	Ethyl alcohol	CO_2	7.1	293	—	Agglomerates	Wubbolts (79)
Acetaminophen	Ethyl alcohol	CO_2	6.2	298	Orifice diameter, 67	Crystals, 200 μm	Wubbolts (99)
Ascorbic acid	—	CO_2	6.2 / 11.5	298 / 318	Orifice diameter, 67	Large crystals Aggregated particles, 1–5 μm	Wubbolts (99)
Albumin	1,1,1,3,3,3 Hexa-fluoro-2-propanol	CO_2	6.7–13.4	323–333	Ultrasonic nozzle	1–10 μm	Niu (100)
Amoxicillin	Dimethyl sulfoxide	CO_2	15	313	Orifice diameter, 60	Film	Reverchon (101)
Amoxicillin	N-Methylpyrrolidone, dimethyl sulfoxide	CO_2	12–15	308–323	Orifice diameter, 60	Spherical particles $0.2 < d_p < 1.2$ μm	Reverchon (102)
Amoxicillin	N-Methylpyrrolidone	CO_2	15	313	Tube-in-tube injector Orifice diameter, 500	Spherical particles $0.3 < d_p < 1.2$ μm	Reverchon (103)
Ampicillin sodium salt	Dimethyl sulfoxide	CO_2	15	313	Orifice diameter, 60	Tightly networked particles	Reverchon (101)
Ascorbic acid	Ethyl alcohol	CO_2	7.5–10	308–313	Orifice diameter, 100–300	Crystalline products, 1–10 μm	Weber (104)
Aspirin	Ethyl alcohol	CO_2	7.5–10	308–313	Orifice diameter, 100–300	Crystalline products, 1–10 μm	Weber (104)
Beclomethasone-17, 21-dipropionate	Methylene chloride, methanol	CO_2	8.5	313	Orifice diameter, 300	Agglomerated microspheres, $d_p < 5$ μm	Steckel (105)
Betamethasone-17-valerate	Methylene chloride, methanol	CO_2	8.5	313	Orifice diameter, 300	Agglomerated microspheres, $d_p < 5$ μm	Steckel (105)
Budesonide	Methylene chloride, methanol	CO_2	8.5	313	Orifice diameter, 300	Agglomerated microspheres, $d_p < 5$ μm	Steckel (105)

Compound	Solvent	Gas		Temp	Conditions	Result	Reference
Camptothecin	Dimethyl sulfoxide	CO_2	10	308	Orifice diameter, 100	0.5–20 μm	Said (106)
Catalase	Ethyl alcohol–water	CO_2	10.7	308	Laser-drilled platinum disk	Microparticles, 1 μm	Lim (107)
Catalase	Ethyl alcohol–water	CO_2	9	308	Orifice diameter, 20	Spherical or rectangular microparticles	Tom (21)
Chloramphenicol–urea	Ethyl alcohol	CO_2	7.5–15	313	Orifice diameter, 100–300	Different morphologies	Weber (108)
Chlorpeniramine maleate	Methylene chloride	CO_2			Orifice diameter, 100	1–5 μm	Bodmeier (109)
Dexamethasone-21-acetate	Methylene chloride, methanol	CO_2	8.5	313	Orifice diameter, 300	Agglomerated microspheres, $d_p < 5$ μm	Steckel (105)
Flunisolide	Methylene chloride, methanol	CO_2	8.5	313	Orifice diameter, 300	Agglomerated microspheres, $d_p < 5$ μm	Steckel (105)
Fluticasone-17-propionate	Methylene chloride, methanol	CO_2	8.5	313	Orifice diameter, 300	Agglomerated microspheres, $d_p < 5$ μm	Steckel (105)
Fluticasone-17-propionate	Methylene chloride	CO_2	8.5	313	Orifice diameter, 300	Median particle size, 1.7 μm	Steckel (110)
Gentamicin–PLA	Methylene chloride	CO_2	8.5–9	308–311	Ultrasonic spray nozzle	Microspheres, d_p: 1 μm	Meyer (124)
Gentamicin–L-PLA	Methylene chloride	CO_2	8.5–9	308–311	Sonicated spray nozzle	Spheres, $0.2 < d_p < 1$ μm	Falk (125)
Griseofulvin	Methylene chloride, dimethyl sulfoxide	CO_2	15	313	Orifice diameter, 60	Long needles[b]	Reverchon (101)
Griseofulvin	Tetrahydrofuran, methylene chloride	CO_2	9.65	308	Orifice diameter, 75	Fiber, acicular crystals, spherical particle, $0.13 < d_p < 0.52$ μm	Chattopadhyay (111)
Griseofulvin	Chloroform, methylene chloride	CO_2 and others	6.71 10.11	293 309	Orifice diameter, 30, 50, and 100	Platelike crystals, needle-shaped crystals	Sarkari (112)

TABLE 3 Continued

Substrate	Solvent	Antisolvent	P (MPa)	T (K)	Nozzle/capillary dimension (μm)	Morphology/particle size[a]	Reference
Griseofulvin–L-PLA	Chloroform, methylene chloride	CO_2 and others	6.71 10.11	293 309	Orifice diameter, 30, 50, and 100	Spherical particles with or without agglomeration	Sarkari (112)
Griseofulvin	Tetrahydrofuran, methylene chloride	CO_2	9.65	308	Orifice diameter, 75	Needlelike crystals, nanoparticles	Gupta (113)
Hydrocortisone	Dimethyl sulfoxide	CO_2	10	308	Orifice diameter, 100	$d_p < 0.6$ μm	Said (106)
Hydrocortisone–poly(vinyl pyrrolidone)	Ethyl alcohol	CO_2	13.3	313		Microparticles containing crystalline hydrocortisone	Corrigan (126)
Hydroquinone	Acetone	CO_2	5.55–6.1	310	Orifice diameter, 67	Micrometer-sized crystals	Wubbolts (114)
Ibuprofen	Dimethyl sulfoxide	CO_2	10	308	Orifice diameter, 100	0.5–1 μm	Said (106)
Indomethacin	Methylene chloride	CO_2			Orifice diameter, 100	1–5 μm	Bodmeier (109)
Insulin	Ethyl alcohol–water	CO_2	13.3	308	Laser-drilled platinum disk	Needlelike and globular microparticles	Lim (107)
Insulin	Ethyl alcohol/water	CO_2	9	308	Orifice diameter, 20	Microspheres and thick needles	Tom (21)
Insulin	Dimethyl sulfoxide Dimethylformamide	CO_2	8.6	298–308	Orifice diameter, 30	Microparticles, $d_p < 4$ μm	Yeo (22)
Insulin	Dimethyl sulfoxide	CO_2	9.06–14.2	301–319	Orifice diameter, 30 or 50		Winters (115)
Insulin	1,1,1,3,3,3-Hexa-fluoro-2-propanol	CO_2	6.7–13.4	323–333	Ultrasonic nozzle	1–10 μm	Niu (100)
Insulin (bovine)–polyllactide glycolide)–polyethylene glycol)		CO_2	10		Orifice diameter, 50	Microparticles	Elvassore (127)

Substance	Solvent	Fluid	Pressure (MPa)	Temperature (K)	Nozzle	Product	Reference
Lobenzarit disodium	Ethyl alcohol–2-propanol	CO_2	12 and 8	278–333	Orifice diameter, 100 and 600	Nano- and microparticles, 0.21 μm–1.8 μm	Amaro-Gonzalez (116)
Lysozyme	Dimethyl sulfoxide	CO_2	7.34–11.5	299.6–318	Orifice diameter, 30 or 50	Microparticles, 1–5 μm	Winters (115)
Lysosyme	Dimethyl sulfoxide	CO_2	9.65	310	Orifice diameter, 75	Micro- and nanoparticles	Gupta (113)
Lysozyme	Dimethyl sulfoxide	CO_2	9.65	310	Orifice diameter, 75	Micro- and nanoparticles	Chattopadhyay (117)
Lysozyme	Methylene chloride	CO_2	6.25–14.1	279–309	Orifice diameter, 50–1020	Micro- and nanoparticles	Sze Tu (118)
Methylprednilosone	Tetrahydrofuran	CO_2, ethane				Microparticles	Schmitt (119)
Naloxone–L-PLA	Methylene chloride	CO_2	8.5–9	308–311	Sonicated spray nozzle	Spheres, $0.2 < d_p < 1$ μm	Falk (125)
Naltrexone–L-PLA	Methylene chloride	CO_2	8.5–9	308–311	Sonicated spray nozzle	Spheres, $0.2 < d_p < 1$ μm	Falk (125)
Naproxen–PLA	Acetone	CO_2	6.7	298	Orifice diameter, 180	Microcomposite particles	Chou (128)
Paracetamol	Ethyl alcohol	CO_2	7.5–10	308–313	Orifice diameter, 100–300	Crystalline products, 1–10 μm	Weber (104)
Paracetamol–ascorbic acid	Ethyl alcohol	CO_2	7.5–15	313	Orifice diameter, 100–300	Microcomposites	Weber (108)
para-Hydroxybenzoic acid	Methyl alcohol	CO_2	6.25–14.13	279–313	Orifice diameter, 180	Crystalline particles Platelets 3 μm long	Sze Tu (118)
Prednisolone	Methylene chloride, methanol	CO_2	8.5	313	Orifice diameter, 300	Agglomerated microspheres, $d_p < 5$ μm	Steckel (105)
Rifampicin	Dimethyl sulfoxide	CO_2	9–18	313	Orifice diameter, 60	Nano- and microparticles $P \le 9$–11 Mpa: $2.5 < d_p < 5$ μm $P \ge 12$ Mpa: $0.4 < d_p < 1$ μm	Reverchon (120)
Saccharose	Ethyl alcohol	CO_2	7.5–10	308–313	Orifice diameter, 100–300	Crystalline products, 1–10 μm	Weber (104)

TABLE 3 Continued

Substrate	Solvent	Antisolvent	P (MPa)	T (K)	Nozzle/capillary dimension (μm)	Morphology/particle size[a]	Reference
Salbutamol	Dimethyl sulfoxide	CO_2	9.5–15	313	Orifice diameter, 60	Rodlike particles: L, 1–3 μm; d_p, 0.2–0.35 μm	Reverchon (121)
Salbutamol	Ethyl alcohol–water	CO_2	14.5	323	Orifice diameter, 60	Very large particles	Reverchon (121)
Soy lecithin	Ethyl alcohol	CO_2	8–11	308	Orifice diameter, 150	Microspheres, d_p 1–40 μm	Magnan (122)
Soy lecithin	Ethyl alcohol	CO_2	8–12	303–323	Orifice diameter, 150	Microspheres, d_p 1–40 μm	Badens (123)
Tetracosactide/L-PLA	Methylene chloride, methanol	CO_2				Microspheres: mean volume diameter, 10 μm	Bitz (129)
Tetracycline	N-methyl 2-pyrrolidone	CO_2	15	313	Orifice diameter, 60	Spheres	Reverchon (101)
Tetracycline	Tetrahydrofuran	CO_2	9.65	308	Orifice diameter, 75	Nanoparticles	Gupta (113)
Tetracycline	Tetrahydrofuran	CO_2	9.65	308	Orifice diameter, 760	Micro- and nanoparticles	Gupta (113)
Triamcinolone acetonide	Methylene chloride, methanol	CO_2	8.5	313	Orifice diameter, 300	Agglomerated microspheres, d_p < 5 μm	Steckel (105)
Trypsin	Dimethyl sulfoxide	CO_2	7.34–13.6	299.6–319.5	Orifice diameter, 30 or 50		Winters (115)
Urea	Ethyl alcohol	CO_2	7.5–10	308–313	Orifice diameter, 100–300	Crystalline products, 1–10 μm	Weber (104)

[a] d_p, pattern difference.
[b] Some as long as millimeters.

In 2002, Reverchon (120) micronized rifampicin from DMSO solutions. Indeed, this antibiotic is quite insoluble in water, and its dissolution in biological liquids is difficult. The micronization of this drug may reduce the therapeutic dosage and the drug toxicity. The particles precipitated by Reverchon are tight aggregates of nanoparticles. Different particle morphologies were obtained depending on the operating pressure (see Table 3). The effect of solution concentration was also investigated, and an increase in concentration led to an increase in mean particle size and an enlargement of particle size distribution. There was no degradation of the antibiotic following supercritical treatment.

Reverchon also performed pilot-scale micronization of amoxicillin from N-methylpyrrolidone solutions in a semicontinuous plant (103). The amoxicillin yield was about 90%, and the particles obtained were suitable for aerosol and/or injectable formulations.

3.2.1. Modified SAS Process

Chattopadhyay and Gupta (111) used the SAS process modified by enhanced mass transfer (SAS-EM) to produce nanoparticles of the antifungal agent griseofulvin. This antibiotic is orally administrated, and its preparation must be designed for maximum absorption. An improvement in absorption from the gastrointestinal tract can be obtained by the reduction of the particle size. In the SAS-EM process, the solution jet is deflected by a surface vibrating at an ultrasound frequency that atomizes the jet into small droplets. Two advantages were claimed: an enhancement of the mass transfer and less agglomeration. Different sizes and morphologies can be obtained by varying the vibration intensity of the deflecting surface. The particles are approximately 10-fold smaller than those obtained from the conventional SAS process, and there is no significant solvent effect on the particle size and morphology. It is worth noting that earlier work with griseofulvin precipitated from methylene chloride or DMSO by means of the SAS process led to the formation of long millimeter-sized needles (101). Chattopadhyay and Gupta (117) succeeded in micronizing lysozyme by the same SAS-EM technique. Nanometer- and micrometer-sized particles of lysozyme were formed, and the biological activity of the protein was retained during the processing. Once again, the particles obtained are up 10-fold smaller than those obtained by the conventional SAS process. Gupta and Chattopadhyay (113) patented the application of the SAS-EM process to lysozyme, tetracycline, griseofulvin, and fullerenes. An earlier patent (106) proposed a method for enhancing mass transfer rates during the solution dispersion.

Many pharmaceutical drugs are in the form of salts, with relatively high solublity in water and very low solubility in organic solvents. Thus, the application of the antisovent process to aqueous solutions of organic salts requires the use of cosolvents to increase the solubility of the antisolvent fluid in the aqueous phase. Amaro-Gonzalez et al. (116) micronized lobenzarit salts from aqueous solutions with ethyl alcohol and 2-propyl alcohol as cosolvents. The process conditions were optimized on the basis of the maximum solubility of water at saturation in the ternary water–antisolvent–cosolvent, and individual crystals were obtained. The influence of the nozzle diameter on the particle size was studied. The particle size decreased when the nozzle diameter increased, and the morphology was strongly modified. Actually, the particles obtained with the smallest orifice consist of agglomerated nano-crystals, while the biggest orifice led to the formation of individual needles without any agglomeration. In the former case, the resulting crystals indicate the predominance of a nucleation mechanism due to high supersaturation of the droplets, followed by agglomeration.

3.2.2. Composites

Production of several composite particles of poly (L-lactic acid) L-PLA loaded with drugs has been described in literature.

Bitz and Doelke (129) formed tetracosactide–L-PLA composite particles. The authors compared the ASES process with two other processes of micronization and focused on the residual content of solvent, the morphology, the particle size distribution, the drug loading, and the encapsulation efficiency. It appeared that the ASES process gives particles with a size understood to be between the particle sizes obtained with the two other processes. The mean volume diameter is of about 10 μm. The ASES process resulted in the highest yield and a good entrapment efficiency. The residual concentration of methylene chloride in the end product was near or above the legal limit using ASES, while the concentration was below the limit using spray drying process.

Falk (125,130), Meyer (124), and their colleagues used a sonicated spray nozzle to form particles of gentamicin–L-PLA from methylene chloride solutions. Spherical particles were obtained, with diameters ranging from to 0.2 to 1 μm. These authors used hydrophobic ion pairing to solubilize gentamicin in a solvent compatible with L-PLA. The resulting precipitate was a homogeneous dispersion of the ion-paired drug in the L-PLA micro-spheres. The release of the ion-paired drugs into phosphate-buffered saline displayed minimal burst effects and occurred by means of matrix-controlled diffusion (125). Composite particles with similar size were formed with naloxone and naltrexone, also from methylene chloride solutions (125). In 1998 Randolph et al. (131) patented the preparation of pharmaceutical powders containing drug entrapped in a biocompatible polymer; precipitation involved spraying the solution through a sonicated orifice.

Griseofulvin and L-PLA were coprecipitated from chloroform or methylene chloride solutions (112). The precipitation with compressed CO_2

is compared with the precipitation with CO_2-philic liquid solvents. Similar microparticle morphologies were obtained in both cases.

In 2002 the coprecipitation of hydroquinone and poly(vinyl pyrrolidone) (PVP) from ethyl alcohol solutions was reported by Corrigan and Cream (126). As for tetracosactide–L-PLA composites particles earlier (129), the characteristics of hydrocortisone–PVP composites prepared by the antisolvent process were compared with those of corresponding systems prepared by spray drying and conventional solvent evaporation/coprecipitation methods. Pure cortisone samples prepared by the coprecipitation or by the antisolvent process gave products having crystal structure similar to that of the original material. The morphology of crystalline hydrocortisone samples obtained by Corrigan was similar to that obtained by Velaga et al. (132), who used the SEDS process. Hydrocortisone–PVP composites also resulted in microcrystalline powder, while amorphous particles were formed by spray drying, which offers good enhancement of drug dissolution and solubility characteristics. Coprecipitation with PVP may increase the initial and intrinsic dissolution rates of hydrocortisone; however, the systems prepared by the antisolvent process did not dissolve more rapidly than those processed by other systems.

3.3. Future Development

All the successful work cited proves that undoubtedly these antisolvent processes will continue to be used in the future for precipitation or coprecipitation of drugs. The mild conditions of temperature and the quite low level of pressure make these processes very attractive. The use of a solvent may remain a problem because of potential disadvantages. The physicochemical properties of the substrates to be recrystallized are predominant factors in the choice of a solvent; and these materials are not always the easiest to remove from the end product, nor the least toxic.

Future research will probably deal mainly with controlling powder size, morphology, and polymorphism. There is no doubt that more promising advances will be made.

4. SOLUTION-ENHANCED DISPERSION BY SUPERCRITICAL FLUIDS (SEDS) PROCESS

4.1. The Process

The SEDS process appeared not long after the other antisolvent processes just described, and it was first developed and patented by Bradford Particle Design (15). The general principle is the same as that for SAS process: supercritical CO_2 and the organic solution are introduced cocurrently in the reactor, and precipitation of the solid is due to the antisolvent effect. The essential difference is related to the way of introducing the different phases; in SEDS, a coaxial nozzle is used. The liquid solution is introduced in the inner channel while CO_2 flows in the outer tube; however; the inverse process was also proposed. Whatever the configuration used, the geometry given to these two channels results in a premixing chamber located just above the injection point in the supercritical medium. This premixing favors turbulence, hence the dispersion of the organic solution in the CO_2 and the efficiency of the mass transfer. It normally leads to small particles with nanometric or micrometric size. This process was used for many substrates that are difficult to manufacture by the RESS process. Compared with the other antisolvent techniques, SEDS is distinguished by two specific features. First the coaxial introduction of $scCO_2$ and liquid solution with a high velocity of CO_2, leads to turbulent flow and to improved conditions for mixing and particle formation. Furthermore, the use of coaxial introduction fixes the composition of solvent, $scCO_2$ and solid material from the mixing point, which should induce uniform conditions for particle formation (see Chapter 3, by Baldyga, Shekunov, and Henczka, this volume).

4.2. Application of the SEDS Process to the Formation of Pharmaceutical Products

The SEDS process has been used for formulating a number of pharmaceutical products. The first and most simple design of the coaxial nozzle consists of introducing the supercritical fluid (+cosolvent) into the center of the nozzle and filling the outer tube with the organic solution with the substrate to be precipitated. This design was implemented for forming very fine particles of salmeterol xinafoate (15,133,134); Table 4 gives a more detailed description. This design was modified to permit the processing of water-soluble drugs, which is not possible with conventional processes because water is not soluble in the most widely used supercritical fluid (i.e., CO_2). The ability of the SEDS process to process drugs of this kind was demonstrated and applied (135). In this specific case, a triaxial nozzle was used for introducing three streams:

> An aqueous solution of the drug
> An organic solvent that is rather miscible with water and with $scCO_2$
> The $scCO_2$

This design (Figure 6) was used successfully to micronize sugars or proteins (16,136,137). More generally, this design allows dissolution of the active substance in a solvent that is not soluble in supercritical CO_2.

This process yielded significant improvements for formulating drugs (138):

Morphological form. Mainly influenced by the cosolvent used and the working pressure. If these parameters are well controlled, it is possible to obtain particles with suitable habit and no agglomeration.

Polymorphism purity. Pressure and temperature influence this property. For instance, salmeterol xinofoate, an antiasthmatic drug, was processed to obtain the right polymorph with a particle size of 5 μm, which is suitable for respiratory delivery (133). More generally, the SEDS process can accurately control these two parameters.

Crystallinity. This is an important property because the rate of dissolution is faster for amorphous forms than for crystalline forms. Until now this property has not been thoroughly studied.

One drawback of this process is the difficulty of finding good operating conditions good enough to produce drug particles without any trace of solvent. However, some examples were reported of micronized powder without any traces of solvent (see, e.g., Ref. 139).

Bradford Particle Design patented another way to perform the SEDS. In this method, nitrogen was the antisolvent, while the substrate to be micronized was solubilized in CO_2. Salicylic acid was processed by means of this approach (140); Table 4 provides additional examples (15,16,132–134,136, 137,141–156).

5. PARTICLES FROM GAS-SATURATED SOLUTIONS (PGSS) AND RELATED METHODS

Though not termed PGSS by its authors, the technique patented in 1982 by Graser and Wickenhaeuser (157) is very similar to those filed later. This patent describes a method for conditioning fine particles of organic pigments. The process is operated in "benign" organic solvents under pressure and at high temperature. In some cases, the operating pressure can be below the critical pressure of the solvent. It is suggested that at the end of recrystallization, the mixture be forced through a valve by means of an inert gas such as nitrogen.

Later on, a number of patents were filed (158–165) that described techniques for spraying different compounds onto various substrates. In these processes the solid of interest is suspended in a liquid carrier, and a compressed gas is added to the suspension. Then, the mixture is sprayed onto the surface to be coated. In some cases the compressed gas plays two roles, acting as a propellant and also as a solvent. Most of these processes were patented by

Table 4 SEDS Experiments for Pharmaceutical Products

Product	Solvent	Supercritical antisolvent	P (MPa)	T (K)	Morphology / particle size (μm)	References
Acetaminophen	Ethanol	CO_2	–	–	6–8	Gilbert (141)
Acetaminophen	Dichloromethane	CO_2	15	323	+ L-PLA	Hanna (134)
Acetaminophen	Ethanol	CO_2	8–25	313–353	5–20	Bristow (142)
Acetaminophen	Ethanol	CO_2	25–30	308–313	10–40	Shekunov (143)
Albumin	Water, ethanol	CO_2	–	–	0.05–0.5	Bustami (144)
Antibody Fab fragment	Water, ethanol	CO_2	–	–	–	Sloan (145)
Aspartame	Methanol, acetone	CO_2	10	308	+ Ethyl cellulose / needles	Hanna (146)
Carbamazepine	Dichloromethane, methanol	CO_2	8–25	318–358	–	Edwards (147)
Ibuprofen	Methanol	CO_2-N_2	15	313	14	Hanna (140)
Insulin	Water, ethanol	CO_2	–	–	0.05–0.5	Bustami (144)
Insulin	Ethyl acetate	CO_2-N_2	13–16	308–313	+ DL-PLG / 10–60	Ghaderi (148)
Insulin	Acetone	CO_2	9–18	313–363	–	Velaga (132)
Ketoprofen		CO_2-N_2	20	323	Fine fluffy white powder	Hanna (140)
Lactose	Methanol	CO_2	15–30	323–363	3–10.5	Palakodaty (149)
Insulin	Water, methanol	CO_2	15–27	323–343	–	Hanna (16,136,137)
Lysozyme	Ethanol	CO_2	20	328	0.78	Sloan (150)
Insulin	Dimethyl sulfoxide	CO_2	8–15	313–323	1–5	Moshashaée (151)
Insulin	Ethanol	CO_2	–	–	0.5–5	Gilbert (141)
Insulin	Ethanol	CO_2	–	–	0.05–0.5	Bustami (144)
Maltose	Water, ethanol	CO_2	25	343	Spongy spheres	Hanna (16)

Compound	Solvent	Gas		Temp	Notes / size	Reference
Nicotinic acid	Ethanol	CO_2	—	—	0.4–0.75	Hanna (133)
Nicotinic acid	Methanol	CO_2-N_2	20	338	—	Hanna (140)
Nicotinic acid	Ethanol	CO_2	—	—	—	Rehman (152)
p-HBA	Ethanol	CO_2	—	—	+ PLGA, PLA	Sze Tu (153)
Plasmid DNA pSVb	Ethanol	CO_2	—	—	—	Sloan (145)
Protein (β-lactamase)	Water, ethanol	CO_2	—	—	—	Hanna (16,136,137)
Quinine sulfate	Ethanol	CO_2	10	308	+ Ethyl cellulose / needles	Hanna (146)
RhDNase	Ethanol	CO_2	—	—	0.05–0.5	Bustami (144)
Salbutamol	Methanol, acetone	CO_2	10	333	0.5	Hanna (133)
Salycylic acid	Methanol, dichloromethane	CO_2-N_2	20	323–338	Fine fluffy white powder	Hanna (140)
Salmeterol xinofoate	Acetone, ethanol	CO_2	10–30	318–368	1–20	Hanna (15)
Salmeterol xinofoate	Acetone, methanol	CO_2	25–30	318–363	Platelet, needles / 4–19	Hanna (133)
Salmeterol xinofoate	1-Butanol	CO_2	15	323	—	Hanna (134)
Salmeterol xinofoate	Methanol	CO_2	25	313–363	—	Tong (154)
Salmeterol xinofoate	Acetone, n-hexane	CO_2	15	333	—	Hanna (137)
Sodium cromoglicate	Methanol	CO_2	—		0.1–2	Jaarmo (155)
Sucrose	Water, ethanol	CO_2	25	333	—	Hanna (16,137)
Sulfathiazole	Acetone, methanol	CO_2	20	353–393	2–20	Kordikowski (156)
Trehalose	Water, ethanol	CO_2	25	343	—	Hanna (16,137)
Trypsin	Ethanol	CO_2	20	328	1.53	Sloan (150)

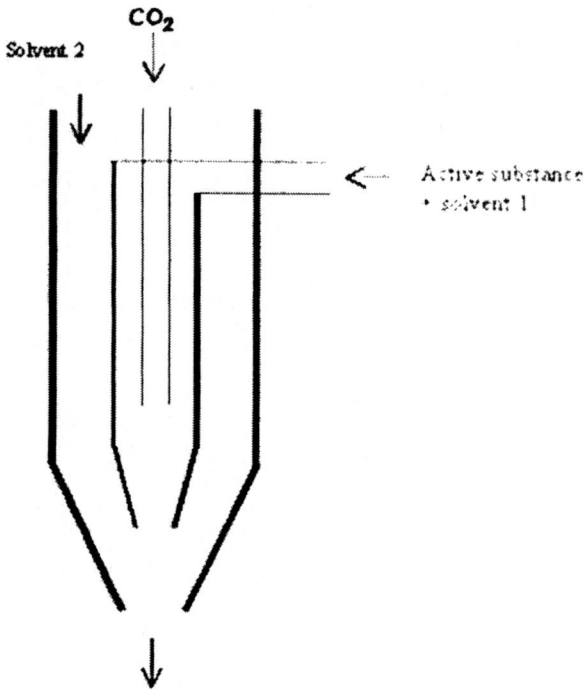

FIGURE 6 Triaxial-channel nozzle for the SEDS process.

Union Carbide and dealt with paint or adhesive applications. In addition to yielding smooth and regular coatings, these techniques are claimed to dramatically reduce the amount of solvent (dispersion medium) released in the atmosphere; this decrease can be as high as 70%.

The so-called PGSS process was described by Weidner et al. in a series of patents filed beginning in 1994 (166–168). Unlike the RESS process, in this technique the compressed gas is dissolved in the material(s) to be treated; thus, the PGSS process takes advantage of a much higher solubility of gases in liquids or solids than that of solids or liquids in compressed gases at same conditions. When the material to be treated is a solid, it is first melted, and the compressible gas is added to the molten material until saturation is reached; the solution thus formed may typically contain 5 to 50 wt % of the compressed gas. The temperature of the solution is preferably adjusted to around 50 K above or below the melting point of the solid under atmospheric pressure. The mixture formed is discharged through a nozzle or other expansion device. The operating conditions of the decompression are tuned in such a way that nearly all the compressed fluid turns to gas, facilitating its further separation from the particles. The evaporation and/or the Joule–Thomson effect results in noticeable cooling of the mixture, which causes the temperature to fall below the melting point of the material, which then precipitates. Additionally, it must be noted that the melting point of the solid is decreased owing to its pressurization with dense gases. Thus, for example, the melting point of glycol 1-stearic ester, which is at 348 K under atmospheric pressure, is lowered to 331 K in the presence of a carbon dioxide atmosphere of 15 MPa. The same value of 331 K is reached with propane under a pressure of 2 MPa.

An almost identical method called the SCAMP process was developed by Ferro in a series of patents (169–171); this process is claimed to produce powder coatings, new polymers, pigments, polymer additives, and temperature-sensitive biomaterials.

The apparatus for carrying out the PGSS process is described in Figure 7. The solid or solids to be treated must first be melted in a feed vessel (3) and sucked in the autoclave (5). A compressible fluid stored in a tank (1) is fed to the autoclave by means of a high pressure pump (2). To enhance the mass transfer between the liquid and gaseous phases, the liquid is drawn off at the bottom of the autoclave and recycled to the top by means of a high pressure circulating pump (4). The expanded liquid phase so formed is then sprayed into a tower (6) through suitable devices (nozzle, capillary, valve, orifice, etc.). The spray tower is designed in such a way that particles with an equivalent diameter equal or greater than 10 μm deposit and are collected at the bottom of the tower (7). The gaseous phase issued from the spray tower is conveyed to a cyclone (8), which allows for the removal of particles with diameter above 1 μm (9). The smallest particles leave the cyclone suspended in the gas stream, which is operated by an electrostatic precipitator. Subject to an intense electric field (20 kV) solid particles gather on the central wire. Periodically the solid is shaken off and collected at the bottom of the precipitator (11) while the purified gas is extracted at the top (12) and recycled to the autoclave. Examples presented in the patent show that propane is effective as a compressed gas for processing materials such as mixtures of

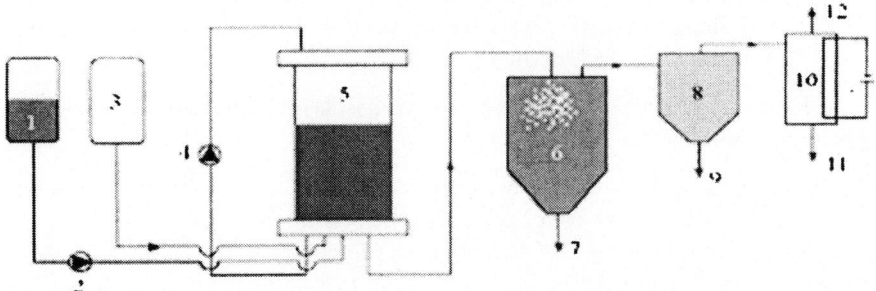

FIGURE 7 Schematic flow diagram of the PGSS process; for 1–12, see text.

mono-, di-, and triglycerides. CO_2 was used to micronize various materials, including monoglycerides of stearic acid, a mixture of citric acid and poly (ethylene glycol), a polyether, and the drug nifedipine [dimethyl-1, 4-dihydro-2,6-dimethyl-4-(2,-nitrophenyl)-3, 5-pyridinedicarboxylate].

The technique is claimed to present a number of advantages over other methods:

Pressure is considerably lower than in the RESS, but of the same order of magnitude and even higher than in antisolvent processes.

The requirement for compressible fluid is smaller. Typically the mass ratio of gas to solid is of 1:10. As a consequence, equipment size can be reduced.

No additional solvent is necessary, and thus no residual traces remain in the end product; this feature is particularly interesting for processing pharmaceutical agents.

The PGSS process seems to be very promising, since in addition to the characteristics just listed, other advantages can be taken into account.

The PGSS can be operated in continuous mode, while antisolvent techniques are discontinuous or semicontinuous. In the particular case of polymers, PGSS allow operation at significantly reduced temperature. Indeed, polymers are melted or swollen at a temperature significantly lower than their melting or glass transition point (see this volume, Part 2, Chapter 10 by Kazarian).

The flexibility of the process makes it applicable to suspensions of active molecule(s) in a polymer or other carrier substances to form composite microspheres after decompression.

Because of the high solubility of compressed gas in liquid solution, the particles obtained from solution have a high concentration of the substance(s).

The concept is simple, since the solid and carrier to be micronized do not have to be soluble in the SCF; this favors the application of PGSS to a wide range of products. Polymers and generally highly viscous compounds, including waxes, resins, and sticky materials can, be preferentially treated by the PGSS process.

However, besides this wide range of assets, the PGSS suffers at least from two limitations:

The particle size cannot be easily controlled, and in many cases micronization leads to a broad particle size distribution.

In the particular case of thermally labile compounds, the temperature required to melt the solid is so high that the material may be damaged. Many drugs may decompose before melting. The patent of Weidner highlights this point and suggests adding a solid auxiliary substance that forms a solution with the solid of interest. The auxiliary substance is chosen to form a low-melting eutectic with the solid to be treated. This answer may be useless when a drug of high purity is required, and indeed only a few references related to drug particle design can be found in literature; these examples are summarized in Table 5 and in Refs. 172 to 178.

Finally, the balance between advantages and drawbacks of the method is positive, since unlike the antisolvent techniques still under development, the PGSS is already used on a large scale; several plants are running with capacities of some hundred kilograms per hour.

6. CO_2-ASSISTED NEBULIZATION AND BUBBLE DRYING (CAN-BD)

A new method referred as CAN-BD, patented by Sievers et al. (179,180), allows the production of micrometer-sized particles of water-soluble compounds (179,180). According to the authors, the invention provides "a process for forming fine particles of compounds which are soluble in fluids, preferably non gaseous fluids, immiscible with the pressurized or supercritical fluid." This process applies particularly to substrates that are significantly soluble in water. Three basic steps are involved in the method.

> The substance of interest is substantially dissolved or suspended in a nongaseous fluid (generally water) to form a solution or suspension.
> The solution or suspension is mixed with a nongaseous fluid (generally a compressed gas). It is suggested that carbon dioxide is a suitable compressible fluid, since it does not solubilize most of the drugs and is also one of the most soluble gases in water.
> The emulsion so formed is rapidly decompressed through a suitable device, whereby a gas-borne dispersion of fine particles is formed.

This aerosolization method produces particles less than 3 μm in diameter. The process can also be applied to form small droplets when the substance to be treated is a liquid, and it may be operated in a continuous mode.

The first patent (179) claims that a wide range of products can be processed by the CAN-BD method: various drugs including antibiotics, enzymes, and DNA, and liposomes, resins, and polymers. Figure 8 provides an illustration of the process. The compressed gas in tank (1) and the liquid solution of the active substance (2) are mixed in a heated low-dead-volume tee (4). At the outlet of the tee, the mixture expands through a restrictor (5) to form a fine aerosol.

TABLE 5 PGSS Experiments for Pharmaceuticals Products

Substrate	Compressible fluid	Pressure (MPa)	Temperature (K)	Particle size (µm)	Observations	References
Catalase enzyme–PLGA	CO_2	—	—	—	—	Mandel (172)
Mixture of mono-, di-, and triglycerides	CO_2	8–20	338–473	15–50	—	Weidner (173)
Nifedipine	CO_2	20	443	10	Mass ratio (gas/solid) 0.1:1	Weidner (167)
Nifedipine	CO_2	10–20	448–463	15 at least	Degradation of the material	Sencar-Bozic (174)
Nifedipine–PEG 4000	CO_2	12–19	323–343	—	Mass ratio (nifedipine/PEG) 1:4	Sencar-Bozic (174)
Nifedipine	CO_2	10–20	438–458	15–30	Nozzle diameter, 0.25 and 0.4 mm	Knez (175)
Felodipine	CO_2	20	423	42	Nozzle diameter, 0.4 mm	Knez (175)
Felodipine–PEG 4000	CO_2	—	—	—	Mass ratio (felodipine/PEG) 1:4	Knez (175)
Nifedipine	CO_2	10–20	438–458	15–30	Nozzle diameter 0.4 mm	Kerc (176)
Felodipine	CO_2	20	423	42	Nozzle diameter 0.25 and 0.4 mm	Kerc (176)
Fenofibrate	CO_2	19	338–353	—	Failed to micronize: agglomeration	Kerc (176)
Fenofibrate–PEG 4000	CO_2	—	—	—	Nozzle diameter 0.25; no modification of the dissolution profile	Kerc (176)
PEG[a] (MW 1500/4000/ 8000/35000)	CO_2	15–25	303–363	170–500	Nozzle diameter, 0.4–1.0 mm	Weidner (177)
Drug–PLGA	CO_2	—	305	—	Mass ratio (CO_2/PEG): 0.17–0.7	Mandel (178)

[a] MW 1500/4000/8000/35,000.

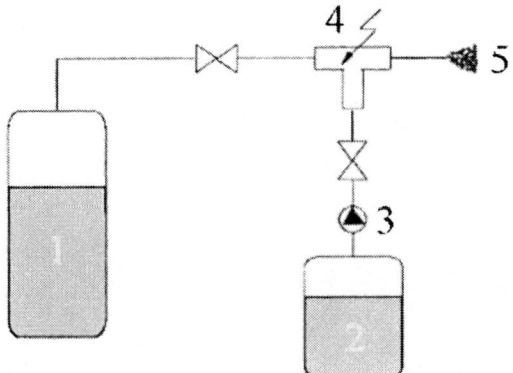

FIGURE 8 Schematic flow diagram of the CAN-BD process; for 1–5, see text.

The role played by the saturated gas is not yet fully understood. The contact time between the gas and the liquid solution being very short, it is unlikely that the former dissolves significantly in the liquid, in which case it would essentially act as a propellant, and this technique would not differ from the classical aerosol method. However according to Sievers, the droplets formed are acidic, which would imply a more or less important dissolution of the gas in the liquid. Nevertheless, unlike the PGSS, the compressible gas is far from dissolving to saturation in the liquid solution.

It is noteworthy that owing to the depressurization of the CO_2, a sharp temperature decrease takes place both in the tee and in the restrictor, which also must be heated to avoid plugging. This method seems to have a lot of advantages, since it is based on a very simple concept and may be applied to a wide range of compounds. The literature provides examples of products treated by the technique (181,182): organic salts, amino acids (e.g., alanine), glucose, peptides, proteins (e.g., glutathione, bovine serum albumin), horseradish peroxidase, trypsinogen, and lactase dehydrogenase.

The method is also claimed to successfully deposit thin films or coatings.

7. CONCLUSION

This overview clearly shows that supercritical fluids offer a wide range of techniques for producing ultrafine particles of pharmaceuticals; Table 6 presents a summary of these methods. Several features make them very attractive:

Particles are in most cases solvent free.
Some of these methods are fully exempt from the use of traditional organic solvents, the others allow a significant reduction of their use.

TABLE 6 Micronization Processes: Summary of Main Features

Process	Solvent	Precipitation driving force	Typical pressure (MPa)	Typical temperature (K)	Typical particle size (μm)	Main advantages	Main drawback	Pharmaceutical examples
RESS	CO_2	Pressure decrease	20–30	310–400	0.2–3	Simplicity (if no cosolvent), no organic solvent use	High pressure; low solubility of molecules; high fluid comsuption	Benzoïc acid, cholesterol, griseofulvin
GAS	Organic solvent	Antisolvent effect	6–10	298–333	0.1–100	Polar molecules; low P and T	Batch; organic solvent use; broad size distribution; Transient conditions	Acetaminophen, Insulin, sulfothiazole
PCA, SAS, ASES	Organic solvent	Antisolvent effect + solvent evaporation	7–15	298–333	0.2–10	Polar molecules Low P and T Controlled particle size	Organic solvent use (small quantities) Separation of solvent and fluid Semicontinuous process	Amoxicillin, griseofulvin, insulin, lysozyme
SEDS	Organic solvent	Antisolvent effect + solvent evaporation	10–30	308–363	0.05–10	Polar molecules, small particles, polymorphism control	Organic solvent use (small quantities); separation of solvent and fluid; semicontinuous process	Acetaminophen, insulin, salmeterol-xinofoate
PGSS	Organic solvent	Pressure decrease	8–20	323–460	15–50	Low P; no organic solvent; low fluid consumption; continuous process	Broad size distribution; high temperature	Nifedipine, felodipine
CAN-BD	Water	Pressure decrease	—	—	0.3–1	Molecules soluble in water	Sharp temperature decrease at depressurization; residual solvent	Glutathione, tobramycin

A broader set of operating parameters than in conventional methods leads to better control of morphology and particle size.

Micrometer and submicrometer sizes can be reached in mild conditions of temperature and without mechanical stress.

Some recent work (139,147,156) has reported that in addition to morphology and particle size, polymorphism can be controlled via a judicious choice of operating parameters.

This wide range of advantages should not induce one to forget that some points remain to be examined before this promising technology becomes an industrial reality.

The foregoing description of the patent literature, shows that the situation may appear slightly complex in some specific cases, since different authors claim priority for publishing original paper on almost identical methods; this could constitute a limitation to their development.

Furthermore, regarding the technical aspect, improvements are needed in various areas, including particle recovery and scale-up.

Particle recovery is currently performed in metallic or porous polymeric baskets; this method is satisfactory for bench scale studies whose aim is, in most cases, limited to studying morphology and particle size. At the industrial scale, this solution may prove to be inadequate. Besides such conventional separation techniques as cyclones, scrubbers, and electrostatic precipitators, different approaches were recently explored (183,184).

The significant advantage of supercritical technology as a procedure for forming ultrafine particles is now clearly demonstrated by a multitude of laboratory-scale studies. To become a commercially viable route, this technology must now face its next challenge: translation to the industrial scale. Undoubtedly, scale-up is the domain that henceforth needs the main research activity. A successful process redesign at large scale requires a better understanding of basic phenomena involved in the particle formation process. This includes thermodynamics, heat and mass transfer kinetics, nucleation and particle growth rates, and hydrodynamics. Nevertheless many research groups are presently focusing on these areas, and hence, effective answers to these pending problems may be available in the near future.

REFERENCES

1. Hannay JB, Hogarth J. On the solubility of solids in gases. Proc R Soc Lond 1879; 29:324–326.
2. Chang CJ, Randolph AD. Precipitation of micron size organic particles from supercritical fluids. AIChE J 1989; 35(11):1876–1882.
3. Subramaniam B, Rajewski RS, Snavely K. Pharmaceutical processing with supercritical carbon dioxide. J Pharm Sci 1997; 86:885–890.

4. Bungert B, Sadowski G, Arlt W. Separations and material processing in solutions with dense gases. Ind Eng Chem Res 1998; 37:3208–3220.
5. Reverchon E. Supercritical antisolvent precipitation: its application to microparticle generation and products fractionation. Proceedings of the 5th Meeting on Supercritical Fluids, Nice, France 1998; 1:221-236.
6. Reverchon E. Review: supercritical antisolvent precipitation of micro- and nanoparticles. J Supercri Fluids 1999; 15(1):1–21.
7. York P. Strategies for particle design using supercritical fluid technologies. Pharm Sci Techno Today 1999; 2:430–440.
8. Subra P, Jestin P. Powder elaboration in supercritical media: comparison with conventional routes. Powder Techno 1999; 103:2–9.
9. Marr R, Gamse T. Use of supercritical fluids for different processes including new developments—a review. Chem Eng and Process 2000; 39:19–28.
10. Thiering R, Dehghani F, Foster NR. Current issues relating to antisolvent micronization techniques and their extension to industrial scales. J Supercrit Fluids 2001; 21:159–177.
11. Jung J, Perrut M. Particle design using supercritical fluids: literature and patent survey. J Supercrit Fluids 2001; 20:179–219.
12. Krober H, Teipel U. Materials processing with supercritical antisolvent precipitation: process parameters and morphology of tartaric acid. J Supercrit Fluids 2002; 22:229–235.
13. Reverchon E. Particle design using supercritical fluids. Proceedings of the 7th Meeting on Supercritical Fluids, Antibes, France 2000; 1:3–20.
14. Reverchon E, Della Porta G. Supercritical fluids-assisted micronization techniques. Low-impact route for particle production. Pure Appl Chem 2001; 73(8):1293–1297.
15. Hanna MH, York P. Method and apparatus for the formation of particles. World patent WO 95/01 221, 1995.
16. Hanna MH, York P. Method and apparatus for the formulation of particles. World patent WO 96/00 610, 1996.
17. Knez Z, Weidner E. Precipitation of solids with dense gases. In: Bertucco A, Vetter G, eds. High Pressure Process Fundamentals and Applications. Amsterdam: Elsevier, 2001:587–611.
18. Gallagher PM, Coffey MP, Krukonis VJ, Klasutis N. Gas antisolvent recrystallization: new process to recrystallize compounds insoluble in supercritical fluids. In: Penniger JML, ed. Supercritical Fluid Science and Technology. ACS Symposium Series 406. Washington, DC: American Chemical Society, 1989:334–354.
19. Dixon DJ, Johnston KP, Bodmeier RA. Polymeric materials formed by precipitation with a compressed fluid antisolvent. AIChE J 1993; 39:127–139.
20. Mawson S, Kanakia S, Johnston KP. Coaxial nozzle for control of particle morphology in precipitation with a compressed fluid antisolvent. J Appl Poly Sci 1997; 64:2105–2118.
21. Tom JW, Lim G, Debenedetti PG, Prud'homme R. Applications of supercritical fluids in the controlled release of drugs. ACS Symp Ser 1993; 514:238–257.

22. Yeo SD, Lim G, Debenedetti PG, Bernstein H. Formation of microparticulate protein powders using a supercritical fluid antisolvent. Biotechnol Bioeng 1993; 41:341–346.
23. Bleich J, Muller BW, Wassmus W. Aerosol solvent extraction system: a new microparticle production technique. Int J Pharm 1993; 97:111–117.
24. Weber M, Thies MC. Understanding the RESS process. In: Ya-Ping Sun, ed. Supercritical fluid technology in materials science and engineering. New York: Marcel Dekker, 2002:387–437.
25. Kwauk X, Debenedetti PG. Mathematical modelling of aerosol formation by rapid expansion of supercritical solutions in converging nozzle. J Aerosol Sci 1993; 24:445–456.
26. Debenedetti P, Tom JW, Yeo SD, Lim GB. Application of supercritical fluids for the production of sustained delivery devices. J Controlled Release 1993; 24:27–44.
27. Kim J, Paxton T, Tomasko D. Microencapsulation of naproxen using rapid expansion of supercritical solutions. Biotechno Prog 1996; 12:650–661.
28. Mishima K, Matsuyama K, Uchiyama H, Ide M. Microcoating of flavone and 3-hydroxyflavone with polymer using supercritical carbon dioxide. 4th International Symposium on Supercritical Fluids, 11–14 May, Sendai, Japan, 1997: 267–270.
29. Krukonis VJ. Supercritical fluid nucleation of difficult-to-comminute solids. Annual Meeting, American Institute of Chemical Engineers Journal, San Francisco, November 1984:140–149.
30. Matson DW, Fulton RC, Petersen RC, Smith RD. Rapid expansion of supercritical fluid solutions: solute formations of powder, thin films and fibers. Ind Eng Chem Res 1987; 26:2298–2306.
31. Petersen RC, Matson DW, Smith RD. The formation of polymer fibers from the rapid expansion of supercritical fluid solutions. Polym Eng Sci 1987; 27: 1693–1697.
32. Coffey MP, Krukonis VJ. Supercritical Fluid Nucleation. An Improved Ultrafine Particle Formation Process. Phasex Corp., Final report to the National Science Foundation, 1988.
33. Pace G, Mishra K, Henriksen B, Krukonis VJ, Godinas A. Processes to generate submicron particles of water-insoluble compounds. World patent WO 99/65469, 1999.
34. Kröber H, Teipel U, Krause H. Formation of submicron particles by rapid expansion of supercritical solutions. GVC-Fachausschus "High Pressure Chemical Engineering," 3–5 March, Karlsruhe, Germany, 1999:247–250.
35. Franck SG, Ye C. Small particle formation and dissolution rate enhancement of relatively insoluble drug using rapid expansion of supercritical solution (RESS) process. Proceedings of the 5th International Symposium on Supercritical Fluids, 8–12, April, Atlanta, 2000.
36. Merrifiel D, Valder C. Process and apparatus for producing particles using a supercritical fluid. World patent WO 00/37169, 2000.
37. Young TJ, Mawson S, Johnston KP, Henriksen IB, Pace GW, Mishra AK.

Rapid expansion from supercritical to aqueous solution to produce submicron suspensions of water-insoluble drugs. Biotechnol Prog 2000; 16:402–407.

38. Helfegen B, Hils P, Holzknecht Ch, Türk M, Schaber K. Simulation of particle formation during the rapid expansion of supercritical solutions. Aerosol Sci 2001; 32:295–319.

39. Türk M, Hils P, Helfegen B, Schaber K, Martin HJ, Wahl MA. Micronization of pharmaceutical substances by the rapid expansion of supercritical solutions (RESS): a promising method to improve bioavailability of poorly soluble pharmaceutical agent. J Supercrit Fluids 2002; 22:75–84.

40. Chen X, Zhao T, Li Y, Yu W. Study on preparation of α- asarone fine particles via RESS. Huaxue Gongcheng (Chem Eng) 2001; 29:12–14.

41. Sievers RE, Hybertson B, Hansen B. European patent EP 0627910B1, 1993.

42. Tavana A, Randolph AD. Manipulating solids CSD in a supercritical fluid crystallizer: CO_2–benzoic acid. AIChE J 1989; 35:1625–1630.

43. Berends EM, Bruinsma OSL, VanRosmalen GM. Nucleation and growth of fine crystals from supercritical carbon dioxide. J Crys Growth 1993; 128:50–56.

44. Cihlar S, Türk M, Schaber K. Submicron particles of organic solids by rapid expansion of supercritical solutions. J Aerosol Sci 1999; 30:355–356.

45. Türk M. Formation of small organic particles by RESS: experimental and theoretical investigations. J Supercrit Fluids 1999; 15:79–89.

46. Domingo C, Wubbolts FE, Rodriguez-Clemente R, van Rosmalen GM. Solid crystallisation by rapid expansion of supercritical ternary mixtures. J Crys Growth 1999; 198:760–766.

47. Helfegen B, Türk M, Schaber K. Theoretical and experimental investigation of the micronization of organic solids by rapid expansion of supercritical solutions. Powder Technol 2000; 110:22–28.

48. Peters S, Steiner K, Stoller H, Weidner E. Process for the manufacture of a pulverous preparation. European patent EP 1 097 705 A2, 2001.

49. Chang CJ, Randolph AD. Precipitation of micronsize organic particles from supercritical fluids. AIChE J 1989; 35:1876–1882.

50. Reverchon E, Taddeo R. Morphology of salicylic acid crystals precipitated by rapid expansion of a supercritical solution. In: Reverchon E, Schiraldi, eds. I Fluidi Supercritici e Le Loro Applicazioni. 20–22 June, Ravello, Italy, 1993: 189–198.

51. Reverchon E, Donsi G, Gorgoglione D. Salicylic acid solubilization in supercritical CO_2 and its micronization by RESS. J Supercrit Fluids 1993; 6:241–248.

52. Subra P, Boissinot P, Benzaghou S. Precipitation of pure and mixed caffeine and anthracene by rapid expansion of supercritical solutions, Proceedings of the 5th meeting on Supercritical Fluids. Vol. 1. Perrut M, Subra P, ed. 23–25 March, Nice, France, 1998:307–312.

53. Frederiksen L, Anton K, van Hoogevest P, Keller HR, Leuenberger H. Preparation of liposomes encapsulating water-soluble compounds using supercritical carbon dioxide. J Pharm Sci 1997; 86:921–928.

54. Kröber H, Teipel U, Krause H. Manufacture of submicron particles via expansion of supercritical fluids. Chem Eng Technol 2000; 23:763–765.

55. Godinas A, Henriksen IB, Krukonis VJ, Mishra KA, Pace GW, Vachon GM. Process to generate submicron particles of water insoluble compounds. US patent 0089852, 1998.
56. King M, Robertson J. Treatment of a substance with a dense fluid (e.g., with a supercritical fluid). World patent WO 9858722, 1998.
57. Mishima K, Matsuyama K, Yamauchi S, Izumi H, Furudono D. Novel control of crystallinity and coating thickness of polymeric microcapsules of medicine by cosolvency of supercritical solution. Proceedings of the 5th International Symposium on Supercritical Fluids, 8–12, April, Atlanta, 2000.
58. Reverchon E, Della Porta G, Taddeo R, Pallado P, Stassi A. Solubility and micronization of griseofulvin in supercritical CHF_3. Ind Eng Chem Res 1995; 34:4087–4091.
59. Peirico NM, Matson HA, Gomez de Azevedo E. Production of drug biocompatible polymer microsize composites by RESS with supercritical CO_2. Proceedings of the 5th Meeting on Supercritical Fluids. Vol. 1. Perrut M, Subra P, eds. 23–25 March, Nice, France 1998:13–318.
60. Charoenchaitrakool M, Dehghani F, Foster NR, Chan HK. Micronization by rapid expansion of supercritical solutions to enhance the dissolution rates of poorly water-soluble pharmaceuticals. Ind Eng Chem Res 2000; 39:4794–4802.
61. Furuta S, Rousseau RW, Teja AS. Production of fine particles by rapid expansion of amino acid solutions at high temperatures and pressures. American Institute of Chemical Engineers, Annual Meeting, Miami, 1992:706–712.
62. Brodin A, Frank S, Ye C. Method of preparing solid dispersions. International patent WO 01/66091A1, 2001.
63. Larson KA, King ML. Evaluation of supercritical fluid extraction in the pharmaceutical industry. Biotechnol Prog 1986; 2:73–82.
64. Mohamed RS, Debenedetti PG, Prud'homme RK. Effects of process conditions on crystals obtained from supercritical mixtures. AIChE J 1989; 35:325–332.
65. Mohamed RS, Halverson DS, Debenedetti RK, Prud'homme RK. Solids formation after the expansion of supercritical mixtures. In: Johnston KP, Penninger JML, eds. Supercritical Fluid Science and Technology. ACS Symposium Series 406. Washington DC: American Chemical Society, 1989:355–378.
66. Stahl E, Quirin KW, Gerard D. High Pressure Micronizing in Dense Gases for Extraction and Refining. Berlin: Springer-Verlag, 1988, Chapter V4.
67. Loth H, Heemgesberg E. Properties and dissolution of drugs micronized by crystallization from supercritical gases. Int J Pharm 1986; 32:265–267.
68. Subra P, Debenedetti P. Application of RESS to several low molecular weight compounds. In: Rudolf von Rohr P, Trepp C, eds. High Pressure Chemical Engineering. Amsterdam: Elsevier Science BV, 1996:49–54.
69. Domingo C, Berends E, van Rosmalen GM. Precipitation of ultrafine organic crystals from the rapid expansion of supercritical solutions over a capillary and a frit nozzle. J Supercrit Fluids 1997; 10:39–55.
70. Alessi P, Cortes A, Kikic I, Foster NR, Macnaughton SJ, Colombo I. Particle

production of steroid drugs using supercritical fluid processing. Ind Eng Chem Res 1996; 35:4718-426.

71. Ohgaki K, Kobayashi H, Katayama T. Whisker formation from jet of supercritical fluid solution. J of Supercrit Fluids 1990; 3:103–107.

72. Krukonis VJ, Gallagher PM, Coffey MP. US patent 5 360 478, 1991.

73. Gallagher PM, Krukonis VJ, Coffey MP. US patent 5 389 263. 1992.

74. Gallagher PM, Coffey MP, Krukonis VJ, Hillstrom WW. Gas antisolvent recrystallization of RDX Formation of ultrafine particle of a difficult-to-comminute explosive. J Supercrit Fluids 1995; 5:130 142.

75. Magnan C. Study of phospholipids micronization by supercritical fluids. Ph.D thesis. Université d'Aix-Marseille III, 1988.

76. Kordikowski A, Schenk AP, Van Nielsen RM, Peters CJ. Volume expansions and vapour–liquid equilibria of binary mixtures of a variety of polar solvents and certain near-critical solvents. J Supercrit Fluids 1995; 8:205 216.

77. Berends EM, Bruinsma OSL, de Graauw J, van Rosmalen GM. Crystallization of phenanthrene from toluene with carbon dioxide by the GAS process. AIChE J 1996; 42:431–439.

78. Kitamura M, Yamamoto M, Yoshinaga Y, Masuoka H. Crystal size control of sulfathiazole using high pressure carbon dioxide. J Cryst Growth 1997; 178:378 386.

79. Wubbolts FE, Kerach C, Van Rosmalen GM. Semibatch precipitation of acetaminophen from ethanol with liquid carbon dioxide at a constant pressure. Proceedings of the 5th Meeting on Supercritical Fluids, Nice, France, 1998; 1:249–255.

80. Yeo SD, Debenedetti PG, Radosz M, Schmidt HW. Supercritical antisolvent process for substituted para-linked aromatic polyamides: phase equilibrium and morphology study. Macromolecules 1993; 26:6207–6210.

81. Debenedetti PG, Tom JW, Yeo SD, Lim GB. Application of supercritical fluids for the production of sustained release delivery devices. J Controlled Release 1993; 24:27–44.

82. Thiering R, Charoenchaitrakool M, Tu LS, Dehghani F, Dillow AK, Foster NR. Proceedings of the 5th Meeting on Supercritical Fluids, Nice, France. 1998; 1:291–296.

83. Muller M, Meier U, Kessler A, Mazzotti M. Experimental study of the effect of process parameters in the recrystallization of an organic compound using compressed carbon dioxide as antisolvent. Ind Eng Chem Res 2000; 39(7):2260–2268.

84. Warwick B, Dehghani F, Foster NR, Biffin JR, Regtop H. Micronization of copper indomethacin using gas antisolvent processes. Ind Eng Chem Res 2002; 41(8):1993–2004.

85. Thiering R, Dehghani F, Dillow A, Foster NR. The influence of operating conditions on the dense gas precipitation of model proteins. J Chem Technol Biotechnol 2000; 75:29–41.

86. Liu Z, Wang J, Song L, Yang G, Han B. Study on the phase behavior of cholesterol–acetone-CO_2 system and recrystallization of cholesterol by antisolvent CO_2. J Supercrit Fluids 2002; 24:1–6.

87. Dillow AK, Dehghani F, Foster N. Production of polymeric support materials using a supercritical fluid gas antisolvent. Proceedings of the 4th International Symposium on Supercritical Fluids, Sendai, Japan, 1997, 247–250.

88. Benedetti L, Bertucco A, Pallado P. Production of micronic particles of biocompatible polymer using supercritical carbon dioxide. Biotechnol Bioeng 1997; 53:232–237.

89. Bertucco A, Pallado P, Benedetti L. Formation of biocompatible polymer microspheres for controlled drug delivery by a supercritical antisolvent technique. Proceedings of the 3rd Symposium on High Pressure Chemical Engineering, Zürich, Switzerland, 1996:217–222.

90. Moneghini M, Kikic I, Voinovich D, Perissutti B, Filipovic-Grcic J. Processing of carbamazepine–PEG 4000 solid dispersions with supercritical carbon dioxide: preparation, characterization, and in vitro dissolution. Int J Pharm 2001; 222:129–138.

91. Pallado P, Benedetti PL, Callegaro L. Nanospheres comprising a biocompatible polysaccharide. US patent 6 214 384, 2001.

92. Kikic I, Alessi P, Cortesi A, Eva F, Fogar A, Moneghini M, Perissutti B, Voinovich D. Supercritical antisolvent precipitation processes: different ways for improving the performances of drugs. 4th International Symposium on High Pressure Process Technology and Chemical Engineering, Venice, 2002.

93. Warwick B, Dehghani F, Foster NR, Biffin JR, Regtop HL. Synthesis, purification, and micronization of pharmaceuticals using the GAS antisolvent technique. Ind Eng Chem Res 2000; 39:4571–4579.

94. Elvassore N, Bertucco A, Caliceti P. Production of insulin-loaded poly (ethylene glycol)/poly(lactide) (PEG/PLA) nanoparticles by gas antisolvent techniques. J Pharm Sci 2001; 90(10):1628–1636.

95. Yeo SD, Kim MS, Lee JC. Recrystallization of sulfothiazole and chlorpropamide using the supercritical fluid antisolvent process. J Supercrit Fluids. In press.

96. Bertucco A, Lora M, Kikic I. Fractional crystallization by GAS antisolvent technique: theory and experiments. AIChE J 198; 44(10):2149–2158.

97. Bothun GD, White KL, Knutson BL. GAS antisolvent recrystallization of semicrystalline and amorphous poly(lactic acid) using compressed CO_2. Polymer 2002; 43:4445–4452.

98. Ventosa N, Sala S, Veciana J. DELOS process: a crystallization technique using compressed fluids. 1. Comparison to the GAS crystallization method. J Supercrit Fluids. In press.

99. Wubbolts FE, Bruinsma OSL, Van Rosmalen GM. Dry-spraying of ascorbic acid or acetaminophen solutions with supercritical carbon dioxide. J Crys Growth 1999; 198/199:767–772.

100. Niu F, Rajewski R, Snaveley WK, Subramanian B. World patent WO 0 235 941, 2002.

101. Reverchon E, Della Porta G. Production of antibiotic micro- and nanoparticles by supercritical antisolvent precipitation. Powder Technol 1999; 106: 23–29.

102. Reverchon E, Della Porta G, Falivene MG. Process parameters and mor-

phology in amoxicillin micro and submicro particles generation by supercritical antisolvent precipitation. J Supercrit Fluids 2000; 17(3):239–248.

103. Reverchon E, De Marco I, Caputo G, Della Porta G. Pilot scale micronization of amoxicillin by supercritical antisolvent precipitation. J Supercrit Fluids. In press.

104. Weber A, Weiss C, Tschernjaew J, Kummel R. Gas antisolvent crystallization—from fundamentals to industrial applications. Proceedings of High Pressure Chemical Engineering, Karlsruhe, Germany, 1999:235–238.

105. Steckel H, Thies J, Muller BW. Micronizing of steroids for pulmonary delivery by supercritical carbon dioxide. Int J Pharm 1997; 152(1):99–110.

106. Said S, Rajewski RA, Stella V, Subramanian B. World patent WO 9 731 691, 1997.

107. Lim GB, Debenedetti PG, Prud'Homme RK. European patent EP 0 542 314, 1993.

108. Weber A, Tschernjaew J, Kummel R. Coprecipitation with compressed antisolvents for the manufacture of microcomposites. Proceedings of the 5th meeting on Supercritical Fluids Materials and Natural Products Processing, Nice, France, 1998:243–248.

109. Bodmeier R, Wang H, Dixon DJ, Mawson S, Johnston KP. Polymeric microspheres prepared by spraying into compressed carbon dioxide. Pharm Res 1995; 12(8):1211–1217.

110. Steckel H, Muller BW. Metered-dose inhaler formulations of fluticasone-17-propionate micronized with supercritical carbon dioxide using the alternative propellant HFA-227. Int J Pharm 1998; 46(1):77–83.

111. Chattopadhyay P, Gupta RB. Production of griseofulvin nanoparticles using supercritical CO_2 antisolvent with enhanced mass transfer. Int J Pharm 2001; 228:19–31.

112. Sarkari M, Darrat I, Knutson L. Generation of microparticles using CO_2 and CO_2-philic antisolvents. AIChE J 2000; 46(9):1850–1859.

113. Gupta R, Chattopadhyay P. US patent US 2 002 000 681, 2002.

114. Wubbolts FE, Bruinsma OSL, De Graauw J, Van Rosmalen GM. Continuous gas antisolvent crystallization of hydroquinone from acetone using carbon dioxide. Proceedings of the 4th International Symposium on Supercritical Fluids, Sendaï, Japan, 1997:63–66.

115. Winters MA, Knutson BL, Debenedetti PG, Sparks HG, Przybycien TM, Stevenson CL, Prestrelski SJ. Precipitation of proteins in supercritical carbon dioxide. J Pharm Sci 1996; 85(6):586–594.

116. Amaro-Gonzalez D, Mabe G, Zabaloy M, Brignole EA. Gas antisolvent ceystallization of organic salts from aqueous solutions. J Supercrit Fluids 2000; 17:249–258.

117. Chattopadhyay P, Gupta RB. Protein nanoparticles formation by supercritical antisolvent with enhanced mass transfer. AIChE J 2002; 48(2):235–243.

118. Sze Tu L, Dehghani F, Foster NR. Micronisation and microencapsulation of pharmaceuticals using a carbon dioxide antisolvent. Powder Technol 2002; 126:134–149.

119. Schmitt WJ, Salada MC, Shook GG, Speaker SM. Finely divided powders by carrier solution injection into a near or supercritical fluid. AIChE J 1995; 41:2476–2486.

120. Reverchon E, De Marco I, Della Porta G. Rifampicin microparticles production by supercritical antisolvent precipitation. Int J Pharm 2002; 243:83–91.

121. Reverchon E, Della Porta G, Pallado P. Supercritical antisolvent precipitation of salbutamol microparticles. Powder Technol 2001; 114:17–22.

122. Magnan C, Badens E, Commenges N, Charbit G. Soy lecithin micronization by precipitation with a compressed fluid antisolvent—influence of process parameters. J Supercrit Fluids 2000; 19:69–77.

123. Badens E, Magnan C, Charbit G. Microparticles of soy lecithin formed by supercritical processes. Biotechnol Bioeng 2001; 72:194–204.

124. Meyer JD, Falk RF, Kelly RM, Shively JE, Withrow SJ, Dernell WS, Kroll DJ, Randolph TW, Manning MC. Preparation and in vitro characterization of gentamicin-impregnated biodegradable beads suitable for treatment of osteomyelitis. J Pharm Sci 1998; 87(9):1149–1154.

125. Falk R, Randolph TW, Meyer JD, Kelly RM, Manning MC. Controlled release of ionic compounds from poly(L-lactide) microspheres produced by precipitation with a compressed antisolvent. J Controlled Release 1997; 44:77–85.

126. Corrigan OI, Crean AM. Comparative physicochemical properties of hydrocortisone–PVP composites prepared using supercritical carbon dioxide by the GAS antisolvent recrystallization process, by coprecipitation and spray drying. Int J Pharm 2002; 245:75–82.

127. Elvassore N, Vezzu K, Bertucco A, Cecchi A, Caliceti P. Protein loading in biodegradable polymeric microparticles produced by compressed gas antisolvent techniques. Proceedings of the 4th International Symposium on High Pressure Process Technology and Chemical Engineering, Venice, Italy, 2002.

128. Chou YH, Tomasko DL. GAS crystallization of polymer–pharmaceutical composite particles. Proceedings of the 4th International Symposium on Supercritical Fluids, Sendai, Japan, 1997:55–57.

129. Bitz C, Doelker E. Influence of the preparation method on residual solvents in biodegradable microspheres. Int J Pharm 1996; 131:171–181.

130. Falk RF, Randolph TW. Process variable implications for residual solvent removal and polymer morphology in the formation of gentamycin-loaded poly(L-lactide) microparticles. Pharm Res 1998; 15(8):1233–1237.

131. Randolph TW, Shefter E, Manning MC, Falk RF. US patent US 5 770 559, 1998.

132. Velaga SP, Ghaderi R, Carlfors J. Preparation and characterisation of hydrocortisone particles using a supercritical fluids extraction process. Int J Pharm 2002; 231:155–156.

133. Hanna M, York P. Method and apparatus for the formation of particles. European patent WO 98/36825, 1998.

134. Hanna M, York P. Method of particles formation. World patent WO 01/03821 A1, 2001.

135. Palakodaty S, York P, Pritchard J. Supercritical fluid processing of materials

from aqueous solutions: the application of SEDS to lactose as a model substance. Pharm Res 1998; 15:1835–1843.

136. Hanna M, York P. Method and apparatus for the formation of particles. US patent 5851453, 1998.

137. Hanna M, York P. Method and apparatus for the formation of particles. US patent 2002/0073511 A1, 2002.

138. Palakodaty S, Sloan R, Kordikowski A, York P. Pharmaceutical and biological materials processing with supercritical fluids. In: Ya-Ping Sun, ed. Supercritical Fluid Technology in Materials Science and Engineering. New York: Marcel Dekker, 2002:439–490.

139. Beach S, Latham D, Sidgwick C, Hanna M, York P. Control of the physical form of salmeterol xinofoate. Org Process Res Dev 1999; 3:370–376.

140. Hanna M, York P. Method and apparatus for the formation of particles. World patent WO 99/59710, 1999.

141. Gilbert R, Palakodaty R, Sloan R, York P. Particle engineering for pharmaceutical applications—process scale-up. Proceedings of the 5th International Symposium on Supercritical Fluids, Atlanta, 2000.

142. Bristow S, Shekunov T, Shekunov BYu, York P. Analysis of the supersaturation and precipitation process with supercritical CO_2. J Supercrit Fluids 2001; 21:257–271.

143. Shekunov BY, Baldyga J, York P. Particle formation by mixing with supercritical antisolvent at high Reynolds numbers. Chem Eng Sci 2001; 56:2421–2433.

144. Bustami RT, Chan HK, Dehghani F, Foster NR. Generation of protein microparticles using high pressure modified carbon dioxide. Proceedings of the 5th International Symposium on Supercritical Fluids, Atlanta, 2000.

145. Sloan R, Tservistas M, Hollowood ME, Sarup L, Humphreys GO, York P, Ashraf W, Hoare M. Controlled particle formation of biological material using supercritical fluids. Proceedings of the 6th Meeting on Supercritical Fluids, Chemistry and Materials. Poliakoff M, George MW, Howdle SM, ed. Nottingham, England, 1999:169–174.

146. Hanna M, York P. Particle formation method and their products. US patent 2002/0073511 A1, 2002.

147. Edwards AD, Shekunov BY, Kordikowski A, Forbes RT, York P. Crystallisation of pure anhydrous polymorphs of carbamazepine by solution enhanced dispersion with supercritical fluids (SEDS). J of Pharma Sci 2001; 90:1115–1124.

148. Ghaderi R, Artursson P, Carlfors J. A new method for preparing biodegradable microparticles and entrapment of hydrocortisone in DL-PLG microparticles using supercritical fluids. Eur J Pharm Sci 2000; 10:1–9.

149. Palakodaty S, York P, Hanna M, Pritchard J. Crystallisation of lactose using solution enhanced dispersion by supercritical fluids (SEDS) technique. Proceedings of the 5th Meeting on Supercritical Fluids. Vol 1. Perrut M, Subra P, eds. Nice, France, 1998:275–280.

150. Sloan R, Hollowood ME, Humphreys GO, Ashraf W, York P. Supercritical

fluid processing: preparation of stable protein particles. In: Proceedings of the 5th Meeting on Supercritical Fluids. Perrut M, Subra P, eds. Vol 1. Nice, France, 1998:301–306.

151. Moshashaée S, Bisrat M, Forbes RT, Nyqvist H, York P. Supercritical fluid processing of proteins. I Lysosyme precipitation from organic solution. Eur J Pharm Sci 2000; 11:239–245.

152. Rehman M, Shekunov BY, York P, Colthorpe P. Solubility and precipitation of nicotinic acid in supercritical carbon dioxide. J Pharm Sci 2001; 90:1570–1582.

153. Sze Tu L, Dehghani F, Dillow AK, Foster NR. Applications of dense gases in pharmaceutical processing. Proceedings of the 5th Meeting on Supercritical Fluids. Perrut M, Subra P, eds. Vol 1. Nice, France, 1998:263–269.

154. Tong HH, Shekunov BY, York P, Chow AHL. Characterisation of two polymorphs of salmeterol xinafoate crystallized from supercritical fluids. Pharm Res 2001; 18:852–858.

155. Jaarmo S, Rantakyla M, Aaltonen O. Particle tailoring with supercritical fluids: production of amorphous pharmaceuticals particles. Proceedings of the 4th International Symposium on Supercritical Fluids. Arai K, ed. Sendai, Japan: Tohoku University Press, 1997:263–269.

156. Kordikowski A, Shekunov T, York P. Polymorph control of sulfathiazole in supercritical CO_2. Pharm Res 2001; 18:685–688.

157. Graser F, Wickenhaeuser G. Conditioning of finely divided crude organic pigments. US patent 4 451 654, 1982.

158. Bok HF, Hoy KL, Nielsen KA. Methods and apparatus for obtaining a feathered spray when spraying liquids by airless techniques US patent, 5 057 342, 1989.

159. Glancy CW, Busby DC. Precursor coating compositions. European patent 388 915, 1990.

160. Lee C, Hoy KL, Donohue MD. Supercritical fluids as diluents in liquid spray application of coatings US patent, 4 923 720, 1990.

161. Taylor JW, Argyropoulos JN, Lear JJ. Monodispersed acrylic polymers in supercritical, near supercritical and subcritical fluids. European patent 506 067, 1992.

162. Nielsen KA. Methods and apparatus for spraying solvent-borne compositions with reduced solvent emission using compressed fluids and separating solvent. US patent, 5 290 604, 1994.

163. Nielsen KA, Glancy CW. Precursor coating compositions suitable for spraying with supercritical fluids as diluents. US patent, 5 509 959, 1996.

164. Nielsen KA, Argyropoulos JN, Wagner BE. Method for producing coating powders catalysts and drier water-borne coatings by spraying compositions with compressed fluids. US patent, 5 716 558, 1998.

165. Kishimoto Y. Supercritical fluid adhesive and production of planar space heating equipment using the same. Japanese patent, 5 132 656, 1991.

166. Weidner E, Knez Z, Novak Z. Slovenian patent, 940 079, 1994.

167. Weidner E, Knez Z, Novak Z. Process for preparing particles or powders. PCT patent application, European patent 0 744 990, 2000.

168. Weidner E, Knez Z, Novak Z. Process for the production of particles or powders. US patent 6 056 791, 1997.

169. Mandel FS, Green CD, Scheibelhoffer AS. Method of preparing coating materials. US patent 5 399 597, 1995.

170. Mandel FS, Green CD, Scheibelhoffer AS. Method of preparing coating materials. US patent 5 548 004, 1996.

171. Mandel FS. Mixing system for processes using supercritical fluids. US patent 5 993 747, 1997.

172. Mandel FS, Don Wang J, Mc Hugh MA. Pharmaceutical material production via supercritical fluids employing the technique of particles from gas saturated solutions (PGSS). Proceedings of the 7th Meeting on Supercritical Fluids, Antibes, France, 2000; 1:35-46.

173. Weidner E, Knez Z, Novak Z. PGSS (particles from gas-saturated solutions), new process for powder generation. Proceedings of the 3rd International Symposium on Supercritical Fluids, Strasbourg, France, 1994; 3:229-234.

174. Sencar-Bozic P, Srcic S, Knez Z, Kerc J. Improvement of nifedipine dissolution characteristics using supercritical CO_2. Int J Pharm 1997; 148:123-130.

175. Knez Z. Micronization of pharmaceuticals using supercritical fluids. Proceedings of the 7th Meeting on Supercritical Fluids, Antibes, France, 2000, 1:21-26.

176. Kerc J, Srcic S, Knez Z, Sencar-Bozic P. Micronization ofh drugs using supercritical carbon dioxide. Int J Pharm 1999; 182:33-39.

177. Weidner E, Steiner R, Knez Z. Powder generation from poly(ethyleneglycols) with compressible fluids. Proceedings of the 3rd Symposium on High Pressure Chemical Engineering, Zürich, Switzerland, 1996:223-228.

178. Mandel FS, Don Wang J. Manufacturing of specialty materials in supercritical fluid carbon dioxide. Inorg Chim Acta 1999; 294:214-223.

179. Sievers RE, Karst U. Methods and apparatus for fine particle production. US patent 5 639 441, 1997, European patent 0 677 332, 1995.

180. Sievers RE, Karst U. Methods and apparatus for fine particle formation. US patent 6 095 134, 2000.

181. Sievers RE, Karst U, Milewski PD, Sellers SP, Miles BA, Schaefer JD, Stoldt CR. Formation of aqueous small droplet aerosols assisted by supercritical carbon dioxide. Aerosol Sci Technol 1999; 30(1):3-15.

182. Villa JA, Sievers RE, Huang ETS. Bubble drying to form fine particles from solutes in aqueous solutions. Proceedings of the 7th Meeting on Supercritical Fluids, Antibes, France, 2000;1:83-88.

183. Perrut M. Method for capturing fine particles by percolation in a bed of granules. US patent 2 002189 454, 2002.

184. Perrut M. Method and device for capturing fine particles by trapping in a solid mixture of carbon dioxide snow type. US patent, 2 002 179 540, 2002.

5

Colloid and Interface Science for CO_2-Based Pharmaceutical Processes

Keith P. Johnston, Sandro R. P. Da Rocha,
C. Ted Lee, Ge Li, and Matthew Z. Yates
The University of Texas at Austin, Austin, Texas, U.S.A.

1. KEY INTERFACES IN SUPERCRITICAL FLUID TECHNOLOGY

Compressed CO_2 has unique solvent properties that may be used advantageously in various pharmaceutical processes (1). The high volatility of CO_2 can be exploited to provide rapid, energy-efficient drying of pharmaceuticals. Carbon dioxide plasticizes many biodegradable and biocompatible polymers, facilitating mass transport inside the polymers during impregnation or microencapsulation processes. It is also advantageous to use compressed CO_2 because it is nontoxic and nonflammable. Since it is nonpolar and has weak van der Waals forces, both polar and nonpolar nonvolatile molecules are often insoluble. Because of this solubility limitation, only a small number of reaction, materials formation, and/or separation processes may be carried out in a homogeneous CO_2-based phase. To overcome this limitation, many of the key advancements in supercritical fluid science and technology have utilized heterogeneous systems with CO_2 at one or more interfaces (2,3). The conjugate phase(s) may be either aqueous or organic solids or liquids. This

TABLE 1 Types of Interface for the Design of New Dispersions and Applications in CO_2–Based Systems

Interface	Process or type of dispersion	Notes
Water–CO_2	Microemulsions 2–10 nm in diameter	Reactions, enzymatic catalysis, separations, materials synthesis
Water–CO_2	W/C and C/W macroemulsions	Reactions, enzymatic catalysis, separations, materials synthesis
Organic–CO_2	Rapid expansion from supercritical solution (RESS)	Nucleation and growth of particles
Organic–CO_2	RESS with a nonsolvent (RESS-N)	Microencapsulation in polymers (e.g., proteins)
Organic–CO_2	Precipitation with a compressed fluid antisolvent (PCA, SAS, SEDS, ASES)	Particle formation and microencapsulation
Organic CO_2	Polymer latexes in CO_2	Formed by synthesis or phase separation
Transfer from Organic–CO_2 to organic–water	Polymers colloids in CO_2 transferred to water	Single "ambidextrous" surfactants or multiple surfactants

Organic–water–CO$_2$	Swelling of polymer latexes in water with CO$_2$	Polymer is water soluble and not soluble in CO$_2$
Organic–water–CO$_2$	Impregnation of swollen polymer latexes in water	Swelling by CO$_2$ plasticizes latex particles allowing for impregnation
Organic–water–CO$_2$	PCA (SEDS) with all three phases	Formation of protein particles from aqueous and ethanol solutions sprayed into CO$_2$
Organic–water–CO$_2$	Rapid expansion from supercritical to aqueous solution	Surfactants stabilize particles produced in expansion and prevent growth
Organic–organic–CO$_2$	PCA of proteins suspended in an organic solvent with dissolved polymer	Protein is encapsulated in polymer without dissolving the protein
Organic–organic–CO$_2$	PCA to form metastable polymer blends	Precipitation fast enough to prevent phase separation of the blend
Inorganic–CO$_2$	Hydrophilic silica dispersions	Polymers adsorbed to silica provide steric stabilization
Inorganic–CO$_2$	Metal nanoparticle suspensions	Synthesized in CO$_2$ by arrested growth precipitation with stabilizing ligands

chapter is organized according to the type of interface, as shown in the summary in Table 1. A variety of single interfaces (e.g., water–CO_2) and double interfaces (e.g., organic–water–CO_2) are addressed. In some cases, the interface changes as a colloid is transferred from one phase to another: for example, a polymer latex may be transferred from CO_2 to water. The phases include solid and liquid organic phases, aqueous phases, and inorganic solids and CO_2 in the gas, liquid, or supercritical fluid states, including cryogenic temperatures down to the sublimation temperature of dry ice. This chapter shows how an emerging understanding of the behavior of these interfaces may serve as a framework for the discovery and development of supercritical fluid technology.

Surfactants have played a key role in the development of colloids in CO_2. A variety of low molecular weight and polymeric surfactants have been designed recently for stabilizing polymer latexes (4), microemulsions (5,6), emulsions (7–9) and inorganic suspensions (10) as described in recent reviews, (3,2,11). Microemulsion droplets are thermodynamically stable and typically 2 to 10 nm in diameter, making them optically transparent, whereas kinetically stable emulsion droplets and latexes in the range of 200 nm to 10 μm are opaque and thermodynamically unstable. The high compressibility of CO_2 makes possible the achievement of large changes in density with small changes in temperature or pressure, which in turn can affect surfactant aggregation behavior (4,12) and colloid stability (2).

The environmentally benign, nontoxic and nonflammable fluids water and carbon dioxide (CO_2) are the two most abundant and inexpensive solvents on earth. Water-in-CO_2 (W/C) or CO_2-in-water (C/W) dispersions in the form of microemulsions and emulsions offer new possibilities in waste minimization for the replacement of organic solvents in fields including chemical processing, pharmaceuticals, and microlectronics for solubilization and separations (e.g., proteins, ions, heavy metals), particle formation, enzymatic catalysis, organometallic catalysis, and synthesis of polymer colloids and inorganic nanoparticles (2,13,11).

Whereas the solvent strength of CO_2 is limited, W/C and C/W microemulsions and emulsions have the ability to function as "universal" solvent media by solubilizing high concentrations of polar, ionic, and nonpolar molecules within the dispersed or continuous water phase. These emulsions may be phase-separated easily for product recovery simply by depressurization, unlike the case for conventional emulsions.

The organic–CO_2 interface plays a central role in physical particle formation processes including rapid expansion from supercritical solution (RESS), precipitation with a compressed antisolvent (PCA, also called SAS and SEDS), impregnation of polymers swollen by CO_2 (see Chapter __), the formation of polymer latexes by polymerization and phase separation, and the formation of polymer films. Because CO_2 is a relatively weak solvent, it is often a good antisolvent that can be tuned by adjusting the pressure to form amorphous and crystalline materials. The rapid transport properties of CO_2 can be advantageous in particle formation technology. For example, in the PCA process, an organic solution is atomized into compressed liquid or supercritical fluid CO_2. The organic solvent diffuses rapidly into the bulk CO_2 phase, while CO_2 simultaneously diffuses into the organic droplets, thereby precipitating the solute. The rate of diffusion in both directions and, thus, the degree of supersaturation are higher than in the case of conventional liquid antisolvents. The large supersaturation and very low interfacial tension between many organic phases and CO_2 lead to high nucleation rates and in many cases materials of submicrometer or micrometer sized. Several studies have shown very low concentrations of residual solvent in the product materials, especially after a CO_2 extraction step upon completion of the spray.

A variety of organic colloids including emulsions and polymer latexes have been dispersed in carbon dioxide in the presence of surfactants (3,13). In most cases, owing to the lower interfacial tension of the former as explained shortly it is easier to form organic-in-CO_2 emulsions than water-in-CO_2, emulsions. Sterically stabilized colloids are stable above the critical flocculation density (CFD) and precipitate below this density. In some cases the CFD occurs at the upper critical solution density of the steric stabilizer, that is, the density at which the stabilizer phase separates from CO_2, as has been shown by theory (14,15) and experiment (16). So-called ambidextrous surfactants have been designed to allow polymer latexes produced in CO_2 to be transferred to an aqueous solution to form a dispersion (17,18).

Various processes utilize multiple interfaces. For example, processes occurring in organic–water–CO_2 interfacial systems can have up to three interfaces (water–CO_2, organic–CO_2, and water–organic). Carbon dioxide may be used to swell polymer latexes in water (19) for impregnation (20) or to modify emulsion polymerization (21). Rapid expansion from supercritical CO_2 to aqueous solution (RESAS) (22) offers a means of achieving greater control of particle size than is possible in conventional rapid expansion from CO_2 to air. Organic–organic–CO_2 interfacial systems may be manipulated to achieve microencapsulation in PCA, or to form metastable polymer blends by PCA (23) or polymerization (24).

Particles may be formed at the inorganic–CO_2 interface. Inorganic suspensions of metals and metal oxides have been formed by reaction or by mixing particles stabilized by surfactants or ligands (25). Semiconductor nanoparticles have been formed through chemical reduction in water-in-CO_2 microemulsions. The vast number of potential interfaces formed by combination of hydrophilic, lipophilic, and CO_2-philic phases will continue to offer many opportunities for discovery and development of new science and practical applications.

2. SOLVENT PROPERTIES OF CARBON DIOXIDE

A property that is useful for describing the solvent strength of CO_2 with regard to solubility behavior is the ratio of polarizability to volume (α/v). The molecular polarizability, a constant for a molecule related to the number and configuration of electrons divided by the molar volume, is a measure of the number and strength of van der Waals interactions in a given volume. This function has been compared for several fluids versus pressure (26), revealing that the van der Waals forces for CO_2 are considerably weaker than those of ethane and ethylene, being more like those of fluoroform. Consequently, the energy interaction parameters between nonpolar aromatic or aliphatic hydrocarbons and CO_2 may be expected to be weaker than those for ethylene and much weaker than those for hexane. This behavior had been shown earlier for the solutes naphthalene, hexamethylbenzene, fluorene, anthracene, phenanthrene, and pyrene (27,28) and will apply to many pharmaceuticals.

While the solvent power of CO_2 has often been compared with that of hexane because the two exhibit similar values of the solubility parameter δ, polymers such as atactic polypropylene and polybutadiene that are quite soluble in hexane are insoluble in CO_2. The van der Waals forces between CO_2 and a nonpolar molecule are weaker than expected from the δ for CO_2, since the δ for pure CO_2 is enhanced by quadrupole–quadrupole interactions between CO_2 molecules. The low polarizability-to-volume ratio for CO_2 reflects weak dispersion forces that lead to very small solubilities of nearly all polymers and drugs in CO_2. An interesting and enlightening exception is the cyclic peptide cyclosporine A. It lacks polar aromatic, carboxylic acid, or hydroxyl groups that tend to enhance solute–solute interactions and lower solubility. At the same time, it contains a large number of carbonyl groups that interact favorably with the quadrupole moment of CO_2 (22). As a result, cyclosporine A has a solubility of over 1 wt % at the moderate pressure of 240 bar at $30\,^\circ C$ (22). The interactions with the quadrupole moment of CO_2 can also raise the solubility of polar polymers containing carbonyl groups, particularly at pressures well over 300 bar, a limiting pressure for many supercritical fluid laboratories (29,30) Sugar pentaacetates have been shown to be unusually soluble in CO_2, owing in part to favorable specific polar interactions with the acetate groups (31).

Spectroscopic probes have been used to characterize the very weak solvent α/v and dipolarity of CO_2, as reviewed elsewhere (1,32). Fourier transform IR spectroscopy (FTIR) has been used to show that CO_2 interacts as an electron acceptor toward carbonyl groups in polymers, based upon splitting of the v_2 bending mode of CO_2 (33). Equilibrium constants have been measured from the v_2 bending mode of CO$_2$ by FTIR spectroscopy for the electron donor–acceptor interactions of CO$_2$ with three Lewis bases: triethylamine, pyridine, and tributylphosphate (TBP) (34). The equilibrium constants were relatively weak compared with those between these bases and Lewis acids such as methanol, water, or phenol. For the strongest interaction, CO$_2$-TBP, the solubility of CO$_2$ in TBP is increased by only up to 28% owing to the specific Lewis acid–base interactions.

3. SOLUBILITY IN CARBON DIOXIDE

The solubilities of nonpolymeric solutes in supercritical fluids such as CO$_2$ can be predicted if the properties that describe the solute–solute interactions are chosen properly (35). For polyaromatic hydrocarbons, examples of appropriate solute properties include the enthalpy of vaporization and the solute molecular volume. To further understand the role of solute–solvent interactions on solubilities and selectivities, it is instructive to define an enhancement factor E as the actual solubility, y_2, divided by the solubility in an ideal gas, with the result $E = y_2 \, P/P_2^{sat}$. This factor is a normalized solubility because it removes the effect of the vapor pressure, providing a means of focusing on interactions in the supercritical fluid (SCF) phase. In carbon dioxide at 35°C and 200 bar, $E \sim 10^7$ for the similar tetracyclic sterols: cholesterol, stigmasterol, and ergosterol, each of which contains a single OH group (35). In fact, enhancement factors typically vary within a narrow range of about one to two orders of magnitude for many types of organic solid in CO$_2$, even though the actual solubilities vary by many orders of magnitude. This means that solubilities in carbon dioxide for a given molecular weight of solute are governed primarily by solute–solute interactions, as described by the solute vapor pressure, and only secondarily by solute–solvent interactions in the SCF phase. Exceptions include bases such as ammonia, primary and secondary amines, and water, which can react with carbon dioxide to form covalent bonds in the form of carbamates and carbonic acid.

O'Neill et al. (27) examined CO$_2$ solubility for a wide variety of homopolymers and block copolymers below 100°C. In general, the solubility of amorphous homopolymers decreases with an increase in cohesive energy density, which may be characterized by its surface tension, as shown in Table 2. For example, poly(fluorooctylacrylates), which have surface tensions σ of only 10 mN/m and are soluble at a level of 10 wt % at 200 bar and 35°C, even at a molecular weight of 10^6. Solubilities are much lower for poly(dimethylsiloxane) ($\sigma = 20$ mN/m) and even lower for hydrocarbons such as poly(propylene oxide) ($\sigma = 30$ mN/m). As the cohesive energy of the polymer decreases, the solute–solute interactions become closer to those of carbon dioxide, favoring solvation as described by regular solution theory. Furthermore, polymers with lower cohesive energy densities have weaker interac-

TABLE 2 Solubility of Polymers in CO_2 as a Function of the Cohesive Energy Density as Characterized by the Surface Tension of the Pure Polymer Melt

Polymer	Surface tension (mN/m) at 20°C	Solubility (wt %, MW) at 3000 psia and 35°C
Poly(1,1-dihydrodecafluorooctyl acrylate)	10	$>>$10 (1 $\times$ 10^6 g/mol)
Poly(dimethylsiloxane), 75,000 g/mol	20	4 (13,000 g/mol)
Poly(2-ethylhexyl acrylate), 34,000 g/mol	30.2	$<<$0.1 (4233 g/mol)
Poly(propylene oxide), 2025 g/mol	31.5	0.5 (2000 g/mol)
Poly(ethylene oxide), 6000 g/mol	42.9	0.4 (600 g/mol)

tions with themselves in a liquid or solid condensed phase, raising their chemical potential and solubility in the CO_2 phase. Polar interactions can further perturb solubility: poly(vinyl acetate) with a more accessible carbonyl group is more soluble than poly(methyl acrylate) even though these polymers are isomers. To develop CO_2-soluble block and graft copolymers without fluorinated segments, Sarbu et al. have synthesized poly(ether–carbonate) copolymers with unusually high solubilities due to the combination of polymer backbone flexibility (as reflected in free volume and glass transition temperature), low cohesive energy density of the propylene oxide monomer, and polarity due to the carbonyl group (30).

Because CO_2 is such a weak solvent, it is an excellent antisolvent. Whether gaseous, liquid, or supercritical, CO_2 is quite soluble in a number of organic solvents at pressures from 10 to 300 bar. The dissolved CO_2 expands the liquid phase, substantially decreasing its cohesive energy density (solvent strength), as has been shown with spectroscopic probes (36). The addition of CO_2 to many organic liquids (with which it is significantly miscible), including ionic liquids, will lower the mixture solubility parameter to a value such that most solutes will precipitate.

4. FUNDAMENTAL PROPERTIES OF INTERFACES IN CO₂-BASED SYSTEMS

A fundamental understanding of colloid and interface science in CO_2-based systems is emerging on the basis of recent studies of interfacial tension and surfactant adsorption measurements (37) along with complementary studies of colloid structure (38) and stability (2,15,16). The interfacial tension γ between a supercritical fluid phase and a hydrophilic or lipophilic liquid or solid, along with surfactant adsorption, play key roles in a variety of processes including nucleation, coalescence and growth of dispersed phases, formation of microemulsions and emulsions, particle and fiber formation, atomization, foaming (39), wetting, adhesion, lubrication, and the morphology of blends and composites (40).

Figure 1 plots values of γ as a function of pressure for the binary CO$_2$–water, CO$_2$–poly(ethylene glycol) (PEG, 0.6 kg/mol), CO$_2$–polystyrene (PS, 1.85 kg/mol),] and CO$_2$–poly(-2-ethylhexylacrylate) (PEHA, 92.5 kg/mol) systems (8,41). In each case, the cohesive energy density of CO$_2$ is lower than that of the second phase. Consequently, γ decreases with an increase in pressure (neglecting the critical region) as the intermolecular forces of CO$_2$ become closer to the force of the other component, as has been predicted theoretically with a gradient model along with lattice–fluid theory (42). As expected, changes are larger at low pressures where the CO$_2$ density variation with pressure is more pronounced. At a given pressure, γ decreases for the above series from CO$_2$–water to CO$_2$-PEHA in the same order as the surface tension of the pure water or organic phase, as expected.

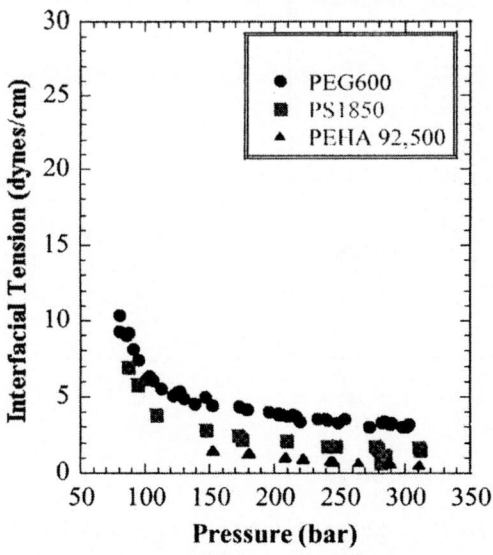

FIGURE 1   Interfacial tension between CO$_2$ and various liquids at equilibrium at 45°C.

Surfactants have been designed to lower γ in CO_2-based systems. The first generation of research involving surfactants in SCFs addressed water-in-oil (W/O) microemulsions and polymer latexes in ethane and propane, as reviewed elsewhere. (43–45). This work provided a foundation for studies in CO_2, which has weaker van der Waals forces (α/v) than ethane. Surfactants with both "CO_2-philic" and "CO_2-phobic" segments have been used to form microemulsions, emulsions, and organic polymer latexes in CO_2.

4.1. Water–CO_2 Interface

4.1.1. Hydrophilic–CO_2–philic Balance: Interfacial Tension and Phase Behavior

The effect of surfactants on the interfacial tension between water and supercritical fluids is a key property for describing emulsions and microemulsions (8), as shown in Figure 2. The x axis may be any formulation variable that influences surfactant partitioning between the phases such as the pressure or temperature. A minimum in γ is observed at the phase inversion point, where the system is balanced with respect to the partitioning of the surfactant

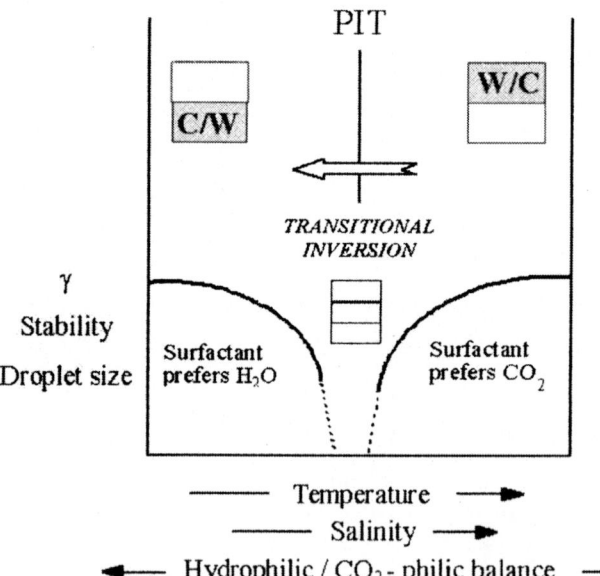

FIGURE 2 Schematic representation of phase behavior and interfacial tension for mixtures of water, CO_2, and nonionic surfactants as a function of formulation variables.

between the phases. Here a balance in the solvation of the surfactant tail by CO$_2$ and the head group by water may lead to interfacial tensions below 1 mN/m. Upon changing any of the formulation variables away from this point—for example, the hydrophilic–CO$_2$-philic balance (HCB) in the surfactant structure—the surfactant will become preferentially soluble in one of the phases. This phase usually becomes the external phase, according to the Bancroft rule. For example, a surfactant with a low HCB such as perfluoropolyether ammonium carboxylate [CF$_3$–(O–CF$_2$–CF(CF$_3$))$_n$–(O–CF$_2$) –COO$^-$NH$_4^+$: PFPE COO$^-$NH$_4^+$, 2500 MW], forms W/C microemulsions in equilibrium with an excess aqueous-rich phase (9). The low HCB results from the favorable solvation of the fluorinated tail by CO$_2$ and the weak tail–tail interactions. The low interfacial tensions and high surfactant adsorption for PFPE COO$^-$NH$_4^+$ explain, in part, the ability to stabilize W/C microemulsions with this surfactant. In contrast, kinetically stable macroemulsions may be formed near the balanced state. However, when too near to the balanced state, the thin film that prevents coalescence of macroemulsions droplets becomes unstable, as explained later. Studies of γ versus HCB have been reported for a homologous series of block copolymers of propylene oxide and ethylene oxide and of poly(dimethylsiloxane) (PDMS) and ethylene oxide (8) as a function of the ratio of the mass fractions of the blocks. These maps of interfacial tension may be used directly to guide the design of surfactant architecture to stabilize either microemulsions or macroemulsions.

4.1.2. Microemulsions of Water and CO$_2$: Fundamentals

A summary of studies of microemulsions in CO$_2$ is given in Table 3. For PFPE COO$^-$ NH$_4^+$ W/C microemulsions, FTIR, UV–visible, fluorescence and electron paramagnetic resonance (EPR) experiments have demonstrated the existence of an aqueous microdomain in CO$_2$ with a polarity approaching that of bulk water (6). From small-angle neutron scattering (SANS) measurements, it is known that the water droplet radius is 3.5 nm, for a molar water-to-surfactant ratio of (30). Examples of surfactants that have been reported to stabilize W/C microemulsions include PFPE COO$^-$ NH$_4^+$, the hybrid hydrocarbon–fluorocarbon surfactant, C$_7$F$_{15}$CH(OSO$_3^-$Na$^+$)C$_7$H$_{15}$, and the surfactant di(1H,1H,5H-octafluoro-n-pentyl) sodium sulfosuccinate (di-HCF$_4$), as well as others listed in Table 3. Organic-in-CO$_2$ microemulsions have also been formed in which the dispersed phase was 600 MW poly(ethylene glycol) (PEG 600) or polystyrene oligomers (41). For microemulsions with low water volume fractions (typically ϕ < 0.02) the absolute amount of dissolved hydrophiles can be somewhat small, and this can limit reaction rates for phase transfer catalysis. Recently, highly concentrated W/C microemulsions ($\phi \sim 0.5$) have been formed (46) providing for much higher solubilization.

TABLE 3 Studies of Water–CO_2 Microemulsions

Surfactant type	Properties or methodology[a]	Ref.
Fluorocarbon–hydrocarbon hybrid sulfate $(C_7H_{15})(C_7F_{15})CHOSO_3^-Na^+$	Phase behavior	5
	SANS of droplet size	76
	Simulation	77
Fluorinated polyether carboxylate $CF_3O(CF_2CF(CF_3)O)_nCF_2COO^-NH_4^+$	Phase behavior, spectroscopy, protein solubilization	6
	spectroscopy, aqueous inorganic reactions	78
	Phase behavior, EPR, time-resolved fluorescence	79
	SANS of droplet size	80
	Fluorescence (pH 3.1–3.5)	47
	CdS nanoparticle synthesis	81
	Buffering pH up to 7	48
	Nucleophilic substitution reactions	82
	Interfacial tension studies	37
	Phase behavior, conductivity (percolation)	46
	Luminescence quenching to probe microenvironment	83

	NMR	84
	SANS (droplet interactions)	49
	TiO$_2$ Nanoparticle synthesis	85
	Rapid expansion to form silver particles	86
Fluorinated perfluoropolyether trimethylammonium acetate	Phase behavior, SANS, FTIR, UV-visible	87
Fluorinated Aerosol OT analogue	Enzymatic catalysis	54
	Phase behavior	88
Methylated acetylenic glycol (nonionic) (Dynol-604)	Phase behavior, UV-visible, lysozyme solubilization	89
Ls-54, C$_{12}$H$_{25}$O(CH$_2$CH$_2$O)$_m$(CH(CH$_3$)CH$_2$))$_n$H	Phase behavior, UV-visible, SAXS, lysozyme solubilization	90
Fluorinated dichain phosphate (C$_6$F$_{13}$-CH$_2$-CH$_2$-O)$_2$-P(O)-O-Na$^+$	Phase behavior, SANS, UV-visible	91
	SANS, simulation	92

[a]SANS, small-angle nuclear scattering; EPR, electron paramagnetic resonance spectrometry; NMR, nuclear magnetic resonance spectrometry; FTIR, Fourier-transform infrared spectrometry; SAXS, small-angle X-ray scattering.

The pH inside microemulsion droplets is typically around 3 owing to the formation of carbonic acid, as determined with fluorescence (47) and absorbance (48) probes. Inorganic and organic bases and buffers, such as NaOH, can be used to control the aqueous pH in microemulsions stabilized by PFPE $COO^-NH_4^+$ from 3 up to 5 to 7.

To form W/C microemulsions, the interfacial tension must be on the order of 1 mN/m or less to overcome the free energy penalty γdA for the large surface area, and the interactions between the droplets must not lead to flocculation. The prevention of this flocculation has been the greatest challenge in designing surfactants for W/C microemulsions. Droplet interactions in water-in-carbon dioxide (W/C) microemulsions formed with PFPE $COO^-NH_4^+$, a perfluoropolyether anionic surfactant, have been studied with electrical conductivity (46), SANS (49), and NMR (50). In Figure 3, the van der Waals attraction constant between droplets is shown to increase with a decrease in CO_2 density. As the solvation of the tails decreases, the interdroplet tail–tail interactions become stronger. At the highest densities, the interaction is weak

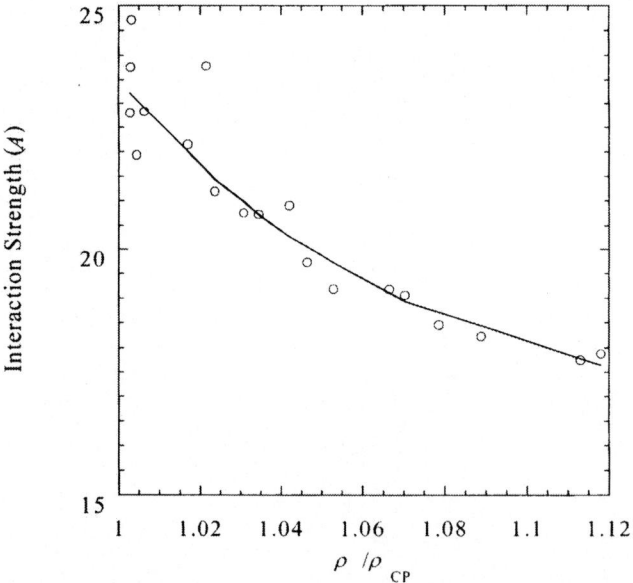

FIGURE 3 Van der Waals interaction strength for W/C microemulsions versus the density of CO_2 divided by the density of CO_2 at the cloud point for temperatures from 15 to 80°C in the one-phase region. The interaction strength at the phase boundary where the droplets flocculate is 21.2 for a hard-sphere fluid with van der Waals attraction. (From Ref. 49.)

enough for the formation of a stable microemulsion, but it is twice as high as that for typical water-in-oil microemulsions as a result of stronger tail–tail interactions resulting from the weak tail solvation by CO_2. Thus, the design of new surfactants for these microemulsions requires overcoming these strong droplet interactions. Recently surfactants with hydrocarbon tails have been designed to form hydrated reverse micelles in CO_2, as shown in Table 3.

4.1.3. Macroemulsions of Water and CO_2: Fundamentals

Stable W/C macroemulsions for either liquid or supercritical CO_2 have been formed with the surfactants (PFPE COO^-)$_2$ Mn$_2$$^+$ (13), PFPE COO^-NH$_4$$^+$ (672–7500 MW) (9), and block copolymer pluronic surfactants composed of poly(propylene oxide) (PPO) and poly(ethylene oxide) (PEO) (51), PDMS and poly(acrylic acid) (PAA), or poly(methacrylic acid) and PDMS and PEO (2). The emulsions may be formed by shear through a 130 μm capillary or commercial homogenizer. The ratio of water to CO_2 has been varied from 9:1 to 1:9 with less than 1 wt % surfactant. W/C microemulsions, on the other hand, contain less than 5 wt % water for this level of surfactant owing to the much higher interfacial area.

In contrast with microemulsions, macroemulsions are stabilized kinetically. The stability of W/C emulsions formed with PDMS-*b*-poly(methacrylic acid) (PDMS-*b*-PMA) and PDMS-*b*-poly(acrylic acid) (PDMS-*b*-PAA) ionomer surfactants has been measured as a function of surfactant architecture, pH, temperature, pressure, and droplet flocculation. Ionization of the PMA or PAA block was manipulated with pH to vary the HCB and, thus, the emulsion stability, as shown in Figure 4. Here the volumes of water and CO_2 were the same. At the balanced state, the interfacial tension passes through a minimum value and an inversion of the curvature of the emulsion occurs (from W/C to C/W or vice versa). Here the Marangoni–Gibbs stabilization becomes weak because the gradients in γ, which otherwise resist drainage of the thin stabilizing film between droplets, are low. Also, it becomes easy to bend and rupture the surfactant monolayer, to allow drainage between droplets or coalescence. As the distance from the balanced state increases, the emulsion stability increases but eventually decreases as the adsorption of surfactant at the interface becomes too low (e.g., at very low pH).

C/W emulsions are easier to form than W/C emulsions because water is a more viscous continuous phase than CO_2. The higher viscosity provides greater shear in droplet disruption and lowers settling or creaming rates. Also, water solvates surfactant head groups more effectively than CO_2 solvates hydrophobic tail groups. Concentrated C/W emulsions have been reported for amphiphiles containing alkylene oxide-, siloxane-, and fluorocarbon-based tails as a function of temperature and salinity (51). Poly(ethylene oxide)-*b*-

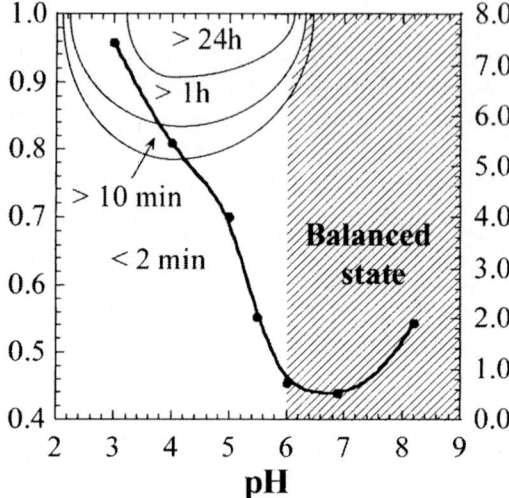

FIGURE 4 Stability contours for W/C emulsions stabilized by PDMS-*b*-PAA (20–0.7 K) in terms of the time for 20% settling. At low pH the surfactant prefers CO_2 and limited adsorption at the interface leads to unstable emulsions. As the pH increases, the PAA groups begin to ionize and adsorption increases, leading to more stable emulsions until the balanced state is reached where rapid coalescence leads to unstable emulsions. (From Ref. 93.)

poly(butylene oxide) (EO_{15}-*b*-BO_{12}) can emulsify up to 70% CO_2 with droplet diameters from 2 to 4 μm, as determined by video-enhanced microscopy.

C/W emulsions may be used as templates in the formation of porous polymers, as shown in Figure 5 (52). Polymerization takes place in the aqueous phase continuous channels between the CO_2 droplets. The CO_2 is vented and the water is removed to form a porous polymer. The median pore diameter on the order of 1 μm reflects the size of the original CO_2 droplets. Polymer foams may be used as adsorbents, as substrates for catalysts, and as scaffolds in biomedical engineering.

As was illustrated earlier in Figure 2, the curvature of an emulsion may be inverted with a formulation variable. The curvature of an emulsion of CO_2 and brine stabilized with a triblock copolymer of PDMS and PEO, PEO-*b*-PDMS-*b*-PEO, was inverted from W/C to C/W by reducing the temperature (53). A "v-shaped" trough was observed in the interfacial tension between the aqueous and CO_2 phases versus temperature at constant CO_2 density, as shown in Figure 6. The minimum in the interfacial tension coincided with the balanced state or phase inversion temperature (PIT), where the curvature of the emulsion inverts. At high temperatures where the surfactant favors CO_2

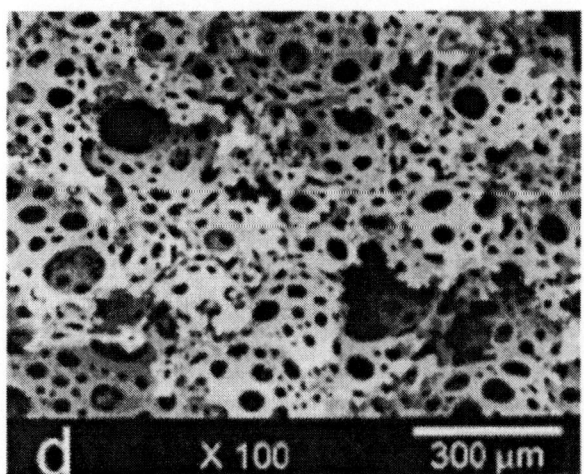

FIGURE 5 Templating of porous polymers in carbon dioxide-in-water emulsions. The polymerization takes place in the continuous phase about CO_2 droplets. (From Ref. 52.)

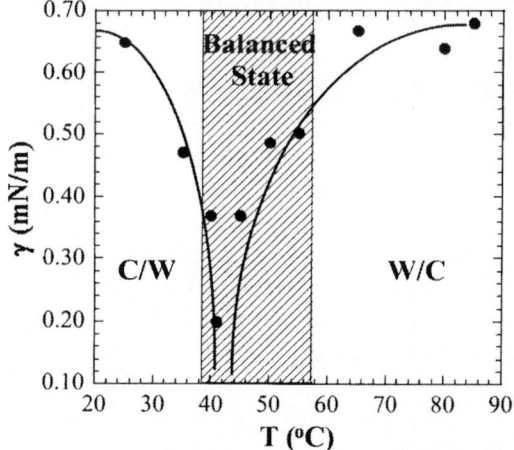

FIGURE 6 Interfacial tension versus temperature at 0.05 wt % PEO-b-PDMS-b-PEO surfactant concentration and a constant CO_2 density of 0.842 g/mL with 0.1 M NaCl. The curvature of the emulsion inverts at the balanced state. (From Ref. 53.)

because hydrogen bonds between water and the EO groups are weakened, the conductivities are low, indicating W/C emulsions. Novel C/W miniemulsions consisting of 200 nm droplets, measured by multiwavelength turbidimetry, were formed by low-shear stirring in the interfacial tension minimum–PIT region and then cooling to 25°C. These droplets were three times smaller than those produced isothermally by high shear at a temperature below the PIT region. The formation of these nanometer-sized droplets is facilitated by the ultralow interfacial tension in the PIT region, which favors droplet nucleation at the onset of inversion to the C/W morphology. Owing to their miniemulsions require about an order of magnitude less surfactant per volume of dispersed phase than microemulsions, smaller surface area. In addition, the droplets settle about an order of magnitude slower than those in macroemulsions owing to the smaller size according to the Stokes equation. This enhanced stability is very important in low viscosity solvents such as CO_2, where settling is unusually rapid.

4.1.4. Applications of Microemulsions and Macroemulsions of Water and Carbon Dioxide

The W/C microemulsions, which have the ability to act as "universal solvents" for solubilizing both polar and apolar molecules, including enzymes, provide a unique medium for enzyme-catalyzed biotransformations. The protein bovine serum albumin (BSA) (M_w = 67 K), has been solubilized in a PFPE $COO^- NH_4^+$-stabilized W/C microemulsion (6). BSA compartmentalized within the water droplets was shown to experience an aqueous environment similar to that of native BSA in buffered solution. Fluoroether-functional surfactants have also been used to extract subtilisin Carlsberg from aqueous buffer and cell culture media into CO_2, with recovery accomplished by depressurization (7). In this case water and CO_2 formed a three-phase system with the surfactant in the middlephase, or a C/W emulsion. Some of the protein was sacrificed to aid the formation of the emulsion. Protein concentration in the emulsion was found to be as high as 19.5 mg/mL. Holmes et al. (54) performed two enzymatic reactions, the lipase-catalyzed hydrolysis of p-nitrophenol butyrate and the lipoxygenase-catalyzed peroxidation of linoleic acid, in W/C microemulsions stabilized by a fluorinated dichained sulfosuccinate surfactant (di-HCF$_4$). The activity of both the enzymes in the W/C microemulsion environment was found to be essentially equivalent to that in a water–heptane microemulsion stabilized by Aerosol OT, a surfactant with the same head group as di-HCF$_4$.

Clarke et al. showed that W/C microemulsions can be used as a new environment for inorganic chemistry by studying well-known inorganic reactions (78). Roberts et al. have recently demonstrated the production of nano-sized metallic copper particles in AOT reversed micelles in compressed propane and supercritical ethane through the reduction of Cu(AOT)$_2$ (55). Also, Ji et al. (56) were able to harvest silver nanoparticles from a W/C microemulsion by reducing silver nitrate. Johnston et al. synthesized semiconductor nanoparticles of cadmium sulfide, in PFPE COO$^-$ NH$_4^+$ stabilized W/C microemulsions (11). At water-to-surfactant ratios of 5 and 10, the excitation energies of the cadmium sulfide particles were 3.86 and 3.09 eV, corresponding to mean particle radii of approximately 0.9 and 1.8 nm, respectively. These radii are consistent with measurements by transmission electron microscopy, as well as the microemulsion droplet diameters from SANS, demonstrating that the size of the microemulsion droplets may be used as templates.

Emulsions of water and CO$_2$ have been used to form liposomes in one step without any organic solvent (57). The liposomes were made by forming a W/C emulsion stabilized with/L-R-dipalmitoylphosphatidylcholine. The pressure was reduced to form the liposomes by a reversed phase evaporation method. Large unilamellar liposomes with diameters of 0.1 to 1.2 μm were formed and used to trap D-(+)-glucose.

4.2. Organic–CO$_2$ Interface

The low interfacial tension between CO$_2$ and many organic phases facilitates the formation of high surface area materials by phase separation.

4.2.1. Rapid Expansion from Supercritical Solution

In rapid expansion from supercritical solution (RESS), nucleation and crystallization are triggered by reducing the solvent density (i.e., solvent strength) through expansion to atmospheric conditions. An interface between an organic precipitate and expanding carbon dioxide forms during depressurization through a nozzle in times less than 1 ms. According to Figure 1, the interfacial tension will increase along the nozzle as the carbon dioxide density decreases. The low interfacial tensions and large supersaturation lead to rapid nucleation rates (58). This process has been used successfully to form a variety of microparticles and microfibrils from polymers, drugs, and inorganic compounds. Most particle sizes reported have been between 5 and 100 μm. From theoretical calculations, it should be possible to form particles as small as 20 to 50 nm (58). The inability to approach the theoretical lower limit is likely due to particle growth resulting from collisions in the free jet. This growth may be inhibited by rapid expansion into aqueous solution containing surfactant stabilizers, as described later.

RESS of polymers in CO$_2$ has been limited by low polymer solubility, such as the system L-poly(lactic acid) (PLA) in carbon dioxide (59). To overcome this limitation, large amounts of cosolvents including ethanol, methanol, and 1-propanol have been utilized in a new process, rapid expansion from supercritical solution with a nonsolvent (RESS-N) to microencapsulate proteins with polymers (60). For low molecular weight solutes and polymers, it has been demonstrated that cosolvents, which cause large increases in solute solubilities in CO_2, need not be good solvents for the solute. In pure form, these cosolvents are nonsolvents for the polymers; thus they are only sparingly soluble in the polymer particles produced during expansion. Since the cosolvent does not swell and soften the polymer product, it is not expected to cause agglomeration. Protein microparticles may be suspended in the polymer–CO_2–cosolvent solution. During RESS, the polymer will precipitate and coat the protein microparticles. For lysozyme, the thickness of coating of polymers including poly(ethylene glycol) and poly(DL-lactide-co-glycolide) (PGLA; MW 5000) may be controlled by changing the feed composition of the polymer.

4.2.2. Precipitation with a Compressed Fluid Antisolvent

The process of precipitation with a compressed antisolvent (PCA) (also called SAS, SEDS, ASES) also takes advantage of interfacial properties. The large supersaturation and very low interfacial tension between many organic phases and CO_2 (Figure 1) lead to high nucleation rates and in some cases submicrometer-sized materials. This process is described in detail in Chapter. In this section, we discuss a particular variation of the PCA process in which both a CO_2 vapor and liquid phase are present in addition to the organic phase. A CO_2 vapor–liquid interface was used to form hollow microspheres of polystyrene from polystyrene in toluene solutions with concentrations above 6 wt % (61). In this case the cell was filled only partially with liquid CO_2, with its equilibrium vapor phase above it. The solution was atomized in the vapor phase, in which a skin formed on the droplets, and the droplets subsequently fell into the liquid phase, where they solidified. Delaying precipitation in the vapor phase caused the formation of hollow microspheres. If the organic solution had been sprayed directly into liquid or dense supercritical fluid CO_2, rapid polymer precipitation at the organic–CO_2 interface would have produced fibers.

For much less viscous 30,000 MW PGLA solutions, a skin did not form in the vapor phase. The particles did not form until the large droplets dispersed on contact with the liquid CO_2 phase, which was agitated to improve mixing. The resulting spherical particles ranged from 3 to 25 μm in diameter. It is shown later that these particles may be used to encapsulate proteins.

PCA may be utilized to produce stable PMMA polymer latexes in CO_2 by spraying the polymer solution in the presence of a steric stabilizer, introduced in either the organic or the CO_2 phase. One of the most successful stabilizers is poly(1,1-dihydroperfluorooctyl acrylate) or copolymers containing this group (62). Dilute PMMA latexes have also been stabilized with block copolymers containing either poly(propylene oxide) (PPO) or poly (butylene oxide) (PBO) stabilizer group(s) and a poly(ethylene oxide) (PEO) anchor group (63). When dissolved PPO-PEO-PPO triblock and PBO-PEO diblock copolymers are introduced with the CO$_2$ feedstream, PMMA particles with diameters between 0.1 and 0.5 μm, are produced with no observed aggregation. The effectiveness of the stabilizer has been described in terms of its concentration and how it partitions between the dispersed phase, the interface, and the CO$_2$ phase to achieve sufficient adsorption on the particles. When the stabilizer is introduced with the solution phase, it does not have to be soluble in CO$_2$ to adsorb at the interface and prevent flocculation. However, steric stabilization of such systems requires solvation of the stabilizer group by CO$_2$.

4.2.3. Polymer Latexes in CO$_2$

Polymer latexes formed in CO$_2$ by phase separation (e.g., PCA) or by dispersion polymerization may be used to encapsulate pharmaceuticals. Polymer latexes coated with steric stabilizers are stable above a certain CO$_2$ density, termed the critical flocculation density(CFD) (16). For acrylate latexes stabilized with block copolymer surfactants based on poly(1,1-dihydroperfluorooctyl acrylate)(PFOA) in CO$_2$, it was shown by light scattering that the CFD corresponds to the upper critical solution density (UCSD) of the PFOA-CO$_2$ system (16). The UCSD is the density in which the homopolymer is miscible with CO$_2$ at all concentrations. The agreement between this result and theory and simulation (15) is likely due to the symmetry in the interactions between the polymer segments and CO$_2$ that results from the low surface tension (cohesive energy) of PFOA. However, 10,000 g/mol PDMS could not stabilize either PMMA latex particles or silica particles in pure CO$_2$ at pressures up to 345 bar at 25°C and 517 bar at 65°C, without stirring (2). In this case, the density is well above the UCSD for the PDMS-CO$_2$ mixture, but below the theta density (i.e., the density at which an infinite molecular weight PDMS becomes soluble in CO$_2$). Further studies are needed to understand the difference between the CFD and UCSD for this system in terms of polymer molecular weight and asymmetry in the polymer–CO$_2$ interactions. There is significant interest in developing nonfluorinated polymer stabilizers for CO$_2$, especially for pharmaceutical applications.

4.3. Transfer of Latexes from Organic–CO$_2$ to Organic–Water Interfaces

A polymer latex in CO$_2$ with an organic–CO$_2$ interface may be transferred into water to form an aqueous latex with an organic–water interface. This transfer may be accomplished through the use of an "ambidextrous" surfactant, that is, a surfactant that is active at both an organic–CO_2 and an organic–water interface (17,18). A second approach is to stabilize a polymer latex in CO_2 with a stabilizer with CO_2-philic tails and then transfer the particles to an aqueous solution containing a separate hydrophilic surfactant stabilizer (18,64). With this concept, it is possible to produce latexes in CO_2 with minimal waste, which can be vented, shipped as dry powder, and resuspended in water to form an aqueous latex for a variety of applications including parenteral administration and the formation of films. Recently, this concept was demonstrated with an ambidextrous surfactant, as shown in Figure 7. The surfactants included a PDMS block, a PMA or PAA block, and in some cases, a third component, methyl methacrylate (MA) or *tert*-butyl acrylate (tBA) units in a random or block architecture (18). The third component is used to increase the adsorption of the stabilizer to the organic polymer latex surface. In CO_2, PMMA particles produced by dispersion polymerization are sterically stabilized by the PDMS segments extending into CO_2. Upon venting the CO_2 and then transferring the surfactant-coated particles to water (or buffered water) at up to 40% by weight, the PDMS block collapses onto the latex surface, and some of the PMA or PAA groups are ionized, producing electrostatic stabilization. The surfactant is ambidextrous in that it stabilizes two different types of interface, each by a different stabilization mechanism. The size of the latex particles did not change upon transfer to water on the basis of measurements by dynamic light scattering.

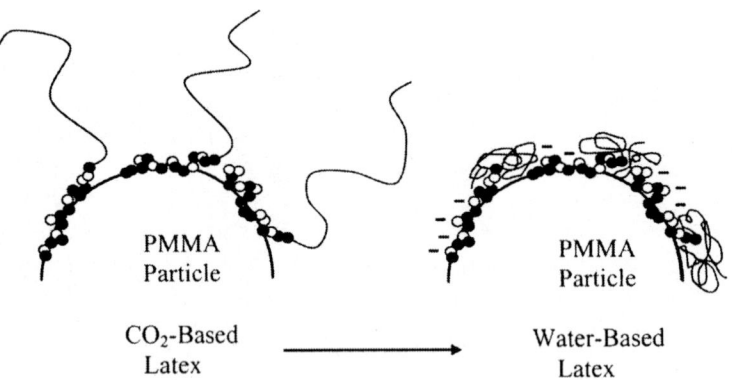

FIGURE 7   "Ambidextrous" surfactant for stabilization of latexes in CO_2 or water: tails: CO_2-philic stabilizer segments; solid circles, anchor group; open circles, an ionizable group. (From Ref. 18.)

4.4. Organic–Water–CO$_2$ Interfacial Systems

4.4.1. Swelling of Polymer Particles

Supercritical CO$_2$ can diffuse quickly into bulk glassy polymers to produce a significant degree of swelling and plasticization, as described in Chapter. This swelling can also be achieved for an aqueous latex in contact with a CO$_2$ phase. The swelling of 50 nm aqueous polystyrene latex spheres in the presence of CO$_2$ was measured as a function of pressure by dynamic light scattering at 25°C and pressures up to 35MPa (19). The latex spheres swelled about 60% more than polystyrene in pure CO$_2$. This enhanced swelling for the latex may be attributed to the adsorption of CO$_2$ at the polystyrene–water interfacial region, which lowers the overall interfacial tension. On the basis of this enhanced swelling, the thickness of this interfacial region containing polystyrene, carbon dioxide, and water was approximately 4 nm, which is on the order of the radius of gyration of the polystyrene chain. Within the experimental time scale of 12 h, flocculation was not present, and the particle diameter returned rapidly to its original value after the release of pressure. However, the latexes flocculated when pressurized for over 36 h.

4.4.2. Impregnation of Latexes

Once a polymer latex has been plasticized by CO$_2$ below the glass transition temperature T_g, it may be impregnated with various water-insoluble or water-soluble solutes that diffuse into the polymer (20). When CO$_2$ is removed, the particles return to the glass state, entrapping the additive. An aqueous latex of monodisperse polystyrene particles was dyed with water-insoluble Sudan Red 7B using CO$_2$ at 25°C and 310 bar. At this condition, polystyrene is a liquid and the increased mobility of the polymer chains greatly enhances the rate of diffusion of dye into the particles. Even though the particles were highly plasticized during the dyeing process, no agglomeration or coalescence of the particles was observed. Two surfactants, poly(ethylene oxide)–b–poly(buty-lene oxide) (PEO-b-PBO) and perfluoropolyether ammonium carboxylate (PFPE COO$^-$NH$_4$) were used to form emulsions of CO$_2$ in water, to increase the amount of dye incorporated in the polystyrene particles. UV–visable analysis showed that the particles produced with PFPE COO$^-$NH$_4$ contained up to 0.46 wt % dye. More recently, the effects of surfactant concentration, particle size, and impregnation time on the impregnation of polystyrene la-texes with several types of dyes has been investigated (65). Both aqueous and ethanol-based latexes may be impregnated with the aid of CO$_2$. This approach holds promise in the development of a CO$_2$-based microencapsulation process that eliminates organic solvents typically used in the solvent evapo-ration or double emulsion processes.

4.4.3. Effect of CO_2 on Emulsion Polymerization

Carbon dioxide may be used to tailor the properties of emulsion polymerization of a water-insoluble monomer in an aqueous solvent by lowering the viscosity of the dispersed organic phase (21). Surfactant-free emulsion polymerization of methyl methacrylate has been conducted in a hybrid CO_2–aqueous medium with the water-soluble initiator potassium persulfate under a varying head pressure of CO_2 (0–350 bar) at $75°C$. In this example there were two continuous phases, namely a lower water-rich phase and an upper CO_2-rich phase, which remained separate with a clear phase boundary. Polymer particles were observed only in the water-rich phase with no visible creaming, and the particle number concentration did not change with CO_2 pressure. At the highest pressures (>280 bar), however, there was a marked decrease in average molecular weight as the lower viscosity appeared to enhance bimolecular chain termination reactions.

4.4.4. PCA(SEDS) at the Water–Organic–CO_2 Interface

The PCA(SEDS) technique is usually used without water. However, it has been modified to allow the dissolution of a water-soluble substance or pharmaceutical protein in an aqueous basic solution prior to spraying into CO_2 (66). A triaxial nozzle was used to introduce the aqueous solution containing the protein, supercritical CO_2, and an organic solvent phase, for example, ethanol separately. The water–organic solvent mixture becomes miscible with the supercritical antisolvent, and micrometer-sized particles of the solute are produced. The contact time between the organic solvent and the pharmaceutical protein is minimized, to reduce denaturation, by using the triaxial nozzle. The mixing of the three flowing streams must be optimized to minimize denaturation and control the particle size.

4.4.5. Rapid Expansion From Supercritical to Aqueous Solution (RESAS)

During RESS, collisions during free jet expansion cause particles at the organic–CO_2 interface to grow . This growth may be inhibited by expanding into an aqueous solution with the use of surfactant stabilizers, a process known as RESAS (22). The surfactants may be introduced into the CO_2 phase, the aqueous phase or both. During the expansion, the nucleated organic particles contact an expanding CO_2 phase and an aqueous phase simultaneously. This process combines aspects of phase separation with interfacial stabilization.

RESAS may be used to enhance the dissolution rate of poorly water soluble drugs (22). By reducing the particle size to increase the interfacial area, dissolution rates in the gastrointestinal tract may be increased. Coating drug particles with polymeric and low molar mass hydrophilic stabilizers to enhance wetting and solvation by intestinal fluids may further increase the dissolution rates. Stable suspensions of submicrometer-sized particles of cyclosporine, a water-insoluble drug, have been produced by RESAS with an aqueous Tween-80 (Polysorbate-80) solution (22). Steric stabilization by the surfactant impedes particle growth and agglomeration. The particles were an order of magnitude smaller than those produced by RESS into air without of the surfactant solution. Concentrations as high as 38 mg/mL for particles 400 to 700 nm were achieved in a 5.0% (w/w) Tween-80 solution. In a related RESS application, nanoparticles of β-carotene (300 nm) were formed by expansion of an ethylene solution, which then flowed into a 10% (w/w) viscous gelatin solution to inhibit postexpansion particle agglomeration (67).

Stable aqueous emulsions of poly(2-ethylhexyl acrylate) (PEHA) were also produced by RESAS from CO$_2$ (68). In this case, a polymer suspension in CO$_2$ was expanded instead of a dissolved solute. A CO$_2$-philic surfactant, Monasil PCA (PDMS-g-pyrrolidonecarboxylic acid), was utilized in dispersion polymerization to form a stable polymer suspension at 65°C and 345 bar. A hydrophilic surfactant, (e.g., SAM 185, Pluronic L61, or Pluronic L62), that is soluble in CO$_2$ and CO$_2$/2-EHA monomer mixtures as well as water was added to CO$_2$ to stabilize the suspension after it had been rapidly expanded through a capillary into aqueous solution. The resulting aqueous emulsion with up to 15.6 wt % polymer content was stable for weeks with an average particle size of 2 to 3 μm. Another approach is to introduce the hydrophilic surfactant in the aqueous phase in addition to the surfactant in the CO$_2$ phase. This approach is more general, since many hydrophilic surfactants are not soluble in CO$_2$. During expansion of the suspension into an aqueous solution, the hydrophilic surfactant—for example, triblock Pluronic copolymers—diffuses to the particle surface to provide stabilization. The resulting aqueous latexes were stable for 100 days for a polymer content reaching 12.7 wt %.

A limitation of RESAS is that few drugs are soluble in CO$_2$. In an alternative process, evaporative precipitation into aqueous solution (EPAS), a drug dissolved in a liquid organic solvent, such as methylene chloride, rather than a supercritical fluid, is sprayed through an atomizer into an aqueous solution containing a hydrophilic stabilizer to produce an aqueous dispersion (69,70). Intense atomization leads to rapid evaporation of the small organic droplets in the aqueous solution. The stabilization of the drug particles with water-soluble stabilizers in the aqueous suspensions facilitates dissolution rates of the final powder after drying.

4.5. Organic–Organic–CO_2 Interface

4.5.1. PCA for Microencapsulation of Pharmaceuticals in Polymers

The PCA method is a promising technique for the preparation of drug-containing microparticles. Potential advantages include the flexibility of preparing microparticles of varying size and morphology, the elimination of surfactants, the minimization of residual organic solvents, low to moderate processing temperatures, and the potential ease of scale-up. Polymer microparticles have been prepared by atomizing organic polymer solutions into a spray chamber containing compressed CO_2 (71–74). The swelling of various pharmaceutically acceptable polymers was investigated to find polymers that did not agglomerate during the spraying process. In general, polymers with low glass transition temperatures agglomerated even at ambient temperatures. Scanning electron microscope characterization showed that the L-PLA microparticles were spherical, smaller than 5 μm, free flowing, and nonagglomerated, with a smooth surface structure (72).

Lysozyme was encapsulated in biodegradable polymer microspheres that were precipitated from an organic solution by spraying the solution into carbon dioxide (75). The polymer, either poly-(L-lactide) (L-PLA) or poly(DL-lactide-co-glycolide) (PGLA), in dichloromethane solution with suspended lysozyme, was sprayed into a CO_2 vapor phase through a capillary nozzle to form droplets, which solidified after falling into a CO_2 liquid phase. By delaying precipitation in the vapor phase, the primary particles became large enough, (5–70 μm) to be able to encapsulate the lysozyme. At an optimal temperature of $-20°C$, the polymer solution mixed rapidly with CO_2, and the precipitated primary particles were so hard such that agglomeration was markedly reduced compared with higher temperatures. This process offers a means of producing encapsulated proteins in poly(DL-lactide-co-glycolide) microspheres without earlier limitations of massive polymer agglomeration and limited protein solubility in organic solvents.

4.6. Polymer Blends by Phase Separation

The effect of dissolved CO_2 on the miscibility of polymer blends and on phase transitions of block copolymers has been measured with spectroscopy and scattering (40). The shifts in phase diagrams with CO_2 pressure can be pronounced. Polymer blends may be trapped kinetically in metastable states before they have time to phase separate. Metastable polymer blends of polycarbonate (PC) and poly(styrene-co-acrylonitrile) were formed with liquid and supercritical fluid CO_2 in the PCA process, without the need for a surfactant. Because of the rapid mass transfer between the CO_2 phase and the solution phase, the blends were trapped in a metastable state before they could phase separate (23). In principle this could be applied to microencapsulate a drug in a polymer blend or composite system.

4.7. Inorganic–CO$_2$ Interface

4.7.1. Silica Suspensions

In addition to polymer latexes, inorganic suspensions may also be formed in CO$_2$. (10) Block copolymers have been shown by turbidimetry to stabilize 1 μm hydrophilic silica dispersions in CO$_2$, with stability decreasing in the order PS-*b*-PFOA > poly(methyl methacrylate-*co*-hydroxyethyl methacrylate)-*g*-PFPE > poly(dimethyl-siloxane)-*g*-pyrrolidone carboxylic acid (PDMS-*g*-PCA) (MW 8500 g/mol, ~2 PCA groups). The decrease in stability with decreasing CO$_2$ density was sharper for higher molecular weight CO$_2$-philic segments, as expected based on the nature of polymer chain collapse.

4.7.2. Metal Nanoparticles

Supercritical fluids may be utilized for the synthesis of metal nanoparticles. Perfluorodecanethiol-stabilized silver nanocrystals were synthesized in supercritical CO$_2$ through arrested precipitation, by reducing silver acetylacetonate with hydrogen in the presence of fluorinated thiol (Figure 8). The CO$_2$ density used during synthesis controls the particle size and polydispersity. At

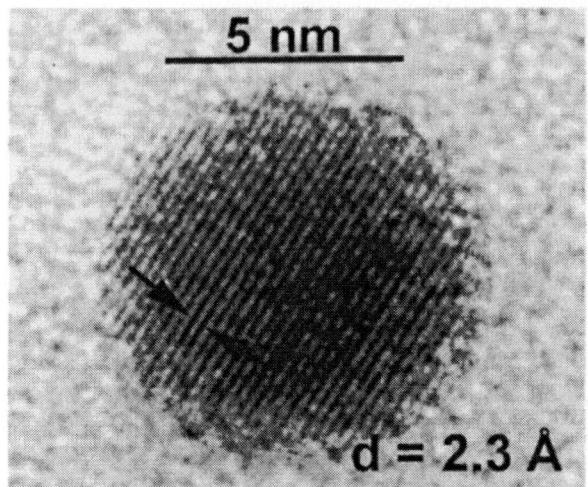

FIGURE 8 Transmission electron micrograph of silver nanocrystals synthesized in supercritical fluid CO$_2$ with perfluoroalkanethiol ligands illustrating the 111 lattice plane. (From Ref. 25.)

high solvent densities ($\rho > 250$ bar), the ligands provide a strong steric barrier that maintains small particles with a 20 Å diameter.

ACKNOWLEDGMENTS

The authors acknowledge financial support from The Dow Chemical Company (Midland, Michigan), partial support from the STC Program of the National Science Foundation under Agreement CHE-9876674, the Welch Foundation, and the U.S. Department of Energy.

REFERENCES

1. Eckert CA, Knutson BL, Debenedetti PG. Supercritical fluids as solvents for chemical and materials processing. Nature 1996; 383:313–318.
2. Johnston KP. Block copolymers as stabilizers in supercritical fluids. Cur Opin Colloid Interface Sci 2000; 5:351–356.
3. Kendall JL, Canelas DA, Young JL, DeSimone JM. Polymerizations in supercritical carbon dioxide. Chem Rev 1999; 99:543–563.
4. McClain JB, Betts DE, Canelas DA, Samulski ET, DeSimone JM, Londono JD, Cochran HD, Wignall GD, Chillura-Martino D, Triolo R. Design of nonionic surfactants for supercritical carbon dioxide. Science 1996; 274: 2409.
5. Harrison K, Goveas J, Johnston KP, O'Rear EA. Water-in-carbon dioxide microemulsions with a fluorocarbon–hydrocarbon hybrid surfactant. Langmuir 1994; 10:3536.
6. Johnston KP, Harrison KL, Clarke MJ, Howdle SM, Heitz MP, Bright FV, Carlier C, Randolph TW. Water-in-carbon dioxide microemulsions: a new environment for hydrophiles including proteins. Science 1996; 271:624.
7. Ghenciu EGRAJ, Beckman EJ, Steele L, Becker NT. Solubilization of subtilisin in CO_2 using fluoroether-functional amphiphiles. Biotechnol Bioeng 1998; 58:572–580.
8. daRocha SRP, Harrison KL, Johnston KP. Effect of surfactants on the interfacial tension and emulsion formation between water and carbon dioxide. Langmuir 1999; 15:419–428.
9. Lee CT, Psathas PA, Johnston KP. Water-in-carbon dioxide emulsions: formation and stability. Langmuir 1999; 15:6781–6791.
10. Calvo L, Holmes JD, Yates MZ, Johnston KP. Steric stabilization of inorganic suspensions in carbon dioxide. J Supercrit Fluids 2000; 16:247–260.
11. Johnston KP, Holmes JD, Jacobson GB, Lee CT, Li G, Psathas P, Yates MZ. Reactions and synthesis in microemulsions and emulsions in carbon dioxide. In: Texter, J, ed. Reactions and Synthesis in Surfactant Systems. New York: Marcel Dekker, 2001:359–372.
12. Johnston KP, McFann G, Lemert RM. Pressure Tuning of reverse micelles for

adjustable solvation of hydrophiles in supercritical fluids. Am Chem Soc Symp Ser 1989; 406:140–164.

13. Johnston KP, Jacobson GB, Lee CT, Meredith C, Da_Rocha SRP, Yates MZ, DeGrazia J, Randolph TW. Microemulsions, emulsions, and latexes. In: Jessop P, Leitner W, eds. Chemical Synthesis Using Supercritical Fluids. Weinheim, Germany: Wiley-VCH, 1999:127–146.

14. Peck DG, Johnston KP. Lattice fluid self-consistent field theory of surfaces with anchored chains. Macromolecules 1993; 26:1537.

15. Meredith JC, Johnston KP. Theory of polymer adsorption and colloid stabilization in supercritical fluids. 2. Copolymer and end-grafted stabilizers. Macromolecules 1998; 31, 5518–5528.

16. Yates MZ, O'Neill ML, Johnston KP, Webber S, Canelas DA, Betts DE, DeSimone JM. Emulsion stabilization and flocculation in CO$_2$. II. Dynamic light scattering. Macromolecules 1997; 30:5060–5067.

17. Yates MZ, Li G, Shim JJ, Maniar S, Johnston KP. Ambidextrous surfactants for water-dispersible polymer powders from dispersion polymerization in supercritical CO$_2$. Macromolecules 1999; 32:1018–1026.

18. Li G, Yates MZ, Johnston KP, Lim KT, Webber SE. Trifunctional ambidextrous surfactants for latexes in supercritical carbon dioxide and water. Macromolecules 2000; 33:1606–1612.

19. Otake K, Webber SE, Munk P, Johnston KP. Swelling of polystyrene latex particles in water by high pressure carbon dioxide. Langmuir 1997; 13:3047–3051.

20. Yates MZ, Birnbaum ER, McCleskey TM. Colored polymer microparticles through carbon dioxide-assisted dyeing. Langmuir 2000; 16:4757–4760.

21. Quadir MA, Snook R, Gilbert RG, DeSimone JM. Emulsion polymerization in a hybrid carbon dioxide/aqueous medium. Macromolecules 1997; 30:6015–6023.

22. Young TJ, Mawson S, Johnston KP, Henriksen IB, Pace GW, Mishra AK. Rapid expansion from supercritical to aqueous solution to produce submicron suspensions of water-insoluble drugs. Biotechnol Prog 2000; 16:402–407.

23. Mawson S, Kanakia S, Johnston KP. Metastable polymer blends by precipitation with a compressed fluid antisolvent. Polymer 1997; 38:2957–2967.

24. Young JLS, Richard J, DeSimone Joseph M. Synthesis of two-stage composite latex particles by dispersion polymerization in carbon dioxide. Polym Prepn (ACS Div Polym Chem) 1999; 40:829–830.

25. Shah PS, Holmes JD, Doty RC, Johnston KP, Korgel BA. Steric stabilization of nanocrystals in supercritical CO$_2$ using fluorinated ligands. J Am Chem Soc 2000; 122:4245–4246.

26. O'Shea K, Kirmse K, Fox MA, Johnston KP. Polar and hydrogen-bonding interactions in supercritical fluids: effects on the tautomeric equilibrium of 4-(phenylazo-1-naphthol). J Phys Chem 1991; 95:7863.

27. O'Neill ML, Cao Q, Fang M, Johnston KP, Wilkinson SP, Smith CD, Kerschner JL, Jureller SH. Solubility of homopolymers and copolymers in carbon dioxide. Ind Eng Chem Res 1998; 37, 3067–3079.

28. Johnston KP, Eckert CA. An analytical carnahan–starling van der Waals model for solubility of hydrocarbon solids in supercritical fluids. AIChE J 1981; 27:773.

29. Rindfleisch F, DiNoia TP, McHugh MA. Solubility of polymers and copolymers in supercritical CO_2. J Phys Chem 1996; 38, 15581–15587.

30. Sarbu T, Styranec T, Beckman EJ. Non-fluorous polymers with very high solubility in supercritical CO_2 down to low pressures. Nature 2000; 405:165–168.

31. Raveendran P, Wallen SL. Sugar acetates as novel, renewable CO_2-philes. J Am Chem Soc 2002; 124:7274–7275.

32. Brennecke JF, Chateauneuf JE. Homogeneous organic reactions as mechanistic probes in supercritical fluids. Chem Rev 1999; 99:433–452.

33. Kazarian SG, Vincent MF, Bright FV, Liotta CL, Eckert CA. Specific intermolecular interaction of carbon dioxide with polymers. J Am Chem Soc 1996; 118:1729.

34. Meredith JC, Johnston KP, Seminario JM, Kazarian SG, Eckert CA. Quantitative equilibrium constants between CO_2 and lewis bases from FTIR spectroscopy. J Phys Chem 1996; 100, 10837–10848.

35. Johnston KP, Peck DG, Kim S. Modeling supercritical mixtures—how predictive is it? Ind Eng Chem Res 1989; 28:1115–1125.

36. Kelley SP, Lemert RM. Solvatochromic characterization of the liquid phase in liquid–supercritical CO_2 mixtures. AIChE J 1996; 42:2047–2056.

37. daRocha SRP, Johnston KP. Interfacial thermodynamics of surfactants at the CO_2–water interface. Langmuir 2000; 16:3690–3695.

38. Wignall GD. Neutron scattering studies of polymers in supercritical carbon dioxide. J Phys-Condens Mat 1999; 11:157–177.

39. Goel SK, Satish K, Beckman EJ. Nucleation and growth in microcellular materials: supercritical CO_2 as foaming agent. AIChE J 1995; 41:357–367.

40. RamachandraRao VS, Watkins JJ. Phase separation in polystyrene–poly(vinyl methyl ether) blends dilated with compressed carbon dioxide. Macromolecules 2000; 33:5143–5152.

41. Harrison KL, daRocha SRP, Yates MZ, Johnston KP, Canelas D, DeSimone JM. Interfacial activity of polymeric surfactants at the polystyrene–carbon dioxide interface. Langmuir 1998; 14:6855–6863.

42. Harrison KL, Johnston KP, Sanchez IC. Effect of surfactants on the interfacial tension between supercritical carbon dioxide and polyethylene glycol. Langmuir 1996; 12:2637–2644.

43. Bartscherer KA, Renon H, Minier M. Microemulsions in compressible fluids—a review. Fluid Phase Equilibria 1995; 107:93–150.

44. Fulton JL. Structure and reactions in microemulsions formed in near-critical and supercritical fluids. In: Kumar P, ed. Microemulsions: Fundamentals and Applied Aspects. New York: Marcel Dekker, 1999:629–650.

45. McFann GJ, Johnston KP. Supercritical microemulsions. In: Kumar P, ed. Microemulsions: Fundamental and Applied Aspects. New York: Dekker, 1999: 281–307.

46. Lee CT, Bhargava P, Johnston KP. Percolation in concentrated water-in-carbon dioxide microemulsions. J Phys Chem B 2000; 104:4448–4456.

47. Niemeyer ED, Bright FV. The pH within PFPE reverse micelles formed in supercritical CO$_2$. J Phys Chem 1998; 1998:1474–1478.
48. Holmes JD, Ziegler KJ, Audriani M, Lee CT, Bhargava PA, Steytler DC, Johnston KP. Buffering the aqueous phase pH in water-in-CO$_2$ microemulsions. J Phys Chem B 1999; 103:5703–5711.
49. Lee CT, Johnston KP, Dai HJ, Cochran HD, Melnichenko YB, Wignall GD. Droplet interactions in water-in-carbon dioxide microemulsions near the critical point: a small-angle neutron scattering study. J Phys Chem B 2001; 105:3540–3548.
50. Nagashima K, Lee CT, Xu B, Johnston KP, DeSimone JM, Johnson CS. NMR studies of water transport and proton exchange in water-in-carbon dioxide microemulsions. 2002.
51. DaRocha SRP, Psathas PA, Klein E, Johnston KP. Concentrated CO$_2$-in-water emulsions with nonionic polymeric surfactants. J Colloid Interface Sci 2001; 239:241–253.
52. Butler RD, Cait M, Cooper Andrew I. Emulsion templating using high internal phase supercritical fluid emulsions. Adv Mater 2001; 13:1459–1463.
53. Psathas PA, Janowiak M, Garcia-Rubio LH, Johnston KP. Formation of carbon dioxide-in-water miniemulsions using the phase inversion temperature method. Langmuir 2002; 18:3039–3046.
54. Holmes JD, Steytler DC, Rees GD, Robinson BH. Bioconversions in a water-in-CO$_2$ microemulsion. Langmuir 1998; 14:6371–6376.
55. Cason J, Roberts C. Metallic copper nanoparticle synthesis in AOT reverse micelles in compressed propane and supercritical ethane solutions. J Phys Chem B 2000; 104:1217–1221.
56. Ji M, Chen X, Wai CM, Fulton JL. Synthesizing and dispersing silver nanoparticles in a water-in supercritical carbon dioxide microemulsion. J Am Chem Soc 1999; 121:2631–2632.
57. Otake K, Imura T, Sakai H, Abe M. Development of a new preparation method of liposomes using supercritical carbon dioxide. Langmuir 2001; 17:3898–3901.
58. Weber M, Russell LM, Debenedetti PG. Mathematical modeling of nucleation and growth of particles formed by the rapid expansion of a supercritical solution under subsonic conditions. J Supercrit Fluids 2002; 23:65–80.
59. Tom JW, Debenedetti PG, Jerome R. Precipitation of poly(L-lactic acid) and composite poly(L-lactic acid)–pyrene particles by rapid expansion of supercritical solutions. J Supercrit Fluids 1994; 7:9–29.
60. Mishima K, Matsuyama K, Tanabe D, Yamauchi S, Young TJ, Johnston KP. Microencapsulation of proteins by rapid expansion of supercritical solution with a nonsolvent. AIChE J 2000; 46:857–865.
61. Dixon DJ, Luna-Barcenas G, Johnston KP. Microcellular microspheres and microballoons by precipitation with a compressed fluid antisolvent. Polymer 1994; 35:3998.
62. Mawson S, Johnston KP, Betts DE, McClain JB, DeSimone JM. Stabilized polymer microparticles by precipitation with a compressed fluid antisolvent. I. Poly(fluoro acrylates). Macromolecules 1997; 30, 71–77.

63. Mawson S, Yates MZ, O'Neill ML, Johnston KP. Stabilized polymer micro-particles by precipitation with a compressed fluid antisolvent. 2. Poly(propylene oxide) and poly(butylene oxide)-based copolymers. Langmuir 1997; 13:1519–1528.

64. Shim J-J, Johnston KP. Aqueous latexes formed from polymer/CO_2 suspensions. 2. Hydrophilic surfactants in water. Ind Eng Chem Res, 2002. Submitted.

65. Liu HY, M.Z. Development of a carbon dioxide-based microencapsulation technique for aqueous and ethanol-based latexes. Langmuir 2002; 18:6066–6070.

66. Palakodaty S, York P, Pritchard J. Supercritical fluid processing of materials from aqueous solutions: the application of SEDS to lactose as a model substance. Pharm Res 1998; 15:1835–1843.

67. Chang CJR, A.D. Precipitation of microsize organic particles from supercritical fluids. AIChE J 1989; 35:1876–1882.

68. Shim J-J, Yates MZ, Johnston KP. Latexes formed by rapid expansion of polymer/CO_2 suspensions into water. 1. Hydrophilic surfactant in supercritical CO_2. Ind Eng Chem Res 2001; 40:536–543.

69. Chen X, Young TJ, Sarkari M, Williams RO, Johnston KP. Preparation of cyclosporine A nanoparticles by evaporation precipitation into aqueous solution. Int J Pharm 2002; 242:3–14.

70. Sarkari M, Brown J, Chen X, Swinnea S, Williams RO, Johnston KP. Enhanced drug dissolution using evaporative precipitation into aqueous solution. Int J Pharm 2002; 243:17–31.

71. Randolph TW, Randolph AD, Mebes M, Yeung S. Sub-micron sized biodegradeable particles of poly(L-lactic acid) via the gas antisolvent spray precipitation process. Biotechnol Prog 1993; 9:429–435.

72. Bodmeier R, Wang H, Dixon DJ, Mawson S, Johnston KP. Polymeric microspheres prepared by spraying into compressed carbon dioxide. Pharm Res 1995; 12:1211–1217.

73. Jung J, Perrut M. Particle design using supercritical fluids: literature and patent survey. J Supercrit Fluids 2001; 20:179–219.

74. Reverchon E. Supercritical antisolvent precipitation of micro-and nano-particles. J Supercrit Fluids 1999; 15:1–21.

75. Young TJ, Johnston KP. Encapsulation of lysozyme by precipitation with a vapor-over-liquid antisolvent. J Pharm Sci 1999; 88:640–650.

76. Eastoe J, Bayazit Z, Martel S, Steytler DC, Heenen RK. Droplet structure in a water-in-CO_2 microemulsion. Langmuir 1996; 12:1423–1424.

77. Salaniwal S, Cui ST, Cummings PT, Cochran HD. Self-assembly of reverse micelles in water/surfactant/carbon dioxide systems by molecular simulation. Langmuir 1999; 15:5188–5192.

78. Clarke MJ, Harrison KL, Johnston KP, Howdle SM. Water in supercritical carbon dioxide microemulsions: spectroscopic investigation of a new environment for aqueous inorganic chemistry. J Am Chem Soc 1997; 119:6399–6406.

79. Heitz MP, Carlier C, deGrazia J, Harrison K, Johnston KP, Randolph TW, Bright FV. The water core within perfluoropolyether-based microemulsions formed in supercritical carbon dioxide. J Phys Chem 1997; 101:6707.

80. Zielinski RG, Kline SR, Kaler EW, Rosov N. A small angle neutron scattering study of water in carbon dioxide emulsions. Langmuir 1997; 13, 3934–3937.

81. Holmes JD, Bhargava PA, Korgel BA, Johnston KP. Synthesis of cadmium sulfide Q-particles in water-in-CO$_2$ microemulsions. Langmuir 1999; 15:6613–6615.

82. Jacobson GB, Lee CT, Johnston KP. Organic synthesis in water/carbon dioxide microemulsions. J Org Chem 1999; 64:1201–1206.

83. Pandey S, Baker GA, Kane MA, Bonzagni NJ, Bright FV. O$_2$ quenching of ruthenium(II) tris(2,2'-bypyridyl)2+ within the water pool of perfluoropoly-ether-based reverse micelles formed in supercritical carbon dioxide. Languir 2000; 16:5593–5599.

84. Fremgen DE, Smotkin ES, Gerald RE, Klinger RJ, Rathke JW. Microemulsions of water in supercritical carbon dioxide: an in-situ NMR investigation of micelle formation and structure. J Supercrit Fluids 2001; 19:287–298.

85. Lim KT, Hwang HS, Lee M-S, Lee GD, Hong S-S, Johnston KP. Formation of TiO$_2$ nanoparticles in water-in-CO$_2$ microemulsions. J Chem Soc Chem Commun 2002; 14:1528–1529.

86. Sun Y-P, Atorngitjawat P, Meziani MJ. Preparation of silver nanoparticles via rapid expansion of water in carbon dioxide microemulsion into reductant solution. Langmuir 2001; 17:5707–5710.

87. Lee CT, Psathas PA, Zielger KJ, Johnston KP, Daib HJ, Cochran HD, Melnichenko YB, Wignall GD. Formation of water-in-carbon dioxide microemulsions with a cationic surfactant: a small-angle neutron scattering study. J Phys Chem B 2000; 104:11094–11102.

88. Liu Z, Erkey C. Water in carbon dioxide microemulsions with fluorinated analogues of AOT. Langmuir 2001; 17:274–277.

89. Liu J, Han B, Li G, Zhang X, He J, Liu Z. Investigation of nonionic surfactant dynol-604 based reverse microemulsions formed in supercritical carbon dioxide. Langmuir 2001; 17:8040–8043.

90. Liu J, Han B, Wang Z, Zhang J, Li G, Yang G. Solubility of Ls-36 and Ls-45 surfactants in supercritical CO$_2$ and loading water in the CO$_2$/water/surfactant systems. Langmuir 2002; 18:3086–3089.

91. Keiper JS, Simhan R, DeSimone JM, Wignall GD, Melnichenko YB, Frie-linghaus H. New phosphate fluorosurfactants for carbon dioxide. J Am Chem Soc 2002; 124:1834–1835.

92. Senapati S, Keiper JS, DeSimone JM, Wignall GD, Melnichenko YB, Frie-linghaus H, Berkowitz ML. Structure of phosphate fluorosurfactant based reverse micelles in supercritical carbon dioxide. Langmuir 2002; 18:7371–7376.

93. Psathas PA, daRocha SRP, Lee CT, Johnston KP, Lim KT, Webber S. Water-in-carbon dioxide emulsions with PDMS-based block copolymer ionomers. Ind Eng Chem Res 2000; 39:2655–2664.

6

Production of Powders for Respiratory Drug Delivery

Boris Y. Shekunov
Ferro Corporation, Independence, Ohio, U.S.A.

1. INHALED DOSAGE FORMS

The last decade created both the motivation and the economic opportunity to develop new technology for respiratory delivery of dry powders. This development was stimulated by several factors including the phaseout of chlorofluorocarbons (CFCs) for pressurized metered dose inhalers (pMDIs), the inadequacy of current inhalation systems in terms of efficiency and reproducibility, and ever-growing requirements related to device ergonomics and ease of use. There are expanding markets for certain classes of drugs for conditions such as asthma and chronic obstructive pulmonary disease (COPD), and emerging new drugs (e.g., for diabetes, osteoporosis, pain) with new targets and absorption mechanisms are requiring new drug delivery approaches. Preparation of powders suitable for inhalation and loaded with biomolecules is of high interest for gene therapy and vaccination. Composite drug–polymer particles can find new applications for sustained or controlled-release formulations in respiratory delivery. Finally, inhalation technology gives new opportunities to realize the full potential of existing drug substances for life cycle management products.

The technology has primarily focused on two principal approaches: engineering of novel, more efficient inhaler devices and improvement of the drug formulation technology. Dry powder inhalers (DPIs) of the first generation commonly exhibit a relatively low efficiency of 10 to 20%, measured as the fine particle fraction (FPF, also called the respirable fraction). In addition, many DPIs do not show consistency in the emitted dose (1). One explanation for this is that the dose delivery from DPIs is sensitive to the inhalation flow rate. This inefficiency is also associated with some formulation problems such as particle adhesion and static charge, leading to material retention in the delivery device, reduction of the uniformity of metering of individual doses, and insufficient deaggregation of particles in the airflow. Therefore, a variety of second-generation inhalers were designed and tested in clinical studies, making use of active dispersion of respiratory powders, electronic synchronization with inspiratory flow, and sophisticated actuation control systems intended to overcome the dependence of dose consistency on the patient's respiratory capability and compliance (1). However, the reliability and practicality of these complex and costly devices have been questioned, and these concerns have precluded their regulatory approval. Thus, more sophisticated formulations are being developed in very simple, cost-effective DPIs (2,3).

The production of particles for efficient respiratory delivery still represents a significant challenge to the pharmaceutical industry. The reason is that micrometer-sized particles can exhibit remarkably different physical properties depending on the type of inhalation system, the manufacturing conditions, and the quality and consistency of the powders produced. The requirement for more consistent particle engineering was probably the most important driving force in the development of supercritical fluid (SCF) technology. This chapter reviews the formulation and particle technology of inhalation powders, current applications, and potentials. Process optimization, as well as analytical and materials science stemming from SCF technology, are considered in a well-defined study of salmeterol xinafoate, an asthma drug.

2. FACTORS AFFECTING PARTICLE DEPOSITION IN THE LUNGS

Although inhalation is probably one of the oldest modes of drug delivery, nature has designed the respiratory tract to prevent deposition of particulate matter in the lungs. The respiratory tract is an efficient anatomical barrier for most particulate mass with aerodynamic diameter d_A above approximately 5 μm. This limit is defined by the mechanism of inertial impaction and sedimentation in the upper airways. Particles below this size can be distributed deep into the smaller airways, and this penetration correlates well with a good clinical response for local treatment (4). In many applications, such local delivery allows a tenfold reduction in therapeutic dose compared with systemic delivery, thereby reducing side effects and ensuring good patient compliance. The treatment of asthma and bronchitis with bronchodilators, corticosteroids, anticholinergic agents, and sodium cromoglycate has been particularly successful. Particle fractions with d_A in the range 1 to 2 µm are the most efficient for deposition into the capillary-rich alveolar air spaces. This is the target for the systemic delivery of drugs that are less efficient or less convenient when delivered by other routes. Submicrometer particles (<0.5 µm) are likely to be exhaled, although this rarely represents an issue because of the very strong aggregation of nanoparticles in dry powder formulations.

The overall efficiency of any inhalation system is a product of the fractions of emitted dose, dose delivered to lung, and lung bioavailability. For example, the bioavailability of some therapeutic proteins can be below 10%, thus making efficient deposition even more important. In vivo clinical studies on the deposition of different dosage forms, calling for kinetic evaluation and assays or direct monitoring of the lungs by means of gamma scintigraphy, are very laborious, costly, and impractical to use in formulation studies. Alternatively, a number of methods are available for screening the inhalation behavior of drug particles and assessing the FPF in vitro. The fraction of particles to be delivered to defined stages of a cascade impactor or impinger, the FPF, usually corresponds to the cutoff aerodynamic diameters in the range of 1 to 5 µm. Inertial impaction techniques, described in pharmacopeia and accepted by regulatory agencies such as the U.S. Food and Drug Administration (FDA), are primarily intended for quality control, where the focus is on the relative measures of product performance. The results obtained in an impactor test will depend on the particle aggregation, the airflow rate and the inhaler design.

A systematic, scientifically driven approach to formulation requires understanding of the fundamental physical parameters involved in the powder aerosolization and deposition. The most important of these parameters and their effects on respiratory formulations are listed in Table 1. The "ideal" powders for respiratory formulation should be noncharged, nonadhesive, nonhygroscopic, and easily dispersible. The desirable product characteristics include high FPF, high dose consistency, and independence of the type of device and the inhalation flow rate. It is expected that powders will show good rheological behavior (flow properties) when loaded into a DPI container, blister, or capsule. The requirement for physical and chemical stability implies that storage must not have a significant effect on particle size distribution or dose content uniformity. The powders should also be eco-

TABLE 1 Powder Properties and Their Effect on Respiratory Formulation

Powder characteristics	Effect on formulation
Yield, recovery, manufacturing complexity	Process economics and capital cost *(developability)*
Process temperature, pressure, solvent, pH, additives	Physical and chemical stability
Solid state, crystallinity, polymorphism, hygroscopicity, impurities	Storage, shelf life *(developability)*
Particle size and shape distribution, particle porosity/particle density	Aerosolization behavior, in vitro and in vivo deposition profile
Surface energy, adhesion, electrostatics, bulk density, flow properties, cohesiveness	Powder handling, inhaler filling, dose metering, storage, shelf life, dose consistency *(bioavailability/ safety/efficacy/consistency)*
Coformulation/blending	Dose uniformity *(consistency)*
Composition/coating	Modified or extended release *(efficacy, compliance)*

nomically produced at large scale, with suitable particle technology available at the front end of the product development process.

3. COMPETITIVE PARTICLE TECHNOLOGIES AND FORMULATIONS

Drugs are rarely crystallized directly to meet the criterion of respirable particle size range. Therefore, additional processing steps such as recrystallization, filtering, drying, and micronization or high energy (jet) milling are required. In addition, granulation, solid state, or surface conditioning may be required before or after the micronization stage (4,5). This processing sequence provides only limited opportunity for control over particle characteristics such as size, shape, and morphology, and it introduces uncontrolled structural variations (decreased crystallinity, polymorphism) and surface modifications (increased surface free energy, adhesion, cohesiveness, and charge), which may have an adverse effect on dry powder formulation and may even render the formulation completely ineffective. For example, in the jet milling process, the starting material undergoes many impact events inside the mill before a significant quantity of required particle size fraction is achieved and separated from the larger particles. Such processing can be time-consuming and inefficient for soft ductile organic pharmaceuticals.

This is not to say however that mechanical micronization is a redundant process. Despite all its drawbacks, jet milling has been a reliable technology perfected over decades. All currently marketed dry inhalation products consist of micronized drug in either agglomerated or blended form (2). In addition, a number of formulation strategies have been successfully developed to enhance the inhalation performance of micronized powders. Thus, reduced particle agglomeration was achieved by coformulation with a surface active material (6,7); by coformulation with fine lactose particles and low density additives such as L-leucine (8), or by modification of the surface roughness (9). As alternatives to lactose, different forms of carrier such as trehalose, mannitol, and materials that are endogenous to the lungs such as albumin and dipalmitoylphosphatidylcholine (DPPC) have been considered. These excipients have shown some advantages for powder aerosolization or powder stability and may eventually replace lactose in some formulations (10,11). Similar principles can be applied for powders intended to be formulated as suspensions for pMDIs. With regard to the potential use in the pMDIs containing CFC-free propellants, the production of particles with enhanced surface properties is very important to prevent aggregation and sedimentation of suspensions in these devices, especially since most of the widely used surfactants have poor solubility in the alternative hydrofluoroalkane (HFA) propellants (12).

The major advantage of another fairly common process, spray drying, is direct production of porous or hollow particles. A sprayed drug solution or emulsion consists of excipients serving as blowing agents to produce porous or hollow structures and also as stabilizers of the active compound in formulation. This approach has attracted the most attention because of its application to proteins and peptides that are susceptible to mechanical micronization and/or to drugs of a narrow therapeutic index, which require very high FPF (2,13). Porous particles are very effective for aerosol drug delivery because of their low particle density. Reduction of interparticle interactions and a corresponding increase of FPF to above 50% have been observed for such formulations (2,13). The technique has already been developed on a large scale for phase III clinical studies. However, spray drying has some disadvantages. Processing inlet temperatures required for sufficient solvent removal typically exceed 100°C (10,14) and may be as high as 220°C for some aqueous solutions (8). Therefore, there is always a concern for the chemical or physical stability of the active ingredients and formulation excipients. For example, the biochemical stability afforded to an active protein molecule can be at expense of the physical stability of a sugar excipient (4), since most excipients precipitate as an amorphous material and may recrystallize spontaneously upon storage into a more thermodynamically stable form, particularly when this process is accelerated by moisture sorption. Similar problems may exist for low molecular weight drug compounds. By spray drying a suspension of nanoparticles, it is possible to obtain aggregates of crystalline material of respirable size. However, this makes the whole process very complex and expensive, since special micronization techniques are required to produce nanoparticulates (1), perhaps adding yet another challenge, namely material loss, during manufacturing. Table 2 compares three particle technologies.

Particle size reduction and coformulation of thermolabile proteins can be done by means of spray freeze-drying (14–16). In this method, a solution containing protein and excipients is sprayed into liquid nitrogen, whereby the frozen droplets are prepared for second lyophilization stage. Spray freeze-drying also tends to produce porous particles with enhanced aerodynamic properties. Because spray freeze-drying is a two-stage process, it is more time-consuming and costly than spray drying, although it may offer an increased production yield (14). Very low temperatures or an abrupt decrease in temperature due to the liquid nitrogen may also result in degradation of proteins, especially for small particles having a large specific surface area (15), or the excipient physical instability produced may deteriorate the aerosol performance of spray freeze-dried powders (14).

To conclude this brief review of existing competitive particle technologies and formulation methods, it is necessary to mention a rapidly growing area of sustained-release/controlled-release formulations for respiratory delivery. Although many current inhalation drugs such as bronchodilators are intended for immediate release, the onset of the effects of most drugs delivered to the respiratory tract is very rapid, and duration is often short-lived as the drug particles are quickly removed from the lung by several types of clearance mechanism (17). For example, if dissolution is the rate-limiting factor for the pulmonary absorption of a drug that is poorly soluble in water, the therapeutic effect of the drug may be reduced and undesired side effects increased, since more drug must be delivered. This situation can be improved by using coformulation with mucoadhesive and water-soluble polymers (18). From another perspective, composite particles (e.g., biodegradable polymer microspheres) can provide selective controlled or prolonged release and increased drug permeability, combined with protection against enzymatic hydrolysis (19). Respirable composite particles can be produced by spray drying (18) or by the water–oil–water double emulsion–solvent evaporation method (19). Emulsion technology is also the most common method for preparing composite polymer particles in the other (oral, injectable) drug delivery applications. Neither technique is perfect because polymer microspheres tend to plasticize and agglomerate at elevated temperatures during spray drying, whereas the evaporation emulsion method is very slow, is difficult to scale up, and may also require a separate drying stage. A significant manufacturing disadvantage of both techniques is that organic solvents used to dissolve

TABLE 2 Comparison of Different Particle Formation Techniques

Factors	Micronization (milling)	Spray drying	SCF precipitation
Process	Secondary	Direct	Direct
Typical product description	Relatively dense powders formed by irregularly shaped, rough-surfaced crystalline particles	Low density powders formed by composite porous or hollow amorphous particles	Low density powders formed by loosely packed, regularly shaped crystalline particles
Major advantage area for respiratory applications	Established, proven	Good particle aerodynamics	Pure, stable, noncohesive powders
Typical composition	Pure small molecules	Biological or small active molecules with excipients	Pure small molecules or composites with polymers
Potential for coformulation	Low	High	High
Current control over physical form	Adequate for small molecules; poor for biomolecules	Poor for small molecules; adequate for biomolecules	Excellent for small molecules and adequate for biomolecules
Batch consistency	Poor	Adequate	Good
Particle size control	Adequate	Good	Good
FPF target	30%	80%	60%
Current industrial scale	Commercial	Small manufacturing	Pilot
Current cost	Low	Moderate	More expensive than the other two technologies
Potential for development	Low	Moderate	High

polymers are difficult to remove, and hence meeting the regulatory requirements is more difficult.

4. SCF TECHNOLOGY

4.1. Advantages of Supercritical Fluids in Pharmaceutical Manufacturing

Supercritical fluids, compressed gases or liquids in a state above their critical pressure and temperatures, offer several fundamental advantages as solvents for pharmaceutical manufacturing. The most important is their ability for efficient extraction and separation, facilitating a potentially clean and recyclable process. The high compressibility and variable solubility power of SCFs can be utilized for direct production of pure and composite particles for respiratory delivery, combined with selective precipitation, impurity separation, and control of crystalline forms. Carbon dioxide is the most important supercritical medium because of its relatively low critical temperature ($31.1°C$) and pressure (73.8 bar), and its low toxicity. Potentially, SCFs should reduce manufacturing complexity, energy needs, and solvent requirements, thus providing a more environment-friendly and benign process than conventional particle formation techniques. All these potentials must be carefully explored, with emphasis on the formulation advantages of the supercritically produced powders.

4.2. Small Molecules

Most of the proof-of-concept studies to date have focused on small-molecule asthma drugs. These drugs are currently the most important class of inhalation medicine and conveniently, commercial inhaler devices and formulations with micronized materials are available for benchmark comparison.

Steckel and coworkers have demonstrated the production of particles in the inhalation size range for eight steroid compounds, including budesonide and fluticasone (12,20). The antisolvent technique employed (aerosol solvent extraction system, ASES) enabled simultaneous particle precipitation and coating by means of physiological surfactants such as phosphatidylcholine or lecithin to improve particle wettability. Some pure and coated particulate products, suitable for formulation into pMDIs, were obtained. Further comparative study of micronized and SCF-processed fluticasone (20) proved the feasibility of formulating this drug with HFA-227 propellant. Coated particles produced with SCF showed a significant increase in FPF over jet-milled products. This delivery efficiency was also comparable to a solution-formulated, CFC-based commercial inhaler (20).

Another antisolvent precipitation technique, the patented SEDS system (see Chapters 3 and 4 of this volume), offers the advantage of very fast, homogeneous nozzle mixing of supercritical antisolvent and drug solution, leading to consistent production of micrometer-sized powders. In vitro assessment of salbutamol sulfate, prepared by micronization and the SEDS process, was carried out by Feeley et al. (21–23). These investigators emphasized powder behavior in different inhalation devices. Drug powders that had been blended with a commercial inhalation grade α-lactose monohydrate and, in certain cases pure drug material alone, were tested with an Andersen type of cascade impactor. An improvement of FPF and total emitted dose, and a decrease in the amount of drug deposited in the throat and preseparator, were observed for the SCF-processed powders compared with both hand-filled micronized samples and the available marketed devices. In addition, the low bulk density (< 0.1 g/cm^3) of the SCF-processed powders provided an opportunity to produce, for drug-alone formulations, in vitro deposition profiles that were comparable to those for drug–lactose formulations. The enhanced performance of the SCF-processed particles was explained by noting that reduced surface energy leads to enhanced downstream processing behavior with respect to blending, powder flow, and content uniformity. Similarly, improved deposition of SCF materials in vitro was observed for different batches of salmeterol xinafoate (24,25), fenoterol hydrobromide (26,27), and some forms of terbutaline sulfate (28). Analysis of hydrocortisone particles by means of a multistage liquid impinger (29) showed that the delivered dose increased by about 30 to 40% for SCF-processed materials in comparison to micronized particles. The improved discharge of SCF-processed particles from both devices was again explained by these particles' low surface adhesion to the inhaler reservoir. Recent work with salbutamol sulfate (30) showed that the SEDS-produced powders gave more consistent delivery in most cases with, however, a lower FPF. In addition, the FPF and emitted dose for the SEDS-produced powders were relatively independent of the flow rate. A moderate FPF, between 16 and 24% (28.3 L/min flow rate) and 22 and 31% (4 kPa differential pressure or 53.5 L/min flow rate), was obtained for these powders and is attributed to the relatively large mean aerodynamic diameter $\bar{d}_A$ of the supercritically produced powders ($2.1 < \bar{d}_A < 6.5$ μm), compared with $\bar{d}_A < 2$ μm for the micronized salbutamol.

Figure 1 compiles all available data (12,23–30) representing FPF values of supercritically produced and micronized materials versus $\bar{d}_A$. The most important fact learned from this diagram is that in vitro deposition is not influenced by particle size exclusively. Thus, particles obtained by using supercritical CO_2 have typically larger aerodynamic diameters than micronized materials but show significantly higher FPF. Clearly, the large deviation

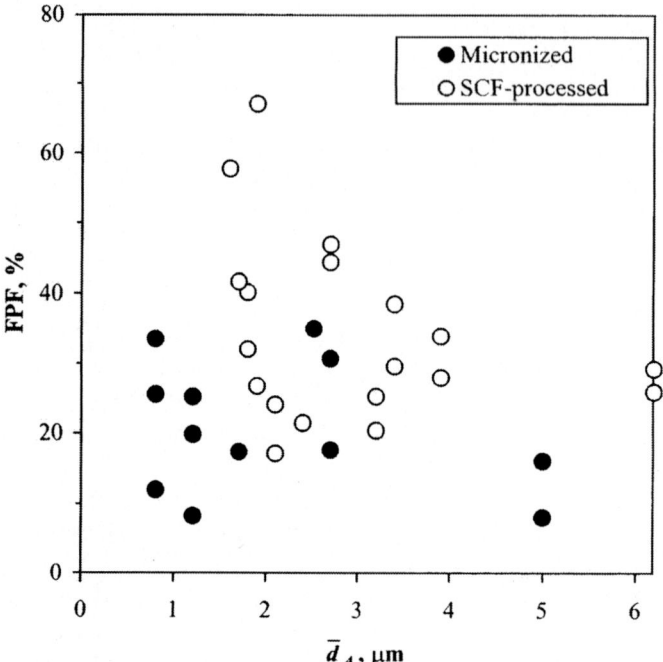

Figure 1 Compilation of data collected from Refs. 12 and 23 to 30 showing fine particle fraction (FPF), versus volume mean aerodynamic diameter $\bar{d}_A$, for micronized and SCF-processed powders investigated. Values of $\bar{d}_A$ obtained by using the AeroSizer time-of-flight technique when available (23,24,26,30) or recalculated from the mean volume (geometric) particle diameters $\bar{d}_V$, obtained from laser diffraction and scanning electron microscope data (12,25,29) by applying the conversion algorithm developed by Shekunov et al. (24).

of data in Figure 1 is a result of the different types of formulation, the various inhalers, and the different flow rates used. For example, a relatively low FPF (i.e., <20%) for pure drug particles of hydrocortisone (29) could be explained by the intrinsic acicular morphology of hydrocortisone crystals, which impeded particle aerosolization. In contrast, values above FPF 45% were observed for a better-tuned SEDS process (23,24) or coated formulations (12). It is also evident from Figure 1 that the full potential of supercritically produced powders for deep-lung delivery (maximal FPF and deposition at the lower impactor stages) has not yet been achieved. Thus, the smallest $\bar{d}_A$ obtained for these powders was about 1.6 μm. Further decrease of the $\bar{d}_A$

below this level would certainly improve the deposition profile. Approaches to the particle size reduction are discussed in Section 5.

As suggested by the studies just cited, interparticle interactions or particle morphology played the dominant role in the comparative in vitro studies. It should be noted that theoretical predictions for FPF are usually based on the ideal case of nonagglomerated, uncharged, spherical particles. In reality, the dispersion of aggregates is crucial for achieving a high and consistent respirable fraction. The surface energy of particles has been recognized to be a very important factor affecting the aggregation–deaggregation forces between the formulation ingredients or between particles and inhaler contact surfaces (22–28). For example, results obtained for several inhalation compounds (27) show lower surface energetics for SCF-processed particles than for micronized powders, in terms of lower dispersive component of the surface free energy γ_s^D and the specific components of surface free energies of adsorption ΔG_A obtained from inverse gas chromatographic (IGC) analysis. More quantitative assessment of particle interactions can be done on the basis of the Hildebrand solubility parameter δ (31).

Control over the physical form of materials produced constituted the major topic of several other studies. The original work on precipitation of salmeterol xinafoate (5,32) showed that two different polymorphic forms of salmeterol can be separated by means of SCF precipitation using methanol or acetone solutions. The relative composition of the two polymorphs could be controlled by varying the working conditions of temperature and pressure. A remarkable degree of crystallinity, stability, and polymorphic purity for SCF-processed salmeterol was further confirmed in detailed thermodynamic, structural, and IGC studies (31,33,34). For flunisolide anhydrate (35), several polymorphic forms, including two new forms, were isolated. A highly crystalline single form of budesonide was also obtained, similar to the results reported by Steckel et al. (12). Significant improvement of crystallinity in comparison to micronized samples was obtained for nicotinic acid (36), used as a model compound in a process optimization study, and also for fenoterol hydrobromide (26). Probably the most interesting phenomenon was observed in work with terbutaline sulfate (26,28). Different crystal forms of terbutaline sulfate were consistently produced, among them two polymorphs, a hydrate and an amorphous material, as well as particles that differed in degree of crystallinity, size, and morphology. A correlation between the FPF, powder flow and particle crystallinity showed that a semicrystalline batch exhibited the best aerosolization behavior. This is an unexpected result because the crystalline bulk structure can usually be associated with the ordered surface and lower specific surface energy (31,33,34,37). It was suggested that the semicrystalline particles underwent surface conditioning in CO_2–ethanol mixtures by which the particle surfaces of terbutaline sulfate relaxed into a more stable and less energetic state (28). A similar mechanism was proposed for SCF treatment of amorphous metastable regions of preformed inhalation particles (38).

Finally, it is worth noticing that antisolvent precipitation with SCF offers a single-step process for coformulations of drugs with excipients, producing homogeneous blends or coatings of respiratory formulations. The example of physiological surfactants (12,20) was discussed earlier. Similarly, solid dispersions or mixtures were prepared to modify the solid state, surface structure of particles obtained (39) or to increase the dissolution rate (40). Usually, a sufficient additive concentration can be achieved by selecting a suitable solvent mixture.

4.3. Processing of Biological Compounds

Molecules of proteins, peptides, and other biological compounds require special consideration in formulation and processing, with a detailed assessment of their chemical and physical stability in solution and solid state form. Methods involving SCF facilitate particle formation at near-ambient temperatures and eliminate the need for liquid–vapor interfaces, which may result in significant loss of biological activity during spray-drying. However, since water and CO_2 are poorly miscible, any precipitation with CO_2 in aqueous solutions requires addition of an organic solvent, which is likely to be a protein denaturant, although CO_2 may dilute the solvent and reduce the damaging influence of the solvent. The effect of CO_2 itself is often beneficial for the recovery of the secondary structure of protein molecules that have been denatured or altered to show less active folded conformation (41).

While some generally robust proteins such as insulin and lysozyme processed in a variety of organic solvents almost completely recovered their natural structure and function after rehydration (42–47), other proteins such as albumin, rhDNase, trypsin (43,46), and DNA molecules (48) showed degradation. Loss of biological activity was also pronounced when the denaturing effect of an organic solvent was combined with an increased processing temperature, as in the case for ribonuclease particles obtained above 50°C (49). Thus biomolecular instability in SCF processing is compound specific. It also depends on the precipitation regime or miscibility between the SCF and solution phases. For example, when a protein solution is prepared in a polar organic solvent such as dimethyl sulfoxide (DMSO) (43,44,49) or in a halogenated alcohol (45), the solvent and the CO_2 become completely miscible at high pressure, and particle formation occurs in the CO_2-rich phase. As a result of the fast precipitation in such systems, submicrometer-sized particles and high yield are achieved. However, prolonged contact with the solvent may cause perturbation of the protein structure (43,44,49). Similarly, in the ternary CO_2–ethanol–water system, a fraction of ethanol above 20 mol % is required for fast water extraction (46,47,49). Decrease of ethanol concentration usually gives rise to micrometer-sized, nonagglomerated particles with a low yield.

Some proteinaceous drugs can be solubilized by means of a hydrophobic ion-pairing (HIP) approach (50). This may help to overcome the problem of loss of secondary and tertiary structure by selecting the more protein-friendly organic phase. It is usually possible to sustain the water content in SCF antisolvent precipitation within the range of 3 to 7%, comparable to the stable form of lyophilized powders. Addition of stabilizing excipients (e.g., disaccharides, surfactants, buffers) is also possible, although both the excipients and the precipitation regime must be carefully selected to avoid solid phase separation between drug and excipient molecules without intimate molecular bonding. Such coprecipitation is likely to improve the processibility (48,51) and the stability (51) of a bioactive compound in a formulation.

There are very few in vitro aerosol studies of the SCF-processed protein powders. A detailed analysis has been carried out for the model compounds lysozyme and albumin, and for insulin (46). All these proteins precipitated as spherical nanoparticles (0.1–0.5 μm primary size) and agglomerated into micrometer-sized clumps, with a characteristic size between 3 and 11 μm volume median diameter as measured by laser diffraction. The FPF was approximately 65, 40, and 20% for lysozyme, albumin, and insulin, respectively, at a flow rate of 60 L/min in a Dinkihaler device. In such systems, aerodynamic dispersion and deposition are governed by the strength and size of the agglomerates. Therefore, the best results were obtained for lysozyme batches with less agglomerated particles showing the smallest mean agglomerate size. This conclusion is likely to be valid for many nanoparticulate powders because of the very strong particle interactions.

Recently, in vivo analysis was carried out on chitosan–plasmid DNA complex powders obtained by means of antisolvent precipitation (51). The powders were prepared with mannitol as a carrier. The chitosan complexation not only suppressed the degradation of the bioactive agent (pCMV-Luc) during SCF precipitation, but also increased the yield and the luciferase activity in mouse lungs in comparison to both pCMV-Luc without chitosan or pCMV-Luc solutions with or without chitosan. The findings of this study revealed that the gene powder produced with the cationic polymer is a promising delivery system to the lungs and showed clear advantages over gene solutions for pulmonary delivery. Such benefits included higher luciferase activity, higher local drug concentration, and longer action time. Although no detailed aerodynamic study was reported, the powders consisted of aggregates in the size range between 3 and 13 μm.

An entirely different processing approach named "CO_2-assisted nebulization with bubble-drying" (CAN-BD) has been applied to a range of biological substances such as lysozyme, lactate dehydrogenase, trypsinogen, anti-CD4 antibody, and α_1-antitrypsin (52–54). The solubility of supercritical or compressed CO_2 in water or organic solvents is utilized in CAN-BD to generate small droplets or bubbles. In a way, it is a spray-drying process that relies on hot or warm air or nitrogen to evaporate the solvent. The advantages of this method lie in its simplicity, the possibility of processing water-soluble compounds without the use of organic solvents and the relative ease of coformulation with stabilizing excipients. In comparison to commercial spray drying, CAN-BD may offer a reduction in processing temperature, which the authors claim to be between 25 and 65°C inside the drying chamber (53,55). Particles in the respirable size range were produced by means of this technique for both biological and low molecular weight compounds. To achieve the desired protein stability and to retain biological activity, pH buffers and appropriate stabilizing excipients such as disaccharide sugars or surfactants (e.g., Tween 80) were added during coformulation (53).

4.4. Composite Particles

The production of respiratory particles for sustained or controlled drug release is a new and virtually unexplored field of SCF processing. Formation of such microparticles by means of the antisolvent SCF techniques is likely to succeed either for structurally similar compounds (39) or for other molecules with affinity to each other that can form ordered solid solutions, molecular dispersions, or amorphous mixtures by strong hydrogen bonding or complex formation (see Chapter 8, this volume). In the event of solid phase separation or recrystallization, nonhomogeneous drug dispersion or partial coating could occur, leading to a characteristic "burst" drug release, as revealed in dissolution studies. An example of successful coprecipitation is microencapsulation of budesonide into poly(L-lactic acid) (PLLA) in an amorphous form (56). Another solid solution of theophylline in ethylcellulose was produced at drug loadings up to 35% (57). The HIP approach was used with PLLA to precipitate gentamicin acid and naltrexone (50). In the HIP process, ionic species can be directly solubilized in nonaqueous solutions by means of the pairing of charged molecules with oppositely charged surfactants. This approach makes it possible to obtain true solutions of drug complexes and polymer molecules in neat organic solvents. It is likely that such complexes prevent drug recrystallization. For example, ion-paired drugs were loaded more efficiently than non-ion-paired preparations into a PLLA matrix and showed uniform release profiles (50). The HIP approach was also used to obtain the prodrug isoniazid for subsequent encapsulation into microspheres for antituberculosis inhalation therapy (58).

Supercritical CO_2 is a very efficient plasticizing agent for polymers (see Chapter 8, this volume). Unfortunately, this property often leads to strong particle aggregation for glassy and even some semicrystalline polymers, such as PLLA (59,60). The relatively low glass transition temperature T_g of most linear polymers makes the antisolvent precipitation of discrete particles a very difficult task. Particle aggregation in an SCF can be significantly reduced by using highly cross-linked polymers, which are less susceptible to swelling and plasticization. A novel method of particle formation from cross-linked polymers has been developed by using simultaneous compressed antisolvent precipitation and photoinitiated polymerization (CAPP) (61). Thus a bimodal particle size distribution in the range between 0.5 and 200 μm was obtained for spherical particles of diacrylated poly(ethylene glycol) (PEG). A large fraction of particles with number-weighted diameters was observed between 1 and 3 μm. This distribution was explained by citing different mechanisms of polymerization in the liquid and gaseous phases (60). However, particle agglomeration during synthesis cannot be ruled out. Mechanical creep studies showed that discrete particles were formed for monomer–polymer conversion above 75%. This technique provides an additional tool for particle size control through photopolymerization kinetics.

The high solubility of SCFs in polymers, lipids, and some other materials (typically between 5 and 40%) is the fundamental reason for particle aggregation when such particles remain in a prolonged contact with CO_2 during antisolvent experiments. However, this phenomenon can also result in lower viscosity, depressed melting point, and reduced T_g, which are beneficial for enhanced processibility of composite materials during mixing and homogenization. The expansion of melts under reduced pressure may produce fine particles, often with a microporous structure (62,63). This process is known as "particles from gas-saturated solutions" (PGSS) (see Chapter 4) and can be applied for the preparation of suspensions of bioactive materials in polymeric or other carrier substances. Particle formation enables us to achieve homogeneous mixtures, coatings, and encapsulation and also occurs at a reduced processing temperature, which is particularly important for the physical and chemical stability of many bioactive materials and drugs. For example, studies of ribonuclease demonstrated that the activity of this protein was preserved after PGSS and that a constant release rate from PLLA polymer could be obtained (63). Microporous composites are excellent candidates for dry powder respiratory delivery and may compete in this way with particles produced using spray drying or spray freeze-drying. Both expansion conditions and solubility parameters must be controlled to achieve high porosity and pores in the nanometer size range. The additional processing advantage is that because the particles are produced at reduced pressure, there is no danger of particle agglomeration.

Particles of very high porosity and surface area can also be produced from some preformed materials such as aerogels (64). Aerogels are prepared by means of a gel method from modified organic carriers such as derivatized mannitol and trehalose. The matrices are than saturated with a therapeutic agent, dried with supercritical CO_2, and micronized by means of a jet mill. The density of the aerogel can be as low as 0.003 g/cm^3. In principle, the respiratory particles produced have the benefit of relatively large volume diameter and therefore may show reduced interparticle interactions and better aerosolization than solid particles.

5. PARTICLE SIZE REDUCTION

The most important prerogative for any dry powder respiratory formulation is production of nonagglomerated particles with a high FPF. SCF antisolvent precipitation is a convenient and direct way to obtain such particles. This section illustrates an optimization strategy based on a case of salmeterol xinafoate (SX) materials processed by SEDS. Although the general principles can be valid for any of the antisolvent precipitation methods differentiated by mixing method, SEDS is a technique that lends itself well to mathematical modeling and offers greater flexibility with process control because of the well-defined turbulent nozzle mixing and precipitation mechanism (see Chapter 3).

A general approach to prediction and optimization of SEDS has been developed by Shekunov, Baldyga and their colleagues (65–67) and also is discussed in this volume in Chapter 3 by Baldyga et al. A variety of parameters can be manipulated, including pressure, temperature, supersaturation, solvent system, solution concentration, nozzle flow velocity, and mixing regime. The requirement for consistent crystal form, reduced processing temperature, low organic solvent uptake, and high yield usually lead to a precipitation regime at or above the mixture critical pressure at which the solvent and SCF are completely miscible (65,67). In the case of SX, the most stable pure polymorphic forms (5,34) were produced at pressures above 150 bar, where most common organic solvents are completely miscible with CO_2.

The first factor to be considered for particle size reduction is the supersaturation profile. For very low concentrations of the nonvolatile solutes in antisolvent crystallization, supersaturation during the initial precipitation stage is characterized by the maximum attainable supersaturation, σ. This corresponds to the ideal case of a fresh feed of CO_2–solvent–solute being completely mixed on a molecular level and is calculated as (66,67)

$$\sigma = \ln \frac{c_A f_A}{c_0(f_A + f)} \qquad (1)$$

where f_A and f are the flow rates (mol/s) of the solution and CO_2 respectively, and c_A is the molar concentration of a drug in the feed solution. The equilibrium solubility, c_0, in the modified CO_2–solvent mixture can be determined prior to crystallization experiments by means, for example, of an on line dynamic solubility method (68). The dependence of σ on the relative rate of solution to CO_2 flow would typically have a maximum that reflects competition between the antisolvent and dilution effect of CO_2 (66). Such dependence for SX is shown in Figure 2 as a function of the solvent mole fraction. According to the homogeneous nucleation model (Chapter 3), this maximum also corresponds to the minimum volume mean diameter $\overline{d}_V$ and the narrowest particle size distribution (PSD) of particles produced. An important conceptual question is whether the resulting solvent–CO_2 mixture remains in the supercritical state, which is a more advantageous phase for obtaining pure and consistent dry product. In most applications that require

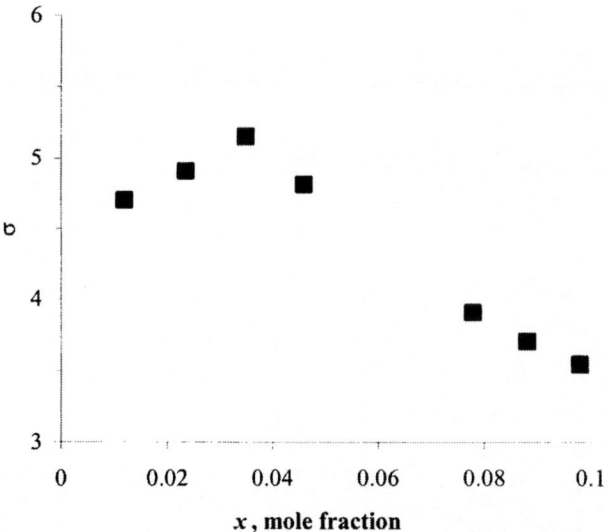

FIGURE 2 Dependency of supersaturation σ of salmeterol xinafoate on the relative methanol–CO_2 flow rate, expressed through the methanol mole fraction x, at temperature 313K, and pressure 200 bar. Solution concentration is 3% w/v.

both high supersaturation and high yield, the mole fraction of the organic solvent is typically below its level at the mixture critical point, and hence, the mixed phase remains supercritical.

The solvent system selected may be an important factor influencing the supersaturation profile, product yield, size, and morphology of the particles produced. Thus in the experiments with SX, methanol, acetone, and tetra-hydrofuran solvents were tested. Although all these solvents are suitable for antisolvent precipitation, methanol was selected for further studies because of its high yield (> 95%) and suitable particle size distribution for respiratory delivery ($\bar{d}_V$ < 5 μm) achieved.

The other process control parameter is the solution concentration, c_A. From Eq. (1) it is to be expected that supersaturation progressively increases with c_A, leading to the precipitation of smaller particles. In reality, the particle formation process is always a competition between particle generation and particle agglomeration. The mean volume diameter of stable particle agglomerates (i.e., the agglomerates that cannot be dispersed aerodynamically without particle attrition) increases in highly concentrated solutions. Therefore, there is a practical limit, in addition to the solubility of the drug, that determines the optimum concentration corresponding to the smallest particle size. As shown in Figure 3, the minimum $\bar{d}_V$ is achieved at an intermediate

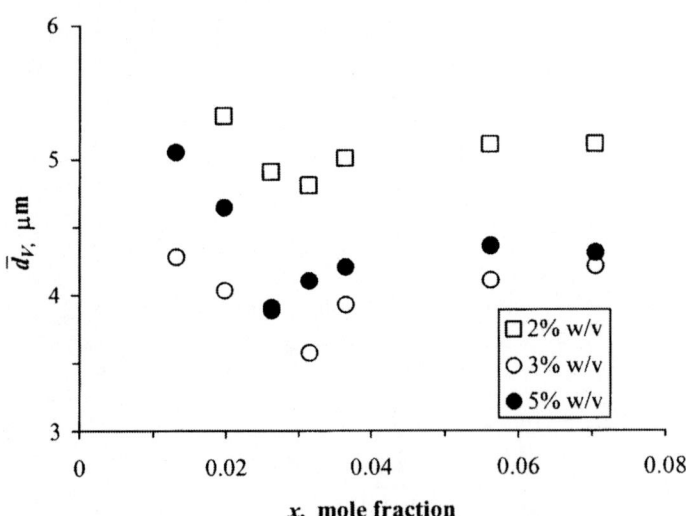

FIGURE 3 Dependency of the volume mean diameter, $\bar{d}_V$, of salmeterol powders produced at different solution concentrations as a function of the relative methanol–CO_2 flow rate, expressed in methanol mole fraction, at temperature 40C, and pressure 200 bar.

solution concentration, 3% w/v. Therefore, for a given nozzle and vessel geometry, there is an optimal concentration of the feed solution as well as an optimum solution feed rate.

Clearly, the particle formation process is very compound specific. For some materials, slow mixing can be adequate to produce micrometer-sized particles; for others, a faster mixing regime is required. For any precipitation process, two major material constants—the nucleation time constant τ_N and the growth time constant, τ_G—define the mean particle size (66). The materials with small τ_N and relatively large τ_G can easily form micro- or even nano-sized particles. As a general rule, the larger the τ_N, the less sensitive particle formation process is to mixing; also, it is more difficult to produce small particles. In addition, even for very fast homogeneous mixing, recirculation and dilution effects outside the nozzle may negatively affect the mean diameter and size distribution of the particles produced (66).

In particular, optimization of the nozzle geometry was found to be the crucial step in achieving the correct size range of SX powders: τ_N determined for this material by optical methods appeared to be about 3 ms: that is, more than an order of magnitude larger than the corresponding constant for acetaminophen, the benchmark compound for SEDS mechanistic studies (66). This relatively slow nucleation probably is attributable to the larger molecular mass and more complex crystal structure of salmeterol. As a result, the particles produced by using a short nozzle (or no nozzle) are outside the respiratory range. Increasing of the nozzle length (Figure 4) decreased the particle size. The optimum nozzle length was found to be about 100 mm, as a compromise between the acceptable mean particle size and possibility of nozzle blockages by relatively large particles formed inside the nozzle mixing chamber. The broken line in Figure 4 shows the size of the particles emerging from the nozzle to be almost proportional to the nozzle length. It was found experimentally that nozzles longer than 100 mm could not provide a stable flow regime because of these blockages.

Thus variation of such major process parameters as pressure, temperature, solution flow, and concentration, combined with fine-tuning by means of different nozzles, enabled the formation of the required crystal form with a respirable particle size. Increase of SCF flow and solution flow velocity should always reach the point at which mixing is faster than precipitation (although this may not always happen in practice). However, as clearly seen from the asymptotic behavior in Figure 4, the theoretical minimal volume diameter for SX is about 3 μm. This limit is dictated by the mechanism of antisolvent precipitation and the competition between the particle nucleation and growth (see Chapter 3). Similarly, the lower limit on particle size can be observed for any material depending on the specific nucleation and growth constants at given pressure and temperature. Fortunately, precipitation with supercritical

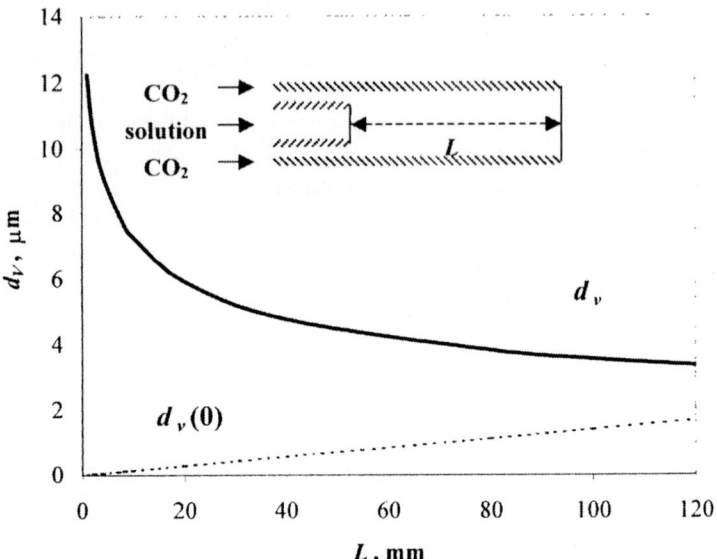

FIGURE 4 The dependence of mean volume diameter of SX on the nozzle length is shown for the final product (d_V) and for particles produced at the nozzle orifice, $d_V(0)$. The computation was performed by using a mixing–precipitation model (66) for the coaxial nozzle shown schematically.

CO_2 is typically fast enough to produce particles in the lower micrometer size range.

6. PHYSICAL PROPERTIES AND AEROSOLIZATION BEHAVIOR OF SCF-PROCESSED POWDERS

A rational formulation of respiratory powders requires understanding of the basic material properties related to aerodynamic particle behavior and adhesion, combined with reliable methods for solid state and particle size analysis. The present section is based on a comparative study of micronized and supercritically produced SX with the goal of distinguishing the physical criteria making SCF-processed material preferable choice for DPI formulation.

A key issue for the enhancement of aerosolization performance of SCF-processed powders is that these materials typically consist of anisometric particles, usually platelets or prisms, packed in a loose "fluffy" structure. Therefore, any data interpretation and size analysis of such particles require a shape correction for the aerodynamic diameter, which also depends on the airflow regime which varies in the human respiratory system, in the inhaler device, or any particle sizing instrument. This ambiguity with respect to aerodynamic diameter can be resolved by using a very general approach wherein the particle drag coefficient C_d is defined for particles having volume-equivalent particle diameter d_V and particle sphericity φ (24,68). A semi-empirical correlation is valid within a wide range of the particle Reynolds number, Re, in the Stokesian ($Re < 1$) and ultra-Stokesian ($1 < Re < 1000$) flow regimes:

$$d_A \cong \frac{1}{\chi} \frac{\rho_c}{\rho_0} d_V \tag{2}$$

$$\chi = \frac{C_d(Re, \varphi)}{C_d(Re)} \tag{3}$$

The aerodynamic diameter d_A, is the diameter of spheres of unit density ρ_0, which reach the same velocity as nonspherical particles of density ρ_c in the air stream; $C_d(Re)$ is calculated for calibration particles of diameter d_A, and $C_d(Re, \varphi)$ is calculated for particles with diameter d_V and sphericity φ. Sphericity is defined as the ratio of the surface area of a sphere with equivalent volume to the actual surface area of the particle determined, for example, by means of specific surface area measurements (24). The aerodynamic shape factor χ is defined as the ratio of the drag force on a particle to the drag force on the particle volume-equivalent sphere at the same velocity. For the Stokesian flow regime and spherical particles ($\varphi = 1$, $\chi = 1$), the drag coefficient is $C_d = 24/Re$ (69). Therefore, Eq. (2) leads to the Stokes aerodynamic diameter well known in the aerosol literature: $\bar{d}_A (Stokes) \cong d_V \sqrt{(\rho_c/\rho_0)}$.

Thus, Eqs. (2) and (3) enable us to convert the volume-equivalent particle size distribution into aerodynamic-equivalent distribution for any flow regime. The characteristic mean diameters are $\bar{d}_V$ (volume geometric) and $\bar{d}_A$ (volume aerodynamic). Although $\bar{d}_A$ is numerically close to the mass-median aerodynamic diameter (MMAD) often used for the aerosols, it is better defined for asymmetrical distribution, so often observed for respiratory powders.

Figure 5 shows typical scanning electron micrographs for micronized (M-SX) and supercritically produced (S-SX) salmeterol particles. Table 3 summarizes the most important physical parameters for these powders included in the present dispersion model. Although the materials are similar in platelet morphology, S-SX consists of thinner particles with much smaller sphericity than the micronized sample. Less than 1% density difference between these crystals is too small to produce any noticeable changes in the aerodynamic behavior. By contrast, a relatively large value of the aerody-

FIGURE 5   Particle images obtained using scanning electron microscope for (a) micronized and (b) supercritically produced salmeterol xinafoate. The magnification is about five times higher for micronized particles.

TABLE 3   Most Important Physical Properties of Micronized (M-SX) and Supercritically Produced (S-SX) Powders

Property and method[a]	M-SX	S-SX
φ (BET adsorption and particle size analysis)	0.46	0.31
ρ_c, g/cm^3 (X-ray powder diffraction)	1.256	1.244
ρ_B, g/cm^3 (bulk)	0.121	0.094
χ (AeroSizer)	2.1	3.3
$\bar{d}_V$, μm (laser diffraction)	1.70	3.55
$\bar{d}_A$, μm (AeroSizer™ time-of-flight)	1.2	1.6
$\bar{d}_A$, μm (Andersen cascade impactor)	6.5	5.1
FPF,% (Andersen cascade impactor)[b]	25.15	57.80
δ_C, MPa$^{0.5}$ (IGC)	31.305	25.619
δ_A, MPa$^{0.5}$ (IGC)	33.353	29.162
θ (IGC)	0.60	0.52

[a] Particle sphericity, φ; aerodynamic shape factor, χ; crystal density, ρ_c; aerosolized powder bulk density, ρ_B; volume mean diameter, $\bar{d}_V$; aerodynamic mean volume diameter, $\bar{d}_A$; total (Hildebrand) solubility parameters of the cohesive (drug–drug), δ_C, and adhesive (drug–carrier), δ_A, interactions.
[b] Obtained for the drug–lactose formulations (DMV Pharmatose 325M α-lactose monohydrate) delivered via the 13 cu/mm Clickhaler device at 49 L/min flow rate (24).

namic shape factor χ for S-SX particles means a relatively small aerodynamic diameter in comparison to its volume diameter. Although M-SX has a smaller aerodynamic diameter, S-SX powder performed much better in the cascade impactor test, showing more than twofold increase in FPF (Table 3). This is attributed to the better dispersibility of S-SX aggregates, which is probably the main cause of enhanced aerosolization performance for the other SCF-processed powders shown in Figure 1.

Dispersibility of powders in the airflow is defined by the balance of forces generated by the mechanical stresses within the dispersion device and the interparticulate forces required to separate the primary particles simultaneously (24). The mechanism of dispersion is very complex and may involve dispersion by acceleration and by shear flow, as well as by impaction or other mechanical forces.

It is hardly possible to anticipate all the factors involved for different devices or even for different flow regimes. However, it is feasible to carry out a comparative study of different powders for the same device, considering the fluid energy dissipated per unit volume of the flow and taking as an adequate measure of such dissipation the magnitude of the viscous turbulent stress τ_S. This stress can be directly compared with the theoretical tensile strength of the aggregate, σ, (24). Of course, τ_S varies within the flow cross section and also depends on the flow geometry and Re. However, if the dispersion is governed by the aerodynamic forces in the viscous subrange below the Kolmogorov scale of turbulence, as it is within the typical DPI flow rates, the shear stress generally depends on the flow velocity in a similar manner, while showing different numerical coefficients (70,71). Therefore, for comparative purposes, the following expression can be used for the viscous shear stress τ_S:

$$\tau_S = \mu \left(\frac{\varepsilon}{\nu}\right)^{1/2} \tag{4}$$

Here ε is the mean turbulent energy dissipation rate in a flow that can be calculated from the Kolmogorov theory (24) and ν is the kinematic air viscosity. The stress acting on the solid particles and aggregates is proportional to τ_S (N/m^2) and to the aerodynamic shape factor χ (24). Now it is possible to obtain an experimental plot showing the mean size of particle aggregates versus τ_S (Figure 6). Clearly, the aerosolization behavior of the salmeterol materials is very different. At low turbulent shear stresses, S-SX powders consistently produce a large fraction of primary particles in the respiratory size range, whereas M-SX powders consist of stable aggregates outside the 5 μm range, which cannot be dispersed at low stresses. It has been shown that the dispersion process in passive DPI devices usually occurs in the lower region of turbulent stresses in Figure 6 (24,70).

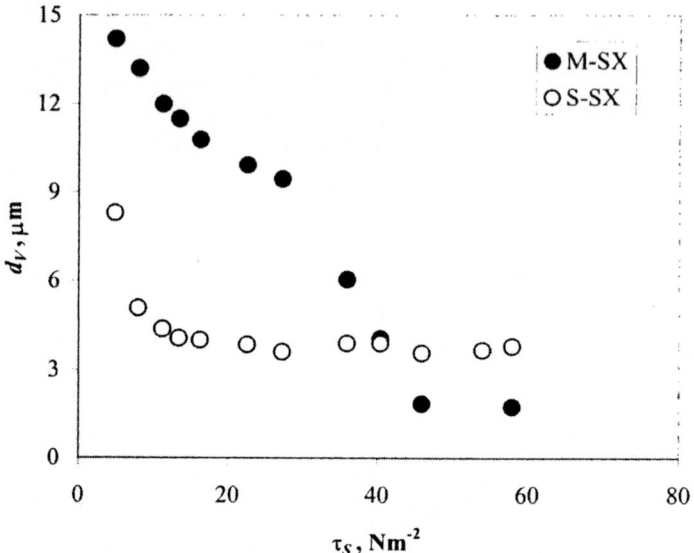

FIGURE 6 Dependence of the volume mean particle diameter $\bar{d}_V$, of the dispersed micronized (M-SX) and supercritically produced (S-SX) salmeterol powders on the magnitude of the aerodynamic turbulent stress $\tau\{\{_S$, calculated according to Eq. (4).

The next step is to predict the aggregate strength, σ, on the basis of known bulk powder properties and compare this parameter with the aerodynamic stress. Based on the model proposed by Kendall and Stainton (72), the balance of forces can be approximated (24) by

$$\chi\tau_S = \sigma = K\phi^4\frac{W_C}{d_V} \tag{5}$$

where $\phi = \rho_B/\rho_c$ is the packing fraction expressed through the ratio between the bulk powder density and particle crystal density correspondingly, and W_C is the work of cohesion of pure drug particles. It is assumed here that the particles have the same size and that the mean curvature of the nonspherical particles rises in proportion to the volume diameter. The K is a numerical coefficient characteristic of a dispersion device.

Thus Eq. (5) allows the measurement of the powder dispersion in relation to a reference material, in our case the micronized powder.

Several important conclusions can be drawn from this simple equation. First, the strong dependence of σ on ϕ (or bulk density, Table 3) given by the quadratic power (72) is explained by the loose aggregate structure that contains fewer contacts, which are therefore weak. Second, the aggregate strength decreases with an increase of the particle volume diameter d_V. It is well known that porous respiratory particles for a given aerodynamic diameter have larger volume diameter than solid particles and therefore are readily dispersible. However, in the case of crystalline salmeterol particles, reduced interparticle interactions come not from the particle density but from the shape factor. The small sphericity of S-SX also translates into a larger volume diameter for a given aerodynamic diameter. Moreover, the large aerodynamic shape factor of S-SX (Table 3) means that aerodynamic forces are more efficient in dispersing these particles than micronized powder.

The strength of the aggregates is proportional to the work of cohesion (drug–drug interactions) or the work of adhesion (drug–lactose) interactions. Theoretically (73), it is assumed that all materials are separated by a plane having a thickness that corresponds to the equilibrium distance at zero potential energy and that each site of one surface interacts with a continuum of sites on another, so it is possible to use the total (Hildebrand) solubility parameters δ_C and δ_A, to predict, respectively, the relative strengths of interaction σ_C (cohesive) or σ_A (adhesive) (73):

$$\sigma_C = 0.25\delta_C^2 \tag{6}$$

$$\sigma_A = 0.25\theta\delta_C\delta_A \tag{7}$$

where θ is the interaction parameter, which is also determined by IGC (73). These total (Hildebrand) solubility parameters can be calculated by using the Hansen component parameters for dispersive forces, polar interactions, and hydrogen bonding by means of a multiple linear regression analysis as described previously by Tong et al. (25,31). The work of cohesion or adhesion, W_C or W_A (J/m^2), is directly proportional to the corresponding value of the strength of the interaction, σ_C (cohesive) or σ_A (adhesive) (24).

Data in Table 3, and Eqs. (6) and (7), show that both cohesive and adhesive interactions are weaker, by about 25%, for S-SX particles. Experimentally, low adhesion of S-SX was observed when a triboelectrification method was used (34). The fraction of adhered material was several times smaller for S-SX powder than for M-SX in both the Turbula™ mixer and cyclone separator tests. In addition, S-SX particles exhibited significantly less (between one and two orders of magnitude) accumulated charge than the micronized powder before and after mixing and also for the nonadhered drug in a cyclone separator. These data point to superior (low energy, low charge) surface properties of the supercritically produced powder. Equation (5) and the data in Table 3 can be used to make a theoretical prediction of the aggregate strength of the pure drug particles. The experimentally measured ratio of aerodynamic stresses required for aggregate dispersion (Figure 5) is $\tau_S(\text{M-SX})/\tau_S(\text{S-SX}) > 7$. This is in qualitative agreement with a theoretical prediction made by means of Eq. (5), which gives a ratio of about 11.4. More accurate determination is subject to analysis of the bulk density in different aerosolization stages and assessment of the aggregate defect structure and interparticle contact area and, possibly, taking into account the nonequilibrium phenomena during dispersion (72).

Adhesive interaction (drug–lactose carrier) is stronger than cohesive interaction (drug–drug) for both M-SX and S-SX (Table 3). This property may encourage particles to adhere to the lactose carrier. However, the tensile strength of an aggregate also depends on the particle diameter. The relatively large size of lactose crystals used in DPI formulations ($\bar{d}_A \approx 50\ \mu\text{m}$) means that the average (tensile) strength of drug–lactose aggregates may decrease, competing with the increased adhesion energy. This phenomenon could be pictured as large lactose crystals surrounded by smaller SX aggregates where the cohesive forces dominate. Separation of such aggregates from lactose requires less aerodynamic stress than separation of individual particles. From the experimental viewpoint, the large $\bar{d}_V$ values obtained in the impactor test (Table 3) are explained by noting that the aggregates of pure drug particles are not dispersed below the preseparation impactor stage, or, in the case of M-SX, down to the third impactor stage (24). This is an indication that the drug particles are able to aggregate themselves within the ordered drug–lactose units. The carrier is less important for dispersion of supercritically processed salmeterol, as shown by the FPF data for drug-alone and drug–lactose formulations obtained by Tong (25); the aerosol performance of supercritically processed salmeterol was only marginally improved by the carrier. The same conclusion was reached by Feeley and coworkers when they analyzed salbutamol materials (23). These observations point out that drug-alone powders composed of supercritically produced materials may well find applications in DPIs.

Equation (5) predicts that the strength of aggregates can be influenced strongly, by packing. There is of course a concern that the aerosolization behavior of low bulk density powders may deteriorate with storage. Although there is no experimental work reporting assessments of this phenomenon for supercritically produced powders, S-SX and other powders tested were readily fluidized after prolonged storage and transportation. The ratio of the tapped bulk density to the aerosolized bulk density (also called the Hausner ratio) can be relatively high for some supercritically produced materials; however, this is because the small particle sphericity leads to loose packing, close to the theoretical random packing for aerosolized powders. Strong interparticle interactions for very cohesive powders can also decrease the strength of the aggregates (72,74). However, this interpretation is contradicted by the reduced cohesive and adhesive energy values measured by IGC and by the reduced adhesion to foreign surfaces of supercritically produced particles (34).

7. APPLICATION OF NEW ANALYTICAL TECHNIQUES

The complex behavior of respiratory powders and the need for a more systematic approach to their formulation call for new analytical approaches. An important analytical task is to discriminate between the specific components of the surface energy, the nature and statistics of the interparticulate forces, the contact area, and the average distance between nonspherical particles. These challenges can be addressed, for example, by using a combination of atomic force microscopy (75) and bulk powder measurements (74), which can provide a comprehensive description of particle aggregation processes on the micro and macro scales.

Another group of phenomena consists of the dynamic behaviors of particles during and after the aerosolization process. The dynamic quality of in situ data is much superior to the ex situ results obtained by means of standard particle size analysis. An example of new optical cell compatible with most DPIs is shown in Figure 7a. Measurements can be made in parallel with an Andersen cascade impactor, allowing a direct comparison between the time-averaged aerodynamic diameter or MMAD given by the impactor results, and the volume particle size distribution reported by laser diffraction, in relation to the emitted dose and FPF (76). The flow rate within the measurement zone can be carefully controlled, and an inhaler device can be mounted either before or after a standard U.S. Pharmacopeia throat. Rapid data acquisition at speeds of up to 2500 Hz are possible (or one full PSD measurement every 0.4 ms), and this speed can further be increased by optimizing the processing electronics. This time-resolved analysis can discriminate the diffraction patterns produced by different sections of an aerosol cloud such as primary drug particles, aggregates, and carrier particles. Figure 7b illustrates this point. The mean size of micronized salmeterol powder released from the inhaler device over a period of 0.5 s remained relatively constant during the release of the powder, with a mixture of lactose and drug particles being observed at all times. By contrast, under the same measurement conditions, drug particles produced from supercritical fluids separate relatively easy, as shown by the appearance of fine, dispersed drug particles in the starting measurement zone (0–0.07 s). In this case, the cloud of pure drug particles moves faster than the coarse lactose particles, suggesting increased deagglomeration of the fine drug particles. During the midpoint of the spray plume (0.07–0.4 s) a mixture of lactose carrier and fine drug particles was observed. Finally, the tail of the spray plume (0.4–0.45 s) contained again

a

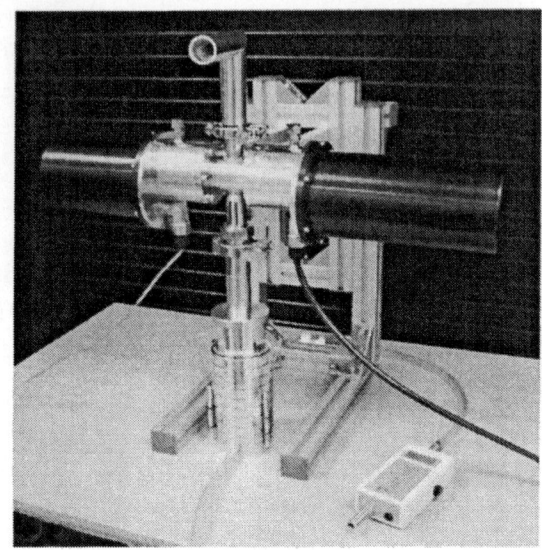

b

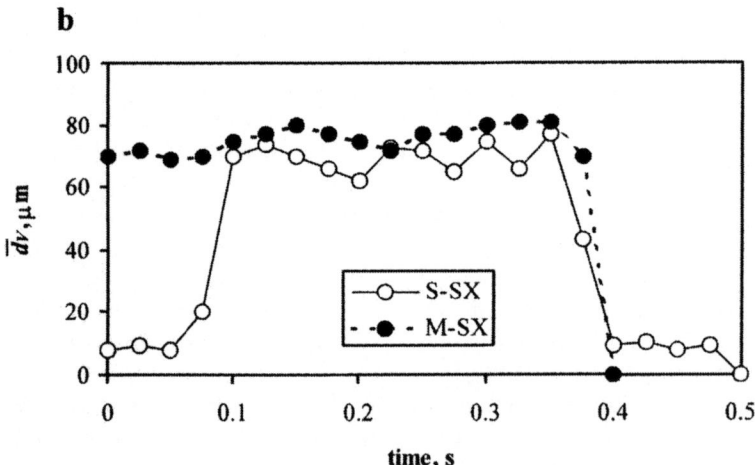

FIGURE 7   (a) Inhalation measurement cell with Malvern Spraytec laser diffraction analyzer, combined with Andersen cascade impactor (76). (b) Time-history showing the mean particle size measured during the actuation of the Clickhaler containing the micronized (M-SX) and supercritically produced (S-SX) salmeterol powders in formulation with lactose and at the airflow rate 49 L/min.

mainly drug particles. It is believed that these particles were delayed by a secondary aggregation–deaggregation process within the inhaler.

In addition, the cumulative size distribution obtained for pure micronized and supercritically produced powders, combined with computation of the aerodynamic diameter according to Eqs. (2) and (3), shows that FPF of supercritically produced particles is about 50%, twice as much as FPF for micronized powder and close to the dispersion efficiency found with lactose (76). The correct values for the total emitted dose of drug particles also were confirmed. In all cases, the state of dispersion for the different formulations was correctly predicted and the results obtained showed good agreement with the cascade impactor measurements made on the same samples. Therefore laser diffraction has distinct advantages over impactor techniques with respect to the speed and reproducibility of measurements and can be used as a complementary method for analysis of aerosol behaviour of different formulations.

8. ADVANTAGES OF USING SUPERCRITICALLY PRODUCED POWDERS IN RESPIRATORY FORMULATIONS

It can be summarized from the experimental data and modeling presented that the main factors responsible for superior in vitro performance of supercritically produced powders are as follows:

> Enhanced dispersibility at low airflow rates or aerodynamic turbulent stresses
> Relatively low surface energy, leading to reduced particle adhesive and cohesive interactions
> Possibility of using drug powders without coformulation with carriers or excipients
> Advantage of consistent production in a stable crystalline form

In particular, the loose packing of particles promotes an open powder structure that is less adhesive and flows and disperses more readily. There is a strong correlation between the interaction parameters derived by IGC and the in vitro data that play an important role in the prediction of aerosol performance of dry powder inhalation formulations. Enhanced dispersibility is particularly important for DPI devices, where performance strongly depends on powder deaggregation at relatively low dispersion forces. Clearly, high turbulence is preferable for dispersion, but it inevitably leads to high pressure differentials, which may prevent many devices from functioning correctly. In addition, low dispersion forces for supercritically produced powders mean that the performance of such materials may be independent of the specific geometry or flow rate of a given DPI. Therefore, combined with the advantage of a more consistent crystalline form, supercritically produced materials offer an attractive alternative to micronized powders in respiratory applications.

9. STRENGTHS AND BOUNDARIES OF THE SCF TECHNOLOGY

This chapter emphasized the role of a mechanistic understanding of the particle formation process and powder behavior for respiratory drug delivery. Supercritical fluid particle technology has clear fundamental advantages based on the unique solvent properties of SCFs. In the SCF process, particle formation, coating or encapsulation, surface treatment, and purification can be achieved in one processing unit. The enclosed environment in SCF precipitation systems makes possible the industrial production of toxic, potent, or sensitive (in terms of heat, light, or shear) compounds. An additional benefit is the sterilization properties of SCF reported in the literature (see Chapter 10 by Knutson and Sarkari, this volume), giving rise to microbial inactivation beneficial for the production according to cGMP (current good manufacturing practices) (see also Chapter 15 by Clavier and Perrut).

These potential benefits of SCF technology in general should be explored with caution, taking into account that this technology is very compound specific and that no particular SCF method is universal. Different particle formation techniques with SCF are complementary and can be used for different material classes. In a way, supercritical fluid technology can be considered to be a versatile "toolbox" rather than a single and universal solution. Indeed, no other particle technology can offer so many different processing approaches. For example, challenges encountered by the antisolvent methods with particle size reduction and production of composite particles for some materials can be resolved by using different expansion techniques. Processing of aqueous solutions with supercritical CO_2 may solve the problem of protein instability in contact with organic solvents.

The enhanced physical properties for supercritically produced salmeterol (e.g., high crystallinity, polymorphic purity, powder uniformity) correlate well with the enhanced dispersion and flow behavior of this powder. However, as shown in some other investigations (12,28), particle size and shape, surface energy, and crystallinity may have to be optimized separately, using all process parameters available, perhaps including a coformulation step. For some compounds with low solubility in water, a compromise must be found between improved bioavailability and physicochemical stability.

Some materials can exhibit negative solid state transformations—for example, formation of different unstable solvate forms or polymorphs and acicular particles. For respiratory applications, the challenge is to achieve many formulation goals with a single-step process.

No doubt new uses of supercritical fluids or compressed gases will be found, particularly in the areas of composite and porous particles, processing of liposomes and emulsions, bioactive materials, and coformulation of different drugs and excipients to increase pulmonary deposition or therapeutic efficiency. Although this chapter was concerned with respiratory drug delivery, particles produced by using SCF techniques can certainly be prepared for the transmucosal nasal delivery, for example, of antimigraine drugs, antibiotic treatment of sinus infections, pain management, or delivery of medicines to the brain via immediate release systems or mucoadhesive particles for controlled release. These applications are yet to be explored. Process modification and optimization should be able to achieve the monodisperse particle size and shape distribution for lung targeting. The standards must be set high to meet the challenge of competitive particle technologies and more economical productions, combined with readiness to scale up for a variety of materials and powder formulations.

REFERENCES

1. Newman SP, Busse WW. Evolution of dry powder inhaler design, formulation, and performance. Respir Med 2002; 96:293–304.
2. Duddu SP, Sisk SA, Walter YH, Tarara TE, Trimble KR, Clark AR, Eldon MA, Elton RC, Pickford M, Hirst PH, Newman SP, Weers JG. Improved lung delivery from a passive dry powder inhaler using an engineered PulmoSphere® powder. Pharm Res 2002; 19:689–695.
3. Edwards DA, Ben-Jebria A, Langer R. Recent advances in pulmonary drug delivery using large, porous inhaled particles. J Appl Physiol 1998; 84:379–385.
4. Malcolmson RJ, Embleton JK. Dry powder formulations for pulmonary delivery. Pharm Sci Technol Today 1998; 1:394–398.
5. Beach S, Latham D, Sidgwick C, Hanna M, York P. Control of the physical form of salmeterol xinafoate. Org Process Res Dev 1999; 3:370–376.
6. Staniforth JN. Pre-formulation aspects of dry powder aerosols. Proceedings of the Conference on Respiratory Drug Delivery, Phoenix, AZ 1996; V:65–74.
7. Kawashima Y, Serigano T, Hino T, Yamamoto H, Takeuchi H. Surface-modified antiasthmatic dry powder aerosols inhaled intratracheally reduce the pharmacologically effective dose. Pharm Res 1998; 15:1753–1759.
8. Lucas P, Anderson K, Potter UJ, Staniforth JN. Enhancement of small particle size dry powder aerosol formulations using an ultra-low density additive. Pharm Res 1999; 16:1643–1647.

9. Chew NYK, Chan HK. Effect of particle size and surface morphology on the dispersion of albumin powders. Proceedings of the Conference on Respiratory Drug Delivery, Palm Harbor, FL 2000; VII:619–622.

10. Bosquillon C, Lomby C, Preat V, Vanbever R. Influence of formulation excipients and physical characteristics of inhalation dry powders on their aerosolisation performance. J Controlled Release 2001; 70:329–339.

11. Tee SK, Marriot C, Zeng XM, Martin GP. The use of different sugars as fine and coarse carriers for aerosolised salbutamol sulphate. Int J Pharm 2000; 208:111–123.

12. Steckel H, Thies J, Mülller BW. Micronizing of steroids for pulmonary delivery by supercritical carbon dioxide. Int J Pharm 1997; 152:99–110.

13. Ben-Jebria A, Chen D, Eskew ML, Vanbever R, Langer R, Edwards DA. Large porous particles for sustained protection from carbachol-induced bronchoconstriction in guinea pigs. Pharm Res 1999; 16:555–561.

14. Maa YF, Nguyen PA, Sweeney T, Shire SJ, Hsu CC. Protein inhalation powders: spray drying vs spray freeze drying. Pharm Res 1999; 16:249–254.

15. Constantino HR, Firouzabadian L, Hogeland K, Wu C, Beganski C, Carrasquillo KG, Cordova M, Griebenow K, Zale SE, Tracy MA. Protein spray–freeze drying. Effect of atomisation conditions on particle size and stability. Pharm Res 2000; 17:1374–1383.

16. Yu ZC, Rogers TL, Hu JH, Johnston KP, Williams RO. Preparation and characterization of microparticles produced by a novel process: spray freezing into liquid. Eur J Pharm Biopharm 2002; 54:221–228.

17. Adjei AL, Qiu Y, Gupta PK. Bioavailability and pharmacokinetics of inhaled drugs. In: Hickey AJ, ed. Inhalation Aerosols. New York: Marcel Dekker, 1996.

18. Sakagami M, Sakon K, Kinoshita W, Makino Y. Enhanced pulmonary absorption following aerosol administration of mucoadhesive powder microspheres. J Controlled Release 2001; 77:117–129.

19. El-Baseir MM, Kellaway IW. Poly(L-lactic acid) microspheres for pulmonary drug delivery: release kinetics and aerosolisation studies. Int J Pharm 1998; 175:135–145.

20. Steckel H, Müller BW. Metered-dose inhaler formulation of fluticasone-17-propionate micronized with supercritical carbon dioxide using the alternative propellent HFA-227. Int J Pharm 1998; 173:25–33.

21. Feeley JC, York P, Sumby BS, Dicks H, Hanna M. In vitro assessment of salbutamol sulphate prepared by micronisation and a novel supercritical fluid technique. Drug Delivery Lungs 1998; IX:196–199.

22. Feeley JC, York P, Sumby BS, Dicks H. Comparison of the surface properties of salbutamol sulphate prepared by micronisation and a supercritical fluid technique. J Pharm Pharmacol 1998; 50:54.

23. Feeley JC, Gilbert DJ, Palakodaty S, Walker SE, York P. Engineering of particle size distributions for respiratory drug delivery by supercritical fluid processing. Proceedings of the Conference on Respiratory Drug Delivery, Palm Harbor, FL 2000; VII:357–360.

24. Shekunov BY, Feeley JC, Chow AHL, Tong HHY, York P. Aerosolization behaviour of micronised and supercritically-processed powders. J Aerosol Sci. In press.
25. Tong HHY. PhD thesis. Chinese University of Hong Kong, 2003.
26. Rehman MU. PhD thesis. University of Bradford, 2002.
27. Shekunov BY, Rehman M, Chow AHL, Tong HHY, York P. Production of crystalline powders for inhalation drug delivery using supercritical fluid technology. Proceedings of the 6th International Symposium on Supercritical Fluids, April 28–30, 2003, Versailles, France.
28. Rehman M, Shekunov BY, York P, Lechuga D, Miller D, Tan T, Colthorpe P. Optimisation of powders for pulmonary delivery using supercritical fluid technology. Eur J Pharm Sci 2003. Submitted.
29. Velaga SP, Ghaderi R, Carlfors J. Preparation and characterization of hydrocortisone particles using a supercritical fluids extraction process. Int J Pharm 2002; 231:155–166.
30. Richardson CH, Chrystyn H, Won ICK, Walker S. Comparison of in-vitro performance of salbutamol sulphate manufactured by a supercritical fluid process micronized salbutamol in a dry powder inhaler. AAPS PharmSci 2002; 4:T3197.
31. Tong HHY, Shekunov BY, York P, Chow AHL. Influence of polymorphism on the surface energetics of salmeterol xinafoate crystallized from supercritical fluids. Pharm Res 2002; 19:640–648.
32. York P, Hanna M. Particle engineering by supercritical fluid technologies for powder inhalation drug delivery V. Proceedings of the Conference on Respiratory Drug Delivery, Phoenix, AZ, 1996:231–239.
33. Tong HHY, Shekunov BY, York P, Chow AHL. Characterisation of two polymorphs of salmeterol xinafoate crystallized from supercritical fluids. Pharm Res 2001; 18:852–858.
34. Shekunov BY, Feeley JC, Chow AHL, Tong HHY, York P. Physical properties of supercritically-processed and micronised powders for respiratory drug delivery. KONA Powder and Particles 2003; 20:178–187.
35. Velaga SP, Berger R, Carlfors J. Supercritical fluids crystallization of budesonide and flunisolide. Pharm Res 2002; 19:1564–1571.
36. Rehman MU, Shekunov BY, York P, Colthorpe P. Solubility and precipitation of nicotinic acid in supercritical carbon dioxide. J Pharm Sci 2001; 90:1570–1582.
37. Feeley JC, York P, Sumby BS, Dicks H. Determination of surface properties and flow characteristics of salbutamol sulphate, before and after micronisation. Int J Pharm 1998; 172:89–96.
38. Bisrat M, Moshashaee S, Nyqvist H, Dumirbuker M. Composition of matter. US patent application 6,475,524, 2000.
39. Shekunov BY, Bristow S, Chow AHL, Cranswick L, Grant DJW, York P. Formation of composite crystals by precipitation in supercritical CO_2. Cryst Growth Design 2003; 3:603–610.
40. Sethia S, Squillante E. Physicochemical characterization of solid dispersions of

carbamazepine formulated by supercritical carbon dioxide and conventional solvent evaporation method. J Pharm Sci 2002; 91:1948–1957.

41. Villa JA, Sievers RE, Huang ETS. Bubble drying to form fine particles from solutes in aqueous solutions. Proceedings of the 7th Meeting on Supercritical Fluids, Antibes / Juan-Les-Pins, France, 2000:83–85.

42. Yeo SD, Debenedetti PG, Patro SY, Przybycien TM. Secondary structure characterization of microparticulate insulin powders. J Pharm Sci 1994; 83:1651–1656.

43. Winters MA, Knutson BL, Debenedetti PG, Sparks HG, Przybycien TM, Stevenson CL, Prestrelski SJ. Precipitation of proteins in supercritical carbon dioxide. J Pharm Sci 1996; 85:586–594.

44. Moshashaee S, Bistrat M, Forbes RT, Nyqvist H, York P. Supercritical fluid processing of proteins. I. Lysozyme precipitation from organic solution. Eur J Pharm Sci 2000; 11:239–245.

45. Snavely WK, Subramaniam B, Rajewski RA, Defelippis MR. Micronization of insulin from halogenated alcohol solution using supercritical carbon dioxide as an antisolvent. J Pharm Sci 2002; 91:2026–2039.

46. Bustami RT, Chan HK, Dehghani F, Foster NR. Generation of micro-particles of proteins for aerosol delivery using high pressure modified carbon dioxide. Pharm Res 2000; 17:1360–1366.

47. Bustami RT, Chan HK, Foster NR. Aerosol delivery of protein powders processed by supercritical fluid technology. Proceedings of the Conference on Respiratory Drug Delivery VII, Palm Harbor, FL, 2000:611–613.

48. Tservistas M, Levy MS, Lo-Yim MYA, O'Kennedy RD, York P, Humphreys GO, Hoare M. The formation of plasmid DNA loaded pharmaceutical powders using supercritical fluid technology. Biotechnol Bioeng 2001; 72:12–18.

49. Hollowood ME, Humphreys GO, Shekunov BY, Sloan R, York P. Formation of protein particles in supercritical fluids. Proceedings of the 7th International Conference on Crystallization of Biological Macromolecules, Granada, Spain, 1998:209.

50. Falk R, Randolph TW, Meyer JD, Kelly RM, Manning MC. Controlled release of ionic compounds from poly(L-lactide) microspheres produced by precipitation with a compressed antisolvent. J Controlled Release 1997; 44:77–85.

51. Okamoto H, Nishida S, Todo H, Sakakura Y, Iida K, Danjo K. Pulmonary gene delivery by chitosan–pDNA complex powder prepared by a supercritical carbon dioxide process. J Pharm Sci 2003; 92:371–379.

52. Sellers SP, Clark GS, Sievers RE, Carpenter JF. Dry powder of stable formulations from aqueous solutions prepared using supercritical CO_2-assisted aerosolization. J Pharm Sci 2001; 90:785–797.

53. Sievers RE, Quinn BP, Huang ETS, Cape SP, Algarov D, Rinner L, Villa JA. Stabilization, micronization, and coating of proteins, peptides, antibodies and enzymes by CO_2-assisted nebulization with bubble dryer. Proceedings of the Protein Stability Conference, Breckenridge, Colorado, 2003:25.

54. Sievers RE, Milewski PD, Sellers SP, Miles BA, Korte BJ, Kusek MD, Clark GS, Mioskowski B, Villa JA. Supercritical and near-critical carbon-dioxide assisted low-temperature bubble drying. Ind Eng Chem Res 2000; 39:4831–4836.

55. Sievers RE, Huang ETS, Villa JA, Kawamoto JK, Evans MM, Brauer PR. Low-temperature manufacturing of fine pharmaceutical powders with supercritical fluid aerosolization in a bubble dryer. Pure Appl Chem 2001; 73:1299–1303.
56. Martin TM, Bandi N, Shultz R, Roberts CB, Kompella UB. Supercritical fluid technology-derived budesonide and budesonide-pla microparticles for respiratory delivery. AAPS Pharm Sci 2001:3.
57. Wilkins S, Shekunov BY, York P. Theophylline: ethylcellulose co-precipitates formed by solution enhanced dispersion by supercritical fluids (SEDS) and solvent co-evaporation—structural analysis by synchrotron powder X-ray diffraction. AAPS Pharm Sci 2001; 3:1136.
58. Zhou H, Lengsfeld C, Claffey DJ, Ruth JA, Hybertson B, Randolph TW, Ng K-Y, Manning MC. Hydrophobic ion pairing of isoniazid using a prodrug approach. J Pharm Sci 2002; 91:1502–1511.
59. Shekunov BY, Edwards AD. Crystallization and plasticization and of poly (L-lactide) by carbon dioxide. Accepted for publication in Proceedings of the 6th International Symposium on Supercritical Fluids, April 28–30, Versailles, France.
60. Edwards AD, Shekunov BY, Forbes RT, Grossmann JG, York P. The structure and morphology of poly(L-lactide) particles formed by spray-drying and solution enhanced dispersion by supercritical fluids (SEDS). Proceedings of the 18th Pharmaceutical Technology Conference, Utrecht, Netherlands, 1999; 1:37–44.
61. Owens JL, Anseth KS, Randolph TW. Compressed antisolvent precipitation and photopolymerization to form highly cross-linked polymer particles. Macromolecules 2002; 35:4289–4296.
62. Mandel FS, Wang JD. Pharmaceutical material production via supercritical fluids employing the technique of particles from gas-saturated solutions. Proceedings of the 7th Meeting on Supercritical Fluids, Antibes, France, 2000:35–45.
63. Howdle SM, Watson MS, Whitaker MJ, Popov VK, Davies MC, Mandel FS, Wang JD, Shakesheff KM. Supercritical fluid mixing: preparation of thermally sensitive polymer composite containing bioactive materials. J Chem Soc Chem Commun 2001:109–110.
64. Lee K, Gould G. Aerogel powder therapeutic agents. International patent publication WO 02/051389, 2002.
65. Shekunov BY, Hanna M, York P. Crystallization process in turbulent supercritical flows. J Cryst Growth 1999; 198/199:1345–1351.
66. Shekunov BY, Baldyga J, York P. Particle formation by mixing with supercritical antisolvent at high Reynolds numbers. Chem Eng Sci 2001; 56:1–13.
67. Bristow S, Shekunov T, Shekunov BY, York P. Analysis of the supersaturation and precipitation process with supercritical CO_2. J Supercrit Fluids 2001; 21: 257–271.
68. Ganser GH. A rational approach to drug prediction of spherical and nonspherical particles. Powder Technol 1993; 77:143–152.
69. Bristow SC, Shekunov BY, York P. Solubility analysis of drug compounds in supercritical carbon dioxide using static and dynamic extraction systems. Ind Eng Chem Res 2001; 40:1732–1739.
70. French DL, Edwards DA, Niven RW. The influence of formulation on emission,

deaggregation and deposition of dry powders for inhalation. J Aerosol Sci 1996; 27:769–783.

71.   Li WI, Perzl M, Heyder J, Langer R, Brain JD, Englmeier KH, Niven RW, Edwards DA. Aerodynamics and aerosol particle deaggregation phenomena in model oral–pharyngeal cavities. J Aerosol Sci 1996; 27:1269–1286.

72.   Kendall K, Stainton K. Adhesion and aggregation of fine particles. Powder Technol 2001; 121:223–229.

73.   Rowe RC. Interaction of lubricants with microcrystalline cellulose and anhydrous lactose—a solubility parameter approach. Int J Pharm 1988; 41:223–226.

74.   Valverde JM, Ramos A, Castellanos A, Watson PK. The tensile strength of cohesive powders and its relationship to consolidation, free volume and cohesivity. Powder Technol 1998; 97:237–245.

75.   Louey MD, Mulvaney P, Stewart PJ. Characterisation of adhesion lactose carriers using atomic force microscopy. J Pharm Biomed Anal 2001; 25:559–567.

76.   Shekunov BY, Kippax P, Jones L, Rehman M, York P. Analysis of dry powders aerosols using laser diffraction. Proceedings of the American Association of Pharmaceutical Sciences Annual Conference, Denver, AAPS Pharm Sci 2001; 3:108.

7

Control of Physical Forms of Pharmaceutical Substances

Albert H.L. Chow and Henry H.Y. Tong
The Chinese University of Hong Kong, Shatin, Hong Kong

Boris Y. Shekunov
Ferro Corporation, Independence, Ohio, U.S.A.

Predictive control of the physical forms of drug and excipient materials is of great relevance in the contexts of patent protection, process development, and product specification in the pharmaceutical industry (1). Pharmaceutical solids exist broadly as either crystalline or amorphous forms. Perfect crystalline substances are rare, and materials are often characterized by imperfections (defects) of varying degrees in their crystal structures, depending on their formation or treatment history. In addition, certain materials can exist in grossly different crystal forms (e.g., polymorphs, solvates, or hydrates) under defined conditions of pressure, temperature, and/or relative humidity. Being different in free energy and thermodynamic activity, all these structural modifications, be they due to different densities of crystal imperfections or to grossly different crystal structures, are generally distinguishable by their differences in physical properties, such as hardness, melting point, solubility, and dissolution rate, all of which can impact significantly the manufacturing process as well as the in vivo performance of the finished drug products.

283

Consequently, it has always been the goal of pharmaceutical formulators to seek effective means of controlling the physical forms of pharmaceutical raw materials, which is important not only to ensure consistency and predictability of product performance but also to satisfy the related drug regulatory requirements.

The pharmaceutical literature is replete with technical information on the production of polymorphs, solvates/hydrates, and amorphous solids (2). Classical techniques for polymorph production include sublimation, crystallization from solutions or melts, evaporation, vapor diffusion, thermal treatment, precipitation by abrupt change of solution pH or by rapid cooling, thermal desolvation or dehydration, grinding/milling, and use of crystallization promoters or inhibitors. As for hydrates, the preparation methods are relatively straightforward: recrystallization of the material from water or aqueous solvents and exposure of the anhydrous form or its organic solvates to a humid atmosphere are some effective ways of generating crystalline hydrates. For producing noncrystalline or amorphous materials, typical strategies are rapid solidification from melts, grinding/milling, spray drying, lyophilization, rapid desolvation or dehydration, and rapid precipitation by change of solution pH for compounds that exist as either a free acid or a free base.

Recent advances in supercritical fluid crystallization (SFC) technologies have made possible the production of pharmaceutical materials in relatively pure physical forms. By proper adjustment of the operating conditions or parameters, such as pressure, temperature, drug concentration, flow ratio of supercritical fluid to drug solution, and nature of organic solvents, materials of predetermined crystal forms and characteristics can be reproducibly generated. This chapter provides a critical review of the current applications of SFC technologies in controlling the crystal form of pharmaceutical materials. Discussed in the sections that follow are the underlying thermodynamic and kinetic principles governing the material formation in supercritical fluids, together with specific illustrative examples where appropriate.

1. BASIC PRINCIPLES GOVERNING MATERIAL PROCESSING IN SUPERCRITICAL FLUIDS

1.1. Thermodynamic Considerations

As with conventional techniques that utilize liquid solvents at normal atmospheric pressure, material formation in supercritical fluids is subject to thermodynamic considerations involving temperature and pressure. The most discernible feature that differentiates the SFC technologies from all the others is the relatively high compressibility and diffusivity of the supercritical solvent, notably carbon dioxide, which provides an additional dimension for process control and manipulation of the solid state properties.

Of all cases of material production and crystal form control using SFC reported thus far, polymorphs and hydrates (or solvates) are probably the most common. Other materials of special interest have also been the subject of investigation, including amorphous solids, solid dispersions, and solid solutions.

1.1.1. Polymorphism

"Polymorphism" refers to the ability of a substance to exist in more than one crystalline structure, which has different arrangements in space (i.e., different cell dimensions and cell packing) of the same structural unit(s) (i.e., same chemical composition). Polymorphs can display widely different physical and chemical properties, including melting point, spectroscopic behavior, solubility, density, hardness, crystal shape, physical and chemical stability, and dissolution–time profile.

Polymorphism can be classified into two types, enantiotropic and monotropic polymorphism. Each type involves two polymorphic forms or a polymorphic pair, whose interconversion is defined by a characteristic transition temperature. The temperature-dependent solid phase transformation is best understood by referring to the respective energy–temperature diagrams at constant pressure, as depicted in Figure 1 (3).

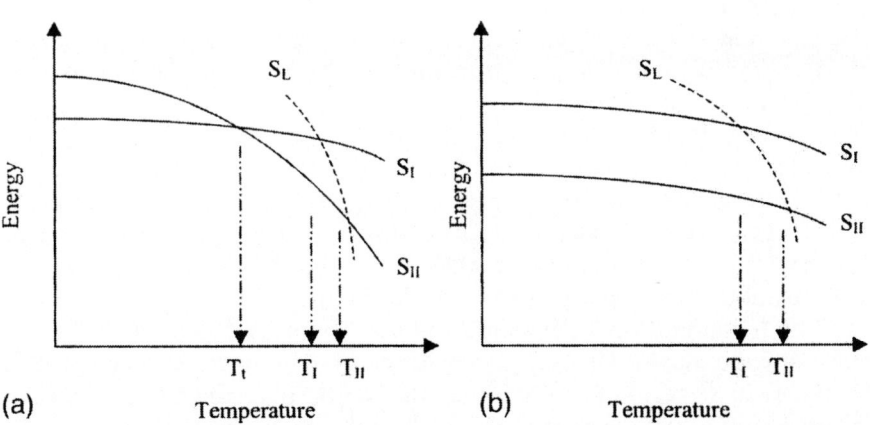

FIGURE 1 Energy–temperature diagrams. (a) For a hypothetical enantiotropic system: T_I and T_{II}, melting points of forms I and II; T_t, transition temperature. (b) For a hypothetical monotropic system: T_I and T_{II}, melting points of forms I and II.

The Gibbs free energy of any crystal at temperature T is given by

$$G_T = H_T - TS_T \tag{1}$$

A plot of G_T versus T at constant pressure according to Eq. (1) yields a curve whose slope (dG_T/dT) equals $-S_T$. In general, the entropy (i.e., the slope of the G_T-T curve) will vary with each polymorph, and therefore, the two curves will intersect at a temperature characteristic of the system under consideration. The phase with the lower free energy at a particular temperature will be the stable phase.

As shown in Figure 1a, the enantiotropic system is characterized by having the transition temperature between the polymorphs below their melting points. Slow cooling of the melt at a rate slow enough to allow maintenance of equilibrium will cause the melt to freeze to the high temperature polymorph S_{II}, which will transform on further cooling to S_I. In contrast, the transition temperature for monotropic polymorphism is above the melting points of the polymorphs (Figure 1b). Slow cooling of the melt as before will yield polymorph S_I, but never S_{II}, which is always obtained as a metastable phase. Thus, the expected transition in this case can only be considered to be virtual because the polymorph melts completely before the transition temperature is reached.

Like temperature, pressure is an important parameter affecting the phase transition of the polymorphs. Figure 2 is a typical pressure–temperature phase diagram of a hypothetical enantiotropic system (4). Each of the two polymorphs has its own S–V sublimation and S–L fusion curves but a common L–V vaporization curve (since they have identical liquid phases). The common L–V curve will normally intersect the two solid–vapor curves (S_I–V and S_{II}–V) above their intersection. Present also in the phase diagram is the S_I–S_{II} transition curve that passes through the intersection between the S_I–V and S_{II}–V curves, forming the S_I–S_{II}–V triple point. This triple point, which represents the solid I, solid II, and vapor phases in equilibrium, is invariably lower than the melting points of the enantiotropes. For each polymorph, there exists a defined range of conditions of temperature and pressure under which it will be the most stable phase, and the conversion from one form to another takes place in a reversible manner.

As for monotropic polymorphism, the common L–V curve will normally intersect the S_I–V and S_{II}–V curves below their intersection (Figure 3) (4). There is no region of stability for the second polymorph (S_{II}), and the melting point of the metastable S_{II} polymorph will invariably be lower than that of the stable form (S_I). Unlike enantiotropic polymorphism, the S_I–S_{II}–V triple point is always higher than the melting point of the stable S_I phase. Only one of the polymorphs remains stable up to the melting point upon heating, and the other polymorph can exist only as a metastable phase, irrespective of

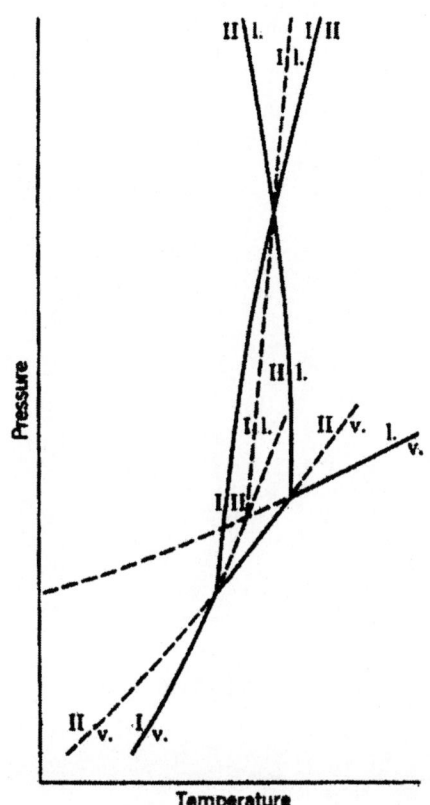

FIGURE 2 Phase diagram of a polymorphic pair exhibiting an enantiotropic relationship. (From Ref. 4, Copyright © 1969 American Pharmaceutical Association.)

both temperature and pressure. For such systems, the transition point can never be attained at atmospheric pressure, and the polymorphic transformation can occur irreversibly in one direction only.

The pressure–temperature phase diagrams also serve to highlight the fact that the polymorphic transition temperature varies with pressure, which is an important consideration in the supercritical fluid processing of materials in which crystallization occurs invariably at elevated pressures. Qualitative prediction of various phase changes (liquid/vapor, solid/vapor, solid/liquid, solid/liquid/vapor) at equilibrium under supercritical fluid conditions can be made by reference to the well-known Le Chatelier's principle. Accordingly, an increase in pressure will result in a decrease in the volume of the system. For most materials (with water being the most notable exception), the specific volume of the liquid and gas phase is less than that of the solid phase, so that

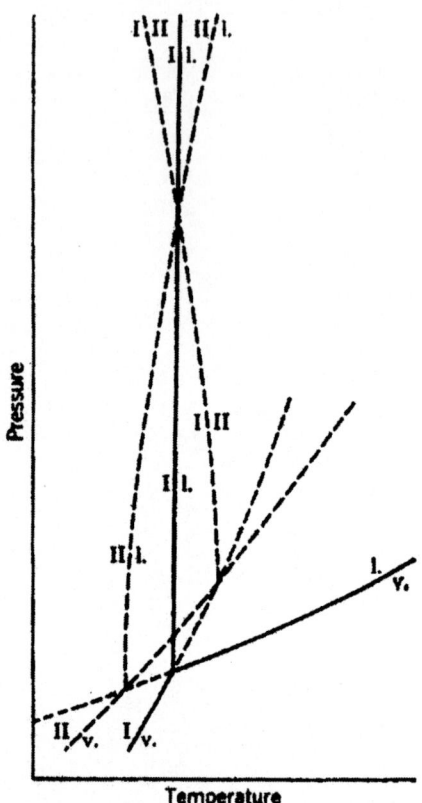

FIGURE 3 Phase diagram of a polymorphic pair exhibiting a monotropic relationship. (From Ref. 4, Copyright © 1969 American Pharmaceutical Association.)

an increase in pressure would have the effect of shifting the equilibria to favor the solid phase. This shift will have the observable effect of raising the melting and boiling points.

If the changes in phase equilibria indeed obey Le Chatelier's principle, the system can be specified quantitatively by the Clausius–Clapeyron equation as follows:

$$\frac{dP}{dT} = \frac{q}{T(v_2 - v_1)} \tag{2}$$

In Eq. (2) q is the quantity of heat absorbed during the transformation of one phase to the other, v_2 and v_1 are the specific volumes of the two phases, and T is the absolute temperature at which the changes occur.

If the phase equilibrium parameters such as melting point, boiling point, and transition temperature at atmospheric pressure are known, the shift of these parameters in response to increasing pressure can be predicted by using the Clausius–Clapeyron equation. For illustration purposes, let us consider two cases involving the transition of the enantiotropic pairs of salmeterol xinafoate, a long-acting antiasthmatic agent, and carbamazepine, an anti epileptic and anticonvulsant drug, under supercritical fluid conditions.

Salmeterol xinafoate is known to exist in two polymorphic forms, forms I and II. Form I is stable, and form II is the metastable polymorph at ambient temperature. The enthalpies of solution (ΔH_{sol}) of forms I and II determined from van't Hoff solubility–temperature plots are 32.1 and 27.6 kJ/mol, respectively, and the transition temperature obtained by linear extrapolation of the van't Hoff plots is 99°C. The enthalpy of polymorphic conversion ($\Delta H_{II \to I}$) calculated from the plots of log solubility ratio of polymorphs versus the reciprocal of absolute temperature is negative (-4.55 kJ/mol) (5). How-ever, the change in molar volume ($\Delta V_{II \to I}$) due to the conversion is positive. Therefore, according to the Clausius–Clapeyron equation,

$$\frac{dP}{dT} = \frac{\Delta H}{T\Delta V} \tag{3}$$

dP/dT should be negative. The calculated dP/dT for the polymorphic transition of salmeterol xinafoate is -12.8 bar/K. Accordingly, an increase in pressure will decrease the transition temperature, which is generally consistent with the trend observed under supercritical fluid conditions. Using the solution-enhanced dispersion by supercritical fluids (SEDS) technique, form I can be crystallized at or below 65°C, while form II is obtainable at or above 70°C at 200 bar (6). Thus, the transition temperature at 200 bar should lie somewhere between 65 and 70°C, which is in line with the expected decrease in transition temperature at elevated pressures, though apparently lower than that predicted (84.1°C) by the equation.

In the case of carbamazepine polymorphs, a similar approach has been adopted to calculate dP/dT for the transition from literature solubility data (7). At least three polymorphic forms of carbamazepine have been identified: form I (trigonal or α form; mp 174–176°C), form II (triclinic or γ form; mp 184–186°C), and form III (monoclinic or β form; mp 189–191°C). Poly-morphs I and III are enantiotropically related with a transition temperature of 71°C (8). The former polymorph has a significantly higher solubility in supercritical carbon dioxide than the latter at 55°C and 350 bar (9). In contrast to the negative value of dP/dT observed for the salmeterol xinafoate polymorphs, Edwards et al. obtained a positive dP/dT value ($+23.7$ bar/K) based on calculations from the carbamazepine monoclinic (β form) and triclinic (γ form) equilibrium curves (7). The transition temperature of the carbamazepine's enantiotropic pair at normal atmospheric pressure determined from linear extrapolation of van't Hoff solubility plots is 71 °C (8). The calculated transition temperature at 250 bar is 81 °C, which accords with the expected increase in transition temperature at higher pressures.

The foregoing examples serve to illustrate the utility of the Clausius–Clapeyron equation in predicting the transition temperature shift at elevated pressures. However, the predicted values do not appear to be comparable to the experimentally determined transition temperatures, and the observed variability of polymorphic transition in the supercritical solvent is too large to be explained by the hydrostatic pressure factor alone. The discrepancy may be ascribed to the strong influence of kinetic factors on the SFC process. In particular, pressure strongly affects the density of the supercritical solvent, which exerts a profound impact on many kinetic crystallization parameters. Since nucleation, and subsequent crystal growth, usually occur within an ultra short period of time, true equilibrium is rarely attainable in material processing in supercritical fluids. Nevertheless, the thermodynamic approach provides useful initial estimates of the transition temperature of the polymorphs and the solubilities of the polymorphs in solutions under supercritical fluid conditions, which will greatly facilitate the subsequent processing development work.

1.1.2. Hydrates

The hydration and dehydration of a crystalline solid may be expressed by the following equilibrium reaction (10):

$$D(s) + nH_2O \rightleftharpoons D \cdot nH_2O(s) \tag{4}$$

The equilibrium hydration constant is given by

$$K_h = \frac{a_3[D \cdot nH_2O_{(s)}]}{a_1[D_{(s)}]a_2[H_2O]^n} \tag{5}$$

where a_1, a_2, and a_3 refer to activities of the anhydrate, water, and hydrate, respectively. It is apparent that the equilibrium is highly influenced by the activity of water in the vapor phase, and hence, the stability of a hydrate relative to an anhydrate (or lower hydrate) depends on the relative humidity (11). A crystalline anhydrate–hydrate system differs from polymorphism in that water activity needs to be specified in addition to pressure and temperature. If both pressure and water activity are kept constant, an energy–temperature phase diagram for a hydrate–anhydrate analogous to that of a polymorphic system can be constructed.

1.1.3. Crystalline and Amorphous Forms

Crystalline materials, including solvates, hydrates, and polymorphs, have defined structures, stoichiometric compositions, and physical properties. In contrast, amorphous materials have no clearly defined arrangement of their structural units and no long-range order; therefore their structure can be viewed as being similar to that of a frozen liquid but without the thermal fluctuations observed in the liquid phase.

Amorphous materials are always in a higher energy state than their crystalline counterparts. Excess free energy and entropy are accrued in the solids as they are converted into the amorphous phase, since solidification occurs rapidly without permitting the basic structural units to attain their lowest energy state. The thermodynamic instability inherent in such a system results in a higher solubility and faster dissolution rate.

Let us consider now the thermodynamic and kinetic factors affecting SFC and crystallinity of the material produced. As shown in Chapter 3, by Baldyga, Henczka, and Shekunov, the SFC process is highly compound specific. Relatively large molecules and molecules with a certain degree of rotational flexibility may form a disordered state. Simpler small molecules can form an almost perfect crystal lattice even when produced at high nucleation and crystal growth rates. Figure 4 illustrates the kinetics leading to generation of crystal defects based on the diffusion–reaction mechanism of crystallization processes (12). If solution supersaturation is defined as $\sigma = (c - c_0)/c_0$, where c and c_0 are the solute concentration and equilibrium concentration

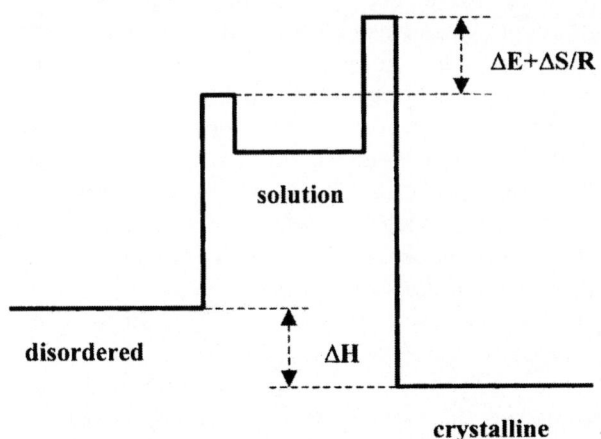

FIGURE 4 Energy diagram for molecular transition between a solution and different solid states.

for a given phase, the ratio of molecular fluxes from solution into a disordered, j, and crystalline, j_0, phases is defined by

$$j/j_0 = \left(1 - \frac{\Delta\sigma}{\sigma_0}\right)\exp\left(-\frac{\Delta H + \Delta E}{RT} - \frac{\Delta S}{R}\right) \tag{6}$$

where $\Delta\sigma$ is the difference of solution supersaturation for the two solid phases, ΔH is the difference of the enthalpies of dissolution, and ΔS and ΔE are the corresponding differences in the entropy and activation energy barriers for solute complexes to be transformed into desolvated molecules in the solid state, which is either perfect crystalline or disordered. In Eq. (6), when $\Delta\sigma > 0$, the preexponential multiplier changes from very small or even negative values at low supersaturation to unity at high supersaturation. The functions ΔH, ΔE, and ΔS are negative and practically independent of supersaturation. As a result, the molecular flux into the crystalline state dominates at relatively low supersaturation. However at high supersaturation, a significant fraction of solute molecules may precipitate into a disordered state. It also follows from Eq. (6) that reduction of the entropy and energy barriers should lead to a less disordered state at the same supersaturation value. A lower entropy barrier typically means a smaller and simpler molecule, whereas lower activation energy reflects weaker solute–solvent interactions. In particular, supercritical solutions can afford a large degree of flexibility in controlling the supersaturation by varying both pressure and temperature. A gradual adjustment of solubility power can be achieved by mixing polar liquid organic solvents such as alcohols with nonpolar supercritical solvents such as carbon dioxide in which the solute activation energy barrier is lower than in the liquid solvents. Thus one of the important advantages of using a supercritical environment for crystallization is the more volatile and nonpolar nature of SCF combined with flexible control of supersaturation. For example, Shekunov et al. showed in a quantitative study (13) involving both thermodynamic and X-ray measurements that the crystallinity of acetaminophen was highest when the crystals were produced in CO_2-rich supercritical solution above the mixture's critical point, whereas less perfect particles (reduction of crystallinity by about 20%) were obtained in an ethanol-rich environment at higher temperatures and at pressures below the mixture's critical pressure.

1.1.4. Solid Solutions

Solid solutions are produced when foreign molecules are incorporated into a host crystal without grossly altering the host's crystal structure. These foreign or guest molecules may occupy the lattice points (i.e., where the structural units of the host crystal are located) to form a substitutional solid solution, or they may intrude the interstices (nonlattice positions) to form an interstitial solid solution. The phase diagram of a solid solution will normally be characterized by limited miscibility of the solute (guest) with the solvent (host) in the solid state, which is a function of both temperature and pressure. In terms of the possibility of formation, a substitutional solid solution is more restrictive than an interstitial solid solution. To be able to form a substitutional solid solution with the host, the guest molecules must meet certain structural and conformational requirements (i.e., isomorphous with the host) in addition to being comparable to the host molecules in size (14).

The sensitivity of the entropy of fusion ΔS_f of a crystalline substance (the host) to the presence of an impurity or additive (the guest) in solid solution has been quantified in terms of a disruption index (d.i.) (15). Specifically, d.i. is given by $(b - c)$ in the equation

$$\Delta S_f = \Delta S_f^\circ - (b - c)\Delta S_{mix}^{id} \tag{7}$$

where ΔS_{mix}^{id} is the ideal entropy of mixing of the host with the guest and ΔS_f° is the entropy of fusion of the pure host. In Eq. (7), b represents the sensitivity of the entropy of fusion of the solid host to the presence of guest, while c depicts the sensitivity of the entropy of fusion of the liquid host to the presence of guest. Because crystals are highly ordered, their structure is readily disrupted by the host molecules, whereas liquids are already highly disordered, and hence the guest molecules have a much smaller effect (i.e., $b \gg c$). Thus, d.i. largely reflects the sensitivity of the disorder of the host crystals to the presence of the guest molecules, over and above that associated with the simple dilution effect, which is reflected in the ideal entropy of mixing, ΔS_{mix}^{id}. Typically, the d.i. value is a function of solvent environment and process conditions. This is particularly relevant to SFC where the concentration of guest molecules and the kinetic adsorption mechanism vary dramatically with the density of supercritical solution and temperature. It has been shown by Shekunov et al. (see also Section 5.2) that the d.i. value $(b - c)$ of the additive p-acetoxyacetanilide in acetaminophen crystals has a typical value between 10 and 25, indicating considerable disruption of the crystal lattice of acetaminophen due to the presence of the additive (13). This disruption effect was shown to be augmented by adsorption of a liquid cosolvent (such as ethanol) during precipitation, and consequently crystallization conditions favoring a lower solvent uptake will also yield a smaller d.i. value.

1.2. Crystallization Kinetics

It is possible to use the computational tools of quantitative thermodynamics to predict the course of an equilibrium process and to determine the most probable outcome or product formed. Unfortunately, classical thermodynamics does not address time-related questions that fall within the realm of kinetics such as when the process will occur and how rapidly it will proceed. To gain an appreciation of the mechanism of material formation, it is relevant to review some of the basic theories governing the kinetics of crystal growth.

The crystallization of different polymorphs and crystal forms is best understood in terms of nucleation, which is often the rate-determining step (16). In the absence of foreign particles for inducing heterogeneous nucleation, spontaneous homogeneous nucleation can be assumed to occur as a first step in the crystallization process. The rate of homogeneous nucleation J can be expressed as

$$J = \Omega \exp\left(\frac{-\Delta G^*}{kT}\right) \tag{8}$$

where Ω is the frequency or preexponential factor, ΔG^* is the Gibbs free energy of activation, k is the Boltzmann constant and T is the absolute temperature.

The free energy of activation ΔG^*, is given by

$$\Delta G^* = \frac{g v^2 \gamma^3}{(\Delta\mu)^2} \tag{9}$$

where g is a geometric shape factor of the nucleus, v is the molar volume, γ is the interfacial energy, and $\Delta\mu$ is the chemical potential difference between crystal and medium (melt, solution, or vapor).

The preexponential factor Ω can be expressed by (17)

$$\Omega = \left(\frac{D}{d^5 N^*}\right)\left(\frac{4\Delta G^*}{3\pi k T}\right)^{1/2} \tag{10}$$

where D is the diffusion coefficient of the solute molecule, d is the interplanar distance in the crystal lattice, and N^* is the number of molecules forming the critical nucleus.

If it is assumed that equilibrium exists among nuclei of all sizes present in the nucleating medium, Eq. (10) will be reduced to:

$$\Omega = \frac{2D}{d^5} \tag{11}$$

All parameters in Eq. (9) are polymorph dependent except for $\Delta\mu$ at the transition temperature, where the Gibbs free energies of the polymorphs are identical. In addition to the ΔG^* term, the d and N^* parameters in Eq. (10) will also be polymorph dependent. The particular polymorph generated would depend on the relative magnitudes of these polymorph-specific parameters, which govern the overall nucleation rate.

The application of Eq. (8) in polymorphic form prediction requires concurrent consideration of the Ostwald's law of successive stages, which states that a phase change will occur stepwise by way of successively more stable phases. In other words, the least stable (and most soluble) modification nucleates and crystallizes first; this then rearranges stepwise to the more stable forms. The phenomenon can be explained by a surface crystallization mechanism, which considers the high energy polymorph to possess a surplus of uncompensated surface bonds, and therefore, a higher surface energy and a lower specific free energy of the crystal growth step in the perpendicular direction. Since the crystal growth rate at the surface normally increases with an increase in surface energy [γ in pre exponential term, Eq. (10)], the nuclei of the metastable polymorph will grow faster than those of the stable form until near equilibrium conditions are reached, whereupon thermodynamics assumes a controlling role. In addition, the number of molecules in the critical nuclei N^* and the interplanar distance d can be smaller [Eq. (10)], and hence the nucleation rate J, higher [Eq. (8)] for the unstable form, similar to the case with amorphous phase explained in Section 1.1.3 and illustrated in Figure 4. Consequently, the crystallization of the least stable form should dominate at high supersaturations. However, it must be noted that this phenomenon is also dependent on surface–solvent interaction, and requires further investigation using both surface measurements and computer modeling.

2. SOLVENT-MEDIATED POLYMORPHIC TRANSFORMATION IN SUPERCRITICAL FLUIDS

Polymorphic transformation can either occur in the solid state per se or be mediated via a solvent, a liquid melt or an interface. From a kinetic standpoint, the solid state transformation is often the slowest of all possible routes of conversion, since it has to overcome a larger activation energy barrier resulting from the highly restricted molecular motion in the solid state. This activation energy is associated with the nucleation and growth of a new crystal form in the parent phase that involves molecular diffusion and structural change at the interface formed between the new crystal phase and the parent phase. On the other hand, recrystallization of the more stable form at the expense of the less stable phase through melt, solution, or interface has the advantage of reducing the total activation energy considerably more than in the conversion in the solid state.

Among the different modes of polymorphic modification, solvent-mediated transformation is of particular relevance to crystal form control in the pharmaceutical industry, since most, if not all, raw materials used in pharmaceutical manufacture are prepared by batch crystallization from solutions. It has been well documented that the solvent type or nature can greatly influence the rate of polymorphic transformation. For instance, the rate of sulfamerazine transformation has been reported to be higher in the solvent yielding higher solubility and lower in the solvent giving lower solubility (18). It is also lower in the solvent with stronger hydrogen bond acceptor propensity.

Solvent-mediated polymorphic conversion in supercritical fluids has been demonstrated for carbamazpine (9) and deoxycholic acid (19). When treated with supercritical carbon dioxide ($scCO_2$) at 55°C and 350 bar for 6 h, a 1:1 mixture of forms I and III of carbamazepine had its form III content increased to 89% (9). The increase in proportion of form III has been ascribed to the dissolution of form I in supercritical carbon dioxide, followed by recrystallization of the stable form III polymorph. Thermal stress alone would not have induced this conversion (20). Similarly, storage of deoxycholic acid at 60°C for one hour inside a vessel pressurized with $scCO_2$ at 12 MPa resulted in the appearance of new X-ray diffraction peaks, indicative of polymorphic conversion (19).

Since compressed carbon dioxide utilized in SFC is usually moisture free, low water activity or relative humidity is an expected condition in most SFC processes unless modified by the addition of water, and polymorphic conversion in such a relatively moisture-free environment will occur via dehydration if a particular hydrate is involved.

Akao et al. investigated the dehydration of trehalose dihydrate to yield form II under supercritical fluid conditions (21). Trehalose form II is a metastable crystalline form of trehalose anhydrate and can be readily converted into the dihydrate by exposure to a moist environment at room temperature (22). Trehalose form III is another anhydrous polymorph. The phase transition behavior, detected by Fourier transform infrared (FTIR) spectroscopy and confirmed by first-derivative euclidean distance analysis (FDE), was found to be dependent on the extraction time, temperature, and pressure of $scCO_2$. At 20 MPa, an increase in temperature from 70°C to 90°C augmented the dehydration rate of trehalose dihydrate. Thus it appears that a temperature higher than 70°C at 20 MPa is required for dehydration of trehalose dihydrate. The polymorphic forms obtained at different temperatures and pressures are summarized in Table 1.

At 70°C, form II was the major product formed at 10 and 20 MPa. However, the initial dihydrate form was kept unchanged during the extraction experiments at 12, 14, and 18 MPa, irrespective of the operating temperature. Akao et al. attributed this anomalous observation to the nonlinear behavior in density and diffusion coefficients of $scCO_2$ with

TABLE 1 Phase Transition of Trehalose Dihydrate Under
Supercritical Fluid Conditions at Different Temperatures

Temperature at $P = 20$ MPa (°C)	Phase transition	Final crystal form
60	No	Dihydrate
70	Yes	Form II
80	Yes	Mixture of forms II and III
90	Yes	Form III

Source: Adapted from Ref. 22.

increasing pressure (22). Interestingly, the authors used maltose monohydrate as a control, which did not show any significant dehydration under the same dehydration conditions employed for trehalose dihydrate. Apparently, the dehydration of maltose monohydrate was kinetically hindered.

Bettini et al. also conducted a comparative study of the solubility and solid state modifications of three pharmaceutical hydrates versus their respective anhydrates in scCO$_2$ (23). They measured the solubilities of theophylline, carbamazepine, and diclofenac sodium in both hydrated and anhydrous forms in scCO$_2$ under dynamic conditions at low CO$_2$ flux. Measurements were performed at 40°C in the pressure range of 100 to 350 bar. Theophylline monohydrate and carbamazepine dihydrate exhibited a substantially higher solubility in scCO$_2$ (by 4- and 1.5-fold, respectively) than their anhydrate counterparts at 40°C for all pressures tested. Both hydrates underwent at least a partial dehydration during the scCO$_2$ treatment, which may be explained by the interaction between water in the crystals and scCO$_2$. Following dissolution in scCO$_2$, carbamazepine recrystallized as the low-melting (177°C) polymorph. Unlike carbamazepine, theophylline dehydration was due to heating only, since it could be effected simply by exposing the drug hydrate at 40°C in air. Although the solubility of diclofenac sodium tetrahydrate has also been found to be higher than that of the anhydrate at both 100 and 300 bar, the solubility values obtained are questionable because, as a result of the presence of water in the crystal structure, the basic sodium salt can interact with the acidic scCO$_2$, yielding the diclofenac acid (mp 156–158°C) and sodium bicarbonate. These observations clearly demonstrate that the three pharmaceutical hydrates exhibit distinctly different behaviors when tested for solubility in scCO$_2$. Nevertheless, the general rule that solvates are more soluble than the corresponding nonsolvated forms in solvents different from but miscible with that entrapped in the crystal structure has been followed for theophylline and carbamazepine. Diclofenac sodium tetrahydrate represents a special case of physicochemical interaction with the acid supercritical fluid (H_2CO_3), being mediated by crystal water.

3. INFLUENCE OF SUPERCRITICAL FLUID PROCESSING ON CRYSTAL FORMS PRODUCED

While fundamental thermodynamic and kinetic principles constitute the basis for the efficient design and operation of any crystallization process, very few studies have specifically considered how these principles can be utilized to control or optimize material processing in supercritical fluids at elevated pressures. Most of the reported SFC studies focus on the influence of operating parameters on material properties. Though not explicitly stated, a change in any of these parameters represents a change in the crystallization conditions in a thermodynamic and/or kinetic sense. For instance, pressure and temperature are representative factors in thermodynamics, while the flow rate of drug solution and supercritical fluid carries an important element in kinetics. It must be noted that most of the reported SFC works are not the result of well-controlled factorial design, and often they involve concurrent changes of more than one operating variable for individual crystallization experiment. Consequently, it would not always be possible to assess and explain the contribution of each operating parameter in relation to the others based on the available literature data. Nevertheless, gross generalization can still be made with regard to the material formation under defined operating SFC conditions. Table 2 summarizes the operating parameters used for processing specific crystal forms of selected drug materials as examples for illustration.

3.1. Influence of Temperature and Pressure on Polymorphism

All SFC processes operate at above the critical temperature (T_c) of supercritical fluids. Temperature is a critical controlling variable of the SFC process based on both thermodynamic and kinetic considerations. First, solubility is a function of temperature, and this will determine the supersaturation ratio or the driving force for the crystallization of individual polymorphs. Second, the kinetics of polymorphic transformation is governed by the Arrhenius law and is also temperature dependent. The rate constant of the conversion is related to the activation energy and the mass transfer process involved (i.e., diffusion, evaporation, or mixing in supercritical fluids).

Flunisolide was first reported to exist in two different anhydrous polymorphic forms (I and II) and as a hemihydrate with markedly different physicochemical properties, as evidenced by FTIR, as well as by powder x-ray diffraction (PXRD), differential scanning calorimetry (DSC), thermogravimetric analysis (TGA), and thermomicroscopy (24). Form I was obtained by heating either form II or the hemihydrate above 230°C. The solubilities of the three crystal forms rank in descending order of form I > hemihydrate > form II. Subsequent studies using the SEDS process have confirmed the existence of two more polymorphic forms (III and IV) for flunisolide (25). It has been observed that with acetone as the solvent of drug solution, increasing vessel temperature will lead to the formation of form I in the form III sample. This suggests that form I is more readily produced at higher temperature in the acetone medium. On the other hand, when methanol is used as the drug solvent, decreasing vessel temperature will cause form III to emerge in the form IV sample. This indicates that form III is more easily formed at lower temperature in the methanol medium. Despite these reported observations, the thermodynamic relationships between the various flunisolide polymorphs have not yet been established. It is interesting to note the effects observed with increasing flow rate of CO_2. At 100 bar and 80°C, increasing the flow rate of CO_2 from 9 mL/min to 18 mL/min while keeping the flow rate of drug solution (acetone and methanol) constant yielded crystals with characteristics similar to those processed at 100 bar and 60°C. A further increase in the flow rate of CO_2 to 25 mL/min produced a material with similar crystal modification but larger particle size.

Employing the same SEDS crystallization technique and keeping the pressure at 200 bar, two enantiotropic polymorphs (I and II) of salmeterol xinafoate have been separately produced in pure forms simply by varying the vessel temperature (6). The vessel temperature can be tightly controlled to within ±2°C. Form I is obtained at 40 to 65°C, while form II is generated at 70 to 90°C. Thus, the transition temperature between the two polymorphs should lie somewhere between 65 and 70°C at 200 bar.

In parallel with the temperature variable, the operating pressure in SFC is maintained above the critical pressure (P_c) of the supercritical fluids. Pressure can exert a significant influence on the SFC process and the resulting material properties, since it is also an important determinant of drug solubility in supercritical fluids. In addition, having a direct bearing on the density (and supersaturation level) of the fluids, pressure can alter the crystallization mechanism and hence the crystal form produced.

The pressure effects have been demonstrated for several drug materials processed by the SEDS technique. With stavudine as an example (26), increasing the pressure to 120 bar while keeping the other conditions constant increases the proportion of form II produced together with form I, suggesting that higher pressures favor the formation of form II. Similarly, for salmeterol xinafoate (SX), increasing the pressure from 50 bar to above 200 bar while

TABLE 2 Summary of the Effects of Vessel Temperature, Vessel Pressure, and Solvent Choice on Crystal Form Produced

Effects	Operating conditions	Operating variables	Physical forms	Possible control mechanism[a]	Ref.
Influence of temperature					
Flunisolide	Flow rate of scCO$_2$: 9–25 mL/min Flow rate of drug solution: 0.3 mL/min Pressure: 100 bar	Drug solvent: acetone Temperature: 80°C Temperature: 60°C Temperature: 40°C	I + III III III	T and K T and K T and K	25
		Drug solvent: methanol Temperature: 80°C Temperature: 60°C Temperature: 40°C	IV III + IV No products	T and K T and K T and K	
Salmeterol xinafoate	Pressure: 200 bar Drug solution: 4.5% SX in methanol Drug solution flow rate: 2.0 mL/min Flow rate of scCO$_2$: 2000 nL/h	Operating temperatures 40–60°C 70–90°C	I II	T T	6
Influence of pressure					
Salmeterol xinafoate	Temperature: 70°C Drug solution: 4.5% SX in methanol Drug solution flow rate: 2.0 mL/min Flow rate of scCO$_2$: 2000 nL/h	Operating temperatures 50–135 bar 135–250 bar	I II	T T	6

Influence of solvent choice

Fomoterol fumarate	Flow rate of $scCO_2$: 18.0 mL/min Temperature: 40°C Pressure: 150 bar Nozzle opening: 0.2 mm	2.0% (w/v) of drug in 99:1 methanol–water mixtures pumped at 0.3mL/min	Totally amorphous formoterol fumarate	K 31
		2.0% (w/v) of drug in methanol pumped at 0.3 mL/min; totally water-saturated CO_2 flushed through the particle-forming vessel after the end of the run, followed by a rinsing period with dry CO_2 equivalent to two volumes of the vessel	Crystalline formoterol fumarate dihydrate	K
Flunisolide	Flow rate of CO_2: 9–25 mL/min Flow rate of drug solution: 0.3 mL/min Pressure: 100 bar	Drug solvent: acetone	I + III (80°C)	T and K 25
		Drug solvent: methanol	III (60 and 40°C)	T and K
			IV (80°C)	T and K
			III + IV (60°C)	T and K
		Unprocessed	II	
Stavudine	Flow rate of $scCO_2$: 9 mL/min Flow rate of drug solution: 0.2 mL/min Temperature: 35°C	Pressure: 120 bar Drug solvent: 1% (w/v) stavudine in 10% water in isopropanol	III	K 26
		5% water in isopropanol	I and II	K

TABLE 2 Continued

Effects	Operating conditions	Operating variables	Physical forms	Possible control mechanism[a]	Ref.
		Pressure: 90 bar			
		Drug solvent: 0.5% (w/v) stavudine in			33
		7.5% water in isopropanolol	I and II	K	
		5.0% water in isopropanolol	I	K	
		2.5% water in isopropanolol	II	K	
Fluticasone propionate	Drug concentration: 0.5% (w/v)	Solvent utilized in drug solution:			
	Pressure: 100 bar	Acetonitrile	I	K	
	Temperature: 75°C	Acetone	I and II	K	
	Flow ratio of drug solution and CO_2: 0.043	Ethyl acetate	I and II	K	
		Methanol	II	K	
Terbutaline sulfate	CO_2 flow rate: 18.0 mL/min	Solvent: pure ethanol	B	T and K	35
		Temperature: 50°C			
		Pressure: 150 bar			
		Solution flow rate: 2.4 mL/min			
		Solvent: methanol–water	Amorphous	T and K	
		Temperature: 45°C			
		Pressure: 250 bar			
		Solution flow rate: 0.2 mL/min			

Solvent: pure water Temperature: 40°C Pressure: 250 bar Solution flow rate: 0.2 mL/min	Hydrate	T and K	
Solvent: pure methanol Temperature: 40°C Pressure: 250 bar Solution flow rate: 1.2 mL/min	A	T and K	38

Influence of flow rate of drug solution and supercritical fluid

Sulfathiazole
Pressure: 200 bar
Temperatures for crystallization: 0–120°C
Flow rate of CO_2: 10 mL/min
Flow rate of drug solution: 0.2–25.6 mL/min

Drug solvent: acetone Drug solution concentration: 1% w/w	I + amorphous, I + IV, and IV produced at different temperatures and flow ratios (Figure 9)	K	
Drug solvent: methanol Drug solution concentration: 1.5% w/w	I, I + III, III, III + IV, and IV resulted at different temperatures and flow ratios (Figure 10)	K	

[a] T, thermodynamically controlled mechanism; K, kinetically controlled mechanism.

keeping the vessel temperature constant at 70°C affords pure form II, whereas pure form I is obtained below 135 bar (6). At 150 bar, form II together with trace amount of form I is produced. Since the crystallization conditions at 70°C and 135 to 150 bar (with methanol as organic modifier) correspond to the supercritical region, it becomes apparent that supercritical conditions are crucial for the generation of form II, while the subcritical region favors the formation of form I. The emergence of form I in trace amount at 150 bar may be due to the momentary fluctuation of local pressure.

Three crystal forms (two polymorphs and one hydrate) of an oxazolidone antimicrobial, RWJ-337813, were processed by means of the SEDS technique at different pressures (80–200 bar) and temperatures (36–120°C) from dimethylformamide (containing 20 and 30 mg/mL of the drug) (27). For the preparation of the hydrate, water was added to the CO_2 stream at up to 16% by volume. The two polymorphs of RWJ-337813 (forms A and B) generated are anhydrous and monotropic to each other, with A as the stable form. Form C is a hemihydrate generated at 200 bar and 40°C with modified CO_2 flow. Form A was produced (as acicular particles) at 200 bar and 120°C, while form B was obtained (as small platelets) at lower pressure and temperature (80 bar and 36°C).

3.2. Influence of Temperature and Pressure on Chiral Separation

Manipulation of the operating pressure (and temperature) in the SFC process also offers a special advantage for indirect chiral separation (resolution) of different enantiomeric forms of drugs via the formation of diastereomeric salts, as exemplified by ephedrine using the SEDS technique (28). Equal molar amounts (0.79 mmol) of (1R,2S)-ephedrine, (1S,2R)-ephedrine and (R)-mandelic acid were dissolved in methanol (40 mL), and the resulting solution pumped (at 0.2 mL/min) together with compressed CO_2 (9 mL/min) in the vessel. Operating temperatures were varied between 35 and 75°C and pressure between 100 and 350 bar. Complete resolution of the (1S,2R)-ephedrinium–(R)-mandelate [(+)-E-(–)-MA] and (1R,2S)-ephedrinium–(R)-mandelate [(–)-E-(–)-MA] was achieved by taking advantage of the solubility differences in different solvent systems and at different operating pressures and temperatures. Thus, when methanol was used as solvent and the process operated at 35°C and 300 bar, (–)-E-(–)-MA was preferentially crystallized while (+)-E-(–)-MA, being more soluble than (–)-E-(–)-MA in CO_2, remained in solution form. Lowering the pressure to 100 bar and replacing methanol with tetrahydrofuran caused (+)-E-(–)-MA to crystallize in $scCO_2$ in preference to (–)-E-(–)-MA. Since (+)-E-(–)-MA is less soluble in tetrahydrofuran, the same solution concentration is more saturated and precipitates more easily when CO_2 is added as an antisolvent.

Racemic mixtures of the two chiral forms can also be obtained from methanol by varying the operating pressure and temperature. Increasing the pressure (or density) of CO_2 while keeping the vessel temperature constant at 35°C results in increased resolution of (–)-E-(–)-MA. The results indicated that (+)-E-(–)-MA is much more soluble in CO_2 than (–)-E-(–)-MA. Hence, at higher pressures, the corresponding density change of the scCO$_2$ is providing more "solvent" for dissolving (+)-E-(–)-MA, thus resulting in higher enantiomeric excess.

In addition to pressure, temperature affects the chiral separation. By maintaining the density constant at 0.713 g/cm^3 through applying a consecutively higher pressure at elevated temperature, a linear decrease in resolution with temperature has been obtained. The loss in resolution can be explained by a reduced solubility difference between the diastereomers with increasing temperature, leading to less efficient resolution. Moreover, the methanol solution of the solute is less saturated at elevated temperatures, which makes crystallization of the diastereomers from the solution more difficult, thereby affording a lower product yield.

As with ephedrine, chiral resolution can be achieved for 2,2'-ninaphthyl-1,1'-diamine (DABN) (29) through diastereomer crystallization. Equal molar quantities (0.19mmol) of (R)-DABN, (S)-DABN, and (R)-camphorsulfonic acid (CSA) were first dissolved in methanol (40 mL). The resulting solution was pumped at low flow rates together with CO_2 (10 mL/

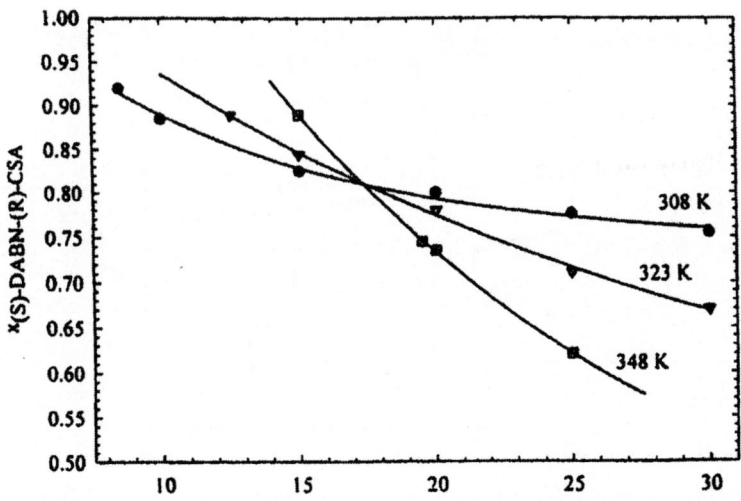

FIGURE 5 Plot of enantiomeric resolution of (S)-DABN–(R)-CSA versus operating pressures at various defined temperatures. (From Ref. 29.)

min) into a crystallization vessel. Operating temperatures and pressures were varied between 308 and 348 K and between 8.5 and 30 MPa, respectively. Both pressure and temperature are important parameters affecting the chiral resolution of (S)-DABN–(R)-CSA. Figure 5 shows the pressure dependence of (S)-DABN–(R)-CSA resolution for a given temperature. Maximum resolution was achieved when the pressure approached the critical pressure of the binary CO_2–solvent system. Figure 5 depicts the dependence of the resolution on temperature for a given pressure. The intersection in Figure 6 indicates that resolution increases with increasing temperature at pressures lower than 17.5 MPa and decreases at higher pressures. In contrast to the pressure dependence, the temperature dependence of resolution exhibited a linear relationship.

3.3. Influence of Solvents on Polymorphism and Hydrate Formation

Much of the literature information concerning the influence of solvent on polymorph crystallization is derived from the conventional crystallization system. The effect of solvent appears to be dictated by kinetics rather than by thermodynamics (30). The solvent may act by selective adsorption to certain crystallographic faces of the polymorphs, thereby inhibiting their nucleation or retarding their growth to the advantage of others. Similar solvent behavior

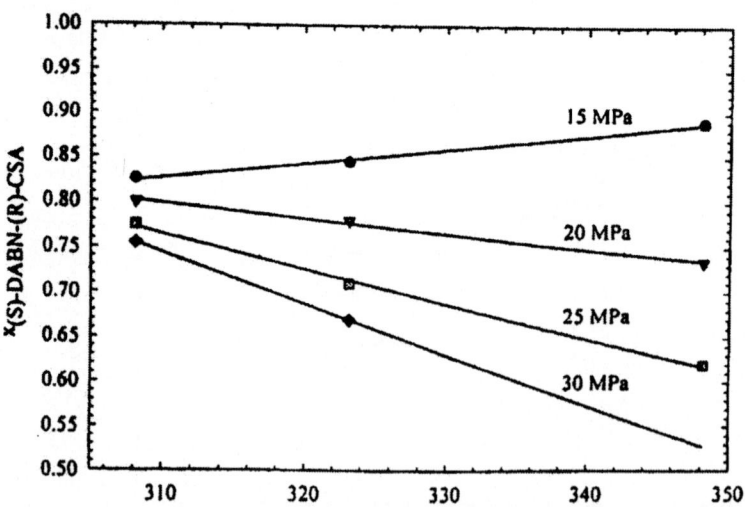

FIGURE 6 Plot of enantiomeric resolution of (S)-DABN–(R)-CSA versus operating temperature at various defined pressures. (From Ref. 29.)

may also be attributed to the SFC process and to the fact that the solvent density can also change the solvent polarity and crystallization mechanism.

In the processing of formoterol fumarate by means of the SEDS technique (31), totally amorphous material was obtained when dry carbon dioxide was used as the antisolvent, while the use of water-saturated carbon dioxide afforded a mixture of amorphous formoterol fumarate and crystalline dihydrate form.

Fluticasone propionate, a steroidal drug, is known to exist in two polymorphic forms, namely, forms I and II. Form I can be readily obtained by crystallization from a variety of solvents. Standard spray-drying techniques can afford only form I. Form II, obtainable only by SFC, has a PXRD pattern completely different from that of form I (32,33). The relative amounts of forms I and II produced can be controlled by using different drug solvents. For example, form I is obtainable from pure acetonitrile while form II can be crystallized from pure methanol. The use of either acetone or ethyl acetate will yield a mixture of forms I and II.

In the case of flunisolide, different polymorphs have also been produced in the SEDS process when either acetone or methanol was used as the drug solvent. With acetone as the solvent, forms I and III were produced, while the use of methanol as solvent afforded predominantly forms III and IV (25).

Stavudine is another drug whose crystal form is sensitive to the particular solvent employed. The drug exists in three different crystal forms (designated as forms I, II, and III) with different solubilities (26). Forms I and II are anhydrous polymorphs, whereas form III is a hydrate and is pseudo-polymorphic with forms I and II. Of these three forms, form I is the most stable and shows no transformation to any of the others. Pure form I can be produced in supercritical fluids from a mixture comprising it and at least one of forms II and III (26). Different polymorphs of the drug can also be crystallized by varying the operating pressure. In addition, the amount of water present in the solvent is an important determinant of the crystal form produced, particularly for the hydrate. The hydrated form III can be crystallized by keeping the pressure at 120 bar and the water content at 10%. Decreasing the water content to 5% will result only in a mixture of forms I and II. At 90 bar, the percentage of water in isopropanol–water mixtures determines the relative proportion of forms I and II.

As mentioned before, carbamazepine displays at least four polymorphs and a dihydrate. SEDS crystallization from dichloromethane yielded α-carbamazepine at low temperature and mixtures of α- and γ-carbamazepine at high temperature. SEDS crystallization from methanol also afforded three pure polymorphs (α-, β-, and γ-carbamazepine) under defined operating conditions of temperature, pressure, and solution flow rates (7). The dependence of crystal form on the SEDS processing conditions is shown in Figures 7

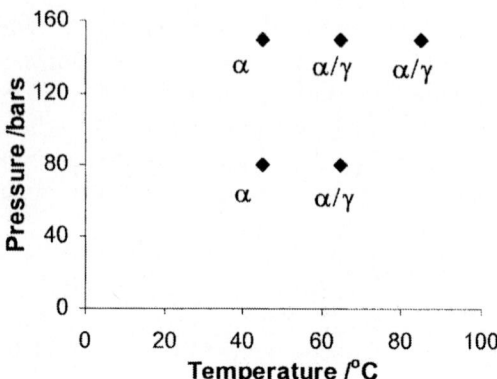

FIGURE 7 Carbamazepine polymorphs processed by SEDS at different operating temperatures and pressures using dichloromethane as solution solvent and a flow rate of 0.5 mL/min. (From Ref. 7, Copyright © 2001 Wiley-Liss Inc. and American Pharmaceutical Association.)

and 8. The α form, the least stable polymorph, was formed at low temperature and high supersaturation level, which is consistent with Ostwald's rule of successive stages and indicative of a process dominated by kinetics rather than thermodynamics. In addition, both dichloromethane and CO_2 are apolar solvents, which favor the formation of α-carbamazepine. However, at higher temperatures but with similar supersaturation level, a mixture of α- and

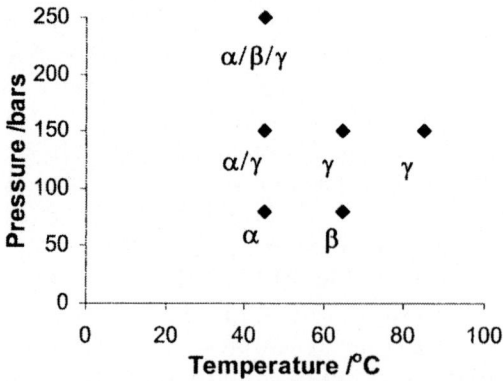

FIGURE 8 Carbamazepine polymorphs processed by SEDS at different operating temperatures and pressures using methanol as solution solvent and a flow rate of 0.5 mL/min. (From Ref. 7, Copyright © 2001 Wiley-Liss Inc. and American Pharmaceutical Association.)

γ-carbamazepine was obtained. On the other hand, pure γ-carbamazepine was generated from methanol at high temperatures and low supersaturation levels. Since α-carbamazepine is a monotrope of γ-carbamazepine, and γ-carbamazepine is an enantiotrope of β-carbamazepine, the physical stability hierarchy depends on the transition temperature of β- and γ-carbamazepine. Equilibrium solubility in propan-2-ol at 25°C of different carbamazepine polymorphs followed the descending order of α form (12.0 mg/mL) > γ form (10.5 mg/mL) > β form (8.5 mg/mL) (34). Pure β-carbamazepine was formed under conditions of partial miscibility, involving transient formation of a methanol-rich phase, from which crystallization occurred slowly.

Depending on the solvent employed, terbutaline sulfate, an antiasthmatic agent, was shown to yield various physical forms when processed by the SEDS technique. Pure methanol solvent afforded form A at 40°C and 250 bar, while aqueous methanol and pure water generated the amorphous form and the hydrate respectively. Form B, the most stable modification, was obtained with pure ethanol at 50°C and 150 bar (35).

3.4. Influence of Flow Rate of Drug Solution and Supercritical Fluid on Polymorphism

As discussed before, the flow rate of drug solution and supercritical fluid is also an important factor governing the crystal form produced because of its relation to the rate at which the supersaturation level builds up in the crystallization vessel. However, the significance of this factor in crystal form control is often not realized in SFC studies owing to the relative dominance of other major operating factors involved such as temperature and pressure.

Sulfathiazole, a potent antibacterial compound, has been reported to exist in five polymorphic forms (36). Crystals prepared by conventional crystallization and SEDS process have been compared (37,38). Conventional crystallization from acetone yields pure form IV, while growth from methanol yields a mixture consisting predominantly of form IV with a small amount of form III. SEDS crystallization from acetone yields amorphous material, form I, and form IV at different temperatures and flow ratios of CO_2 and acetone (Figure 9). Kordikowski et al. commented that the kinetic factor, rather than the operating temperature of the supercritical fluid mixture, is more important for controlling the formation of polymorphs and amorphous form, and this can be accomplished by varying the flow rate ratio of CO_2 and acetone (38). On the other hand, SEDS crystallization from methanol yields forms I, III, and IV. Production of these various polymorphs can also be achieved by simply varying the operating temperature. The three polymorphs are consistent with those reported in the literature and can be separately obtained in pure form by adjusting the temperature for all flow rates of methanol (Figure 10).

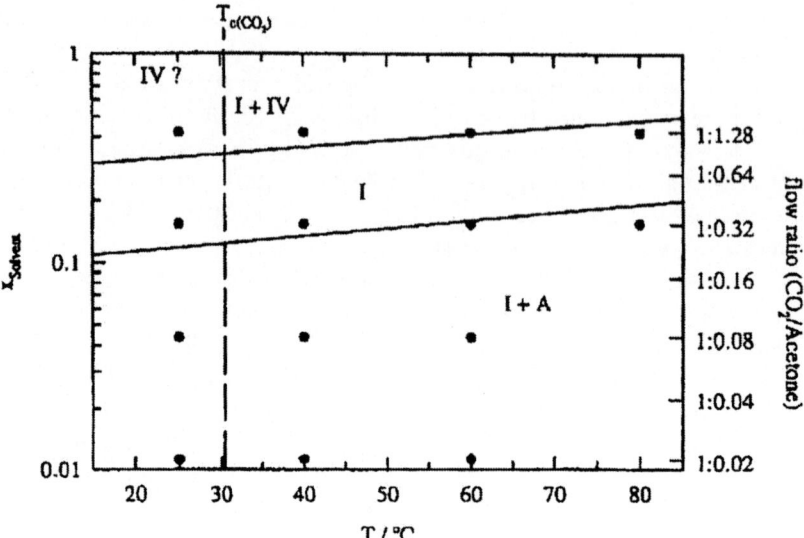

FIGURE 9 Sulfathiazole polymorphs processed from acetone by SEDS at different temperatures and solution flow rates. (From Ref. 38.)

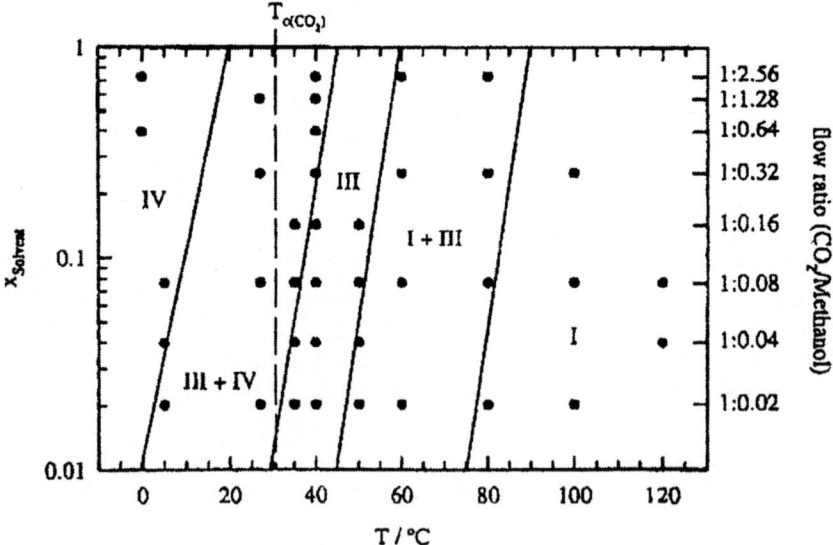

FIGURE 10 Sulfathiazole polymorphs processed from methanol by SEDS at different temperatures and solution flow rates. (From Ref. 38.)

Kordikowski et al. inferred that SEDS crystallization of sulfathiazole polymorphs from methanol is controlled primarily by thermodynamics, since
varying the flow rate of methanol modified the particle size, but not the crystal
form of sulfathiazole (38).

4. FORMATION OF COMPOSITE MATERIALS WITH ADDITIVES

The significant roles of additives in controlling the properties of materials
including the crystal form have been well recognized by material scientists.
Though most of the related information is derived from the conventional
crystallization system, the underlying mechanisms of crystal modification
involving additives should be similar for the SFC process.

Depending on the similarity in size and structure to the host molecules,
guest additive (impurity) molecules may be taken up into the host crystal
lattice to form solid solutions (see Section 1.1.4). In cases of additive molecules that do not satisfy the general incorporation criteria for solid solution
formation (e.g., for large polymers of distinctly different chemical structures),
forced coprecipitation of the additives with the host molecules will lead to the
formation of solid dispersions. Such dispersion systems are highly unstable
thermodynamically and will revert back to the stable state, resulting in phase
separation.

An important practical case during SFC is coprecipitation of structurally similar compounds, for example, reaction by-products, chiral molecules,
additives, and impurities. Such precipitation is important when the aim is to
produce composite crystals or drug–excipient mixtures or, on the contrary, to
separate impurities from a solid product. The similarity of molecular structures ensures that such compounds strongly interact with each other, thus
increasing the likelihood for solid solutions to form. Strong solid–solid
intermolecular interactions typically result in separation problems and in
significant variations of the solid state and particulate properties.

4.1. Influence of Low Molecular Weight Additives and Synthetic Impurities

Physical mixtures (microcomposites) of pharmaceutical compounds, such as
acetaminophen, ascorbic acid, urea, and chloramphenicol, have been
obtained by using antisolvent precipitation, and the effects of pressure on
the separation of these drugs have also been investigated (39).

Physical mixtures of powders have also been produced by using the
rapid expansion of supercritical solution (RESS) process technique (40,41).
Twelve different mixtures of drugs and impurities were evaluated in a study

employing RESS for powder production and $scCO_2$ as the recrystallizing solvent (42). A physical mixture containing 80% drug and 20% impurity was loaded into a reaction vessel in which the solutes were dissolved in $scCO_2$ to form a homogeneous supercritical solution. The SCF region investigated for solute extraction ranged from 35 to 100°C and 1100 to 9000 psi. Nucleation of the solutes was induced by rapid expansion to atmospheric conditions. All the drug–impurity mixtures studied exhibited a general reduction in crystallinity. Extreme cases involving a substantial loss of crystallinity and subsequent amorphous conversion have also been observed with the following three mixture systems: piroxicam–benzoic acid, indomethacin–salicylic acid, and phenytoin–caffeine. Additionally, a number of other significant physical changes were evident in specific mixtures. These include habit modification in salicylic acid–aspirin mixture, polymorphic conversion in tolbutamide–urea mixture, and hydrate formation in theophylline–caffeine mixture. In another RESS study utilizing a similar solvent system, chlorpropamide crystals were doped with different levels (1–50% w/w) of urea, a model impurity (43). The incorporation of urea into the crystal lattice of chlorprop-amide polymorph C brought about a substantial reduction (up to 50%) in the molar heat of fusion, while causing only minor shifts in the melting point. Rapid and uniform nucleation implicit in the RESS process appeared to be responsible for trapping the impurity in the crystal lattice of the drug, thereby causing a lattice strain and hence a reduction in heat of fusion.

In a study investigating the solid solution formation by precipitation in $scCO_2$, acetaminophen (Figure 11; X = OH) was doped with its structurally related synthetic impurity, p-acetoxyacetanilide (PAA) (Figure 11; X = $OCOCH_3$) (13,44). PAA is a synthetic impurity and a suggested prodrug of acetaminophen (45). This study demonstrated that incorporation of PAA into the acetaminophen crystal lattice and modification of the acetamino-phen–PAA crystals are defined by two major competitive factors, namely, the solubility of PAA at a given pressure and temperature (the thermodynamic factor), and the surface molecular structure of the acetaminophen crystals in supercritical solution (the kinetic factor). Accordingly, the uptake of PAA

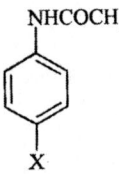

FIGURE 11   Chemical structure of a para-substituted acetanilide.

(Figure 12) followed the trend whereby the amount of incorporated PAA correlated with the solubility of PAA in CO_2. Thus, an increase of the PAA concentration in solution was observed to result in a progressive increase of the PAA content in the precipitated material. However, the amount of incorporated PAA was considerably greater for the samples obtained at 90 bar and 80°C (353 K), even though the solubility of PAA in CO_2 at these conditions is higher than at 90 bar and 40°C (313 K). Even small concentrations of PAA (<5% w/w in acetaminophen) showed a profound effect on the crystal surface structure, which could be explained by a thermodynamic surface transition effect occurring in the CO_2–ethanol solvent at higher temperatures and in the presence of PAA additive. Analysis of this surface transition phenomenon in terms of the dispersive and specific polar interactions derived from inverse gas chromatography data (see Section 5.3) showed that PAA, a weaker proton donor than acetaminophen, decreased the specific polar interactions and increased the dispersive component of the surface free energy of coprecipitated samples, resulting in changes in particle size and shape. A comprehensive thermodynamic and structural analysis done in this work reveals the formation of a secondary acetaminophen–PAA

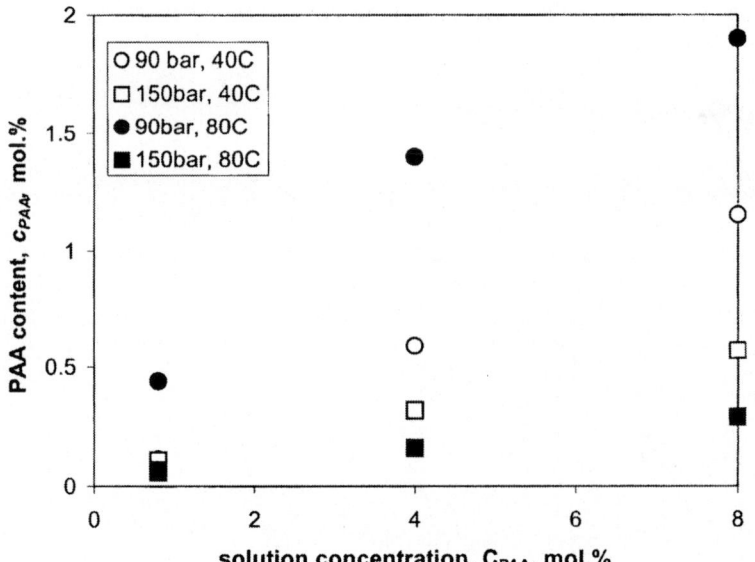

FIGURE 12 Dependence of C_{PAA}, the PAA content in the solid phase, on concentration of additive in solution, C_{PAA}, at various pressures and temperatures. (From Ref. 15, Copyright © 2003, American Chemical Society.)

phase with a deformed crystal lattice, which is characterized by a significant increase of entropy compared with pure acetaminophen crystals that display smaller changes in enthalpy and elastic energy contributions. In conclusion, SFC of acetaminophen–PAA composite crystals illustrates that separation of structurally similar compounds can be achieved thermodynamically by adjusting the density of supercritical fluids where the materials have the greatest solubility difference. However the PAA uptake also depends on the kinetics of PAA adsorption by the surfaces of growing crystals, which needs to be taken into account when one is optimizing the SFC purification or coprecipitation process. PAA exerts a profound effect on the precipitation kinetics, being able to modify the solid state properties of the resulting acetaminophen–PAA particles, as substantiated by quantitative PXRD and computer modeling studies.

4.2. Influence of High Molecular Weight Polymers

Apart from doping with impurities, coformulation with polymers by using supercritical fluid processing to produce materials in different physical forms has been investigated. Crystallinity of a cyclo-oxygenase-2 (COX-2) inhibitor, (Z)-3-[1-(4-chlorophenyl)-1-(4-methanesulfonyl)methylene]-dihydrofuran-2-one) [Cpd (I)], can be regulated by coformulating with polymers in the SEDS process (46). Hydroxylpropyl-cellulose (HPC) and Poloxamer 237 (P-237) [poly(oxypropylene)–poly(oxyethylene) block copolymer] were used. The

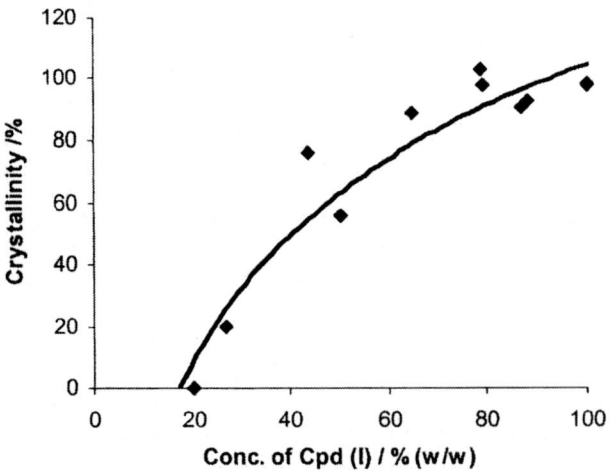

FIGURE 13 Plot of percent crystallinity of SEDS processed compound (Cpd I) versus its proportion in HPC mixtures. (From Ref. 46.)

results for both polymer systems indicate that crystallinity is significantly reduced as the polymer content increases (Figures 13 and 14). The reduction is nearly linear for the P-237 system, but for HPC, a threshold polymer content of at least 20% w/w is required for initiating a drop in crystallinity. For both systems, a 100% amorphous product was achieved at drug loading of 20% w/w or less.

Coprocessing with polymers in supercritical fluids has also been reported for carbamazepine (47). The drug and carriers, poly(ethylene glycol) (PEG) 8000 with either Gelucire 44/14 or vitamin E TPGS NF (d-α-tocopherol PEG 1000 succinate), were dissolved in a minimum amount of methanol. Solid dispersions were produced at around 40°C and 135 bar. The crystal structure of carbamazepine was found to be form I (trigonal or α form) for all polymers studied. Supercritical fluid processing of carbamazepine alone under the same conditions also yielded the same polymorph. Pure carbamazepine processed in supercritical fluids alone was found to have a faster dissolution profile than that of commercial micronised carbamazepine (form III; monoclinic or β form), possibly because of its better wetting property. The supercritically processed coprecipitates were dry and free flowing and had carbamazepine mostly entrapped within the polymers. The presence of form I crystals in these dispersions is consistent with their improved dissolution profiles, since the form I morphology has been shown to display a faster dissolution rate than that of form III.

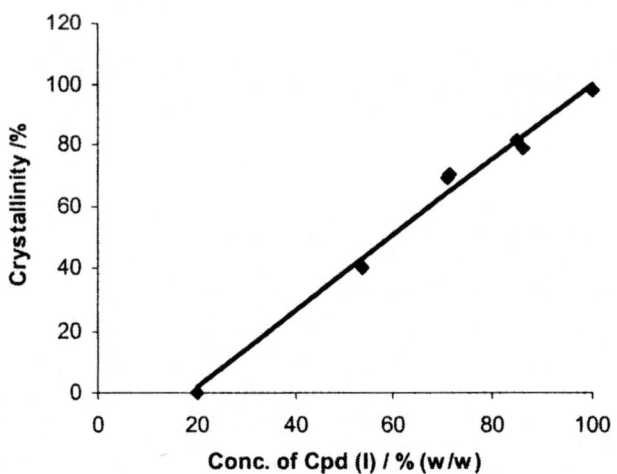

FIGURE 14 Plot of percent crystallinity in SEDS processed compound (Cpd I) versus its proportion in P-237 mixtures. (From Ref. 46.)

5. CHARACTERIZATION OF PHARMACEUTICAL MATERIALS PROCESSED IN SUPERCRITICAL FLUIDS

As discussed in the preceding sections, a significant number of reported studies have attested to the versatility of SFC for preparing a wide range of solid pharmaceuticals, notably, polymorphs, hydrates, and solid solutions. Most of these studies focused on defining the crystallization conditions for particular crystal forms and their structural identification. Few of them have considered the physicochemical and surface properties of the resulting materials in sufficient detail to allow proper assessment of their potential utility in pharmaceutical formulations and to provide a better understanding of the SFC process. Since the properties of materials are intimately related to their growth conditions, such elaborate material characterization will afford an indication on how well the conditions have been controlled during the crystallization process and how the material is formed. A thorough understanding of the kinetic and thermodynamic factors governing SFC is critical for the prediction of the resulting material properties. In the following sections, salmeterol xinafoate (Figure 15) will be presented as an example to illustrate and highlight the advantageous aspects of the SFC process with regard to the control of polymorphic purity, crystallinity, and surface energetics of the material produced.

The salmeterol xinafoate (SX) polymorphs I and II are probably among the very few SFC-processed drug materials that have been intensively characterized. On an industrial scale, SX is produced as polymorph I in a granulated form by adding a hot solution of the drug in 2-propanol to a chilled quench solvent (48). The granulated SX material (GSX) is then subjected to micronization to yield the micronized form (MSX) for subsequent formulation work. While SX form I can be readily obtained by conventional crystallization methods, production of form II has thus far been

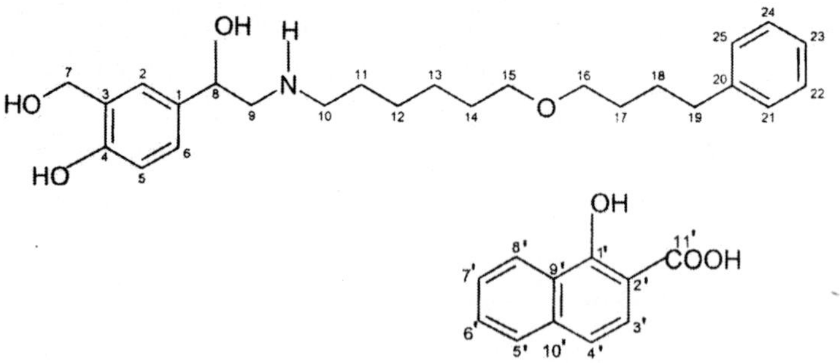

FIGURE 15 Chemical structure of salmeterol xinafoate.

possible only with the SFC technology. Tong et al. conducted a series of physicochemical and surface characterization studies on SX polymorph I (SX-I) and polymorph II (SX-II) crystallized from supercritical fluids under defined conditions and on a reference MSX sample prepared by conventional means (5,49,50). These studies employed a variety of established techniques, including PXRD, FTIR, DSC, TGA, hot stage microscopy (HSM), solid state ^{13}C-nuclear magnetic resonance (NMR) spectroscopy with cross-polarization (CP) and magic angle spinning (MAS), solubility measurements (SMs), Brunauer—Emmett–Teller (BET) nitrogen adsorption, and inverse gas chromatography (IGC) (5,6,50). The salient results of these studies (summarized in Table 3) clearly demonstrate marked differences in crystal structure and physicochemical properties between the SFC-processed form I and form II materials.

Apart from the structure-specific differences in physicochemical properties between the two SX polymorphs shown in Table 3, there exist subtle differences in purity, crystallinity, and surface energetics between the supercritically processed SX-I and conventionally prepared MSX materials, which are elaborated in the sections that follow.

5.1. Polymorphic Purity

Polymorphic purity of solid drug substances is an important parameter for consideration in pharmaceutical formulation. Since different polymorphs or crystalline forms of the same drug exhibit different physical properties, chemical stability, solubility, dissolution rate, and possibly bioavailability, the presence of the alternative (metastable) crystal form(s) may have an adverse impact on the manufacturing and in vivo performance of the drug product.

Polymorphic impurities or contaminants are prone to form if the polymorphic transition temperature can be readily attained under ordinary conditions. The polymorphic conversion could arise during the batch crystallization process when a local rise in temperature causes the transition temperature to be exceeded momentarily, resulting in partial formation of the alternative crystal form. It could also be induced by subsequent processing treatments (e.g., grinding, milling) that generate sufficient heat to cause the temperature to rise above the transition point. In addition to the temperature factor, the presence of trace polymorphic impurity (in embryo or nucleus form) can greatly accelerate the conversion process by lowering the associated activation energy barrier. Regulatory authorities have long recognized the need for limiting polymorphic impurities in pharmaceutical materials. However, the analytical tools available for solid state characterization (e.g., PXRD, DSC) are so low in sensitivity that normally they do not lend

TABLE 3 Solid State Properties of Various Salmeterol Xinafoate Samples

Properties (techniques used)	MSX (reference form I material)	SEDS-processed samples	
		SX-I	SX-II
X-ray diffraction pattern (PXRD)	Similar to SEDS-processed SX-I	A sharp peak at $2\theta = 4.158°$ and multiple peaks at $2\theta = 9-26°$	A sharp peak at $2\theta = 2.853°$, and a few relatively indistinct peaks at $2\theta = 9-26°$
Infrared spectrum (FTIR)	Similar to SEDS-processed SX-I	Obvious band shift in the OH stretching region (at 3308 cm^{-1}) and NH (broad peak at 3000–2273 cm^{-1})	Significant band shift in the OH stretching region (at 3426, 3294 cm^{-1}) and NH (larger broad peak at 3000–2273 cm^{-1})
NMR spectrum (^{13}C-NMR CP/MAS)	Similar to SEDS-processed SX-I	Significant chemical shifts relative to SX-II (≥ 1.50 ppm) for C-11′, C-2′; C-4, C-8, C-7; C-15, C-16 (see Figure 15 for chemical structure)	See comments for SX-I
Thermal behavior and melting point (DSC/TGA)	A melting endotherm at 122.7°C, followed immediately by a large recrystallization exotherm and then by a large melting endotherm at 137.6°C (Figure 16)	A large endotherm at 122.7°C, followed by a small endotherm at 137.6°C (Figure 16)	A single melting endotherm at 137.6°C (Figure 16)
Enthalpy of fusion (DSC)	Comparable to SEDS-processed SX-I	68.3 kJ/mol	42.0 kJ/mol
Entropy of fusion (DSC)	Comparable to SEDS-processed SX-I	172.5 J mol^{-1} K^{-1}	102.3 J mol^{-1} K^{-1}
Enthalpy of solution (SMs)	Comparable to SEDS-processed SX-I	32.3 kJ/mol	27.7 kJ/mol

themselves to the quantification of low levels (<5%) of polymorphic impurity. Thus, the search for a more sensitive technique for such quantification remains a challenge to the formulation scientists.

As reported earlier, the expected polymorphic transition of SX from form I to form II at ~99°C (estimated by solubility measurements) was not observable in DSC without vigorous grinding treatment, a result that can be explained by the presence of a relatively high activation energy barrier against the transition (5). Unlike the conventional MSX sample that underwent rapid transformation to form II in DSC through recrystallization from its melt, supercritically processed SX-I material showed very little of such conversion, as reflected by the much reduced melting endotherm of form II at 137.6°C (Figure 16). GSX, the granulated material used to prepare MSX, also exhibited melting and recrystallization behaviors similar to those of MSX. Despite the observed differences in thermal behavior, all the three SX form I samples (i.e., GSX, MSX, and SX-I) displayed essentially the same powder X-ray diffraction pattern, ^{13}C solid state NMR/CP-MAS spectrum, FTIR spectrum, and solubility–temperature dependence. All these observations suggest that the supercritically processed SX-I is much less prone to poly-

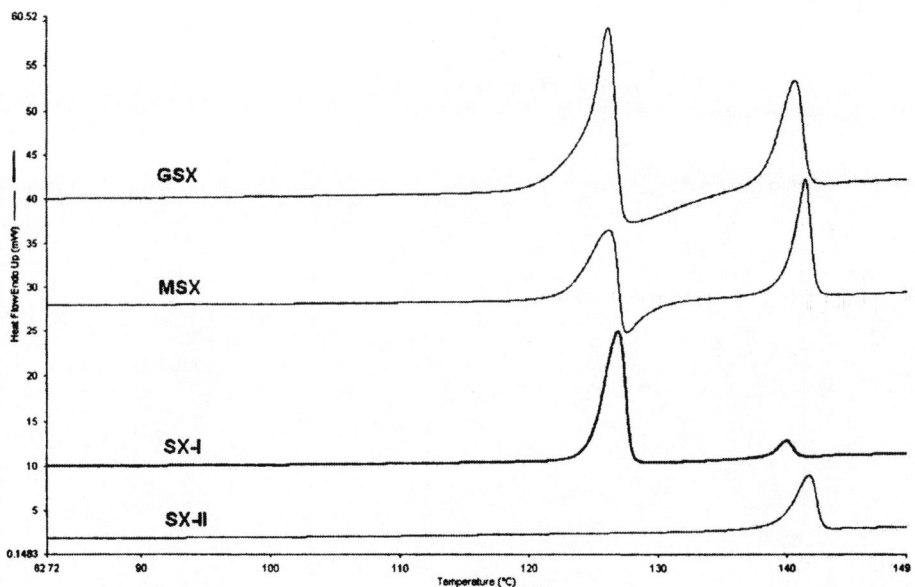

FIGURE 16 DSC profiles of GSX, MSX, SX-I, and SX-II at 10°C/min. (From Ref. 50.)

morphic conversion than are MSX and GSX samples prepared by conventional methods, likely due to its relative freedom from trace form II nuclei.

A method based on quantitative phase analysis and the instantaneous nucleation model has been developed to quantify form II nuclei in both GSX and MSX samples (50). The method involves measurement of the recrystallization rate of form II from the form I melt in DSC at different scanning speeds, and analysis of the data (expressed as α-time curves) by appropriate kinetic models, as explained next.

Assuming in DSC that ΔT, the temperature range of recrystallization, is constant, the time allowed for the recrystallization, t, can be calculated from

$$\Delta T = \beta t \tag{12}$$

where β is the scanning speed.

Equation (12) shows that t is inversely related to β, provided ΔT remains constant. Thus, by measuring the fraction of melt recrystallized (α) at different scanning speeds, an α-time curve can be constructed, as shown in Figure 17. For each form I sample, α was calculated by normalizing the enthalpy of fusion ΔH_f of the second peak (endotherm of form II) obtained at each scanning speed by that of a reference pure form II material (i.e., SFC-processed SX-II) measured at the same scanning speed.

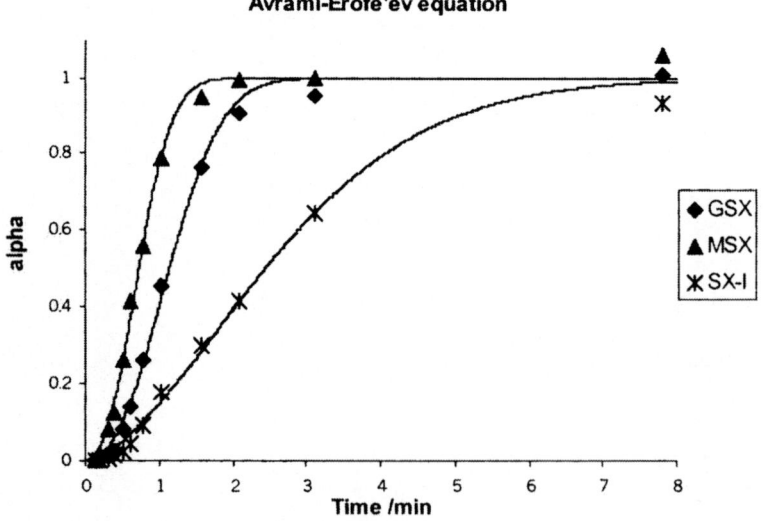

FIGURE 17 Three α-time curves (GSX, MSX, and SX-I) fitted by the Avrami–Erofe'ev rate equation. (From Ref. 50.)

The α-time curves were then fitted to the following classical Avrami–Erofe'ev (AE) rate equation, which has been widely applied to the analysis of nucleation and crystallization kinetics

$$[-\ln(1-\alpha)]^{1/n} = kt \tag{13}$$

where α is the fraction of recrystallized material, k is the crystallization rate constant, t is the crystallization time, and n is the model exponent, normally an integer between 0 and 3, denoting zero-, one-, two-, and three-dimensional nucleation and growth (51).

Subtle differences in the mechanisms of nucleation and crystal growth among the samples can be revealed through comparison of the n and k values obtained. It should be noted that the n values are close to 2 for all the SX samples (data not shown), which is expected because SX particles have a platelet shape and therefore grow predominantly in two dimensions.

The major assumption made in the present quantitative impurity analysis is the model of instantaneous nucleation, which is characterized by extremely rapid onset and is consistent with a relatively small n value (≈ 2 for SX) (52). The homogeneous and heterogeneous nucleation processes can proceed simultaneously. However, the former can occur only in the bulk material of molten SX, whereas the latter is initiated by contact with the surface of the form II seeds present. The number of nuclei formed by heterogeneous nucleation is proportional to the surface area of the metastable phase, and will be proportional to x, the weight fraction of SX-II, if the same specific surface area is assumed for both preexisting nuclei and SX-II particles added (as in physical mixtures). Therefore N_T, the total number of nuclei formed at $x \ll 1$ can be expressed by

$$N_T = (1-x)N_{HON} + xN_{HEN} \tag{14}$$

where N_{HON} and N_{HEN} correspond to the number of nuclei formed per mole (or unit weight) of melt for homogeneous and heterogeneous nucleation, respectively.

To quantify the amount of preexisting nuclei in GSX and MSX, it would be necessary to find a suitable mathematical relationship that relates the (total) rate constant k, computed by the AE model-fitting procedure to the mole fraction of SX-II. Equation (15) follows from the instantaneous nucleation model (52)

$$k = \left(\frac{c_g N_T}{V}\right)^{1/n} G_c \tag{15}$$

where V is the specific molar volume, c_g is the shape factor of SX crystal and G_c is the growth constant.

Combining Eqs. (14) and (15) result in the following expression for k

$$k = [C_1(1 - x) + C_2 x]^{1/n} \tag{16}$$

where C_1 and C_2 are constants related to the homogeneous and heterogeneous contributions, respectively. It should be noted that some other nucleation models, for example, progressive nucleation (52), will also give a relationship similar to Eq. (16) but with different coefficients.

Equation (16) was then employed to construct a calibration curve for calculating impurity level by using physical mixtures of SX-II with SX-I in the appropriate composition range (0–14% w/w in this case) (Figure 18). Since the n values determined for all form I samples (GSX, MSX, and SX-I) and physical mixtures were close to 2, the k values for all SX materials were re-computed by iterative curve fitting with the n parameter fixed at 2. The results are presented in Table 4. The refitted k values at $n = 2$ were subsequently used to compute the coefficients C_1 and C_2, in Eq. (16). Based on their respective k values and $n = 2$, GSX and MSX were found to contain 0.16 and 0.62% w/w of metastable nuclei within their crystal matrices.

The foregoing approach serves well for assessing the polymorphic purity of SX materials and can potentially be applied to other enantiotropic pairs showing similar thermal behavior.

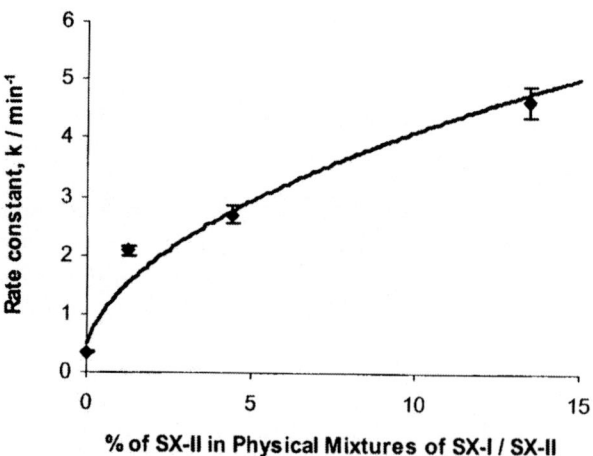

FIGURE 18 Relationship of weight percent of SX-II and k values computed from the Avrami–Erofe'ev equation with $n = 2$. (From Ref. 50.)

TABLE 4 Parameters of the α-Time Curves Fitted by Avrami–Erofe'ev Rate Equation with n Fixed at 2 for All Samples $[- \ln (1 - \alpha)]^{1/2} = k t$

	k $(\text{min}^{-1})^a$	r^2
0.00% of SX-II in SX-I (pure SX-I)	0.348 (0.011)	0.990
1.24% of SX-II in SX-I	2.072 (0.087)	0.937
4.41% of SX-II in SX-I	2.700 (0.144)	0.877
13.47% of SX-II in SX-I	4.645 (0.265)	0.820
GSX	0.724 (0.020)	0.992
MSX	1.135 (0.029)	0.993

[a] Standard deviations are shown in parentheses.
Source: Ref. 50. Reproduced with permission of Plenum Publishing Corporation.

5.2. Crystallinity

Different crystallization processes or process conditions lead to varying degrees of disorder in the form of crystal defects and amorphous regions referred to by the collective term "crystallinity." Such "irregularities" of the crystal structure are often undesirable because of their adverse impact on both the stability and consistency of pharmaceutical ingredients. However, partially crystalline and amorphous forms are much more soluble than highly crystalline drug substances and therefore can be used to promote therapeutic activity in certain cases. Quantification of crystallinity is an important consideration in the controlled modification of drug substances and their solid state stability in dosage forms. According to the U.S. Pharmacopeia, crystallinity is determined by the fraction of completely crystalline material in the mixture (two-state model). The one-state model, which is physically more realistic, incorporates the concept of a gradual decrease of crystallinity with no sharp distinction between the completely crystalline (100% crystallinity) and amorphous (0% crystallinity) states (53). The degree of crystallinity is frequently characterized by DSC or water sorption (%RH) techniques. An alternative approach involves the measurement of changes in entropy, ΔS, between processed and reference samples either from enthalpy of fusion or enthalpy of solution (15). These entropy changes are related to the solid state disorder created by the crystallization process or other processing treatments. However, the most consistent and direct definition of crystallinity is based on PXRD analysis.

Shekunov et al. assessed the crystallinity of SFC-processed SX-I and the reference GSX and MSX materials. High resolution PXRD was applied to determine the characteristic diffraction peaks of SX for distinguishing crystallographic changes (peak shift and peak broadening) of these materials (54). The experiments were conducted at the Synchrotron Radiation Source (SRS) Daresbury Laboratory, which affords an accuracy of 10^{-4} degree and allows determination of the instrumental broadening (i.e., shape of diffraction peaks related to the instrument geometry). The latter is required for absolute determination of the physical diffraction line broadening caused by distribution of domain size and strain in particles. Analysis of the diffraction patterns (Figure 19) showed that peak fitting, background determination, and indexing could be performed more consistently and with higher accuracy by using SFC-produced samples than by using MSX or GSX material. This

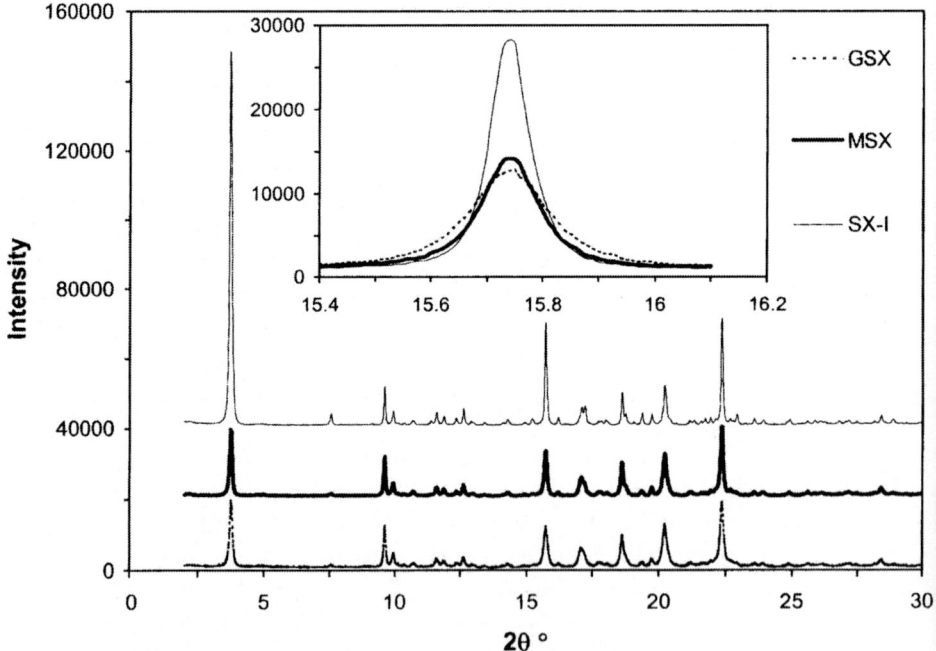

FIGURE 19 X-ray diffraction patterns of salmeterol xinafoate samples: granulated (GSX), micronized (MSX), and SFC-produced (SX-I) obtained at the wavelength 1.4 Å. Insert allows comparison between characteristic $(1\bar{1}3)$ diffraction peaks for these samples. (From Ref. 54.)

observation can be explained by the smaller diffraction line broadening and higher signal-to-noise ratio for the SFC-processed SX-I sample.

Physical broadening of diffraction peaks, a cumulative measure of crystal lattice imperfections, is grouped into size effects (grains, small-angle boundaries, stacking faults) or strain defects (point defects and dislocations). The basis for discrimination between these effects lies in the fundamental principle that the size term does not change with the diffraction angle, whereas the strain does (55). Comparison of the diffraction peaks (Figure 19) indicates that the peak broadening is much smaller for SFC-processed sample. The "double-Voigt" method (55), which approximates the physically broadened line profile as a convolution of Cauchy (C) and Gauss (G) integral breaths, contains the contributions of size (S) and strain (distortion, D) according to the equations:

$$\beta_C = \beta_{CS} + \frac{\beta_{CD}s^2}{s_0^2} \tag{17}$$

$$\beta_G^2 = \beta_{GS}^2 + \frac{\beta_{GD}^2 s^2}{s_0^2} \tag{18}$$

The integral breaths, β, are given in reciprocal angstroms (Å^{-1}) of the wave vector, s:

$$s = \frac{2\sin\theta}{\lambda} \tag{19}$$

$$\beta = \frac{\beta^*\cos\theta}{\lambda} \tag{20}$$

Here, β_{CD}/s_0^2 and β_{GD}/s_0^2 are constant for the whole pattern (taken at zero value of the wave vector s_0 of the first peak). The integral breaths found from the experimental diffraction profiles allow calculation of the surface-weighted (D_S) and volume-weighted (D_V) domain size and a mean-square (Gaussian) strain, ε, which is the total strain averaged over infinite distance:

$$D_S = \frac{1}{2\beta_{CS}} \tag{21}$$

$$D_V = \frac{(EF)\exp(\beta_{CS}^2/\beta_{GS}^2)}{\beta_{GS}^2} \tag{22}$$

$$\varepsilon = \frac{\beta_{GD}}{\left(s_0\sqrt{2\pi}\right)} \tag{23}$$

Here EF denotes a complementary error function related to the background determination (55).

The results of computation using Eqs. (17) to (23) are presented in Table 5. Clearly, SFC-processed SX-I material consists of larger crystalline

TABLE 5 The Integral Breaths of Voigt-Approximated Diffraction Profile, β and Calculated Values of the Surface-Weighted (D_S) and Volume-Weighted (D_V) Domain Size and the Gaussian Strain ε

	SX-I	MSX	GSX
$\beta_{CS},\ \text{Å}^{-1}$	0.11×10^{-2}	0.13×10^{-2}	0.15×10^{-2}
$\beta_{CD},\ \text{Å}^{-1}$	0.30×10^{-5}	0.13×10^{-4}	0.21×10^{-4}
$\beta_{GS},\ \text{Å}^{-1}$	0.39×10^{-3}	0.40×10^{-3}	0.38×10^{-3}
$\beta_{GD},\ \text{Å}^{-1}$	0.63×10^{-4}	0.97×10^{-4}	0.12×10^{-3}
$D_S,\ \text{Å}$	453 ± 44	382 ± 60	326 ± 50
$D_V,\ \text{Å}$	785 ± 60	682 ± 86	606 ± 74
ε	$(0.53 \pm 0.02) \times 10^{-3}$	$(0.83 \pm 0.03) \times 10^{-3}$	$(0.101 \pm 0.006) \times 10^{-2}$

Source: Ref. 54. Reproduced with permission of KONA Editorial Secretariat.

domains (by about 20%) and smaller strain (by about 60%) than both MSX and GSX samples. It should be emphasized that SX-I shows higher crystallinity than both MSX and GSX. Thus the results of this work indicate that "aggressive" crystallization procedures in liquid solvents may also lead to crystal disordering, which is comparable to or even greater than the defects induced by micronization.

A very important factor that influences crystallinity is the presence of impurities and solvent uptake. SFC helps to significantly reduce the amounts of both impurities and solvent incorporated into the solid state. The disruption of crystal lattice was found to correlate with the amount of ethanol present as inclusions or PAA incorporated as solid solution in acetaminophen crystals (13). The observed reduction of the fusion enthalpy, ΔH_f, and increase of the fusion entropy, $\Delta S_f = \Delta H_f / T_m$, were related to the disorder in the crystal lattice. This work has shown that the uptake of ethanol into acetaminophen crystals during processing in supercritical CO_2 was substantial below the mixture's critical pressure where the liquid and vapor phases coexist (see Chapter 3 Baldyga et al.). Above the critical pressure, all compositions of ethanol and CO_2 are completely miscible to form a single homogeneous phase, and the ethanol uptake is substantially reduced with increasing pressure. Similar observations were obtained for the N,N-dimethylformamide (DMF) solvent (56). Rehman showed that different degrees of crystallinity, from amorphous to highly crystalline materials, could be obtained for terbutaline sulfate particles depending on the amount of cosolvent (ethanol) in CO_2 and the regime of crystallization (35). Enhancement of crystallinity for drugs obtained by SFC compared with micronized samples or materials crystallized from liquid solvents was reported for other compounds (25,57–59). However it should be noted that precipitation in SFC is usually associated with very high supersaturations, which tend to reduce the crystallinity according to

Eq. (6). Therefore factors such as level of supersaturation, solvent composition, presence of impurities, pressure, and temperature should all be taken into account when one is assessing the combined effect of the processing conditions on crystallinity. Materials of different degrees of crystallinity can be obtained predictively by changing the regime of SFC.

5.3. Surface Thermodynamic Properties

Another advantageous property of SFC-processed material is its relatively low surface energy, which is closely associated with its high crystallinity or surface-mediated transformation in supercritical fluid media (see Sections 2 and 5.2). As alluded to before, micronization is often applied to particle size reduction of powders prepared by conventional batch crystallization. In addition to the problem of crystallinity reduction, such treatment may generate amorphous domains and induce static charges at the particulate surfaces, leading to increases in surface energy, particle aggregation, and moisture sorption. Of all the techniques available for powder surface characterization, inverse gas chromatography (IGC) is probably the most pragmatic and attractive one that has captured considerable attention in recent years. IGC offers a number of advantages over other techniques: it is nondestructive to the surface, measures only the outermost layers of molecules, and has a resolution down to very small area. IGC is normally conducted at infinite dilution and involves injecting trace vapor of a liquid polar or nonpolar probe into a silanized (surface-deactivated) glass column packed with the sample of interest. The retention time (which reflects the affinity of the probe for the powder surface) measured can be readily converted into various surface energy parameters.

The ability of SFC to regulate particulate surface energy can be illustrated by using SX materials processed in supercritical fluids under defined conditions as an example. In this regard, the SFC-processed SX materials (SX-I and SX-II) and the reference MSX sample have been subjected to surface analysis by IGC at infinite dilution at four different temperatures (30, 40, 50, 60°C) (49). Such measurements at multiple temperatures enable determination of the enthalpic and entropic contributions to the surface free energy of adsorption, which can afford additional insight into the nature and strength of surface interactions. Details on the calculation and interpretation of IGC-derived surface energy data are presented next.

5.3.1. Free Energy of Adsorption and Related Thermodynamic Parameters

The experimental parameter measured in IGC for the adsorption of probes on the stationary phase (SX in this case) inside the glass columns is the retention time of the probes, which can be converted into the retention volume by the relationship (60)

$$V_N = jF(t_r - t_0) \tag{24}$$

where V_N is the net retention volume, j is a correction factor taking into account gas compressibility, F is the carrier gas flow rate, t_r is the retention time of the probe, and t_0 is the void retention time.

The standard free energy of adsorption, ΔG_A, of the probe on the SX sample can be calculated from V_N using the relationship (60)

$$-\Delta G_A = RT \ln \left(\frac{V_N P_0}{S g B_0} \right) \tag{25}$$

where T is the column temperature, R is the gas constant, P_0 is the reference partial pressure of the probe, S is the specific surface area of the sample, g is the weight of sample, and B_0 is the reference bidimensional spreading pressure of the adsorbed probe film on the sample. The reference state of de Boer, where $P_0 = 1.013 \times 10^5$ Pa and $B_0 = 3.38 \times 10^{-4}$ N/m is commonly adopted in the calculation (61).

Equation (25) may also be expressed in the form

$$-\Delta G_A = RT \ln (V_N) + C \tag{26}$$

where C is a constant encompassing the choice of the standard state for ΔG_A and the surface area of the sample.

To a first approximation, the free energy of adsorption is related to the work of adhesion, W_A, by the equation

$$-\Delta G_A = Na W_A + K \tag{27}$$

where N is Avogadro's number, a is the surface area of the probe, and K is a constant that has been introduced to account for the choice of the standard state of ΔG_A.

The enthalpy and entropy of adsorption can be calculated from the surface free energy of adsorption based on the thermodynamic relationship (62)

$$\Delta G_A = \Delta H_A - T\Delta S_A \tag{28}$$

where ΔG_A, ΔH_A and ΔS_A respectively, are the surface free energy, enthalpy, and entropy of adsorption of the probe on the sample, and T is the absolute temperature of columns. Values of ΔH_A and ΔS_A can be obtained from the slope and intercept of the linear plot of $\Delta G_A/T$ against $1/T$, assuming that both ΔH_A and ΔS_A remain invariant over the temperature range of interest.

The surface free energies of adsorption $(-\Delta G_A)$ of MSX, SFC-processed SX-I, and SX-II for both the nonpolar and polar probes were

TABLE 6   Surface Free Energy, Enthalpy, and Entropy of Adsorption of MSX, SX-I, and SX-II

Probes	Samples	$-\Delta G_A$ (kJ/mol at 40°C)[a]	ΔH_A (kJ/mol)[a]	ΔS_A (J k^{-1}mol^{-1})[a]
Pentane	MSX	10.39 (0.17)	−31.33 (0.59)	−67 (2)
	SX-I	10.38 (0.72)	−30.99 (2.42)	−66 (7)
	SX-II	16.15 (0.27)	−16.68 (0.54)	−1 (1)
Hexane	MSX	13.03 (0.12)	−36.58 (0.49)	−75 (2)
	SX-I	12.86 (0.72)	−35.96 (2.05)	−74 (6)
	SX-II	18.46 (0.22)	−22.75 (0.36)	−13 (1)
Heptane	MSX	15.66 (0.11)	−41.91 (0.04)	−84 (0)
	SX-I	15.27 (0.70)	−41.31 (1.35)	−83 (3)
	SX-II	20.77 (0.21)	−29.02 (0.67)	−26 (2)
Octane	MSX	18.33 (0.10)	−47.45 (0.36)	−93 (1)
	SX-I	17.71 (0.70)	−46.71 (1.35)	−93 (3)
	SX-II	23.08 (0.22)	−34.86 (0.41)	−37 (1)
Nonane	MSX	—	—	—
	SX-I	20.22 (0.69)	−51.66 (0.24)	−100 (2)
	SX-II	25.35 (0.20)	−38.57 (1.45)	−42 (5)
Dichloromethane	MSX	—	—	—
	SX-I	11.94 (0.86)	−19.71 (1.79)	−24 (8)
	SX-II	21.10 (0.22)	−20.76 (0.43)	2 (1)
Chloroform	MSX	14.06 (0.17)	−36.72 (0.93)	−72 (3)
	SX-I	13.19 (0.78)	−35.96 (1.74)	−73 (4)
	SX-II	23.16 (0.21)	−25.56 (1.00)	−7 (3)
Acetone	MSX	14.14 (0.13)	−43.59 (1.93)	−94 (7)
	SX-I	13.46 (0.49)	−41.55 (5.84)	−90 (18)
	SX-II	19.95 (0.13)	−24.14 (1.98)	−12 (6)
Ethyl acetate	MSX	16.40 (0.13)	−50.17 (0.58)	−108 (2)
	SX-I	14.96 (0.24)	−45.48 (1.93)	−98 (6)
	SX-II	21.94 (0.21)	−28.14 (0.44)	−19 (2)
Diethyl ether	MSX	12.91 (0.13)	−49.95 (0.36)	−118 (2)
	SX-I	11.65 (0.36)	−38.57 (2.13)	−86 (7)
	SX-II	17.82 (0.21)	−21.15 (0.66)	−10 (2)
Tetrahydrofuran	MSX	16.08 (0.16)	−45.80 (0.51)	−95 (2)
	SX-I	14.77 (0.45)	−43.37 (1.60)	−91 (4)
	SX-II	22.40 (0.20)	−24.28 (0.15)	−5 (1)

[a] Standard deviations are shown in parentheses.
Source: Ref. 49. Reproduced with permission of Plenum Publishing Corporation.

calculated from Eq. (25) and the corresponding enthalpy and entropy of adsorption calculated using Eq. (28). Since the ΔG_A data determined at various temperatures showed the same trend for all three SX samples, only one representative set of data (i.e., data obtained at 40°C) are presented (along with the ΔH_A and ΔS_A data) for illustration in Table 6. Similarly, for the other surface free energy parameters [γ_S^D and ΔG_A^{SP}; Eqs. (30) and (34)] determined at multiple temperatures, only the measurements taken at 40°C are shown in their respective tables (Tables 7 and 8).

As indicated in Table 6, the $-\Delta G_A$ values of the three SX samples obtained for all nonpolar probes in the alkane series followed the order SX-II > MSX ≈ SX-I, suggesting that the adsorption of the alkane probes is energetically similar toward MSX and SX-I, but thermodynamically more favorable on SX-II. Analysis of the temperature dependence of ΔG_A for the SX samples revealed that the MSX and SX-I had statistically equivalent ΔH_A and ΔS_A and hence comparable $-\Delta G_A$, whereas SX-II had a less negative ΔH_A and a much less negative ΔS_A than MSX and SX-I, the net result being a higher $-\Delta G_A$ for SX-II (Table 6). In other words, the more favorable adsorption of the nonpolar probes on SX-II is driven by a considerably reduced loss in surface entropy (disorder or molecular mobility).

The measured ΔS_A values of MSX, SX-I, and SX-II are of the same order of magnitude as the predicted entropy changes (based on ideal gas behavior) for adsorption of heptane, octane, and nonane (-52.4, -53.0, -53.5 J k^{-1} mol^{-1}, respectively, at 20°C) reported in the literature (63). The higher than predicted entropy loss of the alkanes upon adsorption on MSX and SX-I may be explained by the loss of one degree of translational freedom and by a restriction of rotational and vibrational freedom on the surface. In contrast, the entropy loss associated with the adsorption of alkane probes on SX-II was much lower than predicted, suggesting that the adsorbed probe molecules still retain much of their mobility on the surface of SX-II. The less negative ΔH_A obtained with SX-II, which indicated less heat being evolved

TABLE 7 Dispersive Components of Surface Free Energy of Adsorption and Related Thermodynamic Properties of MSX, SX-I, and SX-II

Dispersive energy components	MSX[a]	SX-I[a]	SX-II[a]
γ_S^D (mJ/m^{-2} at 40°C)	38.29 (0.91)	32.48 (0.25)	28.56 (1.11
S_S^D (mJ/m^{-2} K^{-1})	0.253 (0.037)	0.237 (0.033)	0.300 (0.016
H_S^D (mJm^{-2})	117.33 (12.65)	106.67 (10.20)	122.05 (5.97

[a] Standard deviations are shown in parentheses.
Source: Ref. 49. Reproduced with permission of Plenum Publishing Corporation.

TABLE 8 Specific Components of Surface Free Energy of Adsorption of MSX, SX-I, and SX-II

Specific energy components	Dichloromethane[a]	Chloroform[a]	Acetone[a]	Ethyl acetate[a]	Diethyl ether[a]	Tetrahydrofuran[a]
$-\Delta G_A^{SP}$ (kJ/mol at 40°C)						
MSX	—	0.81 (0.05)	4.56 (0.10)	4.00 (0.01)	2.77 (0.04)	3.61 (0.03)
SX-I	2.81 (0.15)	0.15 (0.08)	3.80 (1.17)	2.71 (0.61)	1.49 (0.39)	2.45 (0.28)
SX-II	6.11 (0.17)	4.50 (0.13)	4.47 (0.15)	4.02 (0.11)	1.86 (0.06)	4.42 (0.14)
ΔH_A^{SP} (kJ/mol)						
MSX	—	0.37 (1.13)	−14.02 (2.75)	−14.81 (0.99)	−19.25 (1.11)	−10.31 (0.88)
SX-I	8.38 (2.54)	0.61 (0.41)	−12.30 (8.10)	−10.61 (1.95)	−8.24 (1.17)	−8.36 (1.24)
SX-II	−7.33 (0.04)	−2.25 (0.77)	−9.38 (1.64)	−6.81 (0.53)	−5.11 (0.37)	−2.79 (0.36)
ΔS^{SP} (J k^{-1} mol^{-1})						
MSX	—	4 (4)	−30 (9)	−35 (3)	−52 (3)	−21 (3)
SX-I	36 (9)	2 (9)	−27 (25)	−26 (5)	−21 (3)	−19 (5)
SX-II	330 (578)	317 (536)	311 (562)	325 (577)	319 (569)	332 (565)

[a] Standard deviations are shown in parentheses.
Source: Ref 49. Reproduced with permission of Plenum Publishing Corporation.

through bond formation during the adsorption, may be explained by weaker interactions resulting from a reduced number of nonpolar binding sites and the increased mobility of the probe molecules.

For the polar probes, the $-\Delta G_A$ values of the three SX samples decreased in the following order: SX-II > MSX > SX-I, although MSX showed only a marginally higher $-\Delta G_A$ than SX-I (Table 6). Similar to the case with the nonpolar probes, SX-II afforded a less negative ΔH_A and a substantially less negative ΔS_A than MSX or SX-I (Table 6). These findings again suggest that a much reduced loss in entropy (or increased molecular mobility) rather than a major loss in enthalpy is the driving force for the more favorable adsorption of the polar probes on SX-II, which can be linked to the less stable and more disordered form II structure. However, unlike the case with the nonpolar probes, the ΔH_A and ΔS_A values of MSX for the polar probes exhibited some differences from those of SX-I, although only the data obtained for ethyl acetate and diethyl ether were statistically significantly different. For these probes, MSX had a significantly more negative ΔH_A but only a marginally more negative ΔS_A than SX-I. Thus it would appear that the slightly higher $-\Delta G_A$ is contributed mainly by a higher $-\Delta H_A$, possibly resulting from an increased number and strength of high energy binding sites. As discussed earlier, micronization can increase the surface free energy of SX by introducing structural defects or by exposing more polar groups (i.e., binding sites) at the particle surface (5). The suggested presence of crystal imperfections in MSX has been confirmed by high resolution PXRD (see Section 5.2).

5.3.2. Dispersive Component of Surface Free Energy and Related Thermodynamic Parameters

For adsorption of nonpolar (alkane) probes involving purely dispersive forces, the work of adhesion is given by:

$$W_A = 2(\gamma_S^D)^{1/2}(\gamma_L^D)^{1/2} \qquad (29)$$

where γ_S^D is the dispersive component of surface free energy of the sample and γ_L^D is the dispersive component of surface free energy of the liquid probes.

Combining Eqs. (26), (27) and (29) affords the equation

$$RT \ln V_N = 2aN(\gamma_S^D)^{1/2}(\gamma_L^D)^{1/2} + K_c \qquad (30)$$

where the constant K_c takes into account the choice of the standard state of ΔG_A and the surface area of the sample.

A plot of $RT \ln V_N$ against $a(\gamma_L^D)^{1/2}$ according to Eq. (30) yields a linear slope of $2N(\gamma_S^D)^{1/2}$, from which the dispersive component of surface free energy of adsorption, γ_S^D, can be determined (60).

As with Eq. (28), the sample's surface free energy G_S is related to its surface enthalpy H_S and surface entropy S_S as follows (64):

$$G_S = H_S - TS_S \tag{31}$$

Since the sample remains physically unchanged during the experiment, G_S may be equated to γ_S, and Eq. (31) can be expressed as

$$\gamma_S = H_S - TS_S \tag{32}$$

If only dispersive forces are involved, Eq. (32) may be rewritten as

$$\gamma_S^D = H_S^D - TS_S^D \tag{33}$$

where H_S^D and S_S^D are the dispersive components of surface enthalpy and surface entropy, respectively.

The surface free energies for the dispersive component, γ_S^D, of the MSX, SX-I, and SX-II samples at various temperatures were calculated from Eq. (30), and the corresponding surface enthalpy (H_S^D) and surface entropy (S_S^D) were calculated by using Eq. (33); Table 7 presented the γ_S^D values obtained at 40°C together with the H_S^D and S_S^D data.

The γ_S^D values determined from the whole nonpolar alkane series were the largest for MSX, followed in decreasing order by SX-I and SX-II (Table 7), which was the reverse sequence expressed by the $-\Delta G_A$ values (Table 6). According to Eqs. (27), (29), and (30), the trend of $-\Delta G_A$ should closely parallel that of γ_S^D, provided the constant term K in Eq. (27) (whose magnitude depends on the reference standard state used for ΔG_A) is zero or equivalent for all the SX samples. However, while the plot of $-\Delta G_A$ versus $a(\gamma_L^D)^{1/2}$ (or carbon number) afforded excellent linearity ($R^2 > 0.9999$) for all the SX samples, the intercept value was statistically indistinguishable for MSX and SX-I, but considerably higher for SX-II than for MSX and SX-I (by about -7.1 and -6.1 kJ/mol, respectively, at 30°C) (Figure 20). This substantial additional $-\Delta G_A$ change for SX-II, which is not explicable by purely dispersive interactions, is likely a result of the increased mobility of the probe molecules on the less stable and more disordered SX-II surface, as discussed earlier. Thus although the SX-II showed the largest $-\Delta G_A$, its γ_S^D was the smallest, reflecting a relatively low contribution of the dispersive (nonpolar) forces on a predominantly polar and relatively disordered surface. That MSX has a higher γ_S^D than the SFC-processed SX-I may be explained by the less crystalline or more defective surface of the MSX sample, as alluded to earlier.

Examination of γ_S^D in terms of the enthalpic and entropic contributions revealed that the H_S^D and S_S^D were the highest for SX-II, followed by MSX and then by SX-I (Table 7). This is consistent with the fact that the SX-II is the

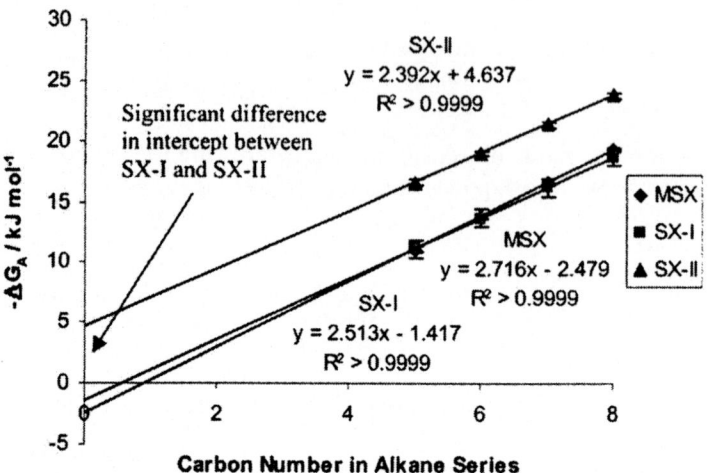

FIGURE 20 Surface free energy of adsorption $-\Delta G_A$, for non-polar probes in alkane series at 30°C. (From Ref. 6.)

metastable polymorph at ambient temperature and has a higher surface enthalpy and entropy than SX-I, while the SFC-processed SX-I is more crystalline than MSX and is characterized by a lower surface enthalpy and entropy. The relatively low γ_S^D observed for SX-II is mainly related to its higher surface entropy.

5.3.3. Specific Interactions and Associated Acid–Base Properties

Polar probes have both dispersive and specific components of surface free energy of adsorption. The specific component of surface free energy of adsorption (ΔG_A^{SP}) is determined by subtracting the dispersive contribution from the total free energy of adsorption, and can be obtained from the vertical distance between the alkane reference line [Eq. (30); Figure 21] and the polar probes of interest according to the following equation (60):

$$-\Delta G_A^{SP} = RT\ln\left(\frac{V_N}{V_N^{ref}}\right) \tag{34}$$

As before, ΔG_A^{SP} is related to enthalpy ΔH_A^{SP} and entropy ΔS_A^{SP} of specific interactions in adsorption by

$$\Delta G_A^{SP} = \Delta H_A^{SP} - T\Delta S_A^{SP} \tag{35}$$

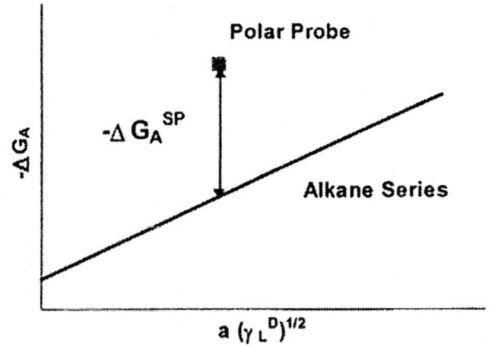

FIGURE 21 Schematic diagram illustrating the calculation of $-\Delta G_A^{SP}$ for a polar probe. (From Ref. 6.)

The determined $-\Delta G_A^{SP}$ values at 40°C together with ΔH_A^{SP} and ΔS_A^{SP} obtained by linear regression of $\Delta G_A^{SP}/T$ against $1/T$ for MSX, SX-I, and SX-II were summarized in Table 8.

The polar probes can be described in terms of the Gutmann electron donor and electron acceptor numbers (DN and AN). DN defines the basicity or electron donor ability of a probe, while AN defines the acidity or electron acceptor ability. The term AN*, introduced by Fowkes (65), is a more appropriate quantity than AN to use, since the former is corrected for the contribution from the dispersive forces. The enthalpy of adsorption for specific interactions can be related to DN and AN* as follows (60):

$$-\Delta H_A^{SP} = K_A DN + K_D AN^* \tag{36a}$$

or

$$-\frac{\Delta H_A^{SP}}{AN^*} = K_A \left(\frac{DN}{AN^*}\right) + K_D \tag{36b}$$

Here K_A and K_D are numbers describing the acid and base characteristics of the solid (SX).

The values of DN and AN* of the liquid probes used can be obtained from literature (66,67). Plotting $-\Delta H_A^{SP}/AN^*$ against DN/AN^* yields a straight line, where K_A and K_D correspond to slope and intercept, respectively. The K_A and K_D values calculated from Eq. (36b) for MSX, SX-I, and SX-II are tabulated in Table 9.

As shown in Table 8, SX-II displayed considerably higher $-\Delta G_A^{SP}$ values than SX-I for all polar probes, reflecting more thermodynamically favorable adsorption of these probes on SX-II. The considerably higher $-\Delta G_A^{SP}$

TABLE 9 Acid and Base Parameters of MSX, SX-I, and SX-II

Acid and base parameters	MSX[a]	SX-I[a]	SX-II[a]
K_A	0.399 (0.039)	0.432 (0.050)	0.102 (0.019)
K_D	5.295 (0.786)	0.213 (0.891)	2.077 (0.267)
K_D / K_A	13.271	0.493	20.363

[a] Standard deviations are shown in parentheses.
Source: Ref. 49.

observed with the adsorption of polar acidic probes (dichloromethane and chloroform) on SX-II suggests that the surface of SX-II exhibits predominantly specific basic interactions. As before, the free energy change was further analyzed in terms of enthalpy and entropy. The ΔS_A^{SP} values determined from ΔG_A^{SP} at various temperatures were highly variable, particularly for the SX-II sample, while the calculated ΔH_A^{SP} data generally showed acceptable consistency (i.e., reasonably low standard deviations) for statistical comparison. The observed data variability with ΔS_A^{SP} and, to a slight degree, ΔH_A^{SP} is mainly due to the involvement of multiple calculation steps in arriving at these data. These steps include the following (a) construction of the alkane reference line based on retention volume data (ln V_N) of the nonpolar probes, (b) subtraction of the ln V_N data of the polar probes from the corresponding ln V_N values for the nonpolar probes on the reference line to afford ΔG_A^{SP}, and (c) linear regression of $\Delta G_A^{SP}/T$ against $1/T$ to obtain ΔH_A^{SP} and ΔS_A^{SP} [see Eq. (35)]. Each step can introduce significant errors, particularly the second step, where the difference of ln V_N measured constitutes a relatively small percentage of the respective ln V_N data. All these errors would be cumulatively reflected in the final results. Thus, while the high $-\Delta G_A^{SP}$ of SX-II appeared to be mostly the contribution of its high ΔS_A^{SP}, the dominance of the entropy factor in the adsorption process could not be statistically verified owing to the considerable variability of the ΔS_A^{SP} data. Despite the difficulty in obtaining consistent ΔS_A^{SP} data for SX-II, the estimation of ΔH_A^{SP} was sufficiently robust to enable reliable comparison of the samples in terms of the nature of the specific interaction (i.e., dominant repulsive or attractive forces for the specific polar groups) involved in the adsorption. As already established, the total enthalpy of adsorption (ΔH_A) of the polar probes on the samples is negative (Table 6). This follows from the fact that the total ΔG_A (Table 6) for any spontaneous adsorption process is negative and the ΔS_A is necessarily negative according to Eq. (28) (Table 6) because the probe molecules in the vapor phase should have a larger entropy than those adsorbed on the crystal surfaces. However, the specific (polar) part of the enthalpy calculated using Eq. (35) can be either negative or positive. As presented in Table 8, both SX-I and SX-II exhibited negative ΔH_A^{SP} for the polar amphoteric (acetone and ethyl acetate) and basic (diethyl ether and tetrahydrofuran) probes, suggesting attractive specific interactions. In addition, the ΔH_A^{SP} of SX-I was more negative than that of SX-II, indicating stronger attractive forces of adsorption for these probes on SX-I. However, for the polar acidic probes, the ΔH_A^{SP} values determined for MSX (with chloroform) and SX-I (with chloroform and dichloromethane) were positive, although the overall free energy change was still negative, implying that the adsorption involves repulsive forces, possibly between similar chemical groups carrying like charges (i.e., between acidic groups in this case).

The foregoing observations suggest that the surface of SX-I involves predominantly acidic forces (from acidic groups), whereas the SX-II surface is characterized mainly by basic forces (from basic groups). Compared with the SFC-processed SX-I, MSX exhibited a higher $-\Delta G_A^{SP}$ for all the polar probes that was attributable to its more negative ΔH_A^{SP}, since the ΔS_A^{SP} values were statistically indistinguishable for all the polar probes (except for the basic diethyl ether probe). The more negative ΔH_A^{SP} values with MSX relative to the SFC-processed SX-I for the various polar probes are indicative of stronger specific acidic and basic forces of interaction (68), possibly resulting from structural defects that expose more acidic and basic functional groups at the surface.

Further information on the strength and relative contribution of acidic and basic forces was obtained from the acid and base numbers, K_A and K_D of the samples [Eq. (36a)]. These numbers, which describe the acid and base properties of the materials, are normally calculated from ΔH_A^{SP} [Eq. (36b)]. As indicated in Table 9, the K_A values were comparable for MSX and SX-I but much higher for SX-I than for SX-II, reflecting relatively weak acidic forces of interactions on the surface of SX-II. On the other hand, the K_D value was the highest for MSX, followed sequentially by SX-II and SX-I, suggesting that the surfaces of MSX and SX-II are dominated by relatively strong basic forces compared with SX-I. In addition, the calculated K_D/K_A ratios (which determines the relative contribution of the basic and acidic properties) of MSX and SX-II were 13.2 and 20.3, respectively, reflecting an overwhelmingly large contribution of basic forces on the surfaces of these samples (largest for SX-II). However, the K_D/K_A ratio of SX-I was less than unity, implying a relative preponderance of the acidic forces on its surface. These findings are consistent with those deduced directly from the ΔH_A^{SP} data.

5.3.4. Summary of IGC Findings for Salmeterol Xinafoate

In summary, SX-II exhibits a more negative ΔG_A and ΔG_A^{SP} (contributed by a much lower entropy loss during the adsorption), a smaller γ_S^D, and a larger

K_D (i.e., stronger basic forces) than SX-I. In contrast, MSX differs from SX-I in that it has a more negative ΔG_A and ΔG_A^{SP} (due to a larger heat loss during the adsorption), a larger γ_S^D, and a larger K_D. Thus, it can be concluded that the metastable SFC-processed SX-II polymorph possesses higher surface free energy and surface entropy than the stable SFC-processed SX-I polymorph, whereas MSX displays higher surface free energy and surface enthalpy than the SFC-processed SX-I material.

6. FUTURE DEVELOPMENT OF SUPERCRITICAL FLUID PROCESSING IN CONTROLLING THE CRYSTAL FORM OF SOLID PHARMACEUTICALS

The ability of SFC to control the crystal form and associated material properties has been clearly demonstrated with a number of pharmaceutical materials. In most cases, the relative composition of the crystal forms produced can be precisely controlled by varying the working conditions of temperature and pressure. Other operating parameters such as nature and type of solvents, and flow rate of solution and supercritical fluids can provide additional control over the crystallization mechanisms and the resulting crystal forms. Unlike conventional powder production by sequential batch crystallization and micronization, which generally results in cohesive, highly charged, and crystallographically defective particles, SFC possesses the capability of affording noncohesive, free-flowing, and highly crystalline powders in a single-step operation. In addition, the flexibility of manipulating an additional operating variable (i.e., pressure/density of the supercritical fluids) allows a wider range of crystallization conditions to be tested for the production of specific materials or crystal forms, which is normally not feasible with conventional batch crystallization technique. All these advantageous aspects of SFC have been clearly demonstrated for the exemplary drug material salmeterol xinafoate.

Apart from the utility of processing pure single substances, SFC technology has proved useful for generating composite materials such as solid solutions or solid dispersions. Crystallinity, surface energy, and bulk thermodynamic properties of composite crystals, as exemplified by host acetaminophen and guest p-acetoxyacetanilide, can all be regulated by controlling the amount of additive uptake and/or by limiting the amount of solvent incorporated through appropriate adjustment of the SFC operating conditions. Thus, it becomes apparent that SFC technology holds good promise for future application in material design and engineering for specific pharmaceutical applications.

Finally, it should be emphasized that a key point of designing low risk, controllable, and predictable formulation processes, as required for pharmaceutical development and manufacture, lies in the understanding of the basic thermodynamic and kinetic factors governing precipitation from supercritical fluids, which is highly dependent on the development of reliable analytical methods for product validation.

REFERENCES

1. Bryn SR, Pfeiffer RR, Stephenson G, Grant DJW, Gleason WB. Solid state pharmaceutical chemistry. Chem Mater 1996; 6:1148–1158.
2. Guillory JK. Generation of polymorphs, hydrates, solvates and amorphous solids. In: Brittain HG, ed. Polymorphism of Pharmaceutical Solids. New York: Marcel Dekker, 1999:183–226.
3. Herbstein FH. Some applications of thermodynamics in crystal chemistry. J Mol Struct 1996; 374:111–128.
4. Haleblian J, McCrone W. Pharmaceutical applications of polymorphism. J Pharm Sci 1969; 58:911–929.
5. Tong HHY, Shekunov BY, York P, Chow AHL. Characterization of two polymorphs of salmeterol xinafoate crystallized from supercritical fluids. Pharm Res 2001; 18:852–858.
6. Tong HHY. Ph.D thesis, The Chinese University of Hong Kong, Hong Kong, 2003.
7. Edwards AD, Shekunov BY, Kordikowski A, Forbes R, York P. Crystallization of pure anhydrous polymorphs of carbamazepine by solution enhanced dispersion with supercritical fluids (SEDS™). J Pharm Sci 2001; 90: 1115–1124.
8. Behme R, Brooke D. Heat of fusion measurement of a low melting polymorph of carbamazepine that undergoes multiple phase changes during differential scanning calorimetry analysis. J Pharm Sci 1991; 80:986–990.
9. Bettini R, Bonassi L, Castoro V, Rossi A, Zema L, Gazzaniga A, Giordano F. Solubility and conversion of carbamazepine polymorphs in supercritical carbon dioxide. Eur J Pharm Sci 2001; 13:281–286.
10. Morris KR. Structural aspects of hydrates and solvates. In: Brittain HG, ed. Polymorphism in Pharmaceutical Solids. New York: Marcel Dekker, 1999:125–182.
11. Bryn SR, Pfeiffer RR, Stowell JG. Solid State Chemistry of Drugs. 2d ed. West Lafayette, IN: SSCI Inc, 1999:233–246.
12. Chernov AA. Modern Crystallography III, Crystal Growth; Springer Series in Solid State Physics. Berlin: Springer-Verlag, 1984.
13. Shekunov BY, Bristow S, Chow AHL, Cranswick L, Grant DJW, York P. Formation of solid solutions by precipitation in supercritical CO_2. Cryst Growth Design 2003; 3:603–610.
14. Kitaigorodsky AI. Molecular Crystals and Molecules. New York: Academic Press, 1973:381–446.
15. York P, Grant DJW. A disruption index for quantifying the solid state disorder induced by additives or impurities. I. Definition and evaluation from heat of fusion. Int J Pharm 1985; 25:57–72.

16. Sato K. Polymorphic transformation in crystal growth. J Phys D Appl Phys
     1993; 26:B77–B84.
17. Nielsen AE. Kinetics of Precipitation. Oxford: Pergamon Press, 1964.
18. Gu CH, JR VY, Grant DJW. Polymorph screening: influence of solvents on the
     rate of solvent-mediated polymorphic transformation. J Pharm Sci 2001; 90:
     1878–1890.
19. Tozuka Y, Kawada D, Oguchi T, Yamamoto K. Supercritical carbon dioxide
     treatment as a new method for polymorph preparation of deoxycholic acid.
     AAPS Pharm Sci 2002; 4(4).
20. Rustichelli C, Gamberini G, Ferioli V, Gamberini MC, Ficarra R, Tomasini S.
     Solid state study of polymorphic drugs: carbamazepine. J Pharm Biomed Anal
     2000; 23:41–54.
21. Akao K, Okubo Y, Inoue Y, Sakurai M. Supercritical CO_2 fluid extraction of
     crystal water from trehalose dihydrate. Efficient production of form II (T_α)
     phase. Carbohydr Res 2002; 337:1729–1735.
22. Akao K, Okubo Y, Asakawa N, Inoue Y, Sakurai M. Infrared spectroscopic
     study on the properties of the anhydrous form II of trehalose. Implications for
     the functional mechanism of trehalose as a biostabilizer. Carbohydr Res 2001;
     334:233–241.
23. Bettini R, Bertolini G, Casini I, Rossi A, Gazzaniga A, Pasquali I, Giordano F.
     Interaction of pharmaceutical hydrates with supercritical CO_2. AAPS Pharm
     Sci 2002; 4(4).
24. Bartolomei M. Solid-state studies on the hemihydrate and the anhydrous forms
     of flunisolide. J Pharm Biomed Anal 2000; 24:81–93.
25. Velaga SP, Berger R, Carlfors J. Supercritical fluids crystallization of
     budesonide and flunisolide. Pharm Res 2002; 19:1564–1571.
26. Bristow S, Cocks PM, Harland R, Gandhi RB. Stavudine polymorphic form I
     process. International patent WO 02/20538 A2, 2002.
27. Gilbert DJ, Palakodaty S, York P, Schultz TW, Shah D. Isolation of the
     polymorphic forms of RWJ-337813 using the solution enhanced dispersion by
     the supercritical fluids (SEDS™) technique. AAPS Pharm Sci 2001; 3(3).
28. Kordikowski A, York P, Latham D. Resolution of ephedrine in supercritical
     CO_2: a novel technique for the separation of chiral drugs. J Pharm Sci 1999; 88:
     786–791.
29. Kordikowski A, York P. Chiral separation using supercritical CO_2. Proceed-
     ings of the 6th meeting on supercritical fluids. Supercritical fluids chemistry and
     materials (Nottingham, UK), 1999.
30. Khoshkhoo S, Anwar J. Crystallization of polymorphs: the effect of solvent. J
     Phys D Appl Phys 1993; 26:B90–B93.
31. Bisrat M, Moshashaee S, Nyqvist HK, Demirbuker M. Crystallization using
     supercritical or subcritical fluids. US patent 6,461,642, 2002.
32. Kariuki BM, Psallidas K, Harris KDM, Johnston RL, Lancaster RW,
     Staniforth SE, Cooper SM. Structure determination of a steroid directly from
     powder diffraction data. J Chem Soc Chem Commun 1999:1677–1678.
33. Cooper SM. Orthorhombic crystalline form of fluticasone propionate and
     pharmaceutical composition thereof. US patent 6,406,718, 2002.

34. Edwards AD, Shekunov BY, Forbes RT, Grossmann JG, York P. Time-resolved X-ray scattering using synchrotron radiation applied to the study of a polymorphic transition in carbamazepine. J Pharm Sci 2001; 90:1106–1114.

35. Rehman MU. PhD thesis. UK: University of Bradford, 2002.

36. Chan FC, Anwar J, Cernik R, Barnes P, Wilson RM. Ab initio structure determination of sulfathiazole polymorph V from synchrotron X-ray powder diffraction data. J Appl Crystallogs 1999; 32:436–441.

37. Kordikowski A, Shekunov T, York P. Crystallization of sulfathiazole polymorphs using CO_2. Proceedings of the 7th meeting on supercritical fluids. Particle design—Materials and natural products processing, Antibes/Juan-les-Pins, France, 2000.

38. Kordikowski A, Shekunov T, York P. Polymorph control of sulfathiazole in supercritical CO_2. Pharm Res 2001; 18:682–688.

39. Webber A, Tschernjaew J, Kümmel R. Coprecipitation with compressed antisolvents for the manufacture of microcomposites. Proceedings of the 5th Meeting on Supercritical Fluids, Nice, France, 1998; 1:243–248.

40. Domingo C, Wubbolts FE, Rodriguez-Clemente R, van Rosmalen GM. Solid crystallization by rapid expansion of supercritical ternary mixtures. J Cryst Growth 1999; 199:760–766.

41. Subra P, Boissinot P, Benzaghou S. Precipitation of pure and mixed caffeine and anthracene by rapid expansion of supercritical solutions. Proceedings of the 5th Meeting on Supercritical Fluids, Nice, France, 1998; 1:307–312.

42. Vemavarapu C, Mollan MJ, Needham TE. Co-crystallization of pharmaceutical actives and their structurally related impurities by the RESS process. AAPS Pharm Sci 2002; 4(4).

43. Vemavarapu C, Mollan MJ, Needham TE. Crystal doping aided by rapid expansion of supercritical solutions. AAPS Pharm Sci 2002; 4(4).

44. Shekunov BY, Bristow S, York P, Chow AHL. Precipitation of acetaminophen and p-acetoxyacetanilide in supercritical CO_2. Proceedings of the 7th meeting on supercritical fluids. Particle Design—Materials and Natural Products Processing. Antibes/Juan-les-Pins, France, 2000.

45. Fairbrother JE. Acetaminophen. In: Florey K, ed. Analytical Profiles of Drug Substances. Vol. 3. New York: Academic Press, 1973:1–109.

46. York P, Wilkins SA, Storey RA, Walker SE, Harland RS. Coformulation methods and their products. International patent WO 01/15664 A2, 2001.

47. Sethia S, Squillante E. Physicochemical characterization of solid dispersions of carbamazepine formulated by supercritical carbon dioxide and conventional evaporation method. J Pharm Sci 2001; 91:1948–1957.

48. Beach S, Latham D, Sidgwick C, Hanna M, York P. P Control of the physical form of salmeterol xinafoate. Org Process Res Dev 1999; 3:370–376.

49. Tong HHY, Shekunov BY, York P, Chow AHL. Influence of polymorphism on the surface energetics of salmeterol xinafoate crystallized from supercritical fluids. Pharm Res 2002; 19:640–648.

50. Tong HHY, Shekunov BY, York P, Chow AHL. Thermal analysis of trace level of polymorphic impurity in salmeterol xinafoate samples. Pharm Res 2003; 20:1423–1429.

51. Michaelson C, Dahms M. On the determination of nucleation and growth kinetics by calorimetry. Thermochim Acta 1996; 288:9–27.
52. Kashchiev D. Nucleation: Basic Theory with Applications. Oxford: Butterworth-Heinemann, 2000.
53. Shekunov BY, York P. Crystallization processes in pharmaceutical technology and drug delivery design. J Cryst Growth 2000; 211:122–136.
54. Shekunov BY, Feeley JC, Chow AHL, Tong HHY, York P. Physical properties of supercritically-processed and micronised powders for respiratory drug delivery. KONA Powder and Particles 2002; 20:178–187.
55. Balzar D, Ledbetter H. Software for comparative analysis of diffraction line broadening. Adv X-ray Anal 1997; 39:457–464.
56. Bristow CS. PhD thesis. UK: University of Bradford, 2003.
57. Rehman MU, Shekunov BY, York P, Colthorpe P. Solubility and precipitation of nicotinic acid in supercritical carbon dioxide. J Pharm Sci 2001; 90:1570–1582.
58. Steckel H, Thies J, Müller BW. Micronizing of steroids for pulmonary delivery by supercritical carbon dioxide. Int J Pharm 1997; 152:99–110.
59. Feeley JC, York P, Sumby BS, Dicks H, Hanna M. In vitro assessment of salbutamol sulphate prepared by micronisation and a novel supercritical fluid technique. Drug Delivery Lungs 1998; IX:196–199.
60. Schultz J, Lavielle L. Interfacial properties of carbon fiber–epoxy matrix composites. In: Lloyd DR, Ward TC, Schreiber HP, Pizana CC, eds. Inverse Gas Chromatography—Characterization of Polymers and Other Materials. Washington, DC: American Chemical Society, 1989:185–202.
61. de Boer JH. The Dynamic Character of Adsorption. 2d ed. Oxford: Clarendon Press, 1968.
62. Gregg SJ. The Surface Chemistry of Solids. London: Whitefriars Press, 1965.
63. Katz S, Gray DG. The adsorption of hydrocarbons of cellophane. I. Zero surface coverage. J Colloid Interface Sci 1980; 82:318–325.
64. Adamson AW, Gast AP. Physical Chemistry of Surfaces. New York: John Wiley & Sons, 1997.
65. Fowkes FM. Quantitative characterisation of the acid–base properties of solvents, polymers, and inorganic surfaces. J Adhesion Sci Technol 1990; 4:669–691.
66. Gutmann V. The Donor–Acceptor Approach to Molecular Interactions. New York: Plenum Press, 1978.
67. Riddle FL, Fowkes FM. Special shifts in acid–base chemistry. 1. Van der Waals contributions to acceptor numbers. J Am Chem Soc 1990; 112:3260–3264.
68. Papirer E, Brendle E, Balard H, Vergelati C. Inverse gas chromatography investigation of the surface properties of cellulose. J Adhesion Sci Technol 2000; 14:321–337.

8

Supercritical Fluid Impregnation of Polymers for Drug Delivery

Sergei G. Kazarian
Department of Chemical Engineering and Chemical Technology,
Imperial College, London, United Kingdom

There has been remarkable progress in the application of supercritical fluids (SCFs) to the processing of pharmaceuticals over the last decade. This has made the publication of this book a timely event, providing a comprehensive overview of the subject. Supercritical fluid technology (1–4) has successfully proved its applicability in the area of polymer processing (5,6) and to the preparation of drug delivery systems (7). It is advantageous to use supercritical CO_2 to prepare such systems because there is no residual solvent in the materials produced, ambient temperature can be used for processing, and it may be possible to obtain, improved performance in drug release applications by altering the materials' polymer and drug morphologies from those achievable by conventional preparation techniques. Supercritical (sc) CO_2 can swell and plasticize glassy polymers, reducing the glass transition temperature. The plasticization of polymers is also accompanied by the swelling of the polymer matrix, with a consequent increase in the free volume of the polymer. Moreover, $scCO_2$ can reduce the melting temperature of semicrystalline polymers. These effects are crucial to the impregnation and modification of polymeric materials. An additional advantage of $scCO_2$ is the possibility of

343

"tuning" solvent properties of CO_2, simply by changing pressure. This tunability is crucial for successful tailoring of materials properties and greatly assists in establishing separation routes.

One of the most important differences between organic solvents that are usually used as polymer plasticizers and $scCO_2$ is that $scCO_2$ is much easier to remove from the polymeric materials when the process is complete. This is because it is a gas at ambient conditions. Small amounts of CO_2 can remain trapped inside the polymer matrix, but they are benign, and pose no danger to the consumer (i.e., a patient who uses a pharmaceutical formulation).

Thus major advantages and reason for using supercritical CO_2 in preparation of polymer–drug formulations are as follows:

> Benign, nontoxic nature of the solvent
> Leaves the product easily, with no solvent residues
> Can extract impurities, such as residual monomer or oligomers.
> "Tunability" of the solvent by varying pressure/temperature
> Plasticizing effect on polymers, allowing them to be processed at lower temperatures

1. EFFECT OF CO_2 ON POLYMERS: INTERACTIONS AND PLASTICIZATION

Most of the beneficial effects of CO_2 on polymers stem from the ability of CO_2 to weakly interact with the functional groups in many polymers. Indeed, if the molecule of CO_2 is viewed as a weak Lewis acid, its interaction with the basic sites in polymers will reduce interchain polymer interactions. This will result in enhanced segmental and chain mobility of the polymer, providing an opportunity to process polymers easily and to enhance impregnation of polymers with doping solutes, such as drug molecules. The evidence of such interactions was obtained spectroscopically by studying interactions between CO_2 and polymers containing basic functional groups (such as carbonyl groups or phenyl rings) (8). This evidence included observation of spectral changes in the infrared spectra of both CO_2 and polymers (9). The spectroscopic evidence for interaction between CO_2 and polymers provided support at the molecular level for suggestions by Koros (10), Sanders et al. (11), and Handa et al. (12) on the mechanism of CO_2-induced plasticization of glassy polymers. The initial spectroscopic evidence of interaction between CO_2 and polymer functional groups (8) included carbonyl groups, fluorine atoms, phenyl rings, and nitrogen atoms. Spectroscopic evidence of specific interactions between CO_2 and carbonyl groups in polymers has recently been supported by ab initio calculations. The evidence of CO_2-induced motion of the polymer segments (9) was also significant because this plasticization was achieved at much lower temperatures than would otherwise normally be needed to induce such mobility. This enhanced polymer segmental mobility can significantly ease the path of diffusing solutes (e.g., drug molecules) into polymer subjected to $scCO_2$. It has been shown that the origin of these phenomena lies in those weak intermolecular interactions between CO_2 and the functional groups in polymers. This led us to suggest that in glassy polymers (a) the effect of $scCO_2$ mimics effect of the heat (13) and (b) $scCO_2$ acts as a "molecular lubricant" (9).

This terminology was also the result of comparing diffusion rates of the azo dye Disperse Red 1 in two systems: poly(methyl methacrylate) heated above its glass and $scCO_2$-swollen PMMA (14,15). The temperature deviations from T_g were similar in these experiments, whereas the rates of diffusion in $scCO_2$-swollen PMMA were higher than those in the heated polymer. However, it is not just the swelling of a polymer matrix that facilitates the mass transport of dye molecules under these conditions. Indeed, it has been shown that the presence of CO_2 in a polymer matrix significantly enhances the diffusivity of water in polymers. Vincent et al. (16) compared the diffusion of water in $scCO_2$-swollen PMMA with that of lyophilically swollen PMMA. It was concluded that free volume theory alone cannot describe the differences between the two systems. The presence of CO_2 in a swollen matrix did play a role in solute mass transport process, presumably by solvating solutes, thus facilitating the diffusion. Based on these observations and the fact that CO_2 is a relatively small molecule compared with most solute molecules (e.g., dyes), the action of $scCO_2$ was described as molecular lubrication (9,15).

It is important to realize that that specific interactions discovered between CO_2 and polymers also explain the high solubility of CO_2 in many glassy polymers—for example, those used in membrane applications (17) or biocompatible polymers (18). This property also provided useful insight to researchers designing $scCO_2$-soluble polymers (19,20). The synthesis of such polymers is driven by the need for inexpensive $scCO_2$-soluble surfactants in dry cleaning applications that use $scCO_2$. These interactions also had an impact on the preparation of scaffolds for tissue engineering via CO_2-assisted foaming of biodegradable polymers (21).

2. USING SUPERCRITICAL FLUIDS TO IMPREGNATE POLYMERIC MATERIALS

Supercritical fluid impregnation is often referred to as a "solvent-free" process because there is essentially no residual solvent to affect the properties of the processed materials. This approach demonstrates great potential for the preparation of new polymeric materials. Pioneering works by Sand (22) and Berens et al. (23) played an important role in stimulating interest in this area. Poliakoff (4) and Howdle (24) and their colleagues successfully used scCO$_2$ to infuse organometallic complexes into polyethylene matrices and applied spectroscopic techniques to analyze these materials (25,26). The latter works have focused on impregnation via solute deposition within a polymer matrix, but it is important to differentiate between the two mechanisms of supercritical fluid impregnation of additives into polymer matrices (27). The deposition approach initially introduced by Berens (23) involves the solubilization of the solute in scCO$_2$ and the exposure of a polymer to such a solution with consecutive depressurizing of the system and removal of scCO$_2$. Indeed, when the high pressure vessel is depressurized, CO$_2$ molecules quickly leave the polymer matrix, leaving the solute molecules trapped inside (Figure 1). Clearly, this approach is particularly effective for solutes that are highly soluble in scCO$_2$. Ferrocene is one example (28,29); also, when glassy polymer matrices are used, the ability of scCO$_2$ to plasticize the polymer matrix contributes to enhanced solute diffusion. A quite different mechanism applies to the impregnation of compounds having very low solubility in scCO$_2$. In such cases the high affinity of these solutes for certain polymer matrices results in the preferential partitioning of a solute in a way that favors the polymer over the fluid phase (Figure 2). This schematic presentation shows that owing to an excess of solid solute, any solute that is impregnated into the polymer phase from the fluid phase is replaced, thus keeping the fluid phase saturated with a solute. The partitioning process of a solute with a low solubility in scCO$_2$ is complete when an equilibrium concentration is achieved in the polymer phase. For example, the high partition (distribution) coefficient of a solute plays a key role in impregnation of many compounds (9,30) that have low solubility in scCO$_2$. Thus, high partition coefficients of polar dye molecules played a role in the success of supercritical fluid dyeing (14). The partitioning mechanism, based on the affinity of a solute for the polymer matrix, has a major advantage over the deposition approach. Indeed, the deposition approach often results in recrystallization of the solute; this is

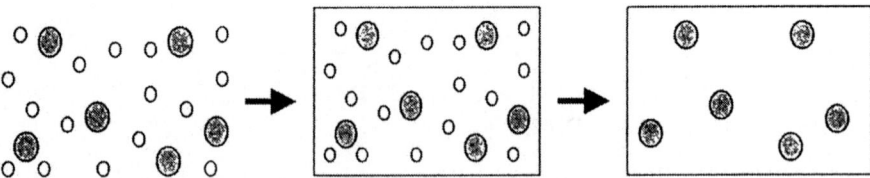

FIGURE 1 Schematic presentation of the impregnation of polymeric materials with drugs that are highly soluble in scCO$_2$: small circles, molecules of CO$_2$; larger dark circles, molecules of drug. Left: solution of drug in scCO$_2$; middle: penetration of solvent and solutes into polymer matrix; right: deposition of drug molecules within the polymer matrix upon depressurizing the system.

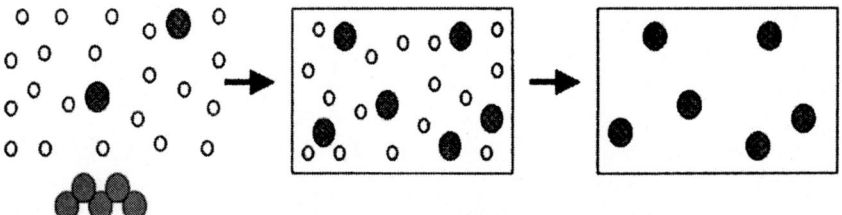

FIGURE 2 Schematic presentation of impregnation with a drug that has a low solubility in scCO$_2$ but a high affinity for a polymer matrix. Left: solution of drug in scCO$_2$ with the excess solid drug at the bottom of the vessel; middle: scCO$_2$ dissolves in polymer, and drug molecules partition into polymer matrix; right: drug molecules remain in polymer matrix after the system has been depressurized, while CO$_2$ leaves the matrix.

because when the vessel is depressurized, the solubility of the solute in scCO$_2$ is suddenly decreased. By contrast, the partitioning mechanism is based on specific interactions between solute and matrix (9) that usually prevent such recrystallization. This approach helps to produce materials with molecularly dispersed solutes interacting with the affinity site inside the polymer matrix. This has important implications for the preparation of systems for drug delivery. The principles of supercritical fluid impregnation for pharmaceuticals are similar to those for the dyeing of polymeric materials with scCO$_2$.

Supercritical fluid dyeing was reviewed in 1999 (15). The high partition coefficients for the azo dyes used in that work were based on specific interactions such as hydrogen bonding between a hydroxyl group of a dye and carbonyl groups in polymers. Fourier transform infrared (FTIR) spectroscopy revealed the evidence of such interaction between Disperse Red 1 and PMMA (9). It is important to remember that interest in supercritical fluid dyeing was motivated by environmental advantages, namely the possibility of replacing water with supercritical fluid in dyeing processes. The difference between the use of supercritical fluid impregnation for dyeing and its pharmaceutical application is that in dyeing the solute needs to be permanently trapped inside the polymer matrix, while in pharmaceutical applications a polymer–drug formulation is designed for drug release either by drug diffusion or via polymer degradation.

A broad (but not exclusive) range of polymers was used in supercritical fluid impregnation; these materials include the following:

Polystyrene (14,23)
Polyethylene (22,24–26,31–33)
Polypropylene (34)

PMMA (8,9,23,30)
Poly (vinyl chloride) (23,35)
Polycarbonate (14,23,31,35)
Polytetrafluoroethylene (35)
Poly (ethylene terephthalate) (15,36–40)
Poly (oxymethylene) (31)
Poly (dimethylsiloxane) (30,41–43)
Polyimides (44,45)

Supercritical fluid impregnation has recently been applied to biode-gradable polymers and is discussed in Section 4. The process of supercritical fluid impregnation largely depends on the solubility of CO_2 in a particular polymer. Dissolved CO_2 can significantly influence mass transport properties in polymer matrix. Berens (23) and other authors have shown that impreg-nation of solutes into polymers is accelerated in $scCO_2$-swollen glassy polymers. Cotton et al. (46) reported an increase in the diffusivities of additives in semicrystalline polypropylene in the presence of $scCO_2$. Dooley and coworkers (47) observed an increase of six orders of magnitude for the diffusivity of ethylbenzene in polystyrene subjected to supercritical carbon dioxide. Paulaitis and coworkers (48), who observed enhanced diffusivity of azobenzene in polystyrene, related this effect to the plasticization phenome-non and subglass transitions. The same investigators have also suggested that solute–CO_2 interactions play a role in the process. This is consistent with the proposal discussed earlier of a CO_2 molecule acting as a "molecular lubri-cant." The tuning of diffusion rates via varying the density of $scCO_2$ allows one to adjust the relative amounts of impregnated solutes.

Another relevant application of supercritical fluids with implications for pharmaceuticals consists of coatings applications. Interest in this area is stimulated by the need to reduce or eliminate emissions of volatile organic solvents during spray painting. The use of $scCO_2$ for coating applications as an alternative solvent was recently reviewed (49). This process is believed to have a number of advantages over conventional powder coating approaches, such as improved film coalescence and quality of the coating. The coating materials are dissolved in $scCO_2$ and subsequently released as an atomized spray through a specially designed nozzle from the spraying head. It is reported that this process produces a highly uniform spray with a narrow droplet size distribution. Unfortunately, the solubility in $scCO_2$ of many compounds used in coating systems is quite low. Therefore, methanol or another organic cosolvent is usually added to the $scCO_2$ solution to increase the solubilities of these compounds.

The alternative to the addition of cosolvents is the use of stabilizers, similar to their use in dispersion polymerization reactions. It was shown that fluorinated acrylate polymers or some block and graft copolymers may act as effective stabilizers in such reactions. Hay and Khan (49) are developing dispersion coatings based on the concept of exploring suitable dispersion stabilizers. This would lead to the elimination of emission of organic solvents in coating processes with $scCO_2$ technology. Applications of this environmentally benign technology range from polymer and paper coatings to biomaterials and pharmaceuticals.

Applications of supercritical fluids for coatings and impregnation of porous and fibrous substrates (e.g., polymer fibers, wood, composite materials) with various chemicals are discussed in a number of articles (50–52). Mandel and Wang (53) reported the use of a solution of polymers diluted with $scCO_2$ for powder-coating applications. They reported that Ferro Corporation has developed the so-called VAMP process used in the production of powder coatings, new polymers and polymer additives, and various biomaterials, with good potential for productions of pharmaceuticals.

3. FOAMING OF POLYMERS WITH $scCO_2$

Plasticization of glassy polymers induced by CO_2 plays an important role in the process of foaming of glassy polymers. The preparation of polymeric foams is aimed at the production of microcellular polymers that have closed cells about 10 μm or less in diameter and cell density of about 10^8 cells/cm^3. Such foamed polymeric materials have applications in many areas including biomedical devices. Pioneering work by Skripov and coworkers (54,55) stimulated the original interest in this field.

Although the $scCO_2$ foaming process was first applied to nonbiodegradable polymers, the principles of the process remain the same when it is applied to biodegradable polymers of biomedical interest. When a polymer is subjected to high pressure gas and the pressure is suddenly decreased or the temperature is rapidly increased, the gas will try to escape from the polymer, causing antiplasticization. This rapid escape of gas can cause nucleation and the growth of bubbles in the polymer. Once a significant amount of gas has escaped, the T_g of the polymer drops and the foamed structure "freezes."

Skripov's work (54,56) in this area began more than 30 years ago with studies of the solubility and diffusion of CO_2 in PMMA. These studies provided the basis for further development of the theory of nucleation kinetics or bubble formation in CO_2-saturated PMMA. The theory of homogeneous nucleation proved to be a satisfactory approximation for describing the kinetics of bubble nucleation in a gas-saturated polymer. Skripov and Blednykh (55) believed that local regions of polymer containing elevated CO_2 concentrations serve as nucleation centers for bubbles. Research on the gas-assisted foaming of glassy polymers has since seen a significant increase in activity. Foaming of PMMA has been studied further by Goel and Beckman, (57,58), while Kumar and Weller (59) studied the foaming of polycarbonate. The importance of Kumar's work is that it questioned the applicability of traditional theories of nucleation. A detailed study of glassy polymers subjected to high pressure CO_2 by Wessling and coworkers (60) suggested that the nucleation mechanism underlying the foaming process is heterogeneous. The significant advance made by Wessling and coworkers was to detect and explain the appearance not only of the porous structure in the polymer film after saturation with CO_2 but also of a dense layer next to the porous layer. They provided a physical explanation and a mathematical model to predict the thickness of this dense layer.

In recent work reporting the effect of the residual oligomer in polystyrene on its foaming with $scCO_2$, the authors have shown that the presence of the oligomer affects the cell size in these foams. It was found that an addition of low molecular weight oligomers to the polymer samples offers a way of controlling the cell structure in polystyrene foams formed by $scCO_2$. This work also questioned the ability of classical nucleation theory to explain the foaming mechanism in these systems. The authors suggest a spinodal mechanism as an alternative route of cell formation.

Another approach to create microcellular materials was demonstrated by Beckman and coworkers (61). They synthesized a number of chemicals soluble in $scCO_2$ or liquid CO_2. These chemicals comprise "monomers" containing one or two urea groups and fluorinated tail groups that enhance the solubility of these compounds in CO_2. When these compounds were dissolved in CO_2, their self-association led to the formation of gels. The removal of CO_2 via depressurization resulted in the formation of foams with cells with an average diameter of less than 1 μm. Beckman and coworkers reported that the bulk density reduction of these foams was 97% compared with the parent materials. A 40-fold expansion of biodegradable polyester foams with $scCO_2$ has also been reported (62).

The foaming of biodegradable polymers with $scCO_2$ is potentially a good route to the preparation of sponge scaffolds for tissue engineering. This approach was used to generate high surface area fibrillar scaffolds that were then used to generate liver tissue. Examples of biodegradable polymers that were foamed for biomedical applications include poly(lactide-co-glycolide) (PLGA) copolymer (63).

Such scaffolds can be used as vehicles for the delivery of bioactive substances (such as proteins, DNA, or drugs). Thus, Nof and Shea (64) recently reported the use of high pressure CO_2 to plasticize PLGA in the preparation of interconnected open-pore scaffolds that release DNA from drug-loaded microspheres. The authors mentioned that although their studies involved the delivery of DNA, the approach could be used to incorporate and release proteins. Hile et al. (65) also described the use of $scCO_2$ for the production of microporous PLGA foams containing encapsulated proteins. Koegler at al. (66) utilized $scCO_2$ to extract residual solvent from PLGA-based biodegradable polymeric devices for drug delivery and tissue engineering. In 2001 Shakesheff and coworkers (67) discussed the release of growth factor from tissue engineering scaffolds.

One of the other uses of CO_2 for polymer foaming is in the preparation of bioabsorbable polymer scaffolds for tissue engineering capable of sustained growth factor delivery. Thus, Mooney and coworkers (21) studied the effects of several processing parameters (such as polymer composition, molecular mass, and gas type) on preparing three-dimensional porous matrices from copolymers of lactide and glucolide. They demonstrated that crystalline polymers (polylactide and polyglycolide) did not produce foamed materials via this approach, while gas treatment of amorphous copolymers produced matrices with porosity up to 95%. This work also demonstrated that only CO_2, among the other gases used (such as nitrogen and helium) created highly porous polymer structures in these copolymers. They explained this by citing possible interactions between CO_2 and the carbonyl groups in these copolymers. Mooney and coworkers (21) also incorporated vascular endothelial growth factor into these porous matrices and studied its release in a controlled manner. This approach shows great potential in preparing scaffolds for tissue engineering. Howdle et al. (68) reported the use of $scCO_2$ to create a diverse range of polymeric composites incorporating thermolabile and solvent-labile guest compounds such as proteins; the work was successful in that the protein function was retained after $scCO_2$ treatment. Early work by scientists from 3M (69) developed an interesting modification of the supercritical fluid polymer impregnation process. In their approach, $scCO_2$ was used as an agent to enhance the diffusivity of water-soluble solutes into various substrates by performing the process in water pressurized with $scCO_2$.

4. INCORPORATION OF DRUG MOLECULES INTO POLYMERS

Supercritical fluids have been used for preparing polymer substrates with drugs. Formulations of drugs with water-soluble polymers are usually prepared to enhance the release rate, and thus the bioavailability of the drugs. This approach is predicated on the very low solubility in water of many solid crystalline drugs. However, if crystallization of the drug incorporated into a polymer matrix is avoided, enhanced dissolution and release are possible, which would increase the therapeutic value of such products. On another hand, in formulations of drugs with high solubility in water, the polymer may moderate the drug release rate by decreasing its fast dissolution process.

Serajuddin (70) provided a comprehensive review of preparations of solid dispersions of drugs in water-soluble polymers.

There are certain requirements for use of polymers in drug delivery (71). Polymers used in drug delivery should be biodegradable, and the products of their degradation should be nontoxic and soluble in body fluids. Also, the polymer must be metabolized within a reasonable period of time. It is also important to distinguish nonenzymatic and enzymatic hydrolysis of polymers. Nevertheless, poly(vinyl pyrrolidone) (PVP), hydroxypropylmethylcellulose (HPMC), and poly(ethylene glycol) (PEG) of different molecular weights remain the most popular polymers for incorporation of drugs with the low solubility in aqueous solutions. Solid dispersions of drugs in these polymers that have been prepared via conventional methods (solvent evaporation, melt extrusion, mechanical mixing) include the following systems: piroxicam-PVP (72), hydrocortisone–PVP (73), indomethacin–PVP (74), frusemide–PVP (75), nifedipine–PVP (76), ibuprofen–PVP (76), oxazepam–PEG (76), and ibuprofen–PEG (77).

Clearly, the opportunity exists to prepare similar systems via supercritical fluid impregnation and to compare the molecular state of the drug molecules in these formulations. Supercritical fluid impregnation of drug molecules into polymer matrices is analogous to the incorporation of dye molecules into polymeric materials as discussed earlier. Therefore, the same two mechanisms of drug incorporation into polymers are present in conventional impregnation processes: deposition in a polymer matrix from a supercritical fluid solution and partitioning into a polymer phase from a supercritical fluid phase. The ability of $scCO_2$ to plasticize amorphous polymers offers the possibility of preparing polymer–drug formulations at lower temperatures than are feasible with some conventional ways of preparation.

The advantage of using $scCO_2$, which can mimic the effect of heat, is particularly important for the impregnation of thermolabile drug molecules. A good basis (78,79) for impregnation studies was provided by an analysis of the solubility of drugs in supercritical CO_2. Kikic reported impregnating PVP with ketoprofen, nimesulide, and piroxicam (80). He reported loading the drug into PVP under different conditions and discussed the limitations and advantages of this method. It is unfortunate that the molecular state of drugs in these formulations has not been assessed in comparison to the state of drugs in analogous formulations prepared via conventional approaches. Elucidation of the molecular state of drugs is needed to understand the mechanism of drug release from such formulations. This has been done recently (27) in an investigation of the molecular state of ibuprofen impregnated into PVP from a supercritical fluid solution. This combination was suitable for the demonstration of supercritical fluid impregnation principles because ibuprofen is soluble (79) in $scCO_2$ in the range from 10^{-5} to 10^{-3} mole fraction in the CO_2

pressure range of 80 to 220 bar, and temperatures of 35 to 45°C. Also, gravimetric studies by Kikic et al. (81) have shown that CO_2 sorption into PVP reaches about 0.1 weight fraction at 100 bar and 40°C.

The studies just cited provided a good basis for a detailed spectroscopic investigation of the PVP–ibuprofen system. In situ spectroscopic attenuated total reflectance (ATR) IR spectroscopy was applied to monitor the process of the impregnation of ibuprofen into PVP (27). The authors demonstrated that ibuprofen was molecularly dispersed in a PVP matrix as a result of supercritical fluid impregnation without formation of a crystallized drug. This was evidenced by comparison of the ATR-IR spectra of the ibuprofen–PVP system with the ATR-IR spectra of the solid drug. For example, Figure 3 shows the spectra of a PVP film, of PVP with ibuprofen impregnated from $scCO_2$ solution, and of solid ibuprofen in the $\nu(C{=}O)$ spectral region. The $\nu(C{=}O)$ of ibuprofen impregnated into PVP absorbs at 1727 cm^{-1}, and this band is shifted to the high wavenumber region in comparison to the corresponding band of solid ibuprofen, which absorbs at 1710 cm^{-1}. This indicates the breakage of interaction (that exists in the crystalline form) between ibuprofen molecules when these molecules are impregnated into PVP from a supercritical CO_2 solution. Shakhtshneider et al. (82) reported a similar shift of the $\nu(C{=}O)$ band of ibuprofen in a solid dispersion in PVP prepared via the

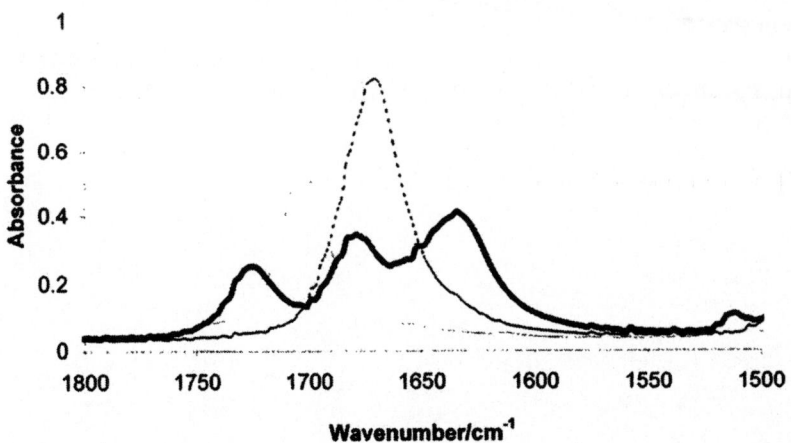

FIGURE 3   ATR-IR spectra in the $\nu(C{=}O)$ spectral region: thin line, spectrum of solid ibuprofen; dashed line, spectrum of PVP film; thick line, spectrum of PVP with impregnated ibuprofen from $scCO_2$ solution. (From Ref. 27, copyright 2002 Elsevier Science.)

conventional method. However, it is important to note that supercritical fluid impregnation of ibuprofen into PVP resulted in the absence of the band corresponding to the solid crystalline drug, while some amount of crystalline drug was present in the formulation prepared via the conventional method. This observation was indicative that the state of ibuprofen impregnated into PVP from an scCO$_2$ solution is that of a molecularly dispersed drug without the presence of crystalline ibuprofen.

Additional evidence for the absence of the crystalline ibuprofen in a system prepared via an scCO$_2$ process stems from Raman spectroscopy (27). Raman spectra of ibuprofen impregnated into PVP from an scCO$_2$ solution have been measured and compared with the Raman spectra of crystalline ibuprofen. These Raman measurements also indicated the absence of a crystalline drug in PVP, indicating that scCO$_2$ impregnation resulted in a molecularly dispersed drug within the PVP matrix. Closer inspection of the spectra in Figure 3 provided further support to this proposal. The spectrum represented by a solid line shows that the ν(C=O) band of molecularly dispersed ibuprofen is accompanied by the appearance of the ν(C=O) band of PVP at 1636 cm^{-1} shifted to the lower wavenumber in comparison to the ν(C=O) band of virgin PVP film at 1682 cm^{-1}. The band at 1636 cm^{-1} is assigned to the absorption of the carbonyl group of PVP that is H-bonded to the hydroxyl group of ibuprofen molecules, as presented in Figure 4. That appearance of the ibuprofen ν(C=O) at 1727 cm^{-1} is accompanied by the shift of ν(C=O) to the lower wavenumbers provides strong evidence that ibuprofen molecules impregnated into PVP are H-bonded to the carbonyl groups of PVP rather than being self-associated (27). Such an interaction was proposed for the ibuprofen–PVP system prepared via the conventional method, but no evidence was obtained for the absence from that system of the crystalline form of ibuprofen.

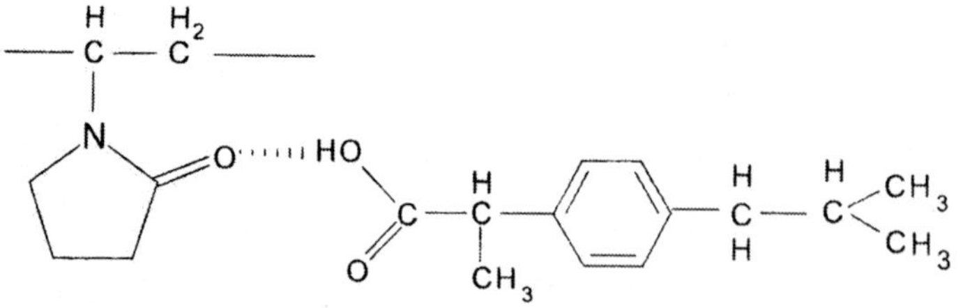

FIGURE 4 Schematic presentation of H-bonding interaction between PVP and an ibuprofen molecule (right) in molecular dispersion of ibuprofen in PVP prepared via supercritical fluid impregnation. (From Ref. 27, copyright 2002 Elsevier Science.)

Spectroscopic analysis of supercritical fluid impregnation of ibuprofen into PVP has shown that essentially all ibuprofen molecules impregnated into PVP are H-bonded to the polymer moieties. This interaction prevents crystallization of drug molecules within the PVP. In addition, the work has shown that competitive interaction of impregnated ibuprofen molecules with the carbonyl groups of PVP prevents CO_2 molecules from interacting with the carbonyl groups of PVP. Indeed, CO_2 molecules are capable of interacting with the basic carbonyl groups in PVP via Lewis acid–base interaction with the lone pair of electrons on the carbonyl oxygen of PVP (27) (Figure 5). Evidence for this interaction was obtained from the splitting of the IR band corresponding to the bending mode of CO_2 in PVP. However, the splitting of the bending band of CO_2 disappeared once ibuprofen had become impregnated into PVP. This observation provided additional evidence that CO_2 interacts with the basic functional groups in polymers: this interaction does not exist once the functional groups have begun to interact with other solutes, such as ibuprofen.

More importantly, such interactions of ibuprofen have an effect on water uptake into PVP at ambient conditions, reducing the amount of absorbed water, presumably because of the unavailability of a certain number of carbonyl groups in PVP to interact with water (27). The significant decrease of moisture uptake into formulations prepared via the $scCO_2$ process is beneficial, since the lower moisture uptake usually improves the stability of polymer–drug formulations. This example shows that supercritical fluid impregnation offers a suitable method of preparation for the molecularly dispersed solid dispersion of the drug in polymers. Thus the key to the successful preparation of polymer–drug formulations that are free of a crystalline drug is to achieve molecular dispersion for the drug within a polymer matrix via H-bonding to ensure the binding of each drug molecule to the functional group of polymers. Thus, this method of impregnation was based on affinity of drug molecules for the polymer matrix. The importance of polymer–drug interactions, polymer architecture, and chemistry for the design of new

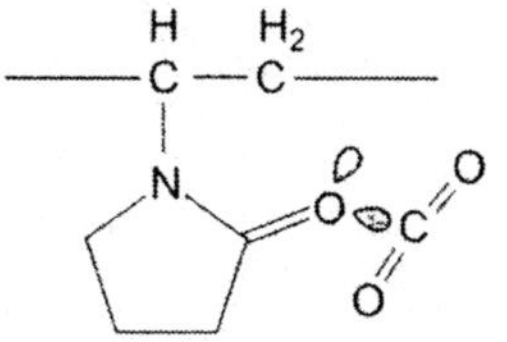

FIGURE 5 Schematic presentation of the interaction of a CO_2 molecule with the lone pair of electrons on the carbonyl group of PVP (in the absence of ibuprofen). (From Ref. 27, copyright 2002 Elsevier Science.)

systems for controlled delivery has also been emphasized in a Monte Carlo simulation study of active ingredient loading (83).

The success of supercritical fluid impregnation of drug molecules into glassy polymers is largely based on the plasticizing ability of CO_2, which results in swelling of the polymer and enhances the diffusion rate of the drug into the polymer. However, in the case of a semicrystalline polymer, such as PEG, high pressure CO_2 is known to be able to reduce the melting temperature of the polymer. Kazarian (84) has applied in situ ATR-IR spectroscopy to the study of the effects of CO_2 on PEG. Figure 6 shows the ATR-IR spectra of PEG film and PEG subjected to high pressure CO_2 at 25°C. Subjecting PEG to 60 bar pressure of CO_2 resulted in dramatic changes, with shifting and broadening of all spectral bands of PEG. It is important to note that similar changes have been observed in the ATR-IR spectrum of PEG when it was heated above its melting temperature of 47°C at ambient pressure. Thus, the spectra presented in Figure 6 demonstrate that high pressure CO_2 is capable of reducing the melting temperature of PEG. Such a reduction of the polymer melting temperature has important implications, since it would allow the viscosity of the polymer to be significantly decreased without additional

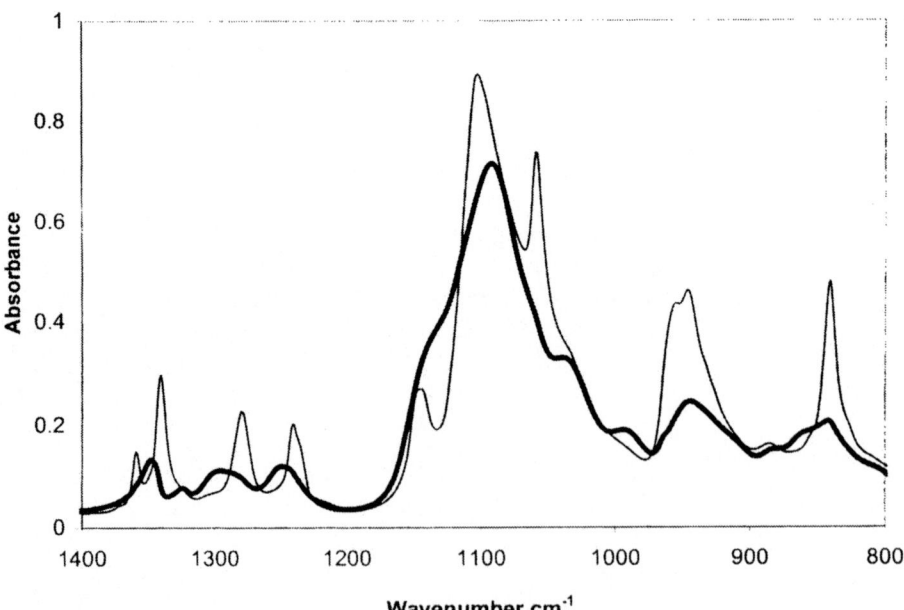

FIGURE 6   ATR-IR spectra of PEG at 25°C (thin line) and during exposure to 60 bar of CO_2 (thick line). (From Ref. 84, copyright Wiley-VCH Verlag GmbH.)

heating, which could make processing easier at milder temperatures. Since PEG is widely used in pharmaceutical applications, $scCO_2$-assisted processing of this polymer at reduced temperatures may offer important benefits. Indeed, the impregnation of bioactive substances into a molten polymer would proceed faster than impregnation into a solid polymer and would also result in homogeneous distribution of a solute in the polymer solution. There is an additional advantage in the case of CO_2-induced melting of a polymer: depressurizing of the system would result in quick escape of the CO_2 molecules from the polymer matrix and the reversion of the polymer to its solid state, thus "freezing" the impregnated drug molecules within the matrix before they are able to recrystallize. Spectroscopic evidence of successful impregnation of ibuprofen into PEG from $scCO_2$ has been obtained (84). The action of CO_2 in that experiment was threefold: to depress the melting temperature of PEG, to dissolve ibuprofen, and to act as a carrier medium of ibuprofen into molten PEG.

Spectroscopy has played an important role in the elucidation of the molecular state of drugs in polymer matrices. Microscopic imaging by means of FTIR spectroscopy has emerged as a powerful method to characterize heterogeneous materials (85). The method relies on the ability of the infrared array detector to simultaneously measure spectra from 4096 different locations in a sample. This imaging approach offers a number of advantages owing to its specificity: it allows one not only to determine the chemical structure but also to distinguish different forms of the same substance and their spatial distribution. We have recently applied FTIR imaging to study the distribution of ibuprofen in PEG and PVP prepared by supercritical fluid impregnation. These studies have unambiguously confirmed uniform distribution of the drug in polymer matrices and the absence of crystalline formation of the drug in all locations in the sample measured with FTIR imaging. We expect to find further applications of FTIR imaging in the analysis of polymer–drug formulations prepared with supercritical fluids.

Foster and coworkers (86) have recently demonstrated an elegant approach to enhancing the dissolution of ibuprofen by reducing the melting temperature. These authors claim that high pressure CO_2 results in the depression of the melting temperature of methyl-β-cyclodextrin (MBCD). The enhancement of dissolution of ibuprofen was achieved by formation of complexes of the drug with MBCD. These complexes dissolved almost instantaneously in water solutions. It was shown that a complex of ibuprofen with MBCD inhibited the formation of the crystalline form of ibuprofen, similar to the effect of the H-bonding of ibuprofen with PVP. Also, similar to the PEG–CO_2 system discussed earlier, Foster and coworkers (86) report that high pressure CO_2 resulted in the reduction of the viscosity of MBCD, which assisted in making possible the loading of a higher amount of ibuprofen. This work has successfully demonstrated that $scCO_2$ processing can be beneficial in the preparation of inclusion complexes with an enhanced dissolution rate. Recently, Foster and coworkers (87) have briefly reviewed different methods of using supercritical fluids to prepare improved drug delivery systems. They discuss the differences between coprecipitation, impregnation, and encapsulation. They define encapsulation as coating of a drug with the thin shell of another material, stating that encapsulation is possible as a continuous domain within a single-layered or multilayered shell, or as a number of small domains throughout a matrix of shell material.

Debenedetti and coworkers (88,89) provided one of the first examples of microencapsulation of a drug in the polymeric matrix. Richard and coworkers (90) provided a recent example of the microencapsulation process when they produced microparticles with the encapsulated model protein that showed sustained release. Foster and coworkers (87) also reported precipitation of copper–indomethacin by PVP with a 96-fold enhancement in the dissolution rate of indomethacin. These examples clearly demonstrated the advantages of using supercritical fluid processing for the preparation of polymer–drug formulations with potentially improved therapeutic properties.

Guney and Akgerman (71) reported a comprehensive analysis of supercritical fluid impregnation of PLGA with 5-fluorouracil for chemotherapy and β-estradiol for estrogen hormone therapy. They also addressed the issue of two different mechanisms in supercritical fluid impregnation: deposition from $scCO_2$ solution and partitioning into polymer from $scCO_2$ solution. They concluded that the process is governed by the solubility of the drug in $scCO_2$ and the equilibrium partitioning between the polymer and $scCO_2$ phases. Nevertheless, they report a satisfactory loading of the drug into polymers despite the low solubility of a drug in $scCO_2$. To elucidate the mechanism of supercritical impregnation for their systems, the authors measured the adsorption isotherm, the drug partition coefficients, and polymer swelling. They have developed a "flow over a flat plate" model that accounts for the polymer swelling and diffusivity, and for the partitioning of drug (71).

Kikic and coworkers (91) outlined the utility of supercritical fluid chromatography in the pharmaceutical field as a tool that can very quickly give useful information on polymer–drug impregnation experiments. Domingo et al. (92) used supercritical fluids to investigate single- and two-solute adsorption processes. They report the results of the experiments of simultaneous adsorption of the mixture of two components (benzoic acid and salicylic acid) on three different matrices (alumina, silica get, and Amberlite). Various process conditions have been studied (duration of the process, pressure and temperature), which allowed the authors to control the uptake of the solute into the various porous adsorbents. The authors rightly speculate that this process could be used in controlled drug delivery and might be used for the separation of structurally similar products such as pharmaceuticals. It should be noted that simultaneous supercritical fluid impregnation of two different solutes had been proposed earlier (15) and was recently realized in the impregnation of two different dye molecules into a PMMA matrix. Domingo et al. (93,94) have also used a chemometric approach based on nonlinear partial least squares modeling to characterize the adsorption process. Their work emphasizes the advantages of supercritical fluid impregnation for impregnation of thermally labile bioactive compounds. In 1999 Domingo and coworkers (95) compared supercritical fluid technology in the preparation of drug delivery systems with conventional techniques.

5. OUTLOOK

Supercritical fluid impregnation offers a clean way to use molecularly dispersed drug molecules in one preparation of polymer–drug formulations. The use of supercritical carbon dioxide as a carrier of drug molecules into a polymer matrix has a number of advantages, such as the plasticizing ability of CO_2 (based on specific interactions between CO_2 and polymer moieties), which enhances both the diffusion rates of drug molecules into the polymer and the ease of solvent removal. This approach compares favorably with conventional methods of preparation of solid polymer–drug dispersions for drugs that have low solubility in water. The mixing of a drug with biopolymers assisted by supercritical fluid (68) is another way to prepare composite materials for biomedical formulations. Although such mixing would not produce molecularly dispersed drugs, it may find utility in medical applications. The full potential of supercritical fluid impregnation of polymeric materials for pharmaceutical applications has yet to be realized (7,87). Indeed, the application of supercritical fluids to polymers in pharmaceuticals and biomedical materials is set to expand.

Earlier we speculated (15) that simultaneous impregnation of two dyes into polymers may provide an unusual way, not easily accessible, to achieve a desired synergism of colors. Similarly, recent research in the preparation of polymer–drug formulations (92) indicates that impregnation of polymers with different compounds may provide an unusual way to prepare novel pharmaceutical products. We have also speculated (6) that potential breakthroughs in supercritical fluid impregnation may result in the formation of particles with complex morphology (e.g., multi layered systems) for controlled-release drug delivery systems. In addition, application of supercritical fluid impregnation appears to be attractive for the preparation of personal care products. For example, bioactive compounds may be impregnated into contact lenses, toothbrushes, bone cements, and so on. Extrusion of biodegradable polymers assisted by supercritical fluids provides an opportunity to combine this process with the impregnation or mixing of bioactive substances. The focus of much recent research (79,96) (as also reflected by other chapters in this book) in the use of supercritical fluid has been on preparing micronized particles containing drugs in ways that will enhance the dissolution rates of pharmaceuticals that are poorly soluble in water. However, it is hoped that an opportunity to prepare molecularly dispersed drugs in biodegradable polymers via supercritical fluid impregnation will offer an additional approach that proves to be suitable for a variety of specific pharmaceutical applications. Discovery of a new drug will require finding ways to prepare novel formulations, and it is hoped that supercritical fluid impregnation will play a role in these developments.

REFERENCES

1. Beckman EJ. Using CO_2 to produce chemical products sustainably. Environ Sci Technol 2002; 36:347A–353A.
2. Hauthal WH. Advances with supercritical fluids [review]. Chemosphere 2001; 43:123–135.
3. Jessop PG, Leitner W. Chemical synthesis using supercritical fluids. Weinheim, Garmany: Wiley-VCH, 1999:480.
4. Poliakoff M, Howdle SM, Kazarian SG. Vibrational spectroscopy in supercritical fluids: from analysis and hydrogen bonding to polymers and synthesis. Angew Chem Int Ed Engl 1995; 34:1275–1295.
5. Cooper AI. Polymer synthesis and processing using supercritical carbon dioxide. J Mater Chem 2000; 10:207.
6. Kazarian SG. Polymer processing with supercritical fluids. Polym Sci Ser C 2000; 42:78–101.
7. Subramaniam B, Rajewski RA, Snavely K. Pharmaceutical processing with supercritical carbon dioxide. J Pharm Sci 1997; 86:885–890.
8. Kazarian SG, Vincent MF, Bright FV, Liotta CL, Eckert CA. Specific intermolecular interaction of carbon dioxide with polymers. J Am Chem Soc 1996; 118:1729–1736.
9. Kazarian SG, Brantley NH, West BL, Vincent MF, Eckert CA. In situ spectroscopy of polymers subjected to supercritical CO_2: plasticization and dye impregnation. Appl Spectrosc 1997; 51:491–494.
10. Koros WJ. Simplified analysis of gas/polymer selective solubility behavior. J Polym Sci Polym Phys 1985; 23:1611.
11. Sanders ES, Jordan SM, Subramanian R. Penetrant-plasticized permeation in polymethylmethacrylate. J Membrane Sci 1992; 74:29–36.
12. Handa YP, Kruus P, O'Neill M. High-pressure calorimetric study of plasticization of poly(methyl methacrylate) by methane, ethylene, and carbon dioxide. J Polym Sci Part B: Polym Phys 1996; 34:2635–2639.

13. Kazarian SG, Briscoe BJ, Lawrence CJ. Supercritical enhanced processing. In: Coates PD, ed. Polymer Process Engineering'99. London: Institute of Materials, 1999:28–36.
14. West BL, Kazarian SG, Vincent MF, Brantley NH, Eckert CA. Supercritical fluid dyeing of PMMA films with azo-dyes. J Appl Polym Sci 1998; 69:911–919.
15. Kazarian SG, Brantley NH, Eckert CA. Dyeing to be clean: use supercritical carbon dioxide. CHEMTECH 1999; 29:36–41.
16. Vincent MF, Kazarian SG, Eckert CA. "Tunable" diffusion of D_2O in CO_2-swollen poly(methyl methacrylate) films. AIChE J 1997; 43:1838–1848.
17. Chiou JS, Barlow JW, Paul DR. Plasticization of glassy polymers by CO_2. J Appl Polym Sci 1985; 30:2633–2642.
18. Kikic I, Lora M, Cortesi A, Sist P. Sorption of CO_2 in biocompatible polymers: experimental data and qualitative interpretation. Fluid Phase Equilibria 1999; 158–160:913–921.
19. Sarbu T, Styranec TJ, Beckman EJ. Design and synthesis of low cost, sustainable CO_2-philes. Ind Eng Chem Res 2000; 39:4678–4683.
20. Sarbu T, Styranec T, Beckman EJ. Non-fluorous polymers with very high solubility in supercritical CO_2 down to low pressures. Nature 2000; 405:165–168.
21. Sheridan MH, Shea LD, Peters MC, Mooney DJ. Bioabsorbable polymer scaffolds for tissue engineering capable of sustained growth factor delivery. J Controlled Release 2000; 64:91–102.
22. Sand ML. US patent, 4,598,006, 1986.
23. Berens AR, Huvard GS, Korsmeyer RW, Kunig RW. Application of compressed carbon dioxide in the incorporation of additives into polymers. J Appl Polym Sci 1992; 46:231–242.
24. Howdle SM, Ramsay JM, Cooper AI. Spectroscopic analysis and in situ monitoring of impregnation and extraction of polymer films and powders using supercritical fluids. J Polym Sci B Polym Phys 1994; 32:541–549.
25. Cooper AI, Howdle SM, Hughes C, Jobling M, Kazarian SG, Poliakoff M, Shepherd LA, Johnston KP. Spectroscopic probes for hydrogen bonding, extraction impregnation and reaction in supercritical fluids. Analyst 1993; 118:1111–1116.
26. Cooper AI, Kazarian SG, Poliakoff M. Supercritical fluid impregnation of polyethylene films, a new approach to studying to equilibria in matrices: the hydrogen bonding of fluoroalcohols to $Cp^*Ir(CO)_2$ and the effect on C—H activation. Chem Phys Lett 1993; 206:175–180.
27. Kazarian SG, Martirosyan GG. Spectroscopy of polymer/drug formulations processed with supercritical fluids: in situ ATR-IR and Raman study of impregnation of ibuprofen into PVP. Int J Pharm 2002; 232:81–90.
28. Kazarian SG. Application of FTIR spectroscopy to supercritical fluid drying, extraction and impregnation. Appl Spectrosc Rev 1997; 32:301–348.
29. Kazarian SG, West BL, Vincent MF, Eckert CA. Spectroscopic method for in-situ analysis of supercritical fluid extraction and impregnation of polymeric matrices. Am Lab 1997; 29(16):18B–18G.

30. Kazarian SG, Vincent MF, West BL, Eckert CA. Partitioning of solutes and cosolvents between supercritical CO_2 and polymer phases. J Supercrit Fluids 1998; 13:107–112.

31. Watkins JJ, McCarthy TJ. Polymerization in supercritical fluids—swollen polymers: a new route to polymer blends. Macromolecules 1994; 27:4845–4847.

32. Ma X, Tomasko DL. Coating and impregnation of a nonwoven fibrous polyethylene material with a nonionic surfactant using supercritical carbon dioxide. Ind Eng Chem Res 1997; 36:1586–1597.

33. Li D, Han BX. Impregnation of polyethylene (PE) with styrene using supercritical CO_2 as the swelling agent and preparation of PE/polystyrene composites. Ind Eng Chem Res 2000; 39:4506–4509.

34. Wang YQ, Yang CH, TD. Confocal microscopy analysis of supercritical fluid impregnation of polypropylene. Ind Eng Chem Res 2002; 41:1780–1786.

35. Muth O, Hirth T, Vogel H. Polymer modification by supercritical impregnation. J Supercrit Fluids 2000; 17:65–72.

36. Saus W, Knittel D, Schollmeyer E. Dyeing of textiles in supercritical carbon dioxide. Textile Res J 1993; 63:135–142.

37. von Schnitzler J, Eggers R. Mass transfer in polymers in a supercritical CO_2-atmosphere. J Supercrit Fluids 1999; 16:81–92.

38. Kazarian SG, Brantley NH, Eckert CA. Applications of vibrational spectroscopy to characterize poly(ethylene terephthalate) processed with supercritical CO_2. Vib Spectrosc 1999; 19:277–283.

39. Drews MJ, Jordan C. The effect of supercritical CO_2 dyeing conditions on the morphology of polyester fibers. Textile Chem Colorist 1998; 30:13–20.

40. Sicardi S, Manna L, Banchero M. Comparison of dye diffusion in poly(ethylene terephthalate) films in the presence of a supercritical or aqueous solvent. Ind Eng Chem Res 2000; 39:4707–4713.

41. Vincent MF, Kazarian SG, West BL, Berkner JA, Bright FV, Liotta CL, Eckert CA. Cosolvent effects of modified supercritical carbon dioxide on cross-linked poly(dimethylsiloxane). J Phys Chem B 1998; 102:2176–2186.

42. Shim JJ, Johnston KP. Adjustable solute distribution between polymers and supercritical fluids. AIChE J 1989; 35:1097.

43. Condo PD, Sumpter SR, Lee ML, Johnston KP. Partition coefficients and polymer-solute interaction parameters by inverse supercritical fluid chromatography. Ind Eng Chem Res 1996; 35:1115–1123.

44. Boggess RK, Taylor LT, Stoakley DM, Clair AKS. Highly reflective polyimide films created by supercritical fluid infusion of a silver additive. J Appl Polym Sci 1997; 64:1309–1317.

45. Rosolovsky J, Boggess RK, Rubira AF, Taylor LT, Stoakley DM, Clair AKS. Supercritical fluid infusion of silver polyimide films of varying chemical composition. J Mater Res 1997; 12:3127–3133.

46. Cotton NJ, Bartle KD, Clifford AA, Dowle CJ. Rate and extent of supercritical fluid extraction of additives from polypropylene: diffusion, solubility, and matrix effects. J Appl Polym Sci 1993; 48:1607–1619.

47. Dooley KM, Launey D, Becnel JM, Caines TL. Measurement and modeling of supercritical fluid extraction from polymeric matrices. In: Hutchenson KW, Foster NR, eds. Innovations in Supercritical Fluids. Washington, DC: ACS Symp Ser American Chemical Society, 1995:269–280.
48. Chapman BR, Gochanour CR, Paulaitis ME. CO_2- enhanced diffusion of azobenzene in glassy polystryrene near the glass transition. Macromolecules 1996; 29:5635–5649.
49. Hay JN, Khan A. Review: environmentally friendly coatings using carbon dioxide as the carrier medium. J Mater Sci 2002; 37:4743–4752.
50. Sahle-Demessie E, Levien KL, Morrell JJ. Impregnating porous solids using supercritical CO_2. CHEMTECH 1998; 12–18.
51. Acda MN, Morrell JJ, Levien KL. Supercritical fluid impregnation of selected wood species with tebuconazole. Wood Sci Technol 2001; 35:127–136.
52. Magnan C, Bazan C, Charbit F, Joachim J, Charbit G. Impregnation of porous supports with active substances by means of supercritical fluids. In: Rudolf von Rohr P, Trepp C, eds. High Pressure Chemical Engineering. Zurich, Switzerland: Elsevier, 1996:509–514.
53. Mandel FS, Wang JD. Manufacturing of specialty materials in supercritical fluid carbon dioxide. Inorg Chim Acta 1999; 294:214–223.
54. Okonishnikov GB, Blednykh EI, Skripov VP. The study of dynamic modulus of PMMA saturated with carbon dioxide. Mehk Polim 1973; 2:370–372.
55. Skripov VP, Blednykh EI. Nucleation kinetics of bubbles in gas-saturated polymethylmethacrylate. Dokl Acad Nauk 1992; 323:326–329.
56. Skripov VP, Supikov MK. Effect on mechanical properties of polymethylmethacrylate by saturation with high-pressure carbon dioxide. Polym Mech 1971; 2:243–246 (in Russian).
57. Goel SK, Beckman EJ. Generation of microcellular polymeric foams II. Cell growth and skin formation. Polym Eng Sci 1994; 34:1148–1156.
58. Goel SK, Beckman EJ. Nucleation and growth in microcellular materials: supercritical CO_2 as foaming agent. AIChE J 1995; 41:357–367.
59. Kumar V, Weller JE. Microcellular polycarbonate I. Experiments on bubble nucleation and growth. ANTEC 1991; 91:1401–1405.
60. Wessling M, Borneman Z, Van den Boomgaard T, Smolders CA. Carbon dioxide foaming of glassy polymers. J Appl Polym Sci 1994; 53:1497–1512.
61. Shi C, Huang Z, Kilic S, Xu J, Enick RM, Beckman EJ, Carr AJ, Melendez RE, Hamilton AD. The gelation of CO_2: a sustainable route to the creation of microcellular materials. Science 1999; 286:1540–1543.
62. Park CB, Liu YJ, Naguib HE. Challenge to forty-fold expansion of biodegradable polyester foams using carbon dioxide as a blowing agent. Cell Polym 1999; 18:367–384.
63. Sparacio D, Beckman EJ. Generation of microcellular biodegradable polymers in supercritical carbon dioxide. Polym Prepr 1997; 38:422–423.
64. Nof M, Shea LD. Drug-releasing scaffolds fabricated from drug-loaded microspheres. J Biomed Mater Res 2002; 59:349–356.
65. Hile DD, Amirpour ML, Akgerman A, Pishko MV. Active growth factor

delivery from poly(D,L-lactide-*co*-glycolide) foams prepared in supercritical CO_2. J Controlled Release 2000; 66:177–185.

66. Koegler WS, Patrick C, Cima MJ, Griffith LG. Carbon dioxide extraction of residual chloroform from biodegradable polymers. J Biomed Mater Res 2002; 63:567–576.

67. Whitaker MJ, Quirk RA. Howdle SM, Shakesheff KM. Growth factor release from tissue engineering scaffolds. J Pharm Pharmacol 2001; 53:1427–1437.

68. Howdle SM, Watson MS, Whitaker MJ, Popov VK, Davies MC, Mandel FS, Wang JD, Shakesheff KM. Supercritical fluid mixing: preparation of thermally sensitive polymer composites containing bioactive materials. Chem Commun 2001; 109–110.

69. Perman C. 3M Company, 5,505.060 US patent, 1996, August 16.

70. Serajuddin ATM. Solid dispersion of poorly water-soluble drugs: early promises, subsequent problems, and recent breakthroughs. J Pharm Sci 1999; 88:1058–1066.

71. Guney O, Akgerman A. Synthesis of controlled-release products in supercritical medium. AIChE J 2002; 48:856–866.

72. Tantishaiyakul V, Kaewnopparat N, Ingkatawor333333nwong S. Properties of solid dispersions of piroxicam in polyvinylpyrrolidone. Int J Pharm 1999; 181:143–151.

73. Raghavan SL, Kiepfer B, Davis AF, Kazarian SG, Hadgraft J. Membrane transport of hydrocortisone acetate from supersaturated solutions: the role of polymers. Int J Pharm 2001; 221:95–105.

74. Taylor LS, Zografi G. Spectroscopic characterization of intercations between PVP and indomethacin in amorphous molecular dispersions. Pharm Res 1997; 12:1691–1698.

75. Doherty C, York P. Mechanisms of dissolution of frusemide–PVP solid dispersions. Int J Pharm 1987; 34:197–205.

76. Forster A, Hempenstall J, Rades T. Investigation of drug/polymer interaction in glass solutions prepared by melt extrusion. Internet J Vib Spectrosc 2001; 5:6–20 [www.ijvs.com].

77. Breienbach J, Schrof W, Neumann J. Confocal Raman-spectroscopy: analytical approach to solid dispersions and mapping of drugs. Pharm Res 1999; 16:1109–1113.

78. Macnaughton SJ, Kikic I, Foster NR, Alessi P, Cortesi A, Colombo I. Solubility of anti-inflammatory drugs in supercritical carbon dioxide. J Chem Eng Data 1996; 41:1083–1086.

79. Charoenchaitrakool M, Dehgani F, Foster NR, Chan HK. Micronization by rapid expansion of supercritical solutions to enhance the dissolution rates of poorly water-soluble pharmaceuticals. Ind Eng Chem Res 2000; 39:4794–4802.

80. Kikic I. Preparation of drug delivery systems through impregnation with supercritical fluids. 5th International Symposium on Supercritical Fluids, Atlanta, 2000.

81. Kikic I, Lora M, Cortesi A, Sist P. Sorption of CO_2 in biocompatible polymers: experimental data and qualitative interpretation. Fluid Phase Equilibria 1999; 158–160:913–921.

82. Shakhtshneider TP, Vasiltchenko MA, Politov AA, Boldyrev VV. The mechano-chemical preparation of solid disperse systems of ibuprofen–polyethylene glycol. Int J Pharm 1996; 130:25–32.

83. Striolo A, Bratko D, Prausnitz JM, Elvassore N, Bertucco A. Influence of polymer structure upon active-ingredient loading: a Monte Carlo simulation study for design of drug-delivery devices. Fluid Phase Equilibria 2001; 183:341–350.

84. Kazarian SG. Polymers and supercritical fluids: opportunities for vibrational spectroscopy. Macromol Symp 2002; 184:215–228.

85. Kazarian SG, Higgins JS. A closer look at polymers. Chem Ind May 20, 2002; 21–23.

86. Charoenchaitrakool M, Dehghani F, Foster NR. Utilization of supercritical carbon dioxide for complex formation of ibuprofen and methyl-β-cyclodextrin. Int J Pharm 2002; 239:103–112.

87. Stanton LA, Dehghani F, Foster NR. Improving drug delivery using polymers and supercritical fluid technology. Aust J Chem 2002; 55:443–447.

88. Debenedetti PG, Tom JW, Yeo SD, Lim GB. Application of supercritical fluids for the production of sustained delivery devices. J Controlled Release 1993; 24:27–44.

89. Yeo S-D, Lim G-B, Debenedetti PG, Bernstein H. Formation of micro-particulate protein powders using a supercritical fluid antisolvent. Biotechnol Bioeng 1993; 41:341–346.

90. Ribeiro Dos Santos I, Richard J, Pech B, Thies C, Benoit JP. Microencapsulation of protein particles within lipids using a novel supercritical fluid process. Int J Pharm 2002; 242:69–78.

91. Cortesi A, Alessi P, Kikic I, Kirchmayer S, Vecchione F. Supercritical fluid chromatography for impregnation optimization. J Supercrit Fluids 2000; 19:61–68.

92. Domingo C, Garcia-Carmona J, Fanovich MA, Llibre J, Rodriguez-Clemente R. Single or two-solute adsorption processes at supercritical conditions: an experimental study. J Supercrit Fluids 2001; 21:147–157.

93. Domingo C, Garcia-Carmona J, Fanovich MA, Saurina J. Study of adsorption processes of model drugs at supercritical conditions using partial least squares regression. Anal Chim Acta 2002; 452:311–319.

94. Domingo C, Garcia-Carmona J, Fanovich MA, Saurina J. Application of chemometric techniques to the characterisation of impregnated materials obtained following supercritical fluid technology. Analyst 2001; 126:1792–1796.

95. Domingo C, Fanovich A, San Roman J, Rodriguez-Clemente R. Possibilities of SCF technology in the preparation of drug delivery systems. Comparison with conventional technologies. 8th Meeting on Supercritical Fluids, Antibes, France, 2000:137–142.

96. Kerc J, Srcic S, Knez Z, Sencar-Bozic P. Micronization of drugs using supercritical carbon dioxide. Int J Pharm 1999; 182:33–39.

9

Formulation of Controlled-Release Drug Delivery Systems

Nagesh Bandi
GlaxoSmithKline, Parsippany, New Jersey, U.S.A.

Christopher B. Roberts and Ram B. Gupta
Department of Chemical Engineering, Auburn University,
Auburn, Alabama, U.S.A.

Uday B. Kompella
Department of Pharmaceutical Sciences, University of Nebraska
Medical Center, Omaha, Nebraska, U.S.A.

Several difficult-to-treat disorders including diabetes and cancer can benefit from controlled-release drug delivery systems. If a drug delivery system can provide some control, whether temporal, spatial, or both, it is considered to be a controlled-release system. The advantages of these products include predictable drug concentrations, reduced dosing frequency, enhanced patient compliance, and cost-effectiveness (1). Controlled-release products can be obtained by coformulating a pharmaceutically active substance with a polymeric excipient that is capable of controlling the drug release. Currently marketed controlled-release products are intended for various routes of administration including parenteral, transdermal, oral, and ocular. While

TABLE 1   Commercial Preparations of Sustained and Controlled Release
Products

Implants

Viadur (leuprolide acetate implant): a nondegradable intramuscularly
injectable system that delivers leuprolide, a luteinizing hormone-releasing
hormone agonist continuously for 12 months as a palliative treatment for
advanced prostate cancer.

Progestasert (progesterone): a nondegradable intrauterine contraceptive
system for localized delivery of progesterone for one year.

Vitrasert (ganciclovir): a nondegradable intravitreal system for localized
delivery of ganciclovir over 6 months for patients with cytomegalovirus
retinitis.

Actisite (tetracycline): a nondegradable 10-day periodontal fiber, for
reduction of pocket depth and bleeding on probing in patients with adult
periodontitis.

Ocusert (pilocarpine): a nondegradable insert for one-week delivery of the
drug for the control of elevated intraocular pressure in pilocarpine-
responsive glaucoma patients.

Zoladex (gosorelin acetate implant): biodegradable polymeric particles
capable of continuously delivering gosorelin for 3 months as a palliative
treatment for advanced prostate cancer.

Microparticles

Lupron Depot (leuprolide acetate for depot suspension): biodegradable
polymeric particles capable of continuously delivering leuprolide for 1, 3,
or 4 months for gynecological or urological disorders.

Liposomes

Doxil (doxorubicin): an anticancer drug for the treatment of metastatic
ovarian cancer in patients with disease that is refractory to both paclitaxel-
and platinum-based chemotherapy regimens and for the treatment of
AIDS-related Kaposi's sarcoma with disease that has progressed.

Transdermal patches

Catapres-TTS (clonidine): Once-weekly product for the treatment of
hypertension.

Duragesic (fentanyl) CII: 3-day patch for the management of chronic pain in
patients who require continuous opioid analgesia for pain that cannot be
managed by lesser means.

Estraderm (estradiol): twice-weekly product for treating postmenopausal
symptoms and preventing osteoporosis.

Transderm Scōp (scopolamine): 2-day patch for prevention of nausea and
vomiting associated with motion sickness.

Testoderm TTS (testosterone) and Testoderm: 1-day patches for replacement
therapy in men with a deficiency or absence of testosterone.

Transderm-Nitro (nitroglycerin) once-daily product for the prevention of
angina pectoris due to coronary artery disease.

TABLE 1 Continued

Oral sustained-release preparations

Alpress LP (prazosin): once-daily extended-release tablet for the treatment of hypertension.

Concerta (methylphenidate HCl) CII: once-daily extended-release tablet for the treatment of attention deficit hyperactivity disorder (ADHD) in patients aged 6 and older.

Covera-HS (verapamil): a controlled-onset, extended-release (COER-24) system for the management of hypertension and angina pectoris.

Ditropan XL (oxybutynin chloride): once-a-day extended-release tablet for the treatment of overactive bladder characterized by symptoms of urinary incontinence, urgency, and frequency.

DynaCirc CR (isradipine): once-daily extended-release tablet for the treatment of hypertension.

Efidac 24 (chlorpheniramine): over-the-counter extended-release tablet providing 24 h relief from allergy symptoms and nasal congestion.

Glucotrol XL (glipizide): extended-release tablet used as an adjunct to diet for the control of hyperglycemia in patients with non-insulin-dependent diabetes.

Sudafed 24 Hour (pseudoephedrine): over-the-counter nasal decongestant for 24 h relief of colds, sinusitis, hay fever, and other respiratory allergies.

Procardia XL (nifedipine) extended-release tablet for the treatment of angina and hypertension.

Volmax (albuterol) extended-release tablet for relief of bronchospasm in patients with reversible obstructive airway disease.

---

the majority of the marketed products including implants, polymeric microparticles, and patches are intended for sustaining the drug delivery, some, such as Doxicil (doxorubicin), are intended for achieving temporal control as well as spatial control of drug delivery. While providing continuous drug release in the blood owing to the use of long-circulating liposomes, Doxicil minimizes drug access to the heart and the associated side effects. Table 1 provides examples of controlled-release products that are currently marketed for various applications. For the purpose of this chapter, the delivery systems can be broadly classified as microscopic or macroscopic. The microscopic systems include the particulate systems, which exist in the micrometer to submicrometer size range including nanospheres, microparticles, microcapsules, and liposomes. The macroscopic systems include porous foams, films, and coated products.

Controlled-release products typically employ one or more polymeric excipients to control drug release. Such polymers include artificial or natural polymers that are biocompatible and/or biodegradable. Examples of the nondegradable polymers include poly(2-hydroxyethyl methacrylate), poly(N-vinyl pyrrolidone), poly(methyl methacrylate), poly(vinyl alcohol), poly(acrylic acid), polyacrylamide, poly(ethylene-co-vinyl acetate), poly(ethylene glycol), and poly(methacrylic acid). Examples of biodegradable polymers include polylactides (PLA), polyglycolides (PGA), poly(lactide-co-glycolides) (PLGA), polyanhydrides, and polyorthoesters (2–4). While the parenteral nondegradable systems require surgical placement and removal, the biodegradable polymers are broken down into biologically acceptable molecules that are metabolized and removed from the body via normal metabolic pathways. To be successfully used in controlled drug delivery formulations, the polymeric material must be chemically inert and free of leachable impurities. It must also have an appropriate physical structure for easy processibility. As described in Chapter 8, such polymeric properties can be tailored using supercritical fluid (SCF) processing. This chapter describes the use of supercritical fluid (SCF) technology for the preparation of polymer-based controlled-release drug delivery systems.

1. DRUG DELIVERY APPLICATIONS OF SCFs

1.1. Supercritical Fluid Processes for Drug Delivery

As described in Chapter 3, several SCF techniques are available for the preparation of drug delivery systems. These include rapid expansion of supercritical solutions (RESS), gas antisolvent recrystallization (GAS), supercritical antisolvent recrystallization (SAS), supercritical antisolvent with enhanced mass transfer (SAS-EM), solution-enhanced dispersion by supercritical fluids (SEDS), supercritical fluid nucleation (SFN), precipitation with compressed antisolvent (PCA), and aerosolized supercritical extraction of solvents (ASES). While RESS and SFN involve the expansion of a supercritical fluid solution of a drug to form drug particles, GAS, SAS, SAS-EM, SEDS, PCA, and ASES use a supercritical fluid as an antisolvent to precipitate particles of a drug dissolved in an organic solvent (5). General RESS and GAS processes are further elaborated in Sections 1.1.1 and 1.1.2.

1.1.1. RESS

RESS is useful for materials that are soluble in CO_2. Unfortunately, CO_2, with no dipole moment and very low polarizability, is a very weak solvent and dissolves very few polymers. Cosolvents such as methanol or acetone can be mixed with SCFs to increase the solvating power of SCFs during RESS. In drug delivery applications, RESS has been used to prepare polymeric films, microparticles, nanospheres, liposomes, and porous foams (Figure 1). A

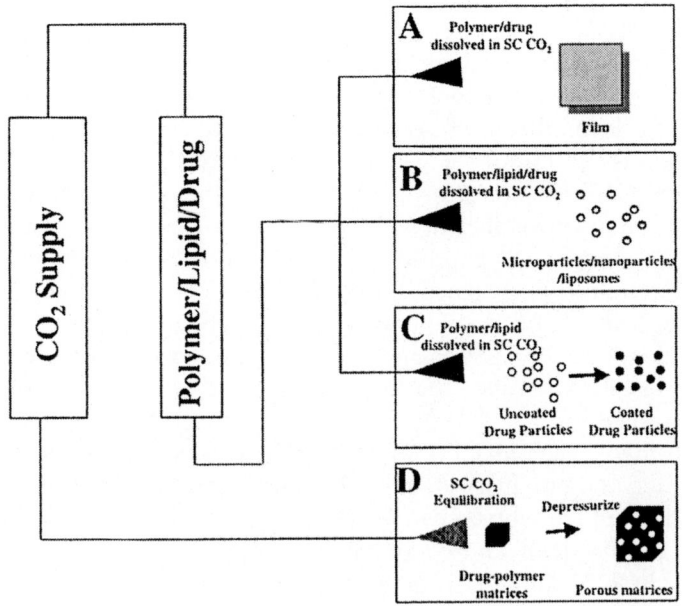

FIGURE 1 Formulation of controlled drug delivery systems using the super-critical fluid–derived RESS process: (A) films, (B) microparticles/nanoparticles/liposomes, (C) coated products including microcapsules, and (D) microporous foams.

typical RESS apparatus consists of a preheater, an extraction unit, and a precipitator. After the solute has been placed in the extraction unit, the compressed and preheated supercritical solvent is passed through this unit. When the solute has dissolved in the supercritical fluid, the supercritical solution is expanded through a capillary or laser-drilled nozzle into a precipitation unit. These nozzles typically have internal diameters between 25 and 150 μm and short lengths to prevent precipitation inside the nozzle due to the rapid pressure drop. Rapid decompression of the solution in the precipitation unit leads to high supersaturation and precipitation of desired solids. High supersaturation ratios and rapidly propagating mechanical perturbations greater than those achieved by thermal perturbations afford RESS the potential to produce small, uniformly sized particles such as microparticles, nanospheres, and liposomes (6–9). Product contamination is minimal since the solvent is a dilute gas after expansion and other solutes need not be introduced to initiate precipitation. Pre- and post-expansion temperature and pressure and nozzle geometry are among the factors that determine the size and morphology of delivery systems precipitated by means of RESS.

1.1.2. GAS

GAS recrystallization uses a dense gas as an antisolvent for precipitating a solute that has been dissolved in an organic solvent (or another SCF). Introduction of a gas into a batch of organic solvent containing a dissolved solute will result in dissolution of the gas (many gases are at least partially miscible with organic solvents), resulting in a lowering of the solvent strength of the liquid phase. Following sufficient reduction in the solvent strength, the liquid solution will no longer be a good solvent for the solute, and nucleation followed by precipitation will take place (Figure 2). This process is termed gas antisolvent recrystallization. For a solute to be processed by GAS, it must be soluble in an organic solvent or an SCF, and it must be reasonably insoluble in antisolvent gas or SCF. Moreover, the solvent must be at least partially miscible with the antisolvent.

Supercritical fluids, such as CO_2, often exhibit weak solvent strength toward many organic and polymeric solutes. This antisolvent or nonsolvent nature of these SCFs for many solutes can further be employed by spraying an organic solution of the drug and/or polymer through a nozzle into a compressed gas or SCF (Figure 2). This process, termed precipitation with

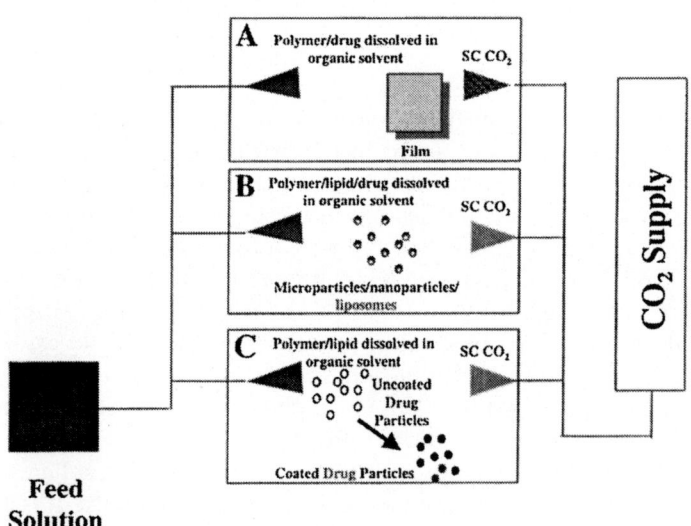

FIGURE 2 Formulation of controlled drug delivery systems using the super-critical fluid–derived GAS process: (A) films, (B) microparticles/nanoparticles/liposomes, and (C) coated products including microcapsules.

compressed antisolvents (PCA), employs either a liquid or supercritical antisolvent. When a supercritical antisolvent is used, the process is often referred to as a supercritical antisolvent (SAS) process or aerosol spray extraction system (ASES). GAS, PCA, SAS and ASES processes offer the flexibility of solvent choice and higher solute throughput than RESS. For this reason, GAS and these related processes were used more often in producing drug delivery systems. These systems can range from films to nanospheres. Various nozzle types, including standard capillary nozzles, ultrasonic atomizers, and coaxial nozzles, have been employed to spray solutions in these processes. The operating pressure, temperature, jet breakup, droplet size, mass transfer rates between the droplet, and antisolvent phase control particle size and morphology. Jet breakup and droplet sizes depend on the nozzle configuration, the spray velocity, and the physical properties of the droplet and antisolvent phases. The rate of antisolvent addition will also influence the size of the final product. With slow antisolvent addition, there is low supersaturation, and relatively few nuclei are formed. Since a large quantity of the solute still remains in the solvent, these nuclei will grow to large crystals. With rapid antisolvent addition, there is high supersaturation, and large numbers of nuclei are formed. The amount of solute remaining in the solvent is low, however, and so the nuclei will not grow much.

2. DRUG DELIVERY SYSTEMS

2.1. Microparticles

Microparticles are polymeric drug delivery systems in the size range of 1 to 1000 μm. By encapsulating drugs in these systems, it is possible to prolong the release of the drug (10–12). Conventional methods for microencapsulation include liquid antisolvent precipitation, spray drying, solvent evaporation, and organic phase separation techniques. Because the polymers are only poorly soluble in water, these methods use large amounts of environmentally hazardous organic solvents, and the high residual organic solvent contents are a major concern based on U.S. Pharmacopeia (USP) specifications. In addition, these methods are likely to produce microparticles with variable size distributions, as well as reduced encapsulation efficiencies. Since particle size and/or morphology and encapsulation efficiency dictate the rate and extent of drug release, precise control of these parameters is desirable. Supercritical fluid technology is likely to overcome some of the limitations of the conventional methods in producing microparticles. This approach minimizes the number of steps in particle formation and also reduces the amount of organic solvent used and the residual content. Several investigators successfully employed SCFs as either solvent or antisolvents in preparing microparticles.

2.1.1. Microparticles Preparation with Supercritical Fluid as a Solvent

Both RESS and different antisolvent techniques have been used for the SCF preparation of various drug and polymeric microparticles. Tom and Debennedetti (8) used a RESS process to produce DL-poly(lactic acid) (DL-PLA) particles of lovastatin, a cholesterol-reducing drug with a molecular weight of 404.55. Owing to differences in the nucleation and growth rates of the drugs and polymers in this study, particles containing embedded lovastatin needles were formed. Although drug microparticles suitable for drug delivery were not formed, this was the first study to explore the coprecipitation of a drug and a polymer from a supercritical solvent. The RESS process has also been used to coprecipitate naproxen, an anti-inflammatory agent (MW 230) with L-poly(lactic acid) (L-PLA) from supercritical carbon dioxide (13). In this study, supercritical carbon dioxide was passed through an extraction vessel containing a mixture of naproxen and L-PLA and then expanded through a 50 μm capillary tube. Microparticles of naproxen in L-PLA were formed at 190 bar and a preexpansion temperature of 114°C. The microparticles produced were spherical, with a diameter of 20 to 90 μm. Table 2 and 3 summarize various drug–carrier particles formed when the RESS process was used.

In addition to low molecular weight drugs, the RESS process is capable of encapsulating macromolecular drugs such as proteins and peptides. Mishima et al. (14) reported a new microencapsulation of lipase (from *Pseudomonas cepacia*) and lysozyme (from chicken egg white) in polymeric particles. The particles were prepared using polymers such as poly(ethylene glycol) (PEG 4000, MW 3000; PEG 6000, MW 7500; PEG 20,000, MW 20,000), poly(methyl methacrylate) (MW 15,000), poly(L-lactic acid) (MW 5000), poly(DL-lactide-*co*-glycolide) (MW 5000), and PEG–poly(propylene glycol) [PPG- PEG triblock copolymer, PEG-PPG-PEG, MW 13,000. The solubility of these polymers in CO_2 was enhanced by using low molecular weight alcohols (methanol, ethanol, and I-propanol) as cosolvents. In this process, known amounts of polymer, protein, and cosolvent were placed in the high pressure vessel containing supercritical CO_2. This mixture was stirred by an agitator rotating at 200 rpm for about 4 h and then kept for more than 1 h without agitation. Prior to the expansion of the polymer solution, it was confirmed by observation through the sapphire window that the entire polymer was dissolved and the proteins dispersed homogeneously. The dissolved polymer–dispersed protein was sprayed through a capillary nozzle toward a target plate. Parameters such as the thickness of the polymer coating the protein, particle size, morphology, and particle size distribution were controlled by changing the feed composition of the polymer. For instance, increasing the feed concentration of PEG-6000 from 10 wt % to 20 wt %

TABLE 2 Polymeric Microparticles Prepared by Using RESS[a]

Polymer	Process/ Solvent[b]	Nozzle i.d. (μm)	Temperature/ Pressure (°C/bar)[c]	Mean size (μm)	Morphology	Ref.
DL-PLA	RESS/sc CO_2		55/200; 82–85; 15–25/0.5	10–20	Irregular	8
PGA	RESS/sc CO_2		55/180–200; 80–82; 15–22/7–9	10–20	Regular (oval/ rectangular)	8
L-PLA	RESS/sc CO_2		55/200–230; 75–100; 15–37/10–15	10–25	Microparticles	8
			55/200–230; 75–100; 49–92/ 10–15		Dendrites	
L-PLA	RESS/sc $CClF_3$		55/114–120; 70–83; 27–32/14–17	2–20	Microparticles (regular)	8
PCL	RESS/sc CDFM	30	100/138; 110; —		Elongated	67
			100/172; 110; —		Powder	
			100/207; 110; —		Powder	
PCL	RESS/sc CDFM	50	90/138; 90; —		Powder	67
			100/138; 110; —		Fiber	
			100/138; 145; —		Spherulite	
PMMA	RESS/sc CDFM	30	100/82.7; 110; —		Fiber	67
			100/110; 110; —		Fiber	
			100/138; 110; —		Fiber	
			100/172; 110; —		Powder	
			100/207; 110; —		Powder	
			70/172; 130; —		Fiber	
PMMA	RESS/sc CDFM	50	70/107; 110; —		Fiber	67
			100/107; 110; —		Powder	
			100/107; 130; —		Fiber	

The RESS process consists of passing the supercritical solvent through an extraction chamber containing the solute followed by expansion into a chamber for precipitation.

Abbreviations: sc, supercritical; CDFM, chlorodifluoromethane; PCL, polycaprolactone; PGA, polyglycolic acid; PLA, polylactic acid; PMMA, poly(methyl methacrylate).

Data given in the following order; extraction temperature and pressure (°C/bar), preexpansion temperature (°C), and expansion temperature/pressure (°C/bar).

TABLE 3 Drug-Containing Polymeric Microparticles Prepared by Using Supercritical Fluid Techniques

| Polymer/Drug[a] | Process/ Solvent[b] | Nozzle i.d. (μm) | Flow rates | | Temperature/ Pressure (°C/bar) | Mean size (μm) | Morphology | Ref. |
			CO$_2$	Solution (mL/min)				
PLA/naproxen	RESS/scCO$_2$	50	—	—	Extraction temperature/ pressure: 60/187–210 Preexpansion temperature: 90–140 Expansion temperature: 29–40	20–90		13
PEG/ lysozyme or lipase	RESS/ethanol	280	20 mL/min	1	35/20 MPa	15–30	Spherical	14
PMMA/ lysozyme or lipase	RESS/ethanol	280	20 mL/min	1	35/20 MPa	15–30	Spherical	
PLA/lysozyme or lipase	RESS/ethanol	280	20 mL/min	1	35/20 MPa	15–30	Spherical	
PLGA/lysozyme or lipase	RESS/ethanol	280	20 mL/min.	1	35/20 MPa	15–30	Spherical	

PEG-PPG-PEG triblock copolymer/lysozyme or lipase	RESS/ethanol	280	20 mL/min	1	35/20 MPa	15–30	Spherical	17
PLA/insulin	GAS/CH$_2$Cl$_2$/DMSO	50	1200–1800 L/h	1	38/105	0.5–2	Spherical and smooth	21
PLGA/estriol (100/110)[b]	ASES/CH$_3$OH + TFE + CH$_2$Cl$_2$	—	11 kg/h	6	34/100	20	Spherical and agglomerated	21
PLGA/BSA (100/114.15)[b]	ASES/CH$_3$OH + TFE + CH$_2$Cl$_2$	—	11 kg/h	6	34/100	16	Spherical and agglomerated	21
Triblock polymer[c]/estriol (100/103.21)[b]	ASES/CH$_3$OH + TFE + CH$_2$Cl$_2$	—	11 kg/h	6	34/100	10	Spherical	21
Triblock polymer[c]/BSA (100/105.6)[b]	ASES/CH$_3$OH + TFE + CH$_2$Cl$_2$	—	11 kg/h	6	34/100	8	Spherical	21

[a] Abbreviations: ASES, aerosol solvent extraction system; BSA, bovine serum albumin; PCA, precipitation with compressed antisolvent; PLA, polylactic acid; PLGA: poly(lactic-co-glycolic acid); PMMA, poly(methyl methacrylate); PPG, poly(propylene glycol); RESS, rapid expansion of supercritical solution; TFE, 2,2,2-trifluoroethanol.
[b] Percent theoretical loading/experimental loading.
[c] Dissolved in dichloromethane by hydrophobic ion pairing with AOT, sodium bis(2-ethylhexyl) sulfosuccinate.
[d] β-poly(L-lactide-co-D,L-lactide-co-glycolide) (62 = 12.5 = 25).
[e] Vibrating nozzle at 120 kHz.

resulted in a particle size increase from 20 μm to 70 μm. However, no information regarding the efficiency of protein encapsulation was provided.

2.1.2. Microparticle Preparation with Supercritical Fluid as an Antisolvent

Most drugs do not exhibit good solubility in supercritical carbon dioxide, and therefore a large number of studies employed CO_2 as an antisolvent. Tables 3 and 4 summarize drug–carrier microparticles formed by using the GAS and related antisolvent approaches. Based on the antisolvent properties of a SCF, Mueller and Fisher (15) first patented a process that uses SCFs to produce polymeric drug microparticles. They used an antisolvent process wherein the drug (clonidine hydrochloride) and the polymer (DL-PLA) in methylene chloride were sprayed into a vessel containing supercritical carbon dioxide. The diffusion of methylene chloride into the supercritical CO_2 caused polymeric drug particles to precipitate. L-PLA particles containing indomethacin and chlorpheniramine maleate were also prepared by means of a PCA process, wherein a methylene chloride solution containing the drug and a polymer were sprayed into a vessel containing supercritical CO_2, which served as an antisolvent (16). The effects of such process variables as temperature, pressure, antisolvent flow rate, and polymer concentration on the morphology and size of L-PLA microparticles were studied. Low polymer concentrations (1%), high CO_2 flow rate and density, and low temperatures (below the glass transition temperature T_g of the polymer) favored the formation of L-PLA microparticles of small diameters and narrow size distribution. The drug-encapsulated microparticles were similar in appearance to pure L-PLA particles but drug-loading efficiency was low (3.73 and 0.73% for chlorpheniramine maleate and indomethacin, respectively, vs 10% theoretical loading), indicating a need for further optimization of the process. The precipitation process can also be used to encapsulate bioactive macromolecules such as insulin. Elvassore et al. (17) used a GAS process to obtain insulin microparticles encapsulated with PLA. To achieve microencapsulation, homogeneous solutions of protein and polymer in dichloromethane or dimethyl sulfoxide (DMSO) were sprayed into supercritical CO_2. The precipitation process resulted in PLA-encapsulated insulin microparticles with an average diameter of 0.5 to 2 μm. The results of the study also indicated that the process did not affect insulin bioactivity. Thus, the SCF process is suitable for preparing particulate carriers of macromolecule therapeutics without affecting their bioactivity.

Several polysaccharides are currently used as drug delivery systems in the form of microparticles, and various methods were used to prepare such microparticles. Besides processing synthetic polymers, the precipitation process can also be used for producing microparticles with natural polymers.

Reverchon et al. (18) produced submicrometer particles of various natural polymers, such as inulin, dextran, and poly(hyaluronic acid) (HYAFF11). Also, Elvassore et al. (19) formulated microparticles of derived from polymers hyaluronic acid (HYAFF11-p100, HYAFF11-p75, HYAFF11-p80, HYAFF302).

The aerosol solvent extraction system has been used by many authors to prepare microparticles (20,21). Bleich et al. (20) tested a variety of biodegradable polymers including DL-PLA, DL-PLGA with lactide glycolide ratios of 50:50 and 75:25, L-PLA and poly(β-hydroxybutyric acid) (PHB) for their suitability to be processed by an ASES process utilizing supercritical carbon dioxide. It was found that both DL-PLA and DL-PLGA (75:25) were extracted by the supercritical carbon dioxide. DL-PLGA (50:50) formed a film at the bottom of the collecting vessel. Both the semicrystalline polymers, DL-PLA and PHB, formed spherical microparticles with mean sizes between 1 and 10 µm (22).

Ghaderi et al. (23) have used solution-enhanced dispersion by supercritical fluids (SEDS) to produce microparticles of DL-PLG, L-PLA, DL-PLA, and PCL. Solutions of these polymers in different organic solvents were sprayed into supercritical carbon dioxide. The size and morphology of the particles formed strongly depended on the solvent used, the polymer concentration, and the temperature, pressure, and density of the antisolvent gas. The authors in this study used a mixture of solvents to dissolve different polymers. For DL-PLGA, they used a mixture of acetone, ethyl actetate, and isopropanol in the ratio of 4:5.6:0.4. These solvents are more environmentally acceptable than other commonly used organic solvents. Also, the addition of a small amount of isopropanol, which is a poor solvent for DL-PLG, helps in attaining higher supersaturation, thereby producing discrete microparticles of this amorphous polymer, which in earlier studies did not form microparticles (16,20). L-PLA formed discrete microparticles with mean diameters from 0.5 to 5.3 µm under various processing conditions (temperature and pressure 35–40°C and 130–185 bar, respectively). L-PLA, a semi crystalline polymer, was more easily processed by the SEDS technique, and also the size distribution was sensitive to changes in the CO_2 density, with particle size increasing (0.5 to 5.3 µm) with decreasing CO_2 density (0.85 to 0.69 g/mL). The amorphous polymers DL-PLA and PCL also formed microparticles with mean diameters ranging from 105 to 216 µm and 27 to 212 µm, respectively, at the different conditions of temperature and pressure tested. Thus, this study demonstrated the applicability of SCF processing to the preparation of microparticles of biodegradable polymers using low amounts of environmentally benign solvents.

Solid lipid particles offer another interesting formulation option for the drug delivery scientist. Solid lipids particles are lipid-based solid colloidal

TABLE 4 Polymeric Microparticles Prepared by Using Supercritical Antisolvent Processes

Polymer/Drug[a]	Process/Solvent[a]	Nozzle i.d. (μm)	Flow rate[b] CO₂	Flow rate[b] Solution (mL/min)	Temperature/ Pressure (°C/bar)	Mean size (μm)	Morphology	Ref.
PLA	ASES/CH_2Cl_2		2–11 kg/h	1, 2.3, 5.5, 8.7, 10	40/90	<10	Agglomerated and spherical	61
PLA	ASES/CH_2Cl_2	40	6.4 kg/h	3	40/90 40/200	2.29 1.76	Spherical	20
PHB	ASES/CH_2Cl_2	40	6.4 kg/h	3	40/90 40/200	5.77 2.23	Spherical	20
PLGA	ASES/TFE + CH_2Cl_2		11 kg/h	6	34/100	20	Agglomerated	21
Triblock polymer	ASES/TFE + CH_2Cl_2		11 kg\h	6	34/100	5	Spherical	21
PLA	GAS/CH_2Cl_2	75	5.34 scfm	0.02–2	31/55.1 31/62 31/75.8 31/82.7 31/96.5	1.43 0.61 0.95 1.40	Irregular film Spherical	68
Polystyrene	PCA/toluene	100	—	—	10–40/ 39.6–224.7 (0.86–0.13)[c]	0.1–20	Uniform microparticles	69
PLA	PCA/CH_2Cl_2	100	—	23/81.6	0/81.6 (0.9049)[c] 1–5 (0.8069)[c] 32/81.6 (0.6850)[c]	<1	Spherical, free flowing, and nonagglomerated	16
Polystyrene	PCA/toluene	50/760[d]	35 mL/min	1	23/—(0.85)[c]	0.1–1	Individual mostly and slightly agglomerated	70

Polymer	System		Flow rate			Particle size	Morphology	Ref.
Polystyrene	PCA/THF	50	35 mL/min	1	0/(0.96)^c	0.1–0.3	Flocculated and slightly agglomerated	70
PAN	PCA/DMF	50/760^d	35 mL/min	1	40/—(0.66)^c	0.1–0.3	Mixture of microparticles; individual fibrils and oriented fibrils	70
PAN	PCA/DMF	50	35 mL/min	1	40/—(0.66)^c	0.1–0.3	Microfibrils	70
L-PLA 1%	PCA/CH$_2$Cl$_2$	50/760^d	35 mL/min	0.5, 1	23/—(0.85)^c	0.5–5	Individual mostly and slightly agglomerated	70
L-PLA 3%	PCA/CH$_2$Cl$_2$	50/760^c	35 mL/min	1	23/(0.85)^c	0.1–1	Individual mostly and slightly agglomerated	70
HYAFF-11	SAS/DMSO	40	12 slpm	1	35 and 40/85	0.4	Spherical and narrow size distribution	71
DL-PLGA (inherent viscosity: 0.63 dL/g)	SEDS/A:E:I (4:5.6:0.4)		21 mL/min	0.13	33/185, 38/130, 38/185	159, 173, 158	Porous and irregular	23
DL-PLGA (inherent viscosity: 0.78 dL/g)					38/130, 33/185, 33/130, 38/185	140, 130, 126, 136	Spherical and discrete	23
DL-PLGA (inherent viscosity: 1.07 dL/g)	SEDS/DM: CH$_3$OH (9:1)		21 mL/min	0.13	34/185	10	Bead network	23
PCL	SEDS/DM:A:I (1.1:6.2:2.7)		21 mL/min	0.13	35/130, 35/185, 40/130, 40/185	106, 189, 27, 212	Large and irregular	23
L-PLA	SEDS/DM:A:I (3.3:6.5:0.2)		21 mL/min	0.13	35/130 (0.77)^c, 35/185 (0.69)^c, 40/130 (0.85)^c	0.72, 0.5, 5.3	Irregular	23

TABLE 4 Continued

Polymer/Drug[a]	Process/ Solvent[a]	Nozzle i.d. (μm)	Flow rate[b]		Temperature/ Pressure (°C/bar)	Mean size (μm)	Morphology	Ref.
			CO$_2$	Solution (mL/min)				
DL-PLA	SEDS/A:E:H (1:7.8:1.2)		21mL/min	0.13	40/185 (0.82)[c]	0.6	Porous and a	23
					35/130	135	mixture of	
					35/185	216	discrete and	
					40/130	178	agglomerated	
					40/185	105		

[a] Abbreviations: A, acetone; ASES, aerosol solvent extraction system; DM, dichloromethane; DMF, N,N-dimethyl-formamide; E, ethanol; GAS, gas antisolvent process; H, hexane; HYAFF-11, hyaluronic acid benzylic ester; I, isopropanol; PAN, polyacrylonitrile; PCA, precipitation with compressed antisolvent; PCL, polycaprolactone; PHB, poly(β-hydroxybutyric acid); PLA, polylactic acid; PLGA, poly(lactic-co-glycolic acid); SAS, supercritical antisolvent process; SEDS, solution enhanced dispersion by supercritical fluids; TFE, 2,2,2-trifluoroethanol; Triblock polymer, β poly(L-lactide-co-D,L-lactide-co-glycolide)(62.5:12.5:25).

[b] Abbreviations: scfm, standard cubic feet per minute; slpm, standard liters per minute.

[c] CO$_2$ density (g/cm^3).

[d] Coaxial nozzle with 50 μm internal diameter for inner nozzle and 760 μm internal diameter for outer nozzle.

carrier systems that are useful for topical, oral, and parenteral administration of drugs (24). Solid lipid particles are biodegradable, scalable for manufacturing, biocompatible, and sterilizable. Solid lipid particles are being investigated for possible use in improving drug bioavailability, protecting drugs from decomposition, and sustaining drug release. Solid lipid particles are also being investigated as vaccine adjuvants. The lipids commonly used for formulating solid lipid particles are triglycerides (tricaprin, trilaurin, trimyristin, tripalmitin, and tristearin), glycerides (Softisan 142), and hard fats (Witepsol W 35, Witepsol H 35, Witepsol H 42, and Witepsol E 85). Drug release can take place either by diffusion or by matrix degradation, which occurs mainly with lipases. The nature of the lipid matrix and the surfactants has been shown to influence the biodegradation of solid lipid particles. Use of triglycerides with long chain fatty acids and/or sterically hindering surfactants like Poloxamer 407 or Poloxamine 908 delay matrix degradation (25). Use of short-chain fatty acid triglycerides and/or bile salts (e.g., cholic acid sodium salt) leads to more rapid degradation (25).

The approaches currently utilized to formulate solid lipid particles include high shear homogenization, high pressure homogenization, and solvent emulsification/evaporation (24). Although these approaches produce solid lipid particles, they suffer from limitations such as wide particle size distribution and high residual organic solvents. The SCF process can overcome these limitations and produce particles with controlled size distribution and reduced residual solvents. Mandel and coworkers have used supercritical carbon dioxide as the solvent in the manufacture of pharmaceutical materials containing polymers and/or lipids and temperature-sensitive biomaterials, in a process called particles from gas saturated solutions (PGSS) (26). In this process, compressed CO_2 is dissolved in a melted drug–carrier mixture and this gas-saturated solution is expanded, leading to supersaturation and precipitation of fine particles (27). The PGSS method offers some advantages over the RESS and GAS processes. Compared with RESS, the consumption of CO_2 in PGSS is reduced. Unlike RESS, PGSS does not require solute solubility in the supercritical CO_2. Unlike the GAS process, the PGSS process does not require any organic solvent. The PGSS process described by Mandel and coworkers was verified on several laboratory, pilot, and commercial scales. The products obtained by means of the PGSS process often displayed superior dispersion properties and possessed unique properties (e.g., porosity, particle size distribution, loading, uniformity) (26).

2.2. Nanoparticles

Nanoparticles and/or capsules are colloidal drug delivery systems whose size ranges from 1 to 1000 nm. Similar to microparticles, nanoparticles also offer the advantages of drug targeting and drug localization. Because they are so small, nanoparticles can afford better access to and entry into remote tissues. Although nanoparticles provide high cellular uptake, their drug-loading capacity limits the efficacy of these systems (28). Low drug-loading efficiencies are often observed with nanoparticles prepared by conventional particle formation methods. This is because a decrease in particle diameter results in an increase in the ratio of surface area to unit volume, thereby increasing the possibility of drug loss by diffusion from the surface toward the continuous medium during particle formation. Because supercritical fluid processes eliminate the number of steps in processing. They are likely to improve encapsulation efficiency. Table 5 summarizes various drug and carrier nanoparticles prepared by using SCF processes.

Chattopadhyay and Gupta (29–31) achieved a significant improvement in the antisolvent process using a SAS-EM process that yields particles of controllable size that are up to 10-fold smaller and have narrower size distributions. In SAS-EM the solution jet is deflected by a surface vibrating at an ultrasonic frequency that atomizes the jet into much smaller droplets. Furthermore, the ultrasound field generated by the vibrating surface enhances mass transfer and, through increased mixing, prevents agglomeration. Particle size is easily controlled by varying the vibration intensity of the deflecting surface, which can be adjusted by changing the power supplied to the attached ultrasound transducer. This new technique was demonstrated by the formation of nanoparticles of different pharmaceuticals such as lysozyme, tetracycline, and griseofulvin (29–31).

Employing external sources of energy to reduce droplet size is a useful approach in preparing nanoparticles. Falk et al. (32) used a PCA technique employing an ultrasonic spray nozzle that vibrates at 120 kHz to produce nanoparticles of gentamicin, naltrexone, and rifampin with L-PLA. The drug polymer microparticles were spherical and between 200 and 1000 nm in diameter. The relative lipophilicities of gentamicin, naltrexone, and rifampin were in the following order: gentamicin < naltrexone < rifampin. The authors used hydrophobic ion pairing to solubilize gentamicin and naltrexone in methylene chloride. This is a process wherein the polar counter ions (sulfate and chloride) of the ionic pharmaceutical agent are replaced with a hydrophobic one (sodium bis-2-ethyl hexyl sulfosuccinate). The more lipophilic rifampin was not solubilzed by hydrophobic ion pairing, and the particles of rifampin had a low loading efficiency (0.5–37.4% experimental vs 5–50% theoretical), and also 70% of the loaded drug was released in a burst effect. Gentamicin, on the other hand, had a high loading efficiency (4.3, 12.2, 17.4, and 24.75% w/w measured vs theoretical values of 5, 10, 15, and 20% w/w, respectively), and the particles prepared had very low burst effects and exhibited linear release kinetics for a period of 7 weeks. Naltrexone, with

TABLE 5 Drug-Containing Polymeric Nanoparticles Prepared by Using Various Supercritical Fluid Techniques

Polymer/Drug[a]	Process/Solvent[a]	Nozzle i.d. (μm)	Flow rates CO$_2$	Flow rates Solution (mL/min)	Temperature/Pressure (°C/bar)	Mean size (μm)	Morphology	Ref.
PLA/Naltrexone	PCA/CH$_2$Cl$_2$	1000[d]	20 mL/min	1	35–38/85–90	0.2–1	Spherical	32
PLA/Gentamicin	Gentamicin		20 mL/min	1				
PLA/Rifampin			20 mL/min	1				
PLA/Gentamicin	PCA/CH$_2$Cl$_2$		15–35 mL/min	1	35/85	<1	Spherical and agglomerated	72
PLA/PEG–insulin	GAS/CH$_2$Cl$_2$/DMSO	50	1200–1500 L/h	1	35/130	0.4–0.6	Spherical and smooth	33

[a] Abbreviations: PLA, polylactic acid; PCA, precipitation with compressed antisolvent; GAS, gas antisolvent; DMSO: dimethyl sulfoxide.

intermediate lipophilicity, produced particles with encapsulation efficiencies in between those of ion-paired gentamicin and rifampin. The particles had a burst effect and exhibited linear release kinetics for up to 3 weeks. Thus, by employing hydrophobic ion pairing, this study overcame the limitation of low applicability of SCF processing to ionic pharmaceuticals.

Encapsulation of therapeutic macromolecules into nanoparticles is a challenge because these agents are water soluble. In addition, factors including size, physicochemical stability, and bioactivity are issues of concern for their encapsulation. Studies by Elvassore et al. (33) indicated that an SCF technique is also capable of encapsulating high molecular weight peptides such as insulin. Elvassore et al. (33) used a GAS CO_2 precipitation technique to formulate (L-PLA nanoparticles) loaded with insulin–poly(ethylene glycol). The polymer was dissolved in methylene chloride and the macromolecule and PEG were dissolved in DMSO. The organic solution (methylene chloride–DMSO mixture) was sprayed into a chamber, wherein carbon dioxide was fed cocurrently from the top of the vessel. The process resulted in high product yield, extensive organic solvent elimination, and maintenance of over 80% of the hypoglycemic activity of the insulin. The nanoparticles presented a mean particle diameter in the range 400 to 600 nm, with more than 90% of the insulin being trapped in the L-PLA nanoparticles. In vitro release studies showed that only a small amount of insulin was released from the preparations. Thus, this study formulated polymeric nanoparticles encapsulating macromolecules without any loss of biological activity. In addition to processing synthetic polymers such as L-PLA, the SCF technique is capable of processing nanoparticles with naturally occurring biodegradable and biocompatible polymers such as hyaluronic acid derivatives. Ethyl hyaluronate nanoparticles were prepared by dissolving the polymer and calcitonin in DMSO at a concentration of 0.1% and 15 IU/mg of the polymer, respectively. When this solution was sprayed into a supercritical CO_2 environment, ethyl hyaluronate nanoparticles encapsulating calcitonin were produced (34). Table 6, briefly describes the application of nano- and microparticles via various routes of delivery.

2.3. Liposomes

Liposomes are lipid-based vesicles useful in delivering conventional as well as macromolecular therapeutic agents (35). The commercially available liposomal formulations include Doxil (doxorubicin) and Ambisome (amphotericin B), which are available for the treatment of tumors and fungal infections, respectively. Egg yolk and soy lecithins that are mainly composed of phospholipids ($\leq$ 95%) are commonly used as raw materials for liposome preparation. Of the two, soy lecithins are generally preferred for reasons of economics. Conventional methods of liposome preparation, including solvent evaporation–hydration, suffer from poor encapsulation efficiencies, especially for hydrophilic compounds. In addition, these methods require high amounts of toxic organic solvents. It has been stated that improvement in the solvent evaporation-hydration method can be achieved by comminution of the lecithin, which occurs during the dissolution and evaporation steps. Yet, the development of the solvent evaporation–hydration on an industrial scale suffers from a major drawback; that is, it cannot yield particles with organic solvent concentration below an acceptable level. To solve this problem, it was suggested that the encapsulation be performed by techniques in which nontoxic solvents such as supercritical fluids are used.

Based on the use of supercritical carbon dioxide, liposomes of soy lecithin were prepared by means of two micronization techniques: the RESS and SAS processes (36). Since phospholipids are only poorly soluble in supercritical CO_2, ethanol was used as a cosolvent. While the RESS process failed to separate the phospholipids from the solvent, resulting in a gel deposit, the SAS process was successful in producing microparticles (14–40 μm) under semicontinuous mode, when CO_2 and the liquid solution were allowed to flow cocurrently. IR spectra of the particles seemed to indicate that they were free of residual solvent. Thus, by employing supercritical fluids, this study demonstrated that the anisolvent process is capable of obtaining liposomes whereas the RESS process failed.

Frederikson et al. (9) modified the RESS method and obtained liposomes encapsulating hydrophilic and hydrophobic drugs. In this study, a laboratory-scale method for the preparation of small liposomes encapsulating a solution of dextran and fluorescein isothiocyanate (FITC), a water-soluble compound, was developed. In this method, phospholipid and cholesterol were dissolved in supercritical carbon dioxide in a high pressure unit, and this phase was expanded with an aqueous solution containing FITC in a low pressure unit. This method used 15 times less organic solvent to get the same encapsulation efficiency as conventional techniques. The length and inner diameter of the encapsulation capillary influenced the encapsulation volume, the encapsulation efficiency, and the average size of the liposomes. In comparison to reverse evaporation of vesicles or dehydration–rehydration of vesicles (37,38), the SCF method minimized the use of toxic combustible solvents. Also, the apolar properties of carbon dioxide facilitated the incorporation of lipophilic markers into liposomes. Thus, this study demonstrated that liposomal carriers of desired sizes and types, with good drug encapsulation efficiency, can be formulated by using RESS-based SCF technologies.

While the formulation method utilized by Frederikson et al. (9) used two chambers, a high pressure and a low pressure chamber, Katsuto et al. (39) developed a one-step method for the preparation of liposomes. In this

TABLE 6 Application of Nano- and Microparticles via Various Routes of Delivery

Route	Particles	Particle size (μm)	Considerations
Ocular	Nano- and microparticles	0.1–1 (intravitreal) (73) 1–5 (subconjunctival) (74)	Particles > 50 μm are irritating upon topical administration and also induce foreign body sensation.
Nasal	Microparticles	10–50 Fraction of particles <10 μm should not be >5%	Particles in the size range of 10–50 μm are suitable for nasal vaccination because they can target the Peyer's patch–like cells and macrophages in the nose associate lymphoid tissue (NALT) (75).
Pulmonary	Microparticles	1–10 Particles < 1 μm are exhaled and particles >10 μm deposit in the upper respiratory tract	Particles with aerodynamic diameters between 2 and 3 μm are ideal for deep lung delivery (76). This aerodynamic diameter depends primarily on the geometric diameter and bulk density (77). Particles with bigger size (~ 10 μm) and lower bulk density aerosolize better and avoid lung macrophage clearance.
Oral	Nano- and microparticles	0.1–100	Microparticles between 1 and 5 μm in diameter cross the intestinal mucosa, whereas microparticles of 5 to 10 μm remain in the lumen or in the Peyer's patches (mucosal immunization) (78). Interestingly, the gastrointestinal uptake of particles < 1 μm is higher than the uptake particles > 1 μm (79).

TABLE 6 Continued

Route	Particles	Particle size (μm)	Considerations
Parenteral			Particles in the micrometer
Intramuscular	Nano- and microparticles	0.6–10	size range can be injected into the
Subcutaneous	Nano- and microparticles	0.2–10	intramuscular or subcutaneous spaces.
Intravenous	Nanoparticles	0.1–0.3	It is recommended that emulsions for the intravenous route have a submicrometer (100–300 nm) droplet size (80). Particles >1 μm block the veins and cause embolism (81).

method, initially the lipid (0.3 wt %) and ethanol (0–15 wt % vs CO_2) were sealed in the view cell and CO_2 (13.74 g) was introduced into the cell. The cell temperature was then raised to 60°C, a temperature higher than the phase transition temperature of the lipid (41°C), and the pressure was kept constant at 200 bar. Aqueous dispersions of liposomes were obtained through emulsion formation by introducing a given amount of water into a homogeneous mixture of supercritical carbon dioxide, L-α-dipalmitoylphosphatidylcholine, and ethanol with sufficient stirring and subsequent pressure reduction. For this, an aqueous D-(+)-glucose solution (0.2 mol/L) was slowly/introduced into the vessel (0.05 mL/min) till the desired amount of solution had been introduced. Transmission electron micrographs indicated that most of the vesicles obtained were large unilamellar liposomes (0.1–1.2 μm). The encapsulation efficiency of D-(+)-glucose was five times higher than that obtained by the conventional method of liposome preparation. The trapping efficiency for cholesterol, an oil-soluble substance, was 63%.

An SCF process was used to prepare liposomes, designated as critical fluid liposomes (CFL), encapsulating hydrophobic drugs such as taxoids, camptothecins, doxorubicin, vincristine, and cisplatin (40). Also, stable paclitaxel liposomes with a size of 150 to 250 nm were obtained. A patent held by Aphios Company (U.S. 5,776,486) on SuperFluids CFL describes a method and apparatus useful for the nanoencapsulation of paclitaxel and campothecin in aqueous liposomal formulations called Taxosomes and Camposomes, respectively. These formulations are claimed to be more effective than commercial formulations against tumors in animals. Table 7 summarizes various liposomal formulations prepared by using SCF processes.

TABLE 7 Liposomes Prepared by Using Various Supercritical Fluid Techniques

Liposome[a]	Process/Solvent[a]	Nozzle i.d. (μm)	Solution flow rate (mL/min)	Temperature/ Pressure	Mean size (μm)	Morphology	Ref.
L-α-DPPC/D- (+)-glucose	Supercritical reversed phase evaporation method/ethanol	—	0.05	60°C/200 bar	0.1–1.2	Spherical	2,39
L-α-DPPC/ cholesterol		—	—	60°C/200 bar			2,39
Soy lecithin	GAS/ethanol	—	—	—	1–40	Spherical and agglomerated	9,36
Soy lecithin	RESS	—	—	—	No liposomes formed		9,36
POPC-CHO/ FITC–dextran	RESS/ethanol	250–1000	—	60°C/25 MPa	200–250 nm	Spherical	9
POPC-CHO/zinc phthalocyanine tetrasulfonic acid	RESS/ethanol	250/1000	—	60°C/25 MPa			9

[a] Abbreviations: CHO, cholesterol; DPPC, dipalmitoylphosphatidylcholine; FITC, fluorescein isothiocyanate; GAS, gas antisolvent; POPC, palmitolyl-oleoylphosphatidyl choline; RESS, rapid expansion of supercritical solution.

2.4. Microporous Foams

Microporous foams serve as an important tool in the field of tissue engineering. Both natural and synthetic polymers have been used as scaffolds for tissue engineering. These scaffolds provide a matrix onto which cells may adhere, and they aid in shaping and defining cell growth in vivo. These scaffolds have been used to form neocartilages, tendons in orthopedic surgery, and human urothelial and bladder muscle structures (41,42). Besides this, several growth factors and/or growth supplements important for tissue growth can be incorporated into these polymeric scaffolds to enhance the tissue generation. These matrices can also sustain the delivery of these growth factors.

One conventional method of polymer scaffold preparation is the solvent casting–salt leaching technique. The first step in this process is to dissolve the polymer (PLLA or PLGA) in chloroform or methylene chloride and then cast it onto a petri dish filled with the porogen. Water-soluble salts such as NaCl or KCl are commonly used as porogens. After evaporation of the solvent, the polymer–salt composite is leached in water for 2 days to remove the porosigen. The resulting scaffold's porosity can be controlled by regulating the amount of salt added, while the pore size is dependent on the size of the salt crystals. With 70 % salt and above, the pores exhibited high interconnectivity. Foams fabricated in this manner have been used extensively with various cell types. With any solvent casting–particulate leaching procedure, organic solvents are used, which in many cases precludes the possibility of adding pharmacological agents to the scaffold during fabrication. Also, the leaching step for water-soluble porogens significantly increases the scaffold preparation time. Use of large amounts of organic solvents and the residual organic solvent in these scaffolds may cause injuries to the interior body organs and/or unwanted side effects. In addition, this process does not provide us with precise control over the cell characteristics of the foam. To circumvent this problem, porous polymeric matrices have alternatively been prepared with carbon dioxide (43–45). This technique involves lowering the glass transition temperature of the polymer under high pressure CO_2 by increasing the number of gas molecules absorbed into the polymer phase (43–45). Upon depressurization, the polymer is suddenly supersaturated with CO_2, causing bubbles to form and grow within the polymer. As the glass transition temperature of the polymer increases following depressurization, the pores generated by CO_2 become permanent, creating microporous foam.

To eliminate the need for organic solvents in the pore-making process, a new technique involving gas as a porogen has been introduced (45). The process begins with the formation of solid disks of PGA, PLLA, or PLGA by means of compression molding with a heated mold. The disks are placed in a chamber and exposed to CO_2 at high pressure and temperature (5.5 MPa; 100°C) for 3 days, at which time the pressure is rapidly decreased to atmospheric pressure. Porosities of up to 93% and pore sizes of up to 100 μm can be obtained with this technique, but the pores are largely unconnected, especially on the surface of the foam. Thus, this fabrication method demonstrated formation of porous foams that requires no leaching step and no use of harsh chemical solvents. The disadvantage is the use of high temperatures, which may alter polymer release properties and also cause unwanted thermal degradation of the active substances.

Hile et al. (46) demonstrated that porous foams incorporating a biologically active agent can be prepared without the use of high temperatures. In their study, using an SCF technique, porous PLGA foams capable of releasing an angiogenic agent, basic fibroblast growth factor (bFGF), were prepared for tissue engineering applications (46). These foams sustained the release of the growth factor. In this technique, first, a homogenous water-in-oil emulsion consisting of an aqueous protein phase and an organic polymer solution was prepared and poured into a longitudinally sectioned and easily separable stainless steel mold. The mold was then placed in a pressure cell and pressurized with CO_2 at 80 bar and 35°C. The pressure was maintained for 24 h to saturate the polymer with CO_2 for the extraction of methylene chloride. Finally, the setup was depressurized for 10 to 12 s, creating microporous foam. Porous biodegradable foams prepared by using an SCF technique are also being investigated as vehicles for delivering specific RNA enzymes, or ribozymes, identified as potential gene therapy agents. PLGA foams prepared by utilizing a depressurization technique with supercritical carbon dioxide as the solvent were used to encapsulate c-*myc* RNA, which shows potential as a gene therapy drug. It is anticipated that these polymer foams will degrade slowly in biological conditions, thereby releasing the encapsulated ribozyme into the body (47).

Several studies have indicated that amorphous polymers are more amenable to porous foam formation than their crystalline counterparts. For instance, biocompatible polymeric foams composed of crystalline poly(vinylidene fluoride) (PVDF) did not result in foam formation, and the product was characterized by poor cell characteristics. However, when PVDF was blended with an amorphous polymer, poly(methyl methacrylate) (PMMA), the process resulted in porous foams with vastly improved morphologies. Morphological characterization of microcellular PVDF/PMMA foams indicated that the cell diameter increased as the PMMA fraction increased. In another study, porous 85:15 PLGA foams were produced by the pressure-quench method using supercritical CO_2 (48). Porous PLGA foams were generated with relative densities ranging from 0.107 to 0.232. Foams showed evidence of interconnected cells with porosities as high as 89%. The cell size ranged from 30 to 70 µm (49). Table 8 summarizes the preparation of porous foams by means of SCF processes.

2.5. Coating

Coating of pharmaceutical dosage forms is a formulation strategy that is commonly employed to achieve the following: (a) masking unpleasant taste and odor, as in sugar-coating, thereby increasing the ease by which the product can be ingested by the patient, (b) protecting the drug from its surrounding environment (gastrointestinal pH), as in enteric-coated products, with a view to improving stability, and (c) modifying the release, as in sustained-release products. Enteric-coated products are obtained by a process called film coating, wherein a thin polymeric film is deposited over solid dosage forms such as tablets. This polymeric film prevents the tablets from dissolving in the stomach and ensures that the drug is released in the intestine. Most often these coatings are made of polymers such as cellulose acetate phthalate. Since these polymers are not soluble in water, the coating process utilizes large amounts of organic solvents (50).

Polymeric coatings applied to particulate material often result in the formulation of microencapsulated products, and this technique is called microencapsulation. Microencapsulation is a modified form of film coating, differing only in the size of the particles to be coated and the methods by which this is accomplished. The process is based on either mechanical methods such as pan coating, air suspension techniques, multiorifice centrifugal techniques, modified spray drying techniques, or physicochemical ones involving coacervation–phase separation, in which the material to be coated is suspended in a solution of the polymer. Polymer phase separation is facilitated by the addition of a nonsolvent, incompatible polymer or inorganic salts, or by altering the temperature of the system. The limitations of these methods include excessive organic solvent usage and temperature effects in the case of spray drying and air suspension techniques. These limitations can possibly be overcome with the use of SCFs.

SCFs can be used to coat drug particles (20 nm–100 µm) with a single or multiple layers of polymers or lipids (51,52). A solution of a coating material in SCF is used at temperature and pressure conditions that do not solubilize the particles being coated. The advantage of supercritical fluids in coating is that the SCF coating process does not use organic solvents. A RESS process for particle coating with a solution of polymer in supercritical CO_2 was studied (52,53). This technique involves extracting the polymer with supercritical CO_2, with or without a cosolvent in an extraction vessel, and then precipitating the polymer onto the surface of host particles in a second precipitation vessel by adjusting the pressure and temperature inside the

TABLE 8 Porous Foams Prepared by Using Various Supercritical Fluid Techniques

Foam[a]	Process/Solvent[a]	Flow rates		Temperature/ Pressure	Mean cell size (μm)	Morphology	Ref.
		CO_2 (mL/h)	Solution (mL/min)				
PVDF	RESS	—	0.05	60/200 60°C/200 bar	—	No foam formation	48
PVDF/ PMMA	RESS	—	—	60°C/200 bar	<10	Formation of microcellular foams	48
PLGA/ bFGF	RESS	15	—	35°C, 80 bar	10–30 μm	Formation of microcellular foams	46
PLGA (85:15)	Pressure quench using supercritical CO_2	—	—	—	30–70 μm	Formation of porous foams (porosity ~90%)	49
PMMA	RESS	—	—	40–80°C/1500– 5000 psi	—	Formation of microcellular foams	82

[a] Abbreviations; bFGF, basal fibroblast growth factor; PLGA, poly(lactic-co-glycolic acid); PMMA, poly(methyl methacrylate); PVDF, poly(vinylidene fluoride); RESS, rapid expansion of supercritical solution.

precipitator to lower its solubility. Techniques including scanning electron microscopy (SEM), energy-dispersive X-ray spectrometry (EDS), energy-dispersive X-ray mapping, and thermogravimetric analysis (TGA) were used to characterize the coatings obtained. Table 9 summarizes various coating applications of SCF processes.

Modifications to the RESS process have also been made for coating particles. In one modified method, fluidized-bed coating for fine particles was performed by means of rapid expansion of a supercritical fluid solution (RESS) (54). Experiments were conducted in a 50 mm diameter circulating fluidized bed with an internal nozzle in the center of the riser. Microspheroidal catalyst particles (average particle size 56 μm) were used as the core particles. Supercritical carbon dioxide solutions of paraffin were expanded through the nozzle into the bed that was fluidized by air. The results of the studies indicated that a stable coating of fine particles was achieved without the formation of agglomerates at room temperature. Several polymers have

TABLE 9 Coating Prepared by Using Various Supercritical Fluid Techniques

Coat[a]	Process/ Solvent[a]	Temperature/ Pressure	Morphology	Reference
HPC–glass beads	RESS		Successful coat formation	53
PVCVA–glass beads	RESS			53
Trimyristin (Dynasan 114)– bovine serum albumin	Novel supercritical coating process	35–45°C/ 200 bar	Successful coat formation	51
Gelucire 50-02– bovine serum albumin	Novel supercritical coating process	35–45°C/ 200 bar	Successful coat formation	51
PLA–iron and silica balloon	GAS		Successful coat formation	55
PEG–iron and silica balloon	GAS		Successful coat formation	
PEG-PPG-PEG– iron and silica balloon	GAS		Successful coat formation	

[a] Abbreviations: GAS, gas antisolvent; HPC, hydroxypropylcellulose; PEG: poly(ethylene glycol); PPG, poly(propylene glycol); PLA, poly(lactic acid); PVCVA, poly(vinyl chloride-*co*-vinyl acetate); RESS, rapid expansion of supercritical solution.

been used for coatings using the supercritical fluid process (55). These include styrene–methyl methacrylate–glycidyl methacrylate triblock copolymer [PS-b-(PMMA co-PGMA), M W 5000;PMMA, MW 15,000; PEG, MW 7500 PLLA, MW 5,000); and PEG-poly(propylene glycol) (PPG-PEG triblock copolymer, MW 13,000]. The results of the studies indicated that changes in the feed composition of the polymer resulted in variations in the thickness of the polymer coating, mean particle diameter, and particle size distribution.

Coatings of polymers have also been performed using CO_2 as an antisolvent. A patent has been issued for the preparation of microcapsules containing active substances coated with a polymer (56). This patent covers the use of SAS process to prepare microcapsules containing an active substance with a substantially polar polymer film. The polar polymers include polysaccharides, cellulose derivatives, acrylic or methacrylic polymers, polymers of vinyl esters, polyesters, polyamides, polyanhydrides, polyorthoesters, and polyphosphazenes. The active substances include analgesics, antiulcer agents, antihypertensives, antipsychotics, antidepressants, antipyretics, antiinflammatory drugs, oligonucleotides, peptides, antibiotics, and vitamins. The patent describes the use of the GAS process to prepare Eudragit L 100, coated microcapsules of bovine hemoglobin. The procedure entails suspending 200 mg of bovine hemoglobin in a solution containing 40 mg of Eudragit L 100 in 54 mL of ethanol. This suspension is mixed with liquid CO_2 at 80×10^5 Pa and 25°C and then treated with supercritical CO_2 at 125×10^5 Pa and 40°C for ethanol extraction and polymer precipitation. After 15 min, the ethanol–CO_2 mixture is evacuated with pressure reduction to 75×10^5 Pa, resulting in the formation of water-resistant microcapsules (230 mg) of size 200 to 300 µm, containing 83.3 wt % hemoglobin.

In addition to polymer coating, the technique has been used to coat with lipids (51). In this study, a novel coating process was used to produce solvent-free microparticles, loaded with a model protein (bovine serum albumin, BSA). Coating material consisted of either trimyristin (Dynasan 114) or Gelucire 50–02, two lipidic compounds having melting points of 45 and 50°C, respectively. The results of the studies indicated that Dynasan 114 forms a discontinuous coating made of crystalline microneedles. This led to particles with an initial burst release of about 70% in 30 min at 37°C. However, a prolonged release of the protein has been achieved over a 24 h period from particles coated with Gelucire 50-02, which produced a more homogeneous, film-forming coating. Furthermore, it was shown that BSA does not undergo any degradation following $scCO_2$ treatment under the supercritical conditions used in the coating process.

In addition to coating particulate systems, SCF processes can be used to coat medical devices for controlled-release applications. A cardiac stent containing the drug RWJ-53308 was coated with PLGA by means of an SCF process (57). The authors' results indicated that the PLGA-coated stent had released approximately 66 μg of RWJ-53308 at the end of an hour, and at the end of 17.5 h, a cumulative amount of approximately 134 μg of drug had been released.

2.6. Films

Polymeric films serve as useful tools in the field of sustained and controlled drug delivery. Polymeric films serve as rate-controlling membranes in transdermal and buccal drug delivery patches. To this end, the marketed (transdermal preparations) include Transderm-Scopo (scopolamine), Duragesic (fentanyl), Estraderm (estrogen), and Catapres-TTS (clonidine). Most of these systems are available as either the matrix or the reservoir type. In the matrix type, the drug is dispersed in the rate-controlling membrane, whereas in the reservoir type, a drug depot or reservoir is separated by a rate-controlling membrane. The rate of drug release from both systems is governed by the thickness of the rate-controlling membrane. Typically, the thickness of the films used in transdermal delivery varies from 50 to 100 μm. The casting and spraying methods currently used for obtaining polymer films in drug delivery produce films with nonuniform thickness, high residual organic solvents, and excessive pore formation. These limitations can be avoided with the use of supercritical fluid technology. Supercritical fluids provide an opportunity to produce films of uniform thickness and reduced residual organic solvent contents. Honda et al. (58) described a process for using supercritical fluid technology to obtain cellulose triacetate films. Briefly, cellulose triacetate was dissolved in methyl acetate in the presence of CO_2 at 100°C and 7.4 MPa, filtered, and cast on a stainless belt to give a film with high tensile strength and no fisheyes.

3. OTHER DRUG DELIVERY APPLICATIONS

3.1. Reduction of Residual Organic Solvents

Conventional methods for microencapsulation include liquid antisolvent precipitation, spray drying, solvent evaporation, and organic phase separation techniques. Because the polymers are only poorly soluble in water, these methods utilize large amounts of environmentally hazardous organic solvents including chloroform, methylene chloride, and ethyl acetate. Formulations with high residual organic solvent content are a major concern from the standpoints of the U.S. Food and Drug Administration and the U.S. Pharmacoseia (USP). This excessive organic solvent usage results in unacceptable levels of residual organic solvent contents, which is a major cause for concern (16). The amount of residual methylene chloride has been found to be 0.3% in particles prepared by spray drying and 2.1% in particles prepared by solvent evaporation (59). Progesterone-loaded microparticles prepared by the solvent evaporation technique contained 1.8 to 4.7% residual solvent (60). The USP limits the amount of organic solvents in pharmaceuticals, and the limit for methylene chloride is 600 ppm. Besides being environmentally unacceptable, methylene chloride is toxic (suspected carcinogen and mutagen), and thus approval for its use can be difficult to obtain. On the other hand, ethyl acetate, with a residual organic solvent limit up to 5000 ppm is considered to be less toxic than methylene chloride. However, ethyl acetate suffers from low acute toxicity associated with local irritation (eye, skin, and respiratory tract) at or above 400 ppm. Through its ability to rapidly extract organic solvents, SCF technology offers a useful tool for preparing drug delivery systems with reduced content of organic solvent. For instance, an aerosol solvent extraction system (ASES) utilizing supercritical gases produced microparticles having a residual organic concentration of less than 30 ppm, a value much below the regulatory limits of the pharmaceutical industry and USP (61,62). Table 10 summarizes the residual solvent (methylene chloride) contents in formulations prepared by using conventional and SCF processes.

3.2. Sterilization of Drug Delivery Systems

In addition to drug delivery system preparation, SCF technology is now making inroads in the field of product sterilization. The most important

TABLE 10   SCF Processing Results in Reduced Residual Methylene Chloride

Particles[a]	Method of preparation	Residual methylene chloride (ppm)[b]	Ref.
Cisplatin–PLA	Solvent evaporation	30,000	83
Progesterone–PLA	Solvent evaporation	18,000–47,000	60
Cinchocaine–PLGA	Solvent evaporation	21,000	59
Cinchocaine–PLGA	Spray drying	3000	59
PLGA–tetracosactide	Solvent evaporation	281–705	84
Indomethacin–PLA	SCF	<30	62
Piroxicam–PLA	SCF	<30	62
Hyoscine–PLA	SCF	<30	62
PLA	SCF	71	61

[a] Abbreviations: PLA, poly(lactic acid); PLGA, poly(lactic-co-glycolic acid).
[b] USP limit for residual methylene chloride is 500 ppm.

consideration before administering controlled- and/or sustained-release preparations in vivo is the sterility of the finished product. The three most common methods of sterilization in use today are ethylene oxide exposure, gamma irradiation, and steam sterilization. Each of these methods has serious limitations for the sterilization of some materials used in medicine, especially thermally and hydrolytically sensitive polymers. Mohr et al. (63) investigated the influence of gamma irradiation (doses ranging from 5.1 to 26.6 kGy) on drug substance, polymer, and microparticles of 17β estradiol: the poly-(D, L-lactide-*co*-glycolide). The drug substance showed excellent stability against gamma irradiation in the investigated dose range, whereas microencapsulated estradiol seems to be converted to conjugation products with the polymer, and to a lesser extent to the degradation product 9,11-dehydroestradiol. The weight-average molecular weight of the PLGA polymers decreased with increasing irradiation dose, while polydispersity indices (M_w/M_n) remained nearly unchanged, compatible with a random chain scission mechanism in lactide/glycolide–copolymer degradation. In vitro drug release studies showed accelerated kinetics with increasing irradiation doses due to dose-dependent polymer degradation.

In another study, the influence of ethylene oxide (ETO) and gamma-irradiation sterilizations was investigated on PLGA scaffolds (Osteofoam) (64). The two sterilization techniques, ETO and gamma, were compared in terms of their immediate and long-term effects on the dimensions, morphology, molecular weight, and degradation profile of the scaffolds. Scaffolds shrank to 60% of their initial volume after ETO sterilization, whereas their molecular weight (M_w) decreased by approximately 50% after gamma irradiation. Thus, both ETO and gamma irradiation posed immediate problems as sterilization techniques for three-dimensional biodegradable polyester scaffolds. Over the 8-week time frame, both the sterilized samples showed morphological and volume changes, with the greatest changes observed for gamma-irradiated samples. The studies cited suggest that the most commonly used sterilization technique, namely gamma irradiation, affects the physicochemical properties of the drug and the polymer, resulting in altered drug release rates.

It has been suggested that high pressure carbon dioxide exhibits microbicidal activity by penetrating into the microbes, thereby lowering their internal pH to a lethal level (20). Dillow et al. (65) described the use of supercritical CO_2 for sterilizing PLGA microparticles (1, 7, and 20 μm). The results of the study, summarized in Table 11 indicated complete sterilization of all microorganisms within 0.6 to 4 h at 205 bar and at temperatures generally between 25 and 40°C. Sterilization occurred in the absence of organic solvents, chemicals, or radiation. In this study, *S. aureus, B. cereus, L. innocua, S. salford, P. vulgaris, L. dunnifii, P. aeruginosa,* and *E. coli* were the

TABLE 11   Supercritical CO_2 Inactivates Various Microorganisms

Microorganism	Temperature ($^\circ$C)	Time (h)	Initial (cfu/mL)	Degree of inactivation (log)
B. cereus	34	0.6	5.1×10^7	2
	34	2	5.7×10^7	1
	60	2	5.2×10^7	5
	60	4	1.8×10^8	8
L. innocua	34	0.6	5.8×10^9	3
	34	0.6	2.1×10^9	9
S. aureus	34	0.6	2.5×10^9	3
	34	0.6	1.2×10^9	7
	40	2	6.7×10^8	6
	40	4	1.9×10^9	9
S. salford	34	0.6	1.5×10^9	3
	34	0.6	1.0×10^9	3
	40	2	6.0×10^8	6
	40	4	2.2×10^9	9
P. aeruginosa	34	0.6	7.4×10^8	6
	40	1.5	2.9×10^8	6
	40	4	2.4×10^8	8
E. coli	34	0.5	6.4×10^8	8
P. vulgaris	34	0.6	9.1×10^8	8
L. dunnifii	40	1.5	6.7×10^4	4

Source: Ref. 65.

model microorganisms. The first three in the list are gram-positive, and the last five are gram-negative bacteria. These species were chosen based on their prevalence in medical contamination and/or their resistance to inactivation. Scanning electron micrographs indicated that there was no gross morphological change in the PLGA spheres before or after SCF sterilization. Fourier transform IR spectroscopy, gas permeation chromatography, and differential scanning calorimetry analysis confirmed the absence of detectable chemical changes in either the PLGA or PLA samples after sterilization compared with untreated samples. In addition, degradation analysis of PLGA indicated that SCF sterilization does not cause unusual degradation behavior in the processed polymers. With further studies in this area, this technique may prove to be a useful method for sterilization of materials and pharmaceutical formulations of many types because of the mild, nonreactive processing conditions, and the ability of supercritical CO_2 to inactivate a wide variety of microorganisms. Although the SCF process was successful against microorganisms, it was much less successful in spores.

3.3. Polymer Purification

The polymers currently used in drug delivery are a mixture of fractions having various molecular weights. Supercritical fluids provide an opportunity to use polymers with narrow molecular weight distribution for drug delivery. The selective extraction of low molecular weight polymers by supercritical fluids had been observed previously where the molecular weight distribution of a polymer in an extraction column changes with extraction time. Saltzman et al. (66) used supercritical propane to separate poly(ethylene-*co*-vinyl acetate) (EVAc), a biocompatible polymer, into narrow fractions with a polydispersity between 1.4 and 1.7. The different EVAC fractions were then used to study the rate of antibody release from a polymer matrix. Greater than 90% of the antibody was released from low molecular weight fractions (M_n 12,500 and 22,900) during the first 5 days, while a high molecular weight fraction (M_n = 253,200) released less than 10% of the antibody during the same time. In another study, impurities from the polymers were removed by contacting the polymer with an extractant under supercritical conditions. The method sharply reduced residual monomer levels in bioabsorbable polymers without altering polymer viscosity or melting point. Their study indicated that extraction of a 92.5:7.5 glycolide–lactide random copolymer (viscosity 1.37

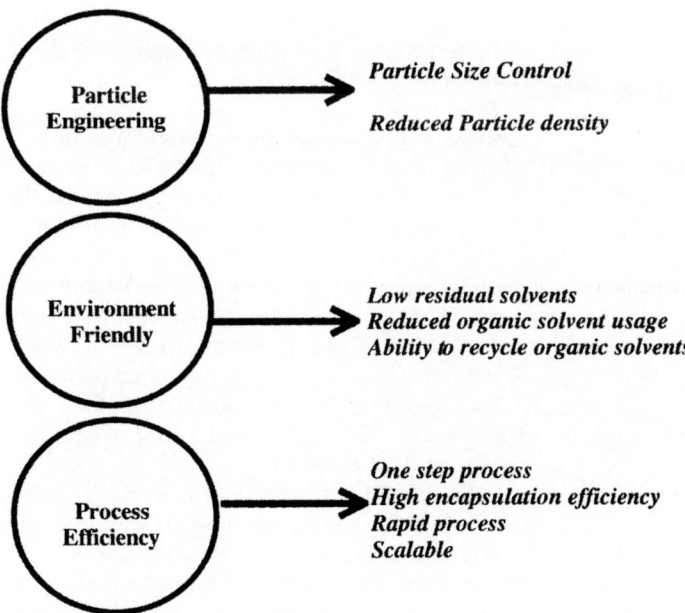

FIGURE 3 Benefits of SCF processing for particle engineering.

dL/g, m.p. 186.5° C) with supercritical CO_2 at 120°C and 350 bar gave a purified sample having viscosity 1.40 dL/g that melted at 186.6°C.

4. CONCLUSIONS

SCF technology can be used in the preparation of drug delivery systems and/ or to improve the formulation properties of certain drug candidates. SCFs

TABLE 12 Some Drug Delivery Companies That Are Adapting SCF Technology

Company	Focus of SCF work
Aphios Corporation	Using proprietary Superfluid CFL technology to prepare liposomes containing hydrophobic drugs. Taxosomes and Camposomes are nanometer-sized liposomes encapsulating taxol and camptothecin, respectively.
Nektar Therapeutics	Developing particles for inhalation with its recently acquired Bradford Particle Technology, UK, a pioneer in the use of supercritical fluids for processing drug powders.
Lavipharm	Preparing particle formulations using the acquired Separex technology. Separex S.A. is a French company known for its supercritical fluid technology for particle engineering.
RTP Pharma, Inc	Generating surface-modified, submicrometer-sized particles of water-insoluble drugs with supercritical fluid technology processes; IDD technology. Acquired Phasex's proprietary SCF- based processes.
Bradford Particle Design PLC, UK, and Bristol-Myers Squibb Company	Developing coformulations of pharmaceutically active drugs and oligomeric and/or polymeric excipients. The coformulations are usually prepared by the solution-enhanced dispersion by supercritical fluids (SEDS) method.
Allied Corporation	Developing coating processes that use supercritical fluid technology.
Thar Technologies	Scaled-up and automated large-scale SAS-EM process for the production of nanoparticles and microparticles of controllable size.
Ferro Corporation	Development of several technology platforms related to production of nano- and microparticles for controlled-release and taste-masking purposes.

TABLE 13 Some Innovations in the Preparation of Drug Delivery Systems by Means of SCF Technology

Innovation	Patent	Ref.
Microparticles/Microcapsules		
Prolonged-release microparticles for injection and preparation method	WO 2001089481 (2001)	85
Method for collecting and encapsulating fine particles	WO 2001049407 (2001)	86
Coformulation of drugs and oligomeric or polymeric excipients	WO 2001015664 (2001)	87
Preparation of microcapsules containing active substances coated with a polymer	9813136 (1998)	56
Encapsulation of active particles by coating of a central core	EP 706821 (1996)	88
Nanoparticles		
SCF process for using surface modifying agents to form aqueous nano- and micro particulate suspensions of poorly soluble compounds	US 6177103 (2001)	89
Nanoparticles comprising a biocompatible polysaccharide	WO 9629998(1996)	34
Nanoparticle preparation with enhanced mass transfer	US 6620351(2003)	97
Liposomes		
Method and apparatus for producing liposomes	WO 2002032564 (2001)	90
Method and apparatus for making liposomes containing hydrophobic drugs	US 5776486 (1998)	40
Foams		
Method and apparatus for molding plastic foams	JP 2002127156 (2002)	91
Open-cell polycarbonate foams and their manufacture by means of supercritical gases	JP 2002144363 (2002)	92
Use of supercritical fluids to obtain porous sponges of biodegradable polymers for pharmaceutical implants or drug delivery systems	WO 9109079 (1991)	93
Coating		
Process for using supercritical carbon dioxide to coat medical devices	WO 2001087368(2001)	57
Methods for particle precipitation and coating of near-critical and supercritical antisolvents	JP 2000256468 (2000)	94
Films		
Environment-friendly method for the manufacture of cellulosic polymer film	JP 2002127166 (2002)	95
Manufacture of polymer solution and films, and apparatus therefor	JP 2000256468 (2000)	58
Deposition of thin films using supercritical fluids	US 4737384 (1988)	96

can be used to formulate drug carrier systems owing to their unique solvent properties, which can be altered readily by slight changes in the operating temperature and pressure. Supercritical fluid technology allows the preparation of drug delivery systems with various morphologies, including microparticles, nanoparticles, films, and foams; it reduces the surface static charge associated with drug–polymer particles, as well as the use of organic solvent and extracts residual organic solvent from drug delivery systems (Figure 3). In recent years, many pharmaceutical and drug delivery companies, some of which are listed in Table 12, have adopted SCF technology to obtain drug delivery solutions. SCF technology is rapidly providing drug delivery innovations, some of which are listed in Table 13.

ACKNOWLEDGMENTS

This work was supported by the NIH grants DK064172 (UBK) and EY013842 (UBK).

REFERENCES

1. Langer R. Drug delivery targets. Nature 1998; 392:5–10.
2. Kost J, Langer R. Responsive polymeric delivery systems. Adv Drug Deliv Rev 2001; 46:125–148.
3. Langer R. Biomaterials in drug delivery and tissue engineering: one laboratory's experience. Acc Chem Res 2000; 33:94–101.
4. Niklason LE, Langer RS. Advances in tissue engineering of blood vessels and other tissues. Transplantation Immunol 1997; 5:303–306.
5. Kompella UB, Koushik K. Preparation of drug delivery systems using supercritical fluid technology. Crit Rev Ther Drug Carrier Syst 2001; 18:173–199.
6. Petersen RC, Matson DW, Smith RD. Precipitation of polymeric materials from supercritical fluid solution: the formation of thin films, powders, and fibers. Polym Prepr (Am Chem Soc, Div Polym Chem) 1986; 27:261–262.
7. Debenedetti PG, Tom JW, Kwauk X, Yeo SD. Rapid expansion of supercritical solutions (RESS): fundamentals and applications. Fluid Phase Equilibria 1993; 82:311–321.
8. Tom JW, Debenedetti PG. Formation of bioerodible polymeric microspheres and microparticles by rapid expansion of supercritical solutions. Biotechnol Prog 1991; 7:403–411.
9. Frederiksen L, Anton K, van Hoogevest P, Keller HR, Leuenberger H. Preparation of liposomes encapsulating water-soluble compounds using supercritical carbon dioxide. J Pharm Sci 1997; 86:921–928.
10. LaVan DA, Lynn DM, Langer R. Moving smaller in drug discovery and delivery. Nat Rev Drug Discov 2002; 1:77–84.
11. Davis SS, Illum L. Polymeric microspheres as drug carriers. Biomaterials 1988; 9:111–115.

12. Cleland JL. Protein delivery from biodegradable microspheres. Pharm Biotechnol 1997; 10:1–43.
13. Kim JH, Paxton TE, Tomasko DL. Microencapsulation of naproxen using rapid expansion of supercritical solutions. Biotechnol Prog 1996; 12:650–661.
14. Mishima K, Matsuyama K, Tanabe D, Yamauchi S, Young TJ, Johnston KP. Microencapsulation of proteins by rapid expansion of supercritical solution with a nonsolvent. AIChE J 2000; 46:857–865.
15. Mueller BW, Fisher W. Manufacture of sterile sustained release drug formulations using liquefied gases. (West) Germany, DE 3744329, 1989.
16. Bodmeier R, Wang H, Dixon DJ, Mawson S, Johnston KP. Polymeric microspheres prepared by spraying into compressed carbon dioxide. Pharm Res 1995; 12:1211–1217.
17. Elvassore N, Bertucco A, Caliceti P. Production of protein-loaded polymeric microcapsules by compressed CO_2 in a mixed solvent. Ind Eng Chem Res 2001; 40:795–800.
18. Reverchon E, De Rosa I, Dela Porta G. Effect of process parameters on the supercritical antisolvent precipitation of microspheres of natural polymers. High Pressure Chem Eng 1999; 6271:251–254.
19. Elvassore N, Baggio M, Pallado P, Bertucco A. Production of different morphologies of biocompatible polymeric materials by supercritical CO_2 antisolvent techniques. Biotechnol Bioeng 2001; 73:449–457.
20. Bleich J, Mueller BW, Wassmus W. Aerosol solvent extraction system. A new microparticle production technique. Int J Pharm 1993; 97:111–117.
21. Engwicht A, Girreser U, Muller BW. Critical properties of lactide-co-glycolide polymers for the use in microparticle preparation by the aerosol solvent extraction system. Int J Pharm 1999; 185:61–72.
22. Bleich J, Kleinebudde P, Mueller BW. Influence of gas density and pressure on microparticles produced with the ASES process. Int J Pharm 1994; 106:77–84.
23. Ghaderi R, Artursson P, Carlfors J. Preparation of biodegradable microparticles using solution-enhanced dispersion by supercritical fluids (SEDS). Pharm Res 1999; 16:676–681.
24. Mehnert W, Mader K. Solid lipid nanoparticles: production, characterization and applications. Adv Drug Deliv Rev 2001; 47:165–196.
25. Olbrich C, Muller RH. Enzymatic degradation of SLN-effect of surfactant and surfactant mixtures. Int J Pharm 1999; 180:31–39.
26. Mandel FS, Wang JD, McHugh MA. Pharmaceutical material production via supercritical fluids employing the technique of particles from gas-saturated solutions. Am Chem Soc, 2001 (PMSE-024).
27. Kerc J, Srcic S, Knez Z, Sencar-Bozic P. Micronization of drugs using supercritical carbon dioxide. Int J Pharm 1999; 182:33–39.
28. Douglas SJ, Davis SS, Illum L. Nanoparticles in drug delivery. Crit Rev Ther Drug Carrier Syst 1987; 3:233–261.
29. Chattopadhyay P, Gupta RB. Production of griseofulvin nanoparticles using supercritical CO_2 antisolvent with enhanced mass transfer. Int J Pharm 2001; 228:19–31.
30. Chattopadhyay P, Gupta RB. Production of antibiotic nanoparticles using su-

percritical CO_2 as antisolvent with enhanced mass transfer. Ind Eng Chem Res 2001; 40:3530–3539.

31. Chattopadhyay P, Gupta RB. Formation of protein micro- and nanoparticles using supercritical antisolvent with enhanced mass transfer. AIChE J 2002; 48: 235–244.

32. Falk R, Randolph TW, Meyer JD, Kelly RM, Manning MC. Controlled release of ionic compounds from poly (L-lactide) microspheres produced by precipitation with a compresed antisolvent. J Controlled Release 1997; 44:77–85.

33. Elvassore N, Bertucco A, Caliceti P. Production of insulin-loaded poly(ethylene glycol)/poly(L-lactide) (PEG/PLA) nanoparticles by gas antisolvent techniques. J Pharm Sci 2001; 90:1628–1636.

34. Pallado P, Benedetti L, Callegaro L. Nanospheres comprising a biocompatible polysaccharide. Italy, WO 9629998, 1996.

35. Storm G, Wilms HP, Crommelin DJ. Liposomes and biotherapeutics. Biotherapy 1991; 3:25–42.

36. Badens E, Magnan C, Charbit G. Microparticles of soy lecithin formed by supercritical processes. Biotechnol Bioeng 2001; 72:194–204.

37. Szoka F Jr, Papahadjopoulos D. Procedure for preparation of liposomes with large internal aqueous space and high capture by reverse-phase evaporation. Proc Natl Acad Sci USA 1978; 75:4194–4198.

38. Kirby CJ, Gregoriadis G. Preparation of liposomes containing factor VIII for oral treatment of haemophilia. J Microencapsul 1984; 1:33–45.

39. Katsuto O, Tomohiro I, Hideki S, Masahiko A. Development of a new preparation method of liposomes using supercritical carbon dioxide. Langmuir 2001; 17:3898–3901.

40. Castor TP, Chu L. Methods and apparatus for making liposomes containing hydrophobic drugs. US Patent 5776486, 1998.

41. Chaikof E, Matthew H, Kohn J, Prestwich GD, Mikos A, Yip C. Bioscaffolds for tissue repair: breakout session summary. Ann NY Acad Sci 2002; 961:112–113.

42. Griffith LG. Emerging design principles in biomaterials and scaffolds for tissue engineering. Ann NY Acad Sci 2002; 961:83–95.

43. Nam YS, Yoon JJ, Park TG. A novel fabrication method of macroporous biodegradable polymer scaffolds using gas foaming salt as a porogen additive. J Biomed Mater Res 2000; 53.

44. Guney O, Akgerman A. Synthesis of controlled-release products in supercritical medium. AIChE J 2002; 48:856–866.

45. Mooney DJ, Baldwin DF, Suh NP, Vacanti JP, Langer R. Novel approach to fabricate porous sponges of poly(D,L,lactic-*co*-glycolic acid) without the use of organic solvents. Biomaterials 1996; 17:1417–1422.

46. Hile DD, Amirpour ML, Akgerman A, Pishko MV. Active growth factor delivery from poly(D,L-lactide-*co*-glycolide) foams prepared in supercritical CO_2. J Controlled Release 2000; 66:177–185.

47. Fontaine PP. Utilizing polymer capsules as a biological delivery system for administering potential ribozyme drugs. 223rd ACS National Meeting, Orlando, FL, 2002.

48. Siripurapu S, Gay YJ, Royer JR, DeSimone JM, Spontak RJ, Khan SA. Generation of microcellular foams of PVDF and its blends using supercritical carbon dioxide in a continuous process. Polymer 2002; 42:5511–5520.
49. Singh L, Kumar V, Ratner BD. Generation of porous microcellular 85/15 poly (DL-lactide-*co*-glycolide) foams using supercritical CO_2 for biomedical applications. Am Soc Mech Eng 2000; 91:29–36.
50. Leopold CS. Coated dosage forms for colon-specific drug delivery. Pharm Sci Technol Today 1999; 2:197–204.
51. Ribeiro Dos Santos I, Richard J, Pech B, Thies C, Benoit JP. Microencapsulation of protein particles within lipids using a novel supercritical fluid process. Int J Pharm 2002; 242:69–78.
52. Chernyak Y, Henon F, Harris RB, Gould RD, Franklin RK, Edwards JR, DeSimone JM, Carbonell RG. Formation of perfluoropolyether coatings by the rapid expansion of supercritical solutions (RESS) process. 1. Experimental results. Ind Eng Chem Res 2001; 40:6118–6126.
53. Yulu W, Dongguang W, Dave R, Pfeffer R, Sauceau M, Letourneau JJ, Fages J. Extraction and precipitation particle coating using supercritical CO_2. Powder Technol 2002; 127:32–44.
54. Tsutsumi A, Nakamoto S, Mineo T, Yoshida K. A novel fluidized-bed coating of fine particles by rapid expansion of supercritical fluid solutions. Powder Technol 1995; 85:275–278.
55. Mishima K, Matsuyama K, Furukawa A, Fujii T, Shida J, Nojiri N, Kubo H, Katada N. Control of polymer coating thickness of microcapsules containing inorganic microparticles using cosolvency of supercritical carbon dioxide. Kagaku Kogaku Ronbunshu 2001; 27:700–706.
56. Benoit JP, Richad J, Thies C. Preparation of microcapsules containing active substances coated with a polymer. France, WO 9813136, 1998.
57. Mehta DB, Corbo M. Process for coating medical devices by using supercritical carbon dioxide. USA, WO 2001087368, 2001.
58. Honda M, Yashima T, Shibue T, Murakami T. Manufacture of polymer solutions and films, and apparatus therefor. Japanese patent 2000256468, 2000.
59. Thoma K, Schluetermann B. Relationships between manufacturing parameters and pharmaceutical-technological requirements for biodegradable microparticles. 5. Relationships between manufacturing parameters, surface properties and degradation characteristics. Pharmazie 1992; 47:368–373.
60. Benoit JP, Courteille F, Thies C. A physicochemical study of the morphology of progesterone-loaded poly(D,L-lactide) microspheres. Int J Pharm 1986; 29:95–102.
61. Ruchatz F, Kleinebudde P, Muller BW. Residual solvents in biodegradable microparticles. Influence of process parameters on the residual solvent in microparticles produced by the aerosol solvent extraction system (ASES) process. J Pharm Sci 1997; 86:101–105.
62. Bleich J, Mueller BW. Production of drug loaded microparticles by the use of supercritical gases with the aerosol solvent extraction system (ASES) process. J Microencapsul 1996; 13:131–139.
63. Mohr D, Wolff M, Kissel T. Gamma irradiation for terminal sterilization of 17-

β-estradiol loaded poly-(D,L-lactide-*co*-glycolide) microparticles. J Controlled Release 1999; 61:203–217.

64. Holy CE, Cheng C, Davies JE, Shoichet MS. Optimizing the sterilization of PLGA scaffolds for use in tissue engineering. Biomaterials 2001; 22:25–31.

65. Dillow AK, Dehghani F, Hrkach JS, Foster NR, Langer R. Bacterial inactivation by using near- and supercritical carbon dioxide. Proc Natl Acad Sci USA 1999; 96:10344–10348.

66. Saltzman WM, Sheppard NF, McHugh MA, Dause RB, Pratt JA, Dodrill AM. Controlled antibody release from a matrix of poly(ethylene-*co*-vinyl acetate) fractionated with a supercritical fluid. J Appl Polym Sci 1993; 48:1493–1500.

67. Lele AK, Shine AD. Morphology of polymers precipitated from a supercritical solvent. AIChE J 1992; 38:742–747.

68. Randolph TW, Randolph AD, Mebes M, Yeung S. Sub-micrometer-sized biodegradable particles of poly(L-lactic acid) via the gas antisolvent spray precipitation process. Biotechnol Prog 1993; 9:429–435.

69. Dixon DJ, Johnston KP, Bodmeier RA. Polymeric materials formed by precipitation with a compressed fluid antisolvent. AIChE J 1993; 39:127–139.

70. Mawson S, Kanakia S, Johnston KP. Coaxial nozzle for control of particle morphology in precipitation with a compressed fluid antisolvent. J Appl Polym Sci 1997; 64:2105–2118.

71. Benedetti L, Bertucco A, Pallado P. Production of microparticles of a biocompatible polymer using supercritical carbon dioxide. Biotechnol Bioeng 1997; 53: 232–237.

72. Falk RF, Randolph TW. Process variable implications for residual solvent removal and polymer morphology in the formation of gentamicin-loaded poly(L-lactide) microparticles. Pharm Res 1998; 15:1233–1238.

73. Merodio M, Irache JM, Valamanesh F, Mirshahi M. Ocular disposition and tolerance of ganciclovir-loaded albumin nanoparticles after intravitreal injection in rats. Biomaterials 2002; 23:1587–1594.

74. Kompella UB, Bandi N, Ayalasomayajula SP. Subconjunctival nano- and microparticles sustain retinal delivery of budesonide, a corticosteroid capable of inhibiting VEGF expression. Invest Ophthalmol Vis Sci 2003; 44:1192–1201.

75. Vajdy M, O'Hagan DT. Microparticles for intranasal immunization. Adv Drug Deliv Rev 2001; 51:127–141.

76. Gonda I. The ascent of pulmonary drug delivery. J Pharm Sci 2000; 89:940–945.

77. Crowder TM, Rosati JA, Schroeter JD, Hickey AJ, Martonen TB. Fundamental effects of particle morphology on lung delivery: predictions of Stoke's law and the particular relevance to dry powder inhaler formulation and development. Pharm Res 2002; 19:239–245.

78. Eldridge JH, Hammond CJ, Meulbroek JA, Staas JK, Gilley RM, Tice TR. Controlled vaccine release in the gut-associated lymphoid tissues. I. Orally administered biodegradable microspheres target the Peyer's patches. J Controlled Release 1990; 11:205–214.

79. Desai MP, Labhasetwar V, Amidon GL, Levy RJ. Gastrointestinal uptake of biodegradable microparticles: effect of particle size. Pharm Res 1996; 13:1838–1845.

80. Floyd AG. Top ten considerations in the development of parenteral emulsions. Pharm Sci Technol Today 1999; 2:134–143.

81. Walpot H. Franke RP, Burchard WG, Agternkamp C, Muller FG, Mittermayer C, Kalff G. [Particulate contamination of infusion solutions and drug additives within the scope of long-term intensive therapy. 1. Energy dispersion electron images in the scanning electron microscope—REM/EDX]. Anaesthesist 1989; 38:544–548.

82. Goel SK, Beckman EJ. Generation of microcellular polymeric foams using supercritical carbon dioxide. II. Cell growth and skin formation. Polym Eng Sci 1994; 34(14):1148–1156.

83. Spenlehauer G, Veillard M, Benoit JP. Formation and characterization of cis-platin-loaded poly(DL-lactide) microspheres for chemoembolization. J Pharm Sci 1986; 75:750–755.

84. Bitz C, Doelker E. Influence of the preparation method on residual solvents in biodegradable microspheres. Int J Pharm 1995; 131:171–181.

85. Dulieu C, Richard J, Benoit JP. Prolonged-release microspheres for injection and preparation method. France, WO 2001089481, 2001.

86. Perrut M. Method for collecting and encapsulating fine particles. France, WO 2001049407, 2001.

87. York P, Wilkins SA, Storey RA, Walker SE, Harland RS. Coformulation of drugs and oligomeric or polymeric excipients. U K, WO 2001015664, 2001.

88. Benoit JP, Rolland H, Thies C, Van De Velde V. Encapsulation of active particles by coating of a central core. France, EP 706821, 1996.

89. Pace GW, Vachon MG, Mishra AK, Henrikson IB, Krukoniz V. SCF process for forming aqueous nano- and micro-particulate suspensions of poorly soluble compounds using surface modifying agents. U S A, US 6177103, 2001.

90. Otake K, Abe M, Sakai H, Imura T, Kaise C, Sakurai M, Arai Y, Kaneko T. Method and apparatus for producing liposome. Japan, WO 2002032564, 2002.

91. Ando T, Tada O, Moriyoshi T, Nakanishi T. Method and apparatus for molding plastic foams. Japan, JP 2002127156, 2002.

92. Saito H, Narushima D, Kawahigashi H, Kanai T. Open-cell polycarbonate foams and their manufacture using supercritical gases. Japan, JP 2002144363, 2002.

93. De Ponti R, Torricelli C, Martini A, Lardini E. Use of supercritical fluids to obtain porous sponges of biodegradable polymers for pharmaceutical implants or drug delivery systems. Italy, WO 9109079, 1991.

94. Subramaniam B, Saim S, Rajewski RA, Stella V. Methods for a particle precipitation and coating using near-critical and supercritical antisolvents. U S A, US 5833891, 1998.

95. Tsujimoto T, Onishi H, Sato T, Yamakawa K, Matsuoka M. Environment-friendly method for manufacture of cellulosic polymer film. Japan, JP 2002127166, 2002.

96. Murthy AKS, Bekker AY, Patel KM. Deposition of thin films using supercritical fluids. U S A, US 4737384, 1988.

97. Gupta RB, Chattopadhyay P. Method of forming nanoparticles and micro-particles of controllable size using supercritical fluids with enhanced mass transfer. USA, US 6620351, 2003.

10

Processing of Biological Materials

B. L. Knutson
University of Kentucky, Department of Chemical and Materials
Engineering, Lexington, Kentucky, U.S.A.

Marazban Sarkari
RX Kinetix, Louisville, Colorado, U.S.A.

1. NEED FOR IMPROVED BIOPROCESSING TECHNIQUES

1.1. Bioprocessing for Drug Delivery

The rapid growth of recombinant biotechnology over the last 20 years has resulted in about 75 protein- and peptide-based drugs available on the market for treatment and more than a 100 in late-phase clinical trials (1). The market for these therapeutic proteins was estimated in 2001 to be $27 billion (2). Global sales of genetically engineered protein drugs were $10.7 billion in 1995 (3) and are forecast to be around $59 billion in 2010 (2). Further, the antibody therapeutics market is expected to reach $7 billion in 2004 (4). In spite of the high demand for protein-based drugs, their processing and delivery remains a challenge. Proteins are easily denatured by the low pH and enzymes present in the gastric environment and, hence, cannot be administered using traditional

411

oral delivery systems. Presently, protein-based drugs are almost exclusively delivered via injectable systems, which suffer from poor delivery efficiencies and low patient compliance. Hence, there is a strong market-driven push toward the development of alternative drug delivery techniques for proteins, including inhaled and controlled or sustained release systems. Further, as increasing applications are found for DNA, RNA, and nucleic acid fragments in gene therapy (5,6) and as vaccines (7), development of novel efficient methods to deliver them are becoming critical.

1.2. Bioprocessing for Sustainable Technology

Far-reaching goals have been established for increasing the role of biobased production in the United States (8), as set forth by the Biomass Technical Advisory Committee. This interagency committee is responsible for advising a broad range of government agencies, including the Environmental Protection Agency (EPA), National Science Foundation (NSF), Department of Energy (DOE), and Department of Agriculture (USDA), on biobased production and biomass conversion. This national initiative includes increasing the use of transportation fuels from biomass from 0.5% in 2001 to 20% in 2030 and increasing the production of chemicals and materials from biobased products from the 5% in 2001 to 25% in 2030 (8). Integral to this biobased production initiative is the concept of the "biorefinery," in which high value-added products (e.g., antibiotics, amino acids, pharmaceuticals) and commodity bioproducts (solvents, ethanol), as well as energy are derived from the starting material. Supercritical fluid (SCF) technology has the potential to contribute to both the production of biocommodities (e.g., the extraction of volatile fermentation products) as well as "high value-added" biobased production (e.g., enzymatic catalysis of chiral pharmaceutical precursors, protein purification and processing, etc.).

The overall processing needs for these high value-added bioproducts include recovery from aqueous broths, fractionation and purification, microparticle formation, microencapsulation, drug delivery applications including controlled and sustained release systems, and stability enhancement. Finally, therapeutic proteins must be formulated in a manner appropriate to their application and mode of delivery. Existing bioprocessing techniques are capable of fulfilling some of these needs, however, yields and residual activities are often very low. Further, the introduction of novel biotechnology products constantly poses new processing challenges. This review explores methods of bioprocessing with SCFs that could meet some of these challenges. It focuses on the effect of SCFs and compressed solvents on nucleic acids, proteins, and whole cells during SCF-based bioprocessing.

2. EFFECT OF CO$_2$ ON PROCESSING OF WHOLE CELLS
AND BIOLOGICALS

Numerous investigators have studied the effect of pressurized CO_2 on cell viability for the sterilization and inactivation of micro-organisms. Mechanisms that have been proposed for the antimicrobial activity of pressurized CO_2 take advantage of its increased solubility relative to atmospheric conditions and include the following (9): cell rupture by the rapid depressurization of high pressure CO_2, the extraction of cell wall lipids, the penetration of pressurized CO_2 into the cellular membrane disrupting membrane properties, and the lowering of the intracellular pH by the dissolution of CO_2 within the cell. Cell inactivation could also occur through the irreversible inhibition of aqueous enzymes related to metabolic function, such as pH-sensitive, intracellular decarboxylases (9). The lowering of aqueous pH through the formation of carbonic acid may also impact the viability of the micro-organism by a reduction in the pH of the medium [Eq. (1)]:

$$H_2O + CO_2 \leftrightarrow H_2CO_3 \leftrightarrow H^+ + HCO_3^- \leftrightarrow 2H^+ + CO_3^{2-} \tag{1}$$

However, lowering of the pH of the media alone cannot describe the inhibition by pressurized CO_2. Acids that do not penetrate the microbial cells as readily as CO_2 (i.e., hydrochloric acid and phosphoric acid) do not inhibit the cells as greatly (10,11). Another possible mechanism of inactivation is the conversion of bicarbonate to carbonate during system depressurization, which may lead to the precipitation of intracellular ions such as calcium and magnesium (9).

The extent and mechanism(s) of microbial inactivation by pressurized CO_2 are dependent on the method of contacting between CO_2 and the cell mass, the depressurization scheme, and the operating temperature, pressure, and phase. The nature of the cell (bacterial, yeast, or fungal, and gram positive or gram negative) and the system water activity also impact the extent of sterilization of microbes (9,12). The performance of CO_2-based sterilization and microbial inactivation in whole cell synthesis, food technology, pharmaceutical processing, and cell disruption for the recovery of intracellular components are described in the following subsections.

2.1. Biosynthesis and Cellular Metabolism in the Presence of CO$_2$

2.1.1. Effect of Pressurized CO$_2$ on Enzymes

The activity of a broad range of suspended purified enzymes and cell extracts in nearly anhydrous supercritical CO_2 has been demonstrated with limited reports of enzyme inactivation (13–15). The rate and selectivity of enzymatic reactions in organic solvents and SCFs may be directly affected by solvent strength (16) or influenced by indirect measures of solvent strength including thermodynamic water activity (a_w) and the solubility of the substrate (17–21). Enzymatic catalysis in SCFs exploits the ability to alter solvent strength with small changes in temperature and pressure in the near-critical region. The versatile solvation characteristics of SCFs, the extreme catalytic specificity of enzymes, and the ability to achieve complete removal of the SCF solvent by depressurization make enzymatic catalysis in SCFs attractive to the pharmaceutical and food industries.

Although some water appears to be necessary for enzyme activity in SCF CO_2 (15), irreversible denaturation of enzymes has been reported for a variety of aqueous enzyme systems in the presence of pressurized CO_2 (22–24). Excess water is thought to be necessary for the denaturation of some proteins by SCF CO_2, although some enzymes experience an increase in activity in the presence of water and pressurized CO_2 (15,22,25). Pressurized in-situ fluorescence spectroscopy studies (26) demonstrate that the change in protein conformation of aqueous bovine trypsin is rapid (occurring in the first several minutes of contacting) and that the enzyme segments experiencing changes are not limited to the active site. The presence of excipients (such as sugars) and buffers increases the resistance of some aqueous enzymes to denaturation (27).

The purposeful denaturation of aqueous proteins by CO_2 for food and pharmaceutical processing is typically explored at temperatures of 35–60°C and CO_2 pressures on the order of 10 MPa. In contrast, hydrostatic pressures of greater than 600 MPa are typically necessary to denature monomeric proteins in the absence of CO_2 (22). Enzyme denaturation by SCF CO_2 in food processing is credited with increasing digestibility and the foaming or emulsifying properties of the treated food (22). Most individual amino acids that were tested (L-glutamic acid, L-methionine, L-leucine, L-alanine, β-alanine, and L-lysine) do not appear to be damaged by CO_2 processing unless they were also susceptible to the corresponding heat treatment in the absence of CO_2 (such as L-glutamine) (22). Arginine was converted into arginine bicarbonate in the presence of humid CO_2 at 80°C and 30 MPa after 6 h. This reaction occurred only in the presence of CO_2 (and not N_2) and is not considered to be significant at the neutral pH associated with food processing (22).

2.1.2. Effect of Near-Atmospheric CO_2 on Microbial Activity

Carbon dioxide is often a substrate or metabolic product in microbial metabolism and, thus, is present in a range of fermentation processes. For instance, CO_2 is a product of decarboxylation reactions, such as the reactions encountered in the production of ethanol (24). Conversely, CO_2 is a reaction substrate in biosynthetic reactions, such as the production of amino acid precursors. Moderate pressure can be used to stimulate nonbarophilic microbes and mammalian cells and affect yield and selectivity by increasing the concentration of dissolved gases (CO_2 and H_2) in the media (28–31).

Partial pressures of CO_2 above ~0.01 MPa have been observed to inhibit a range of bacteria, fungi, and yeasts [as reviewed by Jones and Greenfield (24), Eklund (32), and Dixon and Kell (33)]. This inhibition frequently correlates with the solubility of CO_2 in the fermentation media. High pressure, reduced temperature, and simple versus complex media promote the dissolution of CO_2 in the media and reduce growth and metabolic activity. In contrast, the growth of spore-forming bacteria and fungi is not inhibited by these moderate partial pressures of CO_2 (24).

The mechanism of inhibition in near-atmospheric CO_2 is related to both metabolism and the function of the plasma membrane (24). The permeability of the cellular membrane to dissolved, unhydrated CO_2, a neutral molecule, creates disorder and alters the membrane fluidity even at near-atmospheric pressures. However, unlike typical small "anaesthetic" molecules which alter membrane fluidity, the water permeability of the cell *decreases* upon contact with CO_2 (24). Jones and Greenfield (24) suggest that this unique property of CO_2 inhibition is due to the presence of the bicarbonate ion, which may act on the phospholipid head groups and the proteins near the surface of the membrane to alter the surface charge of the cell.

Some microorganisms may have a greater sensitivity to metabolic effects, rather than membrane effects, at near-atmospheric partial pressures of CO_2 (32). The overall effect of CO_2 partial pressure on metabolism in these organisms is suggested to result from competing effects of carboxylation and decarboxylation activity within the cell and a potential mass action effect (24). At near-atmospheric pressures of CO_2, the relative concentrations of aqueous CO_2 and carbonate ion are adequately buffered within the cell (24). Therefore, a decrease in internal pH is not expected to directly contribute to CO_2 toxicity at these conditions.

2.1.3. Effect of Pressurized CO_2 on Microbial Fermentation

The effect of pressure on biphasic fermentations has been reported for extremely thermophilic organisms (34), supercritical and compressed solvent extractions (35–37), biomethanation of synthesis gases (38), biohydrogenations (39), and bacteria for enhanced oil recovery (40). Fermentation in the presence of compressed and SCF CO_2 has been conducted with a goal of using pressurized CO_2 as an in situ extraction solvent. Extractive fermentation using compressed solvents could prevent toxicity due to the buildup of a fermentation product—in particular, ethanol—and greatly enhance the yield and selectivity to the desired product. Although the solubility of a range of fermentation products may be relatively low, CO_2 has been the first choice of SCF solvents for extractive fermentation largely due to its mild critical temperature and environmental acceptability. In addition, product recovery and solvent regeneration can be achieved through partial depressurization of the compressed solvent. Inhibition studies for several yeasts and bacteria have been used to evaluate the feasibility of extractive fermentation using SCF CO_2 (41–43). However, these investigations have demonstrated that SCF CO_2 negatively impacts the growth, productivity, and viability of yeast and bacterial cells (for a summary, see Ref. 44), which is not surprising based on its effectiveness as a sterilization agent (as reviewed below).

2.1.4. Effect of Other Pressurized Solvents on Microbial Fermentation

Although CO_2 is inhibitory to microbes, compressed hydrocarbon solvents may be appropriate for extractive bioconversions and extractions in biphasic (aqueous-compressed solvent) systems. Our laboratory investigated the metabolic activity of the anaerobic, thermophilic bacteria *Clostridium thermocellum* as a model system (45). Thermophilic bacteria have a distinct advantage over conventional yeasts for ethanol production in their ability to use a variety of inexpensive biomass feedstocks. Extractive fermentation using compressed solvents is an approach to address the end-product toxicity of these bacteria to ethanol and improve the economic viability of biofuel production by thermophilic organisms.

The metabolic activity of nongrowing *C. thermocellum* for the production of ethanol, lactate, and acetate in biphasic systems with compressed solvents (N_2, SCF CO_2, SCF ethane, and liquid propane at 60°C and 7 MPa) was demonstrated (35–37). When *C. thermocellum* was incubated in the presence of nitrogen, the majority of metabolic activity relative to the atmospheric control (>75%) was maintained during the 10-h incubation (35) (Fig. 1). Cell metabolism in the presence of SCF CO_2 was almost completely inhibited relative to the control incubations (35). Incubations with compressed propane or SCF ethane had considerably less activity compared to atmospheric control incubations (39% for ethane; 20% for propane). Activity in these compressed hydrocarbon solvents is consistent with the low biocompatibility of hydrocarbon solvents (e.g., hexane, heptane) to specific yeast and bacteria.

The demonstration of metabolism in compressed biphasic systems allowed us to explore the effect of solvent choice and pressure on biocompatibility. Biocompatibility of liquid solvents is frequently correlated with the *n*-octanol–water partition coefficient (log *P*) of the solvent (46). The log *P* of a substance is defined as the ratio of the molarity of the substance at infinite

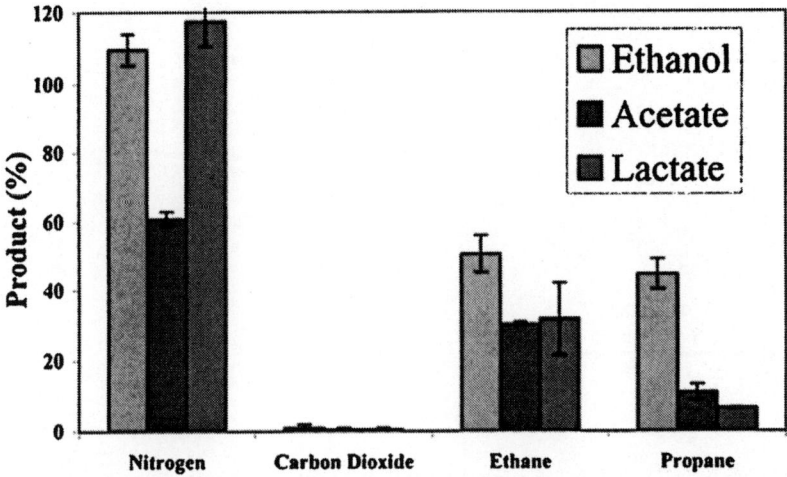

FIGURE 1   Incubation of *C. thermocellum* with compressed solvents at 7.0 MPa. (Adapted from Ref. 45.)

dilution in the two saturated phases. Generally, solvents with log P values less than 2 are not suitable for biocatalysis, whereas solvents with log P values greater than 4 are biocompatible (47).

Interestingly, the activity of *C. thermocellum* in SCF ethane, SCF CO_2, liquid propane, and gaseous propane does not follow the log P trends seen with liquid solvents. Thus, log P may be of limited value when applied to the correlation of metabolic activity in compressed solvents. The traditional definition of log P (25°C and 0.1 MPa) was extended to our incubation conditions (60°C and 7 MPa) using the group contribution associating equation of state (GCA-EOS) (48,49) to calculate the mole fraction of the dissolved compressed fluids in octanol and water. Log P' [$\log_{10}(x_i^O/x_i^W)$] correlated well with the total metabolic activity (ratio of total products formed in the treatment to the total products formed in the control) in liquid hydrocarbon solvents (Fig. 2) (37).

Experiments were also conducted to assess the impact of compressed N_2, ethane, and propane on product selectivity of whole cell biocatalysis (45). Pressurized incubations, with and without a compressed solvent headspace, lead to an increase in the ratio of ethanol to acetate produced by the organism (Fig. 1). These results are consistent with increased dissolution of H_2, a product gas which affects the pathways of acetate production (50), in the fermentation media with increasing pressure. Lactate formation was also decreased in the presence of compressed and liquid solvents (but not nitro-

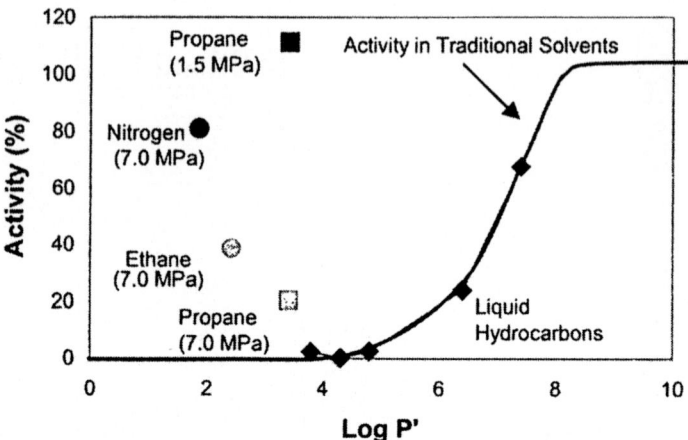

FIGURE 2 Metabolic activity (based on total product formation) as a function of the density-independent (mole-fraction based) octanol–water coefficients of the solvent. (Adapted from Ref. 45.)

gen). After 10 h, incubation with compressed ethane and propane produced 32% and 7% of the lactate in the control, respectively. The reduction in lactate formation was associated with a simultaneous reduction in the rate of cellobiose uptake by the micro-organism. The accumulation of glucose in the fermentation media, a product of cellobiose conversion by an extracellular hydrolase, indicates that extracellular activity is relatively unaffected by the compressed solvents, even SCF CO_2. Pressurized incubations with compressed solvents offer the opportunity to tune product selectivity with solvent choice by causing significant shifts in the fermentation pathways. In addition, compressed hydrocarbon solvents may have clear advantages over CO_2 for the recovery of hydrophobic fermentation products. The ability to alter solvent strength in a single process fluid provides a unique opportunity to examine the physiological and metabolic basis of biocompatibility.

2.2. Microbial Inactivation in Aqueous Systems for Food and Pharmaceutical Processing

Sterilization of aqueous solutions using pressurized CO_2, particularly for food processing, has been examined in a range of studies. Advantages of CO_2 sterilization of heat-sensitive foods include the retention of a fresh taste, the elimination of both pathogenic and spoilage micro-organisms, nutrient preservation, product preservation from oxidation due to the displacement of dissolved oxygen by CO_2, and no residual solvent (CO_2) in the product (51).

Sterilization by pressurized CO_2 has been proposed as an alternative to high pressure treatments (200–700 MPa) for food, pharmaceutical (12), and biomedical (52) applications.

First-order reaction kinetics, with one or two linear stages, have been reported for the sterilization of a range of microbial cells using CO_2 (11,53–55). The presence of water is required for efficient sterilization (12,53,56). For example, Kumagai and coworkers (53) examined the sterilization of *Saccharomyces cerevisiae* in an aqueous solution at 40°C and a pressure range of 4–15 MPa. At low water content, the sterilization rate increased with increasing content of water. At a sufficiently high water content, free water (not bound to the cells) existed in the system and the sterilization rate was almost constant with water content. The authors conclude the presence of dissolved CO_2 in the aqueous-cell environment, and not system depressurization (which occurred at a rate of ~ 8 MPa/h), was responsible for the inactivation of the microbial cells. Cell inactivation kinetics are difficult to compare across laboratories because the aqueous media is frequently not saturated with dissolved CO_2 and the actual concentrations are not measured (44,51).

The mechanisms of CO_2 toxicity at near-atmospheric pressures are amplified at the near-critical and supercritical pressures used to achieve sterilization. For example, Spilimbergo et al. (9) examined the mechanism of inactivation of *Pseudomonas aeruginosa* and *B. subtilis* at 38–54°C and 5.8–20 MPa in a batch process. Total inactivation of the bacteria was observed when exposed to SCF CO_2 at 38°C and 7.4 MPa for 150 s. The mechanism of inactivation is suggested to involve (a) the diffusion of SCF CO_2 into the cells leading to a drop in pH and a subsequent loss of activity of key enzymes and (b) the extraction by SCF CO_2 of intracellular substances, including phospholipids. Further analysis determined a high solubility of CO_2 in model cell membrane phospholipids, suggesting that the enhanced permeability of the membrane in the presence of CO_2 contributed to the inactivation of the cells.

Dillow et al. (12) have reported the inactivation of a number of bacterial species using SCF CO_2 at moderate temperatures. Sterilization of both gram-positive and gram-negative bacteria was examined: *Bacillus cereus, Listeria innocua, Staph. aureus, Salmonella salford, Pseudomonas aeruginosa, Escherichia coli, Proteus vulgaris,* and *Legionella dunnifi.* The method was also recommended for the sterilization of polymers poly(L-lactide) (PLA) and poly(lactide-*co*-glycolide) (PLGA). Aqueous broths, incubated with and without the biodegradable polymers, were treated with SCF CO_2 (30–45°C, 14–21 MPa). All micro-organisms were inactivated when exposed to SCF CO_2 at 25–40°C and 20.5 MPa for 4 h (with the exception of *B. cereus*, which required 6 h) (12). SCF N_2 and near-critical tetrafluoroethane were unable to inactivate the micro-organisms. Only minor cell wall disruption was observed in the scanning electron micrographs of some of the processed gram-negative bacteria (*P. aeruginosa*), indicating that cell rupture did not contribute substantially to the sterilization mechanism.

Fungal spores have high heat resistance, making their treatment with SCF CO_2 attractive (54,57). Similar to whole cells, the sterilization of vegetative fungal cells appears to require the presence of water (53). Conflicting results have been observed for the CO_2-based sterilization of spores in the absence of a thermal treatment, although recent results suggest that heat treatment is necessary to achieve satisfactory inactivation of spores (9). For example, Ballestra and Cuq (54) examined the inactivation of suspended *Bacillus subtilis* spores and *Byssochlomay fulva* ascopores. Inactivation of the spores required a minimum threshold temperature dependent on the micro-organism (80°C for both *B. subtilis* and *B. fulva*). The contribution of CO_2 (5 MPa) to the inactivation was most marked near this threshold temperature and was obscured by the effectiveness of heat treatment at significantly higher temperatures. A 3.5 log reduction of *B. subtilis* spore viability was reported after 1 h of heat treatment (80°C) in the presence of 5 MPa of CO_2. Similarly, Spilimbergo and coworkers (9) demonstrated that exposure of *B. subtilis* spores for 24 h at 75°C and 7 MPa was required for a greater than 7 log reduction of viability. With a goal of reducing the exposure time and process temperature, Spilimbergo et al. (9) developed a pressure cycle analogous to those used in traditional high pressure treatment processes. A 3.5 log reduction of *B. subtilis* spore resulted from a fast cycle treatment (30 min of treatment at 30 cycles per hour, $\Delta P = 8$ MPa) at a moderate temperature (35°C). They further studied the synergistic effect of a pulsed electrical field with high pressure CO_2 for the inactivation of bacterial cells. *E. coli*, *S. aureus*, and *B. cereus* spores suspended in glycerol were treated with PEF (4.5–25 kV/cm, 1–20 pulses) and then with CO_2 at temperatures between 34 and 40°C and pressures of 8–20 MPa for 10 min. *E. coli* and *S. aureus*, for example, were completely inactivated when treated with 10 pulses at 25 kV/cm followed by exposure to SCF CO_2 at 34°C and 20 MPa for 10 min. Lower temperatures and pressures are required for CO_2-based sterilization with PEF than with CO_2 alone (9). Electroporation of the plasma membrane via PEF is a likely mechanism for bacterial inactivation (58).

The inactivation of spores by CO_2 has been described by a single first-order rate constant (54,57). In contrast, a two-step process has been reported for the heat inactivation of spores in the absence of CO_2 (54,59). The presence of CO_2 during heat treatment could increase the rate of spore heat activation, considered to be an initial step in sterilization (59), and/or the rate of spore inactivation to result in first-order sterilization kinetics (54).

Inactivation of spores (a 6 log reduction in *B. subtilis* spores in 1 h) at moderate temperatures (40–55°C) was achieved by Ishikawa and coworkers (60) using a contacting scheme referred to as the SCF CO_2 microbubble method at 30 MPa. CO_2 was introduced into a static aqueous solution containing suspended spores through a 10-μm pore size filter, which promotes saturation of the aqueous phase by CO_2. A 1 log reduction in *Bacillus* spores was observed without the filter (corresponding to 50% saturation), whereas 4 log reduction was observed with the filter (corresponding to 80% saturation) for a 30-min treatment at 40°C and 30 MPa.

Other investigators have also demonstrated the importance of efficient contacting of the pressurized CO_2 and the cell phases (12). The observed rate of cell inactivation can be greatly influenced by slow mass transfer rates of CO_2 in and out of the microbial cells (10). The semicontinuous contacting scheme proposed by Elvassore and coworkers (61) promotes cell inactivation relative to batch processes. Similarly, Sims and coworkers developed a continuous sterilization technique for aqueous easy-to-flow foods using a hollow-fiber membrane contactor to achieve rapid saturation of CO_2 (51). CO_2 saturation of the water was greater than 94% at resident times greater than 2 min at room temperature and 7.5 or 15 MPa. Over this range of conditions, log reduction values of greater than 5 were typically obtained with a residence time of less than 2 min for *Lactobacillus plantarum*.

2.3. Cell Disruption for Recovery of Intracellular Products and Sterilization

Cell lysis is carried out for the extraction of intracellular components and the inactivation of the micro-organism. The advantages of using high pressures and SCFs for cell lysis include preservation of thermally labile molecules, preclusion of the formation of very small particles leading to easier downstream processing, and the possibility of selective protein delipidation when using SCF CO_2 (9,62). Cell lysis is not observed in all antimicrobial investigations of pressurized CO_2 (11) and is favored by rapid decompression processes with high concentrations of dissolved CO_2 and by continuous processing schemes (44).

Lin et al. (55,63) have described a method to lyse baker's yeast cells using subcritical and SCF CO_2 (25–35°C, 7–34 MPa). In a batch process, pressurized CO_2 was allowed to permeate the cells for a fixed time, followed by rapid depressurization. The rapid depressurization resulted in explosive expansion of the CO_2 within the cells, leading to cell rupture and release of the cellular components. SCF CO_2 at 35°C and 20 MPa was the most effective for enzyme release. Higher temperatures were more effective at cell disruption, but they also resulted in lower activity of the released enzymes. Enzyme release from baker's yeast cells was further enhanced with the addition of a cell-wall-lytic enzyme, β-glucuronidase, to the cell mass before pressurization (55).

The effectiveness of rapid decompression for cell (*S. cerevisiae*) inactivation has been described by first-order kinetics with respect to dissolved CO_2 concentration (44). First-order kinetics were valid at both subcritical and supercritical temperatures and pressures, suggesting that CO_2 concentration in the media may be used to predict the degree of sterilization over a range of conditions.

Rapid depressurization of a slurry of cells in the presence of other SCFs, such as SCF N_2O, can also lead to cell disruption and the release of proteins and nucleic acids. Experiments using SCF N_2O and CO_2 were compared to pressurized N_2, which has a limited solubility in water. For *B. subtilis*, the use of SCF N_2O led to improved protein recovery relative to SCF CO_2 (Table 1). SCF N_2O and SCF CO_2 were equivalent for lysing baker's yeast, and CO_2 was very ineffective in lysing *E. coli*. A subsequent investigation of the lysis of baker's yeast through rapid decompression of pressurized CO_2, N_2O, and N_2 (at 40°C and 4 MPa after exposure for 4 h) yielded similar results (57). However, the inactivation was a stronger function of the treatment times than the rate of depressurization. The effectiveness of SCF N_2O as a cell-lysing agent suggests that the dominant mechanism of cell disruption is not specific to SCF CO_2 but is more generally related to compressibility and aqueous solubility of the fluid.

Treatment with near-critical or supercritical fluids has also been proposed for the inactivation of virus particles in biological systems, especially blood products (65). SCF N_2O at 40°C and 17–27 MPa was contacted with Murine-C retrovirus in culture medium and in serum for 5–30 min. Upon

TABLE 1 Protein and Nucleic Acid Yield Obtained After Lysis of Various Micro-organisms Using Various Fluids

Micro-organism	Pressure (MPa)	Time of exposure (min)	Concentration of cells (g dry cell wt/L)	SCF	Protein yield (%)	Nucleic acid yield (%)
E. coli	33.0	25	69	CO_2	5	21
				N_2O	18	50
				N_2	20	26
B. subtilis	34.6	120	93	CO_2	22.7	6.6
				N_2O	41.3	70.7
				N_2	0.0	20.2
S. cerevisiae	30.5	25	68	CO_2	29.6	77.5
				N_2O	26.6	67.0
				N_2	5.5	20.5

Source: Adapted from Ref. 64.

rapid depressurization of the system, the viral titers decreased by 2–4 log units. The temperature of operation had a greater influence in reducing viral titers than the time of contact between the biological fluid and the SCF. The only change in the serum was a lowering of the lactate dehydrogenase (LDH) activity, indicating that this process is relatively benign.

2.4. Using Near-Critical Fluids in the Processing of Sterile Pharmaceutical Solids

The processing of pharmaceutical powders and solids in a noncontaminating, nonoxidizing environment is one of the clear benefits of using pressurized CO_2. In addition, several investigators have developed processes in SCF CO_2 whose sole purpose is the sterilization of solids and dry powders (12,56,66). One of the earliest techniques describes a simple filtration process to obtain solids that are free from microbial contamination (66). The solids (acetylsalicylic acid, ampilicillin sodium, and azlocillin sodium) were dissolved in a SCF (CO_2, N_2O, or Freon 13 + 2% Freon 11) with and without cosolvents at temperatures ranging from 30°C to 50°C and pressures ranging from 10 to 90 MPa. The supercritical solution was passed through a sterile filter that retained all of the micro-organisms. The filtrate was depressurized downstream to obtain a sterile pharmaceutical powder. This process exploits the strong dependence of solid solubility in SCFs on the operating pressure to make powders that are completely sterile.

Supercritical fluid CO_2 was also investigated for sterilization of dry-powder animal blood plasma containing both living micro-organisms and bacterial endospores (56). The initial water content of the plasma powder was 6.8%. No apparent inactivation of the microbes was achieved at 35°C and 20 MPa in a 2-h treatment. However, with the addition of water (initial water content of 16.7%) and/or the use of cosolvents (ethanol or acetic acid), 2–5 log reductions in living cells could be achieved. The need for water to achieve sterilization of the powders is consistent with previous investigations of sterilization at low water content.

3. CO_2 AS A PRESSURE-TUNABLE ANTISOLVENT FOR BIOLOGICAL MOLECULES

Compressed CO_2 is completely miscible with numerous organic solvents and, upon contact, lowers the cohesive energy density of the organic solvent and reduces its solvent power. At the same time, compressed CO_2 is a poor solvent for many solids. These two properties of compressed CO_2 make it a good antisolvent for a variety of substances. Most processes that exploit the antisolvent properties of CO_2 involve, as a first step, the dissolution of the solute in an organic or aqueous–organic solution. When CO_2 is contacted with this solution, it readily diffuses into the solvent and decreases its solvent strength, resulting in rapid precipitation of the dissolved solute. Unlike other liquid nonsolvents, the antisolvent properties of CO_2 are temperature and pressure dependent. Thus, by varying the operating temperature and pressure, the process can be tuned to the characteristics of the precipitated particles.

The primary benefits these antisolvent processes offer include (a) control over particle size by simple manipulation of operating parameters, (b) precipitation from aqueous, organic, or aqueous–organic solutions, (c) mild operating temperatures, (d) single-step operation, and (e) operation in a relatively inert CO_2 environment. Further, certain applications, specifically, the use of proteins for inhalation therapies, require that the final particles be in the 1–5-μm size range, a range easily and consistently achieved using SCF CO_2 antisolvent processes (67,68). Other current methods for protein recovery and size reduction include lyophilization, spray-drying, and mechanical grinding. The salient features of these processes are compared in Table 2 and it is evident that none of them is universally applicable. Extremes of temperature, lack of control over particle size, activity loss during operation, and multistep operation are the main drawbacks of these techniques.

Currently, there exist numerous antisolvent processes based on SCF CO_2. These include, in alphabetical order, ASES (aerosol solvent extraction system) (70), GAS (gas antisolvent) (71), PCA (precipitation using a compressed antisolvent) (72), SAS (supercritical antisolvent) (67), SAS-EM

TABLE 2 Comparison of Techniques for the Formation of Protein Microparticles

	Spray-drying	Lyophilization	PCA
Operating pressure	Near atmospheric (0.1–0.5 MPa)	High vacuum (200–400 mTorr)	Pressurized system (6.5–20 MPa)
Operating temperature	High temperature (75–150°C)	Cryogenic (−40 to +20°C)	Moderately elevated (25–45°C)
Complexity of equipment	Moderate	High (cryogenic system)	High (pressurized system)
Operating cost	Low	High (cryogenics)	High (pressurized system)
Solute requirements	Must not be sensitive to high temperatures	Must not get denatured at cryogenic temperatures	Must not be pressure sensitive or reactive with CO_2
Level of technical development	Well established	Well established	Close to commercialization (72

Source: Adapted from Ref. 69.

(supercritical antisolvent with enhanced mass transfer) (73), and SEDS (solution-enhanced dispersion by supercritical fluids) (68). They vary primarily in their contacting scheme (i.e., the method of introduction of the solvent and antisolvent streams into the precipitation chamber), composition of the antisolvent stream (pure CO_2 or cosolvent enhanced SCF CO_2), and nozzle design. These techniques, along with their various modifications, are described here with relevance to biological systems only. Excellent reviews by Jung and Perrut (74), Kompella and Koushik (75), and Reverchon (76) also address nonbiological systems.

3.1. Historical Perspective

The first reports of particle formation using SCF antisolvents were published in 1987–1988 as patents by Schmitt (77) and by Fischer and Müller (78). Debenedetti et al. (79) reported the first use of SCF CO_2 as an antisolvent for the precipitation of protein microparticles in a 1991 patent. The cocurrent spraying of insulin and catalase solutions (dissolved in ethanol and water in a 9:1 ratio by volume) into a pressurized SCF CO_2 continuum resulted in either globular or needlelike particles that were 1–3 μm in size. Although the patent literature offers considerable insight into the growth and development of the antisolvent technologies for the processing of biological molecules, this review focuses primarily on peer-reviewed publications, which provide more detailed analyses of the antisolvent process variables and their mechanisms.

The first detailed description of protein processing using a SCF antisolvent also examined insulin (67). Insulin was dissolved in pure dimethyl sulfoxide (DMSO) and dimethyl formamide (DMF) solutions and sprayed into cocurrently flowing SCF CO_2. The processed insulin powders were between 1 and 4 μm and had biological activity indistinguishable from the unprocessed insulin.

Although the first reports on protein processing using SCFs were published in 1993 (67), most reports on processing of biological products appeared in 1999 or later. There are several potential reasons for this gap, including the time taken for the SCF antisolvent technology to mature. In the early 1990s, SCF antisolvent technology was under development, with few studies available on the mechanistics of precipitation. Nonbiological systems are easier to handle and study; hence, at that time, considerable effort was focused on studying PCA with model nonbiological systems (76). Further, the early success of Yeo et al. (67) in processing insulin powders (i.e., obtaining free-flowing powders of the exact size required for pulmonary therapy and complete recovery of activity) set a very high standard for an emerging technology that was difficult to match with other biological systems. Another factor hindering the progress of protein precipitation using PCA is the poor solubility of most proteins in organic solvents. The original PCA technique required that the protein be dissolved in an organic solvent and that this organic solvent also be miscible with SCF CO_2. However, few organic solvents can dissolve proteins at concentrations reasonable for PCA processing (1–10 mg/mL) and only a small subset of these can do so without irreversibly denaturing the protein (80–82). Two approaches were developed to circumvent this limitation of the original PCA process. One was to increase the solubility (and stability) of proteins in organic solvents using hydrophobic-ion pairing (83) and reverse microemulsions (84). The other approach was to increase the miscibility of water (the preferred solvent for biologicals) and CO_2 (the most commonly used SCF antisolvent) by using cosolvents, like ethanol, which are miscible with both water and CO_2.

The following subsections examine the CO_2 antisolvent techniques and summarize noteworthy studies. Table 3 provides a comprehensive summary of the numerous biological molecules that have been processed using SCF CO_2-based antisolvent processes.

3.2. Antisolvent Precipitation of Biological Molecules from Organic Solutions

The two main variations of the antisolvent technique are (1) bubbling of SCF CO_2 into a protein solution continuum (generally referred to as GAS) and (2) spraying a protein solution into a SCF CO_2 continuum (generally known as PCA/SAS). The other techniques listed previously are subsets of these.

3.2.1. The Gas Antisolvent Process

The gas antisolvent process is a semibatch process in which the protein is dissolved in a suitable organic solvent and placed in a precipitation vessel (Fig. 3). Pressurized CO_2 is bubbled into this static protein solution. The CO_2 mixes with the organic solvent, reduces its solvent strength, and precipitates the protein. Under isobaric conditions, the expanded organic solvent is drained from the bottom of the vessel through a filter. Pure CO_2 is passed through the vessel to dry the protein particles collected on the filter. The nature of the precipitate depends on the organic solvent, the degree of supersaturation reached before precipitation occurs (i.e., the rate of CO_2 addition and the pressure at precipitation), the temperature, and the concentration of the solute (74).

Winters et al. (87) studied the GAS precipitation of alkaline phosphatase, catalase, lysozyme, ribonuclease, and trypsin from neat DMSO using pressurized gaseous CO_2 at 34°C as the antisolvent. The activity of the reconstituted proteins was measured and was found to decrease with increasing molecular weight (Fig. 4). Thus, although ribonuclease (13.7 kDa) regained

97% of its original activity, alkaline phosphatase (105.1 kDa) recovered only 0.1% of its original activity.

A detailed study by Thiering et al. (102) considered a number of proteins, solvents, and solvent mixtures (Table 3). Insulin was precipitated using SCF CO_2 from DMSO, ethyl acetate, methanol, and ethanol. The particle size ranged between 0.05 and 1.8 μm. Insulin precipitated from solvents in which the solubility of insulin was low (ethyl acetate, methanol, ethanol) had a smaller particle size than insulin precipitated from solvents with a higher insulin solubility (DMSO). Polydisperse particles of myoglobin were precipitated from DMSO solutions (mean particle sizes of 0.03 μm and 0.4 μm), whereas monodisperse myoglobin particles were precipitated from a methanol solution (0.05–0.3 μm).

Lysozyme was precipitated using CO_2 from the following pure solvents and solvent mixtures: DMSO, DMSO–DMF (70:30), DMSO–ethanol (70:30), DMSO–acetic acid (92:8), ethanol–water (85:15, 90:10, 95:5, and 98:2), and methanol (102). For the ethanol–water system, there was little change in the particle size (0.05–1.0 μm) with increasing water content; however, because the antisolvent power of CO_2 with respect to water is limited, it was difficult to obtain dry precipitates at higher water concentrations. Unfortunately, the activity of lysozyme was not reported for this set of experiments. Such data would have helped determine the contribution of the solvent composition to protein denaturation. In the same study, the particle size of lysozyme precipitated from DMSO decreased with temperature (in the range 18–45°C). Further, 95–100% of the activity was recovered from the reconstituted lysozyme particles precipitated at 25°C and 35°C. Increasing the operating temperature to 45°C decreased the recovered activity to 60%.

Recently, Muhrer and Mazzotti (109) reported the GAS precipitation of lysozyme nanoparticles from DMSO solutions. The study involved the modification of the GAS apparatus with the introduction of a mechanical agitator and sparging of the pressurized CO_2 through the impeller shaft. Finally, the scale of operation was much larger than any of the previous studies; 50 mL of solution was processed in a 400-mL jacketed vessel. The precipitated particles recovered activities up to 75% of the unprocessed lysozyme and were 200–300 nm in size. This is the smallest reported size of protein particles using SCF CO_2 antisolvent techniques. There was no significant impact of antisolvent addition rate (range: 5–50 g/min), initial solute concentration (range: 2–8 mg/mL), and temperature (range: 19–35°C) on the final particle size. Particle agglomeration, however, increased with temperature and the initial concentration of lysozyme in the DMSO solution. Further, X-ray powder diffraction measurements showed that these particles were amorphous.

TABLE 3 Summary of Literature on Protein Processing Using SCF CO_2-Based Antisolvent Techniques

	Protein	Process	Solvent
1	α-Chymotrypsin	PCA	Iso-octane
2	α-Chymotrypsin	PCA	Methylene chloride
3	α-Chymotrypsin	PCA	Water
4	β-Lactamase	PCA	Water
5	Albumin	ASES	Water
6	Alkaline phosphatase	GAS	DMSO
7	Antibody fragment 4D5Fab	SEDS	2-propanol
8	Antibody fragment 4D5Fab	SEDS	Ethanol
9	Antibody fragment 4D5Fab	SEDS	Methanol
10	Antibody fragment D1.3Fv	SEDS	Methanol
11	Bovine serum albumin		
12	Bovine serum albumin	ASES	Methanol
13	Catalase	PCA	Ethanol–water (90:10)
14	Deslorelin		
15	Insulin	PCA	1,1,1,3,3,3-Hexafluoro-2-propanol
16	Insulin	GAS	DMSO
17	Insulin	GAS	DMSO
18	Insulin	PCA	DMSO + methanol + methylene chloride
19	Insulin	PCA	DMSO, DMF
20	Insulin	GAS	Ethanol
21	Insulin	PCA	Ethanol–water (90:10)
22	Insulin	GAS	Ethyl acetate
23	Insulin	GAS	Methanol
24	Insulin	PCA	Methanol
25	Insulin	PCA	Methylene chloride + DMSO
26	Insulin	PCA	Methylene chloride + DMSO
27	Insulin	PCA	Pyridine
28	Insulin	PCA	Tetrahydrofuran
29	Insulin	ASES	Water
30	Insulin	GAS	Water
31	Lecithin S75	PCA	Ethanol
32	Lecithins S20, S75, S100	PCA	Ethanol
33	Lipase PS	RESS-N	Insoluble
34	Lysozyme	ASES	DMSO
35	Lysozyme	GAS	DMSO
36	Lysozyme	GAS	DMSO
37	Lysozyme	GAS	DMSO
38	Lysozyme	GAS	DMSO
39	Lysozyme	GAS	DMSO
40	Lysozyme	PCA	DMSO

Cosolvent	Excipient	Major findings	Ref.
	AOT (HIP)	1–10-μm Spheroidal particles	83
	AOT (HIP) + PLA	2–3-μm Spheroidal particles	83
Ethanol		Spheres, 59% activity	69
Ethanol		Particles with substantial activity	85
Ethanol		Spheres/agglomerates, 50–86% activity	86
		0.1±0.2% Activity	87
		12–46% Recovered specific activity	88
		21% Recovered specific activity	88
		20% Recovered specific activity	88
		3% Recovered specific activity	88
	Gelucire	Coating of particles, no activity loss	89
		Encapsulated spheres, high burst effect	90
		Spherical or rectangular 1-μm particles	91
	PLGA	Sustained release particles	92
		No change in protein structure based on ultraviolet, Fourier transform infrared spectra	93
		Fractionation from ribonuclease	87
		Spherical particles, 1.4–1.8 μm	94
		Equivalent to NPH SC, 90% <5 μm	95
		Spheres, activity indistinguishable from original	67
		Spheres, 0.05–0.3 μm	94
		1-μm Particles	79
		Spheres, 0.3–0.7 μm	94
		Spheres, 0.2–0.7 μm	94
	SDS (HIP)	Particles	83
	PEG/PLA	Spheres, 80% in vivo activity	96
		Spheres	96
	SDS (HIP)	Spheroidal particles	83
	SDS (HIP)	Irregular, 1–5-μm particles	83
Ethanol		Spheres/agglomerates, 96–98% activity	86
		NH_3 antisolvent, denatured spheres	94
		Agglomerated spheres	97
		Agglomerated spheres	98
Ethanol	PLA + PEGs	Spheres	99
	PLA	Multiple coaxial nozzles, nanospheres, 12–14% encapsulation efficiency	100
		72% activity	101
		Spherical, 0.05–1.0 μm, 60–100% activity	102
80% DMF		0.1 μm	102
80% Ethanol		0.02–0.04 μm	102
8% Acetic acid		0.05 width × 0.25 length (μm)	102
		Spheres, 44–100% activity	103

TABLE 3   Continued

	Protein	Process	Solvent
41	Lysozyme	PCA	DMSO
42	Lysozyme	GAS	Ethanol
43	Lysozyme	RESS-N	Insoluble
44	Lysozyme	GAS	Iso-octane reverse micelles
45	Lysozyme	PCA	Methylene chloride
46	Lysozyme	ASES	Water
47	Lysozyme	GAS	Water
48	Lysozyme	SEDS	Water
49	Myoglobin	GAS	DMSO
50	Myoglobin	GAS	Methanol
51	Plasmid DNA	PCA	Methylene chloride
52	Plasmid DNA 6.9 kb	SEDS	Water
53	Plasmid DNA 6.9 kb	SEDS	Water
54	rhDNase	ASES	Water
55	Ribonuclease	PCA	Methanol
56	Ribonuclease	PCA	Methanol
57	Ribonuclease A	GAS	DMSO
58	rIgG	SEDS	Aqueous buffer
59	Trypsin	GAS	DMSO
60	Trypsin	PCA	DMSO

Current methods of protein fractionation include chromatography, salt precipitation, and affinity extraction. Based on the difference in precipitation pressures among proteins, the GAS process can also be used to fractionate proteins. Winters et al. (87) fractionated the following protein pairs using GAS precipitation: lysozyme–ribonuclease and alkaline phosphatase–insulin. When only lysozyme or ribonuclease were dissolved in DMSO, their precipitation pressures were 4.8 and 5.3 MPa, respectively. A solution of lysozyme and ribonuclease in DMSO was then subjected to fractional precipitation. The lysozyme activity in the precipitates collected at 4.93, 5.34, 5.82, and 7.07 MPa was 99 ± 5%, 91 ± 1%, 85 ± 2%, and 83 ±1 %, respectively. Thus, successful separation of lysozyme from ribonuclease using pressurized CO_2 as the fractionating agent was demonstrated. Similar fractionation was carried out with alkaline phosphatase and insulin at 4.93 and 5.34 MPa. Sodium dodecyl sulfate–polyacrylamide gel electrophoresis (SDS–PAGE) characterization of the precipitates revealed no insulin in the 4.93-MPa precipitate and only 3% insulin in the 5.34-MPa precipitate.

The reduced pH of aqueous media in the presence of CO_2 [Eq. (1)], not simply the expansion of the solvent, could lead to protein precipitation. The effect of CO_2 as a volatile acid was examined by Hofland et al. (110), who

Cosolvent	Excipient	Major findings	Ref.
		Spheres, 88–100% activity	101
Water (2–15%)		0.05–1.0 μm	102
Ethanol	PLA + PEGs	Spheres	99
	AOT	No change based on ultraviolet spectra	84
	PLGA	Vapor-over-liquid process; particle encapsulation	104
Ethanol		Spheres/agglomerates, 92–95% activity	86
		NH$_3$ as antisolvent; denatured particles	102
Ethanol		Spheres, 95–100% activity	105
		Spherical particles	102
		Spheres, 0.05–0.3 μm	102
	DOTAP (HIP) + PLA	Linear release profile	106
2-Propanol		10% Activity	107
2-Propanol	Mannitol	75% with mannitol	107
Ethanol		Spheres/agglomerates, 14–33% activity	86
	SDS (HIP)	50-μm Spheroidal particles	83
	SDS (HIP) + PEG	Fibers	83
		97 ± 3% Activity	87
Ethanol		38–48% Activity	108
		83 ± 3% Activity	87
		Spheres, 84–98% activity	101

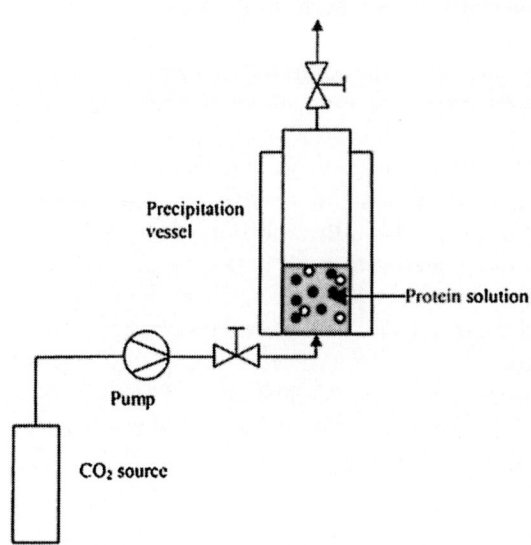

Figure 3 Schematic of the GAS process.

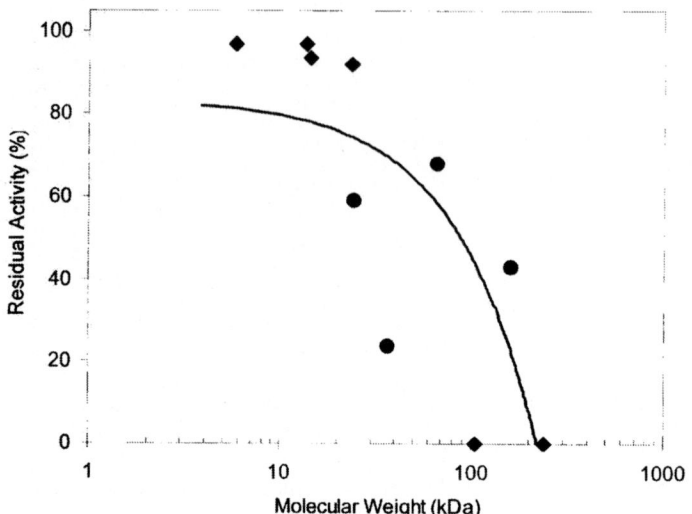

FIGURE 4 Recovered activity of proteins processed by GAS (◆) and PCA (●) as a function of their molecular weight. (Compiled from Refs. 69, 85, and 100.)

studied the precipitation of casein from reconstituted milk. Although the pH of water is reduced to 3.08 in the presence of CO_2 at 25°C and 5.5 MPa, due to the buffering action of the calcium salts present in the milk, the pH of the milk solution was only 4.82 at these conditions. Higher temperatures reduce the solubility of CO_2 in water; hence, the pH of the solutions was a weak function of temperature. Further, the pressure of operation and the protein–CO_2 interaction were deemed to have a minimal effect on the precipitation of casein.

The precipitation and fractionation of soy protein using CO_2 in the pressure range 0.1–4.7 MPa was also examined (111,112). Pressurized CO_2 allowed better control of the solution pH than the addition of mineral acids and prevented local pH overshoots, which can result in poor quality precipitates. Protein precipitates formed by isoelectric precipitation using CO_2 were spherical, whereas those formed using sulfuric acid were irregular (111). The decrease of solution pH to a value close to the isoelectric pH of the proteins was the primary cause of protein precipitation. Volatile acid precipitation is unlike conventional GAS, in which pressures of 5–10 MPa are required to achieve precipitation from an expanded organic solution.

3.2.2. Using Compressed Antisolvents for Precipitation

In the traditional Precipitation using compressed antisolvents (PCA) process (Fig. 5), the biological molecule is dissolved in a suitable organic solvent and

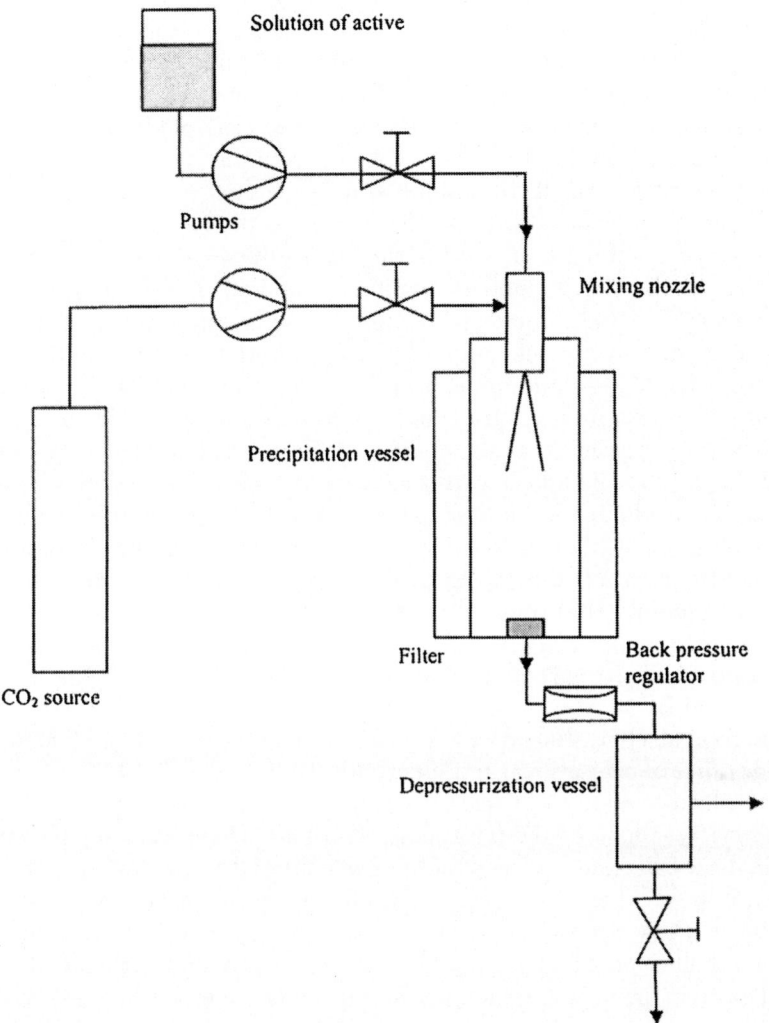

FIGURE 5 Schematic of the PCA process.

sprayed into a CO_2 antisolvent continuum in the precipitation chamber. The organic solvent is expanded by the CO_2 and loses its solvent power, resulting in the precipitation of protein particles. The solvent and the antisolvent continuously flow through the precipitation chamber and the protein particles are collected on a filter at the exit to the chamber. Variations to this scheme include the direction of flow of the antisolvent (cocurrent or countercurrent) to the organic solution (74), coaxial nozzle designs that allow the introduction of solvent and antisolvent as a mixed stream (69), and an intensified breakup of the solvent stream by the application of ultrasonic energy (SAS-EM) (113).

Lysozyme, an inexpensive and robust enzyme, was used in many early studies as a model protein (Table 3). PCA processing of lysozyme formed micron-sized particles that recovered most of their initial biological activity upon reconstitution. Although lysozyme is a good model for preliminary investigations, it is not representative of the therapeutic proteins of commercial significance. Insulin is another commonly studied protein, due to its immense therapeutic and commercial potential. Insulin has been precipitated from a variety of solvents, including ethanol (79), methanol (83), DMSO (67,87,102), 1,1,1,3,3,3-hexafluoro-2-propanol (93), water (86,94), pyridine (114), and tetrahydrofuran (114) (Table 3). Because the biological activity determination of insulin requires study with an animal model, it is not frequently reported. The authors who have conducted animal studies with SCF-processed insulin report biological activities in the range 80–100% of the unprocessed sample. Although recent studies have examined the precipitation of antibody fragments (107), ribonuclease A (83,87), rhDNase (86), rIgG (108), and myoglobin (102), a wider database on protein and peptide particle size and postprocessing activity will have to be developed before PCA is accepted as a standard particle formation technique by the biotechnology industry (108,115).

Winters et al. (101) studied PCA precipitation of lysozyme from DMSO over a wide range of near-critical and supercritical conditions (27–45°C, 7.34–11.5 MPa). The recovered biological activity, upon reconstitution of lysozyme in water, was 88–100% of the unprocessed protein. There were significant changes in the α-helix and β-sheet structures when the lysozyme was dissolved in DMSO; however, the PCA-processed protein showed considerably lower distortions in these secondary structures. An increase in the structural distortion at higher processing temperatures was observed and attributed to increased protein–protein interactions at the higher temperature (45°C). These particles were stored at ambient conditions with no change in their secondary structure or biological activity, indicating their suitability for long-term storage at ambient conditions (116). Similar studies with trypsin resulted in a 69–94% recovered activity (101). Although lysozyme precipitated from DMSO solutions using PCA recovered almost 100% of its activity (101), GAS-processed lysozyme recovered only about 72% of its initial activity (87). A further study of lysozyme precipitation by Moshashaee et al. (103), conducted at conditions similar to Winters et al. (101), highlights the variation seen across research laboratories (Table 4). Lysozyme activities of 44–100% were observed, depending on the conditions of operation. Pressure was determined to be the most important parameter affecting lysozyme activity,

TABLE 4 Comparison of Two PCA Studies on Lysozyme Precipitation from DMSO Solutions

Parameter	Winters et al. (101)	Moshashaee et al. (103)
Temperature (°C)	27–45	40–50
Pressure (MPa)	7.34–11.5	8–15
Concentration (mg/mL)	2.2–6.8	5–10
DMSO flow rate (mL/min)	0.3–2.4	0.2–0.6
CO_2 flow rate (std L/min)	10–21	0.018–0.03
Nozzle diameter (mm)	0.03–0.05	0.2–0.3
Recovered activity (%)	88–100	44–100

although this variable was not determined to be significant by Winters et al. (101).

3.2.3. Modifications to GAS/PCA

One issue deterring the application of antisolvent precipitation to a wider range of biological molecules is the necessity of using organic solvents in GAS/PCA and the concomitant limited solubility of proteins in organic solvents. There are two ways to overcome these limitations: (1) increase the solubility and stability of proteins in organic solvents and (2) increase the mutual miscibility of water and CO_2.

Manning et al. (83) have described hydrophobic ion pairing (HIP) to increase protein solubility and stability in organic solvents. HIP increases the organic solubility of a protein by complexing it with an amphiphilic molecule. The protein is dissolved in a buffered solution (at a pH considerably different from its pI), where it is charged. The polar counterions are stoichiometrically replaced with surfactant molecules by the addition of small quantities of an ionic surfactant, resulting in an ion-paired complex. This protein–surfactant complex is hydrophobic and has improved solubility and stability in organic media. HIP complexes of insulin, subtilisin, α-chymotrypsin, interleukin-4 with SDS, and dioctylsulfosuccinate (AOT) demonstrate increased solubility in organic solvents and show retention of native secondary and tertiary structures (114).

The solution of the HIP-complexed protein in an organic solvent is sprayed using the PCA process to form microparticles. Manning et al. (83) formed microparticles of insulin, α-chymotrypsin, and ribonuclease. Insulin was paired with SDS, dissolved in pyridine, methanol, or THF and sprayed into a SCF CO_2 continuum as in a typical PCA process. Particles precipitated from pyridine were spheroidal and those from THF were 1–5 µm but irregular in shape. α-Chymotrypsin was ion-paired with AOT and sprayed from an iso-octane solution to give <10-μm-diameter spheroidal particles at a precipitation temperature of 34°C and 1-μm irregularly shaped particles at 28°C. Finally, ribonuclease was paired with SDS, dissolved in methanol, and sprayed into SCF CO_2 to give 50-μm spheroidal particles. Unfortunately, no residual activities were reported for any of the proteins, however, retention of activity for a number of HIP-complexed proteins dissolved in organic solvents has been previously demonstrated (114). Thus, the PCA processing of HIP-complexed proteins could potentially eliminate one possible source of activity loss in PCA (i.e., protein denaturation upon dissolution in an organic solvent). The technique has been extensively used to produce microparticles of ionic drugs for controlled-release applications (117).

Another approach developed to increase the solubility of proteins in a bulk aqueous phase is the use of reverse microemulsions. Zhang et al. (84) reported the GAS-based precipitation of lysozyme solubilized in AOT reverse micelles in iso-octane using pressurized CO_2. Comparing the on-line UV-vis (ultraviolet–visible) spectra of processed and unprocessed lysozyme, the authors concluded that the lysozyme was not denatured. The use of reverse micelles to dissolve proteins in a bulk organic phase is a promising variation of the GAS technique. The use of reverse micelles could potentially increase the stability of proteins because they would be in a primarily aqueous local environment until precipitation.

Precipitation of biological molecules using CO_2 can also be achieved by increasing the mutual miscibility of water and CO_2. Aqueous buffers and water are most commonly used to make stable protein solutions. However, the low mutual solubility of water and pressurized CO_2 precludes the use of aqueous protein solutions in SCF CO_2-based antisolvent processes. The poor mutual miscibility of water and SCF CO_2 can be increased by the addition of a cosolvent like ethanol (69,107,118). The three-component phase diagram of water, ethanol, and pressurized CO_2 shows a large single-phase region (118). Thus, another technique developed to increase the range of proteins processed using CO_2 uses ethanol as a cosolvent to aid the precipitation of proteins from aqueous solutions. For successful precipitation, the operating point must lie in the region where water, CO_2, and ethanol all form a single homogeneous phase and the mole fraction of CO_2 should preferably be higher than 0.6 (118).

The one-phase PCA operating region can be achieved by cocurrently spraying an aqueous protein solution into a mixture of CO_2 and ethanol (69,86). α-Chymotrypsin, albumin, insulin, lysozyme, and rhDNase particles have been formed using this technique. Bustami et al. (86) reported the biological activity of the precipitated protein by measuring the monomer content of the final powder. Because aggregation is one of the signs of protein denaturation, the monomer content is an indication of residual protein activity. The residual activities (Table 3) ranged from >96% for lysozyme to 14–33% for rhDNase. Further, lysozyme monomer content was not affected by temperature (in the range of 35–45°C). The monomer content of processed rhDNase was reduced from 33% to 14% with increasing temperature (20–35°C). The monomer content and, hence, residual activity decrease with increasing molecular weight of the protein, a trend similar to that observed by Winters et al. (87) with GAS precipitation from organic solvents (Fig. 4).

Coaxial nozzles have also been used to introduce the aqueous protein solution, ethanol, and CO_2 into the precipitation chamber. This contacting technique, developed at the University of Bradford, is commonly known as Solution-enchanced dispersion by SCFs (SEDS) (85,103,105). To minimize the protein–organic solvent contact time, the protein solution, ethanol, and CO_2 are sprayed through a three-channeled coaxial nozzle (103). The high velocity of the SCF CO_2 flowing in the innermost channel is used to mix the aqueous and organic streams as well as to break them into tiny droplets, facilitating mass transfer. β-Lactamase (85), rIgG (108), lysozyme (105), trypsin, insulin, and antibody fragments Fab and Fv (88) have been precipitated using this technique. Only insulin and lysozyme showed recovered activity greater than 80% of the native protein.

The SEDS process is also scalable, as recently demonstrated by its 10-fold (in terms of flow rates) scale-up of lysozyme precipitation from aqueous solutions (104). Experiments were conducted at the exact same conditions of temperature (55°C), pressure (15 MPa), and protein concentration [5% (w/v) in water] at the laboratory and pilot-plant scale. The pilot-plant-scale flow rates were 180 g/min for CO_2 and 720 mL/h and 36 mL/h for ethanol and lysozyme, respectively. Similar particle morphology and activity (95% of original) were obtained under both conditions. Although Thiering et al. (119) legitimately list the possible hurdles to the scale-up of SCF-based processes, this successful scale-up indicates their potential for the industrial scale micronization of proteins.

3.2.4. Protein Precipitation Based on Antisolvents Other than CO_2

An alternative approach to precipitate solutes from aqueous solutions is the use of a water-miscible compressed antisolvent, such as SCF NH_3 (T_c = 132°C, P_c = 11.3 MPa). However, upon dissolution, NH_3 increases the pH of the aqueous system considerably, thereby denaturing dissolved proteins. Thiering et al. (94) studied the precipitation of insulin and lysozyme from aqueous solutions using SCF NH_3 at 35°C and pressures up to 14 MPa. Lysozyme was observed to precipitate at its pI (11.3), thus by isoelectric precipitation, whereas insulin (pI 5.3) precipitated due to the antisolvent effect of SCF NH_3 [volumetric expansion 370% (v/v)]. For both proteins, the final dried powders were agglomerated and irregular, indicating protein denaturation.

Ziegler et al. (120) have recently reported the control of pH of an aqueous solution contacted with SCF CO_2. Using NaOH and buffers, the system pH was maintained in the 6.2–7.8 range, a range suitable for protein processing. An extension of this study would be to control the pH of aqueous solutions contacted with SCF NH_3, thereby increasing the utility of SCF NH_3 as an antisolvent for aqueous solutions.

3.2.5. Using Supercritical Fluid Antisolvent Techniques in Phospholipid Processing

Phospholipids are biological surfactants that are capable of forming unilamellar and multilamellar vesicles and, hence, are major components of cellular membranes. Purified phospholipids are used to form liposomes which have applications in controlled drug delivery, and the formation of liposomes using SCFs is discussed in Section 5. Microparticles of lecithin (98) have been produced by RESS and PCA. For the RESS process, ethanol was used as a cosolvent to dissolve lecithins in pressurized CO_2. A wide range of temperature, pressure, and ethanol concentrations were explored, but the investigators were unable to form microparticles via RESS (98). The effect of PCA-processed variables on lecithin particle size was also investigated: temperature (30–50°C), pressure (8–12 MPa), lecithin concentration in ethanol (2–25 wt%), and liquid solution flow rate (10–40 mL/h) (98). Spherical, partly agglomerated particles in the 1–50-μm range were obtained. There was no effect of pressure on the particle size, whereas increasing the liquid flow rate decreased the particle size and tightened the particle size distribution. This was attributed to better atomization and, hence, improved mass transfer and high supersaturation (i.e., conditions that favor nuclei formation to particle growth) at the higher liquid flow rates. When analyzed for residual solvent content via infrared spectroscopy, ethanol was not detected in the processed lecithins (97).

3.2.6. Using Supercritical Fluid Antisolvent Techniques in DNA Processing

Recently, plasmid DNA has been considered as a viable therapy to treat diseases like cystic fibrosis, Parkinson's, and diabetes (5,6). Further, DNA vaccines for mucosal administration (7) and intranasal inhalation (121) are currently being developed for a variety of diseases. These developments have increased the interest in developing plasmid DNA powders. PCA, a proven particle formation technology, has been applied to the production of plasmid DNA micropowders. Tservistas et al. (107) precipitated pSVβ, a 6.9-kb plasmid, from an aqueous solution using isopropanol-modified SCF CO_2

using SEDS. The preliminary experiments indicated there was considerable uncoiling of the DNA in the final product, hence, they systematically studied each step of the process for its denaturation ability. Because many aspects of this analysis are applicable to proteins as well, it is presented here in some detail.

1. Contact with an organic solvent: Isopropanol is known to be non-denaturing for DNA; hence, its effect was considered to be negligible.
2. Effect of shear forces: In the SEDS process, DNA is subjected to shear by the pumping apparatus and during flow through the nozzle. The DNA was separately subjected to both of these conditions and no significant decrease in the supercoiled structure was observed. The effect of shear was negligible in this case, probably because of the small size of the plasmid studied.
3. Presence of a stabilizing excipient: Mannitol was used as a stabilizing excipient during these studies. Freeze-dried solutions of pSVβ in mannitol showed > 95% of the original supercoiled structure was retained.
4. Effect of temperature: The plasmid was exposed to 50°C momentarily while in solution and for 2 h as a precipitate. Exposure of unprocessed pSVβ at 20°C and 50°C, in solution, in the dried form, and in solution in the presence of mannitol all resulted in near 100% recovery of the supercoiled portion.
5. Effect of pH: Although the aqueous solution was buffered, exposure of the aqueous solution to large volumes of SCF CO_2 could form carbonic acid, reducing the pH of the solution [Eq. (1)]. When pSVβ was subjected to pH 2.2, it was completely uncoiled within 5 min. Hence, the drop in pH could contribute to the uncoiling. To verify this, pSVβ was SEDS processed while in a heavily buffered aqueous solution. The retention of coiled structure increased to 66%, indicating the pH to be a governing factor in denaturation.
6. Exposure to SCF CO_2: Although this factor was not considered by Tservistas et al., Nivens and Applegate (122) have described a procedure for the extraction of nucleic acids from micro-organisms as a technique for detecting the presence of a micro-organism in an environmental sample. DNA extraction for a number of micro-organisms was examined in a commercial SCF extraction unit. Analysis of the micro-organisms processed at 80°C and 40 MPa showed loss of viability at 89% for *E. coli*, 90% for *Pseudomonas*, 69% for *Sphingomonas*, 74% for *Rhodococcus*, and 79% for *Mycobacterium*. This loss of viability was equivalent or higher than that resulting from the standard treatment of exposure to SDS. Further,

the maximum recovery of DNA for *E. coli* and *Rhodococcus* was obtained after exposure to SCF CO_2 for 30 min at 100°C and 40 MPa. The extracted DNA was analyzed for integrity via agarose gel electrophoresis. The plasmid DNA recovered by SCF extraction at 80°C or 100°C and 40 MPa was of high molecular weight and showed migration similar to control DNA during gel electrophoresis. These results indicate that exposure to SCF CO_2 (in the absence of free water) does not degrade plasmid DNA.

3.2.7. Analysis of Change in Protein Activity Observed During Antisolvent Processing

Figure 4 summarizes the residual activities of proteins obtained after GAS and PCA processing. The only reports of protein activity greater than 80% are for insulin, ribonuclease A, lysozyme, and trypsin. These proteins are relatively small, having molecular weights of 5.8, 13.7, 14.4, and 23.8 kDa, respectively. Other larger proteins have been precipitated in the form of 1–5-μm-sized particles, but none have retained substantial activity. Further, from the limited studies available, we can conclude that the protein particle size is a function of the operating temperature (102) and the activity of the reconstituted protein is inversely related to its molecular weight (Fig. 4). The analysis presented below attempts to describe the interrelated causes of reduction in protein activity during CO_2-based processing.

1.   Contact with an organic solvent: The first step in the GAS and PCA processes is dissolution of the protein in an organic, aqueous, or an aqueous–organic solvent. Substantial data on the activity of proteins dissolved in and suspended in organic solvents exist (for reviews, consult Refs. 82, 123, and 124). Dissolution in organic solvents changes the protein secondary and tertiary structure considerably (67,101). These changes may or may not increase activity. However, there are numerous instances where enzymes dissolved in organic solvents are unfolded but recover complete activity upon reconstitution into aqueous media (82,116).

2.   Effect of temperature: Gilbert et al. (105) compared the activity of lysozyme after spray-drying, PCA processing, and lyophilization and reported that the high temperatures involved in spray-drying can substantially denature proteins. In contrast, the activity of aqueous lysozyme precipitated with CO_2/ethanol did not change significantly. As in this study, most PCA experiments are carried out between 20°C and 50°C—a temperature range in which most proteins are stable. A few studies have shown that the residual protein activity decreases with increasing temperature, especially

above 35°C (86,102,103). Moderate temperature is not a major denaturing influence during the PCA process.

3. Effect of pressure: Hydrostatic pressure above 400–600 MPa can cause unfolding and denaturation of proteins (125) and monomeric globular proteins can be assumed to be stable up to a pressure of 200 MPa (126). The pressures used for PCA and GAS are typically in the range 6–15 MPa and, thus, the operating pressure would not be expected to significantly affect protein activity during antisolvent processing.

4. Effect of pH: Contact of aqueous solutions with pressurized CO_2 lowers the pH of the medium [Eq. (1)]. Exposure to low pH will denature and precipitate some proteins. Although the reduction in antisolvent strength (and not pH reduction) is the major factor that causes protein precipitation during GAS and PCA processing (69), processing at low pH may affect the recovered protein activity.

5. Protein–CO_2 interaction: Although CO_2 is fairly inert, it can react with -NH_2 of lysine residues to form carbamates (127). Carbamate formation can substantially affect the tertiary structure and, hence, activity of proteins. Potential covalent modification of the protein structure can be determined by analyzing the reactive amino acid residues (e.g., free thiol groups and free amino groups) on the protein surface. The environment of tryptophan residues can also be examined for conformational changes in the protein using fluorescence or UV derivative spectroscopy (128). However, a *priori* gauging the protein-specific effect of SCF CO_2 is difficult.

3.2.8. SCF CO_2-Assisted Aerosolization and Bubble Drying

Sievers et al. (129,130) have reported the use of SCF CO_2 as an aerosolization agent to produce microparticles of a number of peptides and proteins, including glutathione, bovine serum albumin (BSA), horseradish peroxidase (HRP), LDH, and rhDNase. In this technique, an aqueous protein solution and SCF CO_2 are fed into the two ends of a low dead volume tee (Fig. 6). The CO_2–aqueous stream is suddenly expanded through a restrictor, releasing the CO_2 dissolved in the aqueous solution and forming an aerosol of fine aqueous droplets. Dry nitrogen was passed through a heated drying tube and the dried protein powders are collected on a filter.

This technique has the advantage of using purely aqueous systems without the addition of organic solvents. No significant loss of activity was observed for glutathione and HRP. Lysozyme, rhDNase, and LDH showed substantial loss of activity when sprayed without any excipients. For rhDNase, buffering the aqueous solution to a pH of 7 with sodium acetate prevented the formation of aggregates, and 99% of the original activity was

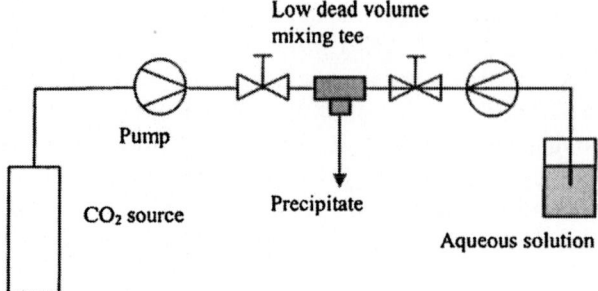

FIGURE 6 Schematic of the bubble dryer technique. (Adapted from Refs. 128 and 129.)

recovered. In the case of lysozyme and LDH, processing in the presence of 10% sucrose and 0.01% Tween-20 increased the regained activity to 100% and 90%, respectively. Drying with heated nitrogen appeared to be the major factor in the loss of activity in the absence of excipients.

3.2.9. Exploiting the High Compressibility of Supercritical Solvents to Comminute Proteins

Rapid depressurization has been used to comminute proteins using SCFs without first dissolving them (131). Particles of insulin (0.5–75 μm) and BSA (1–300 μm) were formed using this method, however, no activities were reported. The fluids studied included propane, Freon-22, CO_2, and N_2. Only CO_2 and N_2 were effective at particle formation. A wide range of temperatures (−173°C to 50°C) was studied, with the smallest particles being formed close to ambient temperatures. The proposed mechanism of particle formation involved weakening of the hydrophobic interactions holding together the secondary and tertiary structure of the protein by the surrounding SCF. This weakening of the protein structure could increase the protein susceptibility to micronization when subjected to flow-induced shear, mechanical impact, and expansion of the SCF in the interstitial spaces within the original protein particle. Finally, low temperatures could also increase the brittleness of the proteins, aiding their breakup.

4. COMPRESSED CO_2 AS A SOLVENT FOR POLYMER PROCESSING

4.1. Polymer Coating of Proteins by Coprecipitation from Organic Solutions

Microencapsulation of therapeutic proteins is a method to preserve activity during sustained delivery as well as a means of controlled release. Current protein microencapsulation methods frequently involve dissolution in water, followed by the formation of a water–oil–water emulsion that is evaporated to form particles. These processes often result in denaturation, aggregation, and cleavage, especially at the oil–water interface (1). The PCA technique can be used to microencapsulate proteins by coprecipitating the dissolved polymer and protein in a single processing step.

Poly(L-lactide) (PLA)-encapsulated insulin has been produced using PCA (95). The PLA was dissolved in methylene chloride and added to a solution of insulin in acidified DMSO or acidified methanol. This solution was sprayed into a CO_2 antisolvent continuum at 37°C and 8.4 MPa. Ninety percent of the microparticles produced were <5 μm. Greater encapsulation efficiencies were obtained using methanol (up to 99% encapsulation) relative to DMSO (49% encapsulation) (132). The tap density of the powders was 0.1–0.2 g/cm^3. When injected in mice, the processed insulin produced no burst effect and suppressed blood glucose levels for more than 48 h.

Elvassore et al. (96) successfully microencapsulated insulin in poly(ethylene glycol) (PEG) of various molecular weights and PLA. Particles of a mean size between 400 and 700 nm were formed at 16–22°C and 13 MPa. The insulin retained more than 80% of its native activity in vivo. Release studies on particles containing PEG with a molecular weight <1900 were free of a burst effect and showed a slow constant protein release for 1500 h. Although these PCA-based microencapsulations were successful, high bursts have been reported for some systems [e.g., BSA in PLGA (90)].

Ion-paired biological molecules have been codissolved with polymers and directly microencapsulated using PCA (83,106). Manning et al. (83) studied the microencapsulation of ion pairs of α-chymotrypsin and AOT and ribonuclease and SDS in PLA and PEG, respectively. The ribonuclease precipitate had a fibrous morphology. No details about the protein activity were reported. HIP processing has also been applied to solubilize plasmid DNA in methylene chloride (106). Because plasmid DNA is negatively charge, it was ion-paired with 1,2-dioleoyl-3-trimethylammonium propane (DOTAP), a cationic lipid. This ion-paired complex was sprayed with PLA to form microspheres with 1% loading of plasmid DNA. Release studies carried out on these microspheres showed a linear release profile with a minimal burst effect.

Young et al. (104) demonstrated the coprecipitation of lysozyme and PLGA using a 'vapor over liquid CO_2' technique, meaning that vapor and liquid CO_2 coexist in the precipitation chamber. Lysozyme suspended in a PLGA solution in methylene chloride was sprayed into the vapor phase. The particle formation process started in the vapor phase (low antisolvent strength) and continued in the liquid phase (higher antisolvent strength) and resulted in microparticles of lysozyme coated with PLGA. The particles were 5–60 μm in diameter and encapsulation was confirmed by elemental analysis via energy-dispersive spectroscopy. Unfortunately, no lysozyme activity or release studies were reported.

4.2. Coating of Proteins Suspended in CO_2–Polymer Solutions

The rapid expansion of a supercritical solution (RESS) technique involves dissolving the solute of interest in a SCF and then rapidly expanding this solution through a nozzle to atmospheric pressure (74). The sudden loss of solvent power results in the precipitation of extremely small particles. The main limitation of this technique is that the solute must be soluble in the SCF phase—a condition that cannot be met by proteins and other polymeric bio-molecules. Hence, RESS cannot be used for their primary particle formation. However, a modification, rapid expansion of a supercritical solution with a nonsolvent (RESS-N) has been used as a microencapsulation technique (99). Various polymers, a cosolvent (ethanol, acetone, or toluene), and lysozyme or PS lipase were loaded in a vessel and pressurized to 35°C and 20 MPa using SCF CO_2. The vessel was agitated to dissolve the polymer in the cosolvent modified CO_2 and to thoroughly disperse the insoluble enzyme. This suspension was depressurized through a heated nozzle and the particles were collected on a plate. The particles were analyzed for size but not for enzyme activity. The particle size, which ranged between 6 and 62 µm, varied with polymer type and increased with polymer concentration in the feed. Because enzyme activities were not reported, it is difficult to judge the effect of RESS-N on enzyme structure. Although the protein was not dissolved in the organic solution, some proteins have been reported to unfold upon contact with organic solvents (81,82). In a similar process, Dos Santos et al. (89) coated BSA particles with Gelucire and trimyristin. The BSA coated with Gelucire showed no structural change and continuous prolonged release over 24 h.

4.3. Using CO_2 as a Polymer Swelling Agent in the Encapsulation of Proteins

The ability of SCF CO_2 to swell and plasticize a broad range of amorphous polymers is well documented (see, for example, Refs. 133 and 134). When the polymer is plasticized, the reduced viscosity allows for the incorporation of insoluble material into the swollen polymer. Subsequent depressurization entraps the insoluble material and creates a porous structure. This property of CO_2 has been exploited to incorporate a variety of inorganics, pharmaceuticals, and biological molecules in biodegradable polymeric matrices for applications in drug delivery and tissue engineering (135–138). Foaming with CO_2 eliminates the need for organic solvents or high temperatures to plasticize the system, making the process well suited for the processing of biological molecules. In addition, CO_2 is easily removed from the polymer foam or composite structure, leaving behind no residual foaming agent. Further, the morphology of the foamed product can be controlled via calculated manipulation of the temperature, pressure, time of exposure, and depressurization process.

Most CO_2-foaming studies related to pharmaceutical and biomedical applications have focused on PLA and PLGA. These amorphous polymers have a relatively low glass transition temperature (below 60°C) (139). The glass transition temperature of these biodegradable polymers is sufficiently reduced by CO_2 gas (at pressures below the critical point) that CO_2-based processing is frequently referred to as gas foaming. Howdle et al. suggest that CO_2 pressures of 20 MPa and near-ambient temperatures (35°C) are sufficient for this SCF mixing process involving these polymers (135). The pore size of the resulting foams ranges from 50 to 100 nm (slow depressurization) and 0.5 to 5 μm (fast depressurization) for the system of PLA (135).

Several studies have examined the formation of biocomposites using gas foaming and the controlled release of biological molecules from these polymeric matrices. Howdle et al. (135) extended CO_2-based polymer foaming to the encapsulation of ribonuclease A, catalase, and β-D-galactosidase in PLA using this technique. Activity of the protein was preserved and release studies showed an initial burst of <10% in the first 2 days, followed by zero-order release for 70 days. Also, 50 wt% catalase was encapsulated in a PLGA matrix monolith by mixing the two in the presence of SCF CO_2 and depressurizing the system. In addition to the production of monoliths, direct atomization of the pressurized polymeric mixture resulted in microparticles with a mean diameter between 24 and 50 μm.

The incorporation and release of vascular endothelial growth factor (VGEF) from amorphous PLGA has also been investigated (138). The matrices were formed by compression molding of the ground polymeric particles, the growth factor, and a salt (NaCl). The resulting disk was equilibrated with pressurized CO_2 and foamed with the rapid release of pressure. NaCl was then leached from the foamed matrix to create an open-pored structure. The released growth factor maintained 90% of its bioactivity and its release was a function of the copolymer system. However, only 28 ± 18% of the initial growth factor was incorporated in the polymer matrix after leaching. To enhance the efficiency of incorporation and demonstrate favorable release kinetics, the VGEF was first immobilized in alginate beads. These beads were then lyophilized and compression molded into a polymeric disk. The incorporation of growth factor was increased (to 55 ± 1%) by the use of alginate beads and released at a steady rate over a 40-day period.

Plasmid DNA has also been incorporated into PLGA copolymer matrices using gas foaming (137). The incorporation efficiency of the DNA into the polymeric matrix (following leaching of the NaCl to form an open-pore structure) was dependent on the ratio of lactide and glycolide and ranged from 49% to 60%. Sustained release of the plasmid DNA was observed over a period greater than 30 days, with retention of the structural integrity of the released plasmid DNA. The content of the supercoiled structure of DNA, however, was reduced from 64% (for unincorporated DNA) to 0% (for DNA released at 28 days). However, the released plasmid DNA was capable of transfecting a high percentage of cultured cells throughout the 28-day in vitro release study. Further, in vivo release studies demonstrated that the transfected cells produced observable physiological effects consistent with the sufficient production of the gene-encoded protein, a response that is not achieved by the direct injection of an equivalent amount of plasmid. The long-term physiological response suggested by this method of controlled release addresses many of the drawbacks of current DNA and protein delivery methods for tissue engineering and gene therapy applications.

An alternative approach to protein encapsulation uses CO_2 to precipitate an amorphous polymer, dry it, and swell it (140). The starting solution is an emulsion consisting of an aqueous phase containing a dissolved protein [in this case, basic fibroblast growth factor (bFGF)] and an organic polymer solution phase (PLGA) dissolved in methylene chloride). A mold inside the pressurized vessel contained the emulsion mixture, and CO_2 flowed through the pressurized vessel (35°C and 8 MPa). Extraction of methylene chloride by SCF CO_2 precipitates and dries the polymer. The dried polymer was then foamed during rapid pressure release. The resulting morphology appears more open and less cellular than the closed cell structures observed in gas foaming in the absence of solvents. Encapsulation efficiency of the protein within the polymeric matrix was 97.5%. Protein release was studied over 30 days and overall protein release was greater than from structures made by a salt-leaching technique. However, the protein activity of bFGF released from the CO_2-foamed matrix is 2% of the total protein released (after day 1), which is half the active protein released from polymeric matrix formed by traditional salt leaching. Significant inactivation may occur due to the interactions of bFGF, which is inherently unstable, with methylene chloride. Inactivation may also be expected based on the contacting of CO_2 with an initial aqueous environment of the proteins. The ability of the 0.01 M phosphate-buffered saline to maintain the aqueous pH during the exposure to CO_2 was not addressed. Finally, additional optimization of the process would be required to lower the residual methylene chloride concentrations to acceptable levels.

The CO_2-based foaming techniques have also been applied to the incorporation of pharmaceuticals and biological molecules in nonbiodegradable polymers. Sproule and coworkers (141) recently demonstrated the CO_2-induced foaming of poly(methyl methacrylate) and the impregnation of a naturally fluorescent protein in powder form at moderate temperatures. For example, the surface modification of polymers using embedded bioactive compounds has the potential to address or control the immune response (141).

Similarly, Luzzi et al. (142) incorporated insulin in ammonium glycyrrhizinate and poly(acrylic acid) by mixing the protein with the matrix material and exposing this mixture to subcritical CO_2 at 20°C and 6.4 MPa for up to 24 h. The microparticles produced by this infusion method contained 5 wt% insulin. The ammonium glycyrrhizinate-based insulin particles when administered intranasally and intravenously to diabetic rabbits could control the blood glucose levels successfully. Further, diabetic rabbits treated intranasally with insulin-infused poly(acrylic acid) particles showed a decrease in blood glucose levels to normoglycemia. Thus, the insulin remained active in both matrices after CO_2 processing.

Finally, the CO_2-based foaming technique has recently been extended to the incorporation of lyophilized whole cells in a polymeric matrix (143). The potential biosensing and biosynthesis applications of whole cell composites are similar to those of purified enzymes. However, whole cells allow for greater complexity in their response based on their ability to perform multiple biochemical reactions in a single system. The viability of lyophilized, cryoprotected *P. aeruginosa* following treatment with pressurized CO_2 (40°C and 16.6 MPa) for 2.5 h was demonstrated (143). This is consistent with previous observations that CO_2 is an ineffective sterilizing agent in low-water-content systems (53). Further, these dry lyophilized powders were directly incorporated in a poly(vinyl acetate) (PVAc) matrix using CO_2-induced polymer swelling (35°C and 12.5 MPa for 2.5 h). Cell incorporation was verified using fluorescent microscope images of the DAPI-labeled bacteria. Rehydration of the *P. aeruginosa* in the PVAc composite led to cell proliferation outside the polymer matrix. This preliminary work suggests that whole cell processing using pressurized CO_2 should be possible in spite of its effectiveness as an antimicrobial agent.

5. LIPOSOME FORMATION VIA SCF PROCESSING

Supercritical fluid CO_2 has been proposed for the encapsulation of hydrophilic and hydrophobic therapeutic agents into liposomes (144–149). Liposomes are closed bilayer structures that enclose an aqueous volume. These vesicles, comprising single or multiple bilayers, are typically formed from phospholipids. On the basis of their biological components and structure, liposomes are model cell membranes and artificial biomembranes. Their similarities to cellular membranes can be exploited for the parental administration of pharmaceuticals, an area of intense research since the 1970s (146). The presence of both hydrophilic regions of the vesicle (in the aqueous core) and hydrophobic regions (in the lipid bilayer) provides for the delivery a broad range of therapeutic agents. Liposomes have been proposed for the delivery of substances having biological activity, including proteins (such as antibodies or enzymes), hormones, and genes for gene therapy [as reviewed by Barenholz (150)].

Drawbacks of the traditional methods for the preparation of liposomes (such as the Bangham method, the organic solvent injection method, reverse phase evaporation, dehydration rehydration vesicle formation) include the use of organic solvent and the multistep preparation techniques that make them unsuitable for industrial scale-up (146,149). The incorporation of water-soluble therapeutics into liposomes is particularly problematic with respect to the volume of organic solvent required to efficiently encapsulate the therapeutic and the subsequent removal of the organic solvent. Additional processing concerns for liposome formulations include the need to maintain a sterile environment throughout the processing and the instability of many of the present liposome products (144).

Liposome formation using SCFs employs contacting and depressurization schemes to take advantage of two key properties of CO_2 and/or pressurized CO_2–aqueous systems: the solubility of phospholipids in cosolvent-enhanced CO_2 and the large change in volume that can be achieved with the expansion of CO_2. In these approaches, the therapeutic agent to be incorporated is dissolved in its respective hydrophilic (aqueous) or hydrophobic (SCF) phase. Frederiksen and coworkers (146,147) have achieved the incorporation of water-soluble compounds in liposomes by simultaneously expanding a dissolved lipid solution in SCF CO_2 and mixing it with an aqueous phase to yield liposomes. These liposomes are primarily unilamellar, with an average size between 20 and 50 nm. A smaller number of multilamellar liposomes (up to ~ 250 nm) are also present. In contrast, Otake and coworkers (149) have added water to a homogeneous phospholipid/ethanol/CO_2 phase. Depressurization results in large unilamellar liposomes (100–1200 nm). Castor (144,145) describes an "injection" method and a "depressurization" method using a variety of SCFs, including CO_2, to solvate the phospholipid. The injection of dissolved lipid/fluid solution into an aqueous phase produces small (< 100 nm), medium (100–400 nm), and large (>400 nm) liposomes, depending on the process variables, including choice of SCF and cosolvent, decompression, nozzle design, and critical fluid density. Unlike the "injection method," the "depressurization method" proposed by Castor (144,145) uses a pressurized aqueous phase in contact with a SCF solution. The volume expansion of the SCF during the rapid depressurization process is believed to create unstable bilayer fragments that are subsequently sealed in the decompression vessel. Uniform size distributions of liposomes (in the range of 60 nm) have been produced using the decompression method.

Liposome formation using SCF CO_2 eliminates or greatly reduces the use of organic solvents. Residual organic solvents or detergents in the liposomes require further removal steps. However, most methods for liposome formation require the solubilization of the lipid components in CO_2. Due to their limited solubility, the addition of a cosolvent may be necessary to achieve sufficient solubilization of the lipid and efficient incorporation of the therapeutic agent. Ethanol is typically used as the cosolvent (at a concentration of 5–10 mol%) because of its low toxicity and its acceptability in injections (146). For example, Otake and coworkers (149) determined that ~ 7 wt% of ethanol was necessary to dissolve ~ 0.3 wt% L-α-dipalmitoylphosphatidylcholine. However, no cosolvent is employed in the formation of lecithin liposomes by the "injection method" (144), where sufficient solubility of chicken egg yolk lecithin is achieved in SCF CO_2 at 60°C and 20 MPa. The solubility experiments are also used to guide the choice of operating conditions to provide for continuous processing, including system cleaning and recycle using CO_2.

These investigations demonstrate the ability to incorporate hydrophilic and hydrophobic compounds in liposomes at greater efficiencies and using less organic solvents than in traditional processing techniques. This advancement in technology offered by SCF CO_2 processing is particularly critical for the incorporation of hydrophilic therapeutic agents. Frederiksen et al. (146) demonstrated the ability to obtain an encapsulation efficiency of 20% of a hydrophilic compound using 1/15 of the alcohol that would have been required with the ethanol injection method. Similarly, Otake and coworkers (149) entrapped a water-soluble solute [D-(+)-glucose] at five times the efficiency of multilamellar vesicles prepared by the traditional Bangham method. The reduced concentration of ethanol suggests important applications of SCF-based liposome formation to proteins and nucleic acids that would otherwise be denatured by organic solvents.

The efficiency of the incorporation of hydrophilic compounds in liposomes is a strong function of the size of the vesicles, the number of bilayers in the vesicle, and the manufacturing techniques (145,146). Castor suggests incorporation efficiencies of 1% for small unilamellar vesicles to 88% for some multilamellar vesicles (145). Hydrophilic drug capture by the passive techniques currently employed in SCF CO_2 processing is related to the volume of water that can be sequestered during liposome formation. Although bimodal distributions of unilamellar vesicles are generally reported for supercritical liposome formation (144–146), evidence for formation of multilamellar and uniform single lamellar liposomes exists. The form of the liposome could greatly impact the incorporation of biological molecules, as suggested by the molecular weight dependence of incorporation of water-soluble markers observed by Frederiksen and coworkers (146). The reduced encapsulation of the water-soluble marker fluorescein isothiocyanate–dextran (FITC–dextran) relative to zinc phthalocyanine tetraculfonic acid (TSZnPc) was attributed to the presence of a small fraction of multilamellar liposomes formed in their SCF CO_2 processing technique. Multilamellar liposomes, particularly those having many bilayers, have a limited capacity for large molecules relative to unilamellar vesicles.

The formation of drug-loaded liposomes using SCF CO_2 has numerous advantages over conventional liposome encapsulation techniques. Solubilization of the lipid components in SCF CO_2 and cosolvent-enhanced CO_2 greatly reduces the organic solvent requirements, thereby reducing the number of processing steps needed for solvent removal. Further, the pressure tunability of SCF CO_2 provides for the recycle of unincorporated therapeutic agents, lipid components, and the SCF. Thus, continuous processing in a sterile environment becomes more feasible. Continuous processing also allows for a more uniform product. In addition, liposome formation using SCFs provides many process variables for the tuning of the final product size and characteristics. The ability to produce uniform product in a continuous process, the high efficiency of encapsulation of water-soluble compounds, and the low organic solvent requirements suggest that SCF CO_2 processing is well suited to the encapsulation of biological molecules in liposomes. Special considerations for the biological processing are the molecular weight dependence of encapsulation efficiency as a function of liposome size and type, which may particularly affect biomacromolecule encapsulation, and the potential need to buffer the aqueous solution in the presence of CO_2. However, the concept of protein encapsulation in SCF-processed liposomes has already been demonstrated for a system of cytochrome-c in SCF alkanes and SCF N_2O (144).

6. RECENT RESULTS AND TRENDS

The recent advances in SCF technology are well suited to meet the needs of the growing biotechnology industry. Recently, microparticles have been designed for very specialized therapeutic applications such as site-specific administration of growth factors and microinjection of therapeutic vaccines. The vast progress made in the processing of biological materials using SCF over the past decade suggest that SCF-based technology may be used to address the current hurdles to implementing these advanced microparticle technologies: the need for microparticles with a very narrow size distribution, high loading efficiencies of the active pharmaceutical, and the aseptic preparation of novel composite materials for targeted and sustained drug delivery. Similar needs exist in the biomedical field. For example, SCF processing has the potential to address the aseptic manufacture of composites for bone and tissue scaffolds.

New SCF-based methods for the materials synthesis of nonbiological molecules, such as the synthesis of Ag, Au, and Cd nanoparticles using biological molecules as templates (151), are expected to transverse the processing of biological molecules.

In addition, biosynthesis and separation using compressed solvents has the potential to enhance biobased production of both value-added and biocommodity chemicals. Metabolic engineering approaches should lead to an improved understanding of the physiological and biological response of both nonbarophiles and extremophiles in the presence of compressed and SCF solvents. Thus, the ability to tune the productivity and optimize the separation of biobased products may be achievable using in situ supercritical fluid processes. Further, these metabolic investigations could elucidate the mechanism of CO_2 inactivation of microbes. This would lead to the broad acceptance of CO_2-based sterilization beyond food technology and into additional critical areas such as the inactivation of viruses in biological products (e.g., blood plasma and blood-derived therapeutics).

Finally, SCF-based processing of biological products is gaining acceptance as a safe, effective, and scalable technology. Advances in SCF processing techniques and applications, initially applied to nonbiological systems, are expected to continue to drive the development of niche applications in biological processing.

REFERENCES

1. Whitaker M, Howdle S, Shakesheff K. Polymeric delivery of protein-based drugs. Business briefing. Pharmatech 2002; 118–123.
2. Datamonitor. Therapeutic proteins: strategic market analysis and forecasts to 2010. Datamonitor 2002.
3. Patton JS. Deep-lung delivery of therapeutic proteins. CHEMTECH 1997; 27:34–38.
4. Das RC, Morrow KJ. Antibody market update. A supplement to D&MD's antibody engineering and antibody therapeutics reports. Drug Market Dev 2002.
5. Varley AW, Munford RS. Physiologically responsive gene therapy. Mol Med Today 1998; 4:445–451.
6. Bowers WJ, Maguire-Zeiss KA, Harvey BK, Federoff HJ. Gene therapeutic approaches to the treatment of Parkinson's disease. Clin Neurosci Res 2001; 1:483–495.
7. Eriksson K, Holmgren J. Recent advances in mucosal vaccines and adjuvants. Curr Opin Immunol 2002; 14:666–672.
8. Roadmap for Biomass Technologies in the United States. Executive Steering Group. Department of Energy 2002.

9. Spilimbergo S, Elvassore N, Bertucco A. Microbial inactivation by high-pressure. J Supercrit Fluid 2002; 22:55–63.

10. Lin H-M, Yang Z, Chen LF. Inactivation of *Leuconostoc dextranicum* with carbon dioxide under pressure. Chem Eng J 1993; 52:B29–B34.

11. Debs-Louka E, Louka N, Abraham G, Chabot V, Allaf K. Effect of compressed carbon dioxide on microbial cell viability. Appl Environ Microbiol 1999; 65:626–631.

12. Dillow A, Dehghani F, Hrkach JS, Foster NR, Langer R. Bacterial inactivation by using near- and supercritical carbon dioxide. Proc Natl Acad Sci USA 1999; 96:10,344–10,348.

13. Aaltonen O, Rantakylä M. Biocatalysis in supercritical CO_2. CHEMTECH April:1991.

14. Kamat S, Beckman E, Russell A. Enzyme activity in supercritical fluids. Crit Rev Biotechnol 1995; 15:41–71.

15. Perrut M. Enzymatic reactions and cell behaviour in supercritical fluids. Chem Biochem Eng Q 1995; 8:25–30.

16. Chaudhary AK, Kamat SV, Beckman EJ, Nurok D, Kleyle RM, Hadju P, Russell AJ. Control of subtilisin substrate specificity by solvent engineering in organic solvents and supercritical fluoroform. J Am Chem Soc 1996; 118:12,891–12,901.

17. Marty A, Chulalaksananukul W, Condoret JS, Willemot RM, Durand G. Comparison of lipase-catalyzed esterification in supercritical carbon dioxide and in *n*-hexane. Biotechnol Lett 1990; 12:11–16.

18. Martins JF, Sampaio TC, Carvalho IB, Nunes de Ponte M, Barreiros S. Lipase catalyzed esterification of glycidol in chloroform and in supercritical carbon dioxide. In: Balny C, Hayashi R, Hereman K, Masson P, eds. High Pressure and Biotechnology. London: Colloque INSERM/John Libbey Eurotext, 1992:411–415.

19. Kamat SV, Iwaskewycz B, Beckman EJ, Russell AJ. Biocatalytic synthesis of acrylates in supercritical fluids: tuning enyme activity by changing pressure. Proc Natl Acad Sci USA 1993; 90:2940–2944.

20. Valivety RH, Halling PJ, Macrae AR. Reaction rate with suspended lipase catalyst shows similar dependence on water activity in different organic solvents. Biochim Biophys Acta 1992; 1118:218–222.

21. Rantakylä M, Aaltonen O. Enantioselective esterification of ibuprofen in supercritical carbon dioxide by immobilized lipase. Biotechnol Lett 1994; 16:825–830.

22. Weder JKP. Influence of supercritical carbon dioxide on proteins and amino acids—an overview. Cafe Cacao The 1990; 34:87–96.

23. Ishikawa H, Shimoda M, Yonekura A, Mishima K, Matsumoto K, Osajima Y. Irreversible unfolding of myoglobin in an aqueous solution by supercritical carbon dioxide. J Agric Food Chem 2000; 48:4535–4539.

24. Jones RP, Greenfield PF. Effect of carbon dioxide on yeast growth and fermentation. Enzyme Microb Tech 1982; 4:210–223.

25. Bauer C, Gamse T, Marr R. Quality improvement of crude porcine pancreatic

lipase preparations by treatment with humid supercritical carbon dioxide. Biochem Eng J 2001; 9:119–123.

26. Zagrobelny J, Bright FV. In situ study of protein conformation in supercritical fluids. Biotechnol Prog 1992; 8:412–423.

27. Tedjo W, Eshtiaghi MN, Knorr D. Impact of supercritical carbon dioxide and high pressure on lipoxygenase and peroxidase activity. Food Chem Toxicol 2000; 65:1284–1287.

28. Dufresne R, Thibault J, Leduy A, Lencki R. The effects of pressure on the growth of *Aureobasidium pullulans* and the synthesis of pullulan. Appl Microbiol Biot 1990; 32:526–532.

29. Klei H, Sundstrom D, Miller J. Fermentation by *Clostridium acetobutylicum* under carbon dioxide pressure. Biotechnol Bioeng Symp 1984; 14:353–364.

30. Takagi M, Ohara K, Yoshida T. Effect of hydrostatic pressure on hybridoma cell metabolism. J Ferment Bioeng 1995; 80:619–621.

31. Macdonald AG, Fraser PJ. The transduction of very small hydrostatic pressures. Comp Biochem Phys A 1999; 122:13–36.

32. Eklund T. The effect of carbon dioxide on bacterial growth and on uptake processes in bacterial membrane vesicles. Int J Food Microbiol 1984; 1:179–185.

33. Dixon NM, Kell DB. The inhibition by CO_2 of the growth and metabolism of microorganisms. J Appl Bacteriol 1989; 67:109–136.

34. Sturm F, Hurwitz S, Kelly R. Growth of the extreme thermophile *Sulfolobus acidocaldarius* in a hyperbaric helium bioreactor. Biotechnol Bioeng 1987; 29:1066–1074.

35. Knutson BL, Strobel HJ, Nokes SE, Dawson K, Berberich JA, Jones CR. Effect of pressurized extractive solvents on ethanol production by the thermophilic bacterium *Clostridium thermocellum*. J Supercrit Fluid 1999; 16:149–156.

36. Berberich JA, Knutson BL, Strobel HJ, Tarhan S, Nokes SE, Dawson KA. Product selectivity shift in *Clostridium thermocellum* in the presence of compressed solvents. Ind Eng Chem Res 2000; 39:4500–4505.

37. Berberich JA, Knutson BL, Strobel HJ, Tarhan S, Nokes SE, Dawson KA. Toxicity effects of compressed and supercritical solvents on thermophilic microbial metabolism. Biotechnol Bioeng 2000; 70:491–497.

38. Ko C, Vega J, Clausen E, Gaddy J. Effect of high pressure on a co-culture for the production of methane from coal synthesis gas. Chem Eng Commun 1989; 77:155–169.

39. Schieche D, Murty M, Kermode R, Bhattacharyya D. Biohydrogenation of fumarate using *Desulfovibrio desulfuricans*: experimental results and kinetic rate modelling. J Chem Tech Biotechnol 1997; 70:316–322.

40. Bubela B. Geobiology and microbiologically enhanced oil recovery. In: Donaldson E, Chilingarian G, Yen T, eds. Microbial Enhanced Oil Recovery. New York: Elsevier, 1989:75–98.

41. van Eijs AMM, Wokke JMP, Ten Brink B. Supercritical extraction of fermentation products. In: Bruin S, ed. Preconcentration and Drying of Food Materials. Amsterdam: Elsevier, 1988:135–143.

42. L'Italien Y, Thibault J, LeDuy A. Improvement of ethanol fermentation under hyperbaric conditions. Biotechnol Bioeng 1989; 33:471–476.
43. Isenschmid A, Marison IW, von Stockar U. The influence of pressure and temperature of compressed CO_2 on the survival of yeast cells. J Biotechnol 1995; 39:229–237.
44. Shimoda M, Cocunubo-Castellanos J, Kago H, Miyake M, Osajima Y, Hayakawa I. The influence of CO_2 concentration on the death kinetics of Saccharomyces cerevisiae. J Appl Microbiol 2001; 91:306–311.
45. Berberich JA. Whole cell biocatalysis in the presence of supercritical and compressed solvents. PhD dissertation. Lexington: University of Kentucky, 2001.
46. Leon R, Fernandes P, Pinheiro HM, Cabral JMS. Whole-cell biocatalysis in organic media. Enzyme Microb Tech 1998; 23:483–500.
47. Laane C, Boeren S, Vos K, Veeger C. Rules for optimization of biocatalysis in organic solvents. Biotechnol Bioeng 1987; 30:81–87.
48. Gros HP, Bottini S, Brignole EA. A group contribution equation of state for associating mixtures. Fluid Phase Equilibria 1996; 116:537–544.
49. Gros HP, Bottini SB, Brignole EA. High pressure phase equilibrium modeling of mixtures containing associated compounds and gases. Fluid Phase Equilibria 1997; 139:75–87.
50. Lamed RJ, Lobos JH, Su TM. Effects of stirring and hydrogen on fermentation products of Clostridium thermocellum. Appl Environ Microb 1988; 54:1216–1221.
51. Sims M, Estigarribia E. Continuous sterilization of aqueous pumpable food using high pressure carbon dioxide. 4th International Symposium on High Pressure Process Technology and Chemical Engineering, Venice, 2002.
52. Bouzidi A, Majewski W, Hober D, Perrut M. Supercritical fluids improve tolerance and viral safety of fresh frozen plaza. Nice: International Society for the Advancement of Supercritical Fluids, 1998:717–721.
53. Kumagai H, Hato C, Nakamura K. CO_2 sorption by microbial cells and sterilization by high pressure CO_2. Biosci Biotechnol Biochem 1997; 61:931–935.
54. Ballestra P, Cuq J-L. Influence of pressurized carbon dioxide on the thermal inactivation of bacterial and fungal spores. Lebensm-Wiss Technol 1998; 31:84–88.
55. Lin H-M, Chan E-C, Chen C, Chen L-F. Disintegration of yeast cells by pressurized carbon dioxide. Biotechnol Prog 1991; 7:201–204.
56. Taniguchi M, Suzuki H, Sato M, Kobayashi T. Sterilization of plasma powder by treatment with supercritical carbon dioxide. Agric Bio Chem 1987; 51:3425–3426.
57. Enomoto A, Nakamura K, Hakoda M, Amaya N. Lethal effects of high-pressure carbon dioxide on bacterial spores. J Ferment Bioeng 1997; 83:305–307.
58. Spilimbergo S, Dehghani F, Bertucco A, Foster NR. Inactivation of bacteria and spores by pulse electric field and high pressure CO_2 at low temperature. Biotechnol Bioeng 2003; 82:118–125.

59. Sapru V, Teixerira AA, Smeradge GH, Lindsay JA. Predicting thermophilic spore population dynamics for uht sterilization processes. J Food Sci 1992; 57:1248–1252.

60. Ishikawa H, Shimoda M, Tamaya K, Yonekura A, Kawano T, Osajima Y. Inactiviation of *Bacillus* spores by the supercritical carbon dioxide microbubble method. Biosci Biotechnol Biochem 1997; 61:1022–1023.

61. Elvassore N, Sartorello S, Spilimbergo S, Bertucco A. Micro-organisms inactivation by supercritical CO_2 in a semicontinuous process. Antibes, France: International Society for the Advancement of Supercritical Fluids, 2000:773.

62. Perrut M. Future trends for supercritical fluids. Applications in the pharmaceutical industry. Proceedings of the 6th Conference on Supercritical Fluids and Their Applications, Maiori, Italy, 2001:1–6.

63. Lin H-M, Yang Z, Chen L-F. An improved method for disruption of microbial cells with pressurized carbon dioxide. Biotechnol Prog 1992; 8:165–166.

64. Castor T, Hong G. Supercritical fluid disruption of and extraction from microbial cells. US patent 5380826, 1995.

65. Castor T, Lander A. Method and apparatus for the inactivation of viruses. US patent 6465168 B1, 2002.

66. Pilz V, Rupp R. Process for rendering solids sterile. US patent 4263253, 1981.

67. Yeo SD, Lim GB, Debenedetti P, Bernstein H. Formation of microparticulate protein powders using a supercritical antisolvent. Biotechnol Bioeng 1993; 41:341–346.

68. York P, Hanna M. Particle engineering by supercritical fluid technologies for powder inhalation drug delivery. Proceedings of Respiratory Drug Delivery V, Phoenix AZ, 1996; 231–239.

69. Sarkari M, Darrat I, Knutson BL. CO_2 and fluorinated solvent-based technologies for protein microparticle precipitation from aqueous solutions. Biotechnol Prog 2003:443–454.

70. Bleich J, Müller BW, Waßmus W. Aerosol solvent extraction system—a new microparticle production technique. Int J Pharm 1993; 97:111–117.

71. Gallagher PM, Coffey MP, Krukonis VJ, Klasutis N. Gas anti-solvent recrystallization: new process to recrystallize compounds insoluble in supercritical fluids. In: Johnston KP, Penniger JML, eds. Supercritical Fluid Science and Technology. Washington, DC: American Chemical Society, 1989:334–354.

72. Dixon D, Johnston KP, Bodmeier R. Polymeric materials formed by precipitation with a compressed fluid antisolvent. AIChE J 1993; 39:127–136.

73. Chattopadhyay P, Gupta RB. Production of griseofulvin nanoparticles using supercritical CO_2 antisolvent with enhanced mass transfer. Int J Pharm 2001; 228:19–31.

74. Jung J, Perrut M. Particle design using supercritical fluids: literature and patent survey. J Supercrit Fluid 2001; 20:179–219.

75. Kompella UB, Koushik K. Preparation of drug delivery systems using supercritical fluid technology. Crit Rev Ther Drug 2001; 18:173–199.

76. Reverchon E. Supercritical antisolvent precipitation of micro- and nanoparticles. J Supercrit Fluid 1999; 15:1–21.

77. Schmitt WJ. Finely divided solid crystalline powders via precipitation into an antisolvent. US patent 5707634, 1998.
78. Fischer W, Müller BW. Method and apparatus for the manufacture of a product having a substance embedded in a carrier. West German patent 3744329, 1987.
79. Debenedetti PG, Lim GB, RK Prud'Homme. Preparation of protein microparticles by supercritical fluid precipitation. US patent 6063910, 1992.
80. Rees E, Singer S. A preliminary study of the properties of proteins in some nonaqueous solvents. Arch Biochem Biophys 1956; 63:144–159.
81. Jackson M, Mantsch H. Halogenated alcohols as solvents for proteins: FTIR spectroscopic studies. Biochim Biophys Acta 1992; 1118:139–143.
82. Stevenson CL. Characterization of protein and peptide stability and solubility in non-aqueous solvents. Curr Pharm Biol 2000; 1:165–182.
83. Manning MC, Randolph TW, Shefter E, Falk R. Solubilisation of pharmaceutical substances in an organic solvent and preparation of pharmaceutical powders using the same. US patent 5770559, 1998.
84. Zhang H, Lu J, Han B. Precipitation of lysozyme solubilized in reverse micelles by dissolved CO_2. J Supercrit Fluid 2001; 20:65–71.
85. Hanna M, York P. Method and apparatus for the formation of particles. US patent 6063138, 2000.
86. Bustami R, Chan HK, Dehghani F, Foster N. Generation of micro-particles of proteins for aerosol delivery using high pressure modified carbon dioxide. Pharm Res 2000; 17:1360–1366.
87. Winters M, Frankel D, Debenedetti P, Carey J, Devaney M, Przybycien T. Protein purification with carbon dioxide. Biotechnol Bioeng 1999; 62:247–258.
88. Sarup L, Tservistas M, Sloan R, Hoare M, Humphreys GO. Investigation of supercritical fluid technology to produce dry particulate formulations of antibody fragments. Trans Inst Chem Eng 2000; 78:101–104.
89. Dos Santos IR, Richard J, Pech B, Thies C, Benoit JP. Microencapsulation of protein particles within lipids using a novel supercritical fluid process. Int J Pharm 2002; 242:69–78.
90. Engwicht A, Girreser U, Müller BW. Critical properties of lactide-co-glycolide polymers for the use in microparticle preparation by the aerosol solvent extraction system. Int J Pharm 1999; 185:61–72.
91. Tom JW, Lim G-B, Debenedetti PG, Prud'homme RK. Applications of supercritical fluids in the controlled release of drugs. ACS Symp Ser 1993; 514:238–257.
92. Koushik K, Bandi N, Kompella UB. Respiratory delivery of a peptide drug: degradation pathways, transport mechanisms, and sustained release large porous microparticles prepared using supercritical fluid technology. AAPS Pharm Sci 2002; 4, (Abstract R6117).
93. Snavely WK, Subramaniam B, Rajewski RA, Defelippis MR. Micronization of insulin from halogenated alcohol solution using supercritical carbon dioxide as an antisolvent. J Pharm Sci 2002; 91:2026–2039.
94. Thiering R, Dehghani F, Dillow A, Foster NR. Solvent effects on the controlled

dense gas precipitation of model proteins. J Chem Technol Biotechnol 2000; 75:42–53.

95. Blonder JM, Burt DG, Cheng YS, Webster LA, Lin B, Gonzales K, Etter J, Rosenthal GJ. Improved bioavailability and safety of polymeric insulin. Toxicologist 2001;60:1785.

96. Elvassore N, Bertucco A, Caliceti P. Production of insulin-loaded poly(ethylene glycol)/poly(l-lactide) (PEG/PLA) nanoparticles by gas antisolvent techniques. J Pharm Sci 2001; 90:1628–1636.

97. Magnan C, Badens E, Commenges N, Charbit G. Soy lecithin micronization by precipitation with a compressed fluid antisolvent—influence of process parameters. J Supercrit Fluids 2000; 19:69–77.

98. Badens E, Magnan C, Charbit G. Microparticles of soy lecithin formed by supercritical processes. Biotechnol Bioeng 2001; 72:194–204.

99. Mishima K, Matsuyama K, Tanabe D, Yamauchi S, Young T, Johnston K. Microencapsulation of proteins by rapid expansion of a supercritical solution with a nonsolvent. AIChE J 2000; 46:857–865.

100. Sze-Tu L, Dehghani F, Foster NR. Micronisation and microencapsulation of pharmaceuticals using a carbon dioxide antisolvent. Powder Technol 2002; 126:134–149.

101. Winters M, Knutson BL, Debenedetti P, Sparks H, Przybycien T, Stevenson C, Prestrelski S. Precipitation of proteins in supercritical carbon dioxide. J Pharm Sci 1996; 85:586–594.

102. Thiering R, Dehghani F, Dillow A, Foster NR. The influence of operating conditions on the dense gas precipitation of model proteins. J Chem Technol Biotechnol 2000; 75:29–41.

103. Moshashaee S, Bisrat M, Forbes RT, Nyqvist H, York P. Supercritical fluid processing of proteins. I. Lysozyme precipitation from organic solution. Eur J Pharm Sci 2000; 11:239–245.

104. Young T, Johnston K, Mishima K, Tanaka H. Encapsulation of lysozyme in a biodegradable polymer by precipitation with a vapor-over-liquid antisolvent. J Pharm Sci 1999; 88:640–650.

105. Gilbert DJ, Palakodaty S, Sloan R, York P. Particle engineering for pharmaceutical applications—a process scale up. 5th International Symposium on Supercritical Fluids, Atlanta, GA, 2000.

106. Patel M, Anchordoguy TJ, Randolph TW, Manning MC. Polymeric microspheres in gene delivery: a DNA release and polymer degradation study. AAPS PharmSci 2001; 3, (Abstract W4758).

107. Tservistas M, Levy MS, Lo-Yim MYA, O'Kennedy RD, Humphreys GO, York P, Hoare M. Controlled particle formation of biological material using supercritical fluids. Biotechnol Bioeng 2001; 72:12–18.

108. Nesta DP, Elliott JS, Warr JP. Supercritical fluid precipitation of recombinant human immunoglobulin from aqueous solutions. Biotechnol Bioeng 1999;67:457–464.

109. Muhrer G, Mazzotti M. Precipitation of lysozyme nanoparticles from dimethyl sulfoxide using carbon dioxide as antisolvent. Biotechnol Prog 2003; 19:549–556.

110. Hofland GW, van Es M, van der Wielen LAM, Witkamp G-J. Isoelectric precipitation of casein using high-pressure CO_2. Ind Eng Chem Res 1999; 38:4919–4927.

111. Hofland GW, de Rijke A, Thiering R, van der Wielen LAM, Witkamp G-J. Isoelectric precipitation of soybean protein using carbon dioxide as a volatile acid. J Chromatogr B 2000; 743:357–368.

112. Thiering R, Hofland G, Foster N, Witkamp G-J, van der Wielen LAM. Fractionation of soybean proteins with pressurized carbon dioxide as a volatile electrolyte. Biotechnol Bioeng 2001; 73:1–11.

113. Chattopadhyay P, Gupta RB. Protein nanoparticles formation by supercritical antisolvent with enhanced mass transfer. AIChE J 2002; 48:235–244.

114. Manning MC, Matsura JE, Kendrick BS, Meyer JD, Dormish JJ, Vrkljan M, Ruth JR, Carpenter JF, Shefter E. Approaches for increasing the solution stability of proteins. Biotechnol Bioeng 1995; 48:506–512.

115. Maa Y-F, Prestrelski SJ. Biopharmaceutical powders particle formation and formulation considerations. Curr Pharm Biot 2000; 1:283–302.

116. Yeo SD, Debenedetti P, Patro SY, Przybycien T. Secondary structure characterization of microparticulate insulin powders. J Pharm Sci 1994; 83:1651–1656.

117. Falk R, Randolph TW, Meyer JD, Kelly RM, Manning MC. Controlled release of ionic compounds from poly (L-lactide) microspheres produced by precipitation with a compressed antisolvent. J Controlled Release 1997; 44:77–85.

118. Amaro-Gonzalez D, Mabe G, Zabaloy M, Brignole E. Gas antisolvent crystallization of organic salts from aqueous solutions. J Supercrit Fluids 2000; 17:249–258.

119. Thiering R, Dehghani F, Foster NR. Current issues relating to anti-solvent micronisation techniques and their extension to industrial scales. J Supercrit Fluids 2001; 21:159–177.

120. Ziegler KJ, Hanrahan JP, Glennon JD, Holmes JD. Producing "pH switches" in biphasic water–CO_2 systems. J Supercrit Fluids ASAP 2003.

121. McCluskie MJ, Millan CLB, Gramzinski RA, Robinson HL, Santoro JC, Fuller JT, Widera G, Haynes JR, Purcell RH, Davis HL. Route and method of delivery of DNA vaccine influence immune responses in mice and non-human primates. Mol Med 1999; 5:287–300.

122. Nivens D, Applegate B, Method for nucleic acid isolation using supercritical fluids. US patent 5922536, 1999.

123. Zaks A, Klibanov A. Enzymatic catalysis in nonaqueous solvents. J Biol Chem 1988; 263:3194–3201.

124. Klibanov A. Improving enzymes by using them in organic solvents. Nature 2001; 409:241–246.

125. St. John RJ, Carpenter J, Randolph T. High pressure fosters protein refolding from aggregates at high concentrations. Proc Natl Acad Sci USA 1999; 96: 13,029–13,033.

126. Webb JN, Carpenter JF, Randolph TW. Stability of subtilisin and lysozyme under high hydrostatic pressure. Biotechnol Prog 2000; 16:630–636.

127. Kamat SV, Beckman EJ, Russell AJ. Enzyme activity in supercritical fluids. Crit Rev Biotechnol 1995; 15:41–71.

128. Giessauf A, Gamse T. A simple process for increasing the specific activity of porcine pancreatic lipase by supercritical carbon dioxide treatment. J Mol Catal B—Enzymol 2000; 9:57–64.

129. Sievers RE, Karst U, Milewski PD, Sellers SP, Miles BA, Schaefer JD, Stoldt CR, Xu CY. Formation of aqueous small droplet aerosols assisted by supercritical carbon dioxide. Aerosol Sci Technol 1999; 30:3–15.

130. Sellers SP, Clark GS, Sievers RE, Carpenter J. Dry powders of stable protein formulations from aqueous solutions prepared using supercritical CO_2-assisted aerosolization. J Pharm Sci 2001; 90:785–797.

131. Castor T. Method for size reduction of proteins. US patent 6051694, 2000.

132. Etter J. Particulate drug-containing products and method of manufacture. PCT/US00/34436, 2000.

133. Condo PD, Paul RD, Johnston KP. Glass transitions of polymers with compressed fluid diluents: Type II and type III behavior. Macromolecules 1994; 27:365–371.

134. Shieh Y-T, Su J-H, Manivannan G, Lee PHC, Sawan SP, Spall WD. Interaction of supercritical carbon dioxide with polymers. II Amorphous polymers. J Appl Polym Sci 1996; 59:707–717.

135. Howdle SM, Watson MS, Whitaker MJ, Popov VK, Davies MC, Mandel FS, Wang JD, Shakesheff KM. Supercritical fluid mixing: preparation of thermally sensitive polymer composites containing bioactive materials. J Chem Soc Chem Commun, 2001:109–110.

136. Shine AD, Gelb J Jr. Microencapsulation process using supercritical fluids. US patent 5766637, 1998.

137. Shea LD, Smiley E, Bonadio J, Mooney DJ. DNA delivery from polymer matrices for tissue engineering. Nature Biotechnol 1999; 17:551–554.

138. Sheridan MH, Shea LD, Peters MC, Mooney DJ. Bioadsorbable polymer scaffolds for tissue engineering capable of sustained growth factor delivery. J Controlled Release 2000; 64:91–102.

139. Engelberg I, Kohn J. Physio-mechanical properties of degradable polymers used in medical applications: a comparative study. Biomaterials 1991; 12.

140. Hile DD, Amirpour ML, Akgerman A, Pishko MV. Active growth factor delivery from poly(D,L-lactide-co-glycolide) foams prepared in supercritical CO_2. J Controlled Release 2000; 66:177–185.

141. Sproule T, Lee JA, Lannutti JJ, Tomasko DL. Bioactive polymers via supercritical fluids. Proceedings of the 8th Meeting on Supercritical Fluids, Bordeaux, 2002:431–436.

142. Luzzi LA, Needham TE, Hossein Z, Dondeti P. Method of incorporating proteins or peptides into a matrix and administration thereof through mucosa. US patent 6190699, 2001.

143. Shah JN. Immobilization of whole bacterial cells in amorphous polymers using supercritical carbon dioxide. Master's thesis. Lexington: University of Kentucky, 2003.

144. Castor TP. Methods and apparatus for making liposomes. US patent 5554382, 1996.
145. Castor TP, Chu L. Methods and apparatus for making liposomes containing hydrophobic drugs. US patent 5776486, 1998.
146. Frederiksen L, Anton K, van Hoogeves P, Keller HR, Leuenberger H. Preparation of liposomes encapsulating water-soluble compounds using supercritical carbon dioxide. J Pharm Sci 1997; 86:921–928.
147. Frederiksen L, Anton K, van Hoogevest P. Process for the preparation of a liposome dispersion under elevated pressure contents. US patent 5700482, 1997.
148. Bridson RH, Al-duri B, Santos RCD, Mcallister SM, Robertson J, Alpar HO. The production of liposomes containing hydrophilic and hydrophobic drugs using supercritical fluid technology. AAPS Pharmsci 2002; 4 (Abstract R6260).
149. Otake K, Imura T, Sakai H, Abe M. Development of a new preparation method of liposomes using supercritical carbon dioxide. Langmuir 2001; 17:3898–3901.
150. Barenholz Y. Liposome application: problems and prospects. Curr Opin Colloid Sci 2001; 6:66–77.
151. Whaley SR, English DS, Hu EL, Barbara PF, Belcher AM. Selection of peptides with semiconductor binding specificity for directed nanocrystal assembly. Nature 2000; 405:665–668.

11

Asymmetric Catalysis in Supercritical Fluids

Philip G. Jessop
Department of Chemistry, University of California, Davis,
California, U.S.A.

Synthesis of chiral pharmaceuticals with catalyst recovery and enhanced rate and selectivity is possible, for some reactions, with the use of liquid or supercritical CO_2 as a substitute for traditional organic solvents. Supercritical CO_2 ($scCO_2$) is an appropriate medium for the synthesis of pharmaceuticals because it is nontoxic, unlikely to be left in the product, relatively inert, inexpensive, and environmentally benign. Because of its unusual physical properties, it can also cause higher rates for certain reactions involving reagent gases. Improved selectivities have also been observed for some syntheses. Supercritical CO_2 and other supercritical fluids (SCFs) are used industrially as solvents for several reactions (1) but not yet for pharmaceutical synthesis. This chapter will describe the potential that SCFs and especially $scCO_2$ have for asymmetric catalysis in the pharmaceutical industry, starting with a description of general aspects of reactions in SCFs, a survey of reports of asymmetric synthesis using catalysts in SCFs, a summary of related techniques, and a discussion of the outlook. The studies described herein show definitively that the advantages of SCFs are significant and realizable for asymmetric syntheses of practical importance. It remains up to the pharmaceutical industry to decide whether to take advantage of these discoveries.

A supercritical fluid is defined as a pure compound or mixture that is at a temperature and pressure above the critical point but below the pressure required to condense it into a solid (1). A SCF is neither a gas nor a liquid, but shares the properties of both states. Like a gas, it fills the available space, but like a liquid, it has (at least under some conditions) enough density to be capable of acting as a solvent.

Reactions in SCFs have been studied for almost 150 years (2), starting with Daubrée's 1857 syntheses of minerals in scH_2O (3). The first simple organic reactions in SCFs were reported by Villard (4,5) in 1896–1898. Industrial use of supercritical conditions include the syntheses of ammonia (since 1913), low-density poly(ethylene), methanol, higher alcohols, quartz, and fluoropolymers (1). Some of these processes are performed on an enormous scale, demonstrating that the cost of pressurizing SCFs does not make their use uneconomical even for low-value products. Despite this long-lasting industrial interest in SFRs (supercritical fluid reactions), academic chemical research was limited to a few isolated studies until a series of articles about scC_2H_4 polymerization started appearing in the 1960s (6). The first reported examples of asymmetric catalysis in a SCF were enzymatically catalysed reactions described by three groups in 1992 (7–9).

The choice of the most appropriate SCF for use as a reaction medium depends on the requirements of the reaction. Normally, a SCF with a critical temperature 5–50 K below the anticipated reaction temperature is chosen, although the use of near-critical liquids should not be rejected—these liquids have most but not all of the advantages of SCFs. Supercritical water, because of its very high critical temperature (374°C) (10), is never used for asymmetric synthesis. Supercritical CO_2, with a critical temperature of 31°C, is ideal for reactions that can be performed near room temperature. Preparations requiring lower temperatures could theoretically be performed in liquid CO_2 or in low-temperature SCFs such as krypton (−63.8°C) (11), methane (−82.6°CC) (12,13), and argon (−122.5°CC) (11,14), although catalysis in low-temperature SCFs has never been investigated perhaps because of concerns about the decreased strength of steel at those temperatures. Polar, low-temperature SCFs (i.e., CO, NF_3, ONF_3) are not practical because of toxicity reasons. Considerations other than temperature can also dictate the choice of SCFs. If one of the reagents is a gas with a suitable critical temperature, then it can be used as both reagent and SCF solvent; this technique is used industrially in the preparation of low-density poly(ethylene), which is performed by catalysis in scC_2H_4. On the other hand, if an inert SCF is desired, then $scCO_2$ and scC_2H_6 are the most obvious choices for moderate temperatures. Note that $scCO_2$ is not always inert; it can react with alkanols, alkoxides, primary and secondary amines, metal amides, metal hydrides, metal alkyls, hydroxides, and water. Partially fluorinated methanes and ethanes are interesting for academic investigations because of their higher and variable dielectric constants (15–17), but they are not appropriate for industrial processes because of their high cost and environmental impact. Supercritical nitrous oxide is strongly oxidizing and therefore should generally be avoided for safety reasons (18,19).

Among the many advantages of SCFs (20) is the high solubility of reagent gases. Permanent gases such as H_2, CO, and O_2 are completely miscible in all proportions with SCFs. This can lead to increased rates of reactions if the rates are first or higher order with respect to the permanent gas (21,22). However, the increase in rate is associated more with the elimination of mass transfer limitations between gas and liquid phases than it is with simply having a higher concentration of the gas in the reaction phase. Elimination of mass transfer problems could also theoretically lead to changes in selectivities. Rate or selectivity improvements, however, must be balanced against phase behavior considerations. The permanent gases act as antisolvents, causing a reduction in the solubility of reagents or catalysts in the SCF. This solubility drop increases with increasing concentration of the permanent gas. For example, an attempt to dissolve 3 mL of MeOH in 31 mL of $scCO_2$ at 31°C in the presence of 100 bar H_2 required enough CO_2 to bring the total pressure over 300 bar (23), whereas in the absence of H_2, the MeOH would have been completely miscible above 65 bar (24). Thus, the use of high concentrations of permanent gases may lead to problems with insolubility of the reagents or catalysts. In any case, experimentalists in the field must observe their reactions through windows and report the number of phases present during the reaction.

Another advantage of SCFs is the possibility of altered selectivity. Dramatic selectivity changes have been observed due to pressure-induced polarity changes, dilution effects, and reactivity of the SCF. For example, changing the pressure of $scCHF_3$ changes its dielectric constant, which has a direct effect on reaction performance; changes have been observed in the enantioselectivity of enzyme-catalyzed transesterification (25,26) and esterification (27) and homogeneously catalyzed cyclopropanation (28,29). The dielectric constant of $scCHF_3$ at temperatures somewhat above the critical temperature increases with pressure over a range of 1 (at 0 bar) to around 7 (at pressures above 80 bar) (17). The dielectric constant changes because it is a function of the density, which changes with pressure. The dielectric constant of $scCO_2$ is also temperature dependent (15) but over such a small range that it has no effect on reaction selectivity. Increasing the pressure of a SCF also causes a dilution effect, which has been shown to favor intramolecular reactions over intermolecular reactions in a study of diene olefin metathesis (30). Finally, the reaction of $scCO_2$ with NH groups to form carbamates can serve as an in situ protection strategy to allow reactions that are normally not possible in the presence of NH groups (30,31).

Promotion of the advantages of SCFs should always be accompanied by a sobering reminder that these fluids and $scCO_2$ in particular are not very good solvents. Many substrates, reagents and catalysts have low or negligible solubility in $scCO_2$. Generally, solubility in $scCO_2$ is high for nonpolar, volatile, or highly fluorinated compounds, whereas solubility is decreased by the presence of OH, CO_2H, NH, PH, or SH groups, aromatic or unsaturated cyclic groups, or long straight-chain alkyl groups (32). Compilations of solubility data for compounds in SCFs have been published (33–35). Solubility of a compound in an SCF can be increased by the addition of a cosolvent such as methanol to the SCF. Also, note that although rate enhancements, selectivity enhancements, and efficient catalyst recovery have all been observed, they only occur for some reactions and there are no examples of reactions which demonstrate all three advantages simultaneously.

Near-critical phenomena, which can take place in SCFs or liquids at conditions very close to the critical point, are scientifically fascinating and have been greatly studied (20,36). The phenomena, which include deviations of partial molar volumes and volumes of activation to very large positive or negative values, have the potential to strongly affect reaction performance. From the viewpoint of the industrialist, however, these phenomena are unappealing because reaction performance under such conditions is a strong function of the exact pressure and temperature. Processes very near the mixture critical point are likely to be unstable and require very demanding temperature and pressure control.

Experimental methods for performing research in SCF chemistry has been reviewed in a recent monograph (2) which contains chapters on equipment design (37), the use of SCF emulsions (38), and techniques for in situ monitoring of reactions by high pressure nuclear magnetic resonance (NMR) (39), infrared (IR) (40), and other spectroscopic techniques (41). Other areas of exciting developments in synthesis in SCFs include the fields of polymerizations (6,42–46), polymer modifications (45,47–49), and loading of biologically active compounds into polymers (50–52). The present chapter will not include a discussion of these topics or uncatalyzed organic reactions (53) or resolutions (54–56). Although all of these areas have potential pharmaceutical relevance, the present review is concerned with asymmetric synthesis using chiral catalysts and supercritical fluids, with an emphasis on classes of products that have pharmaceutical importance.

1. CATALYSIS IN SCFs

Heterogeneous catalysis in SCFs dates from the experiments of Ipatiev, Haber, and researchers at BASF (1). Although the advantages were not known at that time, SCFs are particularly appropriate as media for heterogeneous catalysis because of the extremely rapid diffusion rates of SCFs in pores (compared to liquid solvents) and the superior ability of SCFs to prevent coking/deactivation of the catalyst (57) by extracting deposited material. The subject of heterogeneous catalysis in SCFs has been reviewed recently (58), although not from the viewpoint of asymmetric catalysis.

Homogeneous catalysis in SCFs has great potential to be a method for large-scale asymmetric synthesis using SCFs. The field of homogeneous catalysis in SCFs dates from the experiments of Ipatiev, who used $AlCl_3$ as a (presumably) homogeneous catalyst for the oligomerization of scC_2H_4 (59), but most of the literature dates from the years since 1994. Enantioselective homogeneous catalysis was first reported by Burk et al. (60) in 1995, as described in Section 2. Homogeneous catalysis in $scCO_2$ requires that the catalysts be soluble in the SCF, but this is often not the case. For example, triphenylphosphine, like most compounds with multiple aromatic rings, is not particularly soluble in $scCO_2$ (61,62), and its complexes have very poor solubility (63). In order to circumvent this problem, other types of phosphines have been used, including trialkylphosphines (21,60,64) and fluoroalkyl-substituted triarylphosphines (62,65), which are significantly more soluble. For cationic complexes, the use of a partially fluorinated counterion, especially triflate and the BArF anion ($[(3,5-(CF_3)_2C_6H_3)_4B]^-$), is an effective strategy (60). Beckman's group published the first highly CO_2-soluble polymer containing only C, H, and O atoms; one could anticipate that the design features which lead to that success could be duplicated in a small molecule such as a triarylphosphine, but there have been no examples published so far. Nevertheless, a large body of literature has appeared in the past few years to show that a range of homogeneously catalysed reactions are possible in SCFs such as $scCO_2$, that some of them are faster than in liquid solvents, and some of them have greater enantioselectivity than can be obtained in liquid solvents. Finally, as will be described in Section 3, SCFs have made new catalyst recovery and recycling schemes possible.

Enzymatic catalysts would, at first glance, be considered inappropriate for use in SCFs because they are traditionally used in aqueous solution. However, the discovery that enzymes can be catalytically active in nonaqueous solvents opened the doors for investigations into their use in SCFs. Since the first reports in 1985 (66–68), enzyme catalysis in SCFs has been reported for many reaction types, including asymmetric catalysis (69–71). The fact that enzymes are insoluble in SCFs does not prevent their use; in fact, it helps because it makes catalyst/product separation easy. There can be problems in the use of $scCO_2$ as a solvent if it inserts into enzyme N–H bonds or if it dehydrates the enzyme; in SCFs, as in nonaqueous liquids, at least a monolayer of water is required. However, in many cases, higher activity has been observed in $scCO_2$ than in organic liquids. It has been argued that higher rates should be expected because diffusion from solvent to enzyme particle surfaces and inside pores is rate limiting in organic solvents (72). Additionally, hydrophobic solvents rather than hydrophilic solvents have been found to allow greater rates of reaction for enzymes, because there is a decreased tendency for the solvent to strip off the necessary monolayer of water (71,73); scCO$_2$ serves in this respect as a hydrophobic solvent. The drop in the pH of the water layer around an enzyme in scCO$_2$, due to dissolution of the CO$_2$ in the water, has been shown to be negligible (69). Further discussion of why enzymes are active in scCO$_2$ can be found in reviews of the topic (69–71).

2. ASYMMETRIC CATALYSIS IN SCFs

Syntheses of commercially important drugs, intermediates, and related compounds by asymmetric catalysis in SCFs have been reported (Table 1). There

TABLE 1 Pharmaceutical and Related Applications for Some of the Chiral Compounds Prepared by Catalysis in SCF

Application	Product	Reaction	Enantiomeric excess (e.e.) (%)	Eq.
Antibiotic stabilizer	Cyclopropanecarboxylic acids derivatives[a]	Cyclopropanation	84	27
Antibiotic	Azirine carboxylate esters[a]	Neber reaction	60	28
Antihypercholesterolemic	2-Methylbutanoate esters[b]	Hydrogenation	89	18
Anti-inflammatory	Ibuprofen[c]	Esterification	92	1
Anti-inflammatory	Ibuprofen[b]	Hydroformylation	93	23
Anti-inflammatory	Naproxen[c]	Esterification	nr[d]	2
Anti-inflammatory	Profen[a]	Hydrogenation	88	S2
Antipruritic	Menthol[c]	Hydrolysis	nr	15
β-Blockers	3-Hydroxyesters[a]	Transesterification	78	6
Calcium antagonist	Diltiazem[b]	Hydrolysis	87	13
Calcium antagonist	Mibefradil[b]	Hydrogenation	84	S2
Choleretic	1-Phenylpropanol	Alkylation	98	29
Intermediate[e]	Citronellol[c]	Esterification	98.9	3
Intermediate[e]	1-Phenylethanol[c]	Transesterification	~100	5, 7
Obesity treatment	Xenical[b]	Hydrogenation	98	19
Rice detoxin	N-Acetyl-L-phenylalanine esters[b]	Transesterification	nr	9
Various	Prostaglandins[b]	Hydrolysis	59	12

[a] Model compound representing a class of relevant structures or intermediates.
[b] Tested reaction produces an intermediate for the pharmaceutical.
[c] Pharmaceutical directly prepared by the reaction.
[d] Not reported.
[e] Intermediate primarily for preparation of fragrances.
[f] Scheme 2.

is no doubt, upon inspection of Table 1, that the field of pharmaceutical synthesis in SCFs is still in its infancy. This is evident from the small number of pharmaceutically active species that have been prepared in these media. The number will increase as academic and industrial laboratories continue to work on these applications. There also needs to be an evaluation, in the research community, of whether it is possible or practical to perform multistep syntheses in SCFs. For "green" multistep syntheses, it is likely that SCFs will be used in concert with ionic liquids, the latter media being used for steps which require more polar conditions. These considerations are likely to be the subject of future experimentation in the field.

2.1. Esterification and Transesterification

Lipase-catalyzed esterifications of racemic carboxylic acids in SCFs have been studied by several groups. The target in all of these studies was the preparation of optically pure anti-inflammatory drugs ibuprofen and naproxen. Rantakylä and Aaltonen reported the kinetic resolution of racemic ibuprofen by esterification catalyzed by immobilized lipase from *Mucor miehei* [Eq. (1)] (8,74,75):

$$(1)$$

The enzyme was not deactivated by $scCO_2$ and was active even in the absence of water, but the optimum rate was found at a water concentration of 0.5–1 mL/L. At 25% conversion, the e.e. of the ester was 92%, in favor of the S ester (e.e. = enantiomeric excess). The same reaction was found by Overmeyer et al. (76) to be catalyzed by a lipase from *Candida antarctica B*, which esterified the R enantiomer, but the enantioselectivity was poor. Wu and Liang (77) found that the enantioselectivity of esterification of racemic naproxen [Eq. (2)] was a strong function of the concentration of alcohol, with the greatest selectivity being obtained at lower alcohol concentrations.

$$(2)$$

The lipase from that study was from *Candida rugosa*.

Kinetic resolution of chiral alcohols by selective esterification has also been performed in $scCO_2$ by several groups. A glass-immobilized lipase from *Candida cylindracea* was found to catalyze the stereoselective esterification of only the S isomer from racemic citronellol (Eq. (3)] in a continuous-flow reactor (78):

$$R = -(CH_2)_7CH=CH(CH_2)_8H \tag{3}$$

The enantioselectivity in $scCO_2$ was 98.9 % at 84 bar and 31°C but below 24% at all other conditions tested. The authors suggested that the high selectivity might be due to a greater aggregation of CO_2 around the enzyme taking place at the near-critical conditions, an explanation that is experimentally difficult to test. Esterification of racemic glycidol [Eq. (4)] was evaluated by Martins et al. (7,79):

$$\tag{4}$$

Because free-enzyme porcine pancreatic lipase had much lower activity in $scCO_2$ than in organic solvents, the authors decided instead to test immobilized lipases. Macroporous resin-supported lipase from *M. miehei* had fair activity but poor enantioselectivity. Porcine pancreatic lipase immobilized on supports had the greatest activity if the support was highly hydrophilic; Sephadex G-25 and Bio-gel P6 were selected. Enantioselectivity was 83% for the (*S*)-gycidyl butyrate at 25–30% conversion, comparable to results in organic solvents. Several chiral alcohols were studied by Cernia et al. (80) as substrates for kinetic resolution by esterification catalyzed by silica-supported lipase from *Pseudomonas* sp. [Eq. (5)]:

$$\tag{5}$$

$R = Ph, Cy, C_6H_{13}$, trans-$CH=CHMe$ and $CH_2CH_2CH_2CH=CMe_2$

Enzyme stability in $scCO_2$ was significantly higher than in any of the organic solvents tested. Although the authors did not mention which enantiomer was formed, the enantioselectivity and the activity were also far higher in $scCO_2$ than in liquid hexane, benzene, or toluene.

Kinetic resolution by transesterification of 3-hydroxyoctanoic acid [Eq. (6), R = vinyl or cyclohexyl] was described by Bornscheuer et al. (9,81,82):

$$\text{(6)}$$

The catalyst was the unsupported lipase from *Pseudomonas cepacia*. The reaction was as selective in $scCO_2$ as it was in hexane but was much slower in $scCO_2$. The S ester formed preferentially, in 46% e.e. (after 63% conversion), leaving the R acid in 78% e.e.* Later experiments compared the effectiveness of the crude enzyme with the epoxy- and Celite-stabilized enzymes (82). All three were stable to exposure to $scCO_2$ and greater rates of reaction were found with the immobilized versions, but the conversions dropped greatly when the catalyst was reused. 3-Hydroxy esters can be used as intermediates for the preparation of optically active pharmaceuticals such as propranolol, a β-blocker (82).

Cutinase enzyme from *Fusarium solani pisi*, immobilized on a 4-Å molecular sieve, catalyzes the transesterification of the R isomer of 1-phenyl-ethanol in the racemic mixture [Eq. (7), R = C_2H_5], giving the R ester and the S alcohol, both in essentially 100% e.e:

$$\text{(7)}$$

The authors, Fontes et al. (83), found that the rate, but not the enantiose-lectivity, was strongly dependent on the water content of the $scCO_2$, reaching a maximum at a water activity of 0.5. The same reaction (but with R = CH_3) was shown by Overmeyer et al. (76) to be essentially completely enantiose-lective when catalyzed by a lipase from *Ca. antarctica B*. The activity was pressure and temperature dependent, giving the greatest rate at 90°C and at 170 bar.

---

* In Fig. 2 of Ref. 81, the stereochemistries of the unreacted starting material and the product seem to be reversed.

(S)-α-Cyano-*m*-phenoxybenzyl alcohol, the active constituent in several pyrethroid insecticides, has been synthesized by a kinetic resolution [Eq. (8)] catalyzed by a lipase:

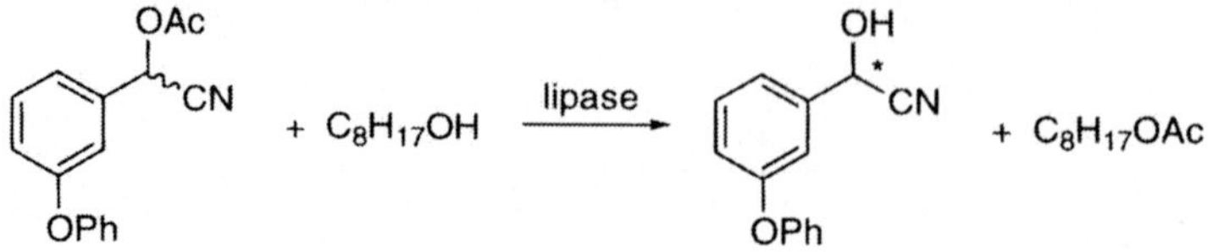

Stephan and colleagues carried out extensive studies to understand the phase behavior of the reaction mixture (84–86).

Enantioselective esterification can be pressure dependent in supercritical fluoroform. For example, the selectivity of the transesterification of *N*-acetyl-phenylalanine ethyl ester with methanol catalyzed by *Subtilisin Carlsberg* was shown by Kamat et al. (26) to be greater at higher pressures than at lower pressures [Eq. (9)]:

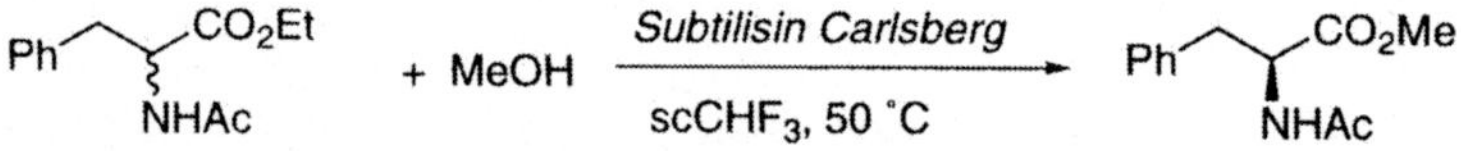

They suggested that at the higher pressures (and therefore higher dielectric constants), the scCHF$_3$ was capable of dissolving more displaced water; this would affect enantioselectivity if one substrate enantiomer displaces more water from the enzyme pocket than does the other enantiomer (26). They also noted that scCO$_2$ inhibited the reaction. *N*-Acetyl-L-phenylalanine esters are used as detoxins in rice production (87). Mori et al. (27) evaluated the effect of scCHF$_3$ pressure on the esterification of racemic phenylethanol [Eq. (10)] with a lipid-coated lipase:

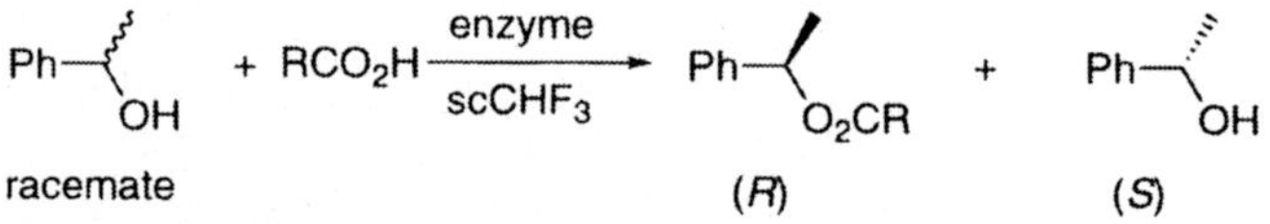

Greater enantioselectivity was found at 60 bar than at higher pressures. The lipid coating of the enzyme was used to enhance the solubility of the enzyme in nonpolar solvents such as alkanes and scCHF$_3$; in fact, it was reported to be soluble in the SCF.

2.2. Hydrolysis

Protease-catalyzed kinetic resolution of N-protected hydrophobic amino acids in scCO$_2$ was described by Chen et al. (88) as a method for the production of unnatural amino acids. With the "alcalase" protease from *Bacillus licheniformus*, carbobenzyloxy-protected racemic amino acid esters (norleucine, α-aminobutyric acid, norvaline, and others) were hydrolyzed at ~34°C, ~80 bar, and 0.5% water [Eq. (11)]:

$$\tag{11}$$

After 20 h, the reaction stopped just short of 50% conversion, giving (for α-aminobutyric acid) the L-acid in 95% yield and >99% e.e. and the D ester in 81% yield and 92% e.e., these results obviously indicating excellent selectivity for the hydrolysis of the L ester only. Additionally, the alcalase was found to have a longer half-life in scCO$_2$ than in a phosphate buffer (pH 8.2) at the same temperature.

A lipase-catalyzed kinetic resolution was described by Parve et al. (89). Bicyclo[3.2.0]heptanol esters were hydrolyzed in scCO$_2$ with the lipase from *Humicola lanuginosa* at 40°C and 200 bar. From the racemic ester in Eq. (12), a 59% isolated yield of the *1R,2S,3S,5R* ester (67% e.e.) and a 3% yield of the *1S,2R,3R,5S* alcohol (92% e.e.) were obtained after 100 h:

$$\tag{12}$$

Although the rate and yield were low, the results are an improvement on the same reaction in water, which does not proceed without added cosolvent. The resolved alcohols are precursors to prostaglandins (90).

An immobilized lipase, from *M. miehei*, was used by Aaltonen's group for the kinetic resolution of *trans*-3-(4-methoxyphenyl)glycidic acid methyl ester [Eq. (13)], a precursor in the preparation of the calcium antagonist drug diltiazem (91):

$$\tag{13}$$

Although both enantiomers reacted, the rate of hydrolysis of the *2S,3R* form was over five times faster than that of the opposite enantiomer. After 53% conversion, the remaining ester had an e.e. of 87%. The selectivity was roughly independent of water concentration. The rate in $scCO_2$ was roughly fivefold greater than that in a water/toluene biphasic medium.

Porcine pancrease lipase was shown by Glowacz et al. (92) to hydrolyze triolein and dioleins in $scCO_2$ [Eq. (14), where $R = -(CH_2)_7CH=CH(CH_2)_8H)$]:

$$(14)$$

$$sn\text{-}1,2\text{-diolein} \quad sn\text{-}1\text{-monoolein}$$

Initially, the enzyme preferentially hydrolyzed the triolein (or 1,3-diolein) at the sn-3 position, but the selectivity decreased over time, possibly due to back-reaction and acyl migration. The enantioselectivity and activity were not functions of the water content of the SCF but were functions of the water content during the preparation of the enzyme. The hydrolysis of rac-1,2-diolein again showed a preference for hydrolysis at the sn-3 position, resulting in an excess of unreacted sn-1,2-diolein in solution.

Michor et al. (93) compared four lipases and an esterase for the resolution of racemic menthol by transesterification [Eq. (15)]:

$$(15)$$

The esterase EP10 was highly selective for esterification of the L-isomer. Supercritical CO_2 was the solvent for the reaction and for the postreaction separation of the acetate from the unreacted D-menthol.

2.3. Hydrogenation

Because of the complete miscibility between H_2 and SCFs such as CO_2 (94) and the resulting potential rate advantage for hydrogenations in such media, there has been considerable interest in hydrogenation reactions in SCFs. Other than a few isolated early studies (95,96), most of the work on catalytic hydrogenation in SCFs has been done since 1994, when the hydrogenation of $scCO_2$ was reported (64). Asymmetric reaction studies were atypically near the forefront, the first example being published in 1995 by Burk et al. (see later in this subsection) (60). Advantages to using SCFs were expected because some hydrogenations have rates that are first order with respect to hydrogen concentration and are limited by the rate of diffusion of H_2 from the gas to the liquid phase. Additionally, enantioselectivities of hydrogenations have been shown to be highly influenced by changes in gas–liquid mass transfer rates (97). Switching from liquid solvents to SCFs increases the H_2 concentration and eliminates gas–liquid mass transfer limitations.

Subsequent research has shown some limitations to the benefit of high hydrogen concentrations. First, catalysts, substrates, and cosolvents are far less soluble in CO_2/H_2 mixtures than they are in $scCO_2$ itself. Another limitation is the possibility of reaction between the CO_2 and the H_2, giving either CO or formates. Carbon monoxide formation has been detected as a problem in heterogeneously catalyzed hydrogenations in $scCO_2$ (98,99). The production of formates, although detected and desired in studies of CO_2 fixation (21), have not been shown to interfere in catalytic hydrogenations of olefins.

Burk et al. (60) described the hydrogenation of six prochiral α-enamide esters to α-amino acid derivatives using cationic [Rh(cod)(EtDuPHOS)]X [where X = $B(C_6H_3$-3,5-$(CF_3)_2)_2$ or O_3SCF_3 and EtDuPHOS is shown in Scheme 1] in methanol, hexane, and $scCO_2$. Greater enantioselectivity was found in $scCO_2$ rather than in organic solvents for two of the six substrates [e.g., Eq. 16]:

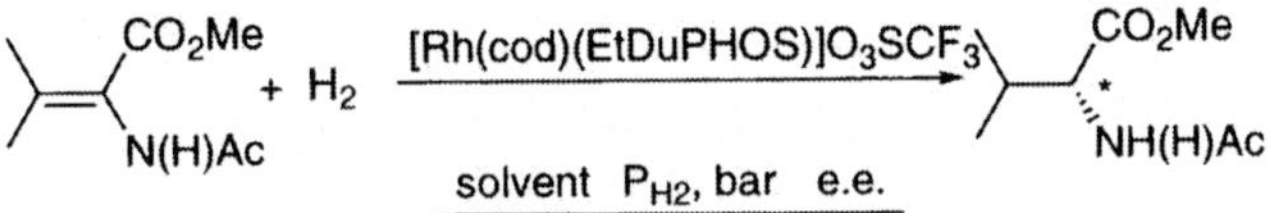

solvent	P_{H2}, bar	e.e.
MeOH	4	67
hexane	4	70
$scCO_2$	14	88

(16)

Although greater solubility in $scCO_2$ was observed with the borate (BArF) anion catalyst precursor than the triflate precursor, the more soluble precursor did not always give the greatest enantioselectivity. This article (60) was the first demonstration that asymmetric homogeneous catalysis could be performed in SCFs and the first demonstration that the technique could lead to greater enantioselectivity than that obtained in traditional solvents.

Development of a more CO_2-philic cationic Rh complex was reported by Lange et al. [Eq. (17), with ligand L^1 in Scheme 1] (100):

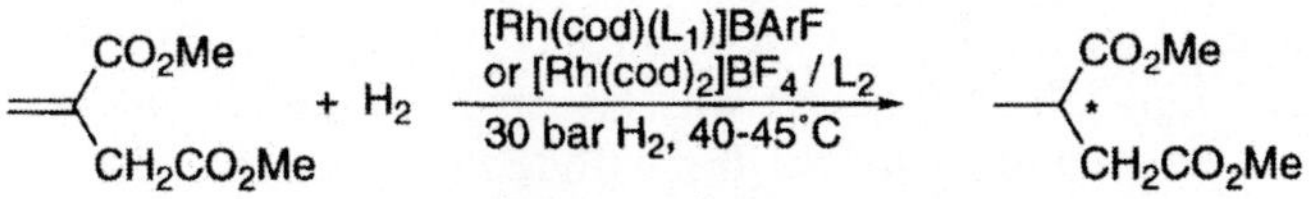

solvent	L	e.e.
MeOH	L^1	8
hexane	L^1	73
$scCO_2$	L^1	72
$scCO_2$	L^3	97
CH_2Cl_2	L^3	99 (at 23 °C)

(17)

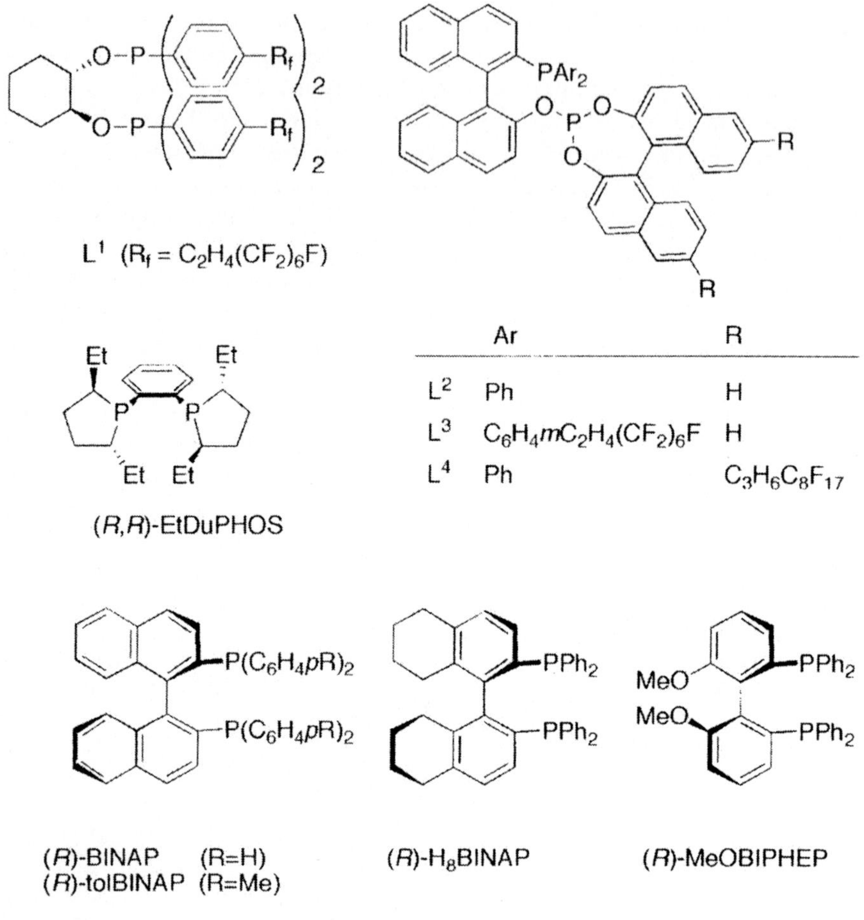

L^1 ($R_f = C_2H_4(CF_2)_6F$)

(R,R)-EtDuPHOS

	Ar	R
L^2	Ph	H
L^3	$C_6H_4mC_2H_4(CF_2)_6F$	H
L^4	Ph	$C_3H_6C_8F_{17}$

(R)-BINAP   (R=H)
(R)-tolBINAP (R=Me)

(R)-H$_8$BINAP

(R)-MeOBIPHEP

SCHEME 1

Although this catalyst precursor was incompletely soluble, it was shown to generate a CO_2-soluble catalytic species active for the hydrogenation of dimethyl itaconate. The rate of hydrogenation was fair in scCO$_2$ but much higher in hexane, whereas the enantioselectivity was comparable in the two solvents. Surprisingly, the selectivity was H$_2$-pressure dependent in hexane but not in scCO$_2$. The same precursor, but with the BF$_4^-$ anion, had very poor selectivity in scCO$_2$. PHIP (*para*-hydrogen-induced polarization) and isotopic labeling showed that the mechanism of hydrogenation in scCO$_2$ was pairwise H$_2$ addition, as in liquid solvents, ruling out the possibility of formate intermediates.

Hydrogenation of dimethyl itaconate but with an in situ catalyst formed from a Rh complex and a fluorinated version of the BINAPHOS ligand [L^3 in Eq. (17) and Scheme 1] gave superior enantioselectivity compared to the earlier ligand; surprisingly, the selectivity and activity were fair, even with the less CO_2-soluble BF_4^- counterion. The same catalyst was also effective for the hydrogenation of a prochiral α-enamide ester using the induced melting technique (Section 3.2) (101).

The question of the effect of H_2 concentration on enantioselectivity is particularly pertinent to asymmetric hydrogenation by ruthenium complexes containing BINAP ligands (Scheme 1). With such catalyst precursors, the enantioselectivity of hydrogenation in methanol is strongly dependent on the concentration of H_2 in the reaction phase (97,102). In practice, the H_2 concentration in the MeOH is a function of the H_2 pressure (102) and the stir rate (i.e., mass transfer limitations exist) (97). For some substrates, the enantioselectivity is greater if there is a greater concentration of H_2 in the liquid phase, whereas for others, the opposite is true. Both classes of substrates have been tested in SCFs.

Substrates that give the greatest enantioselectivity at low H_2 concentrations were first tested in scCO$_2$ by Xiao et al. (103), who investigated the hydrogenation of tiglic acid [Eq. (18)]:

$$
\begin{array}{c}
\text{CO}_2\text{H} \\
\diagup\!\!=\!\!\diagup\!\!\diagdown + H_2 \xrightarrow[\text{Ru(O}_2\text{CMe)}_2\text{(H}_8\text{BINAP)}]{} \quad \diagup\!\!\diagup\!\!\diagdown^{\text{CO}_2\text{H}}_{*}
\end{array}
$$

solvent	P_{H2}, bar	e.e.
scCO$_2$	7	71
scCO$_2$	33	81
scCO$_2$/R$_F$OH	5	89

$$(18)$$

Several optically active esters of the product 2-methylbutanoic acid are important pharmaceutical products, including the antihypercholesterolemics compactin and mevinolin and the vasodilator visnadine (87). A qualitative comparison of three ligands BINAP, TolBINAP, and H$_8$-BINAP (Scheme 1) showed that the partially hydrogenated H$_8$BINAP ligand and presumably the corresponding complex were the most soluble in scCO$_2$. Initial experiments in scCO$_2$ were disappointing in that the e.e.'s obtained were moderate compared to the best that have been obtained in methanol (96%) (104). However, a marked improvement was observed when a fluorinated alcohol $CF_3(CF_2)_6 CH_2OH$ was added to the scCO$_2$ phase; the alcohol may have acted both in the capacity of a cosolvent to increase catalyst solubility and as a proton source for the hydrogenation mechanism. The enantioselectivity of the reaction in scCHF$_3$ was also found to be dependent on the pressure of the SCF, giving higher e.e.'s (89–90%) at scCHF$_3$ pressures above 120 bar but inferior e.e.'s at lower pressures (105).

Atropic acids, representatives of the class of substrates that are hydro-genated with greater enantioselectivity at greater H_2 concentrations, have been tested in $scCO_2$ using tolBINAP catalysts and MeOBIPHEP catalysts. Wang and Kienzle (106) studied the hydrogenation of 2-(4-fluorophenyl)-3-methylbut-2-enoic acid (Scheme 2) using $Ru(OAc)_2((R)$-MeOBIPHEP) cat-alyst, whereas Jessop et al. (23) tested the hydrogenation of atropic acid itself using $Ru(OAc)_2((R)$-tolBINAP) catalyst. Although the miscibility of H_2 with CO_2 was expected to enhance the enantioselectivity obtained in $scCO_2$ compared to liquid MeOH, the results were disappointing; in both cases, the enantioselectivity in $scCO_2$ (with a small amount of MeOH cosolvent) was a few percent lower than that obtained under comparable H_2 pressures in pure MeOH. The enantioselectivity was only slightly better in liquid methanol than in $scCO_2$ with methanol cosolvent. The hydrogenated products belong to the class of 2-arylpropanoic acids which include the important anti-inflammatory drugs ibuprofen and naproxen, which can be prepared by the same reaction (eg., hydrogenation of p-isobutylatropic acid). The product obtained from 2-(4-fluorophenyl)-3-methylbut-2-enoic acid is an intermediate for the prepa-ration of mibefradil, a calcium antagonist (107).

Wang and Kienzle (107) also reported the hydrogenation of a β-ketoester to form an intermediate for the synthesis of Xenical®, a drug used for the treatment of obesity (87). The hydrogenation [Eq. (19), R = an unspecified alkyl chain] was faster and more selective in $scCO_2$ than in the liquid substrate:

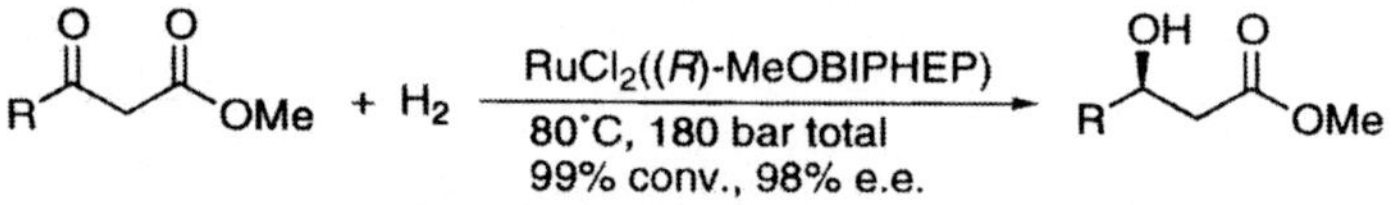

$$\text{(19)}$$

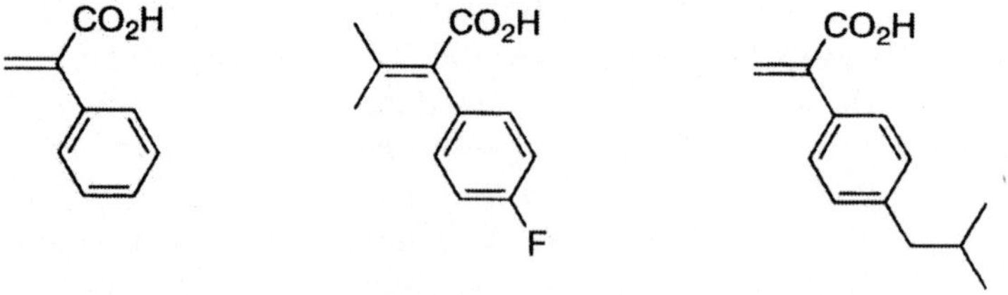

atropic acid   2-(4-fluorophenyl)-3-   p-isobutylatropic acid
                            methylbut-2-enoic acid

SCHEME 2

Baiker and colleagues have extensively studied the heterogeneously catalyzed asymmetric hydrogenation of ethyl pyruvate, giving ethyl D-lactate, in the presence of compressed gases [Eq. (20)] (98,99,108,109):

$$\text{(20)}$$

solvent	e.e.	conv.
toluene	75	100
scCO$_2$ (80 bar)*	28	3	* at 40°C
scC$_2$H$_6$ (60 bar)	74	96

Notably, they found that C_2H_6 was far superior to CO_2 as the compressed gas for this reaction, presumably due to catalyst poisoning by CO formed by CO_2 reduction. Surface coverage by CO was confirmed by in situ Fourier transform infrared (FTIR) spectroscopy. The best results in the presence of ethane were almost identical to those in toluene and superior to those in ethanol. The enantioselectivity in ethane increased with increased H_2 pressure but leveled off at ~60 bar H_2. The mole ratio of H_2: ethane in these experiments was 1 : 3, too high a concentration of H_2 to allow the substrate to dissolve in the supercritical phase. A later phase behavior study (108) showed that much lower H_2 concentrations were needed in order to have a single-phase reaction mixture. Ethyl pyruvate hydrogenation was then investigated in ethane under conditions known to produce a single-phase reaction mixture (108). In general, the switch to single-phase conditions and the resulting elimination of the gas–liquid interface caused an increase in the reaction rate (but only at low H_2 concentrations) and a mild decrease in e.e. A continuous-flow system was developed, taking advantage of the fact that the reaction in ethane was considerably faster than the reaction in the traditional solvent toluene (109).

Alcohol-dehydrogenase catalyzed hydrogenation of acetophenone derivatives [Eq. (21)] was described by Matsuda et al. (110):

$$\text{(21)}$$

Enantioselectivities were high, 97% and above, with the S isomer being the major product. The catalyst was the whole resting cell of the fungus *Geotrechum candidum* immobilized on a water-absorbing polymer.

Imines can also be hydrogenated enantioselectively in scCO$_2$. Leitner's group (111) reported that a cationic iridium complex with the BARF anion

$B(C_6H_3(CF_3)_2)_4^-$, was very active in $scCO_2$ for imine hydrogenation [Eq. (22)]:

$$(22)$$

In fact, the rate was dramatically greater than that in the conventional solvent CH_2Cl_2, making it possible to use significantly lower catalyst loadings.

2.4. Hydroformylation

Hydroformylation could potentially benefit kinetically from the elimination of gas mass transfer limitations when a SCF is the solvent, at least for those cases that are greater than zero order in H_2 or CO concentration. Studies of nonasymmetric hydroformylations in $scCO_2$ have found significant rate enhancements (112,113).

Hydroformylation in SCFs was first studied by Rathke and Klingler from 1991 (114–116), with the first asymmetric examples being reported by Leitner's group in 1998 (117). They noted that enantioselectivity for the hydroformylation of styrene with an in situ catalyst formed from [Rh(CO)$_2$(acac)] and R,S-BINAPHOS [L^2 in Eq. (23) and Scheme 1] was far worse in $scCO_2$ than in a liquid phase under a lower pressure of CO_2 gas:

	in benzene		in scCO$_2$	
L	e.e.	b/l	e.e.	b/l
L^2	94	7	5	6-8
L^3	91	13	92	13
L^4	93	9	74	9

$$(23)$$

The explanation offered was that the $scCO_2$ was able to extract a ligand-free, achiral, active species which then rapidly hydroformylated the styrene with no enantioselectivity. Under the lower pressures of CO_2, no catalyst extraction took place and the ligand-bound Rh active species in the liquid phase performed the catalysis with good, but not excellent, enantioselectivity.

Within a year, the Leitner group completed the difficult synthesis of a BINAPHOS-type ligand with fluorinated "ponytail" groups [L^3 in Eq. (23) and Scheme 1] (118). The new ligand was capable of greater regioselectivity for the desired branched aldehyde, whether or not the solvent was $scCO_2$. In this case, the enantioselectivity was as high in $scCO_2$ as it was in traditional solvents. *Para*-isobutylstyrene was hydroformylated in $scCO_2$ ($d = 0.81$) with excellent selectivity, giving ibuprofen (as the aldehyde) in 93% e.e. and 96% regioselectivity (101,118). Vinylacetate was hydroformylated in higher enantioselectivity in $scCO_2$ than it was in benzene [Eq. (24)]:

solvent	e.e.	b/l
benzene	91	11
scCO$_2$	95	12

$$(24)$$

High pressure NMR spectroscopy was used to detect $RhH(CO)(L^3)$ in $scCO_2$ solution, which indicated the high solubility of the complex and demonstrated that CO_2 did not insert into the Rh—H bond under the reaction conditions.

A BINAPHOS ligand with fluorinated ponytails at the naphthyl positions (ligand L^4 in Scheme 1), prepared by Ojima's group, was obviously difficult to synthesize but disappointing when tested in $scCO_2$ [Eq. (23)], possibly due to racemization (119).

2.5. Oxidation

Although asymmetric oxidations have not yet been performed in SCFs, there is an example of an asymmetric epoxidation of allylic alcohols in liquid CO_2; this reaction requires temperatures lower than the critical temperature of CO_2 (120). Tumas' group (121) reported that Eq. (25), when performed in liquid CO_2 at 0°C, gave the epoxide in 87% e.e., lower than that obtained in CH_2Cl_2:

$$(25)$$

Diastereoselective oxidations and epoxidations have also been reported (122,123) but fall outside the scope of this review. Nevertheless, it is worthwhile noting that a remarkable enhancement of diastereoselectivity was observed when Oakes et al. (123) used $scCO_2$ instead of traditional solvents for the ion-exchange resin-catalyzed oxidation of thioethers such as an S-methyl cysteine methyl ester [Eq. (26)]:

$$(26)$$

No diastereoselectivity was observed in toluene or CH_2Cl_2 but 95% d.e. (diastereomeric excess) was obtained in $scCO_2$ at just under 200 bar and 40°C. The reason for the enormous change in selectivity is not known.

2.6. Ring-Forming Reactions

Cyclopropanation enantioselectivity was found by Wynne et al. (28,29) to be high and pressure independent in $scCO_2$ but lower and pressure dependent in $scCHF_3$ [Eq. (27)]:

$$(27)$$

The pressure dependence was explained by the fact that $scCHF_3$ is a more polar solvent (has a higher dielectric constant) at higher pressures than it is at lower pressures. The enantioselectivity was better in nonpolar CHF_3 than in polar CHF_3, consistent with the observation that the selectivity is better in nonpolar liquid solvents than in polar liquids. The pressure independence of the enantioselectivity in $scCO_2$ is due to the fact that the dielectric constant of that SCF does not vary significantly with pressure. Cyclopropanations of this type can be used in the synthesis of the antidepressant sertraline (124) and of cilastatin, a stabilizer of the antibiotic imipenem (125).

    The preparation of azirine rings using a modified Neber reaction (an asymmetric deprotonation) was studied by Brown and Jessop (unpublished data, 2002). Asymmetric deprotonations are particularly awkward in SCFs because the strong bases typically used react with $scCO_2$ or $scCHF_3$. However, the Neber reaction [Eq. (28)] was found to proceed in high conversion in $scCO_2$ with a chiral organic base, dihydroquinidine, in either stoichiometric quantities or as a catalytic base (with K_2CO_3 as the stoichiometric base):

$$\text{(28)}$$

The e.e. for the stoichiometric reaction was 60% in $scCO_2$ (200 bar with 7 vol% toluene cosolvent) and 70% in liquid toluene (126; Brown and Jessop, unpublished data, 2002). Without toluene cosolvent, the reaction had poor enantioselectivity. Several azirine carboxylic acids and esters related to the product are antibiotics (126).

2.7. Coupling Reactions

Jessop et al. (127) studied the pressure dependence of the aminoalcohol-catalyzed alkylation of benzaldehydes (Eq. (29), R = H or CF_3] in $scCHF_3$ because of the known variability of the dielectric constant of $scCHF_3$ (see Section 2.6):

$$\text{(29)}$$

The enantioselectivity of the alkylation in traditional solvents is greater in nonpolar media than in polar media, but the enantioselectivity in $scCHF_3$ was greater in polar CHF_3 (86–89% e.e. at 59–209 bar) than in nonpolar CHF_3 (77–79% e.e. at 49–52 bar). Calculations by Tucker's group (128) showed that electrostriction at the lower pressures could be ruled out as an explanation, but the investigators were not able to determine the true cause in this very complicated system. The reaction had excellent enantioselectivity in scC_2H_6 (98% e.e.).

A titanium-binaphthol-catalyzed Mukaiyama Aldol reaction [Eq. (30)] was found by Mikami et al. (129) to be more enantioselective in $scCHF_3$ (88%

e.e.) than in $scCO_2$ (72%) or toluene (72%) at 34°C but not as selective as in toluene at 0°C (92% e.e.):

$$(30)$$

The asymmetric hydrovinylation of styrene [Eq. (31)] was found by Wegner and Leitner (130) to be slightly more selective in liquid or supercritical CO_2 (86% e.e. at 1°C) than in CH_2Cl_2 (85%):

$$(31)$$

A strong anion dependence was observed; far better enantioselectivity was observed when BArF was used as the counterion, regardless of solvent.

3. RELATED TECHNIQUES

3.1. Expanded Liquids

Application of subcritical gaseous CO_2 to an organic liquid causes the liquid phase to expand noticeably, due to extensive dissolution of the CO_2 into the liquid phase (131). This expansion is accompanied by a reduction in the liquid phase viscosity, an increase in the solubility of H_2 in the liquid, and an increase in the mass transfer rates from the gas to liquid phase. There is evidence that this can affect the enantioselectivity of reactions in viscous liquids. The enantioselectivity of asymmetric hydrogenation of unsaturated carboxylic acids in a viscous ionic liquid was shown to be strongly affected by CO_2 expansion of the liquid, the enantioselectively being improved for one substrate (atropic acid) and decreased for another (tiglic acid). The results were explained in terms of the solubility and rate of transfer of H_2 gas into the expanded ionic liquid (23). The same effect was not observed in expanded methanol.

3.2. Induced Melts

The presence of gaseous (subcritical) CO_2 above a solid compound can cause the compound to melt at a temperature 20 or more degrees below its normal melting point. This phenomenon can make it possible to perform solventless reactions at temperatures at which the substrate would normally be solid (132). The method, which requires much less CO_2 pressure than supercritical techniques, has been used for nonasymmetric hydrogenations and hydroformylations (132) and for an asymmetric hydroformylation (101).

3.3. Catalyst Recovery and Recycling

Possibly the most exciting area of current SCF research for homogeneous catalysis is the search for industrially viable catalyst recycling techniques. Homogeneous catalysis is, in general, preferable to heterogeneous catalysis for asymmetric synthesis because of the greater enantioselectivity, but non-recoverability of the catalyst is a source of product contamination and expense related to catalyst replacement. Resolved chiral ligands can be more expensive than the precious metals they bind. Use of extremely high substrate/catalyst ratios is not satisfactory if the rate under such conditions is too low due to dilution. Therefore, catalyst recycling is particularly important for asymmetric homogeneous catalysis. It has been shown in the past 2 or 3 years that $scCO_2$ can contribute in several ways to the development of solutions to the problem.

Leitner reported that lower densities of $scCO_2$ can be used to extract product from a reaction mixture without coextraction of a homogeneous catalyst. This process, which he called CESS (catalysis and extraction using supercritical solution), was demonstrated with styrene hydroformylation [Eq. (23), ligand L^3] and imine hydrogenation [Eq. (22)]. The catalyst was used for three cycles before drops in the conversion and selectivity was observed (101,111,133). Sellin and Cole-Hamilton used catalysts specifically chosen for their insolubility in $scCO_2$ to facilitate the catalyst/product separation in a CESS process (134).

Biphasic catalysis can also be used as a means of facilitating post-reaction separation of catalyst from product. In a biphasic catalysis scheme, two mutually immiscible solvents are selected, such that the catalyst is soluble in one and the organic product is soluble in the other. After the reaction, the product-bearing phase can be removed and the catalyst-bearing phase can be reused. Industrially, aqueous/organic biphasic media are used for homogeneously catalyzed hydroformylation (135). There are three corresponding techniques involving a supercritical phase: SCF/aqueous, SCF/IL and SCF/PEG biphasic systems [IL = ionic liquid, PEG = poly(ethylene glycol)]. Of these, only SCF/aqueous biphasic catalysis is currently practiced industrially—in the nonasymmetric hydration of supercritical alkenes (1). No examples of SCF/aqueous biphasic asymmetric catalysis have been reported, although there are nonasymmetric examples (136–139). Although this solvent system is limited by the low pH and the inability of some substrates to dissolve in the aqueous phase, it is obviously inexpensive and environmentally benign. Therefore, asymmetric examples are anticipated in the future.

After the complete insolubility of ionic liquids in scCO$_2$ was reported by Brennecke and colleagues (140–142), it became obvious that the combination of ILs and scCO$_2$ would be ideal for biphasic catalysis. The Jessop group described the enantioselective hydrogenation of atropic acid, *p*-isobutylatropic acid (giving ibuprofen), and tiglic acid [Eq. (18)] in ILs, obtaining e.e.'s up to 95%, the selectivity being strongly dependent on the H$_2$ pressure and choice of IL anion (23,143). After the hydrogenation of tiglic acid, extraction of the product with scCO$_2$ was efficient, and no IL or catalyst was observed in the extracted product. Repeated recycling of the catalyst/IL solution by addition of fresh tiglic acid and H$_2$ lead to 97% conversion of the substrate each cycle, even after five cycles. In SCF/IL biphasic systems, it is not necessary to modify the catalyst with sulfonated or fluorous groups, as there is with aqueous or fluorous biphasic solvent systems, although the use of sulfonated (i.e., charged) phosphines may assist in preventing catalyst leaching into the SCF phase.

A continuous-flow method for asymmetric catalysis in an SCF/IL system was reported by Leitner's group (144), with the hydrovinylation of styrene [Eq. (31)] as the test reaction. The scCO$_2$ solution of styrene and ethylene was continuously bubbled up through a column of ionic liquid containing the catalyst. The enantioselectivity was found to be high (in one of the ILs) and catalyst stability was enhanced due to the fact that there was a constant concentration of substrate in the system; the catalyst was unstable in ILs in the absence of the olefins.

The extension of this method to the enzyme-catalyzed kinetic resolution of 1-phenylethanol by transesterification [Eq. (7)] was published by two groups nearly simultaneously (145,146).

Another promising biphasic catalysis method is the SCF/PEG system, in which catalysis takes place in liquid PEG and the product is subsequently extracted by scCO$_2$, leaving the catalyst in the PEG phase. The PEG-bearing catalyst can be reused repeatedly in this manner. This system has the great advantages that PEG is nontoxic (it is approved as a food additive in the United States), very inexpensive, and able to dissolve organic substrates. This system has only very recently been tested by the Jessop group using a nonasymmetric reaction (147), but further investigations are underway.

Another method for catalyst recovery and recycling is the use of supported, immobilized, or membrane-trapped (148) catalysts. Examples of the use of immobilized enzymes for asymmetric reactions in SCFs have been reported [Eqs. (1), (3)–(6), (13), and (21)]. There have not yet been any examples, to my knowledge, of the use of supported or "heterogenized" complexes or membrane-trapped catalysts for asymmetric reactions in SCFs, although these approaches are likely to be very promising.

4. OUTLOOK AND CONCLUSIONS

The literature summarized here has shown that asymmetric catalysis, whether by enzyme, heterogeneous, or homogeneous catalysis, can be performed in SCFs and that significant advantages can be obtained by using these unusual solvents. Improved enantioselectivity, enhanced rates, and increased catalyst stability have all been observed in various systems. Greater enantioselectivity has been observed in asymmetric hydrogenations [Eqs. (16) and (19)] and a hydroformylation [Eq. (24)]. Examples of greater rates include homogeneously and heterogeneously catalyzed hydrogenations [Eqs. (19), (20), and (22)] and enzyme-catalyzed esterification and hydrolysis [Eqs. (5) and (13)]. Greater stability was observed with a protease catalyst for hydrolysis and a lipase catalyst for esterification [Eqs. (11) and (5)]. These reaction performance enhancements should be sufficient motivation for industry to adopt SCFs as solvents for pharmaceutical syntheses, particularly for hydrogenations and hydroformylations.

The invention of new methods for catalyst recovery appear likely to further increase the attractiveness of $scCO_2$ as a reaction medium, potentially in partnership with a second phase such as water, ionic liquid, or PEG. Given the high price of chiral homogeneous catalysts and the particularly clean separations that can be obtained using the biphasic catalysis techniques described in Section 3.3, one can expect industrial interest in this aspect in particular.

Supercritical fluids used in asymmetric catalysis so far have included CO_2, ethane, and fluoroform. For cost and environmental impact reasons, the use of fluoroform will forever remain restricted to the research laboratory, where it can serve the interests of basic science. Ethane finds use in the laboratory and might find use in industry as a SCF for catalysts that are deactivated by CO_2. However, due to the combined risk of high pressure and high flammability, the use of scC_2H_6 is not expected to be as commercially viable as the use of $scCO_2$. The latter solvent not only is not flammable, but it greatly reduces the explosion and fire risk associated with hydrogenations and oxidations of organic substrates.

The reactions that have so far been tested in SCFs were chosen by researchers to demonstrate the capabilities of the method. These test reactions should not be taken as representative of all the reactions that are possible in SCFs. The field is still young. Similarly, although a number of drug products or intermediates were synthesized in SCFs in the work summarized here, the true industrial potential of the method may lie in syntheses of products that have not yet been prepared in SCFs.

Academic researchers can continue to contribute to this field, not necessarily by testing more and more reactions in SCFs but by exploring and defining which reaction types show enhanced reaction rates and selectivities, by evaluating the use of "heterogenized" or membrane-trapped asymmetric complex catalysts in SCFs, by developing more solvents that can be coupled with $scCO_2$ for biphasic catalysis (such as less expensive, less polar, less viscous, and patent-free ionic liquids), by exploring supercritical and expanded-liquid solvent effects on reaction selectivity, and by developing inexpensive and safe methods for low-temperature reactions in SCFs.

Factors which stand in the way of industrial utilization of SCFs as media for catalysis include the cost of compression, the capital cost of pressurized system construction, and the perception among corporate managers that SCF technology is new and untested. As more and more plants using SCFs as reaction or extraction media come on-line, the perception problem should fade. Industrial experience with ethylene polymerization, ammonia synthesis, and $scCO_2$-based extractions have shown that high pressure processes can be economically successful. The academic community demonstrated that the advantages of SCFs can be realized for asymmetric catalysis. It is now up to the industrial fine chemicals sector to take advantage of these discoveries.

ACKNOWLEDGMENTS

This review includes material based on work supported by the National Science Foundation under Grant No. 9815320. The author also gratefully acknowledges support from the Division of Chemical Sciences, Office of Basic Energy Sciences, Office of Science, U.S. Department of Energy (Grant No. DE-FG03-99ER14986). This support does not constitute an endorsement by DOE of the views expressed in this chapter.

REFERENCES

1. Jessop PG, Leitner W. Supercritical fluids as media for chemical reactions. In: Jessop PG, Leitner W, eds. Chemical Synthesis Using Supercritical Fluids. Weinheim, Germany: VCH–Wiley, 1999:1–36.
2. Jessop PG, Leitner W, eds. Chemical Synthesis Using Supercritical Fluids. Weinheim, Germany: VCH–Wiley, 1999.
3. Daubrée. Sur le métamorphisme et recherches expérimentales sur quelques-uns des agents qui ont fu le produire. Ann Mines 1857; 12(5):289–326.

4. Villard P. Dissolution des liquides et des solides dans les gaz. Seances Soc Fr Phys 1896:234–242.

5. Villard P. The solution of solids and liquids in gases. Chem News 1898; 78:297–298, 309–310.

6. Ehrlich P, Mortimer GA. Fundamentals of the free-radical polymerization of ethylene. Adv Polym Sci 1970; 7:386–448.

7. Martins JF, Sampaio TC, Carvalho IB, da Ponte MN, Barreiros S. Lipase-catalyzed esterification of glycidol in chloroform and in supercritical carbon dioxide. In: Balny C, Hayashi R, Heremans K, Masson P, eds. High Pressure and Biotechnology. London: J. Libbey, 1992:411–415.

8. Aaltonen O, Rantakylä M. Process for producing chiral compounds as pure optical isomers. PCT Int Appl, 1992:14. WI 92 20,812.

9. Bornscheuer U, Capewell A, Scheper T, Meyer HH, Kolisis F. A comparison of enzymatic reactions in aqueous, organic and supercritical phases. Ann N Y Acad Sci 1992; 672:336–342.

10. Levelt Sengers JMH, Kamgar-Parsi B, Balfour FW, Sengers JV. Thermodynamic properties of steam in the critical region. J Phys Chem Ref Data 1983; 12:1.

11. Rabinovich VA, Selover TB. Thermophysical Properties of Neon, Argon, Krypton, and Xenon. Washington, DC: Hemisphere, 1988.

12. Wagner W, de Reuck KM, eds. International Thermodynamic Tables of the Fluid State. Vol. 13. Methane. Oxford: IUPAC/Pergamon Press, 1996.

13. Setzmann U, Wagner W. A new equation of state and tables of thermodynamic properties for methane covering the range from the melting line to 625 K at pressures up to 1000 MPa. J Phys Chem Ref Data 1991; 20:1061.

14. Stewart RB, Jacobsen RT. Thermodynamic properties of argon from the triple point to 1200 K with pressures to 1000 MPa. J Phys Chem Ref Data 1989; 18:639.

15. Kita T, Uosaki Y, Moriyoshi T. Static relative permittivity of some compressed fluids. In: Taniguchi Y, Senoo M, Hara K, eds. High Pressure Liquids and Solutions. Amsterdam: Elsevier, 1994:181–198.

16. Makita T, Kubota H, Tanaka Y, Kashiwagi H. Dielectric constants of refrigerants R-12, R-13, R-22, and R-23. Gakujutsu Koenkai Koen Rombunshu. Refrigeration Tokyo Nippon Reito Kyokui 1976; 52:543.

17. Reuter K, Rosenzweig S. Franck EU. The static dielectric constant of CH_3F and CHF_3 to 468 K and 2000 bar. Physica A 1989; 156:294–302.

18. Hansen BN, Hybertson BM, Barkley RM, Sievers RE. Supercritical fluid transport-chemical deposition of films. Chem Mater 1992; 4:749–752.

19. Raynie DE. Warning concerning use of nitrous oxide in supercritical fluid extractions. Anal Chem 1993; 65:3127–3128.

20. Savage PE, Gopalan S, Mizan TI, Martino CJ, Brock EE. Reactions at supercritical conditions: applications and fundamentals. AIChE J 1995; 41: 1723–1778.

21. Jessop PG, Hsiao Y, Ikariya T, Noyori R. Homogeneous catalysis in supercritical fluids: hydrogenation of supercritical carbon dioxide to formic acid, alkyl formates, and formamides. J Am Chem Soc 1996; 118:344–355.

22. Thomas CA, Bonilla RJ, Huang Y, Jessop PG. Hydrogenation of carbon dioxide catalysed by ruthenium trimethylphosphine complexes: effect of gas pressure and additives on rate in the liquid phase. Can J Chem 2001; 79:719–724.

23. Jessop PG, Stanley R, Brown RA, Eckert CA, Liotta CL, Ngo TT, Pollet P. Comparing neoteric solvents for asymmetric hydrogenation: supercritical fluids, ionic liquids, and expanded ionic liquids. Green Chem 2003; 5:123–128.

24. Chang CJ, Chiu K-L, Day C-Y. A new apparatus for the determination of P-x-y diagrams and Henry's constants in high pressure alcohols with critical carbon dioxide. J Supercrit Fluids 1998; 12:223–237.

25. Kamat SV, Iwaskewycz B, Beckman EJ, Russell AJ. Biocatalytic synthesis of acrylates in supercritical fluids: tuning enzyme activity by changing pressure. Proc Nat Acad Sci USA 1993; 90:2940–2944.

26. Kamat SV, Beckman EJ, Russell AJ. Control of enzyme enantioselectivity with pressure changes in supercritical fluoroform. J Am Chem Soc 1993; 115:8845–8846.

27. Mori T, Funasaki M, Kobayashi A, Okahata Y. Reversible activity changes of a lipid-coated lipase for enantioselective esterification in supercritical fluoroform. Chem Commun, 2001:1832–1833.

28. Wynne D, Olmstead MM, Jessop PG. Supercritical and liquid solvent effects on the enantioselectivity of asymmetric cyclopropanation with tetrakis[1-[(4-tert-butylphenyl)sulfonyl]-(2S)-pyrrolidinecarboxylate]dirhodium(II). J Am Chem Soc 2000; 122:7638–7647.

29. Wynne D, Jessop PG. Cyclopropanation enantioselectivity is pressure dependent in supercritical fluoroform. Angew Chem Int Ed Engl 1999; 38:1143–1144.

30. Fürstner A, Koch D, Langemann K, Leitner W, Six C. Olefin metathesis in compressed carbon dioxide. Angew Chem, Int Ed Engl 1997; 36:2466–2469.

31. Wittmann K, Wisniewski W, Mynott R, Leitner W, Kranemann CL, Rische T, Eilbracht P, Kluwer S, Ernsting JM, Elsevier CL. Supercritical carbon dioxide as solvent and temporary protecting group for rhodium-catalyzed hydroaminomethylation. Chem Eur J 2001; 7:4584–4589.

32. Dandge DK, Heiler JP, Wilson KV. Structure solubility correlations: organic compounds and dense carbon dioxide binary systems. Ind Eng Chem Prod Res Dev 1985; 24:162–166.

33. Fornari RE, Alessi P, Kikic I. High pressure fluid phase equilibria—experimental methods and systems investigated (1978–1987). Fluid Phase Equilibria 1990; 57:1–33.

34. Bartle KD, Clifford AA, Jafar SA, Shilstone GF. Solubilities of solids and liquids of low volatility in scCO$_2$. J Phys Chem Ref Data 1991; 20:713–756.

35. Dohrn R, Brunner G. High-pressure fluid-phase equilibria—experimental methods and systems investigated (1988–1993). Fluid Phase Equilibria 1995; 106:213–282.

36. Fernandez-Prini R, Japas ML. Chemistry in near-critical fluids. Chem Soc Rev 1994; 23:155–163.

37. Fink R, Beckman EJ. High-pressure reaction equipment design. In: Jessop PG,

Leitner W, eds. Chemical Synthesis Using Supercritical Fluids. Weinheim, Germany: VCH–Wiley, 1999:67–87.

38. Johnston KP, Jacobsen GB, Lee CT, Meredith C, Da Rocha SRP, Yates MZ, DeGrazia J, Randolph TW. Microemulsions, emulsions, and latexes. In: Jessop PG, Leitner W, eds. Chemical Synthesis Using Supercritical Fluids. Weinheim, Germany: VCH–Wiley, 1999:127–146.

39. Rathke JW, Klingler RJ, Gerald RE, Fremgen DE, Woelk K, Gaemers S, Elsevier CJ. NMR spectroscopy. In: Jessop PG, Leitner W, eds. Chemical Synthesis Using Supercritical Fluids. Weinheim, Germany: Wiley–VCH, 1999:165–194.

40. Howdle SM, George MW, Poliakoff M. Vibrational spectroscopy. In: Jessop PG, Leitner W, eds. Chemical Synthesis Using Supercritical Fluids. Weinheim, Germany: Wiley–VCH, 1999:147–164.

41. Yonker CR, Linehan JC, Fulton JL. UV, EPR, X-ray and related spectroscopic techniques. In: Jessop PG, Leitner W, eds. Chemical Synthesis Using Supercritical Fluids. Weinheim, Germany: Wiley–VCH, 1999:195–212.

42. Scholsky KM. Polymerization reactions at high pressure and supercritical conditions. J Supercrit Fluids 1993; 6:103–128.

43. Kendall JL, Canelas DA, Young JL, DeSimone JM. Polymerizations in supercritical carbon dioxide. Chem Rev 1999; 99:543–563.

44. Beuermann S, Buback M, Busch M. Free-radical polymerization in reactive supercritical fluids. In: Jessop PG, Leitner W, eds. Chemical Synthesis Using Supercritical Fluids. Weinheim, Germany: Wiley–VCH, 1999:326–350.

45. Cooper AI. Polymer synthesis and processing using supercritical carbon dioxide. J Mater Chem 2000; 10:207–234.

46. Davidson TA, DeSimone JM. Polymerizations in dense carbon dioxide. In: Jessop PG, Leitner W, eds. Chemical Synthesis Using Supercritical Fluids. Weinheim, Germany: Wiley–VCH, 1999:297–325.

47. Yalpani M. Supercritical fluids: puissant media for the modification of polymers and biopolymers. Polymer 1993; 34:1102.

48. Friedmann G, Guilbert Y, Catala JM. Chemical modification of polymers in the presence of supercritical carbon dioxide—Grafting of isocyanato-isopropyl groups onto a poly(ethylene-co-vinyl alcohol) chain. Eur Polym J 2000; 36:13–20.

49. Muth O, Hirth T, Vogel H. Polymer modification by supercritical impregnation. J Supercrit Fluids 2000; 17:65–72.

50. Howdle SM, Watson MS, Whitaker MJ, Popov VK, Davies MC, Mandel FS, Wang JD, Shakesheff KM. Supercritical fluid mixing: preparation of thermally sensitive polymer composites containing bioactive materials. Chem Commun 2001:109–110.

51. Whitaker MJ, Quirk RA, Howdle SM, Shakesheff KM. Growth factor release from tissue engineering scaffolds. J Pharma Pharmacol 2001; 53:1427–1437.

52. Watson MS, Whitaker MJ, Howdle SM, Shakesheff KM. Incorporation of proteins into polymer materials by a novel supercritical fluid processing method. Adva Mate 2002; 14:1802–1804.

53. Oakes RS, Clifford AA, Rayner CM. The use of supercritical fluids in synthetic organic chemistry. J Chem Soc Perkin Trans 2001; 1:917–941.
54. Fogassy E, Acs M, Szili T, Simandi B, Sawinsky J. Molecular chiral recognition in supercritical solvents. Tetrahedron Lett 1994; 35:257–260.
55. Kordikowski A, York P, Latham D. Resolution of ephedrine in supercritical CO$_2$: a novel technique for the separation of chiral drugs. J Pharm Sci 1999; 88:786–791.
56. Simándi B, Keszei S, Fogassy E, Kemény S, Sawinsky J. Separation of enantiomers by supercritical fluid extraction. J Supercrit Fluids 1998; 13:331–336.
57. Tiltscher H, Wolf H, Schelchshorn J. A mild and effective method for reactivation or maintenance of activity of heterogeneous catalysts. Angew Chem Int Ed Engl 1981; 20:892–894.
58. Baiker A. Supercritical fluids in heterogeneous catalysis. Chem Rev 1999; 99:453–473.
59. Ipatiev V, Rutala O. Polymerization of ethylene at a high temperature and pressure in the presence of catalyzers. Berichte 1913; 46:1748–1755.
60. Burk MJ, Feng S, Gross MF, Tumas W. Asymmetric catalytic hydrogenation reactions in supercritical carbon dioxide. J Am Chem Soc 1995; 117:8277–8278.
61. Schmitt WJ, Reid RC. Solubility of paraffinic hydrocarbons and their derivatives in supercritical carbon dioxide. Chem Eng Commun 1988; 64:155–176.
62. Wagner KD, Dahmen N, Dinjus E. Solubility of triphenylphosphine, tris(p-fluorophenyl)phosphine, tris(pentafluorophenyl)phosphine, and tris(p-trifluoromethylphenyl)phosphine in liquid and supercritical carbon dioxide. J Chem Eng Data 2000; 45:672–677.
63. Palo DR, Erkey C. Solubility of dichlorobis(triphenylphosphine)nickel(II) in supercritical carbon dioxide. J Chem Eng Data 1998; 43:47–48.
64. Jessop PG, Ikariya T, Noyori R. Homogeneous catalytic hydrogenation of supercritical carbon dioxide. Nature 1994; 368:231–333.
65. Kainz S, Koch D, Baumann W, Leitner W. Perfluoroalkyl-substituted arylphosphanes as ligands for homogeneous catalysis in supercritical carbon dioxide. Angew Chem Int Ed Engl 1997; 36:1628–1630.
66. Hammond DA, Karel M, Klibanov AM, Krukonis VJ. Enzymatic reactions in supercritical gases. Appl Biochem Biotechnol 1985; 11:393–400.
67. Randolph TW, Blanch HW, Prausnitz JM, Wilke CR. Enzyme catalysis in a supercritical fluid. Biotechnol Lett 1985; 7:325–328.
68. Nakamura K, Chi YM, Yamada Y, Yano T. Lipase activity and stability in supercritical carbon dioxide. Chem Eng Commun 1985; 45:207–212.
69. Kamat SV, Beckman EJ, Russell AJ. Enzyme activity in supercritical fluids. Crit Rev Biotechnol 1995; 15:41–71.
70. Mesiano AJ, Beckman EJ, Russell AJ. Supercritical biocatalysis. Chem Rev 1999; 99:623–633.
71. Aaltonen O. Enzymatic catalysis. In: Jessop PG, Leitner W, eds. Chemical Synthesis Using Supercritical Fluids. Weinheim, Germany: Wiley–VCH, 1999:414–445.

72. Russell AJ, Beckman EJ. Should high diffusivity of SCF increase the rate of enzyme catalyzed reaction? Enzyme Microb Technol 1991; 13:1007.

73. Kamat S, Barrera J, Beckman EJ, Russell AJ. Biocatalytic synthesis of acrylates in organic solvents and supercritical fluids 1. Optimization of enzyme environment. Biotechnol Bioeng 1992; 40:158–166.

74. Rantakyla M, Aaltonen O. Enantioselective esterification of ibuprofen in supercritical carbon dioxide by immobilized lipase. Biotechnol Lett 1994; 16:825–830.

75. Aaltonen O, Rantakylä M. Lipase catalyzed reactions of chiral compounds in supercritical carbon dioxide. Proceedings of the 2nd International Symposium on Supercritical Fluids, Boston, 1991:146–149.

76. Overmeyer A, Schrader-Lippelt S, Kasche V, Brunner G. Lipase-catalysed kinetic resolution of racemates at temperatures from 40°C to 160°C in supercritical CO_2. Biotechnol Lett 1999; 21:65–69.

77. Wu JY, Liang MT. Enhancement of enantioselectivity by altering alcohol concentration for esterification in supercritical CO_2. J Chem Eng Jpn 1999; 32:338–340.

78. Ikushima Y, Saito N, Yokoyama T, Hatakeda K, Ito S, Arai M, Blanch HW. Solvent effects on enzymatic ester synthesis in supercritical carbon dioxide. Chem Lett 1993:109–112.

79. Martins JF, de Carvalho IB, de Sampaio TC, Barreiros S. Lipase-catalyzed enantioselective esterification of glycidol in supercritical carbon dioxide. Enzyme Microb Technol 1994; 16:785–790.

80. Cernia E, Palocci C, Gasparrini F, Misiti D, Fagnano N. Enantioselectivity and reactivity of immobilized lipase in supercritical carbon dioxide. J Mol Catal 1994; 89:L11–L18.

81. Bornscheuer U, Capewell A, Wendel V, Scheper T. On-line determination of the conversion in a lipase-catalyzed kinetic resolution in supercritical carbon dioxide. J Biotechnol 1996; 46:139–143.

82. Capewell A, Wendel V, Bornscheuer U, Meyer HH, Scheper T. Lipase-catalyzed kinetic resolution of 3-hydroxy esters in organic solvents and supercritical carbon dioxide. Enzyme Microb Technol 1996; 19:181–186.

83. Fontes N, Almeida MC, Peres C, Garcia S, Grave J, Aires-Barros MR, Soares CM, Cabral JMS, Maycock CD, Barreiros S. Cutinase activity and enantioselectivity in supercritical fluids. Ind Eng Chem Res 1998; 37:3189–3194.

84. Chrisochoou A, Stephan K, Winkler S, Schaber K. Chem Ing Technol 1995; 67:1153.

85. Chrisochoou AA, Schaber K, Stephan K. Phase equilibria with supercritical carbon dioxide for the enzymatic production of an enantiopure pyrethroid component. 1. Binary systems. J Chem Eng Data 1997; 42:551–557.

86. Chrisochoou AA, Schaber K, Stephan K. Phase equilibria with supercritical carbon dioxide for the enzymatic production of an enantiopure pyrethroid component. 2. Ternary and five-component systems. J Chem Eng Data 1997; 42:558–561.

87. Budavari S, O'Neil MJ, Smith A, Heckelman PE, Kinneary JF, eds. Merck Index. Whitehouse Station, NJ: Merck & Co., 1996.

88. Chen ST, Tsai CF, Wang KT. Resolution of N-protected amino acid derivatives in supercritical carbon dioxide catalyzed by alcalase. Bioorg Med Chem Lett 1994; 4:625–630.

89. Parve O, Vallikivi I, Lahe L, Metsala A, Lille U, Tougu V, Vija H, Pehk T. Lipase-catalysed enantioselective hydrolysis of bicyclo[3.2.0]heptanol esters in supercritical carbon dioxide. Bioorg Med Chem Lett 1997; 7:811–816.

90. Newton RF. Recent syntheses of prostaglandins via polycyclic intermediates. In: Roberts SM, Scheinmann F, eds. New Synthetic Routes to Prostaglandins and Thromoxanes. London: Academic Press, 1982:61.

91. Rantakyla M, Alkio M, Aaltonen O. Stereospecific hydrolysis of 3-(4-methoxyphenyl)glycidic ester in supercritical carbon dioxide by immobilized lipase. Biotechnol Lett 1996; 18:1089–1094.

92. Glowacz G, Bariszlovich M, Linke M, Richter P, Fuchs C, Mörsel JT. Stereoselectivity of lipases in supercritical carbon dioxide. 1. dependence of the regio- and enantioselectivity of porcine pancreas lipase on the water content during the hydrolysis of triolein and its partial glycerides. Chem Phys Lipids 1996; 79:101–106.

93. Michor H, Gamse T, Marr R. Enzyme catalysis in supercritical carbon dioxide: racemate separation of D,L-menthol. Chem Ing Technol 1997; 69:690–694.

94. Tsang CY, Streett WB. Phase equilibria in H_2/CO_2 system. Chem Eng Sci 1981; 36:993–1000.

95. Ipatiev V. Catalytic reactions under pressure at high temperatures. Reduction catalysis. Berichte 1907; 40:1270–1281.

96. Coenen H, Hagen R, Kriegel E. Supercritical extraction and simultaneous catalytic hydrogenation of coal. Fried Krupp Gesellschaft, U.S.A., US 4,485,003, 1984.

97. Sun Y, Landau RN, Wang J, LeBlond C, Blackmond DG. A re-examination of pressure effects on enantioselectivity in asymmetric catalytic hydrogenation. J Am Chem Soc 1996; 118:1348–1353.

98. Minder B, Mallat T, Pickel KH, Steiner K, Baiker A. Enantioselective hydrogenation of ethyl pyruvate in supercritical fluids. Catal Lett 1995; 34: 1–9.

99. Minder B, Mallat T, Baiker A. Enantioselective hydrogenation in supercritical fluids. Limitations of the use of supercritical CO_2. In: von Rohr PR, Trepp C, eds. High Pressure Chemical Engineering: Proceedings of the 3rd International Symposium on High Pressure Chemical Engineering, Zurich, Switzerland, 7–9 October, 1996. Amsterdam: Elsevier, 1996: 139–144.

100. Lange S, Brinkmann A, Trautner P, Woelk K, Bargon J, Leitner W. Mechanistic aspects of dihydrogen activation and transfer during asymmetric hydrogenation in supercritical carbon dioxide. Chirality 2000; 12:450–457.

101. Francio G, Wittmann K, Leitner W. Highly efficient enantioselective catalysis in supercritical carbon dioxide using the perfluoroalkyl-substituted ligand (RS)-3-(HF$_6$)-F-2-BINAPHOS. J Organomet Chem 2001; 621:130–142.

102. Noyori R. Asymmetric Catalysis in Organic Synthesis. New York: John Wiley & Sons, 1994.

103. Xiao J, Nefkens SCA, Jessop PG, Ikariya T, Noyori R. Asymmetric hydrogenation of alpha, beta-unsaturated carboxylic acids in supercritical carbon dioxide. Tetrahedron Lett 1996; 37:2813–2816.

104. Uemura T, Zhang X, Matsumura K, Sayo N, Kumobayashi H, Ohta T, Nozaki K, Takaya H. Highly efficient enantioselective synthesis of optically active carboxylic acids by $Ru(OCOCH_3)_2[(S)-H_8$-BINAP]. J Org Chem 1996; 61:5510–5516.

105. Ikariya T, Noyori R. Organic reactions in supercritical fluids. In: Murahashi S-I, Davies SG, eds. Transition Metal Catalysed Reactions. Oxford: Blackwell Science, 1999:1–28.

106. Wang SN, Kienzle F. The syntheses of pharmaceutical intermediates in supercritical fluids. Ind Eng Chem Res 2000; 39:4487–4490.

107. Crameri YF, Hengartner U, Jenny C, Kienzle F, Ramuz H, Scalone M, Schlageter M, Schmid R, Wang S. Asymmetric hydrogenation vs. resolution in the synthesis of Posicor, a new type of calcium antagonist. Chimia 1997; 51: 303–305.

108. Wandeler R, Kunzle N, Schneider MS, Mallat T, Baiker A. Continuous enantioselective hydrogenation of ethyl pyruvate in "supercritical" ethane: relation between phase behavior and catalytic performance. J Catal 2001; 200:377–388.

109. Wandeler R, Kunzle N, Schneider MS, Mallat T, Baiker A. Continuous platinum-catalyzed enantioselective hydrogenation in 'supercritical' solvents. Chem Commun, 2001: 673–674.

110. Matsuda T, Harada T, Nakamura K. Alcohol dehydrogenase is active in supercritical carbon dioxide. Chem Commun, 2000:1367–1368.

111. Kainz S, Brinkmann A, Leitner W, Pfaltz A. Iridium-catalyzed enantioselective hydrogenation of imines in supercritical carbon dioxide. J Am Chem Soc 1999; 121:6421–6429.

112. Hu YL, Chen WP, Osuna AMB, Stuart AM, Hope EG, Xiao JL. Rapid hydroformylation of alkyl acrylates in supercritical CO_2. Chem Commun, 2001:725–726.

113. Koch D, Leitner W. Rhodium-catalyzed hydroformylation in supercritical carbon dioxide. J Am Chem Soc 1998; 120:13,398–13,404.

114. Rathke JW, Klingler RJ, Krause TR. Propylene hydroformylation in supercritical carbon dioxide. Organometallics 1991; 10:1350–1355.

115. Rathke JW, Klingler RJ. Cobalt carbonyl catalyzed olefin hydroformylation in supercritical carbon dioxide. U.S.A.: US 5,198,589, 1993:8.

116. Klingler RJ, Rathke JW. High pressure NMR investigation of hydrogen atom transfer and related dynamic processes in oxo catalysis. J Am Chem Soc 1994; 116:4772–4785.

117. Kainz S, Leitner W. Catalytic asymmetric hydroformylation in the presence of compressed carbon dioxide. Catal Lett 1998; 55:223–225.

118. Francio G, Leitner W. Highly regio- and enantio-selective rhodium-catalysed

asymmetric hydroformylation without organic solvents. Chem Commun 1999: 1663–1664.

119. Bonafoux D, Hua ZH, Wang BH, Ojima I. Design and synthesis of new fluorinated ligands for the rhodium-catalyzed hydroformylation of alkenes in supercritical CO_2 and fluorous solvents. J Fluorine Chem 2001; 112:101–108.

120. Katsuki T, Sharpless KB. The first practical method for asymmetric epoxidation. J Am Chem Soc 1980; 102:5974–5976.

121. Pesiri DR, Morita DK, Glaze W, Tumas W. Selective oxidation in dense phase carbon dioxide. Chem Commun 1998:1015–1016.

122. Haas GR, Kolis JW. The diastereoselective epoxidation of olefins in supercritical carbon dioxide. Tetrahedron Lett 1998; 39:5923–5926.

123. Oakes RS, Clifford AA, Bartle KD, Petti MT, Rayner CM. Sulfur oxidation in supercritical carbon dioxide: dramatic pressure dependant enhancement of diastereoselectivity for sulfoxidation of cysteine derivatives. Chem Commun 1999:247–248.

124. Corey EJ, Gant TG. A catalytic enantioselective synthetic route to the important antidepressant sertraline. Tetrahedron Lett 1994; 35:5373–5376.

125. Aratani T. Catalytic asymmetric synthesis of cyclopropanecarboxylic acids; an application of chiral copper carbenoid reaction. Pure Appl Chem 1985; 57: 1839–1844.

126. Verstappen MMH, Ariaans GJA, Zwanenburg B. Asymmetric synthesis of $2H$-azirine carboxylic esters by an alkaloid-mediated Neber reaction. J Am Chem Soc 1996; 118:8491–8492.

127. Jessop PG, Brown RA, Yamakawa M, Xiao JL, Ikariya T, Kitamura M, Tucker SC, Noyori R. Pressure-dependent enantioselectivity in the organozinc addition to aldehydes in supercritical fluids. J Supercrit Fluids 2002; 24:161–172.

128. Parsons DF, Boone BI, Jessop PG, Tucker SC. Electrostriction effects on competing transition states in supercritical fluoroform. J Supercrit Fluids 2002; 24:173–181.

129. Mikami K, Matsukawa S, Kayaki Y, Ikariya T. Asymmetric Mukaiyama aldol reaction of a ketene silyl acetal of thioester catalyzed by a binaphthol–titanium complex in supercritical fluoroform. Tetrahedron Lett 2000; 41:1931–1934.

130. Wegner A, Leitner W. Nickel-catalysed enantioselective hydrovinylation of styrenes in liquid or supercritical carbon dioxide. Chem Commun 1999:1583–1584.

131. Kordikowski A, Schenk AP, Van Nielen RM, Peters CJ. Volume expansions and vapor–liquid equilibria of binary mixtures of a variety of polar solvents and certain near-critical solvents. J Supercrit Fluids 1995; 8:205–216.

132. Jessop PG, DeHaai S, Wynne DC, Nakawatase D. Carbon dioxide gas accelerates solventless synthesis. Chem Commun 2000:693–694.

133. Jessop PG, Leitner W. Metal-complex-catalyzed reactions. In: Jessop PG, Leitner W, eds. Chemical Synthesis Using Supercritical Fluids. Weinheim, Germany: Wiley–VCH, 351–387.

134. Sellin MF, Cole-Hamilton DJ. Hydroformylation reactions in supercritical carbon dioxide using insoluble metal complexes. J Chem Soc Dalton Trans 2000; 11:1681–1683.

135. Cornils B, Herrmann WA, eds. Aqueous-Phase Organometallic Catalysis. Weinheim, Germany: Wiley-VCH, 1998.
136. Bhanage BM, Ikushima Y, Shirai M, Arai M. Heck reactions using water-soluble metal complexes in supercritical carbon dioxide. Tetrahedron Lett 1999; 40:6427-6430.
137. Bhanage BM, Ikushima Y, Shirai M, Arai M. Multiphase catalysis using water-soluble metal complexes in supercritical carbon dioxide. Chem Commun 1999:1277-1278.
138. Jacobson GB, Lee CT, Johnston KP, Tumas W. Enhanced catalyst reactivity and separations using water/carbon dioxide emulsions. J Am Chem Soc 1999; 121:11,902-11,903.
139. Bonilla RJ, James BR, Jessop PG. Colloid-catalysed arene hydrogenation in aqueous/supercritical fluid biphasic media. Chem Commun 2000: 941-942.
140. Blanchard LA, Hancu D, Beckman EJ, Brennecke JF. Green processing using ionic liquids and CO_2. Nature 1999; 399:28-29.
141. Blanchard LA, Brennecke JF. Recovery of organic products from ionic liquids using supercritical carbon dioxide. Ind Eng Chem Res 2001; 40:287-292.
142. Blanchard LA, Gu Z, Brennecke JF. High-pressure phase behavior of ionic liquid/CO_2 systems. J Phys Chem B 2001; 105:2437-2444.
143. Brown RA, Pollet P, McKoon E, Eckert CA, Liotta CL, Jessop PG. Asymmetric hydrogenation and catalyst recycling using ionic liquid and supercritical carbon dioxide. J Am Chem Soc 2001; 123:1254-1255.
144. Bösmann A, Franciò G, Janssen E, Solinas M, Leitner W, Wasserscheid P. Activation, tuning, and immobilization of homogeneous catalysts in an ionic liquid/compressed CO_2 continuous-flow system. Angew Chem Int Ed Engl 2001; 40:2697-2699.
145. Lozano P, de Diego T, Carrie D, Vaultier M, Iborra JL. Continuous green biocatalytic processes using ionic liquids and supercritical carbon dioxide. Chem Commun 2002:692-693.
146. Reetz MT, Wiesenhofer W, Francio G, Leitner W. Biocatalysis in ionic liquids: batchwise and continuous flow processes using supercritical carbon dioxide as the mobile phase. Chem Commun 2002:992-993.
147. Heldebrant DJ, Jessop PG. Liquid poly(ethylene glycol) and supercritical carbon dioxide: a benign biphasic solvent system for use and recycling of homogeneous catalysts. J Am Chem Soc 2003; 125:5600-5601.
148. van den Broeke LJP, Goetheer ELV, Verkerk AW, de Wolf E, Deelman BJ, van Koten G, Keurentjes JTF. Homogeneous reactions in supercritical carbon dioxide using a catalyst immobilized by a microporous silica membrane. Angew Chem Int Ed Engl 2001; 40:4473-4474.

12

Analytical and Semipreparative Supercritical Fluid Chromatography in Drug Discovery

Terry A. Berger
Berger Instruments, Newark, Delaware, U.S.A.

Chromatography is seldom understood to be a critical technology in drug discovery. Synthetic organic chemists and medicinal chemists are interested in the chemical product (a series of compounds), not the procedures used to obtain pure representations of the product. Nevertheless, most, if not all, synthetic products require analysis and purification.

High-performance liquid chromatography (HPLC) has been almost universally used to determine the purity of newly synthesized compounds. HPLC–mass spectrometry has been used to prove the right compound was synthesized. Semipreparative HPLC has been used to take complex reaction mixtures and produce relatively pure drug candidates.

Supercritical fluid chromatography (SFC) is a close cousin to HPLC, but with enhanced capabilities. SFC is an ideal technology for analysis and purification of small drug like molecules (1–4).

Supercritical fluid chromatography uses most of the same or slightly modified hardware. The greatest difference is that most of the normal liquid used as the mobile phase in HPLC is replaced with a liquefied gas, like carbon dioxide. Fluids other than carbon dioxide can be used, but for all practical purposes, SFC means CO_2-based chromatography.

Most HPLC and even gas chromatographic (GC) detectors have been interfaced to SFCs. The evaporative light-scattering detector (5–8) and the nitrogen chemilumenesence detector are two such detectors with some importance in drug discovery.

There has been a >30-year effort to develop robust SFC–MS interfaces (9–17). SFC–MS has become a viable technology. The inherent advantages of SFC over HPLC, make SFC–MS superior to HPLC–MS (18,19).

Most of the history of SFC has involved analytical-scale chromatography. Recent advances have made both analytical and semipreparative chromatography highly desirable. The term "semipreparative" (semiprep) refers to the separation of milligrams to grams of materials to prepare modest amounts of material for testing. It is distinct from preparative-scale chromatography, where the intent is the production of the final product or an intermediate of the final product.

A few people recognized the potential of SFC for larger-scale chromatographic separations years ago, but numerous impediments prevented them from fully exploiting the apparent opportunity (4,20–22). A few examples of very large-scale SFC for commercial separation of specific components in complex mixtures (i.e., Refs. 23–25), including simulated moving-bed chromatographs (24,25), have been developed but are not covered in this chapter. Most of the advantages of SFC at the analytical scale are accentuated at the semiprep scale.

Supercritical fluid chromatography is actually a poor name for this technique because the fluids are often (usually) not supercritical. The distinguishing characteristics of interest are that the fluid acts as a polar solvent while retaining many of the positive attributes of a gas (high diffusivity, low viscosity) (26).

The fluids can be supercritical or "subcritical." To add further confusion, part of the region has also been called "near-critical." Both "subcritical" and "near-critical" describe fluids that are defined as "liquids." However, if an external pressure is removed, they will expand to a gas.

Supercritical fluid chromatography can be considered the rebirth of normal phase chromatography. In the 1970s, reversed phase (r) HPLC nearly completely replaced normal phase (n) HPLC, because it was faster, more repeatable, and more robust and decreased the use of large volumes of toxic, flammable solvent. SFC solves all the problems of nHPLC and actually has superior characteristics compared to rHPLC.

1. ADVANTAGES OF SFC OVER HPLC

The advantages of SFC over HPLC will be briefly summarized in the following subsections and will be discussed in further detail later in the chapter, where specific advances in hardware developments are discussed.

1.1. Speed–High Efficiency Per Time

Diffusion coefficients are three to five times higher in carbon dioxide/methanol mixtures than they are in pure methanol or water. Intermolecular forces are much weaker than in a normal liquid, but the fluid still acts as a polar solvent (due to the methanol). This higher diffusion translates into higher optimum linear velocity. The same efficiency can be achieved in one-third to one-fifth the time. The same work can be done faster, or three to five times more work can be done in the same time. In instances requiring very high throughout, this is an important advantage. Unknown purity screening and high-throughput purification are two places in drug discovery where this higher throughput is important (27).

In chiral separations, there is often both a major increase in optimum speed and a simultaneous increase in column efficiency. It is a common occurrence to increase throughput on chiral columns by 10–25 times compared to nHPLC (28,29).

1.2. Faster Method Development

The fluids equilibrate extremely rapidly. Retention times stabilize in as few as three to five column volumes, which is unusually fast compared to HPLC. Because the optimum linear velocity is 3–5 times higher, SFC tends to allow up to 10 times more experiments per unit time. Method development is all about rapid changes in conditions.

1.3. Greater Utilization of Expensive Stationary Phases

In semipreparative separations, the greatest expense is often the stationary phase. This is particularly true in the separation of enantiomers using chiral stationary phases. Significantly faster chromatography allows the user to use smaller, less costly columns to achieve the same throughput.

1.4. Low Viscosity

The viscosity of SFC mobile phases is also as little as 1/20 that of normal liquids. The higher optimum flow rates can be achieved with lower pressure drops than typical in HPLC. Because the vanDeemter curves for these fluids are flatter (30) and the viscosity of the fluids is lower, operation at much higher velocities is both desirable and practical.

1.5. Cost of Solvents

New gas delivery systems allow carbon dioxide to distill out of cylinders or Dewars just before use. This distillation step dramatically decreases the grade of the CO_2 required, making most industrial grades adequate. Prices of $0.10–

1.00/kg are now reported. Compared to HPLC-grade water or acetonitrile, such pricing is dramatic. As the scale of purification increases, the savings in solvent costs becomes more significant.

1.6. Fraction Size

At the end of the SFC separation, the mobile phase is broken down into two phases: one mostly gas and the other mostly polar modifier. Polar solutes tend to be nonvolatile and stay with the polar modifier. The gaseous phase can usually be vented. If the mobile phase consists of 5% or 10% modifier, 90–95% can be vented without significant effort. In semiprep chromatography, this is a huge practical advantage over HPLC. The time and the energy saved can be enormous, and in some instances, it makes previously impractical separations practical.

1.7. Cost of Fraction Dry-Down

In reversed phase HPLC, the mobile phase is usually water/acetonitrile. Typical "universal" gradients start at 10% and end at 90% acetonitrile. On average, peaks might elute at 50% water/acetonitrile. Optimum velocity on a 20–21.2-mm-inner diameter HPLC column is in the vicinity of 20–25 mL/min. Peaks containing 100 mg are likely to be 0.5–1 min wide, creating fractions up to 25 mL.

In SFC, the optimum flow rate is near 70 mL/min and the average modifier composition might be 25%. At the same efficiency, peak widths are 30% as wide, or up to 0.33 mins. Although, the flow rate is higher in SFC, typically 75% of the flow is carbon dioxide, which dissipates on depressurization. A typical fraction is 3–6 mL. It would take ~8520 cal to dry down a water/acetonitrile fraction compared to 480 cal/fraction in SFC. A combi-chem lab might purify 20,000–200,000 compounds per year, requiring up to 2 MW-h/year. Perhaps more important than the direct cost of the energy to remove the solvent, the time required to remove the solvent is also significantly shorter in SFC, decreasing operator time, space required, and lab operation costs per sample.

1.8. Cost of Disposal

Disposal of toxic wastes like acetonitrile is expensive. Generally, there are limits on how much waste can be stored on-site, putting constraints on waste generation. Small laboratories generally collect waste and periodically have a disposal/reclaimation company haul it away. There is usually a fee for coming and an additional fee that depends on the nature and quantity of waste. In many locations, waste disposal is more expensive than the solvent purchase price.

Supercritical fluid chromatography generates significantly less waste, decreasing the number of disposal trips required. Most of the waste is methanol, which is less toxic than acetonitrile.

1.9. Less Barrel Rolling

Gas delivery systems allow most of the mobile phase to be delivered to the lab through a stainless-steel pipe. Similarly, most of the waste exits the lab through another pipe (CO_2 exhaust gas). This approach requires additional infrastructure but dramatically improves ease of use.

1.10. Green Chemistry

Even though carbon dioxide is released into the atmosphere, SFC is considered "green" because the carbon dioxide has been recycled and it replaces more obnoxious fluids. A combi-chem lab performing HPLC purifications will produce 20 L of toxic mixed waste each working day while producing perhaps 100 pure compounds of up to 100 mg each. An SFC would produce 5 L (<one-fourth the volume) of methanol waste for the same production, but is capable of producing 3.5 times higher throughput.

2. SO WHY ISN'T EVERYONE DOING SFC?

In the past, numerous technological and commercial problems have combined to limit the penetration of SFC in drug discovery. Although a few problems, such as unfamiliarity and added complexity, are inherent, over the last few years, virtually all former limitations have been overcome.

2.1. Unfamiliarity

High-performance liquid chromatography is possibly the most widely applicable analytical separation technique, which has been well established for 30 years. However, many synthetic organic chemists and medicinal chemists still prefer thin-layer chromatography (TLC) to HPLC. SFC is a subset of HPLC. As pointed out below, SFC is less widely applicable, more complex, and more costly than HPLC. Few universities own the equipment or use it in teaching classes. Students tend to stay with the techniques they became familiar with in school when they go on to work in industry.

2.2. Complexity

Supercritical fluid chromatography are more complex that HPLCs. One of the fluids is usually a liquefied gas at high pressure. The instrument cannot use a single high pressure pump and a gradient valve to mix such fluids with normal liquids. Consequently, one always requires at least two high pressure pumps to produce gradients of polar modifier in nonpolar carbon dioxide. Further, the fluid wants to expand to a gas, so an extra device, a back-pressure regulator (BPR), is required to maintain the entire system at high pressures (and densities). Detectors require special cells than are not damaged by this higher pressure.

High-performance liquid chromatography allows direct collection of fractions with minimal hardware. A typical fraction collector might cost as little as a few thousand dollars. In SFC, fraction collection has been extremely problematic. The science and engineering is only now being adequately developed to allow easy use and to avoid aerosol generation, allowing very high efficiency recoveries when the fluid expands.

Finally, the logistics, infrastructure, and the budgeting process for SFC is different from HPLC. All these issues had to be addressed before SFC could become a routine technique. All of these issues have now been addressed.

2.3. Capital Cost/Infrastructure/Consumables and Budgets

Because SFC hardware is somewhat more complex, it is also more costly. In addition, at the semiprep scale, hardware cost should also include a gas delivery system to minimize "barrel rolling." On the other hand, a typical SFC uses much less expensive solvent and has roughly 3.5 times higher throughput. To do the same quality work, in the same time, by HPLC, 3.5 times more chromatographic equipment would be required.

The people who decide what lab equipment to buy often have no input into a number of related budgets. Savings in one budget might not be considered when evaluating increased cost in another. Because HPLC is the established technique, most budget are based on HPLC hardware prices and assume similar infrastructure requirements. Because HPLC is less expensive, SFC is at a budgetary disadvantage, requiring additional sign-offs or increased budgetary commitments.

Infrastructure enhancements, like a gas delivery system, are usually in a different budget from laboratory hardware. The requirement of getting several different managers to coordinate their budgets is an additional stumbling block. Savings associated with lower-cost solvents are usually part of another budget and are not necessarily considered in the hardware purchase decision. Similarly, decreased disposal costs are often under yet another budget.

3. WHAT CAN SFC SEPARATE?

As a general rule of thumb, any compound soluble in methanol or a less polar organic solvent is an ideal candidate for SFC. Some of the functional groups routinely separated by SFC are outlined in Fig. 1 along with the appropriate columns and mobile phases. Any compound requiring an aqueous or aqueous buffered environment is probably a poor choice for SFC. Thus, proteins and very large peptides are poor candidates, although peptides with up to 30 amino acids or containing a number of saccharides have been eluted using SFC mobile phases.

3.1. Solubility of Drugs

There is a universal desire in drug discovery to "increase water solubility" or, actually, to decrease hydrophobicity. Many of the lead compounds produced, particularly by combi-chem, tend to be far too lipophilic. In fact, most small organic molecules are not very soluble in water. Nevertheless, many chemists think that because SFC mobile phases are nonaqueous, their compounds will not be soluble and SFC will not work for them.

The log of the partition coefficient (log P) between octanol and water is a convenient guide to solute polarity and has applicability as a drug candidate. Lapinski's Rules of 5 (31) states that log P should be <5 to avoid toxic buildup in fatty tissues. Log P below 0 means that the drug probably must be injected. For orally administered drugs, the ideal range is log $P = 1.35$ to 2. The average log P of commercial drugs is around 2. This means that for the average commercial drug, 99% partitions into the octanol, not the water. A log P of 5 means that 99.999% partitions into the organic phase. The distribution of 5000 commercial drugs is plotted against log P in Fig. 2. Drugs with log P as low as (at least) -1 have been eluted without apparent difficulty by SFC. Thus, SFC is an appropriate analytical technique for >95% of small druglike molecules. It is uncertain how small log P can be and still be amenable to semiprep purification by SFC (what is the solubility in methanol?)

Log P can be related to log S, where S is the molar concentration of a saturated solution in water. There are a number of different computational approaches, but on a gross level, $\log S \sim = -0.8 -\log P - c \times MP$, where c is a constant and MP is the melting point of the compound, representing the energy required to break its crystalline structure (31). Roughly, log $P = 2$ (average for commercial drugs) is equivalent to log $S \sim -3$. The distribution of water solubilities of 2000 commercial drugs versus log S is plotted in Fig. 3. The vertical lines separate regions where solubilities are defined as "practically insoluble," "soluble," and "freely soluble." It is somewhat surprising that at least 60% of commercial drugs have log $S < -3$. If the average small drug molecule has a molecular weight of 400, this translates into water solubility of <0.1 mg/mL. The Merck Index classifies solubility less than 0.1 mg/mL as "practically insoluble." Perhaps more striking, less than 5% of

Polarity Range of Solute Families

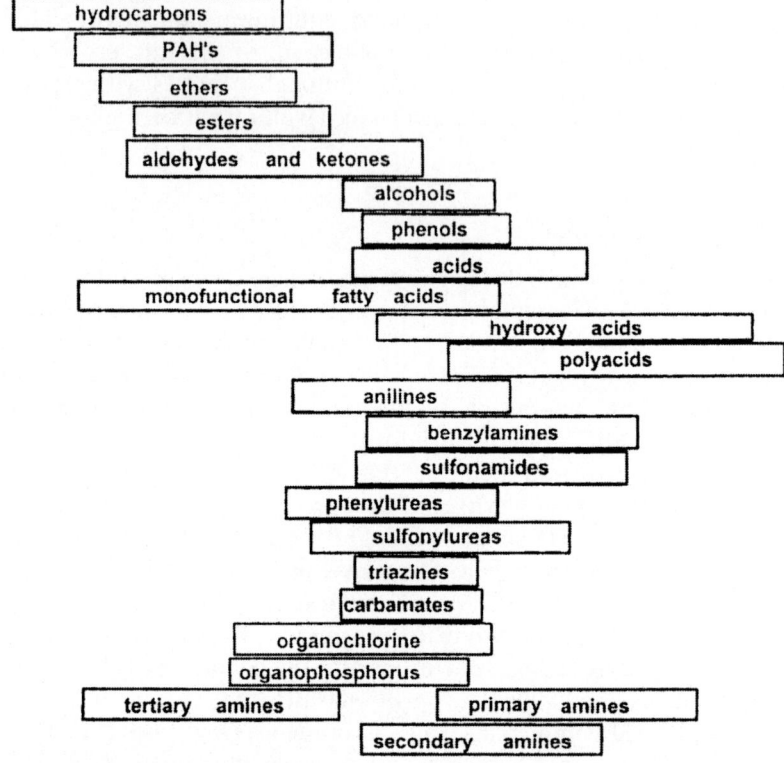

Polarity Ranges of Stationary Phases

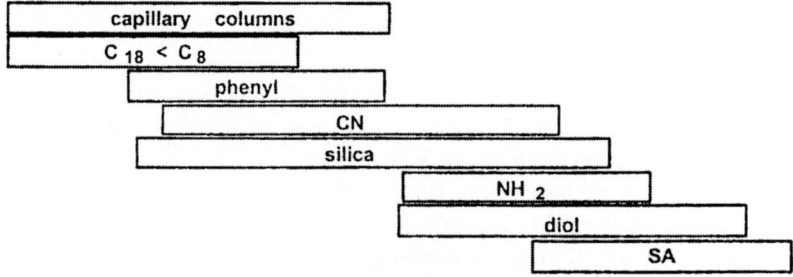

Appropriate Mobile Phase Composition

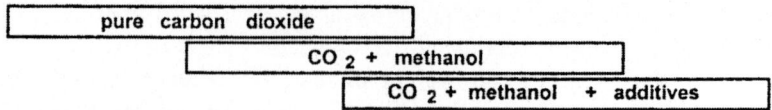

FIGURE 1 Functional groups separated by SFC.

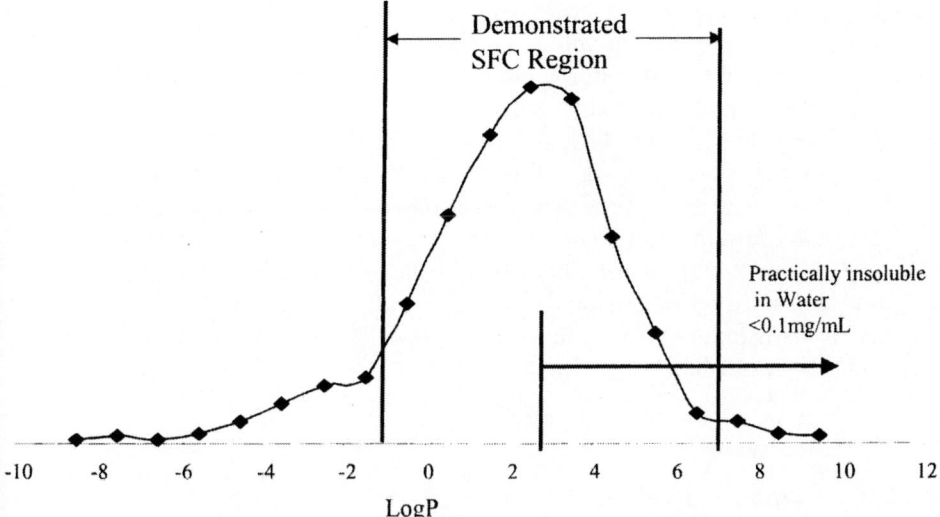

Figure 2 Distribution of log *P* of more than 5000 commercial drugs. (Data from the Actelion Corp. website.)

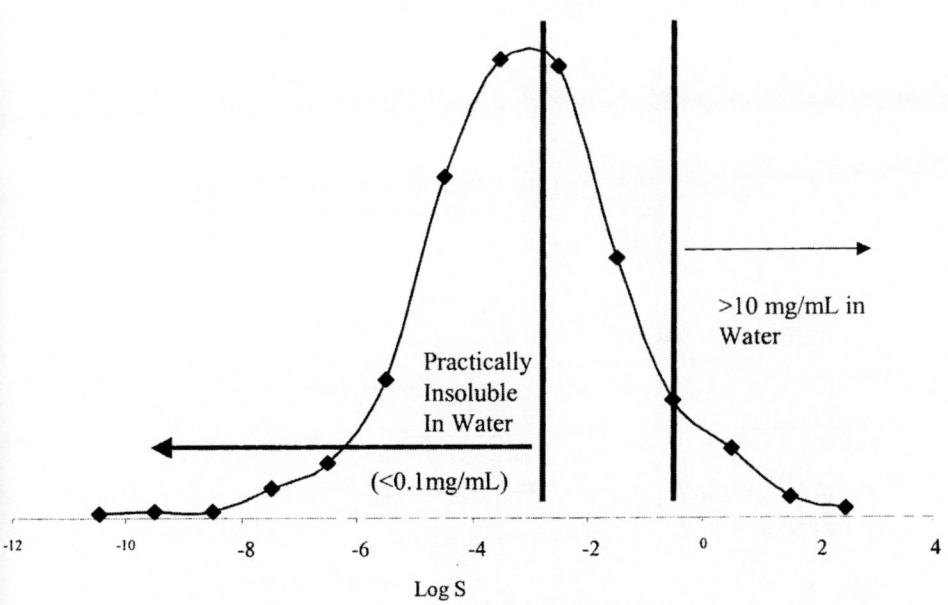

Figure 3 Distribution of water solubility (log *S*) of more than 2000 common drugs. (Data from Actelion Corp. website, assumes MW = 400.)

commercial drugs would be classified as soluble (>10 mg/mL) or freely soluble (>100 mg/mL) in water. Lapinski et al. (31) confirmed the general impression about poor water solubility. They evaluated 1600 phase II drugs for solubility and found 14.2% were soluble to <5 μg/mL in water, which they characterized as having "little or no chance of oral activity."

A recent partial analysis of the 50 best selling drugs (32) in the year 2000 showed that a large fraction were practically insoluble in water, but soluble to freely soluble in methanol. This is not surprising, noting the shape of Figs. 2 and 3. The vast majority of commercial drugs are much *less* polar than water. Methanol is roughly halfway in between water and octanol in polarity. If there is inadequate solubility in water, solubility for the vast majority of drugs should be better in methanol.

3.2. Hardware

3.2.1. Analytical SFC

A schematic of an analytical SFC is shown in Fig. 4. There are two high pressure pumps; one to deliver carbon dioxide and the other to deliver normal liquids as modifier. The modifier pump delivers normal liquids and is, for all practical purposes, an HPLC pump.

The SFC pump is shown connected to a high pressure cylinder. The simplest approach is to use high-purity carbon dioxide and a "dip tube"

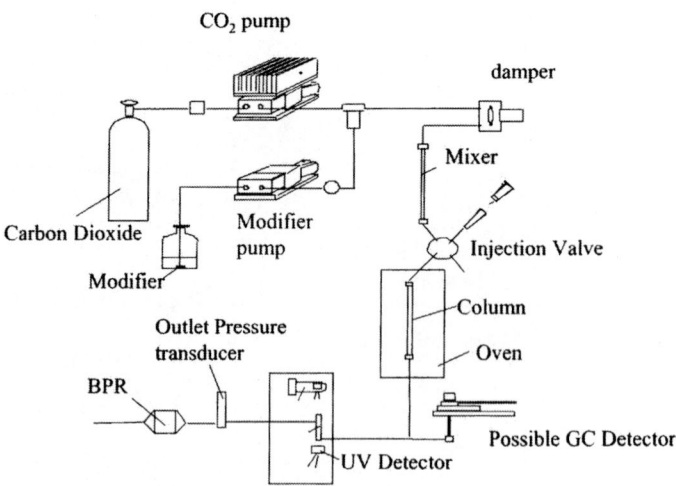

FIGURE 4 Schematic diagram of an analytical packed-column SFC.

cylinder to draw the liquid off the bottom. Cylinders exhibit a pressure near 65 bar at room temperature. Because the inlet to the pump is at high pressure, this pump cannot be used with a switching valve to generate gradients, as is often the case in HPLC.

The carbon dioxide pump has a built-in Peltier chiller and heat exchanger to prechill the fluid entering the pump and chill the pump head itself to a carefully controlled temperature. This pump employs dynamic compressibility compensation.

In the past, it was not uncommon for workers to bolt a heat exchanger onto an HPLC pump and call it an SFC pump. Unfortunately, many people still think that if the fluid is at a temperature and pressure defined as being a liquid, it becomes incompressible. This is emphatically not true. At 5°C, the volume of the fluid can change up to 20% over the normal pressure range of SFC. Decreasing the pressure much below 60 bar will still result in the fluid expanding into a low-density gas.

Most better quality HPLC pumps allow the user to preselect a value for compressibility compensation from a table of values. Each fluid is presumed to have a fixed compressibility. Most commercial HPLC pumps have a limited range of compressibility compensation and do not dynamically change compensation during a run. Although such a pump will deliver some carbon dioxide, the actual flow will depend on the pressure. Achieving an accurate and precise carbon dioxide flow rate is difficult due to the highly compressible nature of the fluid.

Supercritical fluid chromatographic pumps must have both a wide range of compensation and use dynamic compressibility compensation to produce accurate and reproducible flow and composition. Whereas water has a compressibility factor of 75×10^{-6}/bar, methanol is more compressible at 120×10^{-6}/bar. Carbon dioxide has widely varying compressibility from 95 to 395×10^{-6}/bar at 5°C, depending on the pump delivery pressure (column head pressure). The viscosity of pure carbon dioxide is 1/20 the viscosity of pure methanol. During composition programming, the viscosity of the mixed fluid and the column head pressure increases as the modifier concentration increases. Without dynamic compensation, the actual delivery of the carbon dioxide would roll off. The total flow would be less than the set points and the modifier concentration would be more than the set points.

The fluids are mixed together and pass to a standard HPLC injection valve or autosampler. The user should be aware that the injection loop is filled with a dense fluid that will rapidly expand to a gas when the valve is turned to the load position. Autosamplers should keep the needle in the loading port when the valve is turned, to avoid sending a spray of methanol into the lab. The waste line should be placed inside a vented container to trap the methanol.

With manual valves, the user should keep the syringe in the valve and a thumb or finger over the syringe plunger when returning the valve to the load position, to avoid blowing the plunger across the room while spraying sample solvent or modifier.

The column oven tends to be a cross between a GC and an HPLC oven. It is now common to mount a column selection valve in the oven with 6 or even 12 columns, particularly when performing chiral method development.

Columns are generally made of stainless steel and packed with 3–6-μm silica-based particles at 10,000 psi. The typical flow rate for a 4.6-mm-inner diameter column is 3–5 mL/min. Oven temperature is typically 35–60°C, The outlet pressure is generally above 100 bar.

Some applications of SFC outside the pharmaceutical industry employ oven temperatures up to 200°C and use the flame ionization detector (FID) from GC. The FID is not generally used in the pharmaceutical industry because polar solutes will not elute without a modifier, and modifiers give a response in the detector. However, a mass spectrometer is often mounted in the same manner as an FID. A small fraction of the total flow is split off through a tee to a fixed restrictor mounted in the inlet to the MS. The bulk of the flow proceeds through to the back-pressure regulator. Column outlet pressure is controlled by the BPR.

Carbon dioxide is transparent to below 190 nm, making it an ideal fluid for use with ultraviolet (UV) detectors. It can be used with acetonitrile above 195 nm and with methanol above 205 nm. The UV-vis detectors used in SFC are standard HPLC detectors in every way except for high pressure flow cells. Double tapered windows (45° bevels with matching seals) balance the stress and allow cells to be built that withstand more than 600 bar. The operator needs to be sure that the data collection rate is set fast enough to not round off the faster peaks experienced in SFC.

3.2.2. Back-Pressure Regulator

A basic characteristic of what we call SFC is the fact that the fluid wants to expand to a gas. Dropping the pressure on a binary fluid results in the generation of two phases. The presence of a second mobile phase destroys the chromatographic separation. To prevent this, a BPR is placed after the column (and after any dense phase detector like a UV-vis detector), to keep the column outlet pressure high enough to assure there is always a single phase. In general, SF chromatography of small druglike molecules works best between 30°C and 60°C. With carbon dioxide/methanol mixtures, at 35°C, keeping the pressure slightly above 75 bar is adequate to avoid two-phase formation. At 40°C, the pressure should be above 80 bar. At 50°C, the pressure should be >90 bar. At 60°C, the pressure should be >100 bar. Thus, if you keep the column outlet pressure above 100 bar, you will never experience two-phase formation while you separate small druglike molecules. If higher temperatures are used, slightly higher pressures are required. If there is a substantial amount of water added to the mobile phase, slightly higher pressures are also required.

4. EFFECT OF PHYSICAL PARAMETERS ON RETENTION AND SELECTIVITY

Supercritical fluid chromatography has at least one extra control variable compared to HPLC. Many people still associate the name SFC with syringe-pump-based systems in which pure carbon dioxide was used with pressure programming. Modern SFC sometimes still operates in this manner but only for nonpolar solutes. SFC is somewhat different from HPLC. It is useful to discuss the relative effects of different control variable on retention and selectivity. The relative effects of several control variables are represented schematically in Fig. 5.

4.1. Pressure Is a Secondary Control Variable

Retention is a function of pressure (actually density). However, one should expect only modest changes in retention with relatively large changes in pressure under most modes of operation. At the relatively low temperatures (30–60°C) and moderate pressures (100–200 bar outlet) that have been empirically found to give the best results for separating small druglike molecules, the density of carbon dioxide/methanol mixtures is between 0.75

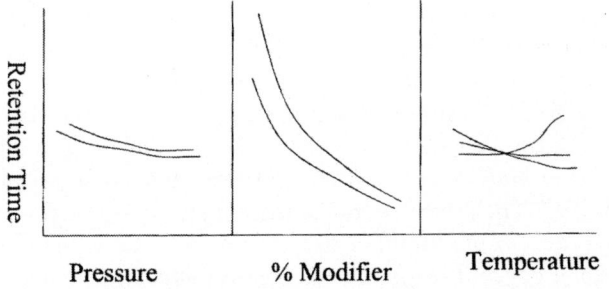

FIGURE 5 Schematic representation of the effect of various control parameters on retention and selectivity. Pressure is a secondary control variable. Modifier concentration has the largest effect on retention but little effect on selectivity. Temperature tends to have only a small effect on retention but a large effect on selectivity.

and 0.95 g/cm^3. Changing pressure (and density) from just above the two-phase region to the maximum available setting is unlikely to change retention of most solutes more than a factor of 2–4 and almost never more than an order of magnitude.

Plotting retention versus pressure for families of related compounds usually results in a series of parallel lines. Two lines almost never cross. Changing pressure is seldom effective in changing selectivity.

4.2. Changing Composition

Pure carbon dioxide is a nonpolar solvent, no more polar than pentane but in a different solvent family (32). Performing chromatography with pure carbon dioxide limits the user to mostly hydrocarbons and molecules with a long hydrocarbon tail like fatty acids. Very small polar molecules like aniline, phenol, or benzoic acid are slightly soluble but tend to give poor peak shapes or do not elute.

Fortunately, carbon dioxide is completely miscible with much more polar solvents. Carbon dioxide is completely miscible with methanol, ethanol, acetonitrile, and most other organic solvents. Because methanol is polar, inexpensive, readily available, and of relatively low toxicity, it has become the modifier of choice in SFC. Many, more exotic fluids have been used. One can include significant amounts of water in the mobile phase if it is added to another fluid like methanol and the temperature and pressure increased, but some of the advantages of SFC are lost.

Changing the composition of the mobile phase has a far more dramatic effect on retention than pressure. It is relatively easy to change retention over several orders of magnitude with moderate changes in composition. As a general rule of thumb, doubling the modifier halves retention. Changing the modifier from 1% to 64% roughly changes the partition ratio by a similar amount. If the partition coefficient were $k = 64$ at 1%, it would typically drop to 32 at 2%, 16 at 4%, and so on until it was $k = 1$ at 64%.

A constant gradient, like 5%/min or 10%/min tends to push all the less retained solutes together and spends excessive time at the end effecting small changes in retention. A "doubling" gradient offers the best trade-off between speed and resolution in SFC. The modifier concentration starts at some low level like 1% and doubles every column hold-up time.

Changing the modifier concentration has the greatest effect in changing retention among all the fluid characteristics. However, for any class of solutes, changing the modifier concentration often has little effect on selectivity. One can think of changing modifier concentration as "more," or "less," of the same thing. The carbon dioxide seems to act as an inert diluent of the modifier. If one plots retention versus percent composition for a group of similar compounds, the result is usually a set of nearly parallel curved lines coming together at higher concentrations. The lines seldom cross.

4.3. Changing Temperature

In drug discovery, we often avoid elevated temperatures because polar organic molecules tend to be thermally labile. Over the modest ranges used, temperature usually has little effect on retention. However, it has been observed that subtle changes in temperature can dramatically affect selectivity or relative retention. It is not uncommon for closely related compounds to show very different responses to 5–10°C changes in temperature. At 35°C, three tricyclic antidepressants coeluted, but at 40°C, they were baseline separated (33).

4.4. Additives

For perhaps 20% of the most polar small druglike molecules, either the peak does not elute or it elutes with severe distortion. The addition of a third component, usually a strong acid or base, alleviates this problem, as shown in Fig. 6. Ion pairing does not work. Ionization suppression seems to be a major mechanism in SFC. To improve the peak shapes of strong bases, another strong base (like dimethylethyl amine) is added. Similarly, to elute a strong acid, another strong acid (like TFA) is added (1).

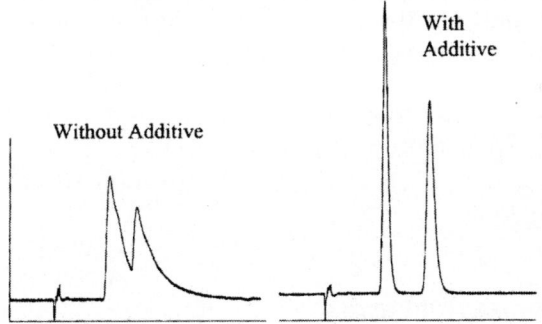

FIGURE 6   Perhaps 20% of the most polar solutes require a additive in the mobile phase to achieve good peak shapes. Propranolol, Chiracel-OD 4.6 × 250 mm, 200 bar, 2 mL/min, 30% methanol without and with 0.5% isopropylamine, 30°C.

4.5. Changing Columns

Perhaps 90% of all rHPLC separations use a C_{18} stationary phase. In SFC, there is a much wider choice. The vast majority of SFC separations have been performed on "totally porous" silica, usually with a bonded organic phase. A full range of achiral stationary phases is available for use in SFC. The most common are silica, cyano, amino, and diol. Several relatively new phases, such as an ethyl pyridine phase, promise to decrease the need for additives in the mobile phase. Almost everything elutes from cyano columns. Diol and bare silica often give the best selectivity for acids and alcohols. Amino is often more retentive but also often yields the best selectivity for amines. Although any of these columns is likely to give a reasonable separation, the use of an automated column selection valve, along with a solvent selection valve, makes it easy to find the one with the best selectivity for a specific application.

Decreasing poor size increases surface area, retention, and loadability as one would expect. There is a tendency to use 60-Å pore size material, even at the semiprep scale, at least with achiral columns.

Virtually all commercial chiral stationary phases have been found to work well in SFC. Some manufacturers have recognized that SFC is becoming a significant part of chiral chromatography and have started packing their columns with that in mind. Columns are most robust when used below the pressure under which they were packed. A few years ago, many manufactures packed their 2-cm-inner diameter columns with larger particles at 200 bar. Warning labels on the columns discouraged many potential users from using those columns in SFC, because the outlet pressure is maintained at 100 bar or above and inlet pressure is always higher. Today, better column hardware allows packing of 2-cm columns at up to 690 bar (10,000 psi).

There are no problems with crushing particles at high pressures. It is the pressure drop, not the absolute pressure, that counts. In fact, the compressible nature of the mobile phases means that the particles are less likely to be damaged by pump pulsations or sudden decompression in SFC than in HPLC.

Very small particle sizes have not been widely exploited. In fact, even with larger-sized particles, it is often difficult to achieve the full-speed potential of SFC because most of the components have been designed for HPLC. Pressure drops tend to be modest.

5. SFC ON THE SEMIPREPARATIVE SCALE

The semiprep schematic in Fig. 7 is similar to the analytical-scale hardware except that the components tend to be larger and extra hardware is needed to collect fractions Peltier cooling is more difficult on a larger scale because it is relatively inefficient. Compressibility compensation is also more difficult because compression of a larger volume creates more heat. Detector flow

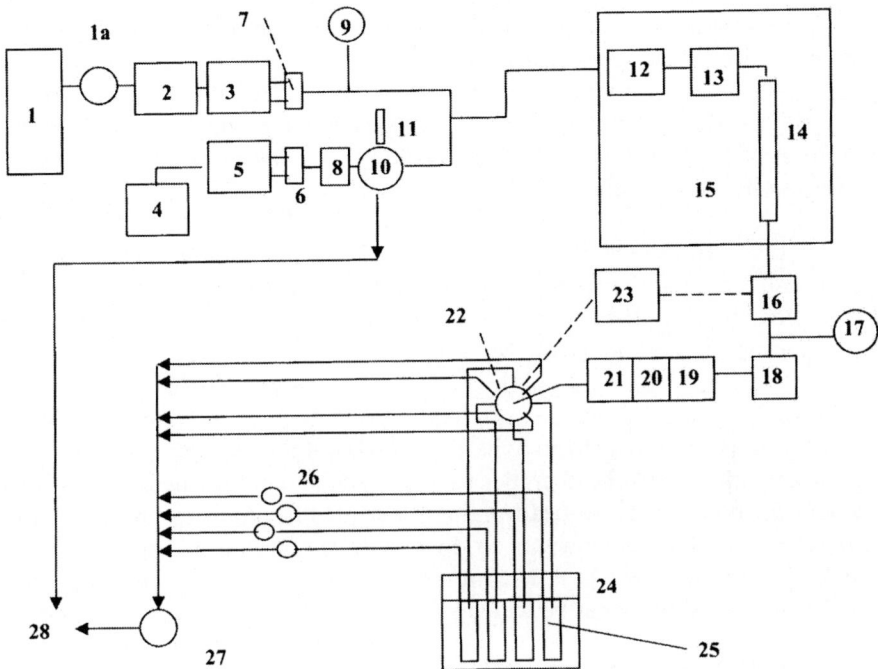

FIGURE 7 Schematic diagram of semipreperative-scale SFC chromatograph: 1: carbon dioxide supply; 1a: regulator; 2: prechiller/heat exchanger; 3: SD-1 Varian pump with (7) 200-mL pump head with special check valves; 4: modifier reservoir; 5: SD-1 modifier pump with (6) 200-mL standard pump heads; 8: check valve; 9: inlet pressure transducer; 10: injection valve; 11: check valve to prevent blowback; 12: mixer; 13: fluid temperature preconditioner; 14: column; 15: column oven; 16: uv detector; 17: outlet pressure transducer; 18: back-pressure regulator; 19: evaporator; 20: restrictor; 21: trim heater; 22: selection valve; 23: peak detector; 24: bank of collection vessels (or cassette); 25: individual collection tubes/bottles; 26: pressure relief valves; 27: waste container, waste vent. The manual cassette can be replaced with an automated cassette fed by a robot holding 128, 25 × 150-mm or 338, 16 × 150-mm test tubes, or with 7 large bottles.

cells need to be shorter to avoid overload. Tubing needs to be larger to avoid excessive pressure drops.

5.1. Pumps

Semipreparative pumps tend to be large, making it more difficult to cool uniformly. They are usually made out of stainless steel with thick, variable-cross-section walls. Stainless steel is effectively an insulator. The compression stroke generates substantial heat. This heat dissipates into the pump head, but there is a finite delay getting heat in and out.

Most modern pumps have two pump heads; one refills while the other delivers. During gradients, the rate of fluid delivery changes significantly. Consequently, the head and fluid temperatures oscillate continuously, but with several different periods.

Low supply pressures or starving the pump by inadequate flow in the supply lines causes part of the fluid aspirated into the pump to gasify. Gas in the pump head requires substantial piston displacement to reliquify without any fluid delivery to the column (i.e., inaccurate flow). The user should ensure that the supply pressure is always adequate.

Pumps are required to perform dynamic compressibility compensation if the user expects to have accurate and precise flow and composition. Most HPLC dumps do not perform dynamic compressibility compensation. If they are used with a pump head chiller for SFC, the actual flow delivered will be dependent on the column head pressure and will deviate substantially from the set point. If the user wants to do repetitive chromatography to collect peaks in time windows, the flow/composition must be accurate, and/or precise or the time windows must be widened.

5.2. Back-Pressure Regulator

When used in semipreparative chromatography, the BPR should have a low "dead" volume to avoid degrading peak shape before collection. Traditional mechanical back-pressure regulators tend to have large unswept or poorly swept volumes on the high pressure side, including large springs wet by the working fluid. Such characteristics make these sorts of device unacceptable for semiprep chromatography. Modern devices tend to be electrically actuated with a large solenoid pushing on a pin or diaphragm. The working fluid is isolated from the solenoid by a sliding seal or the diaphragm. A pressure sensor and circuitry form a feedback loop to the solenoid to control the pressure.

Back-pressure regulators function by pressing a pin or a diaphragm into or over a relatively small hole. The gap between pin and hole is no more than a few tens of nanometers to hold 100–300-bar pressures at flow rates between 1 and 100 g/min of carbon dioxide/methanol. The expansion of the fluid down toward atmospheric pressure occurs in the hole just downstream of the pin or diaphragm. From 50 mL/min (~ 45 g/min) at a density of ~ 0.9 g/cm^3), the fluid expands to up to 25,000 cm^3/min at atmospheric conditions.

5.3. Problems with BPRs

It has been found that repetitive steep modifier gradients and/or high solute loading near saturation can cause erosion/corrosion either on the tip of the pin or in the seat of the BPR when the fluid expands in the throat of the BPR.

This erosion/corrosion has been a concern for many years and has consumed many years of engineering effort. Modern devices use exotic materials to overcome this problem.

6. PEAK COLLECTION: SUPPRESSING AEROSOL GENERATION

A part of this book describes the use of supercritical fluids as an ideal method for making aerosols. There are many patents covering the subject and it is an area of great interest to everyone involved in the design and use of inhalers and other forms for drug delivery. It turns out that this ease of making aerosols is very bad for collecting peaks after chromatography.

Once you make an aerosol, it is difficult to recollect all the material, unless you have a very large surface area, like the lungs. It does one no good to use a very small amount of liquid solvent to separate and purify a compound, then require much larger amounts of liquid to wash out a trap to recover the compound. Various schemes for bubbling the effluent through a column of solvent, high-surface-area trapping material, and so forth have all been shown to be relatively ineffective in trapping an aerosol while negating much of the chromatographic advantage by increasing complexity and/or time required to finish the purification process, after the separation (4,20–22).

For years, many labs used analytical-scale SFCs as miniature prep chromatographs but report only 30–70% recovery. The other 30–70% of each compound was lost into the surrounding air as aerosols! This is inherently unsafe and requires, at a minimum, that the chromatograph be located in a fume hood. Even then, the exposed equipment is likely to be coated with small amounts of each the compounds purified. In drug discovery, this is particularly unsafe because many of the compounds purified will be druglike but with unknown potency and toxicological characteristics. Aerosol formation has been one of the greatest problems preventing acceptance of semipreparative SFC on a wider scale. There are ways to suppress aerosol formation for semipreparative SFC.

6.1. Cyclone Separators: Scaling Down of an Industrial Process

The traditional way to collect fractions in SFC is to use cyclone separators (21,23–25). These large devices allow the fluid to expand in a controlled manner. The pressure is dropped just low enough for the fluid to separate into two phases: one mostly the carbon dioxide, and the other mostly the normal liquid modifer, some dissolved carbon dioxide, and any polar, nonvolatile solutes present. The pressure inside the cyclone separator is set to just low enough to cause the fluid to break down (i.e., 50–60 bar) into these two phases.

The cyclone separator consists, in part, of a large-diameter cylinder with its central axis mounted vertically. A restrictor or BPR introduces the sample into the cylinder through a tube, pointed slightly down from horizontal, parallel to and nearly touching the inner wall. The cylinder acts as an expansion chamber. Its large volume allows the velocity of the expanded fluids to slow down. Heavier, nonvolatile components hit the wall, coalesce, and run down to the bottom, which is often tapered to a smaller diameter. A "gas" exit tube, often on the axis of the cylinder, allows the gas phase to escape out the top of the device. The density of the gas may still be 0.05–0.1 g/cm^3. This fluid is sometimes recompressed and reused if it is pure enough.

The gas escaping from a cyclone separator may still be at 50 or even 60 bar. Because the fluids are always under relatively high pressure, they do not fully expand. Normally, expansion from 200 bar down to 1 bar results in nearly a 500-fold expansion, from 70 mL/min (at density $0.9+$ g/cm^3) to 35,000 cm^3/min (at a density less than 0.002 g/cm^2). Expansion down to only 60 bar may only result in a 10–20-fold expansion. This modest expansion and subsequent modest velocities help minimize aerosol formation.

Cyclone separators can be effective as collection devices in SFC. However, they tend to be best suited for semipermanent applications, like preparative-scale chromatography. They are not user friendly for the separation of a large number of different samples in a short time. Each fraction requires a separate cyclone separator. For modest scale separations, the cyclone separators are bulky, with a very large internal surface area, which is difficult to clean out. Because they are subjected to relatively high pressures, they are usually made of stainless steel. Connections are usually made with large-diameter stainless-steel tubing and the whole apparatus is bolted to some sort of rack mount. One could think of this approaches a scaled-down pilot plant.

6.2. The Berger Separator: Scaling-Up Benchtop Processes

The Berger separator is a relatively new device (34–36) that is dramatically different from cyclone separators. It allows virtually 100% recovery at near-atmospheric pressure without aerosol formation.

The fluid pressure is dropped in multiple stages and heat is gently added to decrease the solubility of carbon dioxide in the modifier, as suggested in Fig. 8. The modifier forms a film on the wall of the larger-diameter tubing, with the gas phase flowing down the middle of the tube. By not dropping the pressure too soon, the carbon dioxide can be stripped out of the liquid phase into the gas phase without excessively disrupting the liquid film. A relatively large amount of thermal energy must be added to counter the cooling caused by the expansion of the fluid, without raising the temperature to levels that

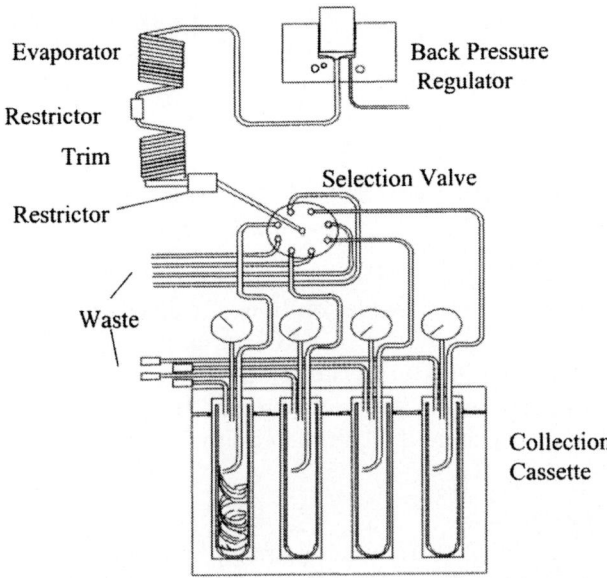

Evaporator

Restrictor

Trim

Restrictor

Waste

Back Pressure Regulator

Selection Valve

Collection Cassette

FIGURE 8 The separator consists of multiple stages of pressure drop and heating. This diagram shows a cassette holding test tubes. Either the user or a robot replaces the tubes after each injection. The robot version, called AutoPrep, supports high-throughput purification and can handle roughly three 96-well plates of samples per day. The larger version, called MultiGram, uses seven 1-L bottles to collect repeatative injections of the same sample.

could damage thermally labile solutes. At 100 mL/min, this heating is ~ 1 kW. The final temperature is maintained below 35°C.

Once the liquid phase is separated from the gas, it is diverted to impinge on the wall of a glass vessel like a simple test tube or collection bottle. The liquid runs down the wall while the gas exits the top. The carbon dioxide expands ~ 500 times. Under typical operating conditions, 25–30 L/min of gas exits out the top of the test tube or bottle. The waste gas is diverted through a waste collection container and then vented outside. A small amount of the modifier, like methanol, is present in this waste stream (due to the vapor pressure of the compound). As with the cyclone separator, the user is never exposed to the effluent of the chromatograph.

The greatest advantage of the Berger separator is its adaptability. The tubing to the collection vessels is $\frac{1}{8}$-inch-outer diameter flexible Teflon, and the apparatus is completely self-cleaning with no carryover. This allows the user to either manually change collection vessels or use a robot. The collection vessels are simple laboratory glassware. Changing to a new sample type takes seconds. The combination of ease of use, very high collection efficiency, and high purity makes high-throughput purification by SFC more practical than before.

7. INFRASTRUCTURE ENHANCEMENTS: GAS DELIVERY SYSTEMS

One might be surprised to find a full section on gas delivery in the instrumentation section, but the lack of reasonable solutions for delivery of pure CO_2 held back the more widespread use of SFC for quite some time. Reasonably priced commercial gas delivery systems designed specifically for SFC have only recently become available. There are subtle problems that make the carbon dioxide delivery more difficult than it appears it should be.

7.1. Purity of Gas

In the early days of SFC and SFE, there was no straightforward means of condensing vapor phase CO_2 near the pump, so cylinders with "dip tubes" were used to draw liquid CO_2 off the bottom of the tank. Commercial grades of CO_2 were called names like "anaerobic" grade, "bone dry", "beverage" grade, and the like. When various groups first started experimenting with SFC and SFE, it was not uncommon for them to contaminate some cylinders with backflow from their high pressure pumps. We once received a cylinder half-full of vegetable oil. In other cases, commercial applications like pressurizing beer in taverns resulted in water contamination or even colloidal rust in some steel cylinders. These and other contaminants were relatively common and caused serious downtime and cleanup issues. Consequently, in the United States, the American Society for Testing Materials (ASTM) established a standard for "Supercritical Fluid Chromatography" Grade carbon dioxide (37), which, subsequently, has been sold by many gas companies, usually packaged in relatively small aluminum cylinders (to avoid colloidal rust from iron cylinders).

7.2. Pure But Expensive

The availability of SFC-grade fluid virtually eliminated the worry over contamination. However, the handling and testing of such relatively small volumes of fluid makes SFC-grade CO_2 quite expensive (\$14–20/L). There is an even more expensive SFC/SFE grade. Further, the smaller aluminum cylinders only contain ~ 12–15 kg of usable fluid. At optimum flow on a 21.2-mm-inner diameter semiprep column, such a cylinder could theoretically last no more than 3.6 h. In reality, the cooling caused by the evaporation of the fluid inside the cylinder is enough to significantly decrease the pressure long before the cylinder is empty. Turning off the flow allows the cylinder to warm up, recovering its initial pressure, but the heat transfer takes hours. Heating blankets are illegal in most jurisdictions because thermal run away could easily rupture a cylinder filled with a liquefied gas.

Ganging a number of dip tube cylinders together spreads out the cooling and allows the user to withdraw a larger fraction of the total fluid. However, dealing with a large number of cylinders under less than optimal conditions is relatively unsatisfying and expensive and raises other safety concerns. Ganged cylinders with dip tubes must each have a check valve to avoid inadvertently filling a new cold cylinder completely with liquid. On subsequent warmup, the cylinder could burst.

7.3. Less Expensive, Pure Gas Alternatives

Gas delivery systems that allow the user to either use gangs of large cylinders without dip tubes or Dewars are now commercially available, as shown in Fig. 9. In either case, the fluid is distilled just before it is supplied to the pump. In the case of high pressure cylinders, the distillation occurs near room temperature. In Dewars, the distillation occurs at much lower temperatures ($\sim 5°C$).

The distillation eliminates virtually all significant contaminants. The vapor phase will not transport colloidal rust from water-contaminated steel cylinders. Greases and oils are not volatile at the temperature of the distillation. Small residuals of nitrogen, oxygen, and even water at the 5–10-ppm level usually do not interfere with analysis or purification. In fact, after a break-in period, the gas delivery system in our laboratory provides significantly lower background noise in SFC–MS than cylinders filled with SFC-grade CO_2.

When the vapor phase is withdrawn, the cooling of cylinders is extreme, whereas carbon dioxide is very poor at conducting heat. The vapor pressure in a single cylinder drops rapidly. At a reduced pressure, fluid traveling through a transfer line from the cylinder to the pump will probably vaporize (the tube is at room temperature) and cause the pump to cavitate after 1–2 h. Gangs of at least three cylinders decrease the rate of cooling and allow the user to draw off much more of the fluid available, because each is cooled at one-third the rate. A single large steel cylinder might contain 20–23 kg of usable fluid. Three ganged cylinders should last at least 14–15 h.

Waiting for a single tank to thaw out or switching back and forth between tanks every 1–2 h is frustrating. Changing a bank of cylinders one or two times a day is still inconvenient. A Dewar can hold >225 kg, which should last >54 h with a single semiprep chromatograph running 24 h/day. The task

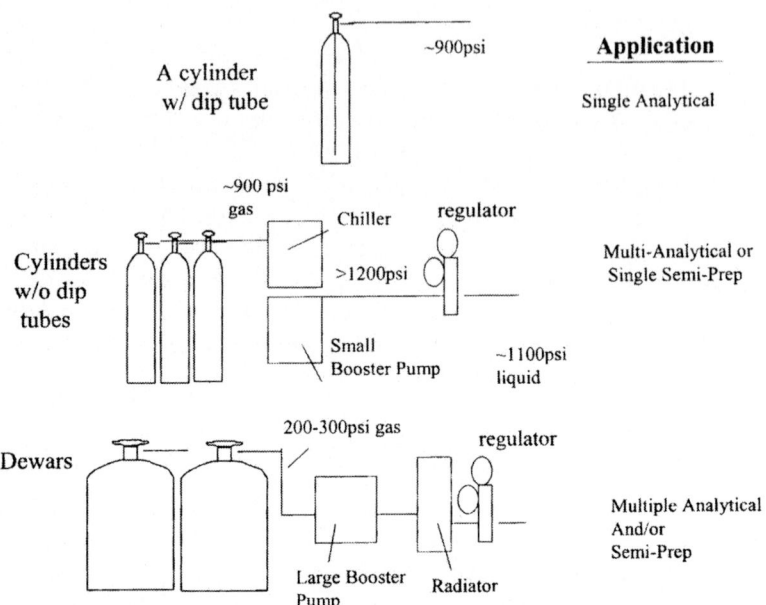

FIGURE 9 There are three practical approaches to gas delivery systems. The simplest is a single aluminum cylinder with a dip tube. Although simple, this approach is expensive and inappropriate for the semiprep scale. Ganged cylinders without dip tubes offer an intermediate scale, with much less expensive solvent and compatibility with a semiprep scale. The Dewar-based system offers the greatest ease of use and least expensive solvent.

of changing over can be delegated. An automatic changeover system can allow the gas supply company to simply change a Dewar every 2–3 days. For larger users, there are very large storage tanks available.

Using either gangs of cylinders without dip tubes or Dewars, a condenser and a pressure booster pump are required in order to ensure that fluid reaching the chromatograph is of uniform high density. For larger systems, redundant pumps ensure a continuous uptime. A full range of accessories are available to change over from nearly empty to full containers: safety sensors, lights, alarms, and so forth. With vapor phase systems, simple pressure-sensing gas changeover systems make it easy to automatically replace empty containers without shutting down the system.

Gangs of large steel cylinders provide fluid at roughly $1.00–1.50/kg in the United States. Dewar-based systems cut that cost to $0.50/kg. A few very large users with fixed storage tanks, filled by a tanker truck, pay as little as $0.10–0.15/kg.

7.4. Decreasing "Barrel Rolling"

In HPLC, a semiprep column might be operated at 25 mL/min, using up solvent and generating waste at a rate of up to 36 L/24-h day, or just under two 20-L containers of solvent in and two 20-L containers out as waste. Each week, the user needs to have available at least 10–14, 20-L containers of new solvent and must dispose of about 10–14 20-L containers of toxic, mixed waste for each chromatograph. Most jurisdictions limit the amount of solvent allowed in a lab at any given time as a fire safety issue. The limit is usually 50 L. A whole lot of "barrel rolling" occurs.

In SFC, gas delivery systems eliminate anywhere from 75% to 95% of this barrel rolling. In many labs, stainless-steel welded pipes are now used to distribute 83 bar (1200 psi) carbon dioxide from gas delivery systems situated up to hundreds of meters away, often in a loading dock area. Automatic changeover systems make the supply transparent to the user. These systems can be fitted with indicator lights in the lab showing the status of the system. Quick disconnect fittings allow chromatographs on carts to be quickly moved around and used. Such distribution systems eliminate most of the "barrel rolling" associated with larger-scale semiprep chromatography. Some solvent handling is still required because the polar modifiers used and the fractions collected are normal liquids.

8. PRACTICAL CONSIDERATIONS IN SEMI PREPARATIVE SFC

8.1. Overload and Focusing

In SFC, one seldom reaches the nonlinear part of the adsorption isotherm for the solute. Instead, the solubility of the solute in methanol or other solvent is reached first.

Saturated solutions in methanol can actually be injected into methanol/carbon dioxide mobile phases (although not recommended) without the solute falling out of solution. This is probably due to the extensive adsorption of the modifier onto the stationary phase (38), creating a form of reservoir of solvent to receive the injected sample.

Peaks are usually broadened more by the use of larger volumes of injection solvent than by excessive solute concentration. Analytical-scale achiral columns tend to have less capacity for polar sample *solvents* than a similar-sized column in rHPLC, because SFC samples are usually dissolved in pure methanol and then injected into a somewhat weaker mobile phase. In HPLC, samples tend to be dissolved in water, an aqueous buffer, or a mixture low in organic modifier which in reversed phase has less elution strength than a solvent like pure acetonitrile.

Injection of more than ~5 µL onto a 4.6-mm-inner diameter SFC column tends to cause peak broadening. Injection volumes greater than 10 µL tend to produce broader but not taller peaks. Chiral columns seem to be more forgiving, perhaps because they generally exhibit much worse efficiency to start with, but direct injections of up to 100 µL tend to produce increasing peak heights with little broadening.

8.1.1. Indirect Injection

In semiprep-scale SFC, directly injecting 1 or 2 mL of methanol into a column with 5% methanol flowing through it results in a local overload of the column with sample solvent (36). This severely broadens peaks. Large volumes of polar solvent distort the peak and shift its retention to shorter times.

Better peak widths and peak shapes are usually obtained if the sample is introduced onto the column as part of a premixed mobile phase. Placing the injection valve in the modifier line instead of the premixed mobile phase line allows the user to slowly add the solute to the mobile phase at a greatly reduced local modifier concentration. Such an approach requires that the solute be strongly retained at the modifier concentration flowing through the column, so that the peak can be focused.

An unretained ideal peak with $k = 0$ and a very small injection volume is likely to have a peak width of 0.032 min at the base. An indirect injection of a much larger volume requires that the loop contents be swept onto the column over an extended time. If the total flow rate is 70 mL/min and the initial concentration is 5% modifier (modifier flow = 3.5 mL/min), a 2-mL loop is swept in 0.285 min. An unretained peak would have a peak width > 0.285 min. Injecting into the modifier line thus artificially broadens a peak compared to the ideal with a small injection volume.

Focusing improves the peak widths. If the solute has a value of $k = 10$ at the initial conditions, solute molecules travel one-tenth as fast as the mobile phase and are roughly focused at a magnification of 10X. If the peak is allowed to travel through the column at this speed, it will emerge nearly as narrow as if a smaller amount of sample solvent were injected faster. Alternately, if the peak is focused into a narrower band, in length, on the column and then washed off with a fast gradient, it will also be narrower in time than the time required to apply it to the column. Injecting into the modifier line allows retention times in analytical chromatography to be scaled directly to the semiprep level.

8.2. Scalability

A number of hardware parameters must be controlled to scale chromatography between analytical and semiprep scale. Large injection volumes should not shift the retention time of peaks in semiprep-scale compared to analytical-scale chromatography.

All of the pumps must be both accurate and reproducible in flow and composition. Accuracy requires careful control of both the temperature of the fluid and the supply pressure, along with comparable dynamic compressibility compensation schemes. Without dynamic compressibility composition, pumps delivering carbon dioxide will deliver different fractions of their set points, depending on the pressure. Total flow and composition will probably deviate from the set points differently at the two different scales. Accurate compressibility compensation assures scalable pumping.

Temperatures and thermal gradients need to be scalable. Larger columns tend to have larger radial thermal gradients than smaller columns. Ensuring that the fluid entering the column is at the oven temperature helps minimize thermal differences between column diameters.

The columns need to have similar pressure drops at the appropriate scaled flow rates. Recall that retention is a (rather weak) function of pressure. Large differences in pressure drop across different sized columns cause noticeable shifts in retention time. A larger pressure drop should cause decreased retention.

8.2.1. Plugging the BPR: Solubility

In semiprep chromatography, the user often injects large amounts of newly synthesized solutes with unknown physical and chemical characteristics. They use the "universal" solvent dimethyl sulfoxide (DMSO) for speed and simplicity. This can result in chromatographic problems.

In HPLC, most solutes are poorly soluble in water and high concentrations of solutes tend to precipitate in front of the column. If the effect is severe, there is a pressure overrange and the system shuts down. The column can usually be cleaned with 100% organic solvents, but the sample is probably lost and the system is down until the column is cleaned or replaced.

In SFC, the solutes tend to travel with the DMSO from the injection port to the head of the column, but are soon separated. Copious amounts of methanol are adsorbed onto the surface of silica-based columns. This adsorbed film acts as a receiving reservoir that helps solubilize incoming solute molecules. However, after the column, there is no longer any large reservoir of solvent and some solutes are just not soluble enough in the mobile phase to stay in solution. If significant amounts of solute drop out of solution after the column and before the BPR, they can potentially plug the device.

To recover, the BPR can be "burped" by momentarily dropping the control pressure. If this does not work, the user can set the modifier concentration to 100% for a brief time, like 30 s. In highly automated situations, a few users have included an additional pump that turns on between runs to deliver a few milliliters of DMSO.

9. APPLICATIONS

9.1. Chiral Separations

More than half of small druglike molecules are chiral. The Food and Drug Administration (FDA) requires testing of pure enantiomers. Such testing is most useful early on in drug development. SFC is dramatically superior to HPLC for chiral separations. SFC offers dramatically faster method development and should be the technique of choice for any molecules soluble in organic solvents (i.e., most druglike molecules). Further, unlike capillary electrophoresis, SFC is fully scalable. A method developed at the analytical scale should work equally well at the semiprep level.

In the early stages of drug development, the toxicological and pharmacological effects of the pure enantiomer must be established, using 99.95% pure enantiomers (e.e. > 99.9%). Chromatographic purification often provides a much faster approach to obtaining pure enantiomers than asymmetric synthesis, recrystallizations, or other purification steps. Thus, chromatography, especially SFC, offers by far the easiest path to early testing.

Many well-known chiral chromatographers (39–46) have published articles or reviews about the significant saving in time and effort possible using analytical SFC in place of HPLC, both in method development and in routine work. Most major pharmaceutical companies have appreciable numbers of SFCs for this type of analysis.

Additives can be important for improving peak shapes. Two chromatograms of propranolol, one without and one with an additive in the mobile phase, are presented in Fig. 6. Clearly, the additive dramatically improves the separation.

It is important to compare apples and apples. In one article (47), HPLC with an additive in the mobile phase was compared to SFC without an additive. Further, the quality of the separation was represented by selectivity (α) values and the partition ratio of the second peak (k_2) but not retention time. In SFC, the optimum flow rate is three to five times higher, meaning that at the same value for k_2, the retention time in SFC is one-third to one-fifth as long. In another article (48), some of the same compounds were separated by SFC but with an additive present in the mobile phase.

A superficial comparison of HPLC, with additive, and SFC, without additive, shown on the left of Table 1. Although it appears that in all cases SFC produced a larger selectivity (α), in two of the three cases shown, k_2

TABLE 1   Comparing Apples to Oranges

	HPLC, α (k_2) (47)	SFC, w/o add. (47)	SFC, w/add. (48)
Atenolol	1.24 (13.2)	2.21 (8.30)	2.0 (1.86)
Metoprolol	1.75 (1.83)	2.74 (8.03)	2.58 (1.10)
Propranolol	1.39 (3.05)	1.73 (8.61)	1.40 (1.46)

Note: Reporting only selectivity and the value of the second partition ratio does not provide adequate information when the optimum flow rate of the two techniques is different. Further, when one technique is used with an additive, the other should also be used with an additive.

(meant to represent analysis time) was larger in SFC. The third row to the right of Table 1 summarizes the other workers' data showing both higher or equal selectivity and noticeably smaller values of k_2. Remember that the SFC data was collected at three times higher flow rate. Comparing both techniques with an additive, Atenelol selectivity improved from 1.24 in HPLC to 2.0 by SFC, whereas k_2 dropped from 13.2 to 1.86 (7.1-fold). However, because the flow rate was at least 3 times faster, the SFC chromatograms were actually more than 20 times faster overall. In all cases, the compounds were at least baseline resolved. This analysis, using only α and k_2 does not fully capture the quality of the two separations because retention time and resolution are not listed.

A chromatograph can employ a column selection valve (49) and a solvent selection valve with automation to empirically find the best column, and to a lesser extent, mobile phase, for a particular application. Samples are usually run overnight and the chromatograms evaluated in the morning. After the column giving the best separation is found, it usually only takes a few runs to further develop the method to the level needed for the separation goals.

A chromatogram of *trans*-stilbene oxide is presented in Fig. 10 and demonstrates many of the attributes of modern SFC. The column is not a "fast" column, being 25 cm long and packed with 10 μm particles. However, the separation shows excessive selectivity in less than 2 min.

The chromatography is scalable, at least to the semiprep level. The same packing materials can be used effectively at the larger scale. Numerous separations have allowed the use of samples dissolved at 100 mg/mL or more.

Chiral separations usually involve only a few peaks. To achieve the highest throughput, separations should be developed where the resolution (R_s) is 1.5, with a solute that is no more than 75% saturated in the sample

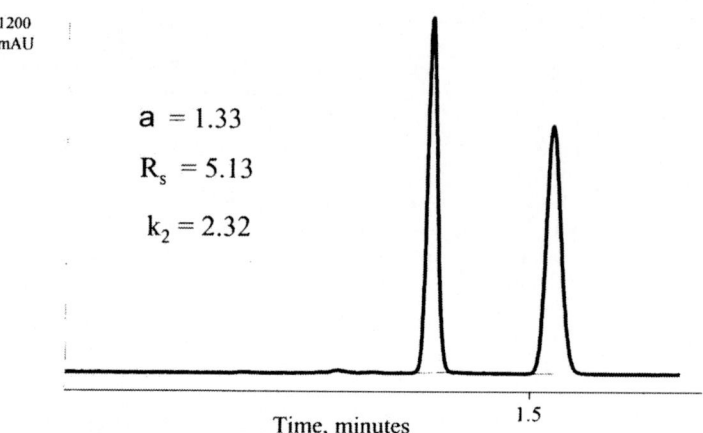

1200
mAU

$a = 1.33$

$R_s = 5.13$

$k_2 = 2.32$

Time, minutes   1.5

FIGURE 10 Chiral separation of *trans*-stilbene oxide showing the speed and efficiency of SFC. Column: 4.6 × 250 mm, 10 μm Chirapak-AD, 5 mL/min, 100 bar, 35°C, 40% MeOH in CO_2.

solvent. It is also desirable to achieve this resolution isocratically, with a partition ratio (k) <1. Such separations will produce pairs of peaks, with each peak 0.2–1 min wide. These injections can be "stacked" (multiple injections are made before the first peaks emerge). With the right injection timing, a continuous series of peaks emerge and the selection valve spends nearly 50% of the time delivering to each of the collection vessels (assuming two isomers). An example of 100 stacked injections is shown in Figure 11. One can use absolute time windows or enabling time windows with either threshold or slope triggering. With an ideal example, 2-mL injections of >100 mg/mL can be injected less than 1.5 min apart, producing >4 g/h of each pure enantiomer, with a 20-mm-inner diameter column.

In another example, a pharmaceutical company required 100 g of a pure enantiomer as an intermediate. An outside service company quoted a price of $100,000 to do the purification. In house, they were limited to a reversed phase approach with very poor solubility (<5 mg/mL), the retention times were over 70 min, and the two peaks were 11 min wide. In SFC, solubility was >100 mg/mL and the two peaks were less than 2 min wide, allowing >100 times the throughput. The result are summarized in Table 2.

The sample is loaded into an appropriate sized vessel and a small syringe pump draws the injection volume into the loop. The loop should be substantially larger than the injection volume to avoid sample loss. Small samples might be placed in a 25–50-mL syringe barrel mounted on top of the

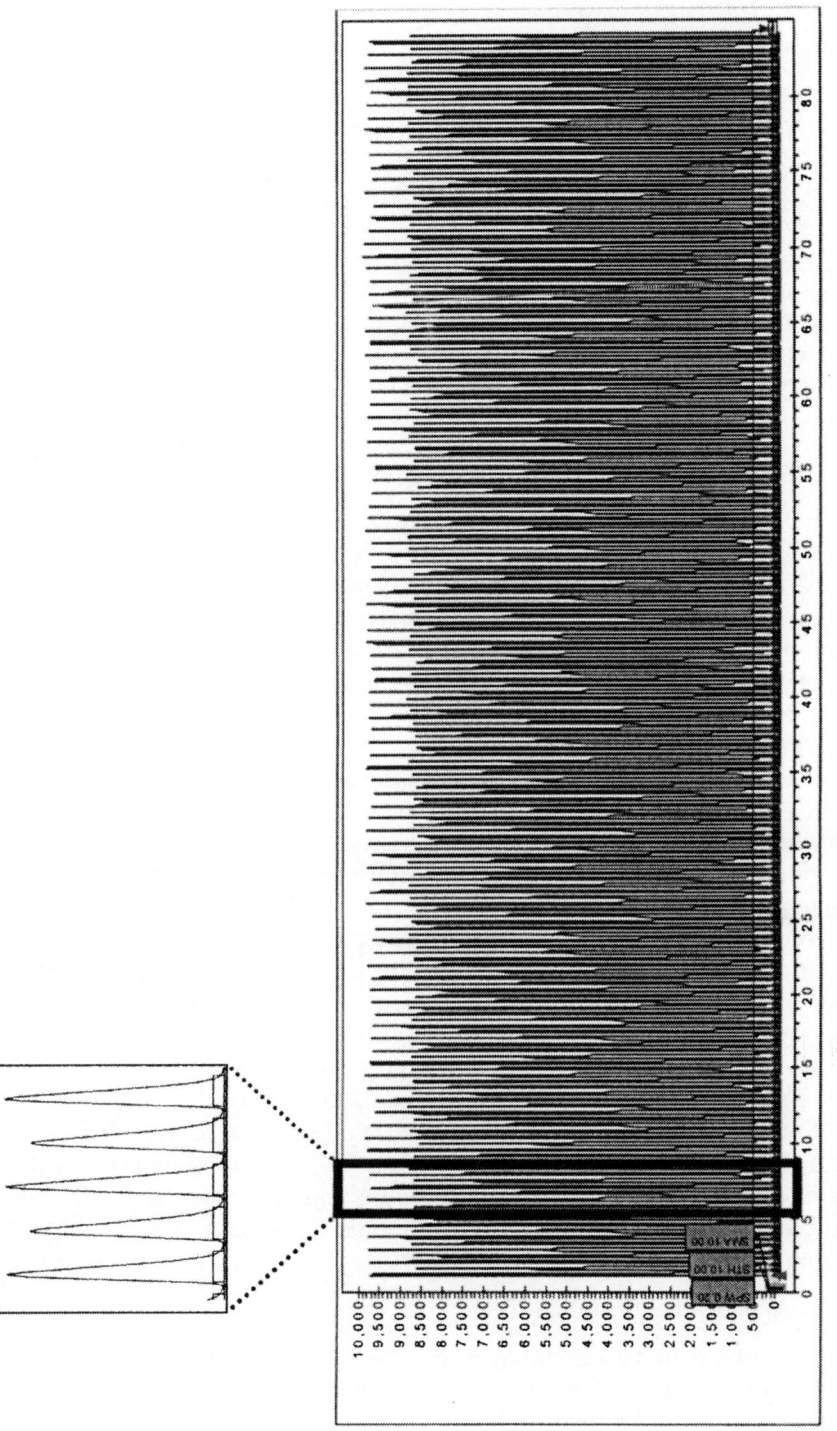

Figure 11 One hundred stacked injections in 85 min. One-milliliter injections at 50-s intervals of 2.5 mg/mL each of ibuprofen and ketoprofen. Column 21.2 × 150 mm, 6 μm Bl Cyano, 20% MeOH in CO_2 at 50 mL/min, 35°C, 100-bar outlet, UV at 220 nm.

TABLE 2 Comparison of Reversed Phase HPLC to SFC for an Actual Industrial Separation of 100 g of a Chiral Intermediate

In Reversed Phase HPLC	In Normal Phase SFC
• 4 months of 8-h days	• 22 h total! (3 working days)
• >$32,000 to buy solvents	• <$600 for solvents
• Peaks recovered in **1200 L (each) of water/acetonitrile**	• Peaks recovered in **10–20 L** methanol!!!
• Requires extensive postchromatographic processing	

injection valve with a Luer adapter. Larger samples might be placed in a 250-, 500-, or even 1000-mL bottle. The collection botles are seldom larger than 1 L because the volume of the fraction from each injection is small. A typical fraction might be 3–5 mL, although much larger volumes are possible (fraction size can be as small as 120–200 mL/h). We have actually collected fractions smaller than the injection volume, although this only occurs at low modifier concentrations.

Peaks can be "time skimmed." It is often not feasible to develop separations with $R_s > 1$. Under these circumstances, the peaks are relatively severely overlapped and recovery of pure fractions will be very limited. Collection at a desired purity requires precise absolute time windows for collection. Such absolute time windows require very good flow precision from the pumps.

With precise absolute time windows, SFC can be used to skim either the leading edge of the first peak or the trailing edge of the last peak and still achieve very high-purity fractions (with degraded recovery), even with relatively poorly overlapped peaks. It is a simple matter to divert the mixed material between the leading-edge and trailing-edge fractions into a separate collection vessel for reinjection later. Unlike HPLC, most of the mobile phase volatilizes are vented, eliminating or minimizing the need to reconcentrate this fraction before reinjection. In general, we do not advise returning this fraction back into the original sample bottle because you will probably dilute the sample and increase the overall separation time. There is also a possibility of contamination.

9.1.1. Using a Laser Polarimeter

An application is under development in which a second detector channel is used to provide additional information, along the lines of using both a belt and suspenders, to ensure that even small amounts of the "wrong" peak do not get accidentally put into the fraction of interest.

The most obvious use for a second channel is in chiral separations using a "chiral" detector that measures the rotation of light. It is often of interest to collect one peak with an enantiomeric excess (e.e.) of >99.9% (corresponding to 99.95% of the mass as one of the enantiomers). Even a small amount of one of the "wrong" peaks negates the intent of the separation. In this instance, the PDR chiral detector is a laser polarimeter. Its output can be set to be ±0.5 V from a $+0.5$-V offset (signal is between 0 and $+1$ V, "zero" is at $+0.5$ V). Right and left rotation results in either a $+$ or $-$ offset. In situations in which the order of peaks gets confused (i.e., a noise spike makes the valve turn at the wrong time), the second detector can detect an error condition. In a series of stacked injections, any situation producing a peak with the wrong sign within a time window can be configured to stop further injections and collect all of the peaks already on-column into a separate container.

At present, the relatively limited dynamic range of laser polarimeters suggests the best semiprep use for such a device is in the mode of error detection. The UV detector can often see the leading edge of a peak long before it becomes visible in the laser polarimeter and has a much higher signal-to-noise (S/N) ratio. The higher S/N ratio allow early, more certain triggering, maximizing sample recovery. For situations where there is no good UV chromophore, the laser polarimeter can become the primary detector.

As in analytical-scale SFC, semipreparative-scale SFC should be the technique of choice for chiral separations of solutes soluble in organic solvents.

9.2. High-Speed Screening

The combination of high-capacity autosamplers and an analytical SFC pumping system allows the collection of purity data on up to eight 96-well plates per day with peak capacity as high as 40. A series of diasteriomers was successfully separated using slightly less aggressive conditions, as shown in Fig. 12.

9.3. SFC–MS

Supercritical fluid chromatography–mass spectrometry is closely related to HPLC–MS except that SFC is "normal" phase and HPLC tends to be used in "reversed" phase mode. In SFC–MS, the mass spectrometer is used in atmospheric pressure ionization (APCI) mode, where there is little fragmentation and the user generally sees a $M + 1$ or $M - 1$ peak (M = molecular weight) is generally prevalent.

Supercritical fluid chromatography–mass spectrometry has been practiced for many years (12–18,50). The mechanical interface between SFC and MS is now very simple, as shown in Fig. 13. A small fraction of the total flow is split off using a tee mounted either just upstream or just downstream of the

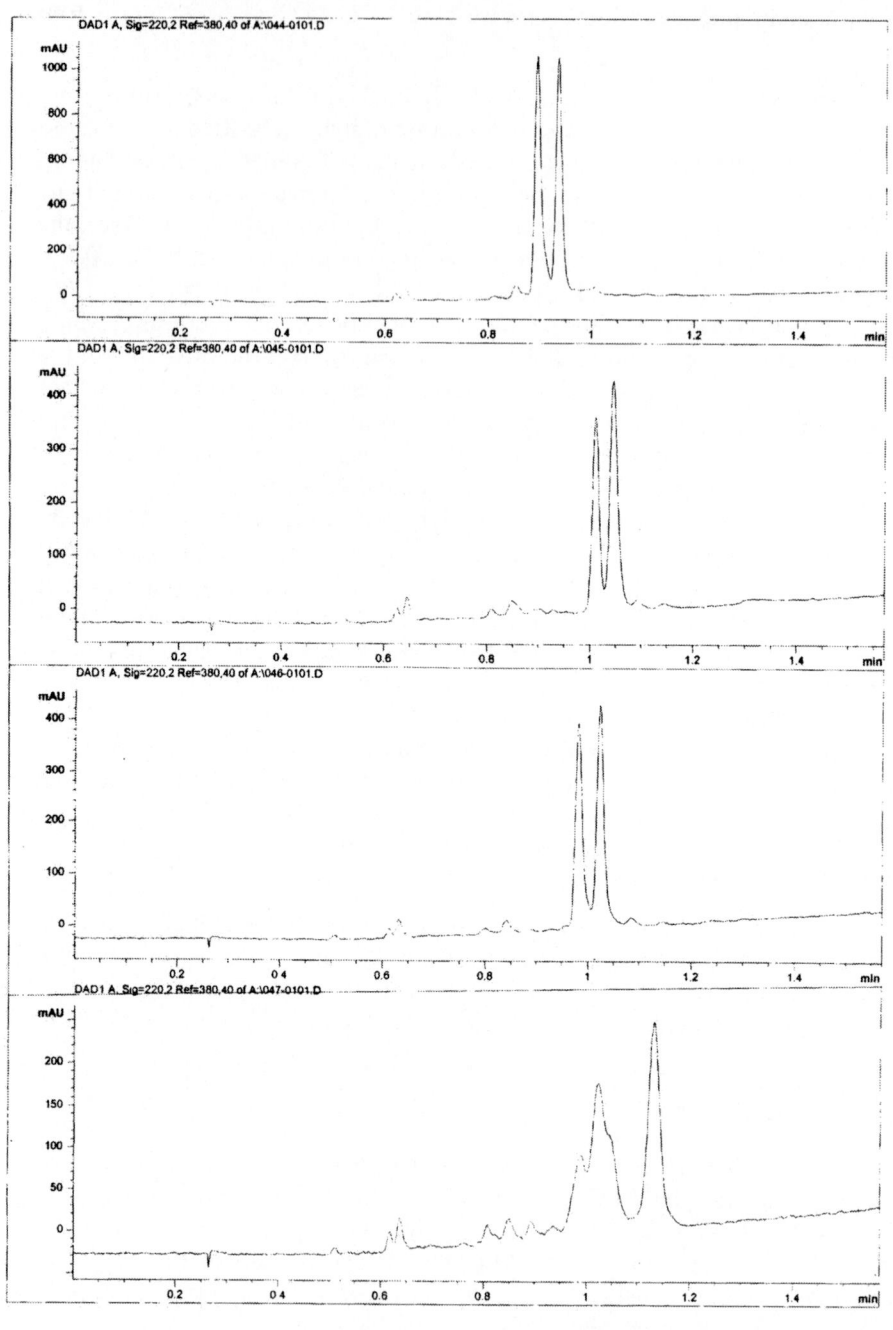

Retention Time, Minutes   1.6 min

FIGURE 12 High-throughput purity analysis showing the results from a series of diasteriomers. Note the high peak capacity. The column was 4.6 × 150 mm packed with 6 μm Cyano.

UV detector. A small-diameter tube is placed in the side arm of the tee and runs to the inlet to the MS. The tubing is usually either PEEK (polyetheretherketone) or fused silica. A smaller-inner-diameter tube penetrates into the source and acts as a restrictor, limiting the flow through this side arm. The fluid pressure drops below the two-phase region somewhere in the transfer line. The small droplets of modifier formed clean the transfer line and rapidly delivers the solute to the MS source.

Restrictor flow is temperature, pressure, and composition dependent, which means that there is a variable, but predictable, split ratio to the MS. A reagent liquid (usually extra methanol) is added with a makeup pump through another tee. This extra methanol flow tends to stabilize the split ratio.

In high-throughput purification, HPLC–MS (and, increasingly, SFC–MS) is usually used to first screen all the samples. Lead generation techniques like combi-chem tend to be relatively inefficient in the automated synthesis of large number of different but related compounds. A private survey of numerous pharmaceutical companies concluded that it is typical to register only 50% of the compounds synthesized. There may be no product, or not enough product, or too many closely related compounds in any one reaction vessel.

Mass spectrometry is used to verify that the proper mass is present. If the solute fragments more than expected or ionizes with a adduct, the MS might not give the information required.

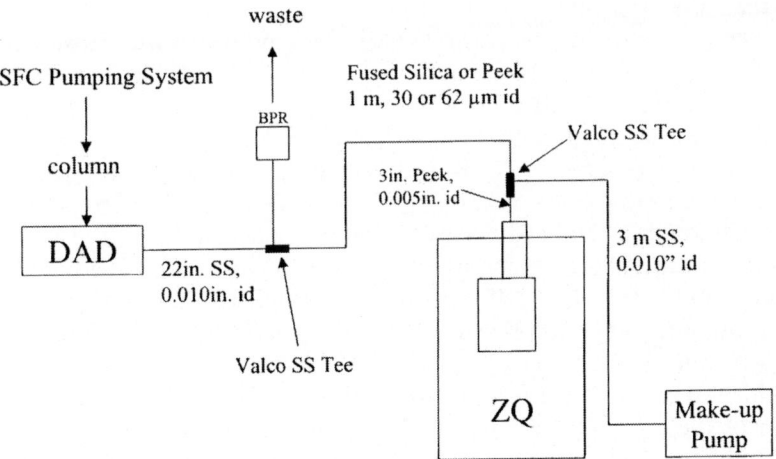

FIGURE 13 Schematic of the simple interface between an SFC and a APCI source one, a Waters ZQ spectrometer.

A UV-vis is diode-array detector is used to indicate the complexity of the sample. Ideally, a single compound is present at high yield. More often, many contaminant peaks are shown. The MS helps determine which peak is of interest, as shown in Fig. 14.

Neither the MS or the UV-vis detector is very good at predicting the amount of material present. A third detector that is mass dependent is sometimes included to quantify the peak of interest. One common detector used in both SFC and HPLC is the evaporative light-scattering detector or (ELSD) (5–8). There are a number of different ELSDs on the market. It is important to use one that is compatible with SFC and is capable of operating during steep composition gradients.

9.4. High-Throughput Purification

Until recently, it was impossible to both rapidly change from one sample to another and collect SFC fractions at near 100% efficiency. Cyclone separators are impractical when each injection involves a different compound, and thousands of compounds are being purified.

With open collection, aerosol generation meant that 30–70% of each sample was lost as an aerosol. In addition to being inefficient, it was also dangerous to the user, because significant quantities of large numbers of compounds with unknown potency were "lost" throughout the system.

The Berger separator does not make aerosols and can collect up to 100% of injected compounds, at the same time allowing a new sample every few minutes. Even though the device does not make aerosols, when the valve actuates there is a momentary "puff" of aerosol generation. Even though this puff represents a negligible amount (<<1 mg) of compound, it was decided to enclose the collection vessels inside a cassette, vented through a waste collection vessel to the outside.

The first embodiment used 25 × 150-mm standard test tubes, compatible with fractions as large as 25–30 mL, containing more than 100 mg of solute. Smaller injections (i.e., 5–10 mg) in up to 1 mL solvent are compatible with 16 × 150-mm test tubes, with a usable volume (~50% full) of 10–12 mL. A few users actually collect close to 100 mg of each compound in these smaller tubes. Shorter tubes (i.e., 16 × 125 mm) are compatible with smaller Genevac dry-down equipment, but decrease the maximum fraction volume allowed in a single tube to 6–10 mL.

A third version employs reusable plastic extenders that allow the expansion of the gas and the separation of the gaseous and liquid phases while collecting into much smaller vials, like the 2-dram vials widely used for storage of libraries. These smaller tubes screw into the bottom of the extender.

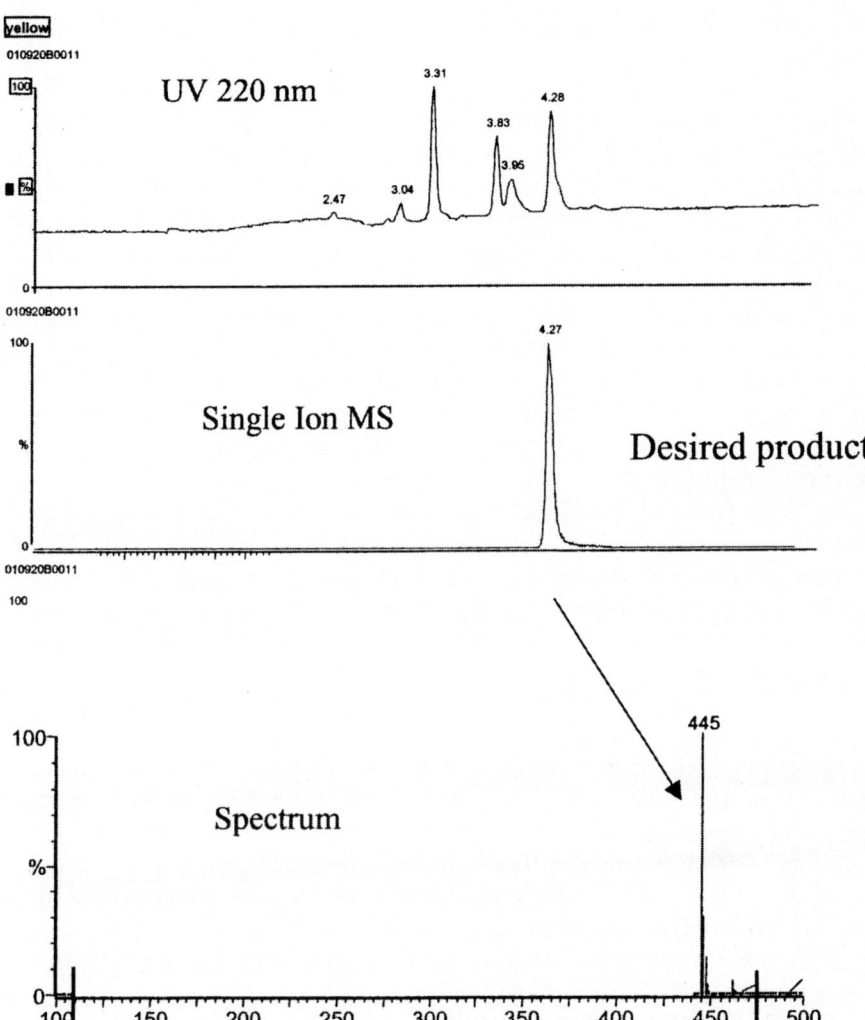

Figure 14 SCF–MS result from a combi-chem sample. Whereas the UV spectrum shows that multiple compounds are present, the mass spectrometer shows that only one of them has the correct mass. Waters ZQ with negative ion APCI interface used.

Both the 16-mm tubes and the smaller vials are mounted in a 4 × 6 plate format.

An enclosed collection cassette has four individual compartments, into which a robot places the various tubes/collection vials, as shown in Fig. 8. After a fraction is collected, the robot returns the fraction to the original position of the tube/vial. This design limits the user to no more than four fractions per injection. However, this hardware is intended to be used with analytical SFC–MS, which identifies the "correct" peak and its retention time. A spreadsheet is created which relates analytical retention time windows to the desired peak. The chromatography is directly scaled to the semiprep level and the retention time windows are passed to the semiprep hardware. In most cases, a single fraction is collected. The analytical SFC–MS is intended as a form of triage to eliminate as many marginal samples as possible to minimize the number of prep runs. With MS tracking, the intent is to collect one peak per injection.

Abbott (51–53) built a custom variant, using the Berger pumping system, peak detection, and separator technology, combined with a two-armed robot. When one arm is positioned to collect a fraction, the other arm moves ahead to preposition for collection of the next sample. The system seals on individual test tubes, mounted in an array, on the robot deck. All of the gas is collected and vented through the standard Berger waste collection and venting hardware.

The Abbott configuration allows the user to collect an unlimited number of peaks for each injection. Such an approach is ideal for collecting peaks from natural products.

An alternative could be to have two MSs, one for analytical triage and the other for mass triggered fraction collection. The analogue input to the peak detector could be a single ion monitoring channel, which would make MS triggered fraction collection trivial. Kassel built another variant performing mass directed fraction collection (54) using the Berger pumping system and separator technology.

10. CONCLUSIONS

Modern SFC offers a multitude of advantage compared to HPLC. In the past, the advantages tended to be at least partially offset by increased complexity and capital cost.

As a technique, SFC has been "crossing the chasm" from early adopters to mainstream acceptance for many years. In the last several years, many larger pharmaceutical companies have begun to treat SFC as a mainstream technology, buying multiple units or a mix of analytical and semiprep units in bulk orders. There is now wide acceptance of the fact that SFC is superior to

HPLC for chiral separations. On the analytical scale, the higher-speed chromatography and faster reequilibration makes method development dramatically faster.

These advantages become more compelling as the scale of the chromatography increases. At the semiprep scale, the lower operating costs can completely pay for the equipment in less than a year. Expensive stationary phases can be utilized at a higher rate, to the point where a much smaller, much less expensive column can produce similar throughput.

The inherent advantages of SFC over HPLC are also becoming more obvious in aspects of drug discovery and development outside of chiral separations. Drug discovery has come to be dominated by combinatorial methods that generate huge numbers of new compounds. These compounds need to be analyzed and often purified. The inherent speed and operating cost advantage of SFC over HPLC means that the former should replace the latter in determining the purity of very large numbers of compounds and in purifying those compounds. The ease of use and scalability of the technique should make it the technique of choice for the majority of chromatographic separations.

Supercritical fluid chromatography and related technologies offer the opportunity for pharmaceutical companies to become more "green" (environmentally friendly). It remains to be see if this aspect will be embraced or simply receive lip service. The financial incentives are very real, however, which is likely to continue the trend toward greater penetration of SFC in drug discovery and development.

REFERENCES

1. Berger TA. Separation of polar solutes by packed column SFC. J Chromatogr A 1997; 785:3–33.
2. Gyllenhaal O, Karlsson A, Vessman J. Applications of packed column SFC in the development of drugs. In: Anton K, Berger C, eds. Supercritical Fluid Chromatography with Packed Columns. Chromatographic Science Series. Vol. 75. New York: Marcel Dekker, 1984, Chap 9.
3. Bargmann N, Caude M. SFC of drugs and related compounds. In: Caude M, Thiebaut D, eds. Practical Supercritical Fluid Chromatography and Extraction. Amsterdam: Harwood Academic Publishers, 1999, Chap 8.
4. Berger TA. Practical advantages of packed column SFC in supporting combinatorial chemistry. In: Unified Chromatography, Parcher JF, Chester TL, eds. ACS Symposium Series. Vol. 748. Washington, DC: American Chemical Society, 2000, Chap 12.
5. Berry AJ, Ramsey ED, Newby M, Games DE. Applications of packed column SFC using light-scattering detection. J Chromatogr Sci 1996; 34:245–253.

6. Herbreteau B, LaFosse M, Morin-Allory L, Dreux M. Analysis of sugars by supercritical fluid chromatography using polar packed columns and light-scattering detection. J Chromatogr 1990; 505:299–305.

7. Camel V, Thiebaut D, Caude M, Dreux M. Packed column subcritical fluid chromatography of underivatized amino acids. J Chromatogr 1992; 605:95–101.

8. Anton K, Bach M, Geiser A. Supercritical fluid chromatography in the routine stability control of antipruritic preparations. J Chromatogr 1991; 553:71–79.

9. Giddings JC, Meyers M, Wahrhaftig AL. Int J Mass Spectrom Ion Physi 1970; 4:9–20.

10. Randal LG, Wahrhaftig AL. Anal Chem 1978; 50:1703–1705.

11. Voorhees KJ, Zaugg SD, DeLuca SJ. Supercritical Fluid Chromatography/Mass Spectrometry. In: White CM, ed. Modern Supercritical Fluid Chromatography. New York: Huthig, 1988, Chap 4.

12. Games DE, Berry AJ, Mylcreest IC, Perkins JR, Pleasance S. Supercritical Fluid Chromatography–Mass Spectrometry. In: Smith RM, ed. Supercritical Fluid Chromatography. RSC Monograph Series. London: Royal Society of Chemistry, 1988, Chap 6.

13. Lane SJ. SFC-MS in the pharmaceutical industry. In: Smith RM, ed. Supercritical Fluid Chromatography. RSC Monograph Series. London: Royal Society of Chemistry, 1988, Chap 7.

14. Matsumoto K, Nagata S, Hattori H, Tsuge S. Development of directly coupled supercritical fluid chromatography with packed capillary–mass spectrometry with atmospheric pressure chemical ionization. J Chromatogr 1992; 605:87–94.

15. Morgan DG, Norwood DL, Fisher DL, Moseley MA III. Directly coupled packed column SFC with APCI/MS. Poster and abstract presented at American Society for Mass Spectrometry. Portland, OR, 1996

16. Pinkston JD. Supercritical fluid chromatography/mass spectrometry. In: Caude M, Thiebaut D, eds. Practical Supercritical Fluid Chromatography and Extraction. Amsterdam: Harwood Academic, 1999, Chap 5b.

17. Harrasch PB, Bente PF, Berger T. Combining mass spectrometry with supercritical fluid chromatography for high-speed pharmaceutical analysis. Business Briefing. Life Sci Technol Spring, 2003.

18. Ventura MC, Farrell WP, Aurigemma CM, Greig MJ. Packed column supercritical fluid chromatography/mass spectrometry for high throughput analysis. Anal Chem 1999; 71:2410–2416.

19. Hoke SH II, Tomlinson JA II, Bolden RD, Morand KL, Pinkston JD, Wehmeyer KR. Increasing bioanalytical throughput using pcSFC-MS/MS: 10 minutes per 96-well plate. Anal Chem 2001; 73:3083–3088.

20. Blum AM, Lynam KG, Nicolas EC. Use of a new Pirkle-type chiral stationary phase in analytical and preparative subcritical fluid chromatography of pharmaceutical compounds. Chirality 1994; 6:302–313.

21. Jusforgues P, Shaimi M, Barth D. Preparative supercritical fluid chromatography: grams, kilograms, tons. In: Anton K, Berger C, eds. Supercritical Fluid Chromatography with Packed Columns. Chromatographic Science Series. Vol. 75. New York: Marcel Dekker, 1998, Chap 14.

22. Coleman K, Boutant R, Verillon F. Practical aspects of preparative supercritical fluid chromatography. Isol Purif 1999; 3:9–19.
23. Lembke P. Production of high purity *n*-3 fatty acid–ethyl esters by process scale supercritical fluid chromatography. In: Anton K, Berger C, eds. Supercritical Fluid Chromatography with Packed Columns. Chromatographic Science Series. Vol. 75. New York: Marcel Dekker, 1998, Chap 15.
24. Nicoud R-M, Clavier J-Y, Perrut M. Preparative SFC: basics and applications. In: Caude M, Thiebaut D, eds. Practical Supercritical Fluid Chromatography and Extraction. Amsterdam: Harwood Academic, 1999, Chap 10.
25. Depta A, Giese T, Johannsen M, Brunner G. Separation of sterioisomers in a simulated moving bed–supercritical fluid chromatography plant. J Chromatogr A 1999; 865:175–186.
26. Berger TA. Packed Column SFC. London: Royal Society, 1995.
27. Berger TA, Wilson WH. High-speed screening of combinitorial libraries by gradient packed-column SFC. J Biochem Biophys Methods 2000; 43:77–85.
28. Stringham RW, Lynam KG, Grasso CC. Application of SFC to rapid method development. Anal Chem 1994; 66:1949–1954.
29. Lynam KG, Nicolas EC. Chiral HPLC vs. chiral SFC: evaluation of long term stability and selectivity of Chiracel OD using various eluents. J Pharma Biomed Anal 1993; 11:1197–1206.
30. Gere DR. Supercritical fluid chromatography. Science 1983; 222:253–259.
31. Lapinski CA, Lombardo F, Dominy BW, Feeney PJ. Experimental and computational approaches to estimate solubility and permeability in drug discovery and development settings. Adv Drug Deliv Rev 1997; 23:4–25; 2001; 46:3–26.
32. Berger TA, Smith J, Fogelman K, Kruluts K. Am Lab News October, 2002.
33. Berger TA. Packed Column SFC. London: Royal Society, 1995:56.
34. Berger TA, Wilson WH. Separation of drugs by packed column supercritical fluid chromatography. J Pharm Sci 1994; 83:281.
35. Berger TA, Fogelman K, Staats T, Bente P, Crocket I, Farrel W, Osonubi M. The development of a semi-preparatory scale supercritical-fluid chromatograph for high throughput purification of 'combi-chem' libraries. J Biochem Biophys Methods 2000; 43:87–111.
36. Inovations in Pharmaceutical Technology.
37. Standard Guide for Purity of Carbon Dioxide Used in Supercritical Fluid Application. ASTM Designation: E 1747-95. West Conshohocken, PA: American Society of Testing Materials, 1995.
38. Strubinger JR, Song H, Parcher JF. High pressure phase distribution isotherms for supercritical fluid chromatographic systems. 2. Binary isotherms of carbon dioxide and methanol. Anal Chem 1991; 63:104–108.
39. Mourier PA, Eliot E, Caude M, Rosset R. Anal Chem 1985; 57:2819.
40. Macaudiere P, Tambute A, Caude M, Rosset R, Alembik MA, Wainer IW. Resolution of enantiomeric amides on a Pirkle-type chiral stationary phase. J Chromatogr 1986; 371:159.
41. Wolf C, Pirkle WH. LC-GC 1997; 15:352–363.

42. Wolf C, Pirkle WH. Enantiomer separation by sub- and supercritical fluid chromatography on rationally designed chiral stationary phases. In: Anton K, Berger C, eds. Supercritical Fluid Chromatography with Packed Columns. Chromatographic Science Series. Vol 75. New York: Marcel Dekker, 1998, Chap 9.

43. Liu Y, Berthod A, Mitchell CR, Xiao TL, Zhang B, Armstrong DW. Super/subcritical fluid chromatography chiral separations with macrocyclic glycopeptide stationary phases. J Chromatogr A 2002; 978:185–204.

44. Sandra P, Medvedovici A, Kot A, David F. Selectivity tuning in packed column SFC. In: Anton K, Berger C, eds. Supercritical Fluid Chromatography with Packed Columns. Chromatographic Science Series. Vol. 75. New York: Marcel Dekker, 1998, Chap 6.

45. Anton K, Eppinger J, Frederiksen L, Francotte E, Berger TA, Wilson WH. J Chromatogr A 1994; 666:395–401.

46. Phinney KW. SFC of drug enantiomers. Anal Chem 2000; 72:A204–A211.

47. Bargmann-Leyder N, Tambute A, Caude M. A comparison of LC and SFC for cellulose- and amylose-derived chiral stationary phases. Chirality 1995; 7:311–325.

48. Biermanns P, Miller C, Lyon V, Wilson WH. Chiral resolution of β-blockers by packed column SFC. LC-GC 1993; 11(10):744–747.

49. Villeneuve MS, Anderegg RJ. Analytical supercritical fluid chromatography using fully automated column and modifier selection valves for the rapid development of chiral separations. J Chromatogr A 1998; 826:217–225.

50. Harrasch PB, Bente PF, Berger T. Combining mass spectrometry with supercritical fluid chromatography for high-speed pharmaceutical analysis. Business Briefing Life Sci Technol, Spring 2003.

51. Olson J, Pan J, Hochlowski J, Searle P, Blanchard D. Customization of a commercially available prep scale SFC system to provide enhanced capabilities. 2002.

52. Hochlowski J, Sauer D, Sowin T. Preparative scale supercritical fluid chromatography in support of HTOS. Presented at High Throughput Organic Synthesis. San Diego, CA, February 2002.

53. Hochlowski J, Searle P, Gunawardana G, Sowin T, Pan J, Olson J, Trumbull J. Development and application of a preparative scale supercritical fluid chromatography system for high throughput purification. Presented at Prep 2001. Washington, DC, 2001.

54. Xu R, Cai Z, Fogelman K, Wikfors R, Worle V, Stublen N, Kassel DB. Mass-directed purification of compound libraries by automated semi-preparative SFC/MS. Oral presentation and poster. American Society for Mass Spectrometry. Orlando, FL, 2002.

13

Drug Extraction

M. D. Luque de Castro, A. Jurado-López, and J. L. Luque-Garcia
Analytical Chemistry Department, University of Córdoba, Córdoba, Spain

After the initial euphoria of the late 1980s and early 1990s, supercritical fluid extraction (SFE) has consolidated as a powerful tool for the preparation of environmental, pharmaceutical, and polymer samples and, especially, food samples (particularly for extracting fat, and lipids in general) (1).

The combined liquidlike solvating capabilities and gaslike transport properties of supercritical fluids (SCF) lead to improved mass transfer and reduced extraction time (2,3). Moreover, the dissolving power of a SCF can be modified by simply changing the applied pressure and/or temperature (2,4). The most frequently used solvent in SFE is supercritical carbon dioxide (CO_2), which has low-critical-temperature (31.1°C) and pressure (73.8 bar) values. It is not hazardous to the environment and it is safer than the common solvents used by analysts in classical extraction procedures. Mixtures of solvents in the supercritical state allow the analyst greater selectivity than conventional extraction techniques. Although CO_2 is nonpolar with a dipole moment of zero, when used in SFE its polarity can be manipulated by adding small amounts of an entraining solvent, such as methanol (5).

Supercritical fluid extraction has so far been applied mostly to solid samples and also, occasionally, to liquid ones (whether as such or following passage through a sorbent that was subsequently placed in the extraction chamber, a procedure also applicable to gaseous samples).

Extraction can be performed in either a static or dynamic mode. In the latter mode, the extractant is continuously flowing through the vessel, whereas the static mode allows the sample to "soak" in it. Problems associated with continuous extraction are restrictor plugging (6) and that dissolved water modifies the SCF, making extraction conditions difficult to predict, whereas static extraction can be inefficient and slow. In practice, a combination of both has been found to be the most effective to achieve maximum efficiency (7).

1. MAIN ASPECTS OF SFE

1.1. The Supercritical Fluid Extractor

All of the SCF extractors, whatever their complexity and cost, share the same essential components, which are schematically depicted in Figure 1 and described in the following subsections.

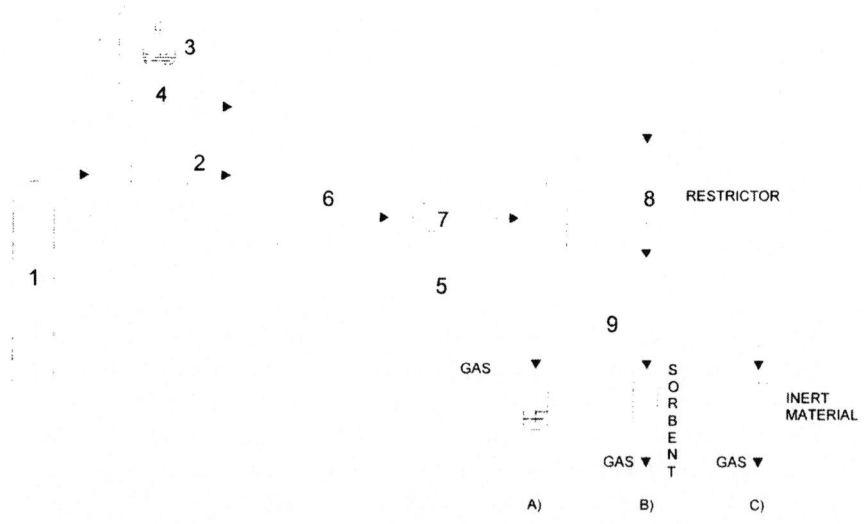

FIGURE 1 Schematic diagram of straightforward SCF extractor: (1) extracting fluid source, (2) extractant propulsion unit, (3) modifier reservoir, (4) modifier propulsion unit, (5) oven, (6) equilibration coil, (7) chamber containing the thimble or sample cell, (8) back-pressure regulator, and (9) collection system (A: bubbling; B: sorption, C: cryogenic trapping). (Reproduced with permission of Elsevier.)

1.1.1. Source of the Extraction Fluid

This unit consists of a steel, aluminum, or stainless-steel cylinder (preferably furnished with a dip tube), a pressurized head, and a device allowing cooling for long enough to ensure equilibration.

1.1.2. Impulsion System

The devices typically used to propel SCFs consist of high pressure pumps of two main types: syringe pumps and dual-piston pumps. The former type has the advantage that it provides a pulseless flow; however, the volume of fluid tha can be used is limited by that of the syringe body. Only a single-impulsion device needs be used in SFE; however, an additional syringe or piston pump may be used to facilitate the addition of modifiers. These substances can also be inserted by adding the entrainer in the pump during loading, using a modifier chamber between the pump and the extraction chamber, soaking the sample in the modifier, or using a commercially available SCF already containing the modifier. Each option has specific advantages and short-comings (8).

The interface between the cylinder and the impulsion system consists of a connecting line, which should be as short as possible, furnished with a series of safety devices, including a bleed valve, which is specially useful to avoid the unwanted presence of gas in the pump when the connecting line is not short enough, a safety valve to avoid pressures above those the system and material used can reasonably withstand, a check valve to minimize contamination of the fluid stored in the cylinder; and filters of small pore size to remove any contaminant particles from the fluid.

1.1.3. Sample Chamber and Cell

The extractor cell or thimble, which should be accommodated in an appro-priate chamber, must meet a series of requirements, namely:

1. It should enable ready, convenient insertion of the sample.
2. It should be easy to shut without the need of any tools.
3. It should be amenable to sealing via software.
4. It should be made of an inert, corrosion-resistant material.
5. It should lend itself to ready, rapid mounting and disassembling from the system.
6. It should encompass a broad range of capacities (the usual range is 1–10 mL).
7. It should be capable of withstanding pressures up to at least 700 bar.

The sample cell can be single or multiple, depending on the specific goal of the extraction process. Thus, a single-cell extractor is usually more appropriate for research purposes, whereas one to nine units are typically used in routine extractions.

An equilibration unit (viz and adequately long coil) is placed between the impulsion system and the sample compartment in the oven to allow the extractant to reach the programmed pressure and temperature conditions before coming into contact with the sample.

1.1.4. Oven

The temperature required for proper development of the extraction process is usually provided by an oven, allowing easy, precise setting and changing of temperatures between room level and 300°C (commercially available equipment encompasses broader ranges).

The thermostat equipment must allow programming of the chamber, restrictor, restrictor outlet, and collection system temperatures—and the impulsion unit to be cooled, if needed. Ovens from gas chromatographs have often been used to construct laboratory SCF extractors. Rapid attainment of the preset temperature and cooling capabilities are desirable, with a view to reducing the extraction time and the interval between successive extractions, respectively.

1.1.5. Back-Pressure Regulator or Restrictor

The depressurization system acts as an interface between the supercritical conditions in the extraction cell and the atmospheric conditions to which the extract is eventually subjected when the extractor is not connected on-line to a chromatographic system for the individual separation of extracted species, with a view to their subsequent detection. A wide variety of commercially available devices for this purpose exists that range from straightforward glass capillaries—the end of which can be readily cut of in the event of clogging—to hand-operated restrictors to computer-controlled units. This is one of the characteristic components of commercial SCF extractors (one that can differ markedly among manufacturers).

1.1.6. Collection System

The unit intended to collect the effluent from the extractor following depressurization is only included in an SFE assembly when the extractor is not coupled on-line to another unit, whether a chromatograph or detector, and is located after the depressurization unit. There are three main ways of collecting solutes, namely solvent bubbling, collection on a sorbent material, and cryogenic trapping.

Solvent bubbling is the simplest way of collecting the effluent from an SCF extractor, as it only requires inserting the restrictor outlet through the septum of a vital containing a few milliliters of solvent. The efficiency of this collection system depends on (a) the solubility of the analyte in the collecting solvent, which is a function of the solvent polarity, (b) the solvent volume used, which may not be close to the saturation value, (c) the size of the bubbles formed, which dictates the exchange interface value, (d) the time of contact with the collector, which depends on the SCF output flow rate and the distance traveled by the bubbles, and (e) the solvent temperature, which has decisive influence on solubility. The different types of collection systems in commercial SCF extractors have been designed with a view to overcoming the adverse effects of the previous factors.

Collection on a sorbent involves the use of a solid material, either in the line or at the restrictor outlet. A number of materials have been (9) and continue to be investigated (10–12) with a view to optimizing the collection of different types of analyte. This collection mode involves and additional step: desorbing the analytes from the sorbent by elution with a small volume of solvent for their subsequent determination or, alternatively, thermal desorption and sweeping by the eluent if an on-line coupled extraction–chromatographic system is being used.

With *cryogenic trapping*, the extraction mixture is cooled down until the SCF expands and the analytes deposit. The trapping temperature to be used depends on whether the analytes are to be isolated from the fluid or this is to be liquefied and the collection vessel sealed in order to avoid losses of analytes through partial crystallization or the formation of aerosols during cooling. When the temperature of the cryogenic trap is very low, the restrictor must be heated in order to avoid the formation of two phases. As with collection on a solid sorbent, an additional preparative step is required.

1.2. Couplings of an SF Extractor

Because SFE is a sample extraction–separation technique, it must precede other steps of the analytical process if the type and content of species of interest in the extract are to be accurately determined. The analytical equipment required to develop the steps following extraction can either be coupled on-line to it or performed off-line. The way the extract should be treated with a view to identifying or quantifying the target analytes depends on its complexity and the type of information required. Thus, the analytes may require cleanup, individual separation, or some other treatment prior to reaching the detector.

The off-line mode is to be preferred when a deep knowledge of the features of the extraction process concerned is required, as this will allow such experimental variables as pressure, temperature, SCF polarity and flow rate, extractor volume and dimensions, extraction time, and sample size to be optimized.

On-line coupling can be used to integrate the extraction step with others of varying complexity ranging from column cleanup (13) to high-performance liquid chromatography–gas chromatography–mass spectrometry (HPLC–GC–MS) (14). On-line coupling methods are especially sensitive as a result of the whole extracted material being transferred to the coupled equipment, which also minimizes potential errors arising from manipulation steps. In addition to increased sensitivity and automatability, direct coupling of SFE to other techniques (particularly chromatographic ones) provides additional benefits derived from the fact that the target species are not exposed to atmospheric oxygen, light, or high temperatures during analysis.

1.2.1. SFE Chromatography

All three types of chromatograph (gas, supercritical fluid and liquid) have been used in combination with SF extractors. The effectiveness of these combined assemblies rests on the availability of an appropriate interface between the two coupled techniques, this role can be assigned to an external accumulator or to a direct connection to the chromatographic column or the retention interface.

SFE–GC. Although gas chromatography can only be applied to thermally stable compounds, it has been widely used in conjunction with SFE for this type of analyte. It is worth emphasizing that the solvent power of CO_2, the most common SCF, is restricted to nonpolar and scarcely polar compounds, so they are suitable for GC. Using these two techniques jointly entails considering the following aspects:

1. Because the supercritical solvent is usually a liquefied gas, its return to the gaseous state on depressurization before reaching the interface causes it to expand by a factor of around 1000.
2. The amount of solute transferred to the column can be diminished by formation of a frozen CO_2 plug in its head. Any water present in the sample may also plug the column restrictor with ice deposits. Both shortcomings can be circumvented using a hot injector as the interface or a cooled thermal desorption injector.
3. The presence of modifiers is occasionally incompatible with an on-line coupled SFE–GC system (15).
4. The restrictor diameter is a key variable in SFE–GC applications (not only with direct injection into the column, the diameter of which is dictated by the flow rate of gas resulting from the depressurized SCF).

Both packed and capillary columns are used with the SFE–GC tandem. The detector is usually of the mass spectrometry, flame ionization, or electron capture type. The most suitable interface for each application will be that allowing the extractant to be removed prior to the column and, hence, to the detector.

The reproducibility of the SFE–GC hyphenated technique is comparable to that of off-line systems (16,17).

SFE–SFC. One of the most immediate advantages of coupling SFE and supercritical fluid chromatography (SFC) is that the former constitutes a highly suitable means for delivering samples to the latter. Because the solvent used to inject the sample is the same as the mobile phase, the principal requisite for effectively coupling two techniques (viz. compatibility between the output of the first system and input of the second) is met. One additional advantage of this hyphenated technique over SFE–GC and SFE–HPLC is the low likelihood of sample components insoluble in the mobile phase reaching the chromatographic column.

The potential of the SFE–SFC couple has been assessed by a number of authors, as this combined technique solves one of the major problems encountered in using SFC for bioanalysis. In fact, many pretreatments for biological samples use polar solvents, which are highly detrimental to the phase systems employed in this technique. This problem is readily solved by the SFE–SFC combination, proper performance of which relies on the following conditions:

1. The extraction chamber volume should be suited to the sample size afforded by the SCF chromatograph.
2. The pressure drop in the SCF during transfer of the extract to the chromatograph should be as low as possible.
3. The chromatograph should be pressurized and equilibrated at the pressure to be used in the chromatographic determination before it receives the extract.

All three types of ordinary SFC column (viz. open-ended, packed-capillaries, and normal-bore models) have been used with the SFE–SFC couple.

Many SFE–SFC applications use a second pump to pressurize the extraction chamber, the first pump being solely employed to effect the chromatographic separation. These systems pose few technical problems and are highly flexible by virtue of the extractor and chromatograph operating independently (in a static or dynamic manner).

SFE–HPLC. The SFE–HPLC combination is a logical extension of the above-described hyphenated techniques aimed at addressing the analysis of extracts inaccessible to GC and SFC owing to the high polarity, molecular weight, or thermal lability of the analytes. Interfacing a SCF extractor to a liquid chromatograph is the most complex operation involved in SFE hyphenated techniques as a result of the difficulty of coupling a preparative technique that produces a gas at the interface with a chromatographic technique that handles a liquid mobile phase.

Selecting the Most Suitable Type of Chromatography for Coupling to SFE. Because all three types of chromatography can be coupled to SFE, the choice in each case should be dictated by the characteristics of the analytes to be isolated and determined. If GC can be used, it is preferred because of its consolidated status and the high sensitivity and flexibility of the detectors with which it is compatible. Analyzing SCF extracts by HPLC provides one special advantage over GC, as most coextracted species contain no chromatographic groups. On the other hand, most organic substances can be sensed with an ordinary GC detector e.g., a flame-ionization detected (FID).

1.2.2. Miscellaneous Combinations

When the target analytes elicit a specific or highly selective response from a given detector, the detector can be coupled on-line to the extractor in order to enhance the determination with increased ease of automation of the overall process and the ability to monitor the extraction kinetics.

As in SFE–chromatography combinations, properly designing the interface between the two techniques is one of the most crucial steps coupling SFE to any other type of technique. These alternative combinations have usually involved detectors and have been aimed at expediting measurements of the extracted species or a monitoring the extraction kinetics. In coupling a SCF extractor to a detector (whether destructive or nondestructive), one should consider the drastic working conditions of the extraction process (i.e., high pressure, temperature and solvent strength, or corrosivity in the SCF) and the requirements of the particular detector technique. The interfaces between an SCF extractor and a molecular spectroscopic detector can be classified as high pressure interfaces and low pressure flow injection. In the first instance, the most common type is the windowed flow cell, used in on-line detection with molecular spectroscopic techniques, although fiber-optic-based flow cells are also used. Low pressure flow injection interfaces have been used as links between the extractor and either a photometric detector, a flowthrough potentiometric sensor or a piezoelectric sensor in dynamic flow injection systems.

Both molecular and atomic detectors have been used in combination with SCF extractors for monitoring purposes. Thus, the techniques used in combination with SFE are infrared spectroscopy, spectrophotometry, fluorescence spectrometry, thermal lens spectrometry, atomic absorption and atomic emission spectroscopies, mass spectrometry, nuclear magnetic resonance spectroscopy, voltammetry, and piezoelectric measurements.

1.3. Variables Influencing SFE

In a SFE process, the analytes are transferred from their host matrix to the SCF, flushed from the extraction cell by the fluid, and collected or sent to a detector or chromatograph for analysis. Unsurprisingly, the performance of the SFE technique is thus affected by a number of variables, the most significant of which are listed in Table 1.

The large number and variety of factors on which SFE performance relies makes optimizing a rather difficult task. Multivariate optimization approaches have been used from the beginning of this technique to both minimize the processing time and increase the extraction efficiency (18,19).

Supercritical fluid extraction variables can be altered with a view to improving the extraction efficiency and/or the selectivity.

Variables can be adjusted in order to increase the extraction rate and/or the maximum amount of analyte that can be extrated. Although increasing the extraction rate expedites the process, it is probably more interesting to improve the overall analyte recovery. In some cases, the analyte can only be recovered to a given extent, however much the extraction is prolonged. The variables on which increased extracted levels rely are the most significant regarding optimization when quantitative extraction is sought.

TABLE 1 Experimental Variables Affecting SFE Efficiency

Properties of the fluid	Dynamic and geometric factors
Nature (polarity)	Extractant flow rate
Pressure (density)	Extraction cell dimensions
Temperature	Extraction time
Presence of a modifier	
Modifier concentration	Sample treatment
	Addition of liquids, solvents,
Properties of the solute	derivatising reagents, etc.
Type of analyte (volatility,	Addition of solids
polarity, molecular weight)	
Concentration	Analyte collection mode
	Solvent bubbling
Properties of the solid	Sorption
Sample size	Cryogenic trapping
Particle size	On-line coupling to a detector
Nature of matrix	or chromatograph
Presence of other	
extractable substances	
Sample conditions	

TABLE 2 Main Applications of SFE to Drug Extraction

Analytes	Source	Modifier	Ref.
β-Blockers	Serum and urine	—	130, 131
β-Carotene	Carrots, paprika	Ethanol	86–89
13-*Cis*-retinoic acid	Pharmaceutical formulations	Methanol	144
Active compounds	Lovage roots	—	23
	Medical plants	—	35, 36, 39
Amphetamines	Hair	Trichloromethane/isopropyl alcohol	103
	Urine	—	135
Anabolic agents	Animal tissues, serum, urine, and tablets	—	104–108, 111–114, 128, 132, 133, 142, 143
Antianxiety drugs	Kava roots, serum	—	25, 128
Anticancer substances	*Salvia miltiorrhiza* Bunge, *Artemisia annua* L., grape and tomato skins, Korean orange peels	—	24, 32, 34, 90, 91, 93, 94
Antiepilepsy drugs	Serum	—	129
Anti-inflammatory drugs	Serum and suppositories	—	128, 154
Antithyroid drugs	Beef muscle	—	109
Baicalein and baicalin	Scutellaria root	—	22
Benzodiazepines	Tablets	Methanol	148
Budesonide	Blood plasma	—	125
Cannabinoids	Cannabis	—	52
Carotenoids	Leaves and algae	—	53, 54
Cocaine	Addict hair	Triethylamine/water or methanol	100–102
Cocaine and its metabolites	Urine	—	134

Compound	Source	Solvent	Ref.
Ephedrine	Ephedra sinica	Diethylamine and methanol	37
Essential oils	Savory, peppermint, and dragonhead	Dichloromethane	69
Flavone	Blood plasma	Methanol	124
Flavor and fragance compounds	Lavender, rosemary, crimson glory, chamomile flowers, rose, peppermint, eucalyptus, and guajava	—	28–31, 58–62
Imidazole antimycotics	Creams	Methanol	155
Indole alkaloids	*Catharanthus roseus*	—	44
Lignans	*Schisandra chinensis* fruits	—	92
Magnolol and honokiol	*Magnolia officinalis*	—	26, 27
Mebeverine alcohol	Blood plasma	Methanol	123
Michellamines A and B	*Ancistrocladus korupensis* leaves	—	33
Misoprostol	Controlled-release drug formulation	Formic acid or methanol	151, 152
Morphine	Dried blood	Methanol/triethylamine	122
Natural antioxidants	Rosemary and tamarind	—	48–51
Nicotine	Moist snuff	—	41
Opiates	Addict hair and urine	Methanol/triethylamine/water	97–99
Oxindole alkaloids	*Uncaria tomentosa*	Methanol	46
Parabens	Cosmetic products	—	157
	Cosmetic products	Acetonitrile	158
Pirrolizidine alkaloids	*Senecio* species	Methanol	45
Prostaglandins	Aqueous solutions	—	136

TABLE 2 Continued

Analytes	Source	Modifier	Ref.
Psoralen, isopsoralen, and pentosalen	Traditional Chinese medicines	Trichloromethane or methanol	149, 150
Purine alkaloids	Roasted coffee beans, mate	—	40, 42
Sulfonamides	Animal tissues	—	117
	Animal tissues	Methanol, ethanol, acetone, acetonitrile	115, 116, 118, 119
Sulfhamethazole and trimethoprim	Septra infusions	—	156
Sunscreen agents	Cosmetic products	—	159, 160
Temazepan	Blood	Ethyl acetate	121
Terpene hidrocarbons	Orange oil, Atractylodes rhizome	—	38, 65, 66
Tipredane	Rodent diet	Ethanol	78
Tropane alkaloids	Plant extracts	—	43
Vitamins	White barley, cereal products animal feeds, milk powder, liver, serum, and ointments	—	55, 56, 71, 84, 85, 120, 127, 137–141
Xanthene dyes	Lipstick	—	161

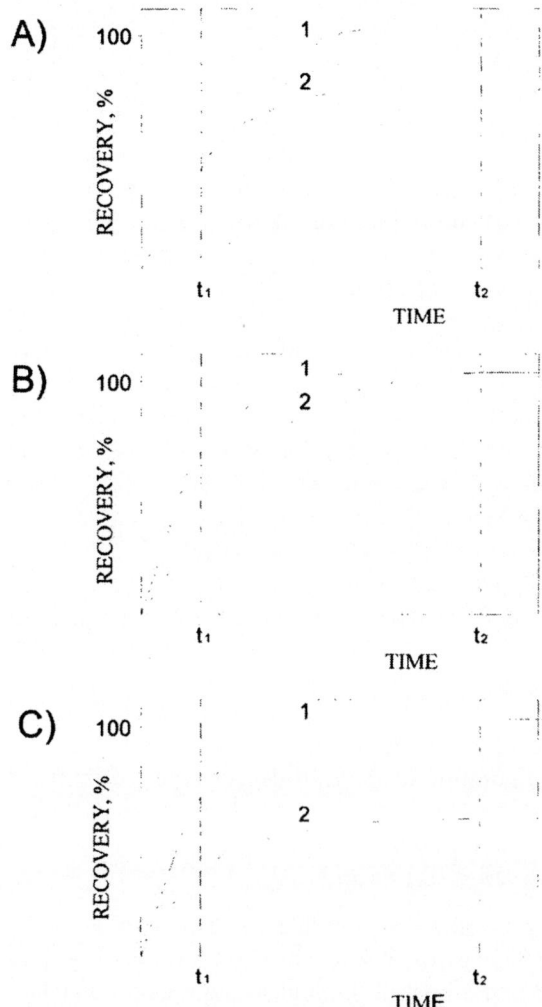

FIGURE 2 Optimization of an SFE process. Study of the influence of an operating variable at two different values (1 and 2). (A–C) Graphs of the extraction kinetics in three different situations. The importance of using appropriate values for time t_1 and/or t_2 is apparent. (Reproduced with permission of Springer-Verlag.)

The variables of an SFE process can be optimized by determining the amount of analyte extracted over a preset interval at different values of the variable concerned or by running the whole cumulative extraction-versus-time curve for each variable value considered. The former choice is more expeditious but, as shown in Figure 2, may lead to spurious conclusions. Thus, if measurements are made at a present time t_1 shorter than that required for maximum efficiency (t_2 or t_{max}), then the process will involve comparing extraction rates, so no information on the extraction efficiency will be obtained. Conversely, if measurements are made at a time $t > t_{max}$, then one will obtain information on the extraction efficiency but not about the extraction rate. In the situations of Figures 2A and 2B, using t_1 or t_2 only would lead to errors of opposite sign (20). On the other hand, under the conditions of Figure 2C, any conclusion drawn from t_1 would be basically correct. Therefore, in studying the effect of a given variable, one should always determine the recovery at both a relatively short time and a long enough one.

2. APPLICATIONS

Supercritical fluid extraction (SFE) has been applied to a broader range of samples. Generally, the applications of SFE have been developed as a faster and less solvent-intensive alternative to traditional extraction schemes. Environmental and food analysis are the main field of applications of SFE (21); however, a wide range of applications have been focused on the extraction of drugs and active compounds from different types of matrices that are commented below.

2.1. Applications of SCF to Drug Extraction

The application of SCF to drug extraction has been divided depending on the type of sample to be extracted. The use of SCF for the extraction of drugs and active compounds from plants and foodstuff, biological tissues, and pharmaceutical preparations are described in the following subsections.

2.1.1. Extraction from Plants and Foodstuff

Supercritical fluid extraction has been applied to plants and food samples with different objectives. Regarding plants, it has been applied for the extraction of active compounds used in the elaboration of pharmaceutical preparations or for the extraction of essential oils, widely used in the pharmaceutical industry as well. In the former case, SFE has been used as a suitable tool for the monitoring of dosage levels of drugs administrated to animals through their feeds and also for the extraction of active compounds from food, thus allowing characterization of some properties of the food such as the vitamin content.

Extraction of Active Compounds from Plant Materials. A variety of natural products present in plant materials, especially high-value-added products, can be processed with supercritical fluids. Because many of these have pharmacological activity, a natural extension of this application is to the field of drug analysis.

A number of natural products have been extracted by SFE. Roots are a typical matrix from which several compounds are extracted. This is the case for the extraction of baicalein and baicalin from Scutellaria root (22), active compounds from lovage roots (23), tanshinone IIA from *Salvia miltiorrhiza* bunge (24), and kava lactones from kava roots (25). In all of these cases, the use of pure CO_2 as extractant was the best option, as the extraction efficiency did not increase using organic-solvent-modified CO_2 as extractant.

Several compounds have also been extracted from flowers such as magnolia (26,27), lavender (28), crimson glory fresh flowers (29), and chamomile (30), for which SFE both at the analytical and preparative scale offered considerable advantages over the traditional methods such as steam distillation, Soxhlet extraction, and maceration. Using CO_2, extractions were performed in a shorter time (30 min by SFE versus 6 h by Soxhlet and 3 days by maceration) and under mild conditions, thus minimizing degradation of thermolabile components (e.g., matricine) and increasing the yield of volatile analytes (31).

Relevant drugs have been isolated using SFE as the active principles of medicinal plants such as artemisinin and artemisinic acid used in the treatment of malaria (32), michellamines A and B—known as the novel atropisomeric pair of anti-human immunodeficiency virus (HIV)-cytopathic alkaloids (33), taxanes as anticancer drugs (34), different Chinese herbal medicines (35,36), and ephedrine, which was selectively extracted from *Ephedra* sinica using methanol diethylamine as modifiers (37). SFE has been on-line coupled with SFC for the determination of actractylon in *Atractylodes rhizome*, which is a useful drug prescribed in many Japanese and Chinese herbal medicines (38). Other arrangements have been performed with different purposes; such is the case of the coupling of SFE to an uterotonic bioassay (39). This approach allowed the isolation of the active compounds as well as the determination of their uterotonic activity.

As stated earlier, a wide range of compounds of different nature has been isolated from plant matrices using SFE; however, the family of compounds more usually extracted by SFE has been alkaloids. Alkaloids of greatest interest for health and consumer reasons are caffeine (40) and nicotine (41), as well as purine alkaloids (42), cocaine (43) and indole alkaloids (44), which have been extracted using pure CO_2. Meanwhile, CO_2-modified methanol has been used for the extraction of pirrolizidine alkaloids (45) and oxindole alkaloids (46). For the extraction of alkaloids from Amaryllidaceae plants, different CO_2 and N_2O modifiers such as methanol, ethanol, NH_3, and so forth were investigated; supercritical N_2O with a mixture of NH_3–methanol as the modifier provided better recoveries than pure or modified supercritical CO_2 (47).

Other compounds related with the pharmaceutical industry extracted from plant materials using SFE have been the following: natural antioxidants from rosemary (48–50) and tamarind (51), needing, in the last case, 5 h for quantitative removal of the target analytes; cannabinoids from cannabis (52), previous sonication of the sample during 30 min in $CHCl_3$; carotenoids from leaves (53) and algae (54), providing results similar to those obtained by conventional liquid extraction; and vitamins from white barley (55) and cereals (56), obtaining quantitative extraction in only 20 min.

Extraction of Essential Oils from Plants.   Essential oils are aromatic substances widely used in the perfume industry, the pharmaceutical sector, and the food and human nutrition field. They are mixtures of more than 200 compounds that can be grouped basically into two fractions: a volatile fraction, which constitutes 90–95% of the whole oil, and a nonvolatile residue, which constitutes the remaining 5–10%. The isolation, concentration, and purification of essential oils have been important processes for many years, as a consequence of the widespread use of these compounds. The common methods used are mainly based on solvent extraction and steam distillation. SFE has been used for the extraction of essential oils from plants, in an attempt to avoid the drawbacks linked to conventional techniques (57). Such is the case with the extraction of flavor and fragrance compounds, such as those from rose (58), rosemary (59), peppermint (60), eucalyptus (61), and guajava (62). The on-line coupling of the extraction and separation–determination steps (by SFE–GC–FID) has been proposed successfully for the analysis of herbs (63) and for vetiver essential oil (64).

Supercritical fluid extraction is also a suitable technique for enhancing the quality of essential oils obtained by conventional extraction methods, by means of fractionation and deterpenation. Thus, the separation of citrus oils into different clssses of substances by supercritical CO_2 has been widely investigated. Temelli et al. reported a method for the extraction of terpene hydrocarbons from cold-pressed Valencia orange oil with supercritical CO_2, using both static and dynamic flow approaches (65). Another article has reported the SFE of terpenes from cold-pressed orange oil in a temperature range from 40°C to 70°C and pressures from 83 to 124 bar (66). The determination and elimination of psoralens from lemon peel oil by SFE has also been conducted (67). The procedure included the increase of CO_2 density in successive steps.

Supercritical fluid extraction applied to the isolation of essential oils has been mainly compared with hydrodistillation (68). In most cases, SFE provided results similar to those obtained by hydrodistillation, but in shorter times. The recoveries obtained for the extraction of essential oils from aromatic plants such as savory, peppermint, and dragonhead applying 15 min of static extraction with CH_2Cl_2 as the CO_2 modifier followed by 15 min of dynamic extraction with pure CO_2 were similar to those obtained by hydrodistillation in 4 h (69). The differences between SFE and hydrodistillation are not only the extraction time but also the extract composition. The oil extracted from eucalyptus by hydrodistillation showed high concentrations of 1,8-cineole, α- and β-pinene, and terpinen-4-ol; meanwhile, the extracts obtained by SFE had lower amounts of these compounds but higher amounts of allo-aromadendrene and globulol (70).

Extraction of Drugs from Animal Feeds. In chronic toxicity studies, administration of drug substances to laboratory animals may involve incorporating the drug into the animal feed (71). The determination of the drug substance level in feed matrices is necessary to monitor dosage levels, verify dose uniformity throughout the feed mix, and confirm the drug stability. Animal feeds are complex mixtures of proteins, lipids, glucicide, cellulose, and mineral matter (72), and if coextracted, the components can interfere in the determination of the analyte of interest. Various means of sample preparation have been investigated to extract drug substances from feed prior to determination. These methods include liquid–solid extraction (73), solid phase extraction (74), Soxhlet extraction (75), and liquid–liquid extraction (76).

Schneiderman et al. (71) was the first to report on the applicability of using SFE as a quantitative method for extracting drugs from animal feed matrices (77). They reported the extraction of menadione (vitamin K_3) from spiked rat chow using supercritical CO_2 for 20 min. Menadione was determined in the extract without any cleanup using HPLC.

In 1991, the SFE of a novel corticosteroid, tipredane, from rodent diet was reported (78). The authors investigated the extraction of the tipredane with and without the presence of ethanol as modifier, obtaining the best results with a 10:1 CO_2–ethanol mixture. A comparison of the optimized SFE conditions with more conventional extraction procedures such as Soxhlet and ultrasonic agitation was performed. The average recovery obtained with SFE was 85% as compared with 92% obtained by ultrasound agitation and 94% by Soxhlet. The main advantage of SFE was the shorter extraction time needed and no preconcentration step required prior to HPLC analysis.

Other drugs such as a hypolipidermic drug, 4-trifluoromethyl-2-biphenyl carboxylic acid (79), fluconazole (80), SC-52151 (an experimental HIV protease inhibitor drug) (81), propanolol, and tamoxifen (82) have also been extracted from animal feeds using supercritical CO_2. However, the potential use of a SCF other than supercritical CO_2 for extraction of pharmaceutical compounds was demonstrated by Sauvage et al. (72) The comparative behavior of supercritical nitrous oxide and CO_2 was studied for the extraction of a halogenated aromatic phenoxy derivative of an aliphatic alkane (HAPA) from a dog feed and a halogenated aromatic phenoxy derivative of urea (HAU) from a rodent feed. Extractions were carried out with pure SCFs with high solvating power and fluids modified with a polar modifier (methanol and acetonitrile). The results indicated that 0.7% (v/v) methanol-modified nitrous oxide gave the best recoveries. The extraction time of HAU from rodent feed was reduced by a factor of 2, with a recovery of 94%. For the extraction of HAPA, the total extraction time required was only a few minutes compared to the 70-min extraction time by the classical method.

Extraction of Active Compounds from Food.   Vitamins are the group of compounds more usually extracted from foods using SFE (83). A method for the analysis of the natural contents of vitamins A and E in milk powder based on SFE, a miniaturized alkaline saponification procedure, and HPLC was proposed by Turner and Mathiasson (84). Modifications of the sample matrix, the combination of static and dynamic extraction modes, and the effect of changes in extraction parameters such as temperature, flow rate, time, collection solvent, and collection temperature were optimized, obtaining recoveries of 99% and 96% for vitamins A and E, respectively. Another method for the determination of vitamins A and E based on the coupling of SFE–enzymic hydrolysis–HPLC has also been proposed providing recoveries between 79% and 152% (85).

Vitamin precursors, such as β-carotene, have also been extracted with supercritical CO_2 from carrots (86,87) and paprika (88). In all cases, ethanol was used as the modifier, obtaining efficiencies similar to those provided by conventional methods but in times around six times shorter. β-Carotene has also been extracted using supercritical CO_2 involving continuous monitoring of the extracted analyte using pulse thermal lens spectrometry (PTLS), thus reporting, for the first time, the behavior of this technique in SCFs (89).

Other compounds such as resveratrol from grape skin (90), lycopene from tomato skin (91), lignans from *Schisandra chinensis* (92), and perillyl alcohol from Korean orange peels (93,94) have been isolated using SFE, obtaining efficiencies 30 times higher than those provided by the conventional solvent extraction methods.

2.1.2. Extraction from Biological Tissues and Fluids

The rapid and accurate measurement of ultratrace levels of drugs and their metabolites in biological samples plays a major role in the pharmaceutical development process. These measurements of drugs and metabolites provide information about the mechanism of action in pharmacology and toxicology studies, as well as for clinical development (95).

Human and Animal Tissues. Sachs and Uhl demonstrated for the first time the use of supercritical fluids in the extraction of drugs in hair by means of a mixture of CO_2 and ethyl acetate, although the extraction efficiency and the method reproducibility remained inferior to those obtained by other conventional techniques (96). Edder et al. extracted opiates from hair with CO_2/ methanol/triethylamine/H_2O at 250 bar and 40°C for 30 min. The extracts were analyzed by CG–MS using a solution of nalorphine as the internal standard. The opiates monitored were morphine, codeine, 6-monoacetylmorphine (6-MAM), and ethylmorphine. The recovery was 93.5% using hair spiked with ^{125}I-labeled morphine. The linear range was 0.5–2 ng/mg for codeine, ethylmorphine, 6-MAM, and morphine. The detection limit was 100 pg/mg. The RSD for five replicates was in the range 3–12% (97). The same experimental conditions were used for the extraction and quantitation of drugs from addict hair and urine; thus, RSDs were 5–12%, detection limits were 0.03–0.05 ng/mg, and the linear range was 0.3–4 ng/mg (98). Cirimele et al. (99) included a cleanup step with supercritical CO_2 at 100°C, a TENAX trap, and a modifier solution of methanol/triethylamine/water (2:2:1). The elution was done with $CHCl_3$. The eluate was evaporated to dryness and the residue treated with NO-bis(trimethylsilyl)-trifluoroacetamide containing 1% of trimethylchlorosilane. The derivative was analyzed by GC–MS using deuterated compound as internal standards. The RSDs were 13%, 17%, and 14% (eight replicates) for codeine, morphine, and 6-MAM, respectively; corresponding detection limits were 0.3, 0.2, and 0.1 ng/mg with recoveries of 61%, 53%, and 96%, respectively. The method was also applied to the extraction of cocaine and cannabinoids from hair. A method based on SFE and immunoassay for the screening of cocaine in hair was described (100). Hair was subjected to SFE with CO_2 modified with triethylamine and water at 110°C and 405 bar, with static extraction for 10 min followed by dynamic extraction for 20–30 min. The extracted components were collected in methanol, which was evaporated, and reconstituted in methanol for radioimmunoassay (RIA) using a solid phase ^{125}I kit. [^{125}I]Benzoylecgonine (major cocaine metabolite) reagent and the sample were added to tubes with immobilized cocaine antibodies, and after 2 h incubation, the bound radioctivity was measured using a gammacounter. Methanolic cocaine standards were used as calibrators. RSDs varied between 1.1% and 7% at 0.5–100 ng/mL cocaine. An RIA cutoff value for distinguishing between negative and positive results was fixed. The results agreed with those obtained by GC–MS. In a later article, the dependence of the SFE behavior of cocaine and benzoylecgonine on the chemical nature of the matrix and the manner in which the target drug analytes are incorporated into or on the matrix was investigated (101), with extraction conditions similar to those reported in Ref. 100. The recovery of cocaine from Teflon wool, filter paper, drug-fortified hair, and drug user hair was studied using a variety of CO_2/modifier mixtures. Optimal results were obtained with triethylamine/water. All samples were silylated to produce TMS derivatives of benzoylecgonine. The results showed that the way the analyte is incorporated into the matrix determines the required conditions for SFE; this dependence suggests the potential of SFE for distinguishing drug present in hair because of passive environmental exposure from drug present because of to active drug use. Cocaine can be extracted from human hair with the use of supercritical CO_2 modified with methanol (102). The extraction conditions were studied by experimental design to characterize the effects of pressure, temperature, and percent methanol on the recovery of cocaine from hair, the optimum values being 304 bar, 145°C, and 10%, respectively. The samples were extracted both statically (5–25 min) and dynamically (10–100 min). Extracts for cocaine analysis were transferred to GC autosampler vials, evaporated to dryness under nitrogen, and reconstituted with 20% methanol in ethyl acetate. Recovery of cocaine from fortified hair by SFE was determined by GC–MS to be two times the amount detected by GC–MS following acid hydrolysis, and the extraction times were much shorter (70–80 min) than for existing classical techniques (24 h).

Amphetamines can be determined in hair by a GC–MS method following SFE (103). Hair was washed sequentially with sodium docecyl sulfate (SDS), CH_2Cl_2, methanol, and water and then dried. The amphetamines were extracted from the hair with 90% CO_2 and 10% $CHCl_3$/isopropyl alcohol at 262 bar and 70°C as the extractant in the dynamic extraction mode for 30 min. The amphetamine were derivatized and determined by GC–MS. Mephentermine was used as an internal standard. The calibration graphs were linear from 0.02 to 20 ng/mg for all of the amphetamines studied. The detection limits were 0.02–0.1 ng/mg, the RSDs (five replicates) were 11–28%, and the recoveries were 71–84%.

Regarding animal tissue, SFE has been widely applied to the extraction of veterinary drugs, mainly anabolic agents. Trimethoprim, hexoestrol, stilboestrol, and dienoestrol were extracted from spiked (10 mg/kg) pulverized freeze-dried pig kidney using liquid CO_2 at 75°C and 302 bar as the SCF. An HPLC precolumn (20 mm × 2 mm) was used as an extraction cell. The extraction stage took 8 mins. The analytes were then absorbed on a column in which a SCF chromatographic separation took place. The determination was achieved by simultaneous ultraviolet (UV) and quadrupole MS–MS detection, using sulfadimidine as a retention index marker (104). An on-line solid phase extraction (SPE) column fitted into a Teflon (PTFE) sleeve assembly was developed by Maxwell et al. for the SFE of anabolic steroids from chicken liver (105). The assembly comprised a 1- or 3-mL SPE column fitted into a PTFE sleeve packed into a SFE vessel together with the sample matrix. The SPE column packing depended on the solute nature. After SFE, the SPE column was removed and the solutes recovered in eluting solvent prior to analysis. Homogenized sample spike with nortestosterone, testosterone, and methyltestosterone was extracted with supercritical CO_2 at 40°C. The SFE vessel was statically heated for 10 min. The on-line SPE column was retrieved and the solutes dissolved in a 3-mL mobile phase. A portion was analyzed by HPLC on a C-18 column with methanol/water as the mobile phase and detection at 254 nm. The determination of androsterone in pig fat could be carried out using GC–MS following SFE (106). The samples consist of sow back fat spiked with 1 µg/g androsterone. The extraction took place, first statically for 10 min with CO_2 at 60°C and 329 bar and then dynamically for 30 min at 40°C and 115 bar. The analytes were trapped on a cartridge and then extracted with cyclohexane containing 5α-androstan-3-one as the internal standard. The detection limit was 50 ng/g and the RSD was 6.6%. Androstenone and skatole have also been extracted from pork fat (107) using CO_2 a the SCF. The extracts were evaporated to dryness and extracted with methanol/cyclohexane/benzene/water, derivatized and analyzed by GC with thermoionic specific detection. At 40°C, $97 \pm 2\%$ androtenone and $65 \pm 3\%$ skatole were extracted after 5 and 20 min, respectively. The extraction was carried out under different pressures (70, 210, and 280 bar), but this parameter did not influence the extraction efficiency.

The extraction of some anabolic steroids directly from bovine tissue was reported by Huopalahti and Henion (108). SFE was performed with CO_2 at 60°C and 405 bar and anlytes were collected in precooled methanol. The quantitation of the analytes was carried out with HPLC–atmospheric pressure–CIMS. The detection limit was 100 µg/mL. Bovine muscle tissue was also used for the determination of residues of 2-thiouracil, 6-methil-2-thiouracil, 6-propyl-2-thiouracil, and 6-phenyl-2-thiouracil by GC–MS (109). The analytes, extracted with acetonitrile from homogenized beef muscle, were spiked with a surrogate compound, 5-ethyl-2-thiouracil, transferred to a column and methylated. The products were extracted with acetonitrile or supercritical CO_2. 2,4,5-Triclorophenoxyacetate was used as the internal standard.

The addition of small amounts of water to the sample matrix prior to SFE on the recovery of dinitolmide and its metabolites (3-amino-5-nitro-, 5-amino-3-nitro-, and N^5-acetyl-3-nitro-o-toluamide) in chicken liver has different effects depending on the analyte nature (110). Five hundred milligrams of ground liver was spiked with 4 µL of a standards solution containing 0.5 µg/µL of each analyte, held at $-10°C$ for 1 h and then mixed with anhydous Na_2SO_4. Water (500 µL) was added to half of the samples. The extraction was carried out with supercritical CO_2 at 60°C directly onto an Al_2O_3–sand SPE cartridge. The analytes were eluted with a 2.5-mL mobile phase and a portion was analyzed by HPLC, with detection at 254 nm. After a 30-min extraction, recoveries increased from 79%, 62%, and 46%, respectively, for each of the dinitolmide metabolites to 92%, 77%, and 66% after water addition prior to SFE. The recovery of dinitolmide was not affected by water addition.

Clenbuterol has been extracted from liver using simultaneous ion-pair formation/SFE (111). The samples were directly weighed in the extraction cell following the addition of 50–100 µL of clenbuterol and 0.5 mL of ion-pair reagent solution. The samples were extracted with supercritical CO_2 for 30 min of dynamic extraction at 40°C and 383 bar. The recoveries ranged from 12% to 87% with an RSD of 15% (three replicates). A combination of SFE with enzyme immunoassay (EIA) was also described for the determination of clenbuterol residues at microgram per milliliter levels (112). The extraction was performed with CO_2 either at 40°C and 620 bar or at 100°C and 300 bar. The extracted material was trapped on neutral Al2O3, then eluted with aqueous 70% methanol, and, finally, evaporated under nitrogen and dissolved in 0.75 mL of assay buffer for the determination by EIA. The recoveries of 5 µg/mL of clenbuterol added to fresh and dried liver were 29.4 and 58.5 with RSDs (three replicates) of 16.2% and 11.6%, respectively. A similar procedure has been carried out for the determination of clenbuterol and salbutamol in liver after treating the sample with β-glucuronidase/sulfatase and lyophilization (113). A portion of the dried sample was mixed with 2 g Hydromatrix, and 1.5 mL methanol was added. SFE was performed with CO_2 at 100°C and 690 bar for 10 min. The determination limit was 0.1 ng/g and the intra-assay and interassay RSDs (five replicates) at the 5-ng/g level were < 7.5% and < 25%, respectively. The recoveries of clenbuterol and salbutamol were 86–92% and 62–70%, respectively.

Finally, a multianalyte, multimatrix method was developed for the routine determination of steroids in animal tissues (muscle, fat, and skin) (114). After the addition of internal standards and sample pretreatment, the analytes of interest were extracted from the matrix with unmodified CO_2 and trapped on an alumina sorbent placed in the extraction vessel. After extraction, alkaline hydrolysis was performed and the analytes were derivatized. The samples were analyzed by GC–MS. The limit of detection for the different matrix–analyte combinations was 2 µg/kg, the repeatability ranged from 4% to 42% (nine replicates), and the reproducibility ranged from 2% to 39%.

Other substances widely extracted by SFE from animal tissues are sulfonamides, which are administered as antimicrobials. The SFE of sulfamethazine and its metabolites from liver and kidney used CO_2 modified with methanol at 60°C and 405 bar (115). When tetramethylammonium (TMA) hydroxide was included as the ion-pairing reagent, static SFE for 30 min was followed by dynamic SFE for 2 h. The extracts were analyzed by HPLC with detection at 266 nm. The recoveries of sulfamethazine and five of its metabolites ranged between 0% and 97%, and ionic metabolite recoveries increased by up to 72% with the use of TMA. Comparison studies carried out in order to know the best way to collect sulfonamides demonstrated that the in-line collection during the dynamic extraction process yielded higher recoveries than off-line collection in standard SPE columns (116). Three sulfonamides were dynamically extracted with CO_2 at 40°C and 680 bar prior to in-line collection on a neutral alumina sorbent bed and HPLC analysis. The recoveries from chicken liver fortified at the 50- and 1000-μg/mL level ranged from 87% to 89.9% and 71.6% to 96.9%, respectively (117). Combs et al. modified supercritical CO_2 at 40°C and 496 bar with different percentages of methanol, ethanol, acetone, or acetonitrile for the extraction of three sulfonamides from chicken liver, beef liver, and egg yolk. Quantitative recovery of the three analytes was achieved with 20% acetone or acetonitrile, but the latter produced less interference in subsequent HPLC (118). The same authors compared the use of supercritical trifluoromethane with carbon dioxide for the extraction of sulfonamides from various matrices, beef liver among them (119). The effects of fluid pressure and addition of methanol modifier on the SFE of sulfonamides from an inert sand matrix were studied, resulting in that the change on the pressure from 365 to 496 bar had little effect on the recoveries with CO_2, but the addition of 10% methanol gave quantitative recoveries. The extraction efficiencies of sulfamethazine and sulfadimethoxine were higher with CHF_3 than with CO_2. In the extraction from beef liver, improved recoveries were obtained with CHF_3 as compared with CO_2. An increase in the extraction of sulfamethazine and sulfadimethoxine from beef liver higher than 200% was obtained with pure CHF_3.

The extraction of vitamin A and β-carotene from liver was carried out with CO_2 at 80°C and 310 bar in two steps: 1 min of static extraction followed by 40 min of dynamic extraction (120). The extracts were eluted into hexane and measurements were performed by spectrophotometry after reverse phase HPLC.

Biological Fluids. The pharmaceutical analyses of biological fluids provide the information necessary for evaluating the safety, therapeutic effect, and mechanism of action of a variety of drugs (95).

When the matrix is liquid, the samples are usually mixed with an adsorbent material called Hydromatrix.

Whole blood has been the matrix from which temazepam (121) and morphine have been extracted (122). In the first instance, the extractant was CO_2 and ethyl acetate at 65°C and 207 bar. The extraction was monitored at 254 nm and individual separation of the analytes was carried out by HPLC. The recoveries ranged from 80% to 100%. The method can be applied to other benzodiazepines of forensic significance. For the extraction of morphine from dried blood, methanol/triethylamine was added as modifier to 90% CO_2 at 100°C and 241 bar. The extraction was performed in a dynamic mode for 30 min, and the extracts were analyzed by GC–MS. A comparison between SFE and SPE was done, with SFE being the fastest and cleanest of both.

A combination of SFE and SPE as a sample preparation technique for the ultratrace analysis of mebeverine alcohol in plasma has been proposed (123). Plasma containing the analyte was applied to conditioned octadecylsilane cartridges. After washing and drying, the packing was placed in an SFE system. Carbon dioxide with 5% methanol was used as the SCF at 40°C and 355 bar. The extract analysis was carried out by GC–MS. A similar method was used for the extraction of flavone, also from blood plasma, but the determination was performed by HPLC in this case (124). In other applications, plasma spiked with tritiated budesonide to 93 nM was deposited onto a filter paper placed in the extraction vessel, where the extraction was performed with pure CO_2 for 30 min; then, the budesonide in the extract was determined with a liquid scintillation counter. The recoveries were over 80% (125). ^{14}C-Flavone and ^{14}C-ketorolac were extracted from plasma at various extraction pressures and times to test recovery (126). The SCF was CO_2 at 60°C and 300 bar. The recoveries were determined by liquid scintillation counting and ranged from 85% to 98% for ^{14}C-flavone (RSD = 3.8–5.8%, three replicates).

Phylloquinone, menaquinone, and menaphthone have been extracted from serum into supercritical CO_2 and separated with SFC, with detection at 250 nm (127). Serum, supported on celite, was transferred to extraction cartridges and extracted with supercritical CO_2 at 45°C prior to the analysis of benzodiazepines, anabolic agents, and nonsteroidal anti-inflammatory drugs by HPLC (128). Phenobarbital, butalbital, pentobarbital, and thiopental have been extracted from human serum prior to liquid chromatography–negative-ion electrospray tandem mass spectrometry analysis (129). The sample was extracted with supercritical CO_2 at 40°C and 507 bar with static extraction for 5 min followed by dynamic extraction for 30 min. Calibrations graphs were linear for 1–60 μg/mL of each barbiturate and the detection limits were 23, 25, 50, and 225 ng/mL of thiopental, pentobarbital, butalbital, and phenobarbital, respectively. In the determination of the cited analytes in serum spiked at the 10- and 50-μg/mL levels, the recoveries were 94.2–106.9% and the RSDs were 1.14–5.18% (three replicates).

A method for the detection of β-blockers in urine by SPE–SFE was developed, using acetic anhydride and pyridine as acetylation reagents in the SFE vessel (130). The detection was carried out by GC–MS. The same method was later applied to the detection of the same analytes in serum and urine (131). Unmodified CO_2 at 40°C and 272 bar has been used for the multiresidue SFE of methyltestosterone, nortestosterone, and testosterone at low ppb levels from fortified urine (132). The eluates were analyzed either by HPLC or GC–MS and the recoveries ranged from $90.8 \pm 6\%$ to $93.9 \pm 3\%$ at the 12.5-ng fortification level. A study on the effect of increasing pressure on the solubility of testosterone, boldenone, androsterone, etiocholanolone, and epitestorone in pure supercritical CO_2 has been carried out, with testosterone being the most soluble in the SCF (133). The extraction efficiency of steroids from an aqueous saline environment exceeded 95%. On-line SFE–GC–MS was feasible for the quantitative extraction and analysis of steroids from both saline and urine solutions, but he GC column efficiency was reduced owing to the fact that the adsorbent vessel filled with Hydromatrix was not sufficient to trap all of the moisture.

Some abuse drugs have been extracted from urine by SFE [viz. cocaine and its metabolites (134) and amphetamine and methamphetamine (135). In the first instance, the levels measured using SFE showed analyte recovery better than 70% for cocaine, better than 40% for benzoylecgonine, and better than 85% for ecgonine methyl ester from whole blood and urine. The limits of detection and quantitation were 1 and 10 ng, respectively, based on a 200-µL sample. Regarding amphetamine (AP) and methamphetamine (MA), an in situ SFE and chemical derivatization procedure followed by GC–isotope dilution mass spectrometry in urine was described. The mean recoveries achieved were 95% (RSD = 3.8%) for AP and 89% (RSD = 4%) for MA. The calibration graphs were linear within 100–500,000 ng/mL, varying the limits of detection and quantitation from 19 to 50 and from 21 to 100 ng/mL, respectively.

Prostaglandins have also been extracted from aqueous solutions with CO_2 at 35–50°C, being then determined by SFC (136). The sample was introduced by direct injection or on-line SFE. The calibration graph was linear from 5 to 125 µg/mL and the detection limit was 9–60 ng with RSD = 1.5–5.6%.

2.1.3. Extraction from Pharmaceutical Preparations

The application of SCF to the extraction of vitamins has been widely reported. Thus, retinyl palmitate and tocopherol acetate have been extracted from a hydrophobic ointment with supercritical CO_2 at 40°C and 196 bar for 4 min, the extract analysis being performed by SFC (137). The calibration graphs were linear from 0.5 to 2.5 µg and the recoveries were quantitative. On the other hand, water-soluble vitamins can be extracted mixing them with low substituted hydroxypropil cellulose. Portions were placed in a column to which a reversed micellar extractant was delivered (138). Extraction of vitamins A and E and their esters from tablet preparations prior to HPLC was performed in the dynamic mode with CO_2 at 40°C and 253 bar for 15 min (139). Calibration graphs were linear from 0.02 to 0.8 and from 0.005 to 0.2 mg/mL of vitamins E and A, respectively. The corresponding RSDs (six

standard method, recoveries being over 95.6%. Vitamin E extracts have been also analyzed by SFC following SFE with carbon dioxide at 60°C and 405 bar; the features of the method are as follows: calibration graphs linear from 0.05 to 2 mg/mL of vitamin E acetate, with recoveries ranging between 96.4 and 104.5 with RSDs of 0.7–7% (140). The SFE of vitamins D_2 and D_3 was reported by Gámiz-Gracia et al. (141). The extraction recovery was enhanced by direct addition of diethyl ether to the sample contained in the extraction cell. Separation and detection of the analytes was performed off-line by reverse phase liquid chromatography with UV detection. The quantification limit of the method was 4.1 μg for both analytes, with RSDs of 3.8% and 6.3% for vitamins D_2 and D_3, respectively (seven replicates). The recoveries were between 85% and 105%.

Eckard and Taylor reported the modifier and additive effects on the SFE of pseudoephedrine hydrochloride from Suphedrine tablets, achieving a recovery of 82% (142). Also, racemic ephedrine was resolved using supercritical CO_2 (143). The temperatures for crystallization were 35–75°C and the pressures were 100–350 bar. The samples were analyzed using electrophoresis. The experimental error for the resolution was 0.3 mol%. Regarding the extraction of isomers, 13-*cis*-retinoic acid and its photoisomers were extracted from pharmaceutical formulations, with CO_2 modified with 5% methanol at 45°C and 329 bar (144); static and dynamic extraction times of 2.5 and 5 min, respectively, were used. The extracts were analyzed by HPLC. The recovery of 13-*cis*-retinoic acid from spiked placebo forms was 98.9%, 98.9%, 98.8%, and 100% for cream, gel, capsule, and beadlet, respectively, with RSDs (four replicates) ranging from 0.6% to 0.9%; its photoisomers were also extracted, with recoveries of 90.4–92.4% with RSDs (four replicates) of 1.5–3.4%.

Tablets have been the most common matrix for the SFE of active components. Pure supercritical CO_2 has been used in the extraction of megestrol acetate (145), caffeine and vanillin (146), and ibuprofen (147). In the first case, a comparison between two SFE systems was done and various support materials were assessed for their suitability as adsorbents. Methanol was used as the modifier for the extraction of benzodiazepines from tablets or capsules (148) in a static and dynamic mode for 5 and 10 min, respectively, at 65°C and 101 bar. The eluate was analyzed using GC–MS. Psoralen and isopsoralen have been extracted from Baishi pills and Baidianfeng capsules—belonging, to traditional Chinese medicines—using supercritical CO_2 modified with tricloromethane (149). The determination of the analytes was performed by GC. By the standard additions method, recoveries were 97.8–101.2% with RSDs (three replicates) less than 2.01%. In the same way, SFE has been applied to the isolation of pentosalen (also known as imperatorin) from *Cnidium monnieri* cusson powder (a traditional Chinese medicine) (150). This experiment was carried out with a homemade SFE system with the use of replicates) were 3.9% and 1.7%. The SFE method was more efficient than the

CO_2 and methanol, as modifier, at $60\,^\circ C$ and 250 bar for 40 min. The determination was done by HPLC.

A very interesting SFE application consists of the development of a method for assaying the active component in a controlled-release drug formulation. The drug substance in the formulation is the active enantiomer of misoprostol, covalently linked to a polymer, which was cross-likned after the covalent linkage (151). Supercritical CO_2 modified with 5% formic acid, at $75\,^\circ C$ and 334 bar, was an effective extraction medium for cleaving the covalent linkage between misoprostol and the polymer and yielded high recoveries of the prostaglandin content in the samples in a short extraction period. The use of formic-acid-modified CO_2 affected the liberation of misoprostol from the samples. Experiments completed under the same conditions, except that the modifier was 5% methanol, did not generate detectable levels of the analyte by HPLC. Static/dynamic SFE together with CO_2 with an 8.7% methanol modifier gave optimal extractions from sustained-release tablets without strong matrix–drug interactions (152). The average recovery (five replicates) was 98.6% with an RSD of 1.2%. The SFE method used 80% less solvent than the traditional liquid extraction procedure and was accurate and precise. Tablet solutions were subsequently analyzed by SFC and HPLC.

Inverse SFE can be used to analytically isolate a polar analyte from its matrix by extracting the drug carrier, a hydrocarbon-based ointment, thereby leaving behind the analyte of interest. The parameters which play important roles in the outcome of "inverse SFE" are as follows: First, the analyte must be totally insoluble in the SCF; second, the matrix must be soluble in the supercritical fluid; third, a highly efficient washing method must be used to transfer the analyte from the extraction vessel for analysis; fourth, the analyte concentration in the matrix should be higher than 2%; fifth, an assay method with low detection limits for the analyte is advantageous (153). The SCF used in the inverse SFE was 5% methanol-modified CO_2. The working conditions were $55\,^\circ C$ and 304 bar for creams and $60\,^\circ C$ and $450\,^\circ C$ for ointments. Portions of the extracts were analyzed by HPLC. The average recoveries and RSDs of polymyxin B sulfate were 108% (six replicates) and 5% for cream, and 137% (six replicates) and 1.9% for ointment. This extraction method was more effective than current SPE methods for polar compounds. Inverse SFE has also been used in the extraction of acetaminophen from suppositories (154). The optimum conditions for the extraction of the waxy matrix were pure CO_2 at $40\,^\circ C$ and 103 bar. The analyte was removed from the extraction cell with warm water and ultrasonication. The method gave quantitative recoveries. The extraction of imidazole antimycotics with 10% methanol-modified CO_2 from creams has also been reported (155).

Mulcahey and Taylor studied the application of SFE for direct extraction of polar active ingredients (sulfamethazole and trimethoprim) from liquid matrices (septra infusion) (156). The extractions were performed in two ways: (1) extraction of septra infusion ingredients directly using a modified extraction vessel designed to bubble the SCF before exiting the trap and (2) spiking the active component onto celite followed by extraction of the drug with 100% CO_2. Quantitative recovery of both drugs was achieved by the latter method, whereas the former one yielded poor recoveries as a result of restrictor plugging caused by precipitation of sulfamethoxazole when the pH of the solution was lowered by the bubbling of CO_2.

Cosmetic products have been the samples for the SFE of their active compounds. Thus, parabens have been extracted from cosmetics, either with pure supercritical CO_2 (157) or with 0.05% acetonitrile-modified CO_2 (158). SFE of sunscreen agents prior HPLC (159) and micellar electrokinetic capillary chromatography (160) has been carried out, yielding 94.8% and 98.4–101.8% recovery, respectively. An inverse SFE procedure has been developed for the efficient isolation of six xanthene dyes from lipstick matrices (161), which only removed the matrix components, whereas the xanthene colorants remained in the extraction vessel. The target analytes were quantitatively recovered by dispersion of the sample in ethanol under sonication. Finally, cosmetic waxes have been extracted by SFE prior to SFC, using multivariate data analysis for their quantitation (162).

2.2. Strategies for Improving SF Drug Extraction

Although SCFs are widely regarded as "supersolvents," this designation is utterly unrealistic, as one can easily see by comparing their solvent powers with those of conventional liquids. In any case, such features as increased solute diffusivity, decreased viscosity, and increased solvent strength, all of which are easier to control than in liquid solvents, have promoted the use of SCFs for treating solid samples. Most analytical SFEs have focused on CO_2 as the fluid because of its high inertness and purity, low toxicity, and moderate cost. Although CO_2 is an excellent solvent for nonpolar organics, its polarity is occasionally too low to ensure efficient extraction, either because the analytes are poorly soluble or because the extractant is unable to displace them from the active matrix sites. In any case, the potential of CO_2 for analytical SFE has been exaggerated, mainly because of its excellent performance in the extraction of analytes spiked to inert supports. The far from natural behavior of these solid samples has led some to consider supercritical CO_2 the ideal SFE leacher. However, matrix–analyte interactions dramatically reduce its extraction efficiency, the effect increasing with increased polarity of the analytes.

The performance of SFE has been improved since its inception by circumventing the shortcomings that hinder quantitative leaching of polar and ionic species. The ensuing modifications have ranged from the mere alteration of extractant characteristics such as pressure and temperature to the use of complex derivatization sequences and include manipulations such as the formation of ion pair, chelation, micellization, esterification, and formation of organometals, which are used to raise the polarity of the extractant and lower that of the target species. However, only the use of alternative SCFs and the formation of ion pairs has been used for improving drug extraction, as commented on in the following subsections.

2.2.1. Use of an Alternative Supercritical Fluid

One way of improving SFE efficiency is by using a more suitable SF to extract the target analytes. Unfortunately, the choice of fluids other than CO_2 is restricted by the desire to have reasonable critical parameter values and costs, chemical inertness, low toxicity, and little environmental impact. The use of supercritical N_2O has proved to increase the extraction efficiency for a halogenated aromatic phenoxy derivative of an aliphatic alkane (HAPA) from a dog feed and for a halogenated aromatic phenoxy derivative of urea (HAU) from a rodent feed (72). For the extraction of alkaloids from Amaryllidaceae plants, different CO_2 and N_2O modifiers such as methanol, ethanol, NH_3, and so forth were investigated, providing the supercritical N_2O with a mixture of NH_3–methanol as modifiers better recoveries than pure or modified supercritical CO_2 (47).

2.2.2. Ion-Pairing Formation

Neutralizing charged species by the formation of ion pairs is one way of lowering the polarity of ionic compounds and increasing their solubility in low-polar extractants such as supercritical CO_2 when the addition of a polar cosolvent is ineffective. Thus, a quaternary ammonium cation, trimethylphenylammonium, was used as the counterion for the quantitative recovery of sulfonamides with CO_2 at 40°C at 281 bar as the extractant (163). Clenbuterol was also extracted from food matrices (viz. feedstuff, freeze-dried milk, and liver), following ion pairing with 10-camphorsulphonate; supercritical CO_2 was used in the dynamic extraction mode after the addition of the ion-pair-forming reagent to the extraction cell containing the sample. The ammonium salt of the reagent was found to provide better results than its acid form (111).

REFERENCES

1. Erickson B. Fattening up SFE sales. Anal Chem 1998; 70:333A–336A.
2. Hawthorne SB. Analytical-scale supercritical fluid extraction. Anal Chem 1990; 62:633A–642A.

3. Taylor LT. Strategies for analytical SFE. Anal Chem 1995; 67:364A–370A.

4. Bartle KD. In: Smith RM, ed. Supercritical Fluid Chromatography. Cambridge: Royal Society of Chemistry, 1988:2–4.

5. Levy JM, Ritchey WM. Investigations of the uses of modifiers in supercritical-fluid chromatography. J Chromatogr Sci 1986; 24:242–248.

6. Hedrick J, Taylor LT. Quantitative supercritical fluid extraction—supercritical fluid chromatography of a phosphonate from aqueous media. Anal Chem 1989; 61:1986–1988.

7. Kane M, Dean JR, Hitchen SM, Dowle CJ, Tranter RL. Experimental design approach for supercritical-fluid extraction. Anal Chim Acta 1993; 271:83–90.

8. Luque de Castro MD, Valcárcel M, Tena MT. Analytical Supercritical Fluid Extraction. Heidelberg, Germany: Springer-Verlag, 1994.

9. Mulcahey LJ, Taylor LT. Collection efficiency of solid surface and sorbent traps in supercritical fluid extraction with modified carbon dioxide. Anal Chem 1992; 64:2353–2358.

10. Howard AL, Taylor LT. Quantitative supercritical fluid extraction of sulfonylurea herbicides from aqueous matrices via solid-phase extraction discs. J Chromatogr Sci 1992; 30:374–382.

11. Wenclawiak BW, Heemken OP, Sterzenbach D, Schipke J, Theobald N, Weighlt V. Device for efficient solvent collection of environmentally relevant compounds in offline SFE. Anal Chem 1995; 67:4577–4580.

12. Charpentier BA, Sevenants MR, eds. Supercritical Fluid Extraction and Chromatography. Techniques and Applications ACS Symposium Series, Washington, DC: American Chemical Society, 1988:130.

13. Yoo WJ, Taylor LT. Supercritical fluid extraction of polychlorinated biphenyl and organochlorine pesticides from freeze-dried tissue of marine mussel, *Mytilus edulis*. J AOAC Int 1997; 80:1336–1345.

14. Johansen HR, Becher G, Greibrokk TG. Determination of planar PCBs by combining online SFE–HPLC and GC–ECD or GC–MS. Anal Chem 1994; 66:4068–4073.

15. Mauldin RF, Vienneau JM, Werhy EL, Mamantov G. Supercritical fluid extraction of vapour deposited pyrene from carbonaceous coal-stack ash. Talanta 1990; 37:1031–1036.

16. Blanch GP, Reglero G, Herraiz M. Analysis of wine aroma by off-line and on-line supercritical-fluid extraction–gas chromatography. J Agric Food Chem 1995; 43:1251–1258.

17. Modey WK, Mulholland DA, Mahomed H, Raynor MW. Analysis of extracts from *Cedrela toona* (Meliaceae) by on-line and off-line supercritical fluid extraction–capillary gas chromatography. J Microcolumn Sep 1996; 8:67–74.

18. Otero ZK. Optimization of SFE by statistical methods. New Orleans: Pittcon'92, 1992.

19. López-Ávila V, Dodhiwala NS, Beckert WF. Supercritical fluid extraction and its application to environmental analysis. J Chromatogr Sci 1990; 28:468–476.

20. Tena MT, Luque de Castro MD, Valcárcel M. Systematic study of the influence of variables on the supercritical fluid extraction of polyromatic hydrocarbons (PAHs). Lab Rob Autom 1993; 5:255–262.

21. Chester TL, Pinkston JD, Raynie DE. Supercritical fluid chromatography and extraction. Anal Chem 1996; 68:487R–514R.

22. Mishima K, Wada N, Uchiyama H, Nagatani M, Choi WS, Kitazaki H, Takai T. Extraction and separation of bacicalein and baicalin from *Scutellaria* root using supercritical carbon dioxide. Solvent Extr Res Dev Jpn 1996; 3:231–237.

23. Dauksas E, Venskutonis PR, Sivik B. Extraction of lovage (*Levisticum officinale* Koch) roots by carbon dioxide. Effect of CO_2 parameters on the yield of the extract. J Agric Food Chem 1998; 46:4347–4351.

24. Dean JR, Liu B, Price R. Extraction of Tanshinone IIA from *Salvia miltiorrhiza* Bunge using supercritical fluid extraction and a new extraction technique, phytosol solvent extraction. J Chromatogr 1998; 799:343–348.

25. Ashraf-Khorassani M, Taylor LT, Martin M. Supercritical fluid extraction of kava lactones from kava root and their separation via supercritical fluid chromatography. Chromatographia 1999; 50:287–292.

26. Dean JR, Liu B, Price R. Extraction of magnolol from *Magnolia officinalis* using supercritical fluid extraction and phytosol solvent extraction. Phytochem Anal 1998; 9:248–252.

27. Miao HJ, Liu ZL, Li YH. Analysis of magnolol and honokiol in *Magnolia officinalis* Rehd et Wils by supercritical fluid extraction and capillary gas chromatography. Yaowu Fenxi Zazhi 1998; 18:182–185.

28. Walker DFG, Bartle KD, Breen DGPA, Clifford AA, Costiou S. Quantitative method for the analysis of flavour and fragrance components from lavender and rosemary for studying the kinetics of their supercritical fluid extraction. Analyst 1994; 119:2789–2793.

29. Zhang L, Xiang ZM, Bi LJ, Xie ZZ, Chen L. Gas chromatographic–mass spectrometric determination of fragrant components of crimson glory fresh flowers extracted by supercritical carbon dioxide. Sepu 1996; 14:438–440.

30. Vuorela H, Holm Y, Hiltunen R, Harvala T, Laitinen A. Extraction of the volatile oil in chamomile flowerheads using supercritical carbon dioxide. Flavour Fragrance J 1990; 5:81–84.

31. Scalia S, Giuffreda L, Pallado P. Analytical and preparative supercritical fluid extraction of chamomile flowers and its comparison with conventional methods. J Pharm Biomed Anal 1998; 21:549–558.

32. Kohler M, Haerdi W, Christen P, Veuthey JL. Extraction of artemisinin and artemisinic acid from *Artemisia annua* L. using supercritical carbon dioxide. J Chromatogr 1997; 785:353–360.

33. Ashraf-Khorassani M, Taylor LT. Supercritical fluid extraction of michellamines A and B from *Ancistrocladus korupensis* leaves. Anal Chim Acta 1997; 347:305–311.

34. Heaton DM, Bartle KD, Rayner CM, Clifford AA. Application of supercritical fluid extraction and supercritical fluid chromatography to the production of taxanes as anticancer drugs. J High Resolut Chromatogr 1993; 16:666–670.

35. Ma X, Yu X, Zheng Z, Mao J. Analytical supercritical fluid extraction of Chinese herbal medicines. Chromatographia 1991; 32:40–44.

36. Dean JR, Liu B. Supercritical fluid extraction of Chinese herbal medicines: investigation of extraction kinetics. Phytochem Anal 2000; 11:1–6.

37.	Choi YH, Kim J, Yoo KP. Selective extraction of ephedrine from *Ephedra sinica* using mixtures of carbon dioxide, diethylamine and methanol. Chromatographia 1999; 50:673–679.

38.	Suto K, Kakinuma S, Ito Y, Sagara K, Iwasaki H, Itokawa H. Determination of atractylon in *Atractylodes* rhizome using supercritical fluid chromatography on-line coupled with supercritical fluid extraction by the direct induction method. J Chromatogr 1998; 810:252–255.

39.	Sewram V, Raynor MW, Raidoo DM, Mulholland DA. Coupling SFE to uterotonic bioassay: an on-line approach to analysing medicinal plants. J Pharm Biomed Anal 1998; 18:305–318.

40.	Elisabeth P, Yoshioka M, Yamauchi Y, Saito M. Infra-red and nuclear magnetic resonance spectrometry of caffeine in roasted coffee beans after separation by preparative supercritical-fluid chromatography. Anal Sci 1991; 7:427–431.

41.	Sharma AK, Prokopczyk B, Hoffmann D. Supercritical fluid extraction of moist snuff. J Agric Food Chem 1991; 39:508–510.

42.	Saldana MDA, Mohamed RS, Baer MG, Mazzafera P. Extraction of purine alkaloids from mate (*Ilex paraguariensis*) using supercritical CO_2. J Agric Food Chem 1999; 47:3804–3808.

43.	Brachet A, Mateus L, Cherkaoui S, Christen P, Gauvrit JY, Lanteri P, Veuthey JL. Application of central composite design in the supercritical fluid extraction of tropane alkaloids in plant extracts. Analusis 1999; 27:772–778.

44.	Lee H, Hong WH, Yoon JH, Song KM, Kwak SS, Liu JR. Extraction of indole alkaloids from *Catharanthus roseus* by using supercritical carbon dioxide. Biotechnol Tech 1992; 6:127–130.

45.	Bicchi C, Rubiolo P, Frattini C. Off line supercritical fluid extraction and capillary gas chromatography of pyrrolizidine alkaloids in *Senecio* species. J Nat Prod 1991; 54:941–945.

46.	López-Ávila V, Benedicto J, Robaugh D. Supercritical fluid extraction of oxindole alkaloids from *Uncaria tomentosa*. J High Resolut Chromatogr 1997: 20:231–236.

47.	Queckenberg OR, Frahm AW. Supercritical fluid extraction, quickness and selectivity in the analysis of natural products. Pharmazie 1994; 49:159–166.

48.	Bicchi C, Binello A, Rubiolo P. Determination of phenolic diterpene antioxidants in rosemary (*Rosmarinus officinalis* L.) with different methods of extraction and analysis. Phytochem Anal 2000; 11:236–242.

49.	Ibañez E, Cifuentes A, Crego AL, Senorans FJ, Cavero S, Reglero G. Combined use of supercritical fluid extraction, micellar electrokinetic chromatography and reverse phase high performance liquid chromatography for the analysis of antioxidants from rosemary (*Rosmarinus officinalis* L.). J Agric Food Chem 2000; 48:4060–4065.

50.	Tena MT, Valcárcel M, Hidalgo PJ, Ubera JL. Supercritical fluid extraction of natural antioxidants from rosemary: comparison with liquid solvent sonication. Anal Chem 1997; 69:521–526.

51.	Tsuda T, Mizuno K, Ohshima K, Kawakishi S, Osawa T. Supercritical carbon dioxide extraction of antioxidative components from tamarind (*Tamarindus indica* T.) seed coat. J Agric Food Chem 1995; 43:2803–2806.

52. Veress T. Sample preparation by supercritical fluid extraction for the HPLC determination of cannabinoids. LC–GC Int 1997; 10:114–122.

53. Ray KS, Chheda M, Mukhopadhyay M. Performance of conventional and supercritical fluid extraction methods for carotene recovery from non-edible leaves. J Food Sci Technol 2000; 37:514–516.

54. Careri M, Furlattini L, Mangia A, Musci M, Anklam E, Theobald A, von Holst C. Supercritical fluid extraction for liquid chromatographic determination of carotenoids in *Spirulina pacifica* algae: a chemometric approach. J Chromatogr 2001; 912:61–71.

55. Colombo ML, Corsini A, Mossa A, Sala L, Stanca M. Supercritical carbon dioxide extraction, fluorimetric and electochemical high performance liquid chromatographic detection of vitamin E from *Hordeum vulgare* L. Phytochem Anal 1998; 9:192–195.

56. Schneiderman MA, Sharma AK, Locke DC. Determination of vitamin A palmitate in cereal products using supercritical fluid extraction and liquid chromatography with electrochemical detection. J Chromatogr 1997; 765: 215–220.

57. Luque de Castro MD, Jiménez-Carmona MM, Fernández-Pérez V. Towards more rational techniques for the isolation of valuable essential oils from plants. Trends Anal Chem 1999; 18:708–716.

58. Reverchon E, Della-Porta G, Gorgoglione D. Supercritical carbon dioxide extraction of volatile oil from rose concrete. Flavour Fragrance J 1997; 12:37–41.

59. Coelho LAF, Oliveira JV, d'Avila SG, Vilegas JHY, Lancas FM. SFE of rosemary oil: assessment of the influence of process variables and extract characterization. J High Resolut Chromatogr 1997; 20:431–436.

60. Reverchon E, Ambruosi A, Senatore F. Isolation of peppermint oil using supercritical carbon dioxide extraction. Flavour Fragrance J 1994; 9:19–23.

61. Della-Porta G, Porcedda S, Marongiu B, Reverchon E. Isolation of eucalyptus oil by supercritical fluid extraction. Flavour Fragrance J 1999; 14:214–218.

62. Reverchon E, Senatore F. Supercritical carbon dioxide extraction of chamomile essential oil and its analysis by gas chromatography–mass spectrometry. J Agric Food Chem 1994; 42:154–158.

63. Polesello S, Lovati F, Rizzolo A, Rovida C. Supercritical-fluid extraction as a preparative tool for strawberry aroma analysis. J High Resolut Chromatogr 1993; 16:555–559.

64. Sagrero-Nieves L, Bartley JP, Provis-Schwede A. Supercritical fluid extraction of the volatile components from the leaves of *Psidium guajava* L. (guava). Flavour Fragrance J 1994; 9:135–137.

65. Blum C, Kubeczka KH, Becker K. Supercritical-fluid chromatography–mass spectrometry of thyme extracts (*Thymus vulgaris* L). J Chromatogr 1997; 773: 377–380.

66. Hawthorne SB, Krieger MS, Miller DJ. Analysis of flavour and fragrance com-

pounds using supercritical-fluid extraction coupled with gas chromatography. Anal Chem 1988; 60:472–477.

67. Blatt CR, Ciola R. Analysis of vetiver essential oil by supercritical fluid extraction and on-line capillary chromatography. J High Resolut Chromatogr 1991; 14:775–777.

68. Reis-Vasco EMC, Coelho JAP, Palavra AMF. Comparison of pennyroyal oils obtained by supercritical carbon dioxide extraction and hydrodistillation. Flavour Fragrance J 1999; 14:156–160.

69. Hawthorne SB, Riekkola ML, Srenius K, Holm Y, Hiltunen R, Hartonen K. Comparison of hydrodistillation and supercritical fluid extraction for the determination of essential oils in aromatic plants. J Chromatogr 1993; 634:297–308.

70. da Cruz-Francisco J, Jarvenpaa EP, Huopalahti R, Sivik B. Comparison of *Eucalyptus camaldulensis* Dehn. Oils from Mozambique as obtained by hydrodistillation and supercritical carbon dioxide extraction. J Agric Food Chem 2001; 49:2339–2342.

71. Schneiderman MA, Sharma AK, Locke DC. Determination of menadione [menaphthone] in an animal feed using supercritical-fluid extraction and HPLC with electrochemical detector. J Chromatogr Sci 1988; 26:458–462.

72. Sauvage E, Rocca JL, Toussaint G. Use of nitrous oxide for supercritical-fluid extraction of pharmaceutical compounds from animal feed. J High Resolut Chromatogr 1993; 16:234–238.

73. Husek P, Rijks JA, Leclercq PA, Cramers CA. Fast esterification of fatty acids with alkyl chloroformates. Optimization and application in gas chromatography. J High Resolut Chromatogr 1990; 13:633–638.

74. Berridge JC, Broad LA. Determination of fluconazole in rodent diet using solid-phase extraction and high-performance liquid chromatography. J Pharm Biomed Anal 1987; 5:523–526.

75. Ozawa H, Tsukioka T. Gas chromatographic determination of sodium monofluoroacetate in water by derivatization with dicyclohexylcarbodi-imide. Anal Chem 1987; 59:2914–2917.

76. Ozawa H, Tsukioka T. Gas chromatographic separation and determination of chloroacetic acids in water by difluoroanilide derivatization method. Analyst 1990; 115:1343–1347.

77. Dean JR, Khundker S. Extraction of pharmaceuticals using pressurised carbon dioxide. J Pharm Biomed Anal 1997; 15:875–886.

78. Euerby MR, Lewis RJ, Nichols SC. Preliminary investigations into the use of supercritical-fluid extraction to extract a novel corticosteroid (tipredane INN) from rodent diet. Anal Proc 1991; 28:287–289.

79. Messer DC, Taylor LT. Development of analytical SFE [supercritical fluid extraction] of a polar drug from an animal food matrix. J High Resolut Chromatogr 1992; 15:238–241.

80. Khundker S, Dean JR, Jones P. A comparison between solid phase extraction and supercritical fluid extraction for the determination of fluconazole from animal feed. J Pharm Biomed Anal 1995; 13:1441–1447.

81. Roston DA, Sun JJ. Supercritical fluid extraction method development for extraction of an experimental HIV protease inhibitor drug from animal feed. J Pharm Biomed Anal 1997; 15:461–468.

82. Williams JR, Morgan ED, Law B. Comparison of supercritical, subcritical, hot, pressurised and cold solvent extraction of four drugs from rodent food. Anal Commun 1996; 33:15–17.

83. Luque-Garcia JL, Luque de Castro MD. Extraction of fat-soluble vitamins. J Chromatogr 2001; 935:3–11.

84. Turner C, Mathiasson L. Determination of vitamins A and E in milk powder using supercritical fluid extraction for sample clean-up. J Chromatogr 2000; 874:275–283.

85. Turner C, King JW, Mathiasson L. On-line supercritical fluid extraction/enzymic hydrolysis of vitamin A esters: a new simplified approach for the determination of vitamins A and E in food. J Agric Food Chem 2001; 49:553–558.

86. Vega PJ, Balaban MO, Sims CA, O'Keefe SF, Cornell JA. Supercritical carbon dioxide extraction efficiency for carotenes from carrots by response surface methodology (RSM). J Food Sci 1996; 61:757–759.

87. Barth MM, Zhou C, Kute KM, Rosenthal GA. Determination of optimum conditions for supercritical fluid extraction of carotenoids from carrot (*Daucus carota* L.) tissue. J Agric Food Chem 1995; 43:2876–2878.

88. Weathers RM, Beckholt DA, Lavella AL, Danielson ND. Comparison of acetals as in situ modifiers for the supercritical fluid extraction of beta-carotene from paprika with carbon dioxide. J Liquid Chromatogr Related Technol 1999; 22:241–252.

89. Amador-Hernández J, Fernández-Romero JM, Ramis-Ramos G, Luque de Castro MD. Monitoring supercritical fluid extraction by thermal-lens spectrometry with pulsed laser excitation. Anal Chim Acta 1999; 390:163–173.

90. Pascual-Martí MC, Salvador A, Chafer A, Berna A. Supercritical fluid extraction of reverastrol from grape skin of *Vitis vinifera* and determination by HPLC. Talanta 2001; 54:735–740.

91. Ollanketo M, Hartonen K, Riekkola ML, Holm Y, Hiltunen R. Supercritical fluid extraction of lycopene in tomato skins. Eur Food Res Technol 2001; 212: 561–565.

92. Choi YH, Kim J, Jeon SH, Yoo KP, Lee HK. Optimum SFE condition for lignans of *Schisandra chinensis* fruits. Chromatographia 1998; 48:695–699.

93. Lee CH, Row KH, Lee YW, Kim JD, Lee YY. Supercritical fluid extraction of perillyl alcohol in Korean orange peel. J Liquid Chromatogr Related Technol 2001; 24:1987–1996.

94. Lee YW, Lee CH, Kim JD, Lee YY, Row KH. Extraction of perillyl alcohol in Korean orange peel by supercritical carbon dioxide. Sep Sci Technol 2000; 35:1069–1076.

95. Yacobi A, Skelly JP, Batra VK. Toxicokinetics and New Drug Development. New York: Pergamon Press, 1989.

96. Sachs H, Uhl M. Opiat-nachweis in Haar-extrakten mit Hilfe von GC/MS/MS und supercritical fluid extraction (SFE). Toxichem Krimtech 1992; 59:114–120.

97. Edder P, Staub C, Veuthey JL, Pierroz I, Haerdi W. Subcritical-fluid extraction of opiates in hair of drugs addicts. J Chromatogr B 1994; 658:75–86.

98. Veuthey JL, Edder P, Staub C. Detection of drug addiction by supercritical-fluid extraction coupled to GC-MS: applications to urine and hair. Analusis 1995; 23:258–265.

99. Cirimele V, Kintz P, Majdalani R, Mangin P. Supercritical fluid extraction of drugs in drug addict hair. J Chromatogr B 1995; 673:173–181.

100. Morrison JF, Chester SN, Reins JL. Supercritical-fluid extraction—immunoassay for the rapid screening of cocaine in hair. J Microcolumn Sep 1996; 8:37–45.

101. Morrison JF, Chesler SN, Yoo WJ, Selavka CM. Matrix and modifier effects in the supercritical fluid extraction of cocaine and benzoylecgonine from human hair. Anal Chem 1998; 70:163–172.

102. Brewer WE, Galipo RC, Sellers KW, Morgan SL. Analysis of cocaine, benzoylecgonine, codeine, and morphine in hair by supercritical fluid extraction with carbon dioxide modified with methanol. Anal Chem 2001; 73:2371–2376.

103. Allen DL, Oliver JS. The use of supercritical-fluid extraction for the determination of amphetamines in hair. Forensic Sci Int 2000; 107:191–199.

104. Ramsey ED, Perkins JR, Games DE, Startin JR. Analysis of drug residues in tissue by combined supercritical-fluid extraction–supercritical-fluid chromatography–mass spectrometry–mass spectrometry. J Chromatogr 1989; 464:353–364.

105. Maxwell RJ, Lightfield AR, Stolker AAM. An SPE column–Teflon sleeve assembly for inline retention during supercritical fluid extraction of analytes from biological matrices. J High Resolut Chromatogr 1995; 18:231–234.

106. Magard MA, Berg HEB, Tagesson V, Jaremo MLG, Karlsson LLH, Mathiasson LJE, Bonneau M, Hansen-Moller J. Determination of androsterone in pig fat using supercritical-fluid extraction and gas chromatography–mass spectrometry. J Agric Food Chem 1995; 43:114–120.

107. Zabolotsky DA, Chen LF, Patterson JA, Forrest JC, Lin HM, Grant AL. Supercritical carbon dioxide extraction of androstenone and skatole from pork fat. J Food Sci 1995; 60:1006–1008.

108. Huopalahti RP, Henion JD. Application of supercritical-fluid liquid chromatography–mass spectrometry for the determination of some anabolic agents directly from bovine tissue samples. J Liquid Chromatogr Related Technol 1996; 19:69–87.

109. Yu GYF, Murby EJ, Wells RJ. Gas chromatographic determination of residues of thyureostatic drugs in bovine muscle tissue using combined resin-mediated methylation and extraction. J Chromatogr B 1997; 703:159–166.

110. Parks OW, Lightfield AR, Maxwell RJ. Effect of sample matrix dehydration during supercritical-fluid extraction on the recoveries of drug residues from fortified chicken liver. J Chromatogr Sci 1995; 33:654–657.

111. Jiménez-Carmona MM, Tena MT, Luque de Castro MD. Ion-pair–supercritical fluid extraction of clenbuterol from food samples. J Chromatogr A 1995; 711:269–276.

112. O'Keeffe MJ, O'Keeffe M, Glennon JD, Lightfield AR, Maxwell RJ. Super-

critical-fluid extraction of clenbuterol from bovine liver tissue. Analyst 1998; 123:2711–2714.

113. O'Keeffe MJ, O'Keeffe M, Glennon JD. Supercritical fluid extraction (SFE) as a multi-residue procedure for beta-antagonists in bovine liver tissue. Analyst 1999; 124:1355–1360.

114. Stolker AAM, Zoontjes PW, van Ginkel LA. The use of supercritical-fluid extraction for the determination of steroids in animal tissues. Analyst 1998; 123:2671–2676.

115. Din N, Bartle KD, Clifford AA, McCormack A, Castle L. Supercritical-fluid extraction of sulphamethazine and its metabolites from meat tissues. J Chromatogr Sci 1997; 35:31–37.

116. Parks OW, Maxwell RJ. Isolation of sulfonamides from fortified chicken tissues with supercritical carbon dioxide and incline adsorption. J Chromatogr Sci 1994; 32:290–293.

117. Maxwell RJ, Lightfield AR. Multiresidue supercritical fluid extraction method for the recovery at low ppb levels of three sulfonamides from fortified chicken liver. J Chromatogr B 1998; 715:431–435.

118. Combs MT, Boyd S, Asraf-Khorassani M, Taylor LT. Quantitative recovery of sulphonamides from chicken liver, beef liver and egg yolk via modified supercritical carbon dioxide. J Agric Food Chem 1997; 45:1779–1783.

119. Combs MT, Ashraf-Khorassani M, Taylor LT. Comparison of supercritical trifluoromethane and carbon dioxide for extraction of sulphonamides from various food matrices. Anal Chem 1996; 68:4507–4511.

120. Burri BJ, Neidlinger TR, Lo AO, Kwan C, Wong MR. Supercritical fluid extraction and reversed-phase liquid chromatography methods for vitamin A and β-carotene. Heterogeneous distribution of vitamin A in the liver. J Chromatogr A 1997; 762:201–206.

121. Scott KS, Oliver JS. Development of a supercritical fluid extraction method for the determination of temazapam in whole blood. J Anal Toxicol 1997; 21:297–300.

122. Allen DL, Scott KS, Oliver JS. Comparison of solid-phase extraction and supercritical fluid extraction for the analysis of morphine in whole blood. J Anal Toxicol 1999; 23:216–218.

123. Liu H, Cooper LM, Raynie DE, Pinkston JD, Wehmeyer KR. Combined supercritical-fluid extraction–solid-phase extraction with octadecylsilane cartridges as a sample-preparation technique for the ultratrace analysis of a drug metabolite in plasma. Anal Chem 1992; 64:802–806.

124. Liu H, Wehmeyer KR. Solid-phase extraction with supercritical-fluid elution as a sample preparation technique for the ultra-trace analysis of flavone in blood plasma. J Chromatogr 1992; 577:61–67.

125. Karlsson L, Jaegfeldt H, Gere D. Supercritical-fluid extraction recovery studies of budesonide from blood plasma. Anal Chim Acta 1994; 287:35–40.

126. Liu H, Wehmeyer KR. Supercritical-fluid extraction as a sample preparation technique for the direct isolation of drugs from plasma prior to analysis. J Chromatogr B 1994; 657:209–213.

127. Hondo T, Saito M, Senda M. Analysis of vitamin K by direct-coupled super-

critical-fluid extraction–supercritical-fluid chromatography. Bunsuki Kagaku 1986; 35:316–319.

128. Simmons BR, Stewart JT. Supercritical-fluid extraction of selected pharmaceuticals from water and serum. J Chromatogr B 1997; 688:291–302.

129. Spell JC, Srinivasan K, Stewart JT, Bartlett MG. Supercritical-fluid extraction and negative-ion electrospray liquid chromatography tandem mass spectrometry analysis of phenobarbital, butalbital, pentobarbital and thiopental in human serum. Rapid Commun Mass Spectrom 1998; 12:890–894.

130. Hartonene K, Riekkola ML. Detection of beta-blockers in urine by solid-phase extraction supercritical-fluid extraction and gas chromatography–mass spectrometry. J Chromatogr B 1996; 676:45–52.

131. Meissner G, Hartonen K, Riekkola LL. Supercritical-fluid extraction combined with solid-phase extraction as sample preparation technique for the analysis of beta-blockers in resum and urine. Fresenius J Anal Chem 1998; 360:618–621.

132. Stolker AAM, van Ginkel LA, Stephany RW, Maxwell RJ, Parks OW, Lightfield AR. Supercritical-fluid extraction of methyltestosterone, nortestosterone and testosterone at low ppb levels from fortified bovine urine. J Chromatogr B 1999; 726:121–131.

133. Ashraf-Khorassani M, Taylor LT. Feasibility of on-line fluid extraction of steroids from aqueous-based matrices with analysis via gas chromatography–mass spectrometry. J Chromatogr Sci 2000; 38:477–482.

134. Allen DL, Oliver JS. The application of supercritical fluid extraction to cocaine and its metabolites in blood and urine. J Anal Toxicol 2000; 24:228–232.

135. Wang SM, Giang YS, Ling YC. Simultaneous supercritical-fluid extraction and chemical derivatization for the gas chromatographic–isotopic dilution mass spectrometric determination of amphetamine and methamphetamine in urine. J Chromatogr B 2001; 759:17–26.

136. Koski IJ, Jansson BA, Markides KE, Lee ML. Analysis of prostaglandins in aqueous solutions by supercritical-fluid extraction and chromatography. J Pharm Biomed Anal 1991; 9:281–290.

137. Masuda M, Koike S, Handa M, Sagara K, Mizutani T. Application of supercritical-fluid extraction and chromatography to assay fat-soluble vitamins in hydrophobic ointment. Anal Sci 1993; 9:29–32.

138. Ihara T, Suzuki N, Maeda T, Sagara K, Hobo T. Extraction of water-soluble vitamins from pharmaceutical preparations using AOT (sodium di-2-ethylexyl sulfosuccinate)/pentane reversed micelles. Chem Pharm Bull 1995; 43:626–630.

139. Scalia S, Ruberto G, Bonina F. Determination of vitamin A, vitamin E, and their esters in tablet preparations using supercritical fluid extraction and HPLC. J Pharm Sci 1995; 84:433–436.

140. Salvador A, Jaime MA, de la Guardia M, Becerra G. Supercritical-fluid extraction and supercritical-fluid chromatography of vitamin E in pharmaceutical preparations. Anal Commun 1998; 35:53–55.

141. Gámiz-Gracia L, Jiménez-Carmona MM, Luque de Castro MD. Determination of vitamins D_2 and D_3 in pharmaceuticals by supercritical-fluid extraction andHPLC separation with UV detection. Chromatographia 2000; 51:428–432.

142. Eckard PR, Taylor LT. Modifier and additive effects in the supercritical-fluid

extraction of pseudoephedrine hydrochloride from spiked-sand and Suphedrine tablets. J High Resolut Chromatogr 1999; 22:469–474.

143. Kordikowski A, York P, Latham D. Resolution of ephedrine in supercritical carbon dioxide: a novel technique for the separation of chiral drugs. J Pharm Sci 1999; 88:786–791.

144. Simmons BR, Chukwumerije O, Stewart JT. Supercritical-fluid extraction of 13-*cis*-retinoic acid and its photoisomers from selected pharmaceutical dosage forms. J Pharm Biomed Anal 1997; 19:395–403.

145. Dean JR, Lowdon J. Application of supercritical-fluid extraction in the pharmaceutical industry: supercritical-fluid extraction of megestrol acetate from a tablet matrix. Analyst 1993; 118:747–751.

146. Anklam E, Mueller A. Extraction of caffeine and vanillin from drugs by supercritical carbon dioxide. Pharmazie 1995; 50:364–365.

147. Khundker S, Dean JR, Hitchen SM. Extraction of ibuprofen by supercritical carbon dioxide. Anal Proc 1993; 30:472–473.

148. Lawrence JK, Larsen AK, Tebbett IR. Supercritical-fluid extraction of benzodiazepines in solid dosage form. Anal Chim Acta 1994; 288:123–130.

149. Chen B, Liu LL, Fang HS, Zhai ZX, Wu YT, Chen W. Determination of psoralen and isopsoralen in Baishi pill and Baidianfeng capsules by capillary GC following supercritical-fluid extraction (SFE). Yaowu Fenxi Zazhi 1999; 19:394–396.

150. Wang QH, Chen R, Zhu DQ, Zhou LM. Applications of supercritical-fluid extraction samples in analytical chemistry. Sepu 1998; 16:344–346.

151. Roston DA, Sun JJ, Collins PW, Perkins WE, Tremont SJ. Supercritical-fluid extraction–liquid chromatography method development for a polymeric controlled-release drug formulation. J Pharm Biomed Anal 1995; 13:1513–1520.

152. Howard AL, Shah MC, Ip DP, Brooks MA, StrodeIII JTB, Taylor LT. Use of supercritical-fluid extraction for sample preparation of sustained-release felodipine tablets. J Pharm Sci 1994; 83:1537–1542.

153. Moore WN, Taylor LT. Analytical inverse supercritical-fluid extraction of polar pharmaceutical compounds from cream and ointment matrices. J Pharm Biomed Anal 1994; 12:1227–1232.

154. Almodóvar RA, Rodriguez RA, Rosario O. Inverse supercritical extraction of acetaminophen from suppositories. J Pharm Biomed Anal 1998; 17:89–93.

155. Bonazzi D, Cavrini V, Gatti R, Boselli E, Caboni M. Determination of imidazole antimycotics in creams by supercritical-fluid extraction and derivative UV spectroscopy. J Pharm Biomed Anal 1998; 18:235–240.

156. Mulcahey LJ, Taylor LT. Supercritical-fluid extraction of active components in a drug formulation. Anal Chem 1992; 64:981–984.

157. Scalia S, Games DE. Determination of parabens in cosmetic products by supercritical-fluid extraction and high-performance liquid chromatography. Analyst 1992; 117:839–841.

158. Wang SP, Chang CL. Determination of parabens in cosmetic products by supercritical fluid extraction and capillary zone electrophoresis. Anal Chim Acta 1998; 377:85–93.

159. Scalia S. Determination of sunscreen agents in cosmetic products by super-

critical-fluid extraction and high-performance liquid chromatography. J Chromatogr A 2000; 870:199–205.

160.   Wang SP, Chen WJ. Determination of p-aminobenzoates and cinnamate in cosmetic matrix by supercritical-fluid extraction and micellar electrokinetic capillary chromatography. Anal Chim Acta 2000; 416:157–167.

161.   Scalia S, Simeoni S. Assay of xanthene dyes in lipsticks by inverse supercritical fluid extraction and HPLC. Chromatographia 2001; 53:490–494.

162.   Li JJ. Quantitative analysis of cosmetic waxes by using supercritical-fluid extraction(SFE)/supercritical-fluid chromatography (SFC) and multivariate data analysis. Chemom Intell Lab Syst 1999; 45:385–395.

163.   Tena MT, Luque de Castro MD, Valcárcel M. Improved supercritical fluid extraction of sulphonamides. Chromatographia 1995; 40:197–203.

14

Development and Potential of Critical Fluid Technology in the Nutraceutical Industry

Jerry W. King
Los Alamos National Laboratory, Los Alamos, New Mexico, U.S.A.

Nutraceuticals, as the name suggests, are ingested substances which combine the benefit of food nutritional requirements while offering some aspect of therapeutic protection to the human body. Such foods and natural substances are called functional foods, designer foods, pharma foods, as well as many less elegant descriptors. Functional foods are similar in appearance to conventional foods, are consumed as part of a normal diet regime, and have demonstrated physiological benefit (i.e., reducing the risk of a disease state). Naturally derived products are purchased to enhance stamina and energy, for weight control, to avoid illness, and to compensate for the lack of exercise. Depending on the definition of a nutraceutical, the market ranges of such products is conservatively estimated to be US$3.15–4.6 billion in the United States and range from US$1.05–1.6 billion in Europe. A broader definition of "functional" food pegs their U.S. market value between US$14.2 and 17.6 billion, and if one assumes that 50% of the food selected for consumption is based on health or medical considerations, then the estimated value of the nutraceutical market expands to US$250 billion (1).

Consumers of nutraceuticals have expressed concern about pesticide or chemical residues, processing technology that contributes to ecological pollution, antibiotics or growth hormones in their foods, and the extensive use of preservatives in the foods they consume. It is for these reasons that technologies incorporating the use of critical fluids become important in the production of nutraceutical ingredients. Critical fluids, such as carbon dioxide ($SC\text{-}CO_2$), $SC\text{-}CO_2$/ethanol mixtures, and subcritical water, are environmentally benign processing agents; leaving no solvent residues in the final products while minimizing the oxidation or degradation of thermally labile components. Even a cursory inspection of the common classes of nutraceutical agents (herbs, specialty oils, plant extracts, specific protein fractions, and antioxidants) suggest a link between the two fields.

Table 1 lists some of the common and popular nutraceutical agents in use today, their application, as well as the use of critical fluid processing for their production. It should be noted that all of the nutraceuticals listed in Table 1 can be processed using critical fluids, however, a "yes" indicates that the actual production of the nutraceutical components has commenced. Indeed, a segment of the production capacity of the over 50 critical fluid processing plants worldwide are devoted to producing products for the nutraceutical market.

TABLE 1 Nutraceuticals and Their Therapeutic Use and Current Production via Critical Fluids

Nutraceutical	Utility	Processed via critical fluids
Saw palmetto	Prostate	Yes
Kava-kava	Anxiolytic	No
Hawthorne	Cardiotonic	No
Ginseng	Tonic	Yes
Garlic	Circulatory	Yes
Ginko biloba	Cognitive	No
St. John's wort	Depression	No
Chamomile	Dermatological	Yes
Echinacea	Colds/flu	Yes
Black cohosh	Gynecological	No
Lutein	Macular degeneration	Yes
Flavanoids	Anticancer	No
Isoflavones	Premenstral syndrome, circulatory	No
Marine oil fatty acids	Circulatory	Yes
Evening primrose	Inflammation	Yes
Phytosterols	Circulatory	Yes
Tocopherols	Antioxidant	Yes
Phospholipids	Cognitive	Yes

Critical fluid processing can be used in several modes for producing nutraceutical ingredients or functional foods. Exhaustive extraction in which SC-CO_2 or a SC-CO_2–cosolvent mixture is used to yield an extract equivalent to those obtained with organic solvent extraction or via pressing/expeling technologies (2), as documented in the recent literature (3). Fractional extraction in which the extraction pressure, temperature, time, or the addition of a cosolvent is varied on an incremental basis is also capable of producing extracts that are somewhat either enriched or depleted in the desired nutraceutical agent (4). Such fluid density-based or cosolvent-assisted extractions frequently yield extracts with considerable extraneous material; indeed, specifically extracting or enriching a desired solute out of a natural product matrix is somewhat akin to "finding a needle in a haystack." The problem is shown in Table 2, where many of the listed naturally occurring oils have been extracted with SC-CO_2; however, the targeted "nutraceutical" components in the right column occur in these SC-CO_2 extracts at very low concentration levels.

To enrich the concentration of desired component(s), researchers have resorted to fractionation techniques utilizing critical fluids. One of the simplest techniques separates the extract with the aid of multiple separators

TABLE 2 Natural Oils Extracted with Critical Fluids and Their Nutraceutical Components

Natural oils	Nutraceutical component
Rice bran	n-6, n-3 Fatty acids
Safflower	Phystosterols
Marine	Tocopherols
Sesame	Carotenoids
γ-Linolemic acid-enriched	Phospholipids
Oat	Tocotrienols
Almond	Oryzanol
Wheat germ	Sesamolin
Amaranth	Glycolipids
Essential	Conjugated fatty acids
Avocado	Lipoproteins
Grape seed	
Macadamia nut	
Kiwi	
Genetically modified oils	

Note: The nutraceutical components listed on the same line as a particular oil does not mean that nutraceutical moiety is associated with that oil.

placed downstream after the main extraction vessel. These separator vessels are held at different combinations of temperatures and pressures (5). Using such an approach, the fractionation of essential oils from waxes and oleo-resins can be accomplished. The use of fractionation columns in which a temperature gradient is imposed on a solute-laden flowing stream of SC-CO_2, either in a batch or countercurrent mode, can also be used to fractionate supercritical fluid extracts. This technique has been used for the production of fish oil concentrates (6), fractionation of peel oil components (7), and mixed-glyceride fractionation (8). The coupling of critical fluids with chromatography on a preparative or production scale offers another alternative route to producing nutraceutical-enriched extracts. These chromatographic-based separations range from simple displacement or elution chromatographic schemes [i.e., for the removal of cholesterol (9)] to the more sophisticated simulated moving bed technology (10), the latter technique perhaps more favored for the purification of pharmacological compounds.

Although there are many examples for processing nutraceutical ingredients using critical fluids, we will discuss here specific methods for producing extracts, fractions enriched in nutraceutical ingredients (i.e., products containing sterols and sterol esters from natural sources). Sterol esters have been clinically evaluated (11) and proven to effective for inhibiting cholesterol absorption and synthesis in the human body. This has resulted in the marketing of two commercial products, Benecol and Take Control, on a worldwide basis (12). Likewise, nutraceuticals containing tocopherols or phospholipids can be readily produced via critical fluid technology, as will be described shortly. In addition, an alternative route for producing "naturally" synthesized nutraceutical components, such as the above sterol esters, or transesterified esters can be accomplished in SC-CO_2, particularly if enzyme catalysts are used. Also amenable to this type of synthesis are the "omega-3" fish oil type of esters, which are highly valued and an article of commerce in the nurtraceutical field.

1. THE RATIONALE FOR USING CRITICAL FLUIDS

The use of critical fluids for the extraction and refining of components in natural products has now been facilitated for over 30 years. Early success in the decaffeination of coffee beans and isolation of specific fractions from hops for flavoring beer, using either supercritical carbon or liquid carbon dioxide, are but two examples of the commercial application of this versatile technology. Critical fluid technology, a term that will be used here to embrace an array of fluids under pressure, has seen new and varied applications which include the areas of engineering-scale processing, analytical, and materials modification.

However, beginning in the early 1990s, an awareness of the potential of critical fluid processing as a viable component of the newly coined term "green" processing arose (13). This, coupled with an increasing consumer awareness of the identity and use of nongreen chemical agents in food and natural products processing, suggested a promising future for the use of such natural solvents as supercritical fluid carbon dioxide (SC-CO$_2$), ethanol, and water. Indeed, labels on food products and ingredients which tout, "naturally decaffeinated" and "nature in its most concentrated form: high pressure extraction with carbon dioxide," have an appeal to the consumer who is aware of food safety issues. Recently, certain nutraceutical products have been labeled as "hexane-free" to alert the general public to the undesirable use of organic chemical-based processing agents.

There is a certain synergy that exists between critical fluid processing and the use of nutraceuticals. Consumer use and acceptance of these natural products is heightened by the appeal that they have been "naturally processed." However, this is not a new development and it is interesting to note that many of today's nutraceutical components have already been extracted or separated using the above-named natural agents. In addition, critical fluid processing has also been used to create healthier and functional foods, such as eggs with a reduced cholesterol content (14), low-fat nut products (15), pesticide-free natural products (16), spice extracts (17), as well as cholesterol reduction in meats (18). There is no doubt that the development of such products coupled with a proper commercial marketing campaign provides a powerful stimulus for the consumer to try these products.

2. A GREEN PROCESSING PLATFORM

As noted previously, SC-CO$_2$ reigns supreme as the principle processing agent in critical fluid technology. However, this situation is changing, particularly with the recognition that SC-CO$_2$ cannot be effectively utilized for all tasks and is a relative poor extraction solvent for polar compounds. For certain applications, the addition of a minimal amount of cosolvent [usually an organic solvent having a higher critical temperature (T_c) than CO$_2$] suffices to improve the extraction of specific components from a natural product matrix. However, the number of GRAS (Generally Recognized As Safe) cosolvents is rather limited (e.g., ethanol, acetic acid). Such cosolvents can be used in conjunction with SC-CO$_2$ under conditions which can favor the formation of a one-phase or multiphase extraction medium, capable of producing the desired end result.

Other processing media which embrace the green processing concept are the use of liquefied gases (e.g., LCO$_2$) and certain liquids under pressure above their boiling point. By the application of external pressure, liquids such as water and ethanol can be used above their boiling point. Such liquids can be used to produce equivalent or superior products compared to those derived by using conventional liquid solvent extraction (19). For example, St. John's wort's active components can be more effectively isolated using critical fluids and/or by using SC-CO$_2$ along with subcritical water (20). Recently, the author and his colleagues have utilized subcritical water between 120°C and 160°C to isolate anthocyanins from natural berry substrates (21). In the processing of natural products, optimization of the extraction temperature is important in order to avoid degradation of the extracted components, such as the anthocyanins. For the extraction of anthocyanins, 120°C proves to be the optimal temperature. LCO$_2$ has also been used at near-ambient or subambient temperatures (22,23), thereby avoiding conditions conducive to thermal degradation of the extracted components.

Subcritical water complements SC-CO$_2$ for the environmentally benign processing of nutraceuticals. Its phase diagram (Fig. 1) is similar to that of CO$_2$, although the critical temperature and pressure of water are much higher (T_c = 374°C, P_c = 221 atm) than those for CO$_2$. However, the region in the phase diagram that is of interest for processing nutraceuticals lies between 100°C and 200°C in the liquid region defined in Figure 1 and at pressures less than 100 atm. Recent studies have shown that subcritical water under these conditions can be effective for the extraction of the natural products, such as cloves (24) and rosemary (25).

Considerable literature exists on the properties and use of water in the superheated state, and its temperature has been shown to be a key parameter

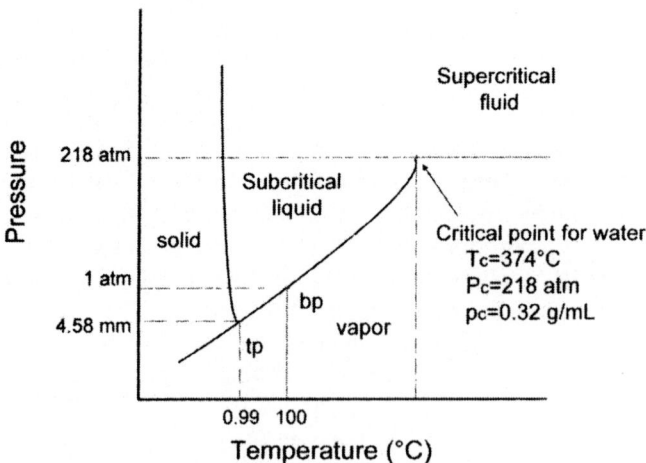

FIGURE 1   Phase diagram of water.

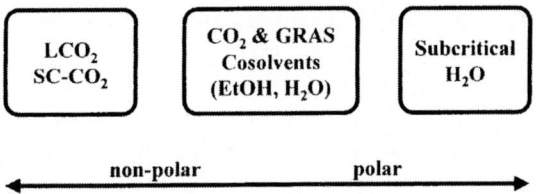

FIGURE 2   "Green" critical fluid processing options.

for regulating the solvent power of superheated water (26). The dielectric constant of water varies inversely with temperature; it varies from 30 to 60 over a temperature range of 100–200°C. Dielectric constants of this magnitude are in the range of those exhibited by polar organic liquids. Therefore, subcritical water can be potentially used as a substitute for less desirable organic solvents, including ethanol, to extract and process natural products.

Figure 2 summarizes the above in terms of offering an "all-natural" approach to processing natural products for nutraceutical ingredients. On one end of the solvent scale lies SC-CO_2 and LCO_2, whereas pressurized water on the other end is available for isolating polar moieties. Several combinations of GRAS cosolvents can be coupled with CO_2 for cases in which this approach proves viable. Sequential processing of a natural substrate by the use of these various fluids is also possible, as suggested by the hypothetical scheme noted in Figure 3 for soybeans as a natural substrate. Here, the nonpolar components, such as carotenoids, triterpenes, or phytosterols, are preferentially removed by CO_2 followed by extraction with a CO_2–cosolvent combination that can remove the more polar components, such as phenolic

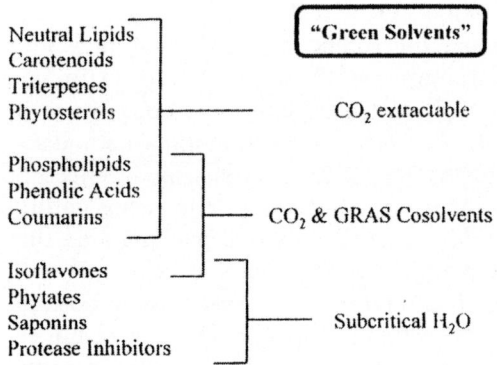

FIGURE 3   Separation scheme for nutraceutical components in soybeans.

acids or coumarins. Finally, after removal of the above components, subcritical water can be applied to isolate the isoflavones, phytates, and so forth. It should be recognized that some of targeted nutraceutical compounds may occur in each of these "green" solvent combinations. An additional benefit of the process depicted in Figure 3 is that left-over residual proteinaceous meal is available for further use, devoid of any solvent residues. This is an appealing extraction and/or fractionation scheme that can be accomplished using the same high pressure processing equipment, as has been noted by King (27).

3. TECHNIQUES FOR ENRICHING AND ISOLATING NUTRACEUTICALS

The development of critical-fluid-based techniques for isolating nutraceutical ingredients requires sequential development in steps of increasing complexity and scale-up. These are summarized as follows:

> Bench-scale supercritical fluid extraction (SFE) evaluation of process feasibility
> Establishment of the need for fractionation
> Pilot-plant evaluation
> Scale-up to plant Stage

Bench-scale evaluation of the feasibility of extracting a nutraceutical is usually accomplished with the aid of a small-scale extractor. Such units are available commercially, but an increased experimental flexibility is achieved by constructing a customized SFE unit. Alternatively, there exists the availability of small-scale SFE equipment, normally intended for analytical uses of critical fluid technology, which can also be used to optimize and assess the extraction of a natural product (28). Both of the above approaches can also incorporate the introduction of a cosolvent into the critical fluid, should this be required in the processing of a nutraceutical source.

Enrichment of a nutraceutical ingredient to a sufficient purity or concentration cannot always be accomplished by using just an extraction step, and this will necessitate the use of a fractionation method, in one of several forms, or even by combining two or more fractionation methods. A combination of mixer/settler modules is one mode of amplifying the effects of single-stage SFE (29). A more popular approach for fractionating natural products is the use of a fractionating tower with internal packing, and that frequently will incorporate a temperature gradient along its length (30). Introduction of the material to be fractionated can be made at the bottom, top, or an intermittent position in the column. Both cocurrent and counter-current operational modes can be enacted with respect to contacting the critical fluid with the substrate whose components need to be separated.

Chromatographic fractionation has also been utilized (31), including the use of simulated moving-bed technology (32). Examples of these chromatographic options will be discussed later.

The pilot-plant evaluation stage incorporates both a scale-up of the previously described approaches as well several other processing options. Although somewhat rare, single-pass (with respect to the extraction fluid) pilot and production plants are known to exist, but plants designed for fluid recycling with integrated heat exchangers are much more the norm. Batch pilot- and production-scale plants have been in use for some time, and several novel approaches have been described which allow the continuous feed of substrates into extraction or fractionation units (i.e., lock hoppers, augers, etc.) (33).

Final scale-up to the production plant stage is a serious undertaking, with respect to mechanical complexity, safety, and economics. Such plants have traditionally been more or permanent in design; however, there is a trend toward increased flexibility (e.g., multiple-use modules for extraction, fractionation, reaction), considering the initial investment in the in common pumps and fluid sources. Portable extraction units have also been constructed and their feasibility demonstrated for the field side processing, thereby permitting extraction of natural products immediately after they are harvested. This avoids degradation during the transport of the nutraceutical ingredient, (i.e., medicinal ingredients in chamomile).

3.1. Examples of Experimental/Processing Equipment and Methods

To describe some of generic approaches mentioned in the previous section for processing nutraceuticals, several examples are provided in this subsection to illustrate the equipment which is appropriate for conducting extractions and fractionations with critical fluids. Figure 4 shows a schematic of a benchtop extraction system that has proven to be very versatile in our laboratories. A fluid source, A (e.g., CO_2), which can be can be either gaseous or liquefied, is, of course, required. Use of liquefied CO_2 has the advantage of being more compatible because this is what is found most frequently in use on scaled-up equipment. A compressor or liquid pump, C, delivers the fluid through a tandem switching valve, SV-1 and SV-2, to a tubular extraction vessel cell that is held at a constant temperature to maintain the fluid in its desired critical state (subcritical or supercritical). The fluid then passes over the material in the extraction vessel and is routed through another switching valve arrangement. In Figure 4, either a micrometering valve, MV, or back-pressure regulator, is used to reduce the pressure on the fluid as it exits the extraction/fractionation stage.

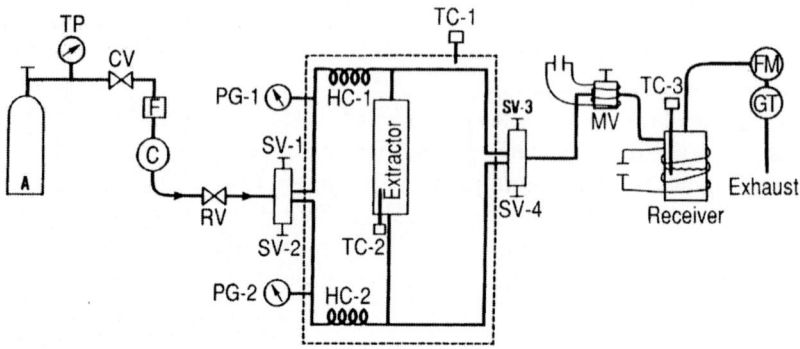

FIGURE 4 Bench-scale supercritical fluid extraction system.

The receiver vessel can be of several formats but will frequently be packed with internal packing to eliminate entrainment of the extracted components forming in the rapidly expanding critical fluid. Flow conditions under ambient conditions are measured with the aid of a flow meter, FM, and fluid totalizing module, GT. This described SFE unit can be changed to accommodate fractionation and reaction experiments as has been shown and documented in our laboratories at a modest cost (34). Such equipment today exists commercially but may be more limited with respect to pressure range, fluid flow rate, and size of the extraction chamber (34).

An increase in processing capability to a mini-pilot-plant stage is shown in Figure 5. This unit incorporates a fluid recycle option which reduces substantially the mass of extraction fluid required for SFE, because the initial charge of extraction fluid is used continuously throughout the extraction. The 4-L extraction vessel permits multikilogram quantities of natural product to be extracted. For this vessel, and others utilized for conducting extractions up 70 MPa (680 atm), a safety factor of 2.5 times the maximum operational pressure is specified when ordering these vessels. Back-pressure or metering valve arrangements used for pressure reduction of the extraction fluid are heated to overcome the attendant Joule–Thomson cooling effect as the depressurized fluid is jettisoned to the collection vessel.

The collection vessel must be of sufficient size to allow separation of the extracted solutes from the depressurized fluid. For these purposes, a 2-L collection vessel is utilized, as shown in Figure 5. By following the vector arrows in Figure 5, one can trace the depressurized fluid flow path back to the compressor, where the fluid is repressurized to continue the extraction. A sorbent-laden column may be inserted in the low pressure side of this scheme for purifying the exhausted fluid of any unwanted odoriferous volatile com-

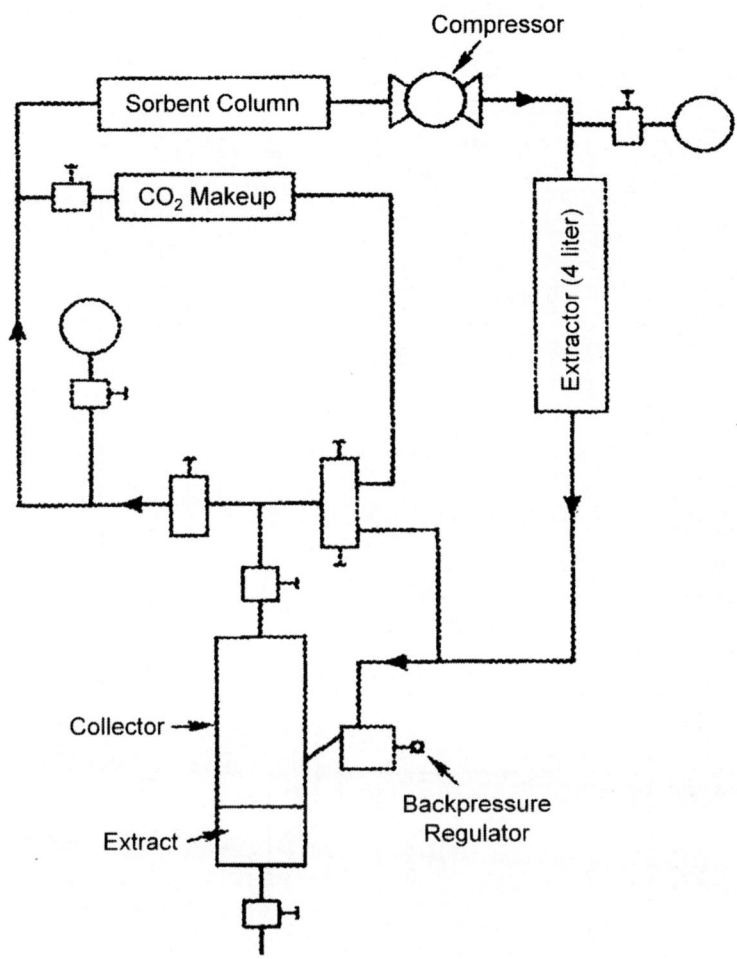

FIGURE 5 Continuous recycle supercritical fluid extractor.

pounds. A provision for makeup of fluid lost upon draining the extract from the receiver vessel is noted in Figure 5 as the "CO_2 makeup."

A larger semicontinuous pilot plant (35) is shown in Figure 6. It consists of three, 4-L extraction vessels, arranged with an appropriate valve sequence to permit several modes of operation. However the principle feature that distinguishes it from the smaller unit noted in Figure 5, is that it allows for semicontinuous processing using the tandem vessel arrangement depicted in Figure 6. Here, carbon dioxide, the extraction fluid, can be routed sequential to one or more of the extraction vessels, which can be operated so that one of

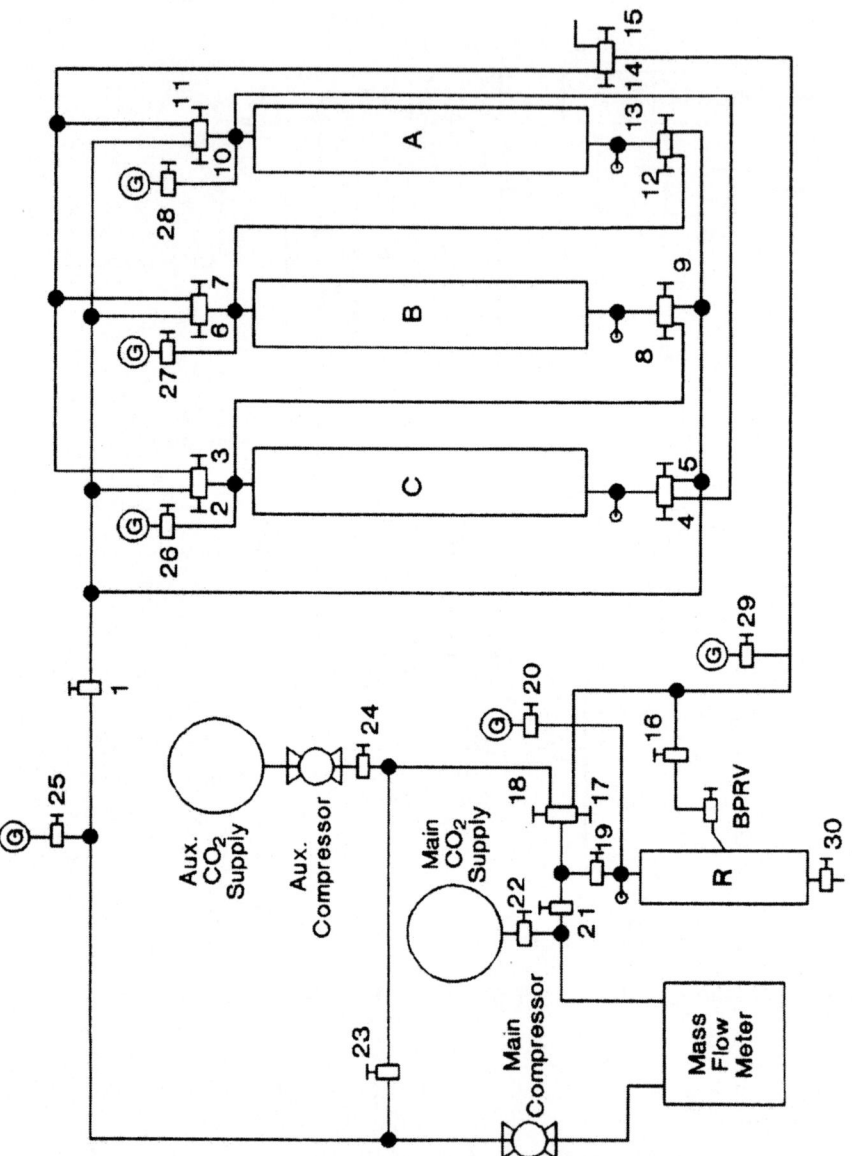

FIGURE 6   Semicontinuous pilot-plant extraction system.

the vessels, A, is being extracted while another vessel, B, is being loaded with product to be processed, and a third vessel, C, is capable of undergoing pressurization/depressurization. That these operations can be accomplished in parallel is also apparent, lending the unit to semicontinuous operation, even for the processing of granular solids.

Other types of pilot plant, including commercial units embracing this principle, are available. Some selected vendors of pilot plants, although this is not an exclusive list, include UHDE, Thar Designs, Applied Separations, Chematur, and Separex.. Likewise, bench-scale equipment for preliminary evaluations are manufactured by such companies as Autoclave Engineers (now called Snap-Tite), Chematur, Nova Swiss, Applied Separations, Nova Sep, Thar Designs, Pressure Products, Inc., Supercritical Fluid Technologies, and Separex.

In the future, extraction and other processing options may be accomplished with the aid of an expeller; that is, the critical fluid component will be introduced into the expeller barrel. This mode of processing not only allows extraction/fractionation to be accomplished on a continuous feed of raw material but also allows the introduction of nutraceutical ingredients through selective dissolution in the fluid, permitting them to be naturally "impregnated" into a product matrix. The addition of CO_2 into the barrel of the extruder, where it becomes a supercritical fluid due to the heat and pressure generated during the extrusion process, not only facilitates solubilization of materials from the substrate being processed but also enhances the fluidity of the potential extract (e.g., a nutraceutical-based oil). Two active companies in this field are Critical Processes Ltd. in North Yorkshire, England and Crown Iron Works in Minneapolis. The former firm is focused on nutraceutical extraction, citrus oil deterpenation, and so forth and the latter company is concerned with CO_2-assisted recovery of oils from seeds.

3.2. Fractionating with Critical Fluids

The ability to fractionate naturally derived materials in a benign way using critical fluids is of particular interest to processors of nutraceutical ingredients. This is due to the fact that to obtain useful extracts, an enrichment or isolation of the more highly purified form of the nutraceutical component increases the value and utility value of the derived product to nutraceutical manufacturers. Hence, approaches to achieve the above result are discussed in this subsection.

One of the important methods of critical fluid fractionation involves the countercurrent separation of phospholipids from a vegetable oil. A system to achieve this end is presented in Figure 7. Here, high pressure CO_2 is fed into a pressure vessel packed with segmented gauze mesh packing (the "refining

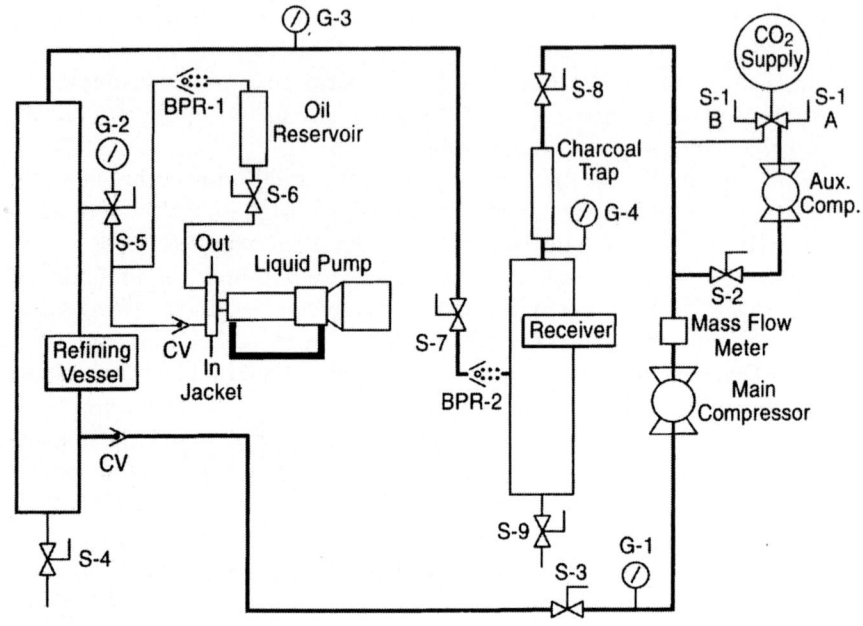

FIGURE 7 Continuous countercurrent refining system.

vessel" in Fig. 7), where it travels upward, contacting soybean oil, which is pumped into the top of the refining vessel using the designated liquid pump. The two media contact one another in the refining vessel, in which oil is solubilized in the SC-CO_2, and the phospholipids being insoluble as the CO_2 descends to the bottom of the refining vessel. The oil can be recovered by lowering the pressure and temperature in the receiver vessel, allowing recycling of the CO_2 back to the main compressor. By using this technique, an extract enriched in lecithin precipitate can be isolated without the use of organic solvents (36).

A somewhat more sophisticated fractionation method involving the use of vertical packed fractionating towers is currently being applied to enrich nutraceutical ingredients from liquid natural feedstocks. The liquid to be fractionated can either be fed concurrently or countercurrently into the fractionating column. There are certain advantages in terms of fractionation efficiency that are provided when using the countercurrent mode and by introducing the liquid feed into the center of column, thus creating extraction and raffination sections. An example of a fractionating column operated in the concurrent mode is shown in Figure 8. For experiments conducted in the author's laboratory, the unit is usually operated in a batch mode, but it can be

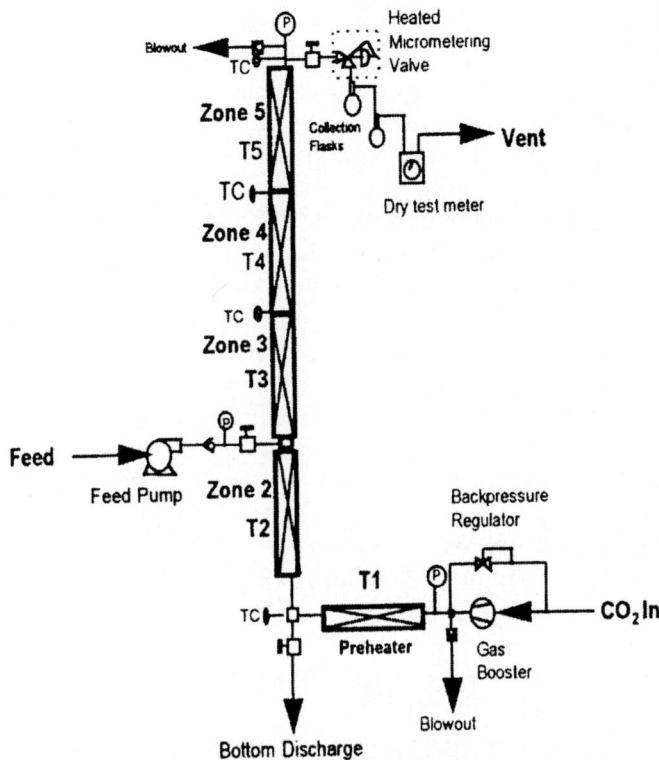

FIGURE 8 Thermal gradient supercritical fluid fractionation column.

made to operate on a semicontinuous basis, using a liquid pump to feed substrate into the column (see in Fig. 8). The components to be separated in the substrate are subjected to a thermal gradient along the length of the fractionating tower, where each of the designated sections of the column have an increasing temperature for the sequence T2, T3, T4, and T5. This allows fractionation of the components in the substrate feed based not only on their relative solubilities in the decreasing CO_2 density gradient but also according to the increase in their respective increasing vapor pressures as they ascend the column. This type of fractionation system has been used to deacidify olive oil, to deterpenate citrus oils, and to fractionate fish oils or butter fat. The recent approach of Clifford et al. (24) involving the deterpenation of citrus oils using subcritical water also applies this fractionation principle.

Chromatographic fractionation using critical fluids as mobile phases has been studied for some time now; however, scaleup from the analytical regime has been less prevalent. Studies at the National Centre for Agricultural

Utilization Research by King and co-workers (31) have demonstrated that preparative supercritical fluid chromatography (SFC) can be coupled advantageously with a selective SFE enrichment stage to yield concentrates rich in nutraceutical ingredients. As shown in the processing scheme in Figure 9, flaked soybeans are initially extracted at a relative low pressure to enrich the components of interest, the tocopherols. This fraction is then moved sequentially on to a sorbent-filled column for further fractionation to yield a tocopherol-enriched extract of nutraceutical value. The advantage of this approach is it allows one to enrich a particularly valuable nutraceutical ingredient from a natural matrix, without contaminating of the remaining matrix with a noxious agent. Enrichment factors relative to the tocopherol content in the original soybean flakes are tabulated in Table 3. Note that these are only modest tocopherol-enrichment factors for application of the single SFE step; however, significant enrichment of the desired components can be obtained by then applying SFC for further fractionation and enrichment of the tocopherols (Table 3).

Recently, using a similar approach, Taylor et al. (37) have been able to produce concentrates enriched in phospholipids (PPLs) for potential use in the nutraceutical industry. Table 4 summarizes the relative amounts of PPLs via initial SFE isolation and in the fractions obtained after SFC. Here, the major component in the starting substrate (soybean oil triglycerides) was initially reduced in the SFE step using neat $SC\text{-}CO_2$, followed by sequential SFE–SFC utilizing $SC\text{-}CO_2$–cosolvent mixtures on the lecithin-containing residue which remains after $SC\text{-}CO_2$ extraction.. By using $SC\text{-}CO_2$/ethanol/water fluid phases, one not only could perform preparative SFC for enriching the PPLs but can also obtain, in certain cases, individual purified PPL moi-

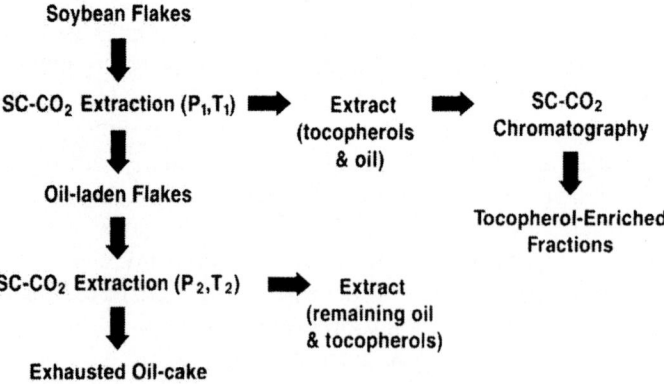

FIGURE 9   Tocopherol-enrichment/fractionation by the SFE–SFC technique.

TABLE 3 Enrichment Factors of Tocopherols from Soybeans by SFE and SFE–SFC

Tocopherol	SFE	SFE–SFC
Alpha	4.33	12.1
Beta	1.83	2.4
Gamma	3.94	15.0
Delta	3.75	30.8

eties. This is achieved by selective density and compositional programming of the SFC fluid phase, coupled with time-based collection of eluent fractions.

Previously, mention has been made of the use of subcritical water as a "green" processing agent. Studies performed in the 1970s by Schultz and Randall (38) showed that useful fractionations of flavor components could be obtained by using LCO_2 in a countercurrent mode with fruit-based aqueous feedstocks. This approach produced fruit flavor essence concentrates enriched in the more hydrophobic components. Recently, a similar method exploiting LCO_2/aqueous phase partitioning has been developed by Robinson and Sims (39) using a microporous membrane to further increase the fractionation efficiency in such systems. This Porocrit fluid fractionation process is depicted in Figure 10 and shows "near"-critical CO_2 being fed into a concentric tube arrangement, countercurrent to the flow of aqueous liquid feedstock, which is being pumped internally through a microporous membrane. Again, an enriched extract exits with the depressurized CO_2, producing an aqueous fruit aroma of the composition described in Table 5. Here, the listed components in the feedstock have been concentrated in many cases, over 100-fold with respect to the starting concentrations. Note also that the

TABLE 4 Relative Amount of PPLs from Soybeans in SFE Isolates and in SFC-Collected Fractions

Phospholipid	SFE[a]	SFC
Phosphatidylethanolamine	16.1	76.8
Phosphatidylinositol	9.2	74.9
Phosphatidic acid	2.8	20.8
Phosphatidylcholine	15.6	55.8

Note: All data in percent of that component.
[a] Relative to other eluting constitutents (oil and unidentified peaks).

Porocritical Membrane Contactor

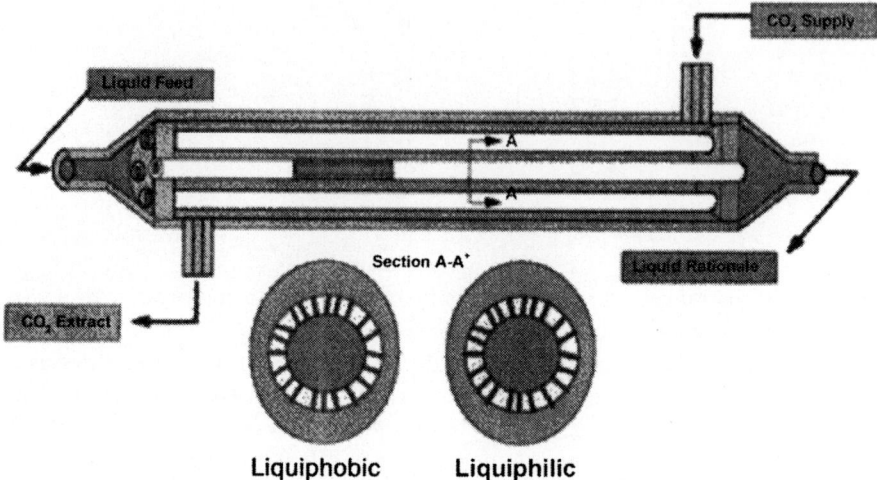

FIGURE 10   Porocrit continuous fluid extraction process.

TABLE 5   Enrichment of Aroma Constituents by the Porocrit Process

Compound	ppm in feed	Conc. factor in extract	% Depletion
Methanol	4,220	3	—
Ethanol	105,000	7	20
1-Propanol	270	82	69
Amyl Alcohol	17	115	81
Hexanol	3	109	95 +
Octanol	5	110	95 +
Z-3-Hexanol	10	127	95 +
α-Terpinenol	10	98	95 +
Terpinen-4-ol	4	106	95 +
Ethyl acetate	29	106	85
Ethyl butyrate	23	106	88
Acetal	37	147	39
ε-2-Hexenal	19	132	86
Hexanal	10	100	84
Octanal	5	156	95 +
Citronellal	3	62	95 +

Porocrit process is very effective at removing the flavor components, as judged by the listed solute depletion data given in the last column of Table 5.

4. ROLE OF ANALYTICAL SCF TECHNOLOGY IN SUPPORT OF NUTRACEUTICALS

The analytical use of critical fluids spans over three decades of endeavor and applications (40,41) and there are many excellent tomes describing activity in this field (42,43). It is not the intention of this review to provide a detailed description of the use of analytical methodology for characterizing nutraceutical products, although such techniques as SFC, can rapidly provide valuable information for the chemist and product formulator (44,45).

There is no doubt that analytical SFC is a logical choice to characterize extracts or fractions obtained by SFE and fractionation, but perhaps of more importance is to describe options that exist for utilizing such analytical instrumentation to assist in optimizing and developing processing methods to produce nutraceuticals for the marketplace. A summary of the information obtainable using analytical-scale techniques and methodology are given in Table 6.

With the advent of automated analytical SFE equipment, it has become possible to rapidly ascertain what extraction or fractionation conditions would be most relevant in scaling up the process. In the United States, analytical SFE instrumentation is produced by such firms as Isco, Applied Separations, Leco, and Jasco. In Europe, analytical-scale SFC equipment is available from Berger Instruments, Thar Designs, Jasco, and Sensar. The equipment is obtained from these vendors can be, if needed, slightly modified to study the conditions that are amenable to processing nutraceuticals. King (34) has provided a interesting review of how lab-constructed equipment can be used for both analytical and process development purposes.

Two examples will be cited that show how analytical SFE instrumentation can be used to obtain information related to the isolation of nutraceu-

TABLE 6 Critical Fluid Analytical Technology — Relevance to Nutraceutical Product Development

To indicate solubility or extractability of a compound
For fractionating a natural product
In support of process development
For analysis of critical-fluid-derived extract
To deformulate a commercial product
To determine required physicochemical data

tical ingredients. Chandra and Nair (46) have studied the extraction of isoflavone components, such as daidzein and genistein, from soya-based products, using a manually operated SFE system. Neat SC-CO$_2$ proved relatively ineffective in removing the isoflavone components from the various soya-based matrices; however, by simply varying the conditions for effecting the extraction, it was found that 20 vol% ethanol in SC-CO$_2$ at 50°C and 600 atm could remove over 90% of the isoflavones in under 60 min extraction time. These survey extractions were performed on sample sizes ranging from 2 to 10 g, saving considerable time and labor in ascertaining what conditions were optimal for the extraction of these components. The need for relatively high pressures and cosolvent to remove the isoflavones from the matrix suggest that subcritical water extraction might be another alternative for isolating these nutraceutical components.

An automated SFE option has been utilized by Montanari et al. (47) and Taylor et al. (37,48) to develop the conditions most amenable to isolating PPL concentrates from deoiled soy meal. In the former case, a set of conditions was tested by programming the microprocessor controller of the automated SFE to run a large number of extractions on common samples at various pressures, temperatures, and cosolvent (ethanol) levels. The results showed that at 70°C and 40.7 MPa with 10 mol% ethanol, phosphatidyl-choline could be extracted preferentially relative to the other phospholipids in the soya meal. By utilizing higher pressures, it was found that the total amount of phospholipids extracted could be substantially increased at the slight reduction of the phosphatidylcholine content in the final extract.

Similarly, an automated SFE unit was also used to ascertain what conditions were necessary to elute and separate phospholipid moieties from extracts obtained from SFE and other extraction methods (37,48). In this case, the PPL-containing extract was layered on top of an adsorbent, such as alumina or silica gel, and elution and separation of the target compounds were assessed by changing the pressure, temperature, quantity, and compo-sition of the cosolvent in the supercritical fluid eluent. As summarized in Table 7, pure SC-CO$_2$ aided in removing the nonpolar (triglyceride) compo-nents from the deposited sample, but the presence of a binary cosolvent with the SC-CO$_2$ proved necessary to effect elution of the PPLs from the sorbent bed. By collecting various fractions during the stepwise elution from the column, while varying the eluent conditions, various phospholipids could be enriched as indicated in Table 7. These experiments could be run in several days and even overnight on the automated SFE module, saving considerable cost and effort before scaling up the SFE–SFC process to a preparative level. Using a similar approach, experiments were run on other natural product matrices in terms of ascertaining whether comminuting the sample prior to

TABLE 7 SFC Fractionation of Lecithin on Silica Gel

Fraction collected[a]	Eluent parameters	Predominate compounds
#1	350 bar, 50°C, CO_2	Triglyceride oil
#2	350 bar, 50°C, CO_2/M	Triglyceride oil
#3	350 bar, 50°C, CO_2/M	
#4	500 bar, 50°C, CO_2/M	Phosphatidylethanolamine
#5	500 bar, 80°C, CO_2/M	Phosphatidylinositol + phosphatidycholine
#6	500 bar, 80°C, CO_2/M	Phosphatidylcholine
#7	500 bar, 80°C, CO_2/M	Phosphatidylcholine

[a] The fraction #2 modifier is 10% ethanol : water (9 : 1). The fractions #3–7 modifier is 25% ethanol : water (9 : 1). M = modifier (cosolvent).

conducting a SFE was necessary. Alternatively, such instrumentation can assist in identifying what part of a natural plant actually contains the targeted components of interest (49).

It was mentioned previously that analytical SFC could be of considerable use in evaluating the content of nutraceutical-containing extracts or feedstocks. This is particularly true if one is looking for a rapid analysis method for quality control of selected ingredients or to monitor qualitative differences in processing and raw materials. The elution order of lipophilic solutes is well known in capillary SFC (50) and permits the separation and identification of key lipophilic nutraceutical components. For example, in Figure 11, the high-resolution separation of fatty acids, squalene, tocopherols, phystosterols, and various glycerides contained in deodorizer distillate is shown using analytical capillary SFC using flame ionization detection (FID) (51). The resultant profile is of a valuable feedstock for nutraceutical components and indicates that its components are soluble in SC-CO_2 (the mobile phase for this analytical SFC separation), as well as revealing valuable information on the chemical composition of the sample to the analytical chemist. This is an excellent assay method considering that the sample preparation is minimal and the analysis time is under 45 min.

On the subject of sample preparation, analytical SFC can save the analyst considerable time, as illustrated by the SFC profile of the composition in a nutraceutical capsule containing sawtooth palmetto berry extract (Fig. 12). In this case, the extract was dissolved and diluted with a minimal amount of hexane and directly injected into the chromatograph. By density programming the CO_2 mobile phase, a high-resolution chromatogram can be facilitated. The complexity of the sawtooth palmetto berry extract is apparent and

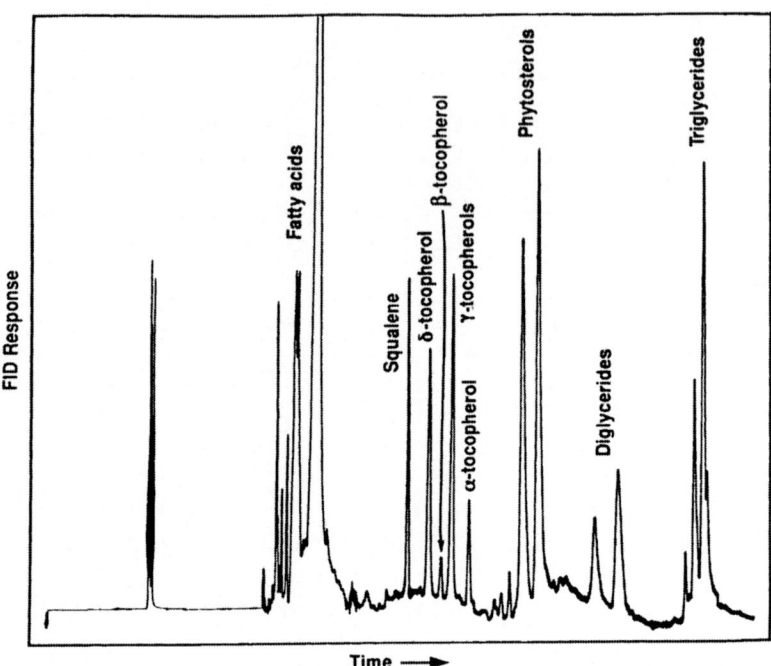

FIGURE 11   Capillary SFC characterization of deodorizer distillate feedstock.

consists of a mixed composition of fatty acids and trace sterol components. Recently, a similar approach has been used by others for characterizing saw palmetto extracts (52).

5.   PRODUCTION PLANTS AND NUTRACEUTICAL PRODUCTS

The purpose of this section is to provide a brief overview of the magnitude and scope of critical fluid technology as applied to the production of industrial products, including nutraceuticals. Information on the use of critical fluid production facilities is be limited due to proprietary constraints; however, production processes using supercritical fluids have existed for over 30 years and critical fluid processing is no longer a novelty industry. These production facilities vary considerably in the magnitude of the operation, ranging from the large Houston-based decaffeination plant of General Foods to smaller extraction facilities focused on the production of noncommodity items. Aside

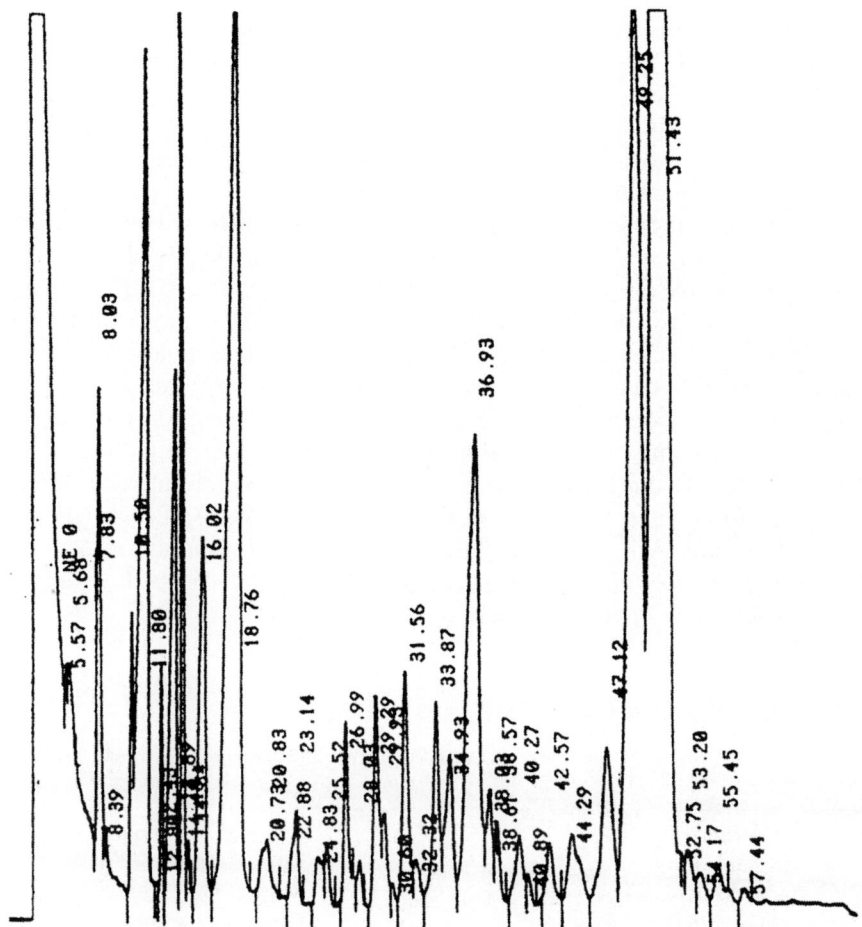

FIGURE 12 Capillary SFC analysis of capsule containing sawtooth palmetto berry extract.

from decaffeination plants, there is a sizable segment of the production facilities devoted to the processing of hops. However, these facilities are devoted to hops processing for only part of the year and potentially have additional capacity which could be devoted to processing of materials having nutraceutical value.

There are over 50 plants (not pilot plants) worldwide devoted to critical fluid processing. Currently, most of these are located in Germany, the United States, France, and Japan. Other nations, such as Britain, Australia, India, and Italy, continue to develop more capacity for critical fluid processing; other nations with a rich litany of natural products will undoubtedly enter the marketplace as users. Table 8 presents a list, which is not inclusive, of many users of critical fluid extraction for food and natural product processing throughout the world. Key players in identified market segments are as follows: decaffeination—General Foods, SKW Trostberg, Kaffee HAG, Hermsen; hops processing—HVG Barth (NATECO$_2$), John Haas, Yakima Chief, Carlton United Breweries, Steiner Hops, English Hops, SKW Trostberg; flavors/spices—Cultor, Quest, Flavex, Norac, US Nutraceuticals, Ogawa, Fuji Flavor, Kobe, Mori Oil Mills, Takeda. Many of the processors listed under the generic heading of flavors/spices process a variety of other natural products such as specialized oils, natural pigments (e.g., Flavex is involved in the processing of ginseng among other moieties). As noted earlier, many of these companies have additional capacity for fulfilling the needs of the nutraceutical market and other firms, on a more diminutive scale, are preparing to enter the market focusing on the processing of niche natural products. Some of the new companies addressing nutraceuticals processing specifically are US Nutraceuticals, KD–Pharma–IQA, Wells Investments Ltd., Aromtech OY, GreenTek 21, and Arkopharma.

TABLE 8   Organizations Processing or Offering Critical Fluid-Derived Products

Flavex (Germany)	Fuji Flavor (Japan)
Hermsen (Germany)	Kobe (Japan)
HVG Barth (Germany)	Mori Oil Mills (Japan)
Kaffe HAG (Germany)	Ogawa (Japan)
SKW Trostberg (Germany)	Takasago (Japan)
KD–Pharma–IQA (Germany–Spain)	Takeda (Japan)
General Foods (United States)	Cultor (France)
John Haas (United States)	HITEX/Separex (France)
Praxair (United States)	Norac (Canada)
Yakima Chief (United States)	Aroma Tech OY (Finland)
Carlton United Breweries (United Kingdom)	Quest (Holland)
English Hops (United Kingdom)	Wells Investment Ltd. (New Zealand)
Steiner Hops (Germany–United Kingdom–USA)	Lavipharm (Greece/USA)
US Nutraceuticals (United States)	GreenTek 21 (South Korea)
Supetrae (Denmark)	Eiffel (Australia)
Ferro Corporation (United States)	Arkopharma (France)

There also exists a cadre of companies that specialize in research and development and take refining up to a certain scale. Among the more prominent ones are Phasex, Praxair, Norac, Marc Sims Inc., Separex–Hitex, Flavex, Critical Processes Ltd., Bradford Particle Design (BPD), CPM Inc., Wells Investments Ltd., and Aphios Corp. These companies and a host of other organizations, consultancies, academia, and government laboratories can be useful sources for technical consultation/information.

What is a typical extraction/processing plant like in terms of scale? This is hard to quantify without picking what might be a "typical" example, but it is worth citing the $NATECO_2$ production facilities in Wolnzach, Germany as an example of a technically sophisticated and diverse operation. Some of its stated capacity and capabilities are as follows:

Production Plant #1: 4×2-m^3 extraction vessels
Laboratory Plant #2: 1000-mL countercurrent column
Production Plant #3: 4×4-m^3 extraction vessels
Production Plant #4: 3×500-L vessels
Pilot Plant #5: 2–200-L countercurrent column

This plant has historically focussed on the processing of hops but has diversified over the years to include the processing of other natural products, thereby providing for maximum utilization of the plant facilities.

To enumerate and discuss all of the nutraceutical or potential nutraceutical products that could be processed by critical fluids is beyond the scope of this review, particularly when one considers that in the past, many of these naturally derived products have been extracted with critical fluids. However, it is worth noting how these extracts are usually obtained in the SFE mode and what generic classes of materials have been extracted. For example, it is rare to obtain an extract without the occurrence of lipid coextractives; hence, the nutraceutical agent will frequently be diluted in a oily matrix. Examples of this type of extract are the previously mentioned tocopherol-containing extracts as well as natural pigments, such as the carotenoids (53). Tocotrienols can also be isolated via SFE as a complex from barley brand; likewise, the newer sterol or steryl ester concentrates can be obtained as mixtures from corn oil, rice brand oil, or saw palmetto sources. Many of these extracts containing other helpful nutraceutical ingredients, such as fatty acids, squalene, and so forth, can act in a synergistic mode with the principle nutraceutical agent.

It is worth noting that many current oil extraction and refining methods remove valuable nutraceutical components from the oil in the name of appearance and flavor. By-products and streams from these milling and refinement steps often contain preconcentrated sources of nutraceuticals (e.g., deodorizer distillate, residual protein meals, fibrous materials). Sources such as corn gluten meal, corn brand fiber, and alfalfa leaf protein concentrate have been extracted with SC-CO_2 successfully for their pigment, sterols, and fatty acid content. The high tocopherol–sterol–squalene content of deodorizer distillate is well known and several schemes employing critical fluids have been reported for fractionating this material to achieve higher–purity materials (54). Another processing agent that is a concentrator for nutraceutical components are the bleaching sorbents used by the vegetable oil processing industry. King and co-workers (55) have shown that very high oil yields can be obtained from clay bleaching earths; however, there is a need for the resultant extract to be analyzed for the enriched content of the nutraceutical components.

Critical fluid extraction has been applied for sometime now in the extraction of speciality oils, such as evening primrose, borage, black currant, and flax. These moieties contain the presence of γ-linoleic acid, a component which has been implicated favorably for the treatment of several medical conditions. Other specialized oils that have also been extracted with SC-CO_2 are wheat germ, avocado, sea buckthorn, sorghum brand/germ oat, and amaranth. Recently, there has been some studies employing critical fluids to obtain oil from fungi or marine sources, like spirulina, which are devoid of cholesterol. Partial deoiling has also been performed using SC-CO_2, more with respect to developing a functional food ingredient that has less fat (oil) or cholesterol content (i.e., low-calorie peanuts). Such deoiling can be done using either SC-CO_2 or propane and still meet the criterion for use in functional food products.

As has been stated previously, PPLs and many of the traditional herbal-medicine-type components are amenable to critical fluid extraction provided that GRAS cosolvents are employed along with SC-CO_2. Pure phosphatidylcholine and phosphatidylserine are finding widespread use, the latter in improving cognitive function; hence, the challenge for those using critical fluid processing for PPL recovery is to develop purification techniques for these naturally derived chemicals. Other nutraceutical agents that can be obtained similarly include extracts from chamomile, paprika, feverfew and gingko biloba (analytical studies), garlic, and ginger. Most of the common spices and mint oils, including a commercially available extract of rosemary, can be obtained via SC-CO_2 extraction.

For the above-mentioned products, critical fluid technology faces stiff competition from molecular (vacuum) distillation techniques, a time-honored technique, although greater selectivity is potentially available utilizing critical-fluid-based methods. This, coupled with the fact that CO_2-derived extracts exhibit, in many cases, extended shelf-lives due to the prophylactic action of the residual, nonoxidative CO_2 atmosphere, as well as micro-organism destruction due to exposure at higher pressures and temperatures, argues for a bright future for critical fluid technology in the nutraceutical marketplace.

6. IMPROVEMENT OF NUTRACEUTICAL FUNCTIONALITY THROUGH PROCESSING WITH CRITICAL FLUIDS

It is obvious from the discussion in the previous sections that many nutraceutical components or compositions are soluble to varying degrees in critical fluid media, particularly SC-CO_2. However, unlike pharmaceutical compounds, which are usually highly purified solutes dissolved in SC-CO_2 and/ or cosolvent mixtures, nutraceutical components and compositions tend to be ill-defined and molecularly complex mixtures, where components often interact synergistically to induce a therapeutic benefit. Testimony to this complexity is provided by the capillary SFC profile of sawtooth palmetto berry extract provided in Figure 12, where the predominate fatty acid mixture along with low levels of sterols help alleviate prostate conditions in men. Obviously, sawtooth palmetto berry extract is quite different than a high-purity pharmaceutical, where perhaps only one or two (i.e., chiral racemates) major components are quantified by chromatography.

As noted at the start of the chapter, the strategy in the nutraceutical or functional food market is to fortify specific foods or natural products with ingredients that have, or are perceived to have, health-promoting benefits. Many of these products are orally ingested and, like pharmacological drugs, can be made effective more rapidly by rapid dissolution within the body. Therefore, it behooves us to consider the benefits of producing nutraceuticals by the processes employed to make small-diameter particles of medicinally active compounds for the pharmaceutical industry using the critical-fluid-based processes described earlier in this book. This can incorporate techniques like GAS (gas anti-solvent), RESS (rapid expansion of a supercritical solution), PCA (precipitation with compressed fluid anti-solvent), and so forth to produce small particles having low polydispersity, or encapsulation of nutraceutical ingredients and impregnation of an active ingredient into a food matrix (i.e., to make a functional food).

Although this would appear to be a good strategy, it is important to consider economics as they relate to the difference between the pharmaceutical and food industries. The food industry tends to be a commodity-driven marketplace with a low return margin on its products. The high hourly production rates characteristic of the food industry are a challenge given the current state of the art of particle formation technology employing supercritical fluids. However, consumers of nutraceuticals and functional foods tend to be willing to purchase these (nutraceutical) more expensive products, so it is not unreasonable to consider the benefit of applying critical-fluid-based techniques to produce ingredients via the methods noted in the previous paragraph. For example, consider the case for winning lecithin, or a more highly purified form of phospholipid concentrate from soybeans. Isolating lecithin from soybean oil or meal can yield a nutraceutical-grade lecithin that sells for $31 a pound. This is a significant mark up versus the price of soybean oil, which might be $0.25 a pound. Hence, when one considers a nutraceutical-grade phospholipid concentrate (concentrated phosphatidycholine or phosphatidyserine), the price is $1500 a pound. Obviously, with these economic incentives, the possibility of applying critical-fluid-based particle production processes may be justified.

The fundamental principles and methodology of producing fine particles, encapsulates, and so forth using critical fluid processing have been covered in depth in previous chapters, and it is not our intention to cover this as we conclude this chapter. However, it is worth considering a few of the possibilities in a generic sense and examining one application of the technology to a specific nutraceutical component or product. Table 9 lists applications of critical fluid technology reported at the 6th International Symposium on Supercritical Fluids in 2003, which have implications for the nutraceutical and/or functional food industry. As one can see, a number of the processes described in previous chapters have been applied to foodstuffs, nutraceutical ingredients or a suitable surrogate (e.g., cholesterol for sterols or sterol esters), or delivery systems that utilize liposomes or biodegradable natural polymers. By controlling the particle size, morphology, or method of delivery (encapsulation), these critical-fluid-based processes can produce a pleathora of materials, as noted in the application column. This includes specific chemicals utilized in nutraceuticals, such as high-antioxidant-containing spices, sterol esters, carotenoids, phospholipids, and fatty acids. This is ample evidence that such processes could impact on the functional food industry, as well as provide new food materials for the integration into the conventional food product chain.

We noted previously the high value and application of phospholipids-containing materials, like lecithin or phospholipid-base liposomes, as functional food ingredients. This is worth examining in greater detail, starting with the historical research of Quirin, Eggers, and Wagner (16,57,58) involving the jet extraction of lecithin from soybean oil. These researchers developed a continuous method for deoiling lecithin obtained from caustic or physical refining–derived lecithin feedstocks that resulted in powdered extracts, somewhat analogous to the lecithin- or phospholipid-based powders noted in Table 9. This process overcame some of the physical difficulties in refining lecithin by SFE with SC-CO_2 or SC-CO_2–csolvent mixtures due to the gelatinous nature of the lecithin feedstock. Further, the method referred to

TABLE 9   Materials Processing Using Critical Fluid Media with Implication or Direct Application in the Nutraceutical or Functional Food Industry

Process	Application	Ref.[a]
Concentrated powder form (CPF), jet dispersion	Dispersion of paprika soup powders	Weidner, pp. 1483–1495
Semicontinuous gas antisolvent process	Cholesterol morphology and precipitation	Subra and Vega, pp. 1629–1634
Rapid expansion of supercritical solution	Phystosterol micronization	Jiang et al., pp. 1653–1658
DELOS crystallization	Crystallization and control of stearic acid morphology	Ventosa et al., pp. 1673–1676
CPF Process	Controlled release of flavors and vitamins	F. Otto et al., pp. 1707–1712
Rapid expansion of supercritical solution	Encapsulation of β-sitosterol in low molecular weight polymer matrix	Turk et al., pp. 1747–1752
Supercritical antisolvent process	Incorporation of cholesterol or proteins in biodegradable matrix	Pellikaan, pp. 1765–1770
PGA and GAS processes	β-Carotene precipitation	Miguel et al., pp. 1783–1788
Particle from gas saturated solution (PGSS)	Lipid micronization of phosphatidylcholine and Tristearin	Elvassore et al., pp. 1853–1858
Supercritical antisolvent process	Biodegradable polymers precipitation studies with	Vega-Gonzalez, pp. 1877–1882

[a] The listed authors and inclusive page numbers all appear in Ref. 56.

as "jet extraction" provided a more environmentally benign process than using propane for deoiling lecithin feedstocks.

The basis of jet extraction is that it facilitates the dispersion of a thin filet of lecithin into a highly turbulent jet of SC-CO_2. This is made possible by the use of two overlapping capillary jets. The lecithin is fed into the inside capillary, and the CO_2 enters into the larger outside capillary encasing the smaller-diameter capillary tube. The compressed CO_2 then mixes with the lecithin extrudate in a mixing chamber under conditions of high turbulence, affecting the deoiling of the lecithin and formation of a powdered product. The physicochemical basis of this separation process in terms of its effect on lecithin viscosity and interfacial tension have been reported by Eggers and Wagner (57,58) and resulted in the conceptualization of a continuous processing plant based on the above principle. Other processing options for lecithin with critical fluids are nicely summarized in Stahl et al.'s book (16).

A device similar in principle to the German researchers has been described by King (34) as shown in Figure 13. Figure 13a shows the general and initial design of the jet extraction system, and Figure 13b provides more critical details. In this laboratory-scale apparatus, the solids collection reservoir and two collector vessels were each 30.5×2.54-cm 316 stainless-steel tubing. The lecithin sample to be extracted is placed into the solids reservoir and extruded into the jet tube assembly with the aid of a nitrogen pressure head. This "pusher" gas flow rate is regulated by a micrometering valve.

As described earlier, the SC-CO_2 interfaces with the lecithin sample in the jet tube assembly (Fig. 13b). It is critical that in the pictured three-way valve that the viscous sample be injected through the 0.16-cm capillary into the larger concentric tube to avoid viscous back-streaming and to assure intimate contact with the SC-CO_2. The solubilized oil components are then routed through the back-pressure relief valve, where the CO_2 decompression occurs, resulting in the precipitation of the oily constituents in the liquid collector. The deoiled lecithin powder then drops into the solids collection vessel. Careful control must be exercised over the relative flow rates of the pushing and extraction fluids so as to maximize the contact time between the lecithin and the extraction fluid. This can also be amplified by using longer extraction chambers, which provide a long contact, or drop time, of the lecithin in the compressed CO_2 atmosphere.

Other notable advancements have been reported in the literature which use critical fluids for producing micron-size phospholipids-based powders as well as the formation of PPL-base liposomes. For example, Castor at Athios Corporation has patented several concepts (the CFL Process) (59,60), which permit the encapsulation of hydrophobic drugs as well as naturally derived drugs, such as taxol. The basis of these patents are that PPLs deposit out at the phase boundary as the PPL–drug aqueous phase (or multilamellar vesicle) undergoes decompression in a critical fluid atmosphere. The resulting liposomes formed in the presence of SC-CO_2 and other alternative fluids showed stability lifetimes exceeding 6 months, and this could be amplified by the inclusion of α-tocopherol in the PLL matrix, which provides prophylactic protection in an extending the lifetime of the resultant liposome.

Similar studies have also been performed by the research group of Charbit in Marseilles (61,62) in which fine PPL particles were formed by decompression using the SAS process. The focus of this research was to develop drug delivery systems, but it would equally applicable to the functional food area. Typically, a 2 wt% solution of soy-derived lecithin is

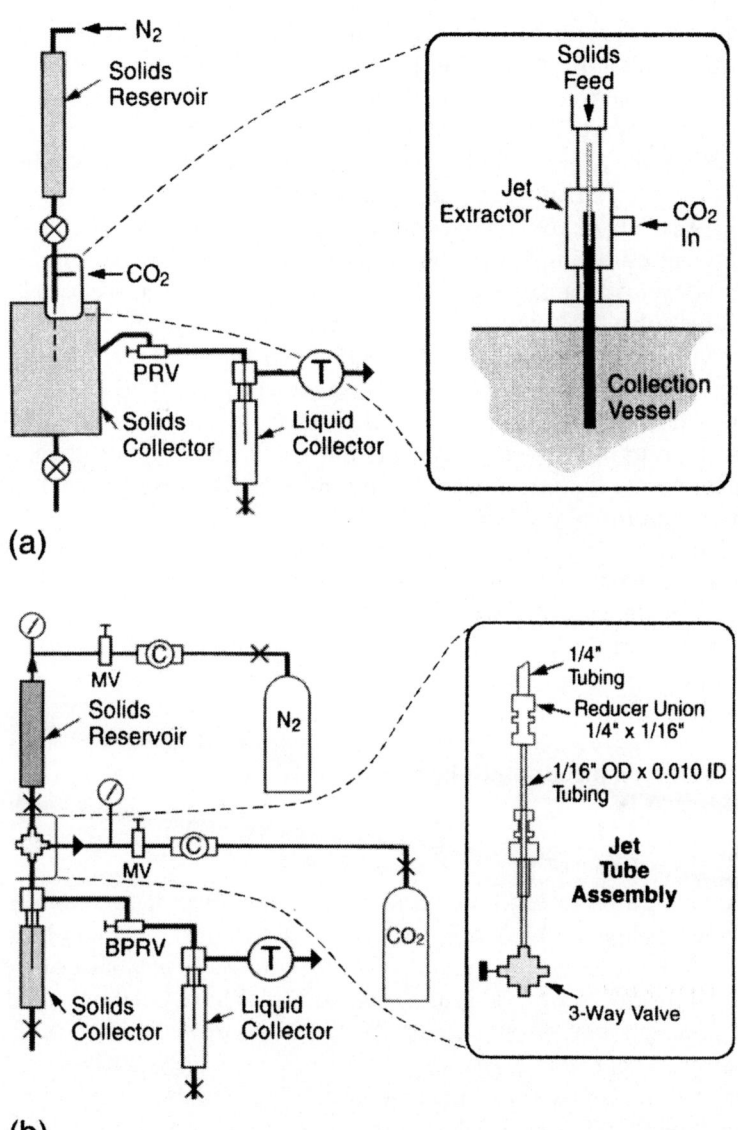

FIGURE 13 Basic schematic of laboratory-scale jet extractor system. PRV = pressure regulating valve; T = flow totalizer. (b) Details of jet extractor for deoiling soya lecithin. BPRV = back-pressure regulating valve; T-flow totalizer.

dissolved in an ethanol solution which is subsequently injected into SC-CO$_2$. Typical precipitation temperatures and pressures were 35°C and 8–11 MPa, respectively. The micronized PPL particles tended to be in the range 15–60μ, were amorphous in nature, and coalescenced when they were exposed to air.

Other schemes have been cited in the literature which employ phospholipids-based materials and supercritical fluids for fine-particle formation or encapsulation. Weber et al. (63) recovered lecithin from egg yolk extracts and induced crystallization by implementing the GAS process. Likewise, Frederiksen et al. (64) developed a new method of preparing liposomes to encapsulate water-soluble substances with the aid of ethanol. In this study, liposomes could be formed from phosphatidylcholine having 40–50-nm dimensions. A mixing process (ESMIC Process) (65) involving supercritical fluids has also reported, which is conducted in a stirred autoclave, to provide embedded pharmaceutical preparations in a lecithin matrix. Final particle size is partially control by the milling process taking place in the stirred autoclave containing the supercritical fluid medium.

In summary, the above studies and processes show the versatility in utilizing critical fluid media for modifying a target nutraceutical for functional food use. In the case of PPLs, the above studies illustrate how they could be prepared for rapid dissolution in food formulations or to encapsulate them for sustained delivery rates, or, alternatively, to use PPL-based liposomes and so on to encapsulate nutraceutical ingredients. Obvious extensions to other target nutraceuticals follow, and application of the technology outside of its traditional niche in the pharmaceutical industry seemed assured.

7. SUMMARY

In summary, we have attempted in this review to provide some understanding of the basic concepts involved in the use of critical fluids and to explain how these fluids are now exploited for the production of nutraceutical and other naturally derived products. Several illustrative examples have been provided of processing concepts and equipment, from the laboratory scale through production plants. There now exists an extensive 30-year history of critical fluid processing technology upon which to draw, replete with many examples of components having nutraceutical value that have been already extracted, fractionated, and reacted in these dense fluids.

In the future, fine-particle production or food material modifications via critical fluid technology will join the other generic processing methods that involve the use of these novel fluids. As noted by King (66), an "all-green" processing platform is emerging involving the use of multiple environmentally benign fluids and combined unit processing applications. Such an integrated critical fluid processing platform will make maximum use of production-scale equipment, which undoubtedly will include facilities for particle production, tailored toward the nutraceutical marketplace.

REFERENCES

1. Koseoglu SS. Overview of nutraceuticals/functional foods. Nutraceuticals and Functional Foods Short Course, College Station, TX, 2000.
2. King JW, List GR, eds. Supercritical Fluid Technology in Oil and Lipid Chemistry. Champaign, IL: AOCS Press, 1996.
3. Mukhopadhyay M. Natural Extracts Using Supercritical Carbon Dioxide. Boca Raton, FL: CRC Press, 2000.
4. King JW. Supercritical fluid processing of cosmetic raw materials. Cosmet Toil 1991; 61:61–67.
5. Revercon E. Fractional separation of of SCF extracts from marjorum leaves: mass transfer and optimization. J Supercrit Fluids 1992; 5:256–261.
6. Krukonis VJ, Vivian JE, Bambara CJ, Nillson WB, Martin RA. Concentration of eicosapentaenoic acid by supercritical fluid extraction: a design study of a continuous production process. In: Flick GJ, Martin RE, eds. Advances in Seafood Biochemistry. Basel: Technomic Publishing Company, 1992:169–179.
7. Reverchon E, Marciano A, Poletto M. Fractionation of peel oil key mixture by supercritical CO_2 in a continuous tower. Ind Eng Chem Res 1997; 36:4940–4948.
8. King JW. Sub- and supercritical fluid processing of agrimaterials: extraction, fractionation, and reaction modes. In: Kiran E, ed. Supercritical Fluids—Fundamentals and Applications. Dordrecht: Kluwer, 2000:451–488.
9. Mohamed RS, Saldana MDA, Socantatype FH, Kieckbusch TG. Reduction in the cholesterol content of butter oil using supercritical ethane extraction and adsorption on alumina. J Supercrit Fluids 2000; 16:225–233.
10. Giese T, Johanssen M, Brunner G. Separation of stereoisomers in a SMB–SFC plant: determination of isotherms and simulation. In: Dahmen N, Dinjus E, eds. Proceedings of the International Meeting of the GVC-Fachausschuss Hochdruckverfahrenstechnik, 1999:283–286.
11. Nicolosi RJ, Ausman LM, Rogers EJ. New Technologies for Healthy Foods & Nutraceuticals. Shrewsbury, MA: ATKL Press, Inc., 1997:84.
12. Fitzpatrick KC. Sterol ester products face US hurdles. INFORM 1999; 10:172–176.
13. Anastas PT, Williamson TC, eds. Green Chemistry. Washington, DC: American Chemical Society, 1996.
14. Froning GW, Wehling RL, Cuppett SL, Pierce MM, Niemann L, Siekman DK. Extraction of cholesterol and other lipids from dried egg yolk using supercritical carbon dioxide. J Food Sci 1990; 55:95–98.
15. Passey CA. Commercial feasibility of a supercritical extraction plant for making reduced-calorie peanuts. In: Rizvi SSH, ed. Supercritical Fluid Processing of Food and Biomaterials. London: Blackie Academic, 1994:223–243.
16. Stahl E, Quirin K-W, Gerard D. Dense Gases for Extraction and Refining. Heidelberg: Springer-Verlag, 1987.

17. Anonymous. CO_2 extracts—food and medicine. Food Market Technol 1997; 11(4):27–31.

18. King JW, Johnson JL, Orton WL, McKeith FK, O'Connor PL, Novakofski J, Carr TR. Effect of supercritical carbon dioxide extraction on the fat and cholesterol content of beef patties. J Food Sci 1993; 58:950–952, 958.

19. Clifford AA, Basile A, Jimenez-Carmona MM, Al-Said SHR. Extraction of natural products with superheated water. Proceedings of the GVC-Fachaussschuss Hochdruckverfahrenstechnik, 1999:181–184.

20. Mannia M, Kim H, Wai CM. Supercritical carbon dioxide and high pressure water extraction of bioactive compounds in St. John's wort. Proceedings of the Super Green 2002 Symposium, 2002:74–78.

21. King JW, Grabiel RD, Wightman JD. Subcritical water extraction of anthocyanins from fruit berry substrates. In: Brunner G, ed. Proceedings of the 6th International Symposium on Supercritical Fluids—Vol. 1, 2003:409–418.

22. King MB, Bott TR, eds. Extraction of Natural Products Using Near-Critical Solvents. London: Blackie Academic, 1993.

23. Brogle H. CO_2 as a solvent: its properties and uses. Chem Ind 1982; 19:385–390.

24. Clifford AA, Basile A, Jiminez-Carmona MM, Al-Saidi SHR. Extraction of natural products with superheated water. Proceedings of the 6th Meeting on Supercritical Fluid Chemistry and Materials, 1999:485–490.

25. Basile A, Jimenez-Carmona MM, Clifford AA. Extraction of rosemary by superheated water. J Agric Food Chem 1998; 46:5205–5209.

26. Hawthorne SB, Yang Y, Miller DJ. Extraction of organic pollutants from environmental solids with sub- and supercritical water. Anal Chem 1994; 66:2912–2920.

27. King JW. Critical fluid options for isolating and processing agricultural and natural products. Proceedings of the Super Green 2002 Symposium, 2002: 74–78.

28. King JW, Eller FJ, Taylor SL, Neese AL. Utilization of analytical critical fluid instrumentation in non-analytical applications. Proceedings of the 5th International Symposium on Supercritical Fluids (CD disk).

29. Pietsch A, Eggers R. The mixer–settler principle as a separation unit in supercritical fluid processes. J Supercrit Fluids 1999; 14:163–171.

30. Nilsson WB. Supercritical fluid extraction and fractionation of fish oils. In: King JW, List GR, eds. Supercritical Fluid Technology in Oil and Lipid. Champaign, IL: AOCS Press, 1996:180–212.

31. King JW, Favati F, Taylor SL. Production of tocopherol concentrates by supercritical fluid extraction and chromatography. Sep Sci Technol 1996; 31:1843–1857.

32. Peper S, Cammerer S, Johannsen M, Brunner G. Supercritical fluid chromatography process optimization of the separation of tocopherol homolgues. In: Kikia I, Perrut M, eds. Proceedings of the 6th International Symposium on Supercritical Fluids, 1999:563–568.

33. Eggers R. High pressure extraction of oilseeds. J Am Oil Chem Soc 1985; 62:1222–1230.

34. King JW. Analytical-process supercritical fluid extraction: a synergistic

combination for solving analytical and laboratory scale problems. Trends Anal Chem 1995; 14:474–481.

35. King JW. Critical fluids for oil extraction. In: Wan PJ, Wakelyn PJ, eds. Technology and Solvents for Extracting Oilseeds and Non-petroleum Oils. Champaign, IL: AOCS Press, 1997:283–310.

36. List GR, King JW, Johnson JH, Warner K, Mounts TL. Supercritical CO_2 degumming and physical refining of soybean oil. J Am Oil Chem Soc 1993; 70:473–477.

37. Taylor SL, King JW, Montanari L, Fantozzi P, Blanco MA. Enrichment and fractionation of phospholipid concentrates by supercritical fluid extraction and chromatography. Ital J Food Sci 1999; 12:65–76.

38. Schultz WG, Randall JM. Liquid carbon dioxide for selective aroma extraction. Food Technol 1970; 24(11):94–98.

39. Robinson JR, Sims M. Method and system for extracting a solute from a fluid using dense gas and a porous membrane. US Patent 5,490,884 (1996).

40. Lee ML, Markides KE, eds. Analytical Supercritical Fluid Chromatography and Extraction. Provo, UT: Chromatography Conferences, 1990.

41. Taylor LT. Supercritical Fluid Extraction. New York: Wiley, 1998.

42. Luque de Castro MD, Valcarel M, Tena MT. Analytical Supercritical Fluid. Heidelberg: Springer-Verlag, 1994.

43. Dean JR, ed. Applications of Supercritical Fluids in Industrial Analysis. London: Blackie Academic, 1993.

44. King JW. Applications of capillary supercritical fluid chromatography–supercritical fluid extraction to natural products. J Chromatogr Sci 1990; 28:9–14.

45. King JW. Capillary supercritical fluid chromatography of cosmetic ingredients and formulations. J Microcol Sep 1998; 10:33–39.

46. Chandra A, Nair MG. Supercritical carbon dioxide extraction of daidzein and genistein from soybean products. Phytochem Anal 1996; 7:259–262.

47. Montanari L, Fantozzi P, Snyder JM, King JW. Selective extraction of phospholipids from soybeans with supercritical carbon dioxide and ethanol. J Supercrit Fluids 1999; 14:87–93.

48. Taylor SL, King JW. Supercritical luid fractionation of phospholipids. Abstracts of the 8th Int. Symposium on Supercritical Fluid Chromatography and Extraction, 1998.

49. Taylor SL, King JW, Snyder JM. Tandem supercritical fluid extraction/ chromatographic studies of the desert botanical species, *Dalea spinosa*. J Microcol Sep 1994; 6:467–473.

50. King JW, Snyder JM. Supercritical fluid chromatography: a shortcut in lipid analysis. In: McDonald RE, Mossoba M, eds. New Techniques and Applications in Lipid Analysis. Champaign, II: AOCS Press, 1997:139–162.

51. Snyder JM, Taylor SL, King JW. Analysis of tocopherols by capillary supercritical fluid chromatography and mass spectrometry. J Am Oil Chem Soc 1993; 70:49–354.

52. DeSwaef SI, Kleibohmer W, Vlietinck AJ. Supercritical fluid chromatography of free fatty acids and ethyl esters in ethanolic extracts of sabal serrulata. Phytochem Anal 1996; 7:223–227.

53. Favati F, King JW, Friedrich JP, Eskins K. Supercritical extraction of carotene and lutein from leaf protein concentrates. J Food Sci 1988; 53:1532–1536.
54. Ssuss D, Brunner G. Countercurrent extraction with supercritical carbon dioxide: behaviour of a complex natural mixture. Proceedings of the GVC-Fachaussschuss Hochdruckverfahrenstechnik-, 1999:189–192.
55. King JW, List GR, Johnson JH. Supercritical carbon dioxide extraction of spent bleaching clays. J Supercrit Fluids 1992; 5:38–41.
56. Brunner G, Kikia I, Perrut M, eds. Proceedings of the 6th International Symposium on Supercritical Fluids, Tome 3, 2003.
57. Eggers R, Wagner H. Extraction device for high viscous media in a high-turbulent two-phase flow with SC-CO_2. J Supercrit Fluids 1993; 6:31–37.
58. Wagner H, Eggers R. Entolung von sojalecithin mit uberkritischen kohlendioxid in hochturbulenter zweiphasestromung. Fat Sci Technol 1993; 95:75–80.
59. Castor TP, Chen L. Method and apparatus for making liposomes containing hydrophobic drugs. US patent 5,776,486, July 7, 1998.
60. Castor TP. Methods and apparatus for making liposomes. US patent 5,554,382, September 10, 1996.
61. Magnan C, Commenges H, Badens E, Charbit G. Fine phospholipid particles formed by precipitation with a compressed fluid anti-solvent. Proceedings of the GVC-Fachaussschuss Hochdruckverfahrenstechnik, 1999:223.
62. Magnan C, Badens E, Commenges N, Charbit G. Soy lecithin micronization by precipitation with a compressed fluid anti-solvent—influence of process parameters. J Supercrit Fluids 2000; 19:69–77.
63. Weber A, Nolte C, Bork M, Kummel R. Recovery of lecithin from egg yolk-extracts by gas anti-solvent crystallization. Proceedings of the 6th Meeting on Supercritical Fluids: Chemistry and Materials. 181–184.
64. Frederiksen L, Anton K, Barrat BJ, VanHoogevest P, Lenenberger H. Proceedings of the 3rd International Symposium on Supercritical Fluids, Tome 3, 1994:235–240.
65. Heidlas J, Zhang Z. New approaches to formulate compounds using supercritical gases. Proceedings of the 7th Meeting on Supercritical Fluids, Tome 1, 2000:167–172.
66. King JW. Critical fluid options for isolating and processing agricultural and natural products. Proceedings of SuperGreen, 2002:61–66.

15

Scale-Up Issues for Supercritical Fluid Processing in Compliance with GMP

Jean-Yves Clavier and Michel Perrut
Lavipharm, East Windsor, New Jersey, U.S.A. and Separex 5,
Champigneulles, France

Although very important research and development (R&D) means are dedicated to applications of supercritical fluids (SCFs) in the pharmaceutical industry, a very limited number of commercial plants are now operating or under construction and few companies have acquired some know-how in process scale-up, especially in SCF formulation and particle design, in compliance with the constraints imposed in this industry [traceability and Good Manufacturing Processes (GMP), sterility, etc].

In this chapter, based on our experience gathered when building tens of SCF plants during the past decade, including some for processing clinical lots in compliance with GMP, we will try to present both the general scale-up rules applicable to large-scale SCF plant construction, operation, and maintenance, the specific scale-up know-how related to pharmaceutical plants, and especially those dedicated to drug formulation and particle design.

1. SCF PROCESSES IN THE PHARMACEUTICAL INDUSTRY

Many SCF processes are now under development in various industries (1–12), including many *pharmaceutical* applications (13):

- Supercritical fluid *extraction* (SFE), *fractionation* (SFF), and *Chromatography* (SFC), for extraction and purification of active substances
- Supercritical fluids as *reaction media* for selective synthesis of active ingredients (6,8,14–16)
- Supercritical fluid *drug formulation* by manufacturing *innovative* therapeutic particles, either of pure active compounds or composites of excipient and active compounds (17,18)

In fact, very few compounds can be used as SCFs for most applications at the commercial scale (see Table 1).

Carbon dioxide is, by far, the most attractive SCF for many reasons: It is inexpensive and abundant at high purity (food grade) worldwide and it is nonflammable, non-toxic, and environment friendly; moreover, its critical temperature ($T_c = 31°C$) permits operations at near-ambient temperature which avoids product alteration and its critical pressure ($P_c = 74$ bar) leads to "acceptable" operation pressure, generally between 100 and 350 bar. In fact, supercritical carbon dioxide behaves as a rather weak "nonpolar" solvent, but its solvent power and polarity can be significantly increased by adding a polar cosolvent that is chosen among alcohols, esters, and ketones. *Ethanol* is often preferred because it is not hazardous to the environment, not very toxic, and available pure at low cost. *Hydro fluorocarbons* (HFCs) are very costly and their specific properties rarely justify their use in the replacement of carbon dioxide.

TABLE 1   Critical Conditions of Usual SCF Solvents

Compound	Molecular weight	P_c (MPa)	T_c (K)	ρ_c (kg/m^3)
Carbon dioxide	44.01	7.38	304.1	468.7
Propane	44.09	4.25	369.8	217.2
Dimethyl ether	46.07	5.24	400.0	255.8
Difluoromethane	52.02	5.83	351.6	430.7
Trifluoromethane	70.01	4.86	299.3	527.6
CF$_3$–CH$_2$F (F134a)	102.00	4.06	374.2	515.0
Water	18.015	22.12	647.3	315.5

Note: P_c = critical pressure; T_c = critical temperature.

Although they do not present the advantages of carbon dioxide, other fluids are also considered for industrial applications in spite of their flammability and related explosion hazards:

- *Light hydrocarbons*, especially liquefied *propane*, which appears to be a much stronger solvent than carbon dioxide vis-à-vis lipids
- *Dimethyl ether*, used as liquefied gas, which behaves as a "polar" solvent able to dissolve a very wide range of compounds, including many polymers (19).

It is also important to notice that SCFs have biocide properties against most micro-organisms (fungi, bacteria, viruses) (20–25) and, even if it cannot be considered as a real sterilizing agent, SCF processes are intrinsically sterile and never increase the bioburden.

2. BASIC RULES FOR SCF PROCESS SCALE-UP IN COMPLIANCE WITH GMP

Prior to dealing with the specific scale-up issues related to SCF extraction/ fractionation and SCF drug formulation plants, we will present the basic background used to build and operate any SCF equipment in compliance with GMP and the safety rules (26) that must be imposed in any case.

2.1. Safety in SCF Fluid Operations

As handling supercritical fluids and liquefied gases presents important hazards, safety must be taken into account at any step of equipment design, building, installation, operation, and maintenance, and a detailed analysis of potential hazards should be conducted for any case.

2.1.1. Mechanical Hazards

Both design standards and official tests that are enforced by state agencies, in combination with strict inspection procedures, limit this hazard to quasizero, especially on large-scale units. No need to mention that equipment shall *never* be modified without enforcing the same standards and obtaining the manufacturer consent. However, some issues are often underestimated, especially on R&D multipurpose equipment:

Plugging. Most solid–fluid equipment use baskets closed by filter disks; on large-scale units, plugging causes disk rupture, but on small-scale equipment, the disk may not break on depressurization and compressed CO_2 remains in the basket that may be brutally ejected and/or explode. Thus, we strongly recommend being extremely prudent when processing material that could lead to plugging, foaming, or expanding (polymers, "sticking" materials, highly viscous extracts, etc.) and to open valves on the autoclave top *and* bottom.

Tubing Connection Rupture. Double-ring connections, commonly used on most small-scale equipment, are safe and reliable when the screwing procedure is strictly followed. Otherwise, the rings may slip on pressurization. We recommend always verifying the proper sealing of the rings and to not use double-ring fittings on tubings larger than 1/4 in. and/or for pressures over 300 bar, for which secure high-pressure screwed connections should be preferred.

Metal Fatigue and Fragilization. Large-scale pressure vessels are commonly built using carbon steel covered by an internal stainless-steel cladding. The life duration of high pressure vessels is linked to the number of pressurization/depressurization cycles, commonly 10,000 to 20,000 cycles, depending on their design, that must include a fatigue calculation and any operation must be stopped when the limit is reached. Moreover, carbon steel may be subject to phase transition and become brittle when the temperature decreases below $-20°C$. Because adiabatic CO_2 depressurization leads to very low temperatures, hot-fluid circulation in the autoclave jacket and a "controlled" depressurization rate are required. Another hazard might also appear in case of perforation of this cladding, as CO_2 (in the presence of water) may corrode the carbon steel; thus, a strict inspection must be made frequently to detect any cladding damage.

2.1.2. Thermodynamic Hazards

Dry Ice. Carbon dioxide handling often leads to a drastic temperature decrease and plugging of filters and tubings by solidified water, products, or dry ice itself. This plugging might be dangerous when occurring in basket-sintered disks, captors tubing, or the vent line that must be always "overdimensioned" and built carefully.

BLEVE. The "boiling liquid expanding vapor explosion" characterizes the physical explosion of a liquefied gas/SCF that is brutally depressurized to atmospheric pressure in case of pressure vessel rupture or opening. In fact, this hazard is almost only to be feared in the case of metal weakening cause by a intense fire around the vessel(s). It is for this reason that it is recommended to install fire detectors which could order immediate depressurization of the whole plant in the case of fire.

2.1.3. Chemical Hazards

Flammable Fluids, Cosolvents, Products. Explosion-proof equipment, buildings, and procedures must be enforced when flammable fluids are handled, especially light hydrocarbons and dimethyl ether. Explosive atmosphere sensors have to be connected to high-power fans and to fluid reservoirs stopvalves. N_2O must be avoided in processing flammable products, as it may behave as a comburant and lead to explosion.

Corrosion. This hazard must be evaluated prior to processing any material. Supercritical water oxidation leads to extreme corrosion rates and special alloys are required.

2.1.4. Biological Hazards

Asphyxia. Carbon dioxide buildup in closed rooms could lead to asphyxia in humans. It is for this reason that all possible CO_2 emissions must be collected into an "overdimensioned" vent line, ensuring a good dispersion of the gas in the outside atmosphere. Moreover, it is highly recommended to install CO_2 detectors in the equipment room, but also in any connected room (especially underground passages and cellars, where CO_2, which is heavier than air, may accumulate) for action with high-power fans and for the information of operators.

Chemical and Biochemical Toxicity. Handling any cosolvent or raw material or fluid that presents a danger in terms of chemical or biological hazard must lead to drastic care, as SCF equipment work at a high pressure with possible leaks at any moment. Fluid leakage often leads to aerosol formation that may be easily inhaled. In particularly dangerous cases, it is necessary to isolate the equipment in a closed room or box (Fig. 1) with remote control, and environment must also be protected from effluent treatment.

2.1.5. Safe Operation

We would stress the fact that a key for a safe and reliable operation of SCF equipment consists of *very cautious training of the operators.* Moreover, automation is required with a high degree of *redundancy* taking into account all potential issues (fire, electrical power or intrument air failure, computer or intrument problems, etc.).

2.2. Large-Scale SCF Plant: Design and Operation

Basic chemical engineering for SCF equipment design can be found in a recent book by Brunner (10), but few data are presently available for the calculation of main parts to be assembled into large-scale SCF plants. Moreover, special issues related to GMP compliance are to be taken into account at all steps of design, construction, operation, and maintenance.

2.2.1. General Layout

In order to avoid contamination of the processed products, it is generally required to install the vessel(s) that have to be filled and emptied in contact

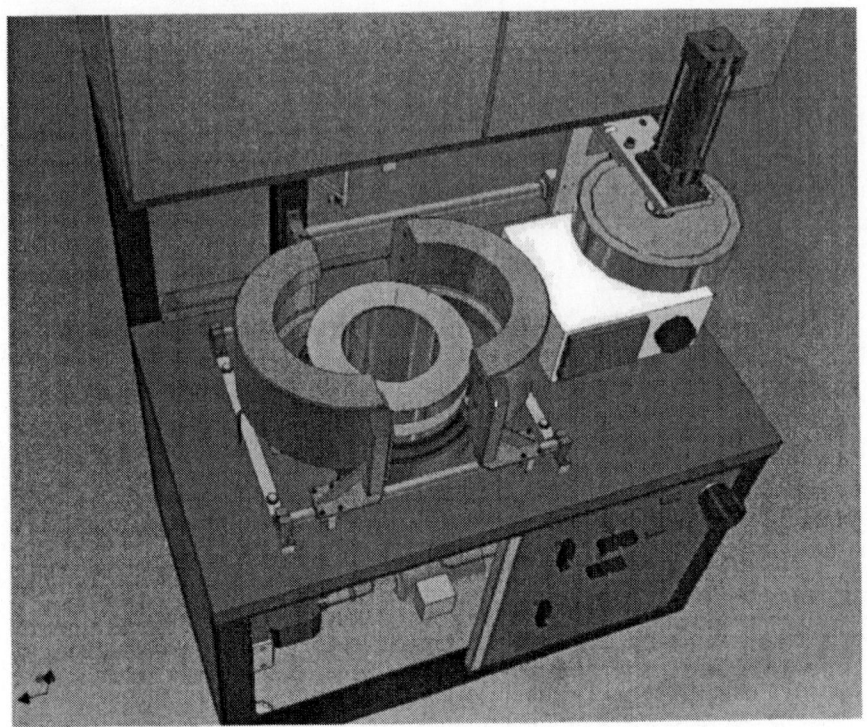

FIGURE 1 Autoclave with clamp-closure system. (Courtesy of Separex.)

with the atmosphere in a "clean" room, the class of which depending on the type of drug and, in certain cases, sterility is demanded. Because a SCF unit is complex and includes various equipment that can generate or accumulate dust and other potential pollutants, we obviously recommend splitting the plant in several rooms and isolating the vessel(s) into and from which the product is transferred in a clean room; the other pieces of equipment can be located outside of this clean room. Because a flammable fluid or organic solvent, or potentially explosive dusts are very often handled, large-scale plants must be built under explosive atmosphere standards; in this case, the utility services shall be located outside the explosion-proof area and the entire automation system located in a remote control room. Of course, worker and environment protections are imposed in all rooms where the process fluids are handled, as a leak may happen and release active substances.

2.2.2. Automation

All large-scale units are designed with complete automation, permitting an easy documentation of what is done during the plant life: cleaning, production, and maintenance, with the edition of detailed run journals and records of all processed substances, all incidents, and plant modifications. This is a basic requirement for complying with both safety standards and GMP.

2.2.3. Heat Exchangers

Double-tube countercurrent heat exchangers are currently used in pilot and semiindustrial plants, but when the fluid flow rate exceeds 500 kg/h, shell-and-tube exchangers are required for providing a sufficient transfer area. According to our experience, we recommend the following:

- Process fluid always on the tube side and heating/cooling medium on the shell side;
- Horizontal SCF heater (when the liquefied fluid is pumped) or cooler (when the fluid is compressed in gas phase), with one pass on the shell side and two passes on tube side, U-tubes being acceptable
- Vertical reheater(s) downward fluid decompression between the high pressure operation and the separators with one pass on the tube side for downward high-speed flow of process fluid in order to avoid extract deposition and plugging
- Vertical condenser (when required to liquefy the fluid prior to recycling by pumping) with one pass on the tube side and downward flow of process fluid so as to avoid plugging by extract entrained from the separators

The designer must also pay attention to the insulation materials used on tubings connected to heat exchangers and utility services inside the clean room, as these materials may accumulate and/or release dust and may be difficult to clean. This is especially difficult to solve for small-scale equipment, although classical solutions using metal envelopes are available for large-scale pipings.

2.2.4. Autoclaves

Most attention is paid to closure systems permitting a fast and easy opening/closure, as all solid processing is realized in the batch mode and requires a great number of repetitive operations to fill and to withdraw the baskets containing the raw materials. First, these closure systems must be very safe and reliable so as to avoid any risk of moving them when the autoclave is pressurized. Generally, a redundancy is proposed through automation and an additional passive system blocking the closure when some pressure remains inside the autoclave. Clamp-closure systems are often preferred, as they are very safe and easy to move automatically (Fig. 1). The remark about insulation materials presented at the precedent paragraph also applies to thermal insulation of all vessels located in the clean room.

2.2.5. Separators

Most SCF plants have separators designed as large empty gravity settlers, but significant savings can be made by using much smaller-volume cyclonic chambers that are of special interest when the density difference between the extract and the fluid is low. In the case where the substrate is very volatile and significantly soluble in the fluid at the separation stage or present at trace concentration, sorption onto an adsorbent bed is required to either purify the fluid prior to recycling (e.g., when pesticides are removed from a raw material) or to recover a valuable fraction of the extract that should have been lost otherwise (like light ends of flavors or fragrances); granular activated carbon is often preferred, as it is inexpensive and has a very large sorption capacity (~1 g/g) and collection efficiency, and it can be easily desorbed by the fluid itself at SCF conditions. For similar applications, distillation of the liquefied fluid has also been proposed. In most cases, it is also valuable to filter the fluid prior to recycling in order to avoid entrainment of raw material or extract fine particles that may accumulate and plug inside the condenser or downward pipings and capacities, resulting in losses of heat transfer efficiency, and finally requiring stoping the plant for a difficult-to-operate cleaning.

2.2.6. Fluid Management

The purity of the fluid contacted with the substrate is obviously a key parameter for preserving the final product from any contaminant. Therefore, much attention must be paid to the fluid supply and delivery to the plant and to the fluid recycle with potential buildup of trace contaminants.

Fluid Supply, Storage, and Delivery. In fact, carbon dioxide is available at high purity (generally 99.5% at least) and most impurities are not at all harmful (mainly moisture and very low concentrations of nitrogen, oxygen, methane, and light hydrocarbons). However, nonvolatile organic pollutants must be detected and evaluated prior to using the fluid in a pharmaceutical application, as this may indicate contamination with lubricants during the manufacture process, although the presence of harmful polycyclic aromatic hydrocarbons is highly improbable. For a plant operating in compliance with GMP, traceability shall be required from the fluid supplier, with systematic analysis of each lot delivered to the storage.

The storage, consisting of a cryogenic vessel (capacity varying from 4 to 10 tons) where carbon dioxide is stored in the liquid state at about 1.8 MPa, may be a source of contamination if not properly designed, cleaned, and operated. Stainless-steel tanks are preferred to the classical carbon steel vessels, as CO_2 induces the formation of rust in the presence of moisture and filtering submicron rust particle from liquid CO_2 is not easily feasible.

However, the most important potential source of contamination is the pump or compressor required to pump the liquefied CO_2 to the plant. A nonlubricated pump must be used instead of lubricated volumetric compressor. Moreover, it is strongly recommended to design the compressor system and the delivery line so that no contamination could occur from any other plant and no reverse flow could happen from the plant toward the storage. Because the operating pressure inside the plant is generally far higher than the pressure in the storage, experience shows that such reverse flow is not easy to avoid when problems happen during plant operation (one-way valve or compressor check-valve leakage, control system failure, etc.). The complete purge and validated cleaning of the storage and delivery line shall be operated between the processing of different products to avoid any risk of cross-contamination.

Fluid Recycling. In all large-scale plants, the fluid is recycled for obvious economical and ecological reasons. This means that the separation of solute(s) from the fluid prior to its recycling must be as perfect as possible to avoid deposition throughout the recycle loop, especially onto the colder parts (condenser, subcooler, liquefied gas reservoir), that may lead to plugging and stoping the unit for a long time. Moreover, some problems, unknown at the laboratory or pilot scale, may appear, as some tiny impurities may accumulate in the fluid phase (water, inert gases, pollutants, etc.). Two fluid cycles can be used:

- *Liquefied gas pumping*: Membrane pumps are preferred for flow rates up to 1000 kg/h and piston-plunger pumps are used for larger capacities. It is important to note that pump check valves present a significant pressure drop requiring a subcooling of the liquefied gas of at least 3°C below the boiling temperature at the inlet pressure (~40–50 bar for CO2) to avoid cavitation in the pump head that must also be cooled. However, some processes require very large fluid flow rates (>1000 kg/h), under a small pressure drop, that are delivered by centrifugal pumps.
- *Gas compression*: It is also possible to compress the gas through a compressor. However, these equipments are generally considered more expensive than pumps and more costly in maintenance, although savings can be made because no refrigeration machine or condenser is required. Moreover, some problems may occur if heavy or waxy materials are entrained in the gas phase to the compressor.

Regarding compliance to GMP, the main issue is related to avoid trace-contaminant buildup in the recycle loop. In fact, this risk is rather limited, as the makeup flow rate for replacing the fluid consumption is generally higher than 10% of the recycle flow rate. However, a periodic check and complete venting of the installation are to be practiced.

Fluid Disposal. When active substances are processed, no direct fluid disposal is acceptable and the depressurized gas can be vented to the atmosphere only after treatment. In most cases, the active substance that may be carried out by the gaseous effluent is solid and filtering is efficient. In rare cases, when the active substance is present in the form of an aerosol, water scrubbing is required. It is to be noted that the vent line is a basic tool of safety and, in no case, shall effluent venting in an emergency be hindered. It is for this reason that any effluent treatment unit must be designed for an emergency release flow rate and can be bypassed in case of plugging.

2.2.7. Construction Constraints

Supercritical fluid equipment must comply with both high-pressure standards and pharmaceutical standards. It is generally not possible to satisfy both demands, although it is not easy to find high pressure pieces compatible with the clean-in-place (CIP) and electro-polished internal surfaces (for thick-wall tubings) requirements. Any kind of threads on the fittings, on the connections, and on the vessel opening system should be avoided when possible because of cleaning issues. When CIP is not possible, all parts must be designed so that they can be dismounted quickly. Also, frequently some minor pieces in contact wth the process fluids have to be changed to avoid potential contamination; similarly, all lubricants present on valve stems must be replaced by pharma-accepted ones.

Obviously, major attention must be paid to ease further cleaning: All welds should be smooth (orbital welding is required for all tubing assembly) and dead volumes minimized. Moreover, it is extremely important to design the whole pipeworks so that a *total* drainage can be easily operated during the cleaning phases. Cleaning validation is a key issue of the design and remains a very time-consuming operation. The careful design of any tube, vessel, and piece of fitting is essential and must be done so that all the necessary clean checks are possible when imposed by the validation method. Straight pipes, no dead-volume fittings, and access to any internal parts of the equipment need to be check during design reviews.

2.2.8. Operation of Large-Scale Units

Large-scale SCF units are operated with the same rules as other pharmaceutical plants according to a validated plan (standard operation procedures) with the help of the control system. We draw, attention to some specific issues in the following subsections.

Solid Processing. According to our experience, many situations may occur with long-term use:

- Fine particles may cause basket filter disk deformation; this is the most widely encountered problem because most operators do not take enough care regarding the raw material particle size distribution during grinding. We strongly recommend avoiding very fine particles by grinding control and/or screening the raw material prior to filling the baskets; a paper filter may be an aid to prevent this problem. In other cases, the raw material may agglomerate in form at a thick "cake," where drastically reduces the fluid–material contact and extraction efficiency and possibly leads to total plugging and deformation of the sintered disks. It is preferable to prepare pellets with such "sticking" materials; when this is not possible, we recommend blending the raw material powder with an inert granular or fibrous (cellulosic for example) material that will prevent agglomeration. In the worst case, disk plugging may lead to basket deformation and blockage inside the extraction autoclave, with resulting damage to the autoclave wall during withdrawal; this may be avoided by controlling the pressure drop between the inlet and the outlet of the vessel below the value that irreversibly damages the basket and/or the sintered disks.
- Basket deformation may happen due to shocks during handling, and autoclave wall damage may be caused by shock with the basket bottom during introduction. This may lead to a drastic loss of efficiency of the extraction due to fluid bypass between the basket and autoclave walls when the external gasket of the basket is not totally sealing. Thus, it is important to operate with adequate means of basket handling and careful manpower.

Liquid Processing. Liquid processing is much easier to operate because pressure vessels are not often opened and closed. However, tubings or column packing plugging may happen. It is generally a slow process with the deposition of a solid or highly viscous material, which progressively reduces the open section and creates zones no longer swept by the fluid, which generally "catalyzes" the deposition until the moment where the operation is be stopped for cleaning. This phenomenon is not easy to detect and to prevent, because, in most cases, it does not appear at the pilot scale where experiments are generally conducted during short periods. To avoid this, it may be necessary to pretreat the liquid(s) (filtration, etc.), to change the operating conditions (possibly increased temperature), or to operate preventative column cleaning regularly.

2.2.9. Maintenance of Large-Scale Units

Industrial production with SCFs requires a high reliability with drastic safety requirements: This requires *preventative* maintenance because many parts must be inspected and some changed periodically; moreover, a rigorous operatio plan must be enforced to eliminate any risk of deterioration of the basic parts, and safety sensors must be continuously logged. This primarily concerns the high pressure pump(s) [check valves and membrane(s) are highly sensitive to abrasion or perforation by solids], autoclave closure systems and gaskets (to prevent fluid leakage), and baskets (external gaskets to avoid solvent bypass; filter disks plugging to avoid deformation or rupture). Of course, pressure vessels must be inspected and submitted to pressure tests according to official standards. Moreover, the main process valves must be checked often, as they are the key to safe operation during autoclave opening for raw material change. Sensors must be recalibrated periodically, in comparison to traceable reference sensors, and data logging validated. Finally, we stress the fact that maintenance is greatly easd when entrainment of some fraction of the substrate(s) through the fluid recycle loop is avoided and an efficient cleaning is frequently operated.

2.2.10. Cleaning Large-Scale Units

One of the most important issues for operating a large-scale SCF flexible unit is probably cleaning and cleaning *validation* according to GMP, as most parts cannot be opened between each lot manufacture and most high pressure parts—like valves—are not CIP. According to our experience, we propose the following cleaning procedures:

- *Between batches of the same product*: Rinsing the unit with an adequate liquid solvent (chosen as a "good" solvent for the processed material), dismantling and cleaning dead ends, rinsing again with the liquid solvent, with sampling for cleaning validation, drying with air, gaseous nitrogen or CO_2 to eliminate solvent vapor, and rinsing with liquid/supercritical CO_2 that is finally vented to atmosphere in order to eliminate most extracted impurities (mainly liquid solvent).
- *Between different products*: Dismantling the equipment, cleaning each part, swabbing the pressure vessels and all "critical" points, reassembling the equipment, and application of the precedent cleaning procedure.

Cleaning is validated through liquid solvent sample analyses (dry weight and chemical identification of residue) and swab characterization according to the classical technique. It is extremely important to consider the cleaning issue at the very beginning of any SCF equipment design. This influences many choices so as to avoid piping/instruments dead ends and all zones that could not be swept easily by the process and cleaning fluids. For example, we developed very low-volume multitubing/multiinstrument connections and very low-volume high-speed separators in the form of cyclonic chambers. Moreover, adequate parts must be installed to permit an easy rinsing of the whole unit with liquid solvent: The port locations must be carefully determined so that a total drainage is rapidly completed.

3. SCF EXTRACTION/FRACTIONATION SCALE-UP

Supercritical fluid extraction (SFE) processes can be scaled-up from lab-scale or pilot-scale results according to a simple procedure (27,28): At first, small-scale experiments lead to the optimal extraction conditions through a scanning of different pressures, temperatures, solvent ratios, and composition. Then, the scale-up method will depend on the mechanism controlling the extraction:

1. Some extractions are only limited by the extract *solubility* in the fluid, as the solvent exiting the autoclave is saturated in extract. For instance, in the case of lipids extraction, the access to the extract in the matrix is easy.
2. Some extractions are only limited by *diffusion*, especially the internal diffusion (e.g., stripping residual solvents from active particles or pesticide elimination from natural stuffs).
3. Many extractions are limited by both *solubility* and *diffusion*. As shown in Figure 2, the extraction versus time (or solvent to feed mass ratio) can be described as a first phase during which the extraction rate is constant (the solvent exiting the autoclave is saturated in extract at a concentration equal to E_0/S_0) and a second phase during which the extraction rate is decreasing due to diffusion limitation (the rate is the higher when the solvent flow rate is the lower). This is confirmed in Figure 3 for spice extraction (28,29). The extraction was performed on seven typical ground spices at various conditions of pressure and temperature chosen in the most common ranges (250–300 bar, 40–60°C). The first part of the extraction curve is always a right line corresponding to CO2 saturation at the autoclave exit, the slope a being the extract solubility in CO_2 in these extraction conditions. In this first part, the seven-spice extraction curves are almost superposed when the yield percentage (ratio of yield to final yield) is presented versus the product of the extract solubility a by the solvent ratio (CO_2 to feed mass ratio). Then, the curves are not superposed, as diffusion controls the

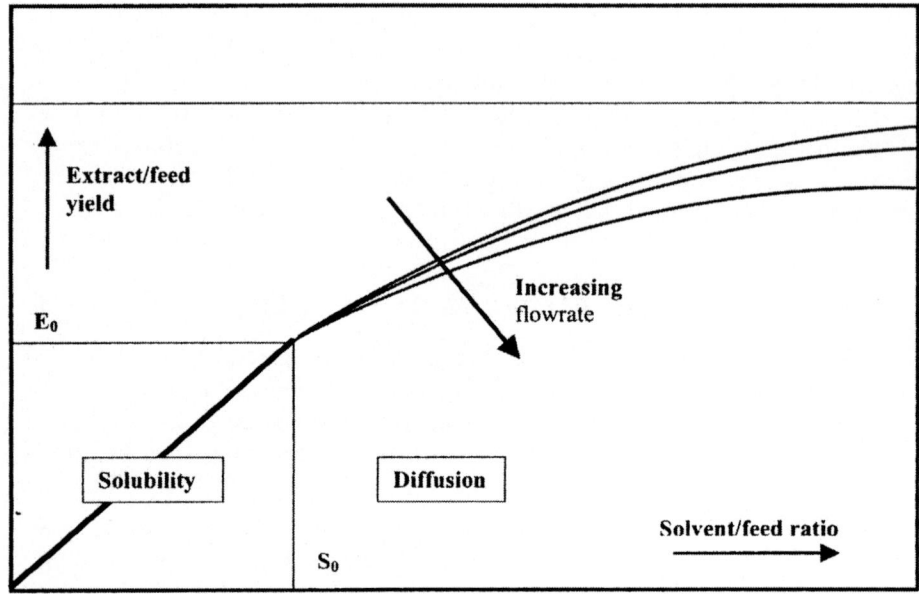

FIGURE 2 Typical extraction curves.

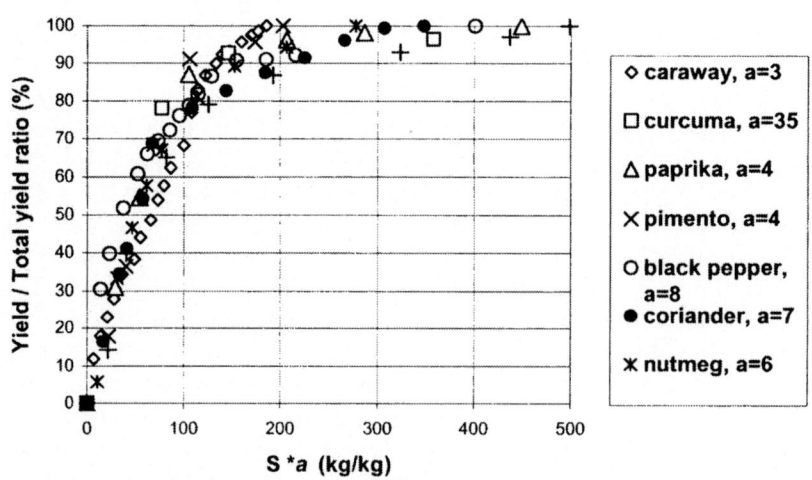

FIGURE 3 Spices extraction by supercritical CO_2. (From Ref. 29.)

extraction rate. Depending on the complexity and on the kinetic limitations of the extraction, different methods are available to design the production unit:

- The easiest method to scale-up experimental data is to keep one or both of the ratios U_s/M_f and M_s/M_f constant, where Us is the solvent flow rate (in kg/h^1), M_f is the feed mass in the extractor (kg), and M_s is the solvent mass required for the extraction (kg).

  - The ratio U_s/M_f is inversely proportional to the residence time of the solvent in the extractor t_r:

$$t_r = \varepsilon \frac{\rho_s M_f}{U_s \rho_f} \tag{1}$$

   where ε is the void fraction of the bed, ρ_s is the specific gravity of the solvent, and ρ_f is the specific gravity of the feed.
   The ratio will be conserved for extraction limited by *internal diffusion* for which the "contact" time of the feed with the solvent is the determinant factor. Therefore, it will be necessary to use very large extractors or to use several extractors in series in order to maximize the contact time, and it is possible to minimize the solvent flow rate and the energy consumption of the plant.
  - The ratio M_s/M_f is the solvent/feed ratio that has to be maintained in the case of extraction limited by the *solubility* of the extract. Thus, for a given plant capacity and, therefore, a given amount of solvent and energy consumption, it will often be possible to reduce the volume and the number of extractors as much as possible. However, the autoclave's volume and dimensions (length H, section S) have to be designed so that the solvent spatial velocity (equal to H/t_r) is low enough to limit the pressure drop and possible channeling through the feed bed.
  - When both *diffusion* and *solubility* control the extraction, both ratios U_s/M_f and M_s/M_f shall be maintained constant.

This method has the definitive advantage of simplicity but does not take into account several important factors (internal diffusion, axial mixing, etc.) and is unable to predict the effect of using a series of autoclaves.

- A refined method integrating these factors requires a numerical simulation that can estimate any configuration and permit industrial-plant optimization. Different models have been proposed in the literature and we built a versatile simulation software allowing the representation of many different systems (28).

Knowing the production requirements, the optimal configuration will be determined. The principal factors are as follows:

- Number of extractors
- Volume and number of shifts/day of the shifts/day of the extractors
- Pump and utilities capacities

Traditional production units are composed of at least two extractors. One is unloaded/loaded while the other one is operated in extraction. Three or more extractor configurations are often preferred in order to reduce the dead times and increase the extraction efficiency. These extractors are connected in series, so that the feed and the solvent are contacted countercurrently. The last extractor of the series is loaded with "new" feed and the first extractor is loaded with feed that has already been contacted the longest time with the solvent and is the next to be unloaded. This "carrousel" implementation allows the reduction of the amount of CO_2 required for a given extraction and, therefore, increased productivity and reduced energy consumption. For a given production capacity, increasing the number of extractors will decrease the energy consumption and the operating costs but increase the investment cost. The extractor volume also depends on the number of shifts/day that can be operated. The economic estimation allows one to decide what is the optimal configuration in each case.

Because the above-described extrapolation and optimization methods concern only the "extraction step" of the process, it is also very important to optimize the other parts of the process. Extract recovery and fractionation, energy management, and improvement of the extractor emptying and loading procedures also have important economic consequences and must be considered in the industrial process design.

Regarding SFF that is performed on a countercurrent column, scale-up is the more difficult because axial mixing in both phases dramatically decreases the fractionation efficiency—in terms of height equivalent to a theoretical plate. At present, no reliable results have been published on column scale-up. We would recommend the following procedure, following what has been done in liquid–liquid extraction:

1. Determination of liquid and fluid holdup curves on the pilot scale in order to estimate the flow rates at which flooding happens in

the given pressure–temperature conditions with the chosen packing (random rings or nonrandom packing). We stress the fact that liquid foaming often happens, especially when the feed is a natural product or is prepared by fermentation, as many biological compounds act as surfactants; similarly, impurities in the form of solid particles or colloids may have the same effect. Because foaming may drastically modify the hydrodynamic pattern and lead to flooding, experimental runs must be performed with a "representative" liquid, not with a synthetic mixture of pure compounds.

2. Scale-up of the column diameter on the basis of these curves, avoiding the zone where hold-up may happen.

3. The column length should be evaluated very prudently because axial mixing increases with the increase in column diameter. Great care must be taken to design the fluid and liquid phase distributors; channeling will be reduced by fractionating the packing in several beds separated by redistribution plates.

4. Regarding reflux, large-diameter columns cannot be operated with the internal reflux caused by a temperature gradient along the column that is operative only for diameters below 15 cm. Thus, an external reflux through a pump reinjected part of the extract must be implemented, which also permits easy control of this parameter.

4. SCF DRUG FORMULATION SCALE-UP

Particle formation processes using SCF are now subjected to an intense R&D effort in the pharmaceutical industry both for formulating water-insoluble (or poorly soluble) and hydrophilic molecules, including fragile biomolecules. In fact, SCF drug formulation is a very large domain leading to very different particles in terms of size, shape, and morphology through several very different processes described in the precedent chapters and in many publications and reviews (17,18). Presently, in spite of very attractive results obtained on the laboratory scale, commercial development is still pending because scale-up requires the solution of some major issues that we will investigate in the following subsections. Hardly any articles were published on this domain, except for a few dealing with antisolvent processes. The first two (30,31) claim success on scale-up from lab to pilot scale, but on a very small scale (a few grams per hour), and another one presents interesting considerations, listing scale-up issues but without tangible results (32). A few brief economic studies (33,34) were also proposed, but based on contestable technical assumptions, especially regarding the fluid/substrate ratio, and no solutions to scale-up issues were presented; the most recent one (34) is the more complete, although it significantly underestimates the equipment cost and does not evaluate the extra cost related to compliance with GMP if a drug is processed. Although the SCF formulation processes are very different, all of them comprise three main steps for which scale-up is difficult: particle generation, particle collection, and fluid purification and recycle (when recycled).

4.1. Microparticle Generation

For most SCF formulation processes leading to microparticles [except batch antisolvent GAS that is rarely used (35) and some coating (36–40) and impregnation processes (41–44)], particles are generated by *atomization* through a nozzle:

- Rapid Expansion of a Supercritical Solution (RESS) (45–47)
- Particle precipitation by Supercritical Anti-Solvent (SAS) according to various implementations like the batch Gas-Anti-Solvent (GAS) (48,49) and the continuous Aerosol Solvent Extraction System (ASES) (32,50–54) or Solvent Enhanced Dispersion by Supercritical fluids (SEDS) (30,31,55)
- Particles from water solution by antisolvent (31,56,57) or nebulization (58–60) or extraction by a polar fluid (54), and from water-in-oil (W/O) emulsion droplets (emulsion drying process (13,61,62)
- Particles from a Gas-Saturated liquid Solution (PGSS) (63–66) or microcapsules from a fluid-expanded suspension (67–69)
- Porous particles impregnated by a substrate by copulverization of the fluid-saturated substrate and the porous excipient, using the Concentrated Powder Form (CFP) process (66,70).

4.1.1. Atomization

On a laboratory scale, the difficulty is to manufacture and operate very small-diameter nozzles (<100 μm) that are often plugged by microparticles. For a given process and processing conditions, the nozzle diameter and shape (length-to-diameter ratio) and the fluid (or liquid) velocity through the nozzle determine the size and shape of the generated particles, but no model is presently available to correlate these process and nozzle parameters and particle properties for any of the processes. Although it looks very simple, RESS is very difficult to describe, as shown in recent detailed contributions (71,72), from which it is not possible to find easy scale-up rules. Similarly, antisolvent modeling was subjected to extended investigations (73–76), which did not lead to practical results yet. In fact, the phenomena are extremely complex because the particles' size and morphology result not only from the atomization itself—and the resulting particle nucleation—but also from their growth by condensation and/or coagulation during their residence time inside the atomization vessel.

Ideally, scale-up should be made in maintaining all parameters constant by replacing one nozzle by several similar nozzles of the same design and dimension through which the fluid or liquid velocity and pressure drop should be kept identical to those through the sole nozzle. However, it is difficult to realize practically an atomization system with several nozzles presenting exactly the same pressure drop and leading to a perfect fluid or liquid distribution as has been experienced in liquid–liquid or gas–liquid contactors; moreover, in such multinozzle systems, the resulting droplets or bubbles do interact with each other and may coalesce, leading to a wide size distribution. In the case where multinozzle systems must be used, it is recommended to perform experiments with a unitary nozzle to optimize the process parameters in order to obtain adequate particles and to scale up by multiplying the nozzle number working in these optimal conditions.

Another route, applicable to the RESS process, consists in using a sintered disk as the fluid distributor, as proposed by Domingo et al. (77), although this system's performance is limited by the imperfect sintered disk porosity distribution that is continuously modified by the unavoidable plugging of the finest pores by nanoparticles generated inside the disk.

Regarding nucleation of solid particles from liquid droplets (antisolvent, nebulization, PGSS, microencapsulation processes), it is obvious that the droplets size is a basic parameter in the process. Some investigators reported results of experimental works supported by theoretical considerations on droplet formation by injection of a liquid into a pressurized fluid using a simple capillary nozzle (78) or a coaxial nozzle (79). Both articles, following previous works on liquid droplets formation into an insoluble liquid or into a gas, suggest the correlation of the experimental results with two dimensionless numbers

- The Reynolds number $Re = \rho U \Phi / \mu$ (representing the inertial force to friction force ratio)
- The Weber number $We = \rho U^2 \Phi / \sigma$ (representing the kinetic energy to capillary energy ratio)

where ρ is the specific gravity of the fluid, U is the relative velocity of the two fluids, Φ is the nozzle diameter, μ is the viscosity of the fluid, and σ is the interfacial tension between the two phases. The larger the Weber number, the greater is the turbulence versus the capillary forces and the smaller are the resulting droplets. Experimental results and modeling show that the diameter of the initial droplets and of the resulting particles increase when the Reynolds number is increased and decrease when the Weber number is decreased (79), and the ratio $(We)^{0.5}/Re$ is proposed to evaluate the zone where a free liquid jet is atomized into the fluid (78). From a practical point of view, these results are not easy to use for scale-up, except by maintaining both Re and We constant: When all other parameters are constant, this leads to the obvious rule of maintaining both the nozzle diameter and the liquid velocity through the nozzle constant, leading to a multinozzle system, as stated earlier. However, our experience showed that the key parameter seems to be the liquid velocity; the nozzle diameter can be scaled up on a certain range without modifying the particle size significantly (80).

Regarding the postnucleation growth of particles, it could be assumed that maintaining the residence time before particle collection would lead to the same final particles when all other parameters are kept constant, especially the fluid/substrate ratio, temperature, and pressure.

4.1.2. Temperature Control

As recently reviewed (81,82), several authors showed that the solid morphology (amorphous and crystalline pattern) is dependent on the temperature at which the particles are formed and maintained during their residence through the atomization vessel, which could explain the very scattered results found in the literature as the temperature distribution changes from one experimental system to another. For the difference with lab-scale equipment, where the vessel thermal inertia in comparison with the heat flux brought by the fluid permits one to easily stabilize the temperature distribution, careful design and operation are required at a large scale in order to obtain reproducible conditions in the atomization vessel that are required to reach reliable particle size and morphology.

In fact, this practically implies that the particle formation and the particle collection cannot be operated in the same vessel!

4.2. Particle Collection

When micronic or submicronic particles are produced, particle collection and harvesting is certainly the most acute issue! Based on the considerable experience gathered in dust collection at any scale (83), different particle collection systems can be proposed depending on the final form desired by the operator: dry powder, dry mixture of active substance and excipient particles, aqueous or solvent suspension.

Basically, four concepts are currently used to collect particles from a gas flow: inertia (gravity settler, impact chambers and cyclones), electrostatic forces, liquid scrubbing, and filtering through porous media. In the present case, we have great doubts about the feasibility of electrostatic precipitation for both technical (use of very high voltage, powder explosivity, reagglomeration, etc.) and economic reasons. Gravity settling is only efficient for large particles (> 100 μm) and requires very large chambers. Impact chambers and high-efficiency cyclones are adapted for medium-size particle (between 10 and 100 μm), but the collection efficiency rapidly decreases below 10 μm and is nearly zero for 1-μm-diameter particles; even wet-wall cyclones (leading to a liquid suspension) have a limited efficiency in this size range. Finally, for micronic and submicronic particles such as those required for pulmonary delivery [e.g., Figure 4 presents the size distribution of insulin particles obtained by emulsion drying (63)], the choice is limited to filtration for obtaining dry particles or to liquid scrubbing for obtaining a suspension.

4.2.1. Particle Collection by Filtration

Bag filters are traditionally used for very fine-particle collection. The filtering media are either woven textile or nonwoven paper or fiber mats. In the present case, the filtering media should be chosen in materials both compatible with the processed fluids (especially for antisolvent processes where the filter is contacted with SCF and organic solvents) and acceptable for drug manufacture. We currently tested various filter types: woven or nonwoven polymer (like PTFE or polyamides) fibers to form bags or paper bags (as for vacuum cleaners), woven stainless-steel fibers to from a disk, or filter paper supported by a sintered disk at the bottom of a basket at a smaller scale; ceramic filters can also be considered favorably.

In spite of its great efficiency in particle collection, filtration has several main inconveniences that must be overcome before reaching reliable production:

- It is batch process and filtration conditions vary with time as the filter cake is building, with correlative pressure drop buildup and

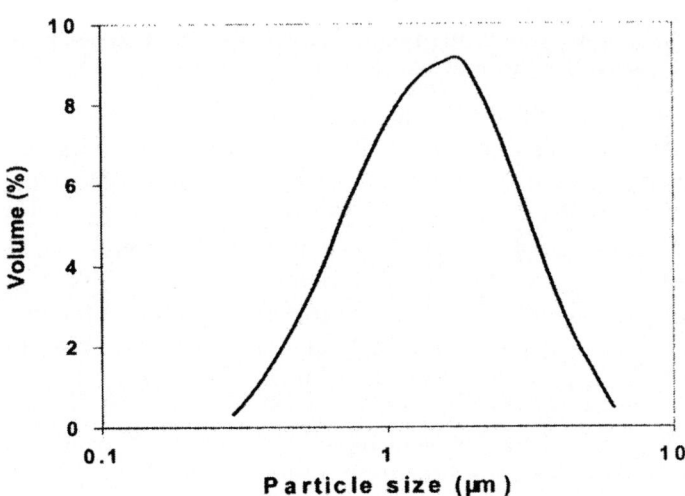

FIGURE 4 Size distribution of insulin particles obtained by emulsion drying. (From Ref. 63.)

particle layer compression that often leads to particle agglomeration, especially if some residual solvent is present.

- The particle residence time on the filter varies and may lead to possible morphology changes (crystal pattern) and nonhomogeneous properties. The worst case happens when the particles are collected before they are completely dried by the fluid (antisolvent or drying processes), leading to particle agglomeration and "sintering" inside the filter cake.
- Particle harvesting is not easy, especially when very fine particles are collected. On a small scale, it is possible to close the bag after deppressurizing the collection vessel and to transport it into a glove box, where the powder can be recovered in an adequate vessel. On a large scale, this system appears to be not applicable in most cases and in situ harvesting is often mandatory.

These issues cannot be solved through a single technique, but several routes are proposed and adapted case by case.

Dual-Filtration Vessels. In order to transform a batch process into a semibatch process, the particle-loaded fluid is directed to one of at least two collection vessels that are operated altenatively in fluid filtration and particle harvesting.

Harvesting from Filter. Several methods can be used for particle harvesting, depending on the particle size and final form. For large-scale operation, it is possible to shake the loaded bag filter and to entrain the released particles by a reverse flow of low pressure neutral gas (carbon dioxide or nitrogen). However, this collection is often difficult again for very fine particles, possibly requiring a second collection/harvesting step.

Deep-Bed Filtration. In the case where the particles are to be mixed into an excipient prior to tableting, it could be interesting to replace the bag filtration by a deep-bed filtration through a bed of excipient, avoiding any further handling of these fine particles (84). The bed depth depends on the excipient particles' size distribution, but for "classical" fine powder, it can be set at a few tens of centimeter. Of course, the solid mixture can be easily removed and must be submitted to an intense mixing for obtaining an homogeneous repartition of the active substance inside the final powder. This very simple method is easily applicable to the RESS process; for antisolvent particle formation, the organic solvent must be carefully stripped from the resulting solid mixture prior to the final powder recovery, because it may adsorb onto the excipient beads at a significant concentration.

Carbonic Snow Collection. Because it is well known that rain droplets or snow crystals formation is controlled by the presence of dust particles in the clouds that serve as nucleation germs, we demonstrated particle collection by capture into large particles of "carbonic snow" generated by depressurization of a stream of liquid CO_2 into the particle-loaded gas stream. Surprisingly, the collection efficiency was very high and the "snow" was easily collected in a cyclone (85). After transfer of this snow in a clean vessel and slow sublimation, it results in a dry powder of micronic particles, showing no agglomeration. This method seems to be of particular interest when particles are formed by RESS, as collection can be made without any filter at the exit of the atomization vessel.

4.2.2. Particle Collection by Scrubbing

Liquid scrubbing is also known as one of the most efficient methods for very fine-particle collection, but they are gathered in the form of a suspension that can be rather used for nasal, pulmonary, or parenteral delivery. The suspension stability is often a delicate issue because no chemical or biological degradation shall occur, nor particle decantation, which requires the use of complex mixtures, including a buffer (especially for biomolecules), preservatives (antimicrobial, antioxidant, etc.), and surfactants. Several systems have been described (37,86–88):

- Direct injection of the particle-loaded gas resulting from RESS into an aqueous solution of a surfactant, especially lecithin (86).
- Coinjection through the same nozzle of the fluid solution containing the substance and a warm aqueous solution supplying the enthalpy required to completely vaporize the fluid without important preheating before the nozzle (87). This process is especially adapted when using a liquefied gas as solvent and/or a thermolabile substance is processed; a surfactant or stabilizing agents can be added either in the fluid phase, or in the aqueous phase, or in both.
- Scrubbing the particle-loaded gas by a liquid solution in which is dissolved a coating agent that will entrap the particles either by coacervation caused by antisolvent precipitation (37) or by deposition caused by coating agent sursaturation caused by solvent extraction by the fluid (88).

4.3. Residual Solvent Stripping

For processes using an organic solvent (antisolvent, emulsion drying, etc.), one major issue is related to the elimination of residual solvent adsorbed onto the formed particles. On the laboratory scale, most investigators collect the particles at the bottom of the atomization vessel onto a filter and propose to percolate pure carbon dioxide after stopping the stream of organic solution; in the rare publications dealing with data on stripping conditions and efficiency and according to our own experience, it appears that very large fluid/substrate ratios are required, often higher than those used for the atomization itself, the worst case being when the substrate is precipitated from an aqueous solution and should be obtained with a low moisture content. Moreover, we think that this method cannot be directly applied on a large scale because fluid percolation through the particle layer collected onto the filter is not satisfactory because of fluid bypass and formation of aggregates with unacceptable concentration of solvent, especially when submicronic or micronic particles are formed.

For these reasons and in order to decrease the fluid/substrate ratio (and the cost), we propose a two-step method for large-scale operation:

- At the end of particle precipitation, a stream of pure CO2 is pumped through the collection vessel in order to eliminate the fluid rich in organic solvent and a part of the solvent present on the particles; this requires a low fluid/particle ratio.
- Then, after depressurization, the particles are harvested and the powder is set into a basket to be installed in an extraction autoclave and submitted to a flow of SCF CO_2. The fluid/substrate ratio can be reduced by changing the fluid direction upward/downward several times to prevent channeling, the fluid being recycled after solvent collection and purification as detailed in the next subsection.

4.4. Fluid Purification and Recycle

For most SCF formulation processes at large-scale, solvent recycle is mandatory whatever are the encountered problems detailed below. Most investigators do ignore this major economic and technical issue as they only work at lab-scale where the fluid can be vented; otherwise, many results would never have been presented as attractive and "promising" breakthroughs as they are certainly useless for any commercial application.

4.4.1. PGSS, CFP, and Microencapsulation Processes

The fluid is carbon dioxide and the ratio fluid/substrate(s) is very small, of an order of magnitude of 1: No need to recycle the fluid that is depressurized to atmosphere and vented (64–70).

4.4.2. RESS

In most cases, the fluid must be recycled. Either it is carbon dioxide and the substrate solubility is low so that a very large fluid/substrate ratio is required, currently of the order of magnitude of 10^3–10^4 or even higher, or it is another fluid with better solvent properties, like propane or dimethyl ether, leading to a lower ratio; however, both demands fluid recycling! In both cases, recycling is difficult and costly because the fluid must be recompressed from the atomization pressure to the dissolution pressure. In fact, this requires that the atomization pressure be much higher than the atmospheric pressure, at the difference that is currently done at the laboratory scale (although the so-generated particles may be significantly different from those obtained when the solution is atomized at a much higher but "realistic" pressure). Another difficulty appears when the SCF is a mixture of two or more components (like carbon dioxide + propane or polar cosolvent). On one hand, the collected particles have to be stripped from residual solvent(s) by pure carbon dioxide, and, on the other hand, the fluid mixture recycling demands a composition control.

4.4.3. Antisolvents

Carbon dioxide is generally used and the particle formation and collection is operated at high pressure, permitting an "easy" recycling of the fluid—after purification. However, similar to RESS, antisolvent particle formation requires very large fluid/substrate ratios, especially when water solutions are processed; it is important to know that the fluid is required not only for the particle formation itself but also for extraction of the solvent residue adsorbed onto the particles after collection. In fact, fluid recycling is difficult because it is almost impossible to eliminate the organic solvent borne by the fluid after particle collection by the classical means used in SFE because the residual solubility of the organic solvent in the depressurized fluid at the recycle pressure (~ 40–50 bar) is far from null. For the antisolvent precipitation itself, the fluid can be recycled as it is, at the condition that the solvent concentration is perfectly constant. We developed an original system to maintain the recycled fluid composition at a constant value, based on the Gibbs' law, that was successfully demonstrated in preparative SFC (89). For solvent stripping from the collected particles, pure carbon dioxide is required and can be recycled only after classical solvent collection by fluid depressurization followed by carbon bed adsorption so as to drastically reduce the solvent content in the fluid and, consequently, inside the stripped particles. In large-scale units, this requires two separate fluid circuits; one for antisolvent precipitation and one for stripping.

4.4.4. Drying Processes

Several SCF processes are being developed for obtaining dry particles from aqueous solutions, but drastic limitations appear at scale-up.

    Antisolvent.   The fluid is a CO_2–ethanol mixture, the alcohol serving as entrainer of water into the fluid: the same recycle system as proposed for the antisolvent process is applicable. However, requires very large fluid/substrate ratios, of the order of magnitude of 10^4–10^5, rendering scale-up economically hazardous.

Emulsion Drying. Supercritical CO_2 extracts the water and organic solvent (a heavy alcohol or ketone) and the same recycle system as proposed for the antisolvent process is applicable. The difference is that, much lower fluid/substrate ratios, of the order of magnitude of 10^2–10^3 are required, with the resulting savings in the recycle loop.

4.4.5. Impregnation and Coating by Deposition

These processes (36,39–44) do not raise any major difficulty for fluid recycle and can be operated with the same systems as currently used in SFE.

4.5. Scale-Up Results

For several years, we have been designing and operating equipment permitting the operation of the various SCF particle design processes:

- Samples below 1 g can be processed on a very small scale on new chemical entities or biomolecules on equipment (X0.1, Fig. 5) pro-

FIGURE 5 Lab-scale equipment (X0.1) inside an isolator. Fluid flow rate: 0.5 kg/h; atomization vessel: 0.01–0.10 L. (Courtesy of Separex.)

tected inside an isolator (glove box) with a sapphire cell and small atomization vessels (~10 mL).

- Larger samples (1–20 g) are processed on a pilot plant (X1, Fig. 6) installed inside a "laminar" hood.
- Clinical lots (10–100 g) can be prepared in a larger pilot plant (X10, Fig. 7) operated inside a clean room, in compliance with GMP; for instance, we recently processed bovine and human insulin in order to obtain inhalable particles (particle size distribution presented in Fig. 4) from an aqueous solution using our emulsion drying process (13,62,63).
- Large lots can be prepared on a commercial-scale plant (X100, Fig. 8), where several tens of kilograms of particles can be obtained daily, either by RESS, antisolvent, or microencapsulation processes.
- For a nonpharmaceutical application, we envisage modifying a large SFE plant (X1000, Fig. 9) for processing at least 1000 kg of powder per day by an antisolvent process.
- Microencapsulation by atomization of a fluid-expanded suspension (66,67) is performed at pilot plants (X1 and X10) with the addition of large low pressure vessels in which the mixture is depressurized and

FIGURE 6 Pilot-scale equipment (X1). CO_2 flow rate: 5 kg/h; atomization vessel: 0.5 L. (Courtesy of Separex.)

FIGURE 7   A GMP pilot plant (X10). CO_2 flow rate: 50 kg/h; atomization vessel: 5 L. (Courtesy of Separex.)

FIGURE 8   A SCF plant (X100). CO_2 flow rate: 500 kg/h; atomization vessel: 50 L. (Courtesy of Separex.)

FIGURE 9   A SCF plant (X1000). CO_2 flow rate: 3000 kg/h; atomization vessel: 500 L. (Courtesy Hitex and Novelect; photo by M. Philippe Baudet.)

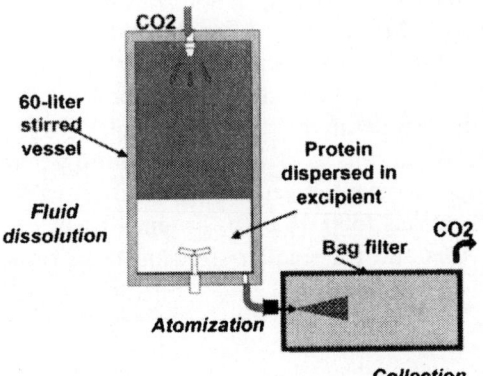

FIGURE 10   Microencapsulation by atomization of a fluid-expanded suspension. Dissolution agitated vessel: 60 L; atomization vessel: 100 L. (From Refs. 67 and 68.)

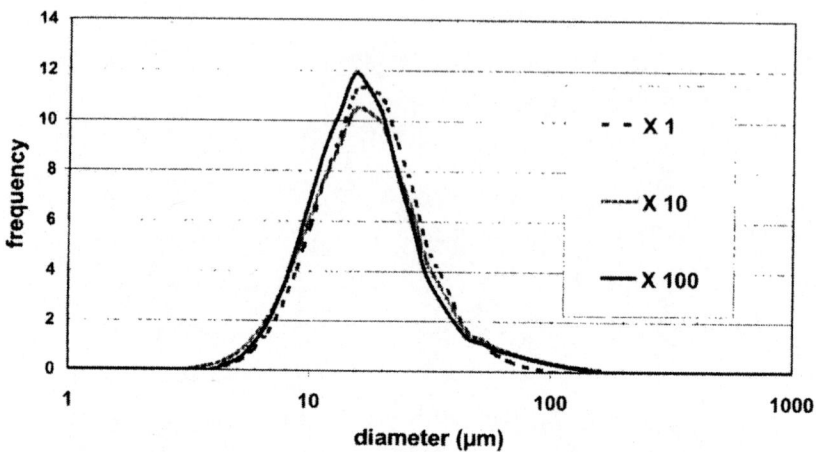

FIGURE 11 Particle size distribution of inulin particles prepared by the ASES process at various scale plants: X1 (2g sample), X10 (20g sample), and X100 (200-g sample). (From Ref. 80.)

the particles collected. Larger lots were produced from a 60-L agitated vessel as presented in Figure 10. No major diffulty appears when scaling up to this size, which corresponds to the commercial scale for most biomolecule encapsulation processes, because as several -kilogram lots can be processed daily.

In order to demonstrate that scale-up can be successfully performed from lab to commercial scale, we performed the atomization of inulin (a polysaccharide extracted from chicory root) from NMP solutions (300 g/L) by antisolvent with supercritical CO_2 (20 MPa, 40°C): After the first test a lab scale (X0.1), we prepared samples in three plants: 2 g in X1, 20 g in X10, and 200 g in X100 (80). As shown in Figure 11, the particle size distributions (by volume) are strictly the same at the three scales in the range for which we want to obtain a "nondusty" powder. Moreover, this work permits us to show that the fluid/substance ratio (~ 50 kg/kg) can be optimized at a much lower value than generally stated in most publications (500–10,000). Extended work is now ongoing on therapeutic molecules and for smaller-sized particles on a large scale.

5. CONCLUSION

Supercritical fluid technology is not yet widespread in the pharmaceutical industry, except for extraction of active compounds from vegetal sources (phytopharmaceuticals/nutraceuticals). Many innovative drug formulations based on SCF technology are now under development, demanding a wide R&D effort because process choice and optimization will be adapted case by case on technical and economic bases as summarized in Table 2. Ironically, this intense R&D work is leading not only to many attractive results but also to many patents that are rendering the intellectual property situation rather complex and may make pharmaceutical companies reconsider entering this technology in their formulation "toolbox" on the short term. More importantly, the process complexity conjugated with the high investment required

TABLE 2 Formation of Neat or Composite Microparticles

Substrate soluble in SCF	Matrix soluble in SCF	Available process	Type of particles produced	Remarks
Yes	—	RESS	Nano/ microparticles	Few substrates/ coatings are soluble in SCF CO_2
Yes	Yes		Microspheres	
Yes	No	Impregnation	Microspheres	Difficult scale-up
No	Yes	Liposome–RESS	Liposomes	No scale-up development?
		RESS fluidized-bed coating	Microcapsules	Few coatings are soluble in SCF CO_2
		Coating deposition	Micro-capsules	Very large fluid ratio
No	No	Antisolvent processes	Nano/ microparticles	Very large fluid ratio
			Microspheres	Difficult scale-up
		Coating coacervation	Microcapsules	To be demonstrated on a large scale
		Fluid-assisted microencapsulation	Microcapsules	Very low CO_2 consumption; easy scale-up
		CPF process	Microspheres	Easy scale-up
No	—	Emulsion drying	Nano/ microparticles	Biological molecules
		Nebulization	Nano/ microparticles	Wide range of water-soluble molecules

has probably delayed the introduction of the SCF formulation techniques in spite of their proven performances and interests for designing innovative drugs.

However, the situation is moving and the "pipeline" is now rich with several formulations to be introduced shortly for registration, especially for manufacturing inhalable and sustained-release particles. We already built three semi-industrial particle-design plants under strict quality assurance and documentation that are presently used for clinical lots manufacture, and scale-up with compliance to GMP seems accessible at present, as shown by this chapter. Moreover, as most promising drugs are based on proteins and peptides delivered by injection, the intrinsic sterility of SCF processes will appear as a major advantage for preferring these environment-friendly solutions.

ACKNOWLEDGMENTS

The authors thank Dr. Jennifer Jung and Dr. Fabrice Leboeuf for fruitful discussions and text reviewing.

REFERENCES

1. Stahl E, Quirin KW, Gerard D. Dense Gases for Extraction and Refining. Berlin: Springer-Verlag, 1987.
2. Rizvi SSH. Supercritical Fluid Processing of Food and Biomaterials. London: Blackie Academic and Professional, 1994.
3. King JB, List TR. Extraction of Natural Products Using Near-Critical Solvents. London: Blackie A&P, 1993.
4. McHugh MA, Krukonis VJ. Supercritical Fluid Extraction: Principles and Practice. 2d ed. Boston: Butterworths-Heinemann, 1994.
5. Perrut M, Subra P. Proceedings of the 5th Meeting on Supercritical Fluids, 1998.
6. Poliakoff M, George MW, Howdle SM. Proceedings of the 6th Meeting on Supercritical Fluids, 1999.
7. Perrut M, Reverchon E. Proceedings of the 7th Meeting on Supercritical Fluids, 2000.
8. Besnard M, Cansell F. Proceedings of the 8th Meeting on Supercritical Fluids. Bordeaux, 2002.
9. Bertucco A. High Pressure in Venice. Chemical Engineering Transactions. 2002; Vol. 2. Milan: AIDIC, 2002.
10. Brunner G. Gas Extraction. Berlin: Springer-Verlag, 1994.
11. Perrut M. Supercritical fluid applications: industrial developments and economic issues. Ind Eng Chem Res 2000; 39:4531–4535.

12. Sun Y-P, ed. Supercritical Fluid Technology in Materials Science and Engineering. New York: Marcel Dekker, 2002.
13. Perrut M. Applications of supercritical fluid solvents in the pharmaceutical industry. In: Marcus Y, ed. Ion Exchange & Solvent Extraction. Vol. 17. New York: Marcel Dekker, 2003.
14. Hutchenson KW. Organic chemical reactions and catalysis in supercritical fluid media. In: Sun Y-P, ed. Supercritical Fluid Technology in Materials Science and Engineering. New York: Marcel Dekker, 2002:87–187.
15. Herkley C. Homogeneous catalysis in supercritical carbon dioxide. In: Sun Y-P, ed. Supercritical Fluid Technology in Materials Science and Engineering. New York: Marcel Dekker, 2002:189–226.
16. Harrod M, van den Hark S, Holmqvist A, Moller P. Hydrogenation at supercritical single-phase conditions. In: Besnard M, Cansell F, eds. Proceedings of the 8th Meeting on Supercritical Fluids. Bordeaux, 2002:133.
17. Kompella UD, Koushik K. Preparation of drug delivery systems using supercritical fluids. Crit Rev Ther Carrier Syst 2001;18, (2):173–199.
18. Jung J, Perrut M. Particle design using supercritical fluids: Literature and patent review. J Supercrit Fluids 2001; 20:179–219.
19. Lemert RL, DeSimone JM. Solvatochromic characterization of near- and supercritical ethane, propane and dimethyl ether. J Supercrit Fluids 1991; 4:186–193.
20. Stahl E, Quirin KW, Gerard D. Non-extractive applications. In: Stahl E, et al., eds. Dense Gases for Extraction and Refining. Berlin: Springer-Verlag, 1987:218–221.
21. Fages J, Marty A, Delga C, Condoret JS, Combes D, Frayssinet P. The use of supercritical CO_2 for bone delipidation. Biomaterials 1994; 15:650–656.
22. Fages J, Mathon D, Poirier B, Autefage A, Larzul D, Jean E, Frayssinet P. Supercritical processing enhances viral safety and functionality of bone alografts. Proceedings of 4th International Symposium on Supercritical Fluids, 1997:383–386.
23. Castor TP, Lander AD, Cosman MD, D'Entremont PR, Pelletier MR. Viral inactivation method using near critical, critical or supercritical fluids. US patent 5,877,005, 1999.
24. Bouzidi A, Majewski W, Hober D, Perrut M. Supercritical fluids improve tolerance and viral safety of fresh frozen plasma. In: Perrut M, Subra P, eds. Proceedings of the 5th Meeting on Supercritial Fluids, 1998:717–722.
25. Splilimbergo S, Elvassore N, Bertucco A. Microbial inactivation by high-pressure. J Supercrit Fluids 2002; 22:55–63.
26. Clavier JY, Perrut M. Safety in Supercritical Operations. In: von Rohr R, Trepp C, eds. High Pressure Chemical Engineering, Process Technology Proceedings. Vol. 12. Amsterdam: Elsevier, 1996:627–631.
27. Clavier JY, Majewski W, Perrut M. Extrapolation from pilot plant to industrial scale SFE: a case study. In: Kikic I, Reverchon E, eds. I Fluidi Supercritici Et Le Loro Applicazioni. Grignano, Italy: Atti del III Congresso, 1995:107–114.
28. Clavier JY, Majewski W, Perrut M. Extrapolation from pilot plant to industrial

scale SFE. In: von Rohr R, Trepp C, eds. High Pressure Chemical Engineering. Process Technology Proceedings. Vol. 12. Amsterdam: Elsevier, 1996:639–644.

29. Majewski W. Natural product SFE: a data and samples bank. In: Perrut M, Subra P, eds. Proceedings of the 5th Meeting on Supercritical Fluids, 1998:545–548.

30. Gilbert DJ, Palakodaty S, Sloan R, York P. Particle engineering for pharmaceutical applications—a process scale-up. Proceedings of the 5th International Symposium on Supercritical Fluids 2000.

31. Palakodaty S, Sloan R, Kordikowski A, York P. Pharmaceutical and biological materials processing with supercritical fluids. In: Sun Y-P, ed. Supercritical Fluid Technology in Materials Science and Engineering. New York: Marcel Dekker, 2002:439–490.

32. Thiering R, Deghani F, Foster NR. Current issues relating to anti-solvent micronisation techniques and their extension to industrial scales. J Supercrit Fluids 2001; 21:159–177.

33. Weber A, Tschernjaew J, Berger T, Bork M. A production plant for gas antisolvent crystallization. In: Perrut M, Subra P, eds. Proceedings of the 5th Meeting on Supercritical Fluids, 1998:281–285.

34. Rantakyla M, Aaltonen O, Hurme M. Cost study of a supercritical antisolvent (SAS) particle production process. In: Bertucco A, ed. High Pressure in Venice. Chemical Engineering Transactions. Vol. 2. Milan: AIDIC, 2002:525–530.

35. Wubbolts FE. Supercritical crystallization. PhD thesis. Veenedaal, Netherlands: Universal Press Science Publishers, 2000:209–226.

36. Benoit JP, Rolland H, Thies C, Van de Velde V. Method of coating particles and coated spherical particles. European patent EP 0 706 821, 1996. World patent WO 96/11055, 1996.

37. Benoit JP, Richard J, Thies C. Method for preparing microcapsules comprising active materials coated with a polymer and novel microcapsules in particular obtained according to the method. French patent FR 2 753 639, 1996; US patent 6,183,783, 2001.

38. Perrut M. Process for encapsulation of fine particles in form of micro-capsules. French patent FR 2 811 913, 2000; World patent WO 02/05944, 2001.

39. Shine AD, Gelb J. Microencapsulation process using supercritical fluids. US patent 5,766,637, 1998.

40. Howdle SM, Watson MS, Whitaker MJ, Popov VK, Davies MC, Mandel FS, Wong JD, Shakesheff KM. Supercritical fluid mixing: preparation of thermally sensitive polymer composites containing bioactive materials. J Chem Soc Chem Commun, 2001:109–110.

41. Majewski W, Perrut M. On-line direct impregnation of natural extracts. In: Perrut M, Reverchon E, eds. Proceedings of the 7th Meeting on Supercritical Fluids, 2000:779–780.

42. Luzzi LA, Needham E, Zia H, Dondeti P. Method of incorporating proteins or peptides into a matrix and administration thereof through mucosa. US patent 6,190,699. 2001.

43. Zia H, Dondeti P, Needham TE. Comparison of nasal insulin powders prepared

by supercritical fluid and freeze-drying techniques. Particulate Sci Technol 1997; 15:214–217.

44. Sproule T, Lee JA, Lannutti JL, Tomasko DL. Bioactive polymers via supercritical fluids. In: Besnard M, Cansell F, eds. Proceedings of the 8th Meeting on Supercritical Fluids. Bordeaux, 2002:431–436.
45. McHugh MA, Krukonis VJ. Supercritical Fluid Extraction: Principles and Practice. 2d ed. Boston: Butterworths-Heinemann, 1994:333–342.
46. Palakodaty S, York P. Phase behavioural effects on particle formation processes using supercritical fluids. Pharm Res 1999; 16:976–984.
47. Weber M, Thies MC. Understanding the RESS process. In: Sun Y-P, ed. Supercritical Fluid Technology in Materials Science and Engineering. New York: Marcel Dekker, 2002:387–437.
48. McHugh MA, Krukonis VJ. Supercritical Fluid Extraction: Principles and Practice. 2d ed. Boston: Butterworths-Heinemann, 1994:342–357.
49. Wubbolts FE. Supercritical crystallization. PhD thesis. Veenedaal, Netherlands: Universal Press Science Publishers, 2000:61–119.
50. Schmitt WJ. Finely divided solid crystalline powders via precipitation into an antisolvent. US patent 5,707.634, 1998.
51. Yeo S-D, Lim G-B, Debenedetti P, Bernstein H. Formation of microparticulate protein powders using a supercritical fluid antisolvent. Biotechnol Bioeng 1993; 41:341–346.
52. Reverchon E. Supercritical antisolvent precipitation of micro- and nano-particles. J Supercrit Fluids 1999; 15:1–21.
53. Thiering R, Deghani F, Dillow A, Foster NR. The influence of operating conditions on the dense gas precipitation of model proteins. J Chem Technol Biotechnol 2000; 75:29–41.
54. Thiering R, Deghani F, Foster NR. Micronization of model proteins using compressed carbon dioxide. Proceedings of the 5th International Symposium on Supercritical Fluids, 2000.
55. Hanna M, Yorks P. Method and apparatus for the formation of particles. World patent WO 95/01221, 1994.
56. Palakodaty S, Sloan R, Kordikowski A, York P. Pharmaceutical and biological materials processing with supercritical fluids. In: Sun Y-P, ed. Supercritical Fluid Technology in Materials Science and Engineering. New York: Marcel Dekker, 2002:439–490.
57. Palakodaty S, York P, Pritchard J. Supercritical fluid processing of materials from aqueous solutions: the application of SEDS to lactose as model substance. Pharm Res 1998; 15:1835–1843.
58. Sievers RE, Karst U. Methods for fine particle formation. US patent 5,639,441, 1997.
59. Sievers RE, Karst U. Methods and apparatus for fine particles formation. US patent 6,095,134, 2000.
60. Sievers RE, Huang ETS, Villa JA, Walsh TR. CAN-BD process for rapidly forming and drying fine particles. In: Besnard M, Cansell F, eds. Proceedings of the 8th Meeting on Supercritical Fluids. Bordeaux, 2002:73–78.

61. Leboeuf F, Jung J, Perrut M. Process of manufacture of very fine particles from at least one water-soluble compound. French patent application FR 01.06403, 2001.

62. Jung J, Leboeuf F, Perrut M. Preparation of inhalable protein particles by SCF emulsion drying. Proceedings 6th International Symposium on Supercritical fluids, 2003:1837–1842.

63. Weidner E, Knez Z, Novak Z. Process for the production of particles or powders. European patent 0 744 992, 1995. US patent 6,056,791, 2001.

64. Knez Z. Micronization of pharmaceuticals using supercritical fluids. In: Perrut M, Reverchon E, eds. Proceedings of the 7th Meeting on Supercritical Fluids, 2000:21–26.

65. Mandel F, Wang JD, McHugh MA. Pharmaceutical material production via supercritical fluids employing the technique of particles from gas-saturated solutions (PGSS). In: Perrut M, Reverchon E, eds. Proceedings of the 7th Meeting on Supercritical Fluids, Antibes, France, 2000:35–46.

66. Perrut M. Process for encapsulation of fine particles in form of micro-capsules. French patent FR 2 811 913, 2000, World patent WO 02/05944, 2001.

67. Jung J, Leboeuf F, Perrut M. Supercritical fluid micro-encapsulation. In: Besnard M, Cansell F, eds. Proceedings of the 8th Meeting on Supercritical Fluids. Bordeaux, 2002:805–812.

68. Heidlas J, Zhang Z. New approaches to formulate biocompounds using supercritical fluids. In: Perrut M, Reverchon E, eds. Proceedings of the 7th Meeting on Supercritical Fluids, 2000:167–172.

69. Weidner E. Powder generation by high pressure spray processes. Proceedings of the GVC-Fschausschub High Pressure Chemical Engineering, 1999:225–230.

70. Weber M, Thies MC. Understanding the RESS process. In: Sun Y-P, ed. Supercritical Fluid Technology in Materials Science and Engineering. New York: Marcel Dekker, 2002:387–437.

71. Weber M, Russell LM, Debenedetti P. Mathematical modeling of nucleation and growth of particles formed by the rapid expansion of a supercritical solution under subsonic conditions. J Supercrit Fluids 2002; 23:65–80.

72. Kikic I, Bertuco A. Thermodynamic and mass transfer for the simulation of recrystallization processes with a supercritical antisolvent. Proceedings of the 4th Italian Conference on Supercritical Fluids, 1997:299–306.

73. Werling JO, Debenedetti P. Numerical modeling of mass transfer in the supercritical antisolvent process. J Supercrit Fluids 1999; 6:167–181.

74. Werling JO, Debenedetti P. Numerical modeling of mass transfer in the supercritical antisolvent process: miscible conditions. J Supercrit Fluids 2000; 18:11–24.

75. Bristow S, Shekunov T, Shekunov BY, York P. Analysis of the supersaturation and precipitation process with supercritical CO_2. J Supercrit Fluids 2001; 21:257–271.

76. Domingo C, Berends E, Van Rosmalen GM. Precipitation of ultrafine organic crystals from the rapid expansion of supercritical solutions over a capillary and a frit nozzle. J Supercrit Fluids 1997; 10:39–55.

77. Czerwonatis N, Eggers R. Disintegration of liquid jets and drop drag coefficient in pressurized nitrogen and carbon dioxide. Chem Eng Technol 2001; 24:619–624.
78. Rantakyla M, Jäntti M, Aaltonen O, Hurme M. The effect of initial drop size on particle size in the supercritical antisolvent precipitation (SAS) technique. J Supercrit Fluids 2002; 24:251–263.
79. Jung J, Clavier JY, Perrut M. Gram to kilogram scale-up of supercritical fluid antisolvent process. Proceedings of the 6th International Symposium on Supercritical Fluids, 2003:1683–1688.
80. Perrut M, Jung J, Leboeuf F. Solid state morphology of particles prepared by a supercritical fluid process. In: Bertucco A ed. High Pressure in Venice. Chemical Engineering Transactions. Vol 2. Milan: AIDIC, 2002:711–716.
81. Perrut M, Jung J, Leboeuf F. Process for preparation of amorphous particles. French patent application FR-02 05046, 2002.
82. Maas JH. Gas–solid separations. In: Schweitzer PA, ed. Handbook of Separation Techniques for Chemical Engineers. New York: McGraw-Hill, 1980:6.3–6.55.
83. Perrut M. Process and equipment for collection of fine particles by percolation through a granular bed. French patent FR 2 803 538, 1999. World patent WO 01/43853, 2001.
84. Perrut M. Process and equipment for fine particles collection by trapping into a solid mixture like carbonic snow. French patent FR 2 802 445, 1999; World patent WO 01/43845, 2001.
85. Pace GW, Vachon GM, Misrha KA, Henriksen IB, Krukonis V. Processes to generate submicron particles of water-insoluble compounds. US patent 6,177,103, 2001.
86. Perrut M, Jung J, Leboeuf F. Process for preparation of a stable suspension of particles in a liquid. French patent application FR-02.04673, 2002.
87. Perrut M. Process for collection and encapsulation of fine particles formed by a process using a fluid at supercritical pressure. French patent FR 2 803 539, 2000. World patent WO 01/49407, 2001.
88. Perrut M, Jusforgues P. Process and equipment of separation using a supercritical fluid. European patent EP 0 254 610, 1988.
89. Clavier J-Y, Allemand A-V. Equipment for SCF processing of hazardous biologically active powders. Proceedings of the 6th International Symposium on Supercritical Fluids, 2003:1843–1848.

Index